Sustainable Development and Biodiversity

Volume 25

Series Editor

Kishan Gopal Ramawat
Botany Department, Mohanlal Sukhadia University, Udaipur, India

This book series provides complete, comprehensive and broad subject based reviews about existing biodiversity of different habitats and conservation strategies in the framework of different technologies, ecosystem diversity, and genetic diversity. The ways by which these resources are used with sustainable management and replenishment are also dealt with. The topics of interest include but are not restricted only to sustainable development of various ecosystems and conservation of hotspots, traditional methods and role of local people, threatened and endangered species, global climate change and effect on biodiversity, invasive species, impact of various activities on biodiversity, biodiversity conservation in sustaining livelihoods and reducing poverty, and technologies available and required. The books in this series will be useful to botanists, environmentalists, marine biologists, policy makers, conservationists, and NGOs working for environment protection.

More information about this series at http://www.springer.com/series/11920

Ajar Nath Yadav · Joginder Singh ·
Ali Asghar Rastegari · Neelam Yadav
Editors

Plant Microbiomes for Sustainable Agriculture

Springer

Editors
Ajar Nath Yadav
Department of Biotechnology
Eternal University
Baru Sahib, Himachal Pradesh, India

Joginder Singh
Department of Microbiology
Lovely Professional University
Phagwara, Punjab, India

Ali Asghar Rastegari
Falavarjan Branch
Islamic Azad University
Isfahan, Iran

Neelam Yadav
Gopi Nath PG College, Veer Bahadur Singh
Purvanchal University
Ghazipur, Uttar Pradesh, India

ISSN 2352-474X ISSN 2352-4758 (electronic)
Sustainable Development and Biodiversity
ISBN 978-3-030-38452-4 ISBN 978-3-030-38453-1 (eBook)
https://doi.org/10.1007/978-3-030-38453-1

This Springer imprint is published by the registered company Springer Nature Switzerland AG
The registered company address is: Gewerbestrasse 11, 6330 Cham, Switzerland

Foreword

The plant microbiomes play important role in plant growth, development, and soil health. The plants and rhizospheric soil are the hotspots harboring microbes playing critical roles in the maintenance of global nutrient balance and ecosystems. The diverse groups of microbes are the key components of soil–plant systems, where they are engaged in an intense network of rhizosphere/endophytic/phyllospheric interactions. The rhizospheric, endophytic, and epiphytic microbes with plant growth-promoting (PGP) attributes have emerged as an important and promising tool for sustainable agriculture. PGP microbes promote plant growth directly or indirectly, by releasing plant growth regulators; solubilization of phosphorus, potassium, and zinc; biological nitrogen fixation or by producing siderophores, ammonia, HCN, and other secondary metabolites which are antagonistic against pathogenic microbes. These PGP microbes could be used as biofertilizers/bioinoculants in place of chemical fertilizers for sustainable agriculture. The aim of the present book is to collect and compile the current developments in the understanding of the rhizospheric, endophytic, and epiphytic microbial diversity associated with plants. The book encompasses current knowledge of plant microbiomes and their potential biotechnological applications for plant growth, crop yield, and soil health for sustainable agriculture. The book will be highly useful to the faculty, researchers, and students associated with microbiology, biotechnology, agriculture, molecular biology, environmental biology, and related subjects.

The present book entitled *Plant Microbiomes for Sustainable Agriculture* is a very timely publication providing state-of-the-art information in the area of microbial biotechnology focusing on microbial biodiversity, plant–microbe interaction, and their biotechnological application in plant growth and soil fertility for sustainable agriculture. The book volume comprises sixteen chapters. Chapter 1 by Subrahmanyam as the lead author describes the biodiversity of rhizospheric microbiomes and their biotechnological applications for plant growth promotion and soil health for sustainable agriculture. Chapter 2 presented by Saleem et al. highlights the culturable endophytic fungal communities associated with cereal crops and their role in plant growth promotion. Chapter 3 by Sujatha et al. describes the genetic diversity of endophytic fungi having phosphate-solubilizing attributes and their ecological significances. Chapter 4 by Berde et al. highlights the opportunities and challenges of endophytic fungi from medicinal plants and their secondary metabolites for diverse applications. Sivakumar et al. describe the biodiversity of epiphytic microbiomes and their ecological significance and biotechnological applications in agriculture and environments in Chap. 5. Chapter 6 by Kumar and Singh deals with the functional attributes of biofilms forming microbial communities, mechanisms of biofilms formation, and role in agriculture. Nalini and Prakash highlight the recent advancements in taxonomical progress in phylum Actinobacteria and biotechnological applications of actinobacterial genera in agriculture in Chap. 7. In Chap. 8, Ashok and his colleagues describe in detail the alleviation of abiotic stresses in plants by epiphytic pink-pigmented methylotrophic bacteria. Pandey et al. highlight the diversity, ecological significance, and biotechnological applications in agriculture of the potassium-solubilizing microbe in Chap. 9. Etesami et al. explain the mitigation of different abiotic stresses in plant by the use of ACC deaminase-producing microbes in Chap. 10. The roles of halophilic microbes for plant growth promotion and alleviation of saline stress in plants have been described by Enespa et al. in Chap. 11. Chapter 12 by Gontia-Mishra deals with the microbial-mediated drought tolerance in plants. Odoh et al. highlight the microbial consortium formulation for use as biofertilizers and their applications under the natural as well as abiotic stress condition in Chap. 13. Singh et al. discuss the current advancements and future challenges in plant microbiome research in Chap. 14. The biotechnological applications of microbes as bioinoculants and biopesticides have been discussed in Chap. 15 by Misra et al. Finally, the conclusion and future visions on plant microbiomes have been given by Ajar Nath Yadav in Chap. 16.

Overall, great efforts have been carried out by Dr. Ajar Nath Yadav, his editorial team, and scientists from different countries to compile this book as a unique and up-to-date source on plant microbiomes for the students, researchers, teachers, and academician. I am sure the readers will find this book highly useful and interesting during their pursuit on plant microbiomes.

H.S. Dhaliwal

Dr. H. S. Dhaliwal
Vice Chancellor
Eternal University
Baru Sahib, Himachal Pradesh, India

Dr. H. S. Dhaliwal is presently the Vice Chancellor of Eternal University, Baru Sahib, Himachal Pradesh, India. He holds Ph.D. in Genetics from the University of California, Riverside, USA (1975). He has 50 years of research, teaching, and administrative experience in various capacities. He is also a Professor of Biotechnology at the Eternal University, Baru Sahib from 2011 to date. He had worked as Professor of Biotechnology at IIT, Roorkee (2003–2011); Founder Director of Biotechnology Centre, Punjab Agricultural University, Ludhiana (1992–2003); Senior Scientist and wheat Breeder-cum-Director at PAU Regional Research Station, Gurdaspur (1979–1990); Research Fellow FMI, Basel, Switzerland (1976–1979); D.F. Jones Postdoctoral fellow, University of California, Riverside, USA (1975–1976). He was elected as Fellow, National Academy of Agricultural Sciences, India, (1992); worked as Visiting Professor, Department of Plant Pathology, Kansas State University, Kansas, USA, (1989) and Senior Research Fellow, CIMMYT, Mexico, (1987). He has many national and international awards such as Cash award from the Federation of Indian Chambers of Commerce and Industry (FICCI) in 1985 and Pesticide India Award from Mycology and Plant Pathology Society of India (1988). He has to his credit more than 300 publications including research papers: 250; reviews: 12; chapters contributed to books: 15; papers presented in meetings, conferences, and abstracted: 105; popular articles: 18; project report/books/bulletins/manuals. His important research contributions are: Identification of a new species of wild diploid wheat *Triticum urartu* and gathered evidences to implicate *it* as one of the parents of polyploid wheat; Team leader in the development of seven wheat varieties, *viz*., PBW 54, PBW 120, PBW 138, PBW 175, PBW 222, PBW 226, and PBW 299 approved for cultivation in Punjab and North Western Plain Zone of India; molecular marker-assisted pyramiding of bacterial blight resistance genes *Xa5, Xa21*, and xa13 and the green revolution semi-dwarfing gene *sd*1 in Dehraduni basmati and development of elite wheat lines biofortified for grain iron and zinc through wide hybridization with related non-progenitor wild *Aegilops* species and molecular breeding. He made a significant contribution to the development of life and epidemiology cycle of *Tilletia indica* fungus, the causal organism of Karnal bunt disease of wheat and development of Karnal bunt-tolerant wheat cultivars. He has been the member/chairperson of several task forces and committees of Department of Biotechnology, Ministry of Science and Technology, Government of India, New Delhi and ICAR, New Delhi. Presently, he is a Member of an Expert Committee of DBT for DBT-UDSC Partnership Centre on Genetic Manipulation of Crop Plants at UDSC, New Delhi (2016 onwards), SAC of NABI of DBT and RAC of IIAB, Ranchi of ICAR.

Preface

The plant microbiomes (rhizospheric endophytic and epiphytic) play important role in plant growth promotion and nutrient recycling. The microbes associated with plant growth-promoting (PGP) attributes have emerged as an important and promising tool for sustainable agriculture. PGP microbes promote plant growth directly or indirectly, either by releasing plant growth regulators; solubilization of phosphorus, potassium, and zinc; biological nitrogen fixation or by producing siderophore, ammonia, HCN, and other secondary metabolites which are antagonistic against pathogenic microbes. The PGP microbes belonged to different phylum of archaea (Euryarchaeota); bacteria (Acidobacteria, Actinobacteria, Bacteroidetes, Deinococcus-Thermus, Firmicutes, and Proteobacteria); and fungi (Ascomycota and Basidiomycota), which include different genera, namely, *Achromobacter, Arthrobacter, Aspergillus, Azospirillum, Azotobacter, Bacillus, Erwinia, Gluconoacetobacter, Methylobacterium, Paenibacillus, Pantoea, Penicillium, Piriformospora, Planomonospora, Pseudomonas, Rhizobium, Serratia,* and *Streptomyces*. These PGP microbes could be used as biofertilizers/bioinoculants at place of chemical fertilizers for sustainable agriculture. The present book on *Plant Microbiomes for Sustainable Agriculture* covers biodiversity of plant-associated microbes and their role in plant growth promotion, mitigation of abiotic stress and soil fertility for sustainable agriculture. This book will be immensely useful to biological sciences, especially to microbiologists, microbial biotechnologists, biochemists, researchers, and scientists of microbial and plant biotechnology. We are thankful to the leading scientists who have extensive, in-depth experience and expertise in plant–microbes interaction and microbial biotechnology took the time and effort to develop these outstanding chapters. Each chapter is written by internationally recognized researchers/scientists so the reader is given an up-to-date and detailed account of our knowledge of the microbial biotechnology and innumerable agricultural applications of plant microbiomes.

We are grateful to the many people who helped to bring this book to light. Dr. Ajar Nath Yadav gives special thanks to his exquisite wife for her constant support and motivation in putting everything together. Dr. Yadav also gives special thanks to his esteemed friends, well-wishers, colleagues, and senior faculty members of Eternal University, Baru Sahib, India.

Baru Sahib, Himachal Pradesh, India	Ajar Nath Yadav
Phagwara, Punjab, India	Joginder Singh
Isfahan, Iran	Ali Asghar Rastegari
Mau, Uttar Pradesh, India	Neelam Yadav

Acknowledgements

All authors are sincerely acknowledged for contributing up-to-date information on the plant microbiomes (epiphytic, endophytic, and rhizospheric microbes), their biodiversity, and application as biofertilizers, biopesticides, and biotechnological implication for mitigation of abiotic stress in plants. The editors are thankful to all authors for their valuable contributions.

The editor, Dr. Ajar Nath Yadav is thankful to his Ph.D. research scholars Ms. Divjot Kour, Ms. Tanvir Kaur, Ms. Rubee Devi, Mrs. Kusam Lata Rana and colleagues for their support and motivation in all my efforts during this project.

All editors would like to thank their families who were very patient and supportive during this journey. Our sincere thanks to the whole Springer team who was directly or indirectly involved in the production of the book. Our special thanks to Dr. Kishan Gopal Ramawat, Dr. Valeria, and Dr. Ineke Ravesloot for the assistance and support.

We are very sure that this book will interest scientists, graduates, undergraduates, and postdocs who are investigating "plant microbiomes" microbial and plant biotechnology.

Contents

Editors and Contributors

About the Editors

Ajar Nath Yadav is an Assistant Professor (Sr. Sacle) in Department of Biotechnology, Dr. Khem Singh Gill Akal College of Agriculture, Eternal University, Baru Sahib, Himachal Pradesh, India. He has 4.5 years of teaching and 10.5 years of research experience in the field of Microbial Biotechnology, Microbial Diversity, and Plant–Microbe Interactions. He obtained doctorate degree in Microbial Biotechnology, jointly from Indian Agricultural Research Institute, New Delhi and Birla Institute of Technology, Mesra, Ranchi, India; M.Sc. (Biotechnology) from Bundelkhand University and B.Sc. (CBZ) from University of Allahabad, India. He has 150 publications, which include 75 research/review papers, 15 books, 58 book chapters, and 02 patents with h-index of 28, i10-index of 64, and 2356 citations (Google Scholar). He has published 111 research communications in different international and national conferences. He has got twelve Best Paper Presentation Awards, one Young Scientist Award (NASI-Swarna Jyanti Purskar), and three certificate of excellence in reviewing awards. He received "Outstanding Teacher Award" in 6th Annual Convocation 2018 by Eternal University, Baru Sahib, Himachal Pradesh. He has a long-standing interest in teaching at the UG, PG, and Ph.D. level and is involved in taking courses in agriculture microbiology, bacteriology, bioprocess engineering and technology, environmental microbiology, industrial microbiology, and microbial biotechnology. He is currently handling two projects, one funded by Department of Environments, Science & Technology (DEST), Shimla entitled "Development of Microbial Consortium as Bio-inoculants for Drought and Low Temperature Growing Crops for Organic Farming in Himachal Pradesh" as Principal Investigator and another funded by HP Council for Science, Technology & Environment (HIMCOSTE) on "Value-added products" as Co-PI. He also worked as an organizing committee member for 7 international conferences/symposia in the related field. Presently, he is guiding 05 scholars for Ph.D. degree and 01 for M.Sc. dissertations. In his credit, ~6700 microbes (archaea, bacteria, and fungi) isolated from diverse sources and ~550 potential and efficient microbes deposited at culture collection National Bureau of Agriculturally Important Microorganisms (NBAIM), Mau, India. He has deposited **2423** nucleotide sequences, **03** whole genome sequences (*Bacillus thuringiensis* AKS47, *Arthrobacter agilis* L77, and *Halolamina pelagica* CDK2), and **02** transcriptome to NCBI GenBank databases: in public domain. He and his group have developed method for screening of archaea for phosphorus solubilization for the first time. He has been serving as an editor/editorial board member and reviewer for more than 35 national and international peer-reviewed journals. He has lifetime membership of Association of Microbiologist in India, Indian Science Congress Council, India and National Academy of Sciences, India. Please visit https://sites.google.com/site/ajarbiotech/ for more details.

Joginder Singh is presently working as Professor at the Department of Microbiology, Lovely Professional University, Punjab, India. Previously, he has worked as Young Scientist at Microbial Biotechnology and Biofertilizer Laboratory, Department of Botany, Jai Narain Vyas University in Department of Science and Technology, Government of India. He is an active member of various scientific societies and organizations including Association of Microbiologists of India, Indian Society of Salinity Research Scientists, Indian Society for Radiation Biology, and European Federation of Biotechnology. He has more than 60 research and review articles in the peer-reviewed journals, edited four book entitled "Arbuscular Mycorrhizal Fungi" and "Microbes: In Action" published by AgroBios, India and "Microbial Bioprospecting for Sustainable Development" and "Fungi and their Role in Sustainable Development: Current Perspectives" published by Springer International Publishing, and authored/co-authored 26 chapters in edited books. He serves as reviewer for many prestigious journals, including Current Research in Engineering, Science and Technology, Science of the Total Environment, Environmental Monitoring and Assessment, Pedosphere, Soil and Sediment Contamination, Symbiosis, International Journal of Phytoremediation, Ecotoxicology and Environmental Safety, Annals of Agricultural Sciences, Annals of the Brazilian Academy of Sciences, and many more. He attended several international and national seminars, symposia, conferences, and chaired technical sessions and presented papers in them.

Ali Asghar Rastegari is currently working as an Assistant Professor in the Faculty of Biological Science, Department of Molecular and Cellular Biochemistry, Falavarjan Branch, Islamic Azad University, Isfahan, I.R. Iran. He has 09 years of research and 13 years of teaching experience in the field of Enzyme Biotechnology/Nanobiotechnology, Biophysical Chemistry, Computational Biology, and Biomedicine. He gained a Ph.D. in Molecular Biophysics in 2009, the University of Science and Research, Tehran Branch, Iran; M.Sc. (Biophysics) in 1994 from Institute of Biochemistry and Biophysics, the University of Tehran; and B.Sc. (Microbiology) in 1990 from the University of Isfahan, Iran. In his honor 36 publications [21 research papers, 02 book, 13 book chapters] in various supposed international, national journals, and publishers. He is editor of 02 books in Elsevier, 02 in CRC press, Taylor & Francis, and 06 in Springer, under the process of publication. He has issued 12 abstracts in different conferences/symposiums/workshops. He has presented 02 papers presented at national and international conferences/symposiums. He has delivered to organizing 03 conference/workshops. In his validity 01, SUB2459130 of *Homo sapiens* apolipoprotein C3 (APOC3), mRNA with a single nucleotide variant and accession: SCV000555657.1 in NM_000040.1(APOC3):c.55C>T (p.Arg19Ter) located on NCBI(ClinVar) database at recorded 2019. He is a reviewer of different international journals including *Journal of Environmental Chemical Engineering* (Elsevier), *Symbiosis Journal* (Springer), Cellulose (Springer), Frontiers in Microbiology, Scientific Report (Springer Nature), Biotechnology Progress (Wiley), Mycosphere, The 2nd International Conference on New Energy and Future Energy System (NEFES2017), The 3rd International Conference on New Energy and Future Energy System (NEFES2018), Nanomedicine Journal, and Iranian Journal of Plant Biology and got 02 Award, Certificate of Reviewing, Awarded September 2017, and In Recognition of the Review made for JECE (Elsevier) and Certificate of Outstanding Contribution in Reviewing, February 2017, In Recognition of the Contributions made to the Quality for JECE (Elsevier). He has a lifetime membership of Iranian Society for Trace Elements Research (ISTER), The Biochemical Society of I.R. IRAN, Member of Society for Bioinformatics in Northern Europe (SocBiN), Membership of Boston Area Molecular Biology Computer Types (BAMBCT), Bioinformatics/Computational Biology Student Society (BIMATICS Membership), Ensemble genome database and Neuroimaging Informatics Tools and Resources Clearinghouse (NITRC).

Neelam Yadav is currently working on microbial diversity from diverse sources and their biotechnological applications in agriculture. She obtained her postgraduation degree from Veer Bahadur Singh Purvanchal University, Uttar Pradesh, India. She has research interest in the area of plant–microbes interaction, and beneficial soil microbiomes from diverse sources. In her credit 35 publications in different reputed international and national journals and publishers. She is editor of 14 books, 02 in Elsevier, 03 in CRC press, and 09 in Springer under the process of publication. She has published 19 research communications in different conferences/symposiums/workshops. She got 02 best paper presentations award and 02 certificates of excellence in reviewing Award. In her credit >1700 microbes (archaea, bacteria, and fungi) isolated from diverse sources and >115 potential and efficient microbes deposited at culture collection. She has deposited 295 nucleotide sequences to NCBI GenBank databases: in public domain. She is Editor/Associate Editor/Reviewer of different international and national journals. She has the lifetime membership of Association of Microbiologists in India, Indian Science Congress Council, India and National Academy of Sciences, India.

Contributors

Ovaid Akhtar Department of Botany, Kamla Nehru Institute of Physical and Social Sciences, Sultanpur, India

Xavier Alexander National Institute of Pharmaceutical Education and Research, Kolkata, India

Kalimuthu Arjunekumar Department of Molecular Microbiology, School of Biotechnology, Madurai Kamaraj University, Madurai, Tamilnadu, India

Ganapathy Ashok Department of Biotechnology, Sree Narayana Guru College, Coimbatore, Tamil Nadu, India

Krishnan Baskaran Department of Biochemistry, Sree Narayana Guru College, Coimbatore, Tamil Nadu, India

Vikrant B. Berde Department of Zoology, Arts, Commerce and Science College, Lanja, Maharastra, India

Pallaval Veera Bramhachari Department of Biotechnology, Krishna University, Machilipatnam, Andhra Pradesh, India

Prem Chandra Department of Environmental Science, School for Environmental Sciences, Babasaheb Bhimrao Ambedkar (a Central) University, Lucknow, Uttar Pradesh, India

Kanhaiya L. Chaudhri Sadasivan Mycopathology Laboratory, Department of Botany, University of Allahabad, Allahabad, India

Mahananda Chutia Central Muga Eri Research & Training Institute, Central Silk Board, Ministry of Textiles, Goverment of India, Lahdoigarh, Jorhat, Assam, India

Shivika Datta Department of Zoology, Doaba College Jalandhar, Jalandhar, Punjab, India

Reena Deshmukh Biotechnology Centre, Jawaharlal Nehru Agriculture University, Jabalpur, Madhya Pradesh, India

Daljeet Singh Dhanjal Department of Biotechnology, Lovely Professional University, Phagwara, Punjab, India

Boniface Barinem Dumpe Faculty of Agriculture, Department of Crop and Soil, University of Port Harcourt, Choba, Rivers State, Nigeria

Ali Ebadi Agriculture Biotechnology Research Institute of Iran (ABRII), AREEO, Karaj, Iran

Enespa Department Plant Pathology, MPDC, University of Lucknow, Lucknow, Uttar Pradesh, India

Hassan Etesami Department of Soil Science, University College of Agriculture and Natural Resources, University of Tehran, Tehran, Iran

Chibuzor Nwadibe Eze Department of Microbiology, University of Nigeria, Nsukka, Enugu State, Nigeria

Iti Gontia-Mishra Biotechnology Centre, Jawaharlal Nehru Agriculture University, Jabalpur, Madhya Pradesh, India

Kuraganti Gunaswetha Department of Microbiology, Kakatiya University, Warangal, Telangana, India

Harbans Kaur Kehri Sadasivan Mycopathology Laboratory, Department of Botany, University of Allahabad, Allahabad, India

Ajay Kumar School of Bioengineering and Biosciences, Lovely Professional University, Phagwara, Punjab, India

Amit Kumar Central Muga Eri Research & Training Institute, Central Silk Board, Ministry of Textiles, Goverment of India, Lahdoigarh, Jorhat, Assam, India

Vijay Kumar Regional Ayurveda Research Institute for Drug Development, Gwalior, Madhya Pradesh, India

Charity Laku Environment and Conservation Unit, Centre for Environment, Human Rights and Development (CEHRD), Legacy Centre, Port Harcourt, Rivers State, Nigeria

Modhurima Misra Department of Bio-Engineering, Birla Institute of Technology, Ranchi, Jharkhand, India

Hareem Mohsin Department of Microbiology and Molecular Genetics, University of the Punjab, Lahore, Pakistan

Monnanda Somaiah Nalini Department of Studies in Botany, University of Mysore, Manasagangotri, Mysuru, Karnataka, India

Guruvu Nambirajan Department of Microbiology, Sree Narayana Guru College, Coimbatore, Tamil Nadu, India

Fatemeh Noori Department of Biotechnology and Plant Breeding, Sari Agricultural Sciences and Natural Resources University, Sari, Iran

Amechi S. Nwankwegu Department of Environmental Science and Engineering, Hohai University, Nanjing, China

Chuks Kenneth Odoh Environment and Conservation Unit, Centre for Environment, Human Rights and Development (CEHRD), Legacy Centre, Port Harcourt, Rivers State, Nigeria;
Department of Microbiology, University of Nigeria, Nsukka, Enugu State, Nigeria

Dheeraj Pandey Sadasivan Mycopathology Laboratory, Department of Botany, University of Allahabad, Allahabad, India

Chanda V. Parulekar Berde Department of Microbiology, Gogate Jogalekar College, Ratnagiri, Maharashtra, India

Harischandra Sripathy Prakash Department of Studies in Biotechnology, University of Mysore, Manasagangotri, Mysuru, Karnataka, India

Jai Prakash Department of Microbiology, School for Environmental Sciences, Babasaheb Bhimrao Ambedkar (a Central) University, Lucknow, Uttar Pradesh, India

Prachiti P. Rawool Department of Microbiology, Gogate Jogalekar College, Ratnagiri, Maharashtra, India

Yasir Rehman Department of Life Sciences, School of Science, University of Management and Technology, Lahore, Pakistan

Narges Reiahi Samani Department of Agronomy and Plant Breeding, Sari Agricultural Sciences and Natural Resources University, Sari, Iran

Ashish Sachan Centre for Life Sciences, Central University of Jharkhand, Ranchi, Jharkhand, India

Shashwati Ghosh Sachan Department of Bio-Engineering, Birla Institute of Technology, Ranchi, Jharkhand, India

Hira Saleem Department of Microbiology and Molecular Genetics, University of the Punjab, Lahore, Pakistan

Kabari Sam Environment and Conservation Unit, Centre for Environment, Human Rights and Development (CEHRD), Legacy Centre, Port Harcourt, Rivers State, Nigeria;
Faculty of Marine Environmental Management, Department of Marine Environment and Pollution Control, Nigeria Maritime University, Warri, Delta State, Nigeria

Sosanka Protim Sandilya Central Muga Eri Research & Training Institute, Central Silk Board, Ministry of Textiles, Goverment of India, Lahdoigarh, Jorhat, Assam, India

Swapnil Sapre Biotechnology Centre, Jawaharlal Nehru Agriculture University, Jabalpur, Madhya Pradesh, India

Ramamoorthy Sathishkumar Department of Molecular Microbiology, School of Biotechnology, Madurai Kamaraj University, Madurai, Tamilnadu, India

Gopal Selvakumar Department of Microbiology, Allagappa University, Karaikudi-03, Tamilnadu, India

Rajaram Shyamkumar Department of Biotechnology, Kamaraj College of Engineering and Technology, Vellakovil, Kallikudi, Madurai, Tamilnadu, India

Sumana Sikdar Biotechnology Centre, Jawaharlal Nehru Agriculture University, Jabalpur, Madhya Pradesh, India

Simranjeet Singh Punjab Biotechnology Incubators, Mohali, Punjab, India;
Department of Biotechnology, Lovely Professional University, Phagwara, Punjab, India;
Regional Advanced Water Testing Laboratory, Mohali, Punjab, India

Joginder Singh Department of Biotechnology, Lovely Professional University, Phagwara, Punjab, India

Satyender Singh Regional Advanced Water Testing Laboratory, Mohali, Punjab, India

Uma Singh Sadasivan Mycopathology Laboratory, Department of Botany, University of Allahabad, Allahabad, India

Natesan Sivakumar Department of Molecular Microbiology, School of Biotechnology, Madurai Kamaraj University, Madurai, Tamilnadu, India

Gangavarapu Subrahmanyam Central Muga Eri Research & Training Institute, Central Silk Board, Ministry of Textiles, Goverment of India, Lahdoigarh, Jorhat, Assam, India

Edla Sujatha Department of Microbiology, Kakatiya University, Warangal, Telangana, India

Rabia Tanvir University Diagnostic Lab (UDL), Department of Microbiology, University of Veterinary and Animal Sciences(UVAS), Lahore, Punjab, Pakistan

Sharad Tiwari Department of Plant Breeding and Genetics, Jawaharlal Nehru Agriculture University, Jabalpur, Madhya Pradesh, India

Chandran Viswanathan Department of Biotechnology, Sree Narayana Guru College, Coimbatore, Tamil Nadu, India

Ajar Nath Yadav Department of Biotechnology, Dr. KSG Akal College of Agriculture, Eternal University, Baru Sahib, Sirmour, Himachal Pradesh, India

Nenibarini Zabbey Environment and Conservation Unit, Centre for Environment, Human Rights and Development (CEHRD), Legacy Centre, Port Harcourt, Rivers State, Nigeria;
Faculty of Agriculture, Department of Fisheries, University of Port Harcourt, Choba, Rivers State, Nigeria

Ifra Zoomi Sadasivan Mycopathology Laboratory, Department of Botany, University of Allahabad, Allahabad, India

Chapter 1
Diversity, Plant Growth Promoting Attributes, and Agricultural Applications of Rhizospheric Microbes

Gangavarapu Subrahmanyam, Amit Kumar, Sosanka Protim Sandilya, Mahananda Chutia and Ajar Nath Yadav

Abstract Rhizosphere harbors potential microbiomes which play a pivotal role in nutrient cycling, enhancing soil fertility, maintaining plant health and productivity. Specific microbiomes that are assembled near roots are considered to be some of the most complex ecosystems on the Earth. Heterogeneous microbial communities of rhizospheric microbiomes considerably vary by soil type, land use pattern, plant species, and host genotype. It is demonstrated that root exudates act as substrates and signaling molecules which are required for establishing plant–rhizobacterial interactions. The present chapter focused on the rhizosphere microbiomes of different agricultural crops, their functions, and possible biotechnological applications for increasing crop production in a sustainable manner. Further, the plant growth-promoting mechanisms of rhizobacteria were highlighted. Although much work has been done on the biocontrol characteristics of rhizospheric bacteria, it has to be considered that soil type, plant species, and the pathogen affect altogether influence the biocontrol efficiency of strain applied against a soil-borne pathogen.

Keywords Bacterial community · Biotechnological application · Microbiome · Plant growth promotion · Rhizosphere

1.1 Introduction

Soil microorganisms play a pivotal role in nutrient cycling, regulating soil fertility, maintaining plant health, and productivity (Wagg et al. 2014). Soil microbial communities are exceedingly complex and consist of various organisms such as bacteria, archaea, fungi, algae, and viruses. Most of these microorganisms largely utilize plant root-derived nutrients such as root exudates and secondary metabolites (Huang et al.

G. Subrahmanyam (✉) · A. Kumar · S. P. Sandilya · M. Chutia
Central Muga Eri Research & Training Institute, Central Silk Board, Ministry of Textiles, Goverment of India, Lahdoigarh, Jorhat 785700, Assam, India
e-mail: subbugangavarapu@gmail.com

A. N. Yadav
Department of Biotechnology, Dr. KSG Akal College of Agriculture, Eternal University, Baru Sahib, Sirmour, Himachal Pradesh, India

A. N. Yadav et al. (eds.), *Plant Microbiomes for Sustainable Agriculture*, Sustainable Development and Biodiversity 25, https://doi.org/10.1007/978-3-030-38453-1_1

2014). Rhizosphere microorganisms are component of microbiomes that assemble near plant roots. Rhizospheric microbiomes are considered to be some of the most complex ecosystems on Earth. It is estimated that one gram of soil contains more than 50,000 different microbial species, but majority of them are uncultivable in nature (Roesch et al. 2007a, b). Beneficial free-living rhizospheric bacteria are generally referred to as plant growth-promoting rhizobacteria—"PGPR". Conceptually, "PGPR" represents beneficial portion of rhizospheric microbiome and can have positive effect on both growth and development of plants by direct or indirect mechanisms.

Chemical compounds that are released by roots apparently modify physical and chemical characteristics of the soil (Mukherjee et al. 2018) and subsequently regulates the diversity and composition of soil microbial community in the rhizosphere (Huang et al. 2014). Moreover, plants may also influence composition of rhizosphere microbial communities by selectively stimulating microorganisms with beneficial traits that are needed for both plant growth and health (Chaparro et al. 2014). For example, *Acidobacteria, Proteobacteria (*mainly *Alpha, Beta,* and *Deltaproteobacteria* classes*), Chloroflexi*, and *Actinobacteria* are enriched in the rhizosphere of *Oryza sativa,* whereas soybean selected a specific microbial community consists of *Bacteroidetes, Acidobacteria, Proteobacteria*, and *Actinobacteria* (Lu et al. 2018; Ding et al. 2019; Yadav et al. 2016b). These microbial populations are found to colonize in the root rhizosphere because of their functional traits and also beneficial to plant nutrient absorption, growth, and disease suppression. In turn, the plant provides root exudates to the microbes which are used as substrates and signaling molecules (Mendes et al. 2013).

Studies revealed that root microbiomes considerably vary by soil type, habitat, land use pattern, plant species, and host genotype (Bouffaud et al. 2014; Fitzpatrick et al. 2018; Lu et al. 2018; Ding et al. 2019; Yadav et al. 2019f). In recent, the relationship between rhizosphere microbial communities and plant genotypes is well studied and the results may lead to increased plant productivity (Bouffaud et al. 2014; Bulgarelli et al. 2015; Pérez-Jaramillo et al. 2017; Leff et al. 2017; Ding et al. 2019). In this chapter, we summarize recent progress made in rhizosphere microbiomes of agriculture crops. We also discuss the importance of rhizosphere microbial communities particularly PGPR and their immense biotechnological values for sustainable production and productivity of agriculture crops.

1.2 Rhizosphere and Root Exudates

The narrow zone of soil surrounding the plant roots and influenced by roots, root hair, and plant-produced exudates is referred to as rhizosphere (Dessaux et al. 2009). There are three distinct interacting systems which are reported in the plant rhizosphere, *viz*., rhizoplane, rhizosphere, and the root itself. Rhizoplane is defined as the root surface including the strongly adhering soil particles. Group of bacteria which are inhabitants of rhizosphere and able to compete in colonizing the root system is known as "rhizobacteria" while the total microbial component (prokaryotes, eukaryotes, and viruses) of rhizosphere is termed as rhizo-microbiome or rhizosphere microbiome. The "rhizobacteria" term was first time introduced by Kloepper and Schroth (1978) to refer the soil bacterial population that competitively colonize the roots and stimulate plant growth, thereby reducing the incidence of diseases in a sustainable manner.

Specific microbiomes that are assembled near roots are proposed to be some of the most complex ecosystems on the Earth (Raaijmakers et al. 2009). Most of these microorganisms utilize diverse array of compounds/nutrients which are derived from plant roots in the rhizosphere (Lu et al. 2018; Yadav et al. 2017b). The chemicals that are released by roots in the soil are known as "root exudates." It was suggested that chemicals secreted by plant roots act as signaling molecules and recruit wide variety of heterogeneous and metabolically active soil microbial populations (Ahemad and Kibert 2014) (Table 1.1).

Most importantly, the exudation of chemical compounds by roots apparently modifies the physical and chemical characteristics of the soil and subsequently regulates the structure and composition of rhizosphere microbial community (Doornbos et al. 2012). Impact of root exudates on bacterial communities in the rhizosphere was extensively reviewed by Doornbos et al. (2012). Further, it is estimated that around five to twenty-one percent of caron (photosynthetically fixed carbon) gets transported to the rhizosphere through the process of root exudation (Doornbos et al. 2012). Therefore, the rhizosphere is redefined by Dessaux et al. (2009) as "any volume of soil selectively influenced by plant roots, root hairs and plant-produced materials."

Table 1.1 Different kinds of compounds in root exudates of plants

Chemical nature	Compounds
Carbohydrates, amino acids, and derivatives	Chlorogenic acid, caffeic acid, cinnamic acid, canavanine, strigolactone 5-deoxystrigol, arabinogalactan proteins, arabinogalactan-like glycoprotein, glucose, fructose, galactose, ribose, xylose, rhamnose, arabinose, oligosaccharides, raffinose, maltose, α-Alanine, β-alanine, asparagines, aspartate, cysteine, cystine, glutamate, glycine, isoleucine, leucine, lysine, methionine, serine, threonine, proline, valine, tryptophan, ornithine, histidine, arginine, homoserine, phenylalanine, γ-Aminobutyric acid and α-Aminoadipic acid
Secondary metabolites and hormones	Benzoxazinoids, flavonoids, strigolactones, and related compounds that mimic quorum-sensing signals
Vitamins	Biotin, thiamine, pantothenate, riboflavin, and niacin
Enzymes	Protease, amylase, acid and alkaline-phosphatase and invertase
Organic acids	Malic acid, oxalic acid, fumaric acid, succinic acid, acetic acid, butyric acid, valeric acid, glycolic acid, erythronic acid, piscidic acid, citric acid, formic acid, aconitic acid, lactic acid, pyruvic acid, glutaric acid, malonic acid, tetronic acid, and aldonic acid

Source Huang et al. (2014), Ahemad and Kibert (2014)

The quality and quantity of the root exudates rely on type of plant species and distinct developmental/physiological status of plants (Kang et al. 2010). Furthermore, root exudates significantly enhance the plant-beneficial microbial–symbiotic interactions in the rhizosphere. These interactions, in turn, affect the rooting patterns, supply of available nutrients, thereby modifying the quantity and/or quality of root exudates. Microbial colonization in/on root tissues is known as root colonization, similarly microbial colonization of the adjoining volume of soil under the influence of the plant root system is defined as "rhizosphere colonization" (Ahemad and Kibert 2014). Compared with the bulk soil, microbial activity and biomass are relatively enhanced in the rhizosphere as a result of root exudation (Ahemad and Kibert 2014; Huang et al. 2014).

1.3 Rhizosphere Microbiome and Its Diversity

Most of the soils contain exceedingly high microbial diversity including bacteria, fungi, algae, viruses, and protozoa. It was reported that one gram of soil contains approximately 9×10^7 bacteria, 2×10^5 fungi, 4×10^6 actinomycetes, 5×10^3 protozoa, and 3×10^4 algae. The rhizosphere which is under influence of root exudates can harbor up to 10^{-11} microbial cells and around 30,000 different prokaryotic species per gram of root (Egamberdieva et al. 2008). Metagenomic analysis of tomato rhizosphere revealed that approximately 3,050 different bacterial species (OTUs at 3% distance cutoff) were associated in the rhizosphere (Tian et al. 2015). The rhizosphere microbiomes are very diverse and can actively interact with plants and mediate distinct agro-ecological process. The rhizosphere microbiome is considerably important in bridging the plant microbiomes and bulk soil and facilitates plant growth promotion by providing nutrition (Pathak et al. 2016) . The rhizobacterial microbiota also improves host plant's health by protecting from phytopathogens and promotes plant growth and fitness in different physiochemical stresses by producing phytohormones (Fig. 1.1). It is imperative to elucidate the assembly, composition, and variation among the microbial communities present in the rhizosphere for understanding the diversity and metabolic functions of the rhizosphere microbiome. This information could be beneficial for sustainable management of plant health and the underlying mechanisms that drive microbiome assembly.

It has been revealed that the rhizosphere, rhizoplane (root surface), endosphere (root interior), and of host plants harbor a distinct microbiome (Edwards et al. 2015). Diversity, distribution, and the composition of the core rhizospheric microbiomes from several plant species such as *Arabidopsis* (Bulgarelli et al. 2012; Carvalhais et al. 2013; Chaparro et al. 2014), and economically important crops, *viz.,* maize (Bouffaud et al. 2014), rice (Edwards et al. 2015; Malyan et al. 2016a, b; Lu et al. 2018; Moronta-Barrios et al. 2018; Ding et al. 2019), barley (Bulgarelli et al. 2015), citrus (Xu et al. 2018), sugar beet (Chapelle et al. 2016), sunflower (Leff et al. 2017), tomato (Tian et al. 2015), French bean (Pérez-Jaramillo et al. 2017), soybean (Mendes et al. 2011, 2014), wheat (Kour et al. 2019d; Verma et al. 2015a,

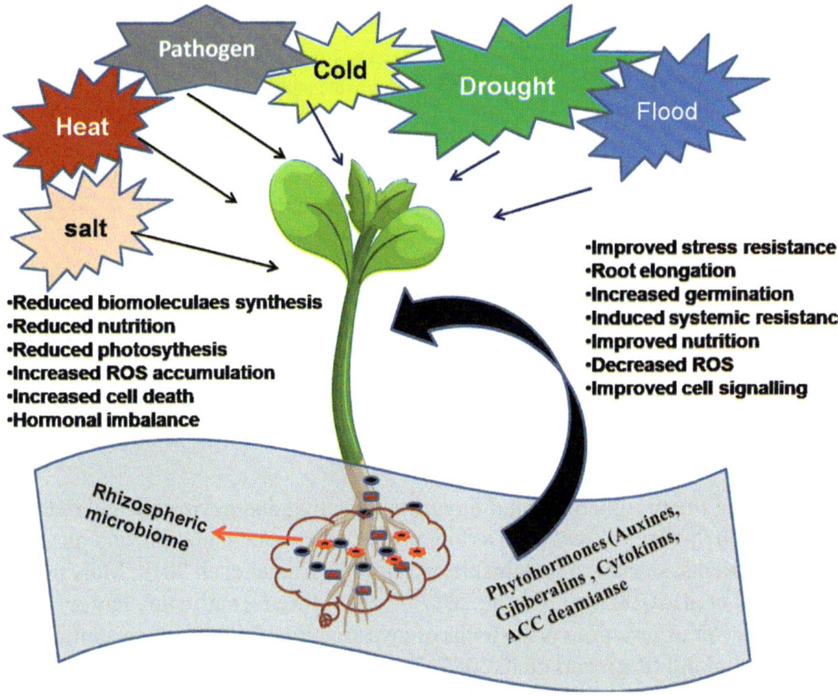

Fig. 1.1 Amelioration of abiotic and biotic stresses in plants by phytohormones produced by rhizospheric microbiome

b, 2016a, b; Yadav 2017a, 2019), and other tropical crop plants (Yadav 2017a, b; Yadav et al. 2019a; Yadav and Yadav 2018) have been established. All these studies have utilized 16S rRNA gene-based high-throughput sequencing analysis for understanding the microbial community dynamics. Although different methodologies have been suggested to explore soil microbial diversity and functions, culture-independent molecular methods are appropriate choice for deciphering diversity of microbiomes in high resolution (Fig. 1.2). Dominant microbial communities and their functions in core rhizospheric microbiomes of different agricultural crops have been extensively summarized in Table 1.2.

1.3.1 Diversity of Rhizospheric Microbiome in Wild Plants

Microorganisms represent the richest gamut of molecular and chemical diversity in nature, as they comprise the simplest yet dynamic forms of life (Yadav et al. 2015). Interest in the exploration of microbial diversity has been spurred by the fact that microbes are essential for life as they perform numerous functions integral to the

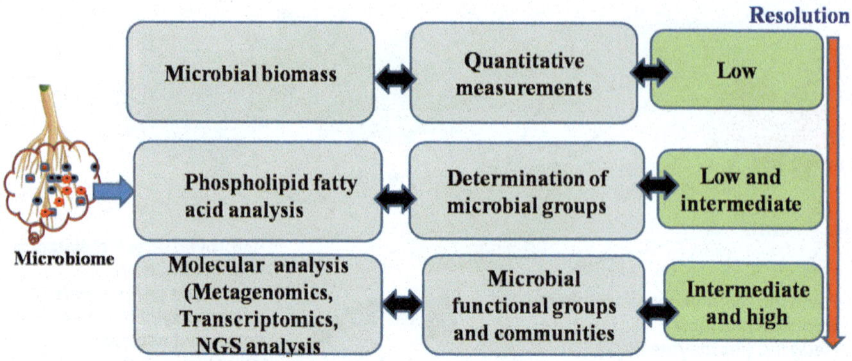

Fig. 1.2 Different methods for elucidation of rhizospheric microbiomes assembly and activity. Molecular methods are preferable choice to establish microbial communities with a higher resolution

sustenance of the biosphere, including nutrient cycling and environmental detoxification, which involve process such as augmentation, supplementation, and recycling of plant nutrients, so vital to sustainable agriculture (Kumar et al. 2019; Malyan et al. 2019; Rana et al. 2018; Yadav et al. 2017a, c, d). More recently, this largely unexplored reservoir of resources is the focus of investigations for innovative applications useful to mankind (Rastegari et al. 2019; Yadav et al. 2019c, d, e).

The distribution and diversity of bacterial community compositions in the rhizosphere microbiomes of six different wild plant species (*Bidens biternata* of the Asterales order, *Ageratum conyzoides*, *Artemisia argyi*, *Euphorbia hirta*, *Viola japonica* of the Malpighiales order, *and Erigeron annuus*) were evaluated by Lei et al. (2019). All the six different wild plant species were grown in the same experimental field. In this study, high-throughput sequencing of 16S rRNA gene targeting the hypervariable V3 and V4 regions was carried out with Illumina MiSeq platform. Comprehensive details for composition and distribution of rhizospheric microbiomes of wild plants have been shown in Fig. 1.2.

Approximately, 3000 OTUs for each rhizosphere sample were obtained. Rhizosphere microbiomes in the six wild plant species were dominated by bacterial phyla Proteobacteria (35%), Acidobacteria (12%), Actinobacteria (11%), Bacteroidetes (10%), Planctomycetes (8%), Chloroflexi (6%), and Verrucomicrobia (6%) and the details have been shown in Fig. 1.3a, b, c, d. *Rhizobiales* (8%) and *Sphingomonadales* (3.5%) orders of class Alphaproteobacteria (15%); *Nitrosomonadales* (4.28% ± 1.24%), and *Burkholderiales* (3%) orders of class Betaproteobacteria (9%); *Myxococcales* (5.5%) order of class *Deltaproteobacteria* (8%); and *Xanthomonadales* (4%) orders of class Gammaproteobacteria (7%) were found to be abundant in phylum Proteobacteria (Fig. 1.3a). Abundant members of phylum Actinobacteria were found to be Acidimicrobiales (4%). Similarly, Subgroup 4 (6%) and Subgroup 6 (4%) were abundant in Acidobacteria phylum.

The core rhizospheric microbiome of wild plant species showed a total of 1,109 operational taxonomic units (OTUs) affiliated to 113 bacterial genera accounting

Table 1.2 Diversity and functions of rhizosphere microbiomes in different agricultural crops

Crop	Major findings on rhizospheric microbiome	Dominant members in the rhizospheric microbiome	Functions	References
Phaseolus vulgaris	Significant associations between rhizobacterial community composition, crop genotype, and specific root phenotypic traits were reported	*Bacteroidetes* and *Verrucomicrobia* in roots of wild beans; *Actinobacteria* and *Proteobacteria* were enriched on roots of modern bean accessions	Growth and health of common bean	Pérez-Jaramillo et al. (2017)
Hordeum vulgare	Host genotype has a significant effect on the diversity of root-associated bacterial communities	*Comamonadaceae, Flavobacteriaceae, Rhizobiaceae*	Traits related to pathogenesis, phage interactions, and nutrient mobilization are enriched in the barley root-associated microbiota	Bulgarelli et al. (2015)
Arabidopsis thaliana	Soil type defines the composition of root-inhabiting bacterial communities. Host genotype determines specific bacterial groups in microbiome of rhizosphere. Plant cell wall features serve as sufficient colonization (up to 40%) of root-associated microbiota	*Proteobacteria, Bacteroidetes Actinobacteria, Acidobacteria, Planctomycetes*	Growth, health, and crop domestication	Bulgarelli et al. (2012)

(continued)

Table 1.2 (continued)

Crop	Major findings on rhizospheric microbiome	Dominant members in the rhizospheric microbiome	Functions	References
Arabidopsis thaliana	Jasmonic acid plant defence pathway may mediate plant–bacteria interactions in the soil and alters the composition of rhizosphere bacterial communities	*Bacillales sp., Bacillus sp., Planococcacea sp.,, Lysinibacillus sp., Pseudomonas sp.,*	Induced systemic resistance (ISR)	Carvalhais et al. (2013)
Beta vulgaris	Invading pathogenic fungus and plant stress responses directly influence the shift in rhizobacterial community in microbiome composition	*Oxalobacteraceae, Burkholderiaceae, Sphingobacteriaceae Sphingomonadaceae*	Antagonistic traits that restrict pathogen colonization and infection	Chapelle et al. (2016)
Helianthus annuus	Plant-associated fungal communities are strongly influenced by host genetic factors than bacterial communities	*Nectriaceae, Olpidiaceae Mortierellaceae, unclassified Pleosporales, Preussia spp., unclassified Thelebolaceae, Fusarium spp., Conocybe spp.*	Native microflora may have symbiotic action	Leff et al. (2017)

(continued)

Table 1.2 (continued)

Crop	Major findings on rhizospheric microbiome	Dominant members in the rhizospheric microbiome	Functions	References
Poaceae crops such as *Zea mays L; Zea mays ssp. Parviglumis; Sorghum bicolor cv. Arprim; (Triticum aestivum L. cv. Fiorina)*	Bacterial community composition in the rhizosphere is different from that in bulk soil. Rhizobacterial community composition differed according to the *Poaceae* genotype. The extent of diversification of eukaryotic hosts can be a significant factor for selection of their associated bacterial compartment	*Rhodospirillales* such as *Azospirillum, Gluconacetobacter, Rhodospirillum Sphingomonadaceae (Alphaproteobacteria); Burkholderiales* such as *Acidovorax, Alcaligenes, Burkholderia, and Hydrogenophaga (Betaproteobacteria); Francisella, Moraxella, Pantoea, Photorhabdus, Pseudomonas, and Xanthomonas (Gammaproteobacteria); Myxococcales* such as *Anaeromyxobacter (Deltaproteobacteria), (v) Megasphaera, Mogibacterium; Bacillales* such as *Bacillus* and *Peanibacillus (Firmicutes); Collinsella*; and *Actinomycetales* such as *Actinomyces, Corynebacterium, Kocuria, and Propionibacterium (Actinobacteria)*	Normal growth and plant health	Bouffaud et al. (2014)

(continued)

Table 1.2 (continued)

Crop	Major findings on rhizospheric microbiome	Dominant members in the rhizospheric microbiome	Functions	References
Citrus crop	Provided a comprehensive taxonomic and functional biogeographical analysis of the citrus rhizosphere microbiome	*Pseudomonas, Agrobacterium, Cupriavidus, Bradyrhizobium, Rhizobium, Mesorhizobium, Burkholderia, Cellvibrio, Sphingomonas, Variovorax and Paraburkholderia,, Ascomycota, Fusarium* and *Hirsutella*	Nutrition acquisition and plant growth promotion	Xu et al. (2018)
Amaranthus albus, Sonchus arvensis, Sisymbrium officinale, Medicago sativa, Solanum dulcamara, and other 25 angiosperm plant species	Evolutionary divergence among host plant species affects the assembly of the endosphere and rhizosphere microbiome. This indicates that the potential role of host-associated microbial communities in mediating interactions between hosts and their biotic and abiotic environment	*Proteobacteria, Actinobacteria, Firmicutes, Verrucomicrobia, Bacteroidetes, Chloroflexi, Acidobacteria*	The root microbiome is associated with drought tolerance across host plant species	Fitzpatrick et al. (2018)
Arabidopsis thaliana	Different developmental stages of plant influence rhizosphere microbiome assemblage. Plant can select a subset of microbes at different stages of development, presumably for specific functions	*Acidobacteria, Actinobacteria, Bacteroidetes, Cyanobacteria*	Disease suppression	Chaparro et al. (2014)

(continued)

Table 1.2 (continued)

Crop	Major findings on rhizospheric microbiome	Dominant members in the rhizospheric microbiome	Functions	References
Oryza sativa	Rhizosphere microbiome is shaped by soil- and plant-related conditions such as geographic location, soil type, rice genotype, oxic and anoxic interface, agricultural management and growth stages	*Proteobacteria* (mainly *Alpha-, Beta- and Deltaproteobacteria* classes), *Acidobacteria, Actinobacteria* and *Chloroflexi* phyla. *Archaeal* communities composed of *Crenarchaeota, Thaumarchaeota* and *Euryarchaeota* phyla	Growth promotion and disease inhibition	Ding et al. (2019), therein references
Triticum aestivum, Hordeum vulgare, Oryza sativa (indica and japonica)	*Triticum aestivum* and *Hordeum vulgare* had shown much stronger selection effects than *Oryza sativa* for the rhizosphere microbial community	*Proteobacteria, Bacteroidetes, Acidobacteria, Planctomycetes, Nitrospirae, Actinobacteria, Verrucomicrobia, Firmicutes, Cyanobacteria, Chloroflexi, Gemmatimonadetes, Ascomycota, Basidiomycota, Zygomycota, Unidentified fungi*	Glycan, limonene, and pinene degradation; Nitrogen and sulfur metabolism; Plant growth promotion	Lu et al. (2018)

(continued)

Table 1.2 (continued)

Crop	Major findings on rhizospheric microbiome	Dominant members in the rhizospheric microbiome	Functions	References
Solanum lycopersicum	Nematode infections were associated with diversity and composition of rhizosphere bacterial populations in tomato plant roots. Root endophytes and rhizobacteria had significantly different community structures and species abundance	*Proteobacteria, Alphaproteobacteria. Sphingomonadales, Rhizobiales*	Degradation of plant polysaccharides; Carbohydrate and protein metabolism and biological nitrogen fixation	Tian et al. (2015)
Glycine max (L.)	Selection of the microbial community in the rhizosphere is based on niche-based processes as a result of the selection power of the plant and other environmental factors	phyla *Actinobacteria, Acidobacteria, Chloroflexi, Cyanobacteria, Chlamydiae, Tenericutes, Deferribacteres, Chlorobi, Verrucomicrobia,* and *Aquificae*	Growth promotion and nutrition	Mendes et al. (2014)

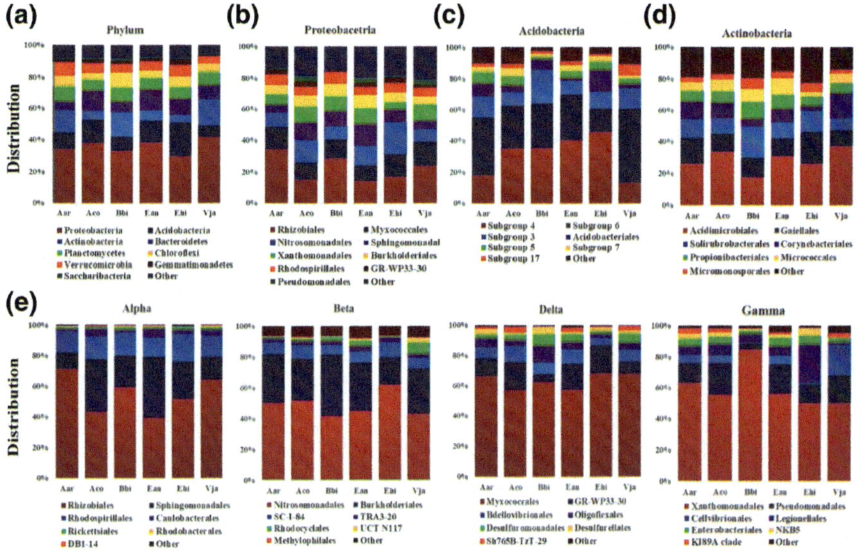

Fig. 1.3 The composition and relative abundance of major bacterial taxa in a typical rhizosphere of six different wild plant species Adapted with permission from Lie et al. (2019)

for more than 70% of the total sequencing data analyzed. The predominant bacterial genera of core OTUs are *Variovorax, Acidibacter, Ferruginibacter, Bradyrhizobium, Blastocatella, Variibacter, Sphingomonas,* and unclassified bacteria (Fig. 1.3e). The predominant bacterial orders were found to be composed of *Xanthomonadales, Rhodospirillales, Rhizobiales, Burkholderiales, Sphingomonadales, Myxococcales, Nitrosomonadales* of Proteobacteria; *Acidimicrobiales* of Actinobacteria; *Subgroup 4* and Subgroup *6* of Acidobacteria.

Variations in microbial community compositions at the order level in the rhizosphere of six different plant species were also demonstrated (Lei et al. 2019). Predominant bacterial group in *E. hirta* rhizosphere is Proteobacteria, while the same group is least represent in *V. japonica* microbiome. Highly enriched *Rhizobiale*s order of Proteobacteria was found in *V. japonica* and *A. argyi.* Predominant members of *Myxococcales* were noticed in *V. japonica* rhizosphere. Abundant members of *Nitrosomonadales* were observed in *E. hirta*. Similarly, higher abundance of Burkholderiales and *Sphingomonadales* was noticed in *E. annuus*. Members of *Xanthomonadales* were dominated in *V. japonica* rhizosphere.

1.3.2 Diversity of Rhizospheric Microbiome in Agriculture Crops

1.3.2.1 Rhizospheric Microbiome of Rice

The structure of microbial communities present in the rice rhizosphere is very complex, dynamic, and diverse (Edwards et al. 2015; Lu et al. 2018; Moronta-Barrios et al. 2018; Ding et al. 2019). Recently, microbiome inhabiting rice roots and rhizosphere is extensively reviewed by Ding et al. (2019). A study taken by Edwards et al. (2015) revealed that endosphere (inside the root compartment), rhizoplane (surface of the root), and rhizosphere of rice had distinct microbiomes. Microbial communities from the rice rhizosphere are established by amplification of the 16S rRNA gene (variable regions V4-V5) followed by high-throughput sequencing using the Illumina MiSeq platform (Edwards et al. 2015). Results indicate that rice endosphere microbial communities had the lowest α-diversity, whereas rice rhizosphere had higher α-diversity. Furthermore, the mean α-diversity was found to be relatively high in the rhizosphere than in the bulk soil (Edwards et al. 2015).

The most dominant bacterial genera of rice rhizosphere is summarized in Fig. 1.4. Bacterial community profiles and their relative abundance are shown in Fig. 1.5 (Lu

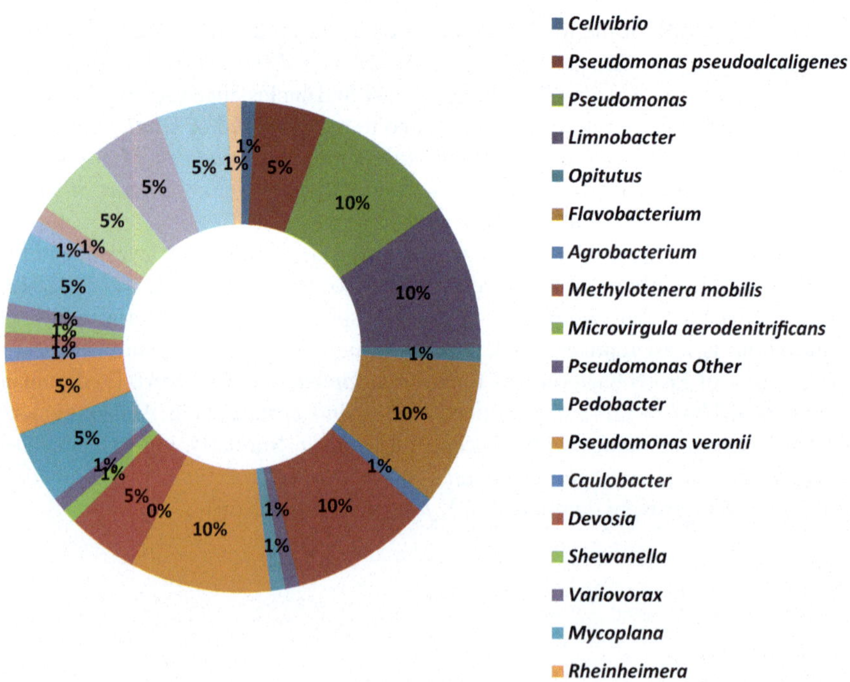

Fig. 1.4 Dominant bacterial genera in the rhizosphere microbiome of rice Adapted from Moronta-Barrios et al. (2018)

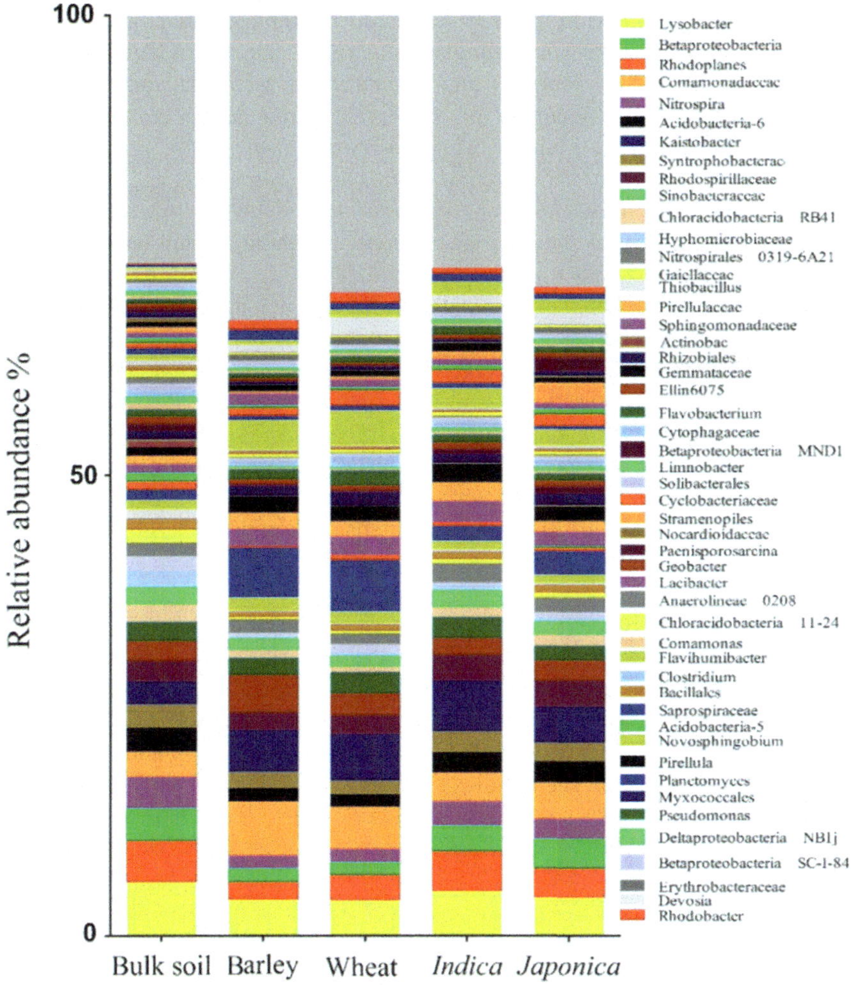

Fig. 1.5 Bacterial community profiling in the rhizospheric microbiomes of *Hordeum vulgare (Barley), Triticum aestivum (Wheat), Oryza* sativa *Indica* and *Japonica* (Rice) Adapted with permission from Lu et al. (2018)

et al. 2018). Bacterial, archaeal, and fungal communities and their relative abundance in the rice rhizosphere have been studied (Ding et al. 2019). Bacterial populations were found to be abundant in the rice rhizosphere (Edwards et al. 2015). The abundance of rhizosphere microbial populations such as bacterial, fungal, and archaeal was twice those that of the bulk soil which is an indication of rhizospheric effect (Ding et al. 2019).

Proteobacteria dominated the microbiome of rice rhizosphere accounting more than 71%. Among Proteobacteria, the most abundant class was Gammaproteobacteria followed by *Betaproteobacteria, Alphaproteobacteria, Deltaproteobacteria*, and

Epsilonproteobacteria (Moronta-Barrios et al. 2018). Representatives of *Epsilonproteobacteria* and *Deltaproteobacteria* classes were not detected in the rice endorhizosphere. *Phyla Verrucomicrobia* and *Bacteroidetes* were abundant across the samples. Representative members of *Nitrospirae and Acidobacteria* were found only in rice-rhizospheric samples (Moronta-Barrios et al. 2018). Bacterial phyla *Proteobacteria, Firmicutes, Chloroflexi, Nitrospirae, Spirochaetes, Fibrobacteres, Planctomycetes, Bacteroidetes, Proteobacteria, Actinobacteria, Verrucomicrobia, Cyanobacteria*, and *Acidobacteria* are the most commonly found bacterial members of the rice rhizosphere (Edwards et al. 2015; Lu et al. 2018; Moronta-Barrios et al. 2018; Ding et al. 2019). Similarly, the most common bacterial genera of rice rhizosphere are as follows: *Pseudomonas sp., Limnobacter, Devosia, Opitutus, Flavobacterium, Shewanella, Caulobacter, Agrobacterium, Pseudomonas veronii, Methylotenera mobilis, Microvirgula aerodenitrificans, Pedobacter, Rhodoferax, Variovorax, Mycoplana, Rheinheimera, Flavisolibacter, Fluviicola, Chryseobacterium, Asticcacaulis, Halothiobacillus, Pleomorphomonas, Sphingobium, Thiobacillus, Bacillus sp., Flavobacterium gelidilacus, Methylophaga,* and *Acidovorax* (Moronta-Barrios et al. 2018). Further details on dominant microbial communities and their functions in rice-rhizospheric microbiomes have been summarized in Table 1.2.

Methanogenic archaea, viz., *Methanobacterium, Methanosarcina, Methanocella*, and *Methanosaeta* were also reported in the rice rhizosphere (Malyan et al. 2016a; Edwards et al. 2015). *Firmicutes, Bacteroidetes, Betaproteobacteria*, and *Chloroflexi* were found to be differentially enriched in rice rhizosphere. The *Betaproteobacterial* OTUs that are enriched in rice rhizosphere belong to mainly *Comamonadaceae* and *Rhodocyclaceae* families (Edwards et al. 2015). Total bacterial count of rice rhizosphere is approximately 5×10^{9} cells g dw^{-1} soil, whereas as archeal members are found to be 2.5×10^{8} cell g dwt^{-1} soil.

1.3.2.2 Rhizospheric Microbiome of Wheat and Barley

Rhizosphere community of wheat was analyzed by comparative metatranscriptomics approach (Hayden et al. 2018). The rhizosphere community of wheat was predominately bacteria. Classes *Gammaproteobacteria, Alphaproteobacteria,* and *Actinobacteria* were dominant in the rhizosphere of wheat and barley. Bacterial families such as *Micrococcaceae*, *Enterobacteriaceae,* and *Pseudomonadaceae* were abundant in the rhizosphere microbiomes (Hayden et al. 2018). Predominant archaeal members in the rhizosphere are affiliated to family *Nitrososphaeraceae* under phylum Thaumarchaeota. *Ascomycota* is the dominant fungal phylum found in the rhizosphere representing more than 72% of total fungal transcripts. Other fungal phyla in the rhizosphere of wheat and barley were affiliated to Basidiomycota (>10%), which includes the genus *Rhizoctonia*, and Glomeromycota (4%) form arbuscular mycorrhizae. Interestingly, fungal families represent a smaller proportion of the total microbial transcripts analyzed in the rhizosphere (Hayden et al. 2018). Rhizosphere community of barley was reported by Lu et al. (2018). Bacterial phyla *Proteobacteria, Bacteroidetes, Acidobacteria, Planctomycetes, Nitrospirae, Actinobacteria, Verrucomicrobia, Firmicutes, Cyanobacteria, Chloroflexi*, and *Gemmatimonadetes* were associated with barley rhizosphere. Further, fungi *Ascomycota, Basidiomycota, Zygomycota*, and Unidentified fungi were distributed in the barley rhizosphere (Hayden et al. 2018).

1.3.2.3 Rhizospheric Microbiome of Soybean

Shotgun metagenomics approach was used to study functional and taxonomic diversities of microbial communities in the rhizosphere of soybean, *Glycine max* (L.) (Mendes et al. 2014). Metagenomic libraries were dominated by bacteria (>95%) followed by eukaryotes (3%) and archaea and virus (1%). *Proteobacteria* was found to be the most abundant phylum in soybean rhizosphere and represented around 47% distribution. Other dominant bacterial phyla in the rhizosphere of soybean were found to be *Actinobacteria* (23%), *Acidobacteria* (5%), and *Firmicutes* (6%) (Mendes et al. 2014). In general, 28% of total sequences found in the soybean rhizosphere were novel and were not affiliated to known bacterial taxa. Results indicate clear differences in microbial community structure among rhizosphere and bulk soil. Overrepresentation of the phyla *Acidobacteria, Actinobacteria, Chloroflexi, Chlamydiae, Cyanobacteria, Deferribacteres, Tenericutes, Chlorobi, Aquificae,* and *Verrucomicrobia* was found in rhizosphere and the results were significant at $P < 0.01$ (Mendes et al. 2014). Similarly, abundance of class *Mollicutes, Bacilli, Clostridia, Epsilonproteobacteria, Gammaproteobacteria, Thermomicrobia,* and *Chlamydiae* was found in the rhizosphere of *Glycine max* (L.).

1.3.2.4 Rhizospheric Microbiome of French Bean

Microbiome of French bean was elucidated by amplification of 16S rRNA (V3–V4 region) followed by high-throughput sequencing performed at Illumina MiSeq platform (Pérez-Jaramillo et al. 2017). Phylum *Proteobacteria* was the dominant member, whereas lower abundance of *Acidobacteria* was noticed in wild bean rhizosphere. The phyla *Verrucomicrobia* and *Bacteroidetes* were predominant in the wild bean rhizosphere. Phylum *Actinobacteria* was found to be more abundant in the modern bean rhizosphere and these results were statistically significant.

Significant increase in the relative abundance of bacterial families *Sphingomonadaceae* and *Rhizobiaceae* was observed in the rhizosphere as compared to the bulk soil. Furthermore, it was noticed that there is a gradual decrease in the relative abundance of the *Chitinophagaceae* and *Cytophagaceae* of the Bacteroidetes phylum in the French bean rhizosphere. Gradual increases in relative abundance of families *Streptomycetae* and *Nocardiodaceae* of *Actinobacteria* and *Rhizobiaceae* of *Proteobacteria.*

1.3.2.5 Microbiome of Maize and Other Members of Poaceae Crops

Rhizospheric microbiome of Poaceae crops such as *Zea mays L; Zea mays ssp. Parviglumis; Sorghum bicolor cv. Arprim; Triticum aestivum L. cv. Fiorina* was established by Bouffaud et al. (2014). The dominant bacterial members of *Microbiome of Poaceae crops* were found to be *Rhodospirillales such as Gluconacetobacter, Rhodospirillum, Azospirillum,* and *Sphingomonadaceae of* class *Burkholderiales; Actinomycetales* such as *Corynebacterium, Actinomyces, Propionibacterium,* and *Kocuria; Acidovorax* of *Alphaproteobacteria; Xanthomonas, Francisella, Pantoea, Moraxella, Pseudomonas*, and *Photorhabdus* of class *Gammaproteobacteria; Burkholderia, Hydrogenophaga*, and *Alcaligenes* of class *Betaproteobacteria; Myxococcales* such as *Anaeromyxobacter* of class *Deltaproteobacteria; Mogibacterium, Bacillales (Firmicutes)* such as *Bacillus* and *Paenibacillus; Megasphaera* and *Collinsella* (Bouffaud et al. 2014).

1.4 Factors Influencing Rhizospheric Microbiome in Agriculture Crops

The rhizosphere microbiomes participate in very important functions suitable for plant growth promotion. The key functions mediated by rhizosphere microbiome include abiotic stress tolerance, nutrient acquisition, and protection against plant pathogen infection. Therefore, understanding the assembly of rhizosphere microbiome and their molecular mechanisms will provide us basic information. This information will be useful to develop soil management practices, designing of healthy rhizosphere microbiome, and introduction of biofertilizers and biological control agents to develop sustainable agricultural strategies. Different factors that are influencing structure, assembly, and function of rhizospheric microbiomes are depicted in Fig. 1.6.

Rhizobacterial community composition in *Phaseolus vulgaris* was influenced by specific root morphological traits and host plant genotype (Pérez-Jaramillo et al. 2017). Impact of host plant genotype on rhizosphere microbial community was mediated by qualitative and quantitative composition of root exudates (Huang et al. 2014; Ahemad and Kibert 2014). Host genotype had a tremendous effect on the composition of root-associated microbial communities in *Hordeum vulgare* (Bulgarelli et al. 2015). Bulgarelli et al. (2012) reported that host genotype and soil type define the diversity of root-inhabiting bacterial communities in *Arabidopsis thaliana*. Plant cell wall properties confer sufficient colonization (40%) of root-associated microbiota in *Arabidopsis thaliana* (Bulgarelli et al. 2012).

Invading fungal pathogens and plant stress response induces a shift in microbiome composition of sugar beet (Chapelle et al. 2016). Rhizosphere microbial community structure varied according to the Poaceae genotype (Bouffaud et al. 2014). Evolutionary divergence among host plants and type of plant species affects the assembly

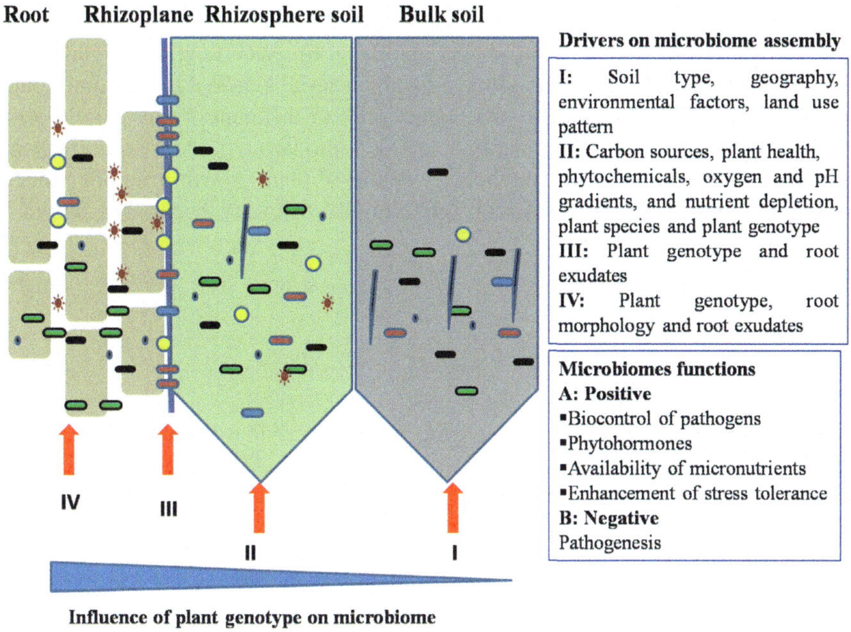

Fig. 1.6 Factors influencing assembly of microbiomes in agriculture crops. Microbial community structures in the four different compartments (I–IV) such as bulk soil, rhizosphere, rhizoplane, and root systems are distinct in nature

of the rhizosphere and endosphere and microbiome (Fitzpatrick et al. 2018). The root microbiome is also associated with drought tolerance of host plants (Fitzpatrick et al. 2018, Kour et al. 2019). Different developmental stages of plant also influence rhizosphere microbiome assemblages (Chaparro et al. 2014). Rhizosphere microbiome of *Oryza sativa* is shaped by plant and soil-related conditions such as soil type, geographic location, rice genotype, oxic–anoxic interface, agricultural management, and growth stages (Ding et al. 2019). Selection of the microbial community in the wheat rhizosphere depends on niche-based processes as a result of environmental factors and the selection power of the plant (Mendes et al. 2014). Further, agricultural management practices and growth stages of host plants exerted much influence on the rice rhizosphere microbiome (Edwards et al. 2015).

1.5 Plant Growth-Promoting Mechanisms of Rhizospheric Microbiome

Rhizobacteria plays a crucial role in growth promotion and immunity of the agricultural crops. These plant growth promoters follow certain mechanisms during the entire sequential process for nutrient mobilization, phytohormones for the growth and development, and chemical agents for defense-related issues of the crops (Suman et al. 2016; Verma et al. 2017; Yadav et al. 2018a, b). According to Mahanty et al. (2016), similar mechanisms are adopted by all bacterial genera during promotion of plant growth, although they are phylogenetically of different origins. The mechanisms behind the scene could mainly be divided into two types, direct mechanisms and indirect mechanisms. Comprehensive details of plant growth-promoting mechanisms of rhizobacteria in different agricultural crops have been summarized in Table 1.3.

1.5.1 Direct Mechanism

The direct mechanisms mainly involve the bacterial activities like phosphate solubilization, nitrogen fixation, secretion of plant hormones, ACC deaminase activities, and siderophore production.

1.5.1.1 *Phosphate Solubilization*

In spite of the large reservoir of phosphorus in soil, a very low amount of it is available to the plants (Ahemad and Kibret 2014). This is because plants utilize them in only two forms: (a) monobasic and (b) dibasic ions (Bhattacharyya and Jha 2012). It has also been reported that due to rapid conversion of phosphorus into insoluble complexes of different metal oxides most of the cultivable soils are deficit of available phosphate (Sandilya et al. 2016). Phosphate fertilizers are mostly applied to the agricultural soils in order to overcome the overall loss. But, continuous use of these chemical fertilizers is harmful to the soil and the environment in vivo. Hence, the importance of biofertilizers having plant growth-promoting traits was raised worldwide.

Native rhizobacteria pays an immense attribute to solubilize the inorganic phosphate so as to make it available for the utilization of various crops or plants (Widawati 2011). Certain bacterial genera, viz., *Bacillus*, *Pseudomonas*, *Azospirillum*, *Achromobacter*, *Acetobacter*, *Acinetobacter*, *Enterobacter*, *Klebsiella*, and *Serratia* are able to solubilize the inorganic form of phosphate to the available form (Kumar et al. 2012; Rana et al. 2019a, b). Besides, the role of bacterial organic acids for cation uptake by the plants is also worth mentioning (Sandilya et al. 2016). Researchers further stated that the bacterial genera belonging to the Proteobacteria and some of the *Firmicutes* and *Actinobacteria* are the most capable of the abovementioned conversion process.

Table 1.3 Plant growth-promoting mechanisms of plant microbiomes

Rhizobacteria	PGP traits	Crops/Plant rhizosphere	References
Bacillus sp. *Burkholderia* sp.	Phosphate solubilizer	*Zea mays*	Oliveira et al. (2009)
Pseudomonas aeruginosa strain MAJ PIA 03 *Bacillus firmus* strain MAJ PSB12	IAA, GA_3, ACC deaminase activity, HCN production, NH3 production, Siderophore production, antagonist, and phosphate solubilizer	*Ricinus communis*	Sandilya et al. (2016, 2017)
Rhodococcus sp. EC35, *Pseudomonas* sp. EAV, and *Arthrobacter nicotinovorans* EAPAA	Phosphate solubilizer	*Zea mays*	Sofia et al. (2014)
Azospirillum brasilense Az39, *Bradyrhizobium japonicum* E109	Phytostimulation	*Glycine max*	Cassan et al. (2009)
P. fluorescens Aur6, *Chryseobacterium balustinum* Aur9	Biocontrol agents	*Oryza sativa*	Lucas et al. (2009)
Bacillus, *Azotobacter*, *Pseudomonas*, and *Acinetobacter*	IAA, NH_3, HCN, Siderophore, phosphate solubilizer, antagonistic activity, nitrate reducer	*Momordica charantia*	Singh et al. (2017)
Pseudomonas putida, *Gluconacetobacter azotocaptans*, *Azospirilum lipoferum*	Phosphate solubilizer, plant hormones, siderophore	*Zea mays*	Mehnaz and Lazarovits (2006)
Sphingobacterium canadense	Phosphate solubilizer, plant hormones, siderophore	*Zea mays*	Mehnaz et al. (2007)
Chryseobacterium palustre, *Chryseobacterium humi*, *Sphingobacterium*, *Bacillus*, *Achromobacter*	IAA, HCN, NH_3, siderophore, ACC deaminase	*Zea mays*	Marques et al. (2010)
Pseudomonas fluorescens strain Psd	Zinc solubilizer, nitrate reducer	*Triticum aestivum*	Sirohi et al. (2015)
Bacillus sp. strain WG4	Antifungal metabolite pyrrolo [1, 2-a] pyrazine-1, 4-dione, hexahydro-3-(phenylmethyl)	*Zingiber officinale*	Jimtha et al. (2016)
Bacillus sp. PSB10	IAA, siderophores, HCN, ammonia	*Cicer arietinum*	Wani and Khan (2010)

(continued)

Table 1.3 (continued)

Rhizobacteria	PGP traits	Crops/Plant rhizosphere	References
Bradyrhizobium sp. 750, *Pseudomonas* sp., *Ochrobactrum cytisi*	Heavy metal mobilization	*Lupinus luteus*	Mehnaz et al. (2010)
Mesorhizobium sp. strain MRC4	IAA, siderophores, HCN, ammonia, exo-polysaccharides	*Cicer arietinum*	Ahemad and Khan (2009, 2010a, b)

1.5.1.2 *Nitrogen Fixation*

Nitrogen being the most important limiting factors, its fixation in nature is an interesting phenomena led by the plant growth-promoting rhizobacteria both in the symbiotic and non-symbiotic or free-living forms (Fagodiya et al. 2017a, b). It has been believed that the free-living nitrogen fixers provide a very lower amount of available nitrogen to the plants in comparison to the symbiotic nitrogen fixers since time immemorial (James and Olivares 1997). The *nif* genes found in the nitrogen-fixing rhizobacteria complete the nitrogenase enzyme by the means of its structural and regulatory proteins responsible for activation of the Fe protein, iron molybdenum, cofactor biosynthesis, and electron donation in case of the former and synthesis and function of the enzyme in the later (Glick 2012). Numerous PGPR genera capable of converting nitrate into nitrite by the catalysis of the nitrate reductase enzyme have also been reported. The most common among them are *Azospirillum*, *Azotobacter*, *Achromobacter*, *Bradyrhizobium*, *Beijerinckia*, and *Rhizobium* (Kour et al. 2019b, c; Yadav et al. 2019b).

1.5.1.3 Phytohormones and ACC Deaminase Enzyme Activity

Major plant hormones such as IAA and GA_3 (Marques et al. 2010; Ahmed and Hasnain 2010 and Khan et al. 2014) along with cytokinin secretion (Liu et al. 2013) by the PGPR's have often being reported by various authors. The IAA secreted by the bacterial population associated with the roots of the agricultural crops could augment the root surface area and length that could pave an easier route for absorption of the soil nutrients by the plants (Ahemad and Khan 2012). Amino acid tryptophan being a major precursor of IAA boosts the level of IAA biosynthesis. Almost five different types of IAA pathways have been reported by Spaepen and Vanderleyden (2011).

The role of GA_3 has also been explained by some authors in the context of plant growth-promoting rhizobacteria. The most important among them are the induction of seed germination and emergence and development of stem, leaf, flower, and fruits (Bottoni et al. 2004). The most common bacterial strains *Bacillus cereus*, *Sphingomonas* sp. LK11 were reported by them to enhance the growth and production of red pepper and tomato. Similarly, some other mechanisms of a plant body, viz., cytokinesis, sensitivity of vascular cambium, and their differentiation and root apical dominance are being conducted by the hormone cytokinin. Root-associated bacteria such as *Azotobacter chroococcum*, *Bacillus megaterium*, and *B. subtilis* were accounted to produce cytokinin thereby enhancing plant growth. On the other hand, synthesis of ethylene by the plant growth-promoting rhizobacteria induces ripening of fruits, opening of flowers, and leaf abscission.

Plants growing under stress are able to withstand the adverse effects of the environment with the due help of these phytohormones (de Garcia et al. 2006). Ethylenes produced in such conditions are called as "stress ethylene" that adds to the existing production of ethylene. However, excessive production of ethylene is a harmful phenomena for the longer development of the roots and in order to check such level of production, PGPR's with the help of 1-aminocyclopropane-1-carboxylate (ACC) deaminase plays a vital role in the early stages of growth which modulates the level of ethylene by hydrolyzing ACC, a precursor of ethylene, in ammonia and a-ketobutyrate (Glick et al. 1998; Marques et al. 2010). Bacteria synthesizing IAA along with endogenous plant IAA could stimulate plant growth or accelerate the amalgamation of the enzyme ACC synthase translating the compound *S*-adenosyl methionine to ACC being the immediate precursor of ethylene in higher plants (Glick 2012). Different kinds of phytohormones and their plant growth-promoting activity in agriculture crops have been summarized in Table 1.4.

1.5.1.4 Siderophore Production

Iron being one of the most important nutrients for all forms of life is found to occur as Fe^{3+} that could most likely form insoluble hydroxides and oxyhydroxides making it nearly impossible for plants and microflora for easy access (Rajkumar et al. 2010). In order to overcome such situations, bacteria secretes siderophores which are iron chelating agents with low molecular mass. According to Glick (2012), siderophores are mostly water soluble and could be divided into extracellular and intracellular siderophores. Siderophore forming Fe^{3+}-siderophore complex on the bacterial membranes gets reduced to Fe^{2+}. These ionic forms of iron are released into the cell from the complex via another mechanism linking both the membrane systems (inner and outer) which may finally lead to the destruction or recycling of the left out siderophore (Rajkumar et al. 2010). Thus, the siderophores prove to be excellent iron solubilizing agents from minerals and other inorganic sources. Pseudomonads, the bacterial genera, are the best-known secretors of siderophores playing an important role in the overall plant growth promotion activities (Sandilya et al. 2017).

Table 1.4 Phytohormones produced by microbiomes

Phytohormone produced	Plant growth-promoting rhizobacteria	Agriculture crop	References
Indole-3-acetic acid (IAA)	*Aeromonas veronii*	*Oryza sativa*	Mehnaz et al. (2001)
	Azospirillum brasilense	*Triticum aestivum* L.	Kaushik et al. (2000)
	Enterobacter sp.	*Saccharum officinarum*	Mirza et al. (2001)
	Enterobacter cloacae	*Oryza sativa*	Mehnaz et al. (2001)
	Pseudomonas sp. RJ10	*Brassica napus*	Sheng and Xia (2006)
	Bacillus sp. RJ16	*Brassica napus*	Sheng and Xia (2006)
	Enterobacter sp.	*Cicer arietinum* L.	Fierro-Coronado et al. (2014)
	Pseudomonas sp, *Bacillus* sp.	*Sulla carnosa*	Hidri et al. (2016)
	Bacillus licheniformis	*Triticum aestivum* L.	Singh and Jha (2016)
	Bacillus subtilis	*Acacia gerrardii* Benth	Hashem et al. (2016)
	Pseudomonas sp.	*Zea mays*	Mishra et al. (2017)
	Enterobacter sp. C1D	*Vigna radiata* L.	Subrahmanyam and Archana (2011)
	Proteus vulgaris JBLS202	*Arabidopsis thaliana*	Bhattacharyya et al. (2015)
Cytokinin	*Pseudomonas fluorescens*	Glycine max L.	De Salamone et al. (2001)
	Pseudomonas fluorescens	*Pinus* sp.	Bent et al. (2001)
	Paenibacillus polymyxa	*Triticum aestivum* L.	Timmusk et al. (1999)
	Micrococcus luteus	*Zea mays*	Raza and Faisal (2013)
	Bacillus subtilis	*Platycladus orientalis*	Liu et al. (2013)
	Arthrobacter sp., *Bacillus* sp., *Azospirillum* sp.	*Glycine max* L.	Naz et al. (2009)
	Proteus vulgaris JBLS202	*Arabidopsis thaliana*	Bhattacharyya et al. (2015)
Gibberellin	*Bacillus* sp.	*Alnus* sp.	Gutierrez-Manero et al. (2001)

(continued)

Table 1.4 (continued)

Phytohormone produced	Plant growth-promoting rhizobacteria	Agriculture crop	References
	Aspergillus fumigatus	*Glycine max* L.	Khan et al. (2011)
	Azospirillum lipoferum	*Triticum aestivum* L.	Creus et al. (2004)
	Phoma glomerata, *Penicillium* sp.	*Cucumis sativus*	Waqas et al. (2012)
	Proteus vulgaris JBLS202	*Arabidopsis thaliana*	Bhattacharyya et al. (2015)
ACC deaminase	*Enterobacter cloacae*	*Brassica napus*	Saleh and Glick (2001)
	Pseudomonas putida	*Vigna radiata* L.	Mayak et al. (1999)
	Pseudomonas sp.	*Zea mays* L.	Shaharoona et al. (2006)
	Methylobacterium fujisawaense	*Brassica* sp.	Madhaiyan et al. (2006)
	Rhizobium leguminosarum	*Pisum sativum*	Ma et al. (2003)
	Achromobacter xylosoxidans, *Acidovorax facilis*	*Brassica juncea* L. Czern	Belimov et al. (2005)
	Bacillus, *Microbacterium*, *Methylophaga*, *Agromyces*	*Oryza sativa*	Bal et al. (2013)
	Enterobacter sakazakii 8MR5, *Pseudomonas* sp. 4MKS8, *Klebsiella oxytoca* 10MKR7	*Zea mays* L.	Babalola et al. (2003)
	Methylobacterium fujisawaense	*Brassica campestri*	Madhaiyan et al. (2006)
	Enterobacter sp. C1D	*Vigna radiata* L.	Subrahmanyam et al. (2018)

1.5.2 Indirect Mechanisms

Plant growth-promoting rhizobacteria has been implemented in various crop fields for their promising capability to work both as biocontrol agents and growth promoters since last two decades. Bacteria secretes various metabolites and chemical agents that makes them wonderful candidates for controlling different crop diseases most of them being originated from fungal sources. According to Bhattacharyya and Jha (2012), PGPRs are able to synthesize different antifungal secondary metabolites such as phenazines, HCN, pyrrolnitrin, 2, 4-diacetylphloroglucinol, viscosinamide, tensin, and pyoluteorin. Availability of bacterial antagonist in the rhizosphere soil may even adapt the plant for developing induced systemic resistance against broad-spectrum bacterial, fungal, and viral pathogens (Lugtenberg and Kamilova 2009). Cyanide is the most dangerous chemical known for its high toxic properties which can well inhibit the pathogens sensitizing agricultural crops. HCN being the secondary metabolite secreted by the PGPRs does not have any pessimistic effect on the host plants, and hence they are frequently used for controlling weeds (Zeller et al. 2007). According to various reports, HCN-producing PGPRs are very helpful in controlling dreaded phytopathogens such as *Pythium ultimum*, *Fusarium oxysporum*, and pathogenic *Agrobacterium*. The mode of action mechanisms involves lysis of fungal cell walls (Maksimov et al. 2011), root colonization (Kamilova et al. 2005), reduction of stress ethylene level (Van Loon 2007), siderophore and antibiotic production (Beneduzi et al. 2012).

Certain genera like *Bacillus* have been best studied for their ability to secrete antimicrobial traits with higher rate of agricultural applicability (Compant et al. 2005). The members of this group of bacteria hold a key role in biocontrol aspects as they could reluctantly replicate at a very faster rate and are mostly resistant to environmental stress (Shafi et al. 2017). They secrete bacillomycins, iturins, and mycosubtilin very much effective against fungal pathogens, particularly *Aspergillus flavus* (Gong et al. 2015). Similarly, Lee et al. (2015) reported almost 99.1% of the antagonistic success in crops fields inoculated with *Bacillus amyloliquefaciens* strain HK34 against *Phytophthora cactorum* in *Lycopersicum esculentum*, *Sclerotium rolfsii*, *Capsicum annuum* var. acuminatum, *Colletotrichum gloeosporioides*, and *Cucumis sativus*.

Apart from that, other bacterial genera like *Pseudomonas* and *Paenibacillus* have also been reported by various authors having antimicrobial properties in both in vitro and in vivo conditions. Although laboratory results may not always be relied under field conditions, PGPR has been reported to be effective in both the conditions in different agricultural cropping systems. That is why they may be termed as multifunctional agents by controlling a wide spectrum range of phytopathogens and a spectacular replacement for chemical fertilizers by enhancing plant growth and overall yield per hectares of cultivated soil further playing a vital role in maintaining ecological balance across the globe (Ahemad and Kibret 2014).

Although much work has been done on the biocontrol characteristics of rhizospheric bacteria, it has to be considered that soil type, plant species, and pathogen affect in rhizosphere competence and/or biocontrol efficiency of applied biocontrol strain against a soil-borne pathogen.

1.6 Biotechnological Applications of Rhizosphere Microbiomes

In the recent past, sustainable technologies have gained lot of momentum to improve quality and yield of agricultural crop production. Nevertheless, still there is uncertainty about success of chemical-based formulations in plant protection management. In general, pests and diseases are mainly controlled by chemical-based pesticides which pose major health risks as well as adverse negative impacts in the ecosystem and environment. In addition to this, indiscriminate use of chemical fertilizers resulted in negative impacts on biodiversity and function of biogeochemical cycles. Most importantly, agricultural practices require novel products according to the demand of farmers and consumers. Therefore, alternative management tools have to be developed on the basis of biological solutions.

The plant rhizosphere hosts a considerable amount of microbiome. Plant growth-promoting rhizobacteria (PGPR) is an integral component of rhizosphere microbiome and is competent to promote plant growth by direct and indirect mechanisms. PGPR also promotes defense against diseases causing organisms using diverse plant-beneficial functions. Therefore, it is anticipated that crop inoculation with suitable PGPR could reduce the use of pesticides and fertilizers in agrosystems. Biotechnological applications of various PGPR inoculants for enhancing crop production were summarized in Table 1.5. Since most of the research information on PGPR comes from rhizosphere microbiome, one can further explore and exploit biotechnical prospects of rhizosphere microbiomes for sustainable agricultural production. We have specially highlighted the production of extracellular lytic enzymes, bioactive metabolites, and volatile organic compounds (VOCs) of rhizosphere bacteria in this section and the details are given extensively in Table 1.6.

1.6.1 Production of Lytic Enzymes by Rhizospheric Bacteria

Rhizosphere bacteria can benefit plant growth indirectly through biocontrol mechanisms which can inhibit the growth and colonization of phytopathogens. This potential antagonism character of biocontrol agent might occur through different mechanisms which include production of extracellular lytic enzymes, secondary metabolites, siderophores, antibiotics, and induction of systemic responses (Saraf et al. 2014, Jadhav and Sayyed 2016; Kour et al. 2019a; Yadav et al. 2016a, 2019f). One of the important mechanisms for biocontrol agent is the production of lytic enzymes which are able to degrade the membrane constituents of phytopathogens, such as proteases (Felestrino et al. 2018), acylases, and lactonases (Combes-Meynet et al. 2011). These hydrolytic enzymes degrade the structural integrity of the pathogen cell wall. Their ability to inhibit phytopathogens makes them to be the preferable choice in biological control process. The application of

Table 1.5 Biotechnological applications of bio-inoculants for sustainable agriculture

SN	Crop	Organism/Biofertilizers	Mode of action	References
1	Canola and lettuce	*Rhizobium leguminosarum*	Early development, Growth promotion	Sneha et al. (2018), Abd El-Lattief (2016)
2	Wheat, oat, barley mustard, seasum, rice, linseeds, sunflower, castor, maize, sorghum, cotton, jute, sugar beets, tobacco, tea, coffee, rubber, and coconuts	*A. chroococcum, A.vinelandii, A. beijerinckii, A. nigricans, A. armeniacus, and A. paspali.*	Nitrogen fixation, produce thiamine and riboflavin, indole acetic acid (IAA), gibberellins (GA) and cytokinins (CK), improves the plant growth by enhancing seed germination and advancing the root architecture, inhibiting pathogenic microorganisms around the root systems of crop plants	Revillas et al. (2000), Abd El-Fattah et al. (2013), Gholami et al. (2009), Mali and Bodhankar (2009), Wani et al. (2013), Bhardwaj et al. (2014)
3	Chickpea lentil, pea, alfalfa and sugar beet rhizosphere, berseem, ground nut and soybean *Cicerarietinum* and *Tigonella foenum-gracecum*	*Rhizobium* inoculants	Increase the grain yields	Patil and Medhane, (1974), Rashid et al. (2012), Ramachandran et al. (2011), Hussain et al. (2002), Grossman et al. (2011), Sharma et al. (2011, 2012a, b), Kumar et al. (2013)
4	Tomato	*Pseudomonas putida*	Early developments, Growth stimulation	Sneha et al. (2018), Abd El-Lattief (2016)
5	Wheat and maize	*Azospirillumbrasilense and A. irakense*	Growth of wheat and maize plants by secrete gibberellins, ethylene, and auxins	Abd El-Lattief (2016); Perrig et al. (2007), Bhardwaj et al. (2014), Sneha et al. (2018)

(continued)

Table 1.5 (continued)

SN	Crop	Organism/Biofertilizers	Mode of action	References
6	Banana Pearl millet	*P. flourescens*	Enhance growth, leaf nutrient contents, and yield	Sneha et al. (2018), Abd El-Lattief (2016)
7	Rice	Cylindrospermum musicola	Nitrogen fixation, liberation of growth-promoting substances and vitamins; increase the root growth	Venkataraman and Neelakantan (1967)
8	Canola	*Azotobacter* and *Azospirillum* spp.	Growth and productivity	Sneha et al. (2018), Abd El-Lattief (2016)
9	Maize crop.	*P. alcaligenes*, *Bacillus polymyxa*, and *Mycobacterium phlei*	Improves the uptake of N, P, and K	Sneha et al. (2018), Abd El-Lattief (2016)
10	Chick pea.	*Pseudomonas*, *Azotobacter*, and *Azospirillum* spp.	Stimulates growth and increases the yield	Sneha et al. (2018), Abd El-Lattief (2016)
11	Wheat	*R. leguminismarum and Pseudomonas* spp.	Enhances the yield and phosphorus uptake	Sneha et al. (2018) Abd El-Lattief (2016)
12	Maize.	*P. putida, P. fluorescens, A. brasilense,* and *A.lipoferum*	Enhances seed germination, seedling growth, and yield	Sneha et al. (2018), Abd El-Lattief (2016)
13	Wheat, maize, and rice	*Azotobacter Azotobacte*r *Alcaligenes, Azospirillum, Bacillus, Enterobacter, Herbaspirillum, Klebsiella, Pseudomonas,* and *Rhizobium*	Improves growth and grain yield	Sridhar (2012), James (2000)
14	Wheat maize, and rice	*Azospirillum*	Synthesis of phytohormones (indole-3-acetic acid, IAA), and regulation of plant hormonal balance by deamination of the ethylene precursor	Abd El-Lattief (2016)

(continued)

Table 1.5 (continued)

SN	Crop	Organism/Biofertilizers	Mode of action	References
15	Rice	*Alcaligenes*, *Azospirillum*, *Bacillus*, *Herba spirillum*, *Klebsiella*, *Pseudomonas*, and *Rhizobium*	N-fixation	Sneha et al. (2018)
16	Chickpea	Co-inoculation (*Pseudomonas* + *Bacillus* strains + effective *Rhizobium* spp.)	Stimulate chickpea growth, nodulation, and nitrogen fixation	Sneha et al. (2018)
17	Chickpea	Co-inoculation (PSB + *Rhizobium* + *Trichoderma)*	Increase sugar, protein, starch contents, nodule weight and seed nitrogen, potassium, phosphorus	Mohammadi (2010, 2011)
18	Rice	Green manure and biofertilizer	Stimulated the growth of plants with more number of tillers and broader leaves, increased leaf area	Shanmugam and Veeraputhran (2000)

these hydrolytic enzymes from rhizospheric origin is a viae solution as they are totally natural and are eco-friendly in nature (Mishra et al. 2019).

Lytic enzymes produced by various microorganisms can hydrolyze polymeric compounds like cellulose, hemicellulose, chitin, and protein of phytopathogens. Extracellular hydrolytic enzymes like chitinases, lipases, proteases, and glucanases are involved in the lysis of fungal cell wall (Neeraja et al. 2010). These enzymes either disintegrate or digest the molecular components of cell wall of fungal phytopathogens. Therefore, this process would be considered as eco-friendly control of soil-borne pathogens in agriculture crops. These enzymes further involve in nutrient cycling by decomposition of organic matter and plant residues in the rhizosphere. It is demonstrated that extracellular lytic enzymes produced by *Myxobacteria* sp. have the ability to suppress fungal plant pathogens (Bull et al. 2002). In an another study, glucanase-producing antagonistic bacteria *Lysobacter sp.* is capable of controlling diseases of *Pythium* sp. *and Bipolaris* sp. (Palumbo 2005). These hydrolytic enzymes rescue plants from biotic stresses and directly contribute in the parasitization of phytopathogens.

Hydrolytic enzymes of rhizospheric microbes were reviewed extensively by Jadhav and Sayyed 2016. Many rhizobacterial microbial species are capable of producing extracellular enzymes and effectively hydrolyze wide variety of polymeric substances like cellulose, hemicellulose, proteins, and chitin of phytopathogens (Jadhav and Sayyed 2016). Microbial strains like *B. subtilis* strains PCL1608 PCL1612, Streptomyces cyaneofuscatus B-49, *Serratia marcescens* strain ETR17,

Table 1.6 Biotechnological prospects of different rhizobacteria suitable for enhanced production of agricultural crops

Biotechnological prospects	Rhizospheric organism	Crop	Growth parameter	References
Hydrolytic enzymes				
β-1,3-glucanase and β-1,4-glucanase activities, Chitinase, oxalate oxidase	*Bacillus sp.*	Bean	Biocontrol	Kumar et al. (2012)
Glucanases, protease	*B. subtilis* strains, PCL1608 and PCL1612	Avocado	Biocontrol	Cazorla et al. (2007)
Chitinase	*Bacillus subtilis* BP 9	Cotton	Biocontrol	Gajbhiye et al. (2010)
Chitinase, β-1,3-glucosidase, cellulose,. Protease	*Streptomyces cyaneofuscatus* B-49, *Streptomyces kanamyceticu* X-4, *Streptomyces rochei* Z-13, *Streptomyces flavotricin*	Cotton, chili pepper, canola	Biocontrol	Xue et al. (2013)
Glucanases, proteases or chitinases,	*B. megaterium B5*, *B. cereus* sensulato *B25*, and *Bacillus* sp. *B35*	Maize	Biocontrol	Ramírez et al. (2016)
β-1,3-, β-1,4-, and β-1,6-glucanases	*Actinoplanes philippinensis*, *Microbispora rosea*, *Micromonospora chalcea*, and *Streptomyces griseoloalbus*	Cucumber	Biocontrol	El-Tarabily (2006)
Chitinase, β-1,3 glucanase (laminarinase) and β-1,4 glucanase (cellulase)	*Pseudomonas fluorescens* and *Bacillus subtilis*	Tomato	Biocontrol	El-Gamal et al. (2016)

(continued)

Table 1.6 (continued)

Biotechnological prospects	Rhizospheric organism	Crop	Growth parameter	References
chitinase, protease, lipase, cellulase	*Serratia marcescens* strain ETR17	Tea	Biocontrol	Purkayastha et al. (2018)
Extracellular proteases and lipases	*Lysobacter enzymogenes*	Cucumber	Biocontrol	Folman et al. (2003)
Antibiotics/secondary metabolites				
DAPG and HCN	*P. fluorescens strains CHA0, Pf-5, Q2-87,*	Tobacco	Biocontrol	Weller (2007)
DAPG and HCN	*P. fluorescens Pf-5*	Cotton	Biocontrol	Nowak-Thompson et al. (1994)
DAPG and HCN	*P. fluorescens*	Rice	Biocontrol	Reddy et al. (2007)
DAPG and HCN	*P. fluorescens Q2-87*	Wheat	Biocontrol	Harrison et al. (1993)
HCN	*Bacillus* sp. BPR7	Bean	Biocontrol	Kumar et al. (2012)
HCN	*Bacillus subtilis BP-9*	Cotton	Biocontrol	Gajbhiye et al. (2010)
Bacilysin and iturin	*Bacillus subtilis* ME488	Cucumber and pepper	Biocontrol	Chung et al. (2008)
Pyrrolnitrin and prodigiosin	*Serratia marcescens* strain ETR17	Tea	Biocontrol	Purkayastha et al. (2018)
Lipopeptides surfactin, fengycin, and/or iturin A	*B. subtilis* strains, PCL1608 and PCL1612	Avocado	Biocontrol	Cazorla et al. (2007)
Viscosinamide, cyclic lipopeptides	*P. fluorescens* DR54	Cotton	Biocontrol	Nielsen et al. (1999)
Antifungal protein	*Bacillus* sp. KM 5	Rice	Biocontrol	Majumdar et al. (2011)
polymyxin B1	*Paenibacillus* sp. strain B2	sorghum	Biocontrol	Selim et al. (2005)

(continued)

Table 1.6 (continued)

Biotechnological prospects	Rhizospheric organism	Crop	Growth parameter	References
Eicosane ($C_{20}H_{42}$) and dibutyl phthalate ($C_{16}H_{22}O_4$), cyclononasiloxane, octadecamethyl-($C_{18}H_{54}O_9Si_9$) and benzoic acid, 2,5-bis(trimethylsiloxy) ($C_{16}H_{30}O_4Si_3$)	*Streptomyces* strain KX852460	Tobacco	Biocontrol	Ashan et al. (2017)
Fungichromin	*Streptomyces padanus* strain PMS-702 i	Cabbage	Biocontrol	Shih et al. (2003)
Pyrrolnitrin	*P. fluorescens* BL915	Wheat	Biocontrol	Chin-A-Woeng et al. (2003)
Oomycin A	*P. fluorescens*	Cotton	Biocontrol	Howie and Suslow (1991)
Phenazine-1-carboxylic acid	*P. fluorescens*	Wheat		Pierson and Pierson (1996)
Volatile organic compounds (VOCs)				
VOCs	*Bacillus subtilis* GB03	Arabidopsis	Vegetative growth promotion, elevated photosynthetic capacity, and iron accumulation	Xie et al. (2009)
VOCs	*Bacillus subtilis* GB03	Arabidopsis	Vegetative growth promotion, plant acquisition of iron, and increased photosynthetic capacity	Zhnag et al. (2009)
2-pentylfuran, aldehydes, alkanes, ketones, and aroma components	*Bacillus megaterium*	Arabidopsis	Fresh weight	Zou et al. (2010)

(continued)

Table 1.6 (continued)

Biotechnological prospects	Rhizospheric organism	Crop	Growth parameter	References
4-nitroguaiacol and quinoline	*Pseudomonas simiae*	Soybean	Salt tolerance, increased shoot and root length, fresh weight and leaf surface area	Vaishnav et al. (2016)
13-Tetradecadien-1-ol, 2-butanone and 2-Methyl-n-1-tridecene	*Pseudomonas fluorescens* SS10	Tobacco	Fresh weight, dry weight	Park et al. (2015)
Dimethylhexadecylamine	*Arthrobacter agilis* UMCV2	*Sorghum*	Fresh weight, shoot length, chlorophyll concentration, and lateral root number	Castulo-Rubio et al. (2015)
Brassinosteroid	*Proteus vulgaris* JBLS202	Arabidopsis	Fresh weight	Bhattacharyya et al. (2015)
Long-chain VOC signaling molecules, acetoin 2,3-butanediol, ethanethiol, isoprene, and acetic acid-butyl ester, tridecane, a C13 hydrocarbon compound	*Pseudomonas fluorescens* 89B-61, *Bacillus amyloliqefaciens* IN937a, *Bacillus subtilis* GB03, and *Paenibacillus polymyxa* E681	*Arabidopsis*	Induced systemic resistance	Lee et al. (2012a)
β-Caryophyllene	*Fusarium oxysporum* MSA 35	Lettuce	Root length, fresh weight, chlorophyll content	Minerdi et al. (2011)
Sesquiterpenes	*Laccaria bicolor*	*Arabidopsi and Populus*	Lateral root development	Ditengou et al. (2015)
α-pinene, (-)-trans-caryophyllene, tetrahydro-2,2,5,5-tetramethylfuran, dehydroaromadendrene, and (+)-sativene	*Cladosporium cladosporioides* CL-1	Tobacco	Fresh weight	Paul and Park (2013)
2,3-butanediol	*Bacillus subtilis*	Pepper	Plant defences/resistance	Yi et al. (2016)

Table 1.7 Mode of action of different lytic enzymes produced by rhizobacteria

S.No	Extracellular lytic enzyme	Mode of action
1	Chitinase	Degradation of chitin involves breakdown of chitin polymer into monomer, random cleavage at internal sites of chitin micro-fibril or progressive release of diacetylchitobiose in a stepwise manner without releasing monosaccharide or oligosaccharides
2	Glucanase	Glucanase can hydrolyze the substrate by sequentially cleaving glucose residues from non-reducing end. The enzyme can also initiate cleaving linkages at random sites along the polysaccharide chain, releasing smaller oligosaccharides
3	Protease	Protease can hydrolyze major proteins of phytopathogenes into small peptide chains, subsequently release their constituent amino acids and thereby destroy capacity of phytopathogen's protein to act on plant cells. Some of the proteases are capable of inactivating extracellular enzymes produced by phytopathogenic fungi
4	Cellulase	Cellulases hydrolyze the β-(1,4) glucosidic linkages in cellulose polymer and play a significant role in recycling this polysaccharide in the rhizosphere. Cellulose chains are composed of complex, rigid, insoluble, crystalline microfibrils. Therefore, complete degradation of cellulose involves a complex interaction between different cellulolytic enzymes such as cellulose/endoglucanases, exo-cellobiohydrolase/exo-glucanases, and β-glucosidases act synergistically to convert cellulose into glucose molecules

Pseudomonas fluorescens, *Serratia marcescens* strain ETR17, and many other antagonistic microbes have a potential to synthesize hydrolytic enzymes for the biocontrol of fungal phytopathogens like *P. ultimum, F. oxysporum, R. solani,* and *S. rolfsii,* (Cazorla et al. 2007; Kumar et al. 2012a, b; Purkayastha et al. 2018, El-Gamal et al. 2016). The mode of actions of extracellular enzymes is given in Table 1.7.

Chitinolytic microorganisms are heavily colonized in plant rhizosphere among which actinobacteria are the most abundant members (Yadav et al. 2018c). Actinobacteria such as *Streptomyces* flavotricini, *Streptomyces kanamyceticu, Streptomyces cyaneofuscatus*, and *Streptomyces rochei* produce chitinases and inhibit the growth of phytopathogen, viz, *Verticillium dahlia* in cotton rhizosphere (Xue et al. 2013). Chitinase-producing *Bacillus thuringiensis* spp. *colmeri* can inhibit the growth of plant pathogenic fungi, including *Rhizoctonia solani*, *Penicillium chrysogenum*, and *Physalospora piricola* (Liu et al. 2010). Biocontrol agent *Bacillus subtilis* inhibits the growth of pathogenic fungi *Fusarium oxysporum* through production of extracellular chitinase (Gajbhiye et al. 2010). Chitinases produced by *Brevibacillus laterosporus* effectively inhibit the growth of phytopathogenic fungi *Fusarium equiseti* (Prasanna et al. 2013). *Lysobacter enzymogenes* showed to inhibit *Pythium aphanidermatum* by producing extracellular protease and lipases (Folman et al. 2003)

Minimal use of chitinase-based fungicides in agriculture crops was associated with the perception that their efficacy will be slowly reduced in the soil environment. Nevertheless, Dahiya et al. (2006) extensively reviewed biotechnological prospects of chitinolytic enzymes and suggested that chitinases can be used as supplementary inputs along with other chemical-based fungicides to enhance their effectiveness against phytopathogenic fungi and reduce the required amount of chemical fungicides. In addition to this, it was shown that the application of mixed consortia containing two different chitinolytic bacteria is more effective in controlling the pathogen. Application of chitinase-producing *Streptomyces* sp. 385, *Paenibacillus* sp. 300, and both together is more effective in controlling cucumber wilt caused by *F. oxysporum* than individual strains applied (Singh et al. 1999). Similar kind of observation was reported by El-Tarabily et al. (2000) wherein growth of fungal pathogen *Sclerotinia* responsible for vegetable rot was effectively controlled by combination of *S marcescens, Streptomyces viridodiasticus,* and *Micromonospora carbonacea* strains. In recent, chitinase, protease, lipase, and cellulose-producing *Serratia marcescens* strain ETR17 showed in vitro antagonism toward nine different root and foliar pathogens of tea (Purkayastha et al. 2018).

Actinomycetes were considered to be strong biocontrol agents against fungal pathogens. This is mainly due to production of different types of antifungal compounds such as antibiotics and extracellular hydrolytic enzymes which includes chitinases and glucanases (Xue et al. 2013; Yadav et al. 2018c). *Streptomyces halstedii, Streptomyces cavourensis* SY224, and *Streptomyces griseus* are known to produce potential antifungal extracellular chitinases, which makes them to be used as biocontrol agents in crop protection strategies (Ki et al. 2012; Gherbawy et al. 2012). *Lysobacter* spp. was reported to be an effective biocontrol agent against soilborne pathogens through production of extracellular enzymes and other metabolites (Folman et al. 2003). *Lysobacter* spp. was abundant in the soil which is suppressive to root pathogen, viz., *Rhizoctonia solani*. Certain antagonistic strains showed in vitro biocontrol activity against *Xanthomonas campestris, R. solani*, and other important phytopathogens such as *Aspergillus niger, Fusarium oxysporum,* and *Pythium ultimum.*

These natural microbial biofungicides will be used as integrated pest management supplement for reduction of negative impact of chemical pesticides on the environment and maintain the sustainable production of agriculture.

1.6.2 Production of Antibiotics

Rhizospheric bacteria produce distinct antimicrobial products to inhibit the growth and colonization of plant pathogens to compete the nutrients present in the rhizosphere. This has become a beneficial trait to the host plant as disease development is significantly reduced by PGPR. Rhizosphere harbors diverse actinomycetes species which have been further exploited for secondary metabolites (Yadav et al. 2018b; Geetanjali and Jain 2016). Actinobacteria is known to produce wide variety of natural antimicrobial products (approximately 10,000 secondary metabolites) (Passari et al. 2015, 2017; Yadav et al. 2018a, b). Production of antibiotics by Actinobacteria was extensively reviewed by Yadav et al. (2018b, c). Application of secondary metabolites producing rhizobacterial isolates against phytopathogens is increasing over the past decade (Yilmaz et al. 2008). A variety of antimicrobial agents such as 2,4-diacetylphloroglucinol (DAPG), pyoluteorin (PRN), phenazine, cyclic lipopeptides, tensin, and pyrrolnitrin (PLT) have been screened and identified from *Pseudomonas* sp., *Arthrobacter* sp., and *Streptomyces* sp., (Weller 2007; Gupta et al. 2015). Details of antibiotics/secondary metabolites producing organisms and their application in different crops have been summarized in Table 1.6.

Rhizospheric soil isolates *Bacillus* sp. S2 and *Pseudomonas fluorescens* S5 were found to exert good antimicrobial activity against multi-drug-resistant clinical pathogens such as *Pseudomonas aeruginosa, Klebsiella pneumonia, Escherichia coli,* and *Staphylococcus aureus* obtained from different samples (Dhore et al. 2014). Thirty *Pseudomonas fluorescens* strains isolated from rice rhizosphere against pathogenic fungi *Sarocladium oryzae, Dreschelaria oryzae, Magnaporthe grisea,* and *Rhizoctonia solani.* Among these, *P. fluorescens* Pf 003 effectively inhibited (62–85%) the mycelial growth in all the pathogenic fungi in dual culture. The antifungal compounds extracted with ethyl acetate from *P. fluorescens* at 5% completely inhibited the pathogens (Reddy et al. 2007). Walia et al. (2013) isolated the bacteria from the tomato rhizosphere for having broad-spectrum antifungal activity against *Sclerotinia sclerotiorum, Rhizoctonia solani,* and *Fusarium oxysporum.*

DAPG, phenazines, PLT, and PRN are considered to be potent antibiotics synthesized by *Pseudomonas* biocontrol agents affiliated to gammaproteobacteria (Table 1.6). In recent, antibiotics-producing *Pseudomonas spp.* has got much attention in biocontrol research, and corresponding genes involved in the expression and regulation of these metabolites are now fully understood (Weller 2007and there in references). For the last 30 years, developments on biocontrol applications of *Pseudomonas* sp. against soil-borne pathogens have been summarized by Weller (2007). *P. fluorescens* strain CHA0 was isolated from tobacco rhizosphere which is naturally suppressive to black root rot of tobacco caused by *Thielaviopsis basicola* (Stutz et al. 1986). *P. fluorescens* CHA0 produces siderophore (pseudobactin), PLT, DAPG, PRN, HCN, salicylic acid, pyoverdine, indoleacetic acid, pyochelin, and other secondary metabolites (Voisard et al. 1994). Antagonistic bacterium *P. fluorescens* F113 isolated from sugar beet was applied in the field for suppression of damping-off of sugar beet infection caused by a pathogen *Pythium ultimum* (Cronin et al. 1997a, b).

Antibiotics such as bacilysin- and iturin-producing *Bacillus subtilis* ME488 suppressed soil-borne pathogens in pepper and cucumber crops (Chung et al. 2008). Secondary metabolites, viz., Pyrrolnitrin and prodigiosin-producing *Serratia marcescens* strain ETR17 *Serratia marcescens* strain ETR17 showed significant level of in vitro antagonistic property against different root and foliar pathogens of tea (Purkayastha et al. 2018). Antifungal lipopeptides such as surfactin-, fengycin-, and iturin-producing *B. subtilis* strains PCL1608 and PCL1612 have shown biocontrol mechanism toward soil-borne pathogen *Fusarium oxysporum* (Cazorla et al. 2007). *Paenibacillus* sp. strain B2 isolated sorghum mycorrhizosphere showed production of antibiotic polymyxin B1 and significantly inhibited the growth of fungal pathogens (Selim et al. 2005). Antifungal peptides-producing *Bacillus* sp. KM 5 isolated from rice rhizosphere showed antagonist activity toward pathogenic fungi *Gibberella fujikuroi, Sclerotium rolfsii* Saccardo, *Fusarium udum, Helminthosporium oryzae, and Rhizoctonia solani* Nees (Majumdar et al. 2011).

1.6.3 Production of Volatile Organic Compounds (VOCs)

Volatile organic compounds are lipophilic low molecular weight (<300 g mol − 1) compounds emitted from microbial metabolic pathways with high vapor pressure and low boiling point. VOCs can act as signal molecules in rhizosphere over short and long distances (Fincheira and Quiroz 2018). It is evidenced that VOCs released from diverse rhizospheric microorganisms, e.g., *Arthrobacter* sp., *Proteus* sp., *Bacillus* sp., *Fusarium* sp., *Pseudomonas* sp., *Alternaria* sp. and *Laccaria* sp., can promote plant growth on a specific "target". Detailed description about chemical nature of VOCs and their functions have been summarized in Table 1.6. Ryu et al. (2003) reported for the first time about the mechanism mediated by volatile organic compounds released by *Bacillus subtilis* GB03 which induced growth on *Arabidopsis thaliana*. This study evidenced that VOCs can modulate stress, growth, nutrition, and health processes in host plants. Some identified VOCs compounds, such as acetoin, β-Caryophyllene 2,3-butanediol, Sesquiterpenes, 2-pentylfuran, and dimethylhexadecylamine, have shown their ability to elicit plant growth at above and below ground biomass (Fincheira and Quiroz 2018; Chung et al. 2016) (Table 1.6).

Few studies indicate that VOCs act as signals and chemical messengers to regulate phytohormone synthesis, metabolic pathways, and nutrition levels. Effects of VOCs for induction of resistance and tolerance in plants are documented, wherein compounds such as 3-pentanol, dimethyl disulfide, 6-pentyl-α-pyrone, and acetoin were reported. VOCs derived by rhizospheric bacteria showed antagonistic activity toward plant pathogen *Rhizoctonia solani* and inhibit mycelial growth (Kai et al. 2007). Certain plant volatiles are proven to induce plant growth promotion through biochemical signals, eliciting local defence reactions known as induced systemic resistance (Chung et al. 2016; Kai et al. 2007). Long-chain VOCs signaling molecules, acetoin 2,3-butanediol, ethanethiol, isoprene, and acetic acid-butyl ester, and tridecane are found to be involved in induced resistance in *Arabidopsis* (Lee et al. 2012a, b). Yi et al. (2016) reported that 2,3-butanediol is produced by a *Bacillus subtilis* isolate involved in plant defense mechanisms. Root exudates of pepper inoculated with the *B. subtilis* were used to challenge various phytopathogens. For example, growth of *Trichoderma* sp (saprophytic fungus) and *Ralstonia solanacearum* (soil-borne pathogen) was inhibited by VOCs. This indicates that VOCs triggered the secretion of root exudates and subsequently acted as a plant defence inducer toward soil-borne fungal and bacterial pathogens.

Volatile organic compounds *such as* dehydroaromadendrene, α-pinene, tetrahydro-2,2,5,5-tetramethylfuran, (-)-trans-caryophyllene, and (+)-sativene-producing *Cladosporium cladosporioides* strain CL-1 showed increased growth parameters in Tobacco crop (Paul and Park 2013). In an another study, rhizospheric isolates such as *Bacillus subtilis* GB03, *Bacillus amyloliquefaciens* IN937, *Pseudomonas fluorescens* 89B-61, and *Paenibacillus polymyxa* E681 produced Brassinosteroid **a** long-chain VOC and signaling molecules such as acetoin 2,3-butanediol, ethanethiol, acetic acid-butyl ester, and isoprene. These VOCs are involved in induced systemic resistance in *Arabidopsis (*Lee et al. 2012a, b). Fresh weight, shoot length, chlorophyll concentration, and lateral root numbers of *Sorghum were significantly increased by dimethylhexadecylamine produced by Arthrobacter agilis* UMCV2. Salt tolerance, increased shoot and root length, fresh weight, and leaf surface area were increased in soybean by VOCs, 4-nitroguaiacol, and quinoline produced by Vaishnav et al. (2016). VOCs of fungal origin also showed increased growth parameters in host plants like lettuce, Arabidopsis, and tobacco. *Fusarium oxysporum* MSA 35 showed production of β-Caryophyllene and increased fresh weight of tobacco in field experiment (Minerdi et al. 2011). Sesquiterpenes synthesized by ectomycorrhizal fungi *Laccaria bicolour* increased the lateral root of *Arabidopsis* (Ditengou et al. 2015). In the same study, it was demonstrated that other ectomycorrhizal ascomycote, *Cenococcum geophilum*, which cannot synthesize Sesquiterpenes does not promote lateral root of *Arabidopsis.* These studies indicate that volatile organic compounds emitted by microorganisms in the rhizosphere are cheaper, effective, efficient, and eco-friendly alternatives for controlling phytopathogens.

Environmentally friendly biotechnological approaches offer the development of PGPR inoculants and their potential application in metal-contaminated systems. Plant growth promotion by PGPR is a result from improved nutrient acquisition or phytohormonal stimulation (Table 1.3). Different mechanisms involved in plant growth promotion were shown in Fig. 1.6. PGPR inoculants were widely used in agriculture, forestry, horticulture, and in environmental restoration/phytoremediation sectors.

1.7 Conclusion and Future Prospects

Although studies have focused on plant microbiome structure and its function under natural and agricultural environments, there have been no significant coordinated efforts to combine and translate research results into practical solutions for farmers. According to Busby et al. (2017), integration of beneficial plant microbiome into agricultural production is one of the ways to assist in achieving these goals. However, this requires large-scale efforts from academic and industry researchers, farmers, and policy-makers to understand and manage complex plant–microbiome interactions under current challenges of the agriculture production.

For achieving this goal, five key research priorities have been identified by Busby et al. (2017). Few research priorities include development of host–microbiome model systems with associated microbial culture collections and reference genomes; characterization and refinement of a model "plant genotype–environment stress–microbiome–management interactions"; elucidation of the role core microbiome and determine functional mechanisms of plant–microbiome interactions. These research priorities may enable us to manipulate agricultural microbiomes and thereby to develop management strategies for increased production and productivity of global agriculture in a sustainable manner. One of the challenges for future research work includes protection and conservation of rhizosphere biodiversity and their potential application in agricultural soils. Sustainable agriculture production may not be possible unless integration of plant germplasm and beneficial microbial species in the current agricultural practices globally.

Exploitation and production of natural drug formulations from microbial species have gained a significant leap during last three decades. Therapeutic applications of anticancerous compounds extracted from actinobacteria have been well addressed (Busi and Pattnaik 2018). The research priority is now shifted toward rhizosphere microbial communities for developing new drugs through high-throughput screening and fermentation techniques. Exploitation of bioprospecting potential of rhizosphere microbiomes is an upcoming new avenue.

Acknowledgements Authors are thankful to Director, CMER&TI, Central Silk Board, Lahdoigarh for valuable guidance and constant encouragement.

References

Abd El-Fattah DA, Ewedab WE, Zayed MS, Hassaneina MK (2013) Effect of carrier materials, sterilization method, and storage temperature on survival and biological activities of Azotobacter chroococcum inoculants. Ann Agric Sci 58:111–118

Abd El-Lattief EA (2016) Use of azospirillum and azobacter bacteria as biofertilizers in cereal crops: a review. Int J Res Eng Appl Sci 6(7):36–44

Ahemad M, Khan MS (2009) Effect of insecticide-tolerant and plant growth promoting Mesorhizobium on the performance of chick pea grown in insecticide stressed alluvial soils. J Crop Sci Biotechnol 12:213–222

Ahemad M, Khan MS (2010a) Ameliorative effects of Mesorhizobium sp. MRC4 on chickpea yield and yield components under different doses of herbicide stress. Pestic Biochem Physiol 98:183–190

Ahemad M, Khan MS (2010b) Improvement in the growth and symbiotic attributes of fungicide-stressed chickpea plants following plant growth promoting fungicide-tolerant Mesorhizobium inoculation. Afr J Basic Appl Sci 2:111–116

Ahemad M, Khan MS (2012) Effect of fungicides on plant growth promoting activities of phosphate solubilizing Pseudomonas putida isolated from mustard (Brassica campestris) rhizosphere. Chemosphere 86(9):945–950

Ahemad M, Kibret M (2014) Mechanisms and application of plant growth promoting rhizobacteria: Current perspective. J King Saud Univ Sci 26(1):1–20

Ahmed A, Hasnain S (2010) Auxin-producing Bacillus sp.: auxin quantification and effect on the growth of Solanum tuberosum. Pure Appl Chem 82(1):313–319

Ahsan T, Chen J, Zhao X, Irfan M, Wu Y (2017) Extraction and identification of bioactive compounds (eicosane and dibutyl phthalate) produced by Streptomyces strain KX852460 for the biological control of Rhizoctonia solani AG-3 strain KX852461 to control target spot disease in tobacco leaf. AMB Exp 7:54

Babalola OO, Osir EO, Sanni AI, Odhaimbo GD, Bulimo WD (2003) Amplification of 1-aminocyclopropane-1-carboxylic (ACC) deaminase from plant growth promoting rhizobacteria in Striga-infested soils. Afr J Biotechnol 2(6):157–160

Bal HB, Das S, Dangar TK, Adhya TK (2013) ACC deaminase and IAA producing growth promoting bacteria from the rhizosphere soil of tropical rice plants. J Basic Microbio 53(12):972–984
Belimov AA, Hontzeas N, Safronova VI, Demchinskaya SV, Piluzza G, Bullitta S, Glick BR (2005) Cadmium-tolerant plant growth-promoting bacteria associated with the roots of Indian mustard (Brassica juncea L. Czern.). Soil Biol Biochem 37:241–250
Beneduzi A, Ambrosini A, Passaglia LMP (2012) Plant growth-promoting rhizobacteria (PGPR): their potential as antagonists and biocontrol agents. Gen Mole Biol 35(4):1044–1051
Bent E, Tuzun S, Chanway CP, Enebak S (2001) Alterations in plant growth and in root hormone levels of lodgepole pines inoculated with rhizobacteria. Can J Microbiol 47(9):793–800
Bhardwaj D, Ansari MW, Sahoo RK, Tuteja N (2014) Biofertilizers function as key player in sustainable agriculture by improving soil fertility, plant tolerance and crop productivity. Micro Cell Fact 13:66
Bhattacharyya D, Garladinne M, Lee Y (2015) Volatile indole produced by rhizobacteriumProteus vulgaris JBLS202 stimulates growth of Arabidopsis thaliana through auxin, cytokinin, and brassinosteroid pathways. J Plant Growth Regul 34:158–168
Bhattacharyya PN, Jha DK (2012) Plant growth-promoting rhizobacteria (PGPR): emergence in agriculture. World J Microbiol Biotechnol 28:1327–1350
Bouffaud ML, Poirier MA, Muller D, Moënne-Loccoz Y (2014) Root microbiome relates to plant host evolution in maize and other P oaceae. Env Microbiol 16(9):2804–2814
Bulgarelli D, Garrido-Oter R, Münch PC, Weiman A, Dröge J, Pan Y, McHardy AC, Schulze-Lefert P (2015) Structure and function of the bacterial root microbiota in wild and domesticated barley. Cell Host Microbe 17(3):392–403
Bulgarelli D, Rott M, Schlaeppi K, van Themaat EVL, Ahmadinejad N, Assenza F, Rau P, Huettel B, Reinhardt R, Schmelzer E, Peplies J (2012) Revealing structure and assembly cues for Arabidopsis root-inhabiting bacterial microbiota. Nature 488(7409):91
Bull CT, Shetty KG, Subbarao KV (2002) Interactions between Myxobacteria, plant pathogenic fungi, and biocontrol agents. Plant Dis 86:889–896
Busby PE, Soman C, Wagner MR, Friesen ML, Kremer J, Bennett A, Morsy M, Eisen JA, Leach JE, Dangl JL (2017) Research priorities for harnessing plant microbiomes in sustainable agriculture. PLoS Biol 15(3):2001793
Busi S, Pattnaik SS (2018) Current status and applications of actinobacteria in the production of anticancerous compounds. In: New and future developments in microbial biotechnology and bioengineering, Elsevier, pp 137–153
CarvalhaisLC Dennis PG, Badri DV, Tyson GW, Vivanco JM, Schenk PM (2013) Activation of the jasmonic acid plant defence pathway alters the composition of rhizosphere bacterial communities. PLoS ONE 8(2):56457
Cassan F, Maiale S, Masciarellia O, Vidal A, Luna V, Ruiz O (2009) Cadaverine production by Azospirillum brasilense and its possible role in plant growth promotion and osmotic stress mitigation. Eur J Soil Biol 45:12–19
Castulo-Rubio DY, Alejandre-Ramírez NA, Orozco-Mosqueda MC, Santoyo G, Macías-Rodríguez L, Valencia-Cantero E (2015) Volatile organic compounds produced by the rhizobacterium Arthrobacter agilis UMCV2 modulate Sorghum bicolor (Strategy II Plant) morphogenesis and SbFRO1 transcription in vitro. J Plant Growth Regul 34:611–623
Cazorla FM, Romero D, Pérez-García A, Lugtenberg BJJ, Vicente AD, Bloemberg G (2007) Isolation and characterization of antagonistic Bacillus subtilis strains from the avocado rhizoplane displaying biocontrol activity. J Appl Microbiol 103(5):1950–1959
Chaparro JM, Badri DV, Vivanco JM (2014) Rhizospheremicrobiome assemblage is affected by plant development. ISME J 8(4):790
Chapelle E, Mendes R, Bakker PAH, Raaijmakers JM (2016) Fungal invasion of the rhizospheremicrobiome. ISME J 10(1):265
Chin-A-Woeng TF, Bloemberg GV, Lugtenberg BJ (2003) Phenazines and their role in biocontrol by Pseudomonas bacteria. New Phytol 157(3):503–523

Chung JH, Song GC, Ryu CM (2016) Sweet scents from good bacteria: case studies on bacterial volatile compounds for plant growth and immunity. Plant Mol Biol 90:677–687

Chung S, Kong H, Buyer JS, Lakshman DK, Lydon J, Kim SD, Roberts DP (2008) Isolation and partial characterization of Bacillus subtilis ME488 for suppression of soil borne pathogens of cucumber and pepper. Appl Microbiol Biotechnol 80(1):115–123

Combes-Meynet E, Pothier JF, Moenne-Loccoz Y, Prigent-Combaret C (2011) The Pseudomonas secondary metabolite 2,4-diacetylphloroglucinol is a signal inducing rhizoplane expression of Azospirillumgenes involved in plant-growth promotion. Mol Plant Microbe Interact 24:271–284

Compant S, Brion D, Jerzy N, Christophe C, Essaid AB (2005) Use of Plant Growth-Promoting Bacteria for biocontrol of plant diseases: principles, mechanisms of action and future prospects. Appl Environ Microbiol 71(9):4951–4959

Creus CM, Sueldo RJ, Barassi CA (2004) Water relations and yield in Azospirillum inoculated wheat exposed to drought in the field. Can J Bot 82:273–281

Cronin D, Moënne-Loccoz Y, Fenton A, Dunne C, Dowling DN, O'Gara F (1997a) Ecological interaction of a biocontrol Pseudomonas fluorescens strain producing 2,4-diacetylphloroglucinol with the soft rot potato pathogen Erwinia carotovora subsp. atroseptica. FEMS Microbiol Ecol 23:95–106

Cronin D, Moënne-Loccoz Y, Fenton A, Dunne C, Dowling DN, O'Gara F (1997b) Role of 2,4-diacetylphloroglucinol in the interactions of the biocontrol pseudomonad strain F113 with the potato cyst nematode Globodera rostochiensis. Appl Environ Microbiol 63:1357–1361

Dahiya N, Tewari R, Hoondal GS (2006) Biotechnological aspects of chitinolytic enzymes: a review. Appl Microbiol Biotech 25:1–10

de Garcia, Salamone IE, Hynes RK, Nelson LM (2006) Role of cytokinins in plant growth promotion by rhizosphere bacteria. PGPR: biocontrol and biofertilization. Springer, Netherlands, Amsterdam, pp 173–195

de Salamone IEG, Hynes RK, Nelson LM (2001) Cytokinin production by plant growth promoting rhizobacteria and selected mutants. Can J Microbiol 47(5):404–411

Dessaux Y, Hinsinger P, Lemanceau P (2009) Rhizosphere: so many achievements and even more challenges. Plant Soil 321:1–3

Dhore M, Barate D, Musaddiq M (2014) Studies on in-vitro anti microbial potential of rhizospheric soil bacteria against multi drug resistant clinical isolates. Ind J Appl Res 4(7):446–449

Ding LJ, Cui HL, Nie SA, Long XE, Duan GL, Zhu YG (2019) Microbiomes inhabiting rice roots and rhizosphere. FEMS Microbiol Ecol. https://doi.org/10.1093/femsec/fiz040

Ditengou FA, Muller A, Rosenkranz M, Felten J, Lasok H, Van Doorn MM, Legué V, Palme K, Schnitzler JP, Polle A (2015) Volatile signalling by sesquiterpenes from ectomycorrhizal fungi reprogrammes root architecture. Nat Commun 6:6279

Doornbos R, Loon L, Bakker PHM (2012) Impact of root exudates and plant defence signaling on bacterial communities in the rhizosphere: a review. Agron Sust Dev 32(1):227–243

Edwards J, Johnson C, Santos-Medellín C, Lurie E, Podishetty NK, Bhatnagar S, Eisen JA, Sundaresan V (2015) Structure, variation, and assembly of the root-associated microbiomes of rice. P Natl Acad Sci 112(8):911–920

Egamberdieva D, Kamilova F, Validov S, Gafurova L, Kucharova Z, Lugtenberg B (2008) High incidence of plant growth-stimulating bacteria associated with the rhizosphere of wheat grown on salinated soil in Uzbekistan. Environ Microbiol 10:1–9

El-Gamal NG, Shehata AN, Hamed ER, Shehata HS (2016) Improvement of lytic enzymes producing Pseudomonas fluorescens and Bacillus subtilis isolates for enhancing their biocontrol potential against root rot disease in tomato plants. Res J Pharm Biol Chem Sci 7(1):1393–1400

El-Tarabily KA (2006) Rhizosphere-competent isolates of streptomycete and non-streptomycete actinomycetes capable of producing cell-wall-degrading enzymes to control Pythium aphanidermatum damping-off disease of cucumber. Bot 84(2):211–222

El-Tarabily KA, Soliman MH, Nassar AH, Al-Hassani HA, Sivasithamparam K, McKenna F, Hardy GESTJ (2000) Biological control of Sclerotinia minor using a chitinolytic bacterium and actinomycetes. Plant Pathol 49(5):573–583

Fagodiya RK, Pathak H, Bhatia A, Kumar A, Singh SD, Jain N (2017a) Simulation of Maize (Zea Mays L.) yield under alternative nitrogen fertilization using infocrop-maize model. Biochem Cell Arch 17:65–71
Fagodiya RK, Pathak H, Kumar A, Bhatia A, Jain N (2017b) Global temperature change potential of nitrogen use in agriculture: a 50-year assessment. Sci Rep 7:44928
Felestrino ÉB, Vieira IT, Caneschi WL, Cordeiro IF, Assis RDAB, de CarvalhoLemes CG, Fonseca NP, Sanchez AB, Cepeda JCC, Ferro JA, Garcia CCM (2018) Biotechnological potential of plant growth-promoting bacteria from the roots and rhizospheres of endemic plants in ironstone vegetation in southeastern Brazil. World J Microbiol Biotechn 34(10):156
Fierro-Coronado RA, Quiroz-Figueroa FR, García-Pérez LM, Ramírez-Chávez E, Molina-Torres J, Maldonado-Mendoza IE (2014) IAA-producing rhizobacteria from chickpea (Cicerarietinum L.) induce changes in root architecture and increase root biomass. Can J Microbiol 60(10): 639–648
Fincheira P, Quiroz A (2018) Microbial volatiles as plant growth inducers. Microbiol Res 208:63–75
Fitzpatrick CR, Copeland J, Wang PW, Guttman DS, Kotanen PM, Johnson MT (2018) Assembly and ecological function of the root microbiome across angiosperm plant species. Proc Natl Acad Sci USA 115(6):E1157–E1165
Folman LB, Postma J, Van Veen JA (2003) Characterisation of Lysobacter enzymogenes (Christensen and Cook 1978) strain 3.1 T8, a powerful antagonist of fungal diseases of cucumber. Microbiol Res 158:107–115
Gajbhiye A, Rai AR, Meshram SU, Dongre AB (2010) Isolation, evaluation and characterization of Bacillus subtilis from cotton rhizospheric soil with biocontrol activity against Fusariumoxysporum. World J Microbiol Biotechnol 26(7):1187–1194
Geetanjali, Jain P (2016) Antibiotic production by rhizospheric soil microflora-a review. Int J Pharm Sci Res 7(11):4304–4314
Gherbawy Y, Elhariry H, Altalhi A, El-Deeb B, Khiralla G (2012) Molecular screening of Streptomycesisolates for antifungal activity and family 19 chitinase enzymes. J Microbiol 50(3):459–468
Gholami A, Shahsavani S, Nezarat S (2009) The effect of plant growth promoting rhizobacteria (PGPR) on germination seedling growth and yield of maize. Int J Biol Life Sci 5:1
Glick BR (2012) Plant growth promoting bacteria: mechanisms and applications. Scientifica 1–5
Glick BR, Penrose DM, Li J (1998) A model for the lowering of plant ethylene concentrations by plant growth promoting bacteria. J Theoretical Biol 190:63–68
Gong AD, Li HP, Yuan QS, Song XS, Yao W, He WJ, Zhang JB, Liao YC (2015) Antagonistic mechanism of Iturin A and Plipastatin A from Bacillus amyloliquefaciens S76-3 from wheat spikes against Fusarium graminearum. PLoS ONE 10(2):e0116871
Grossman JM, Schipanski ME, Sooksanguan T, Drinkwater LE (2011) Diversity of rhizobia nodulating soybean Glycine max (Vinton) varies under organicand conventional management. Appl Soil Ecol 50:14–20
Gutiérrez-Mañero FJ, Ramos-Solano B, Probanza AN, Mehouachi J, Tadeo FR, Talon M (2001) The plant-growth-promoting rhizobacteria Bacillus pumilus and Bacillus licheniformis produce high amounts of physiologically active gibberellins. Physiol Plant 111(2):206–211
Harrison LA, Letendre L, Kovacevich P, Pierson E, Weller D (1993) Purification of an antibiotic effective against Gaeumannomyces graminis var. tritici produced by a biocontrol agent, Pseudomonas aureofaciens. Soil Biol Biochem 25:215–221
Hashem A, Abd Allah EF, Alqarawi A, Al-Huqail AA, Wirth S, Egamberdieva D (2016) The interaction between arbuscularmy corrhizal fungi and endophytic bacteria enhances plant growth of Acacia gerrardii under salt stress. Front Plant Sci 7:1089
Hayden HL, Savin K, Wadeson J, Gupta V, Mele PM (2018) Comparative metatranscriptomics of wheat rhizosphere microbiomes in disease suppressive and non-suppressive soils for Rhizoctonia solani AG8. Front Microbiol 9:859
Hidri R, Barea JM, Mahmoud OM, Abdelly C, Azcón R (2016) Impact of microbial inoculation on biomass accumulation by Sulla carnosa provenances, and in regulating nutrition, physiological and antioxidant activities of this species under non-saline and saline conditions. J Plant Physiol 201:28–41

Howie WJ, Suslow T (1991) Role of antibiotic synthesis in the inhibition of Pythium ultimum in the cotton spermosphere and rhizosphere by Pseudomonas fluorescens. Mol Plant-Microbe Interact 4:393–399

Huang XF, Chaparro JM, Reardon KF, Zhang R, Shen Q, Vivanco JM (2014) Rhizosphere interactions: root exudates, microbes, and microbial communities. Botany 92(4):267–275

Hussain N, Mujeeb F, Tahir M, Khan GD, Hassan NM, Bari A (2002) Effectiveness of Rhizobium under salinity stress. Asian J Plant Sci 1:12–14

Jadhav HP, Sayyed RZ (2016) Hydrolytic enzymes of rhizospheric microbes in crop protection. MOJ Cell Sci Rep 3(5):00070

James EK (2000) Nitrogen fixation in endophytic and associative symbiosis. Field Crops Res 65:197–209

James EK, Olivares FL (1997) Infection and colonization of sugar cane and other graminaceous plants by endophytic diazotrophs. Crit Rev Plant Sci 17(1):77–119

Jimtha JC, Jishma P, Arathy GB, C Anisha, Radhakrishnan EK (2016) Identification of plant growth promoting Rhizosphere Bacillus sp. WG4 antagonistic to Pythium myriotylum and its enhanced antifungal effect in association with Trichoderma. J Soil Sci Plant Nutr 16(3):578–590

Kai M, Effmert U, Berg G, Piechulla B (2007) Volatiles of bacterial antagonists inhibit mycelial growth of the plant pathogen Rhizoctonia solani. Arch Microbiol 187:351–360

Kamilova F, Validov S, Azarova T, Mulders I, Lugtenberg B (2005) Enrichment for enhanced competitive plant root tip colonizers selects for a new class of biocontrol bacteria. Environ Microbiol 7:1809–1817

Kang BG, Kim WT, Yun HS, Chang SC (2010) Use of plant growth-promoting rhizobacteria to control stress responses of plant roots. Plant Biotechnol Rep 4:179–183

Kaushik R, Saxena AK, Tilak KVBR (2000) World J Microbiol Biotechnol 16(6):567–570

Khan AL, Hamayun M, Kim YH, Kang SM, Lee JH, Lee IN (2011) Gibberellins producing endophytic Aspergillus fumigatus sp. LH02 influenced endogenous phytohormonal levels, isoflavonoids production and plant growth in salinity stress. Process Biochem 46:440–447

Khan AL, Waqas M, Kang SM (2014) Bacterial endophytes Sphingomonas sp LK11 produces gibberellins and IAA and promotes tomato plant growth. J Microbiol 52:689–695

Kloepper JW, Schroth MN (1978) Plant growth-promoting rhizobacteria on radishes. In: Proceedings of the 4th international conference on plant pathogenic bacteria. Gilbert-Clarey, Tours, pp 879–882

Kour D, Rana KL, Kumar A, Rastegari AA, Yadav N, Yadav AN, Gupta VK (2019a) Extremophiles for hydrolytic enzymes productions: biodiversity and potential biotechnological applications. In: Molina G, Gupta VK, Singh BN, Gathergood N (eds) Bioprocessing for biomolecules production. Wiley, USA, pp 321–372

Kour D, Rana KL, Yadav AN, Yadav N, Kumar V, Kumar A, Sayyed RZ, Hesham AEL, Dhaliwal HS, Saxena, AK (2019). Drought-tolerant phosphorus-solubilizing microbes: biodiversity and biotechnological applications for alleviation of drought stress in plants. In: Plant growth promoting rhizobacteria for sustainable stress management. Springer, Singapore, pp 255–308

Kour D, Rana KL, Yadav N, Yadav AN (2019b) Bioprospecting of phosphorus solubilizing bacteria from Renuka Lake Ecosystems, Lesser Himalayas. J Appl Biol Biotechnol 7:1–6

Kour D, Rana KL, Yadav N, Yadav AN, Kumar A, Meena VS, Singh B, Chauhan VS, Dhaliwal HS, Saxena AK (2019c) Rhizospheric microbiomes: biodiversity, mechanisms of plant growth promotion, and biotechnological applications for sustainable agriculture. In: Kumar A, Meena VS (eds) Plant growth promoting rhizobacteria for agricultural sustainability: from theory to practices. Springer Singapore, Singapore, pp 19–65. https://doi.org/10.1007/978-981-13-7553-8_2

Kour D, Rana KL, Yadav N, Yadav AN, Singh J, Rastegari AA, Saxena AK (2019d) Agriculturally and industrially important fungi: current developments and potential biotechnological applications. In: Yadav AN, Singh S, Mishra S, Gupta A (eds) Recent advancement in white biotechnology through fungi: Volume 2: perspective for value-added products and environments. Springer International Publishing, Cham, pp 1–64. https://doi.org/10.1007/978-3-030-14846-1_1

Kumar A, Chaturvedi AK, Yadav K, Arunkumar KP, Malyan SK, Raja P, Kumar R, Khan SA, Yadav KK, Rana KL, Kour D, Yadav N, Yadav AN (2019) Fungal Phytoremediation of Heavy Metal-Contaminated Resources: Current Scenario and Future Prospects. In: Yadav A, Singh S, Mishra S, Gupta A (eds) Recent Advancement in White Biotechnology Through Fungi. Fungal Biology. Springer, Cham

Kumar A, Gupta DK, Kumar M (2013) Green manure crops: a boon for agricultural soil. Int J Agri Environ Biotechnol 6:193

Kumar A, Kumar A, Devi S, Patil S, Chandani P, Nagi S (2012) Isolation, screening and characterization of bacteria from rhizospheric soils from different plant growth promotion activities: as in vitro study. Recent Res Sci Technol 4(1):1–5

Kumar P, Dubey RC, Maheshwari DK (2012) Bacillus strains isolated from rhizosphere showed plant growth promoting and antagonistic activity against phytopathogens. Microbiol Res 167(8):493–499

Lee B, Farag MA, Park HB, Kloepper JW, Lee SH, Ryu CM (2012a) Induced resistance by a long-chain bacterial volatile: elicitation of plant systemic defense by a C13 volatile produced by Paenibacilluspolymyxa. PLoS One 7:48744

Lee BD, Dutta S, Ryu H, Yoo SJ, Suh DS, Park K (2015) Induction of systemic resistance in Panax ginseng against Phytophthora cactorum by native Bacillus amyloliquefaciens HK34. J Ginseng Res 39(3):213–220

Lee SY, Tindwa H, Lee YS, Naing KW, Hong SH, Nam Y, Kim KY (2012b) Biocontrol of anthracnose in pepper using chitinase, beta-1,3 glucanase, and 2-furancarboxaldehyde produced by Streptomyces cavourensis SY224. J Microbiol Biotechnol 2(10):1359–1366

Leff JW, Lynch RC, Kane NC, Fierer N (2017) Plant domestication and the assembly of bacterial and fungal communities associated with strains of the common sunflower, Helianthus annuus. New Phytol 214(1):412–423

Lei S, Xu X, Cheng Z, Xiong J, Ma R, Zhang L, Yang X, Zhu Y, Zhang B, Tian B (2019) Analysis of the community composition and bacterial diversity of the rhizosphere microbiome across different plant taxa. Microbiol Open. https://doi.org/10.1002/mbo3.762

Liu D, Cai J, XieCh-Ch Liu Ch, Chen Y-H (2010) Purification and partial characterization of a 36-kDa chitinase from Bacillus thuringiensis spp. colmeri, and its biocontrol potential. Enzyme Microb Technol 46:252–256

Liu F, Xing S, Ma H, Du Z, Ma B (2013) Cytokinin-producing, plant growth-promoting rhizobacteria that confer resistance to drought stress in Platycladus orientalis container seedlings. Appl Microbiol Biotechnol 97:9155–9164

Lu T, Ke M, Peijnenburg WJGM, Zhu Y, Zhang M, Sun L, Fu Z, Qian H (2018) Investigation of rhizospheric microbial communities in wheat, barley, and two rice varieties at the seedling stage. J Agri Food Chem 66(11):2645–2653

Lucas JA, Solano BR, Montes F, Ojeda J, Megias M, Gutierrez Ma-nero FJ (2009) Use of two PGPR strains in the integrated management of blast disease in rice (Oryza sativa) in Southern Spain. Field Crop Res 114:404–410

Lugtenberg B, Kamilova F (2009) Plant-growth-promoting rhizo-bacteria. Annu Rev Microbiol 63:541–556

Ma W, Guinel FC, Glick BR (2003) Rhizobium leguminosarum biovar viciae 1-aminocyclopropane-1-carboxylate deaminase promotes nodulation of pea plants. Appl Environ Microbiol 69(8):4396–4402

Madhaiyan M, Poonguzhali S, Ryu J, Sa T (2006) Regulation of ethylene levels in canola (Brassica campestris) by 1-aminocyclopropane-1-carboxylate deaminase-containing Methylobacterium fujisawaense. Planta 224:268–278

Mahanty T, Bhattacharjee S, Madhurankhi Goswami, Bhattacharyya P, Bannhi Das, Ghosh A, Tribedi P (2016) Biofertilizers: a potential approach for sustainable agriculture development. Environ Sci Pollut Res 24(4):3315–3335

Majumdar K, Razdan M, Aggarwal N, Murali KK, Bhattacharya RC, Dureja P (2011) Isolation and characterization of a potential biocontrol agent Bacillus KM5 from rhizosphere soil of a rice plant. Arch Phytopathol PFL 44(12):1196–1212

Maksimov IV, Abizgil'dina RR, Pusenkova LI (2011) Plant growth promoting rhizobacteria as alternative to chemical crop protectors from pathogens (Review). Appl Biochem Microbiol 47:333–345

Mali GV, Bodhankar MG (2009) Antifungal and phytohormone production potential of Azotobacterchroococcum isolates from Groundnut (Arachis hypogea L.) rhizosphere. Asian J Exp Sci 23:293–297

Malyan SK, Bhatia A, Kumar A, Gupta DK, Singh R, Kumar SK, Tomer R, Kumar O, Jain N, (2016a) Methane production, oxidation and mitigation: a mechanistic understanding and comprehensive evaluation of influencing factors. Science of The Total Environment 572:874–896

Malyan SK, Kumar A, Baram S, Kumar J, Singh S, Kumar SS, Yadav AN (2019) Role of Fungi in Climate Change Abatement Through Carbon Sequestration. In: Yadav A., Singh S., Mishra S., Gupta A. (eds) Recent Advancement in White Biotechnology Through Fungi. Fungal Biology. Springer, Cham

Malyan SK, Kumar SS, Kumar A, Kumar J (2016b) Water management tool in rice to combat two major environmental issues: global warming and water scarcity. In: Kumar S, Beg MA (eds) Environmental concerns of 21st century: Indian and global context, pp 43–58. (ISBN: 978-93-83281-65-7)

Marques APGC, Pires C, Moreira H, Rangel AOSS, Castro PML (2010) Assessment of the plant growth promotion abilities using Zea mays as indicator plant. Soil Biol Biochem 42:1229–1235

Mehnaz S, Baig DN, Lazarovits G (2010) Genetic and phenotypic diversity of plant growth promoting rhizobacteria isolated from sugarcane plants growing in Pakistan. J Microbiol Biotechnol 20:1614–1623

Mehnaz S, Lazarovits G (2006) Inoculation effects of Pseudomonas putida, Gluconabacter azotocaptans and Azospirilum lipoferum on corn plant growth under greenhouse conditions. Microbial Ecol 51:326–335

Mehnaz S, Mirza MS, Haurat J, Bally R, Normand P, Bano A, Malik KA (2001) Isolation and 16S rRNA sequence analysis of the beneficial bacteria from the rhizosphere of rice. Can J Microbiol 47(2):110–117

Mehnaz S, Weselowski B, Lazarovits G (2007) Sphingobacterium canadense sp. nov., an isolate from corn roots. Syst Appl Microbiol 30:519–524

Mendes LW, Kuramae EE, Navarrete AA, Van Veen JA, Tsai SM (2014) Taxonomical and functional microbial community selection in soybean rhizosphere. ISME J 8(8):1577

Mendes R, Garbeva P, Raaijmakers JM (2013) The rhizosphere microbiome: significance of plant beneficial, plant pathogenic, and human pathogenic microorganisms. FEMS Microbiol Rev 37(5):634–663

Mendes R, Kruijt M, De Bruijn I, Dekkers E, van der Voort M, Schneider JH, Piceno YM, DeSantis TZ, Andersen GL, Bakker PA, Raaijmakers JM (2011) Deciphering the rhizosphere microbiome for disease-suppressive bacteria. Science 332:1097–1100

Minerdi D, Bossi S, Maffei M, Gullino M, Garibaldi A (2011) Fusarium oxysporum and its bacterial consortium promote lettuce growth and expansin A5 gene expression through microbial volatile organic compounds (MVOC) emission. FEMS Microbiol Ecol 76:342–351

Mirza MS, Ahmad W, Latif F, Haurat J, Bally R, Normand P, Malik KA (2001) Isolation, partial characterization, and the effect of plant growth-promoting bacteria (PGPB) on micro-propagated sugarcane in vitro. Plant and Soil 237(1):47–54

Mishra S., Goyal D., Kumar A., Dantu P.K. (2019) Biotechnological Applications of β-Glucosidases in Biomass Degradation. In: Yadav A., Singh S., Mishra S., Gupta A. (eds) Recent Advancement in White Biotechnology Through Fungi. Fungal Biology. Springer, Cham

Mishra SK, Khan MH, Misra S, Dixit KV, Khare P, Srivastava S, Chauhan PS (2017) Characterisation of Pseudomonas spp. and Ochrobactrum sp. isolated from volcanic soil. Antonie Van Leeuwenhoek 110:253–270

Mohammadi K (2010) Ecophysiological response ofcanola (Brassica napus L.) to different fertility systems incrop rotation. PhD thesis. Agronomy Department. Tarbiat Modares University, Tehran, Iran, p 354

Mohammadi K, Ghalavand A, Aghaalikhani M, Heidari GR, Sohrabi Y (2011) Introducing the sustainable soilfertility system for chickpea (Cicer arietinum L.). Afr J Biotechnol 10(32):6011–6020

Moronta-Barrios F, Gionechetti F, Pallavicini A, Marys E, Venturi V (2018) Bacterial microbiota of rice roots: 16S-based taxonomic profiling of endophytic and rhizospheric diversity, endophytes isolation and simplified endophytic community. Microorgani 6(1):14

Mukherjee J, Mridha N, Mondal S, Chakraborty D, Kumar A (2018) Identifying suitable soil health indicators under variable climate scenarios: a ready reckoner for soil management. In: Bal S, Mukherjee J, Choudhury B, Dhawan A (eds) Advances in crop environment interaction. Springer, Singapore

Naz I, Bano A, Ul-Hassan T (2009) Isolation of phytohormones producing plant growth promoting rhizobacteria from weeds growing in Khewra salt range, Pakistan and their implication in providing salt tolerance to Glycine max L. Afr J Biotechnol 8:5762–5766

Neeraja C, Anil K, Purushotham P, Suma K, Sarma P et al (2010) Biotechnological approaches to develop bacterial chitinases as a bioshield against fungal diseases. Crit Rev Biotechnol 30:231–241

Nielsen TH, Christophersen C, Anthoni U, Sorensen J (1999) Viscosinamide, a new cyclic depsipeptide with surfactant and antifungal properties produced by Pseudomonas fluorescensDR54. J Appl Microbiol 87:80–90

Nowak-Thompson B, Gould SJ, Kraus J, Loper JE (1994) Production of 2,4-diacetylphloroglucinol by the biocontrol agent Pseudomonas fluorescens Pf-5. Can J Microbiol 40:1064–1066

Oliveira CA, Alves VMC, Marriel IE, Gomes EA, Scotti MR, Carneiro NP, Guimaraes CT, Schaffert RE, So NMH (2009) Phosphate solubilizing microorganisms isolated from rhizosphere of maize cultivated in an oxisol of the Brazilian Cerrado Biome. Soil Biol Biochem 41:1782–1787

Palumbo JD, Yuen GY, Jochum CC, Tatum K, Kobayashi DY (2005) Mutagenesis of beta-1,3-glucanase genes in Lysobacter enzymogenes strain C3 results in reduced biological control activity toward bipolaris leaf spot of tall fescue and Pythium damping-off of sugar beet. Phytopathol 95:701–707

Park Y, Dutta S, Ann M, Raaijmakers J, Park K (2015) Promotion of plant growth by Pseudomonas fluorescens strain SS101via novel volatile organic compounds. Biochem Biophys Res Commun 461:361–365

Pathak H, Jain N, Bhatia A, Kumar A, Chatterjee D (2016) Improved nitrogen management: a key to climate change adaptation and mitigation. Indian J Fertil 12(11):151–162

Patil PL, Medhane NS (1974) Seed inoculation studies in gram (Cicer arietinum) with different strains of Rhizobium sp. Plant Soil 40:221–223

Paul D, Park KS (2013) Identification of volatiles produced by Cladosporium cladosporioides CL-1, a fungal biocontrol agent that promotes plant growth. Sensors (Basel) 13:13969–13977

Pérez-Jaramillo JE, Carrión VJ, Bosse M, Ferrão LF, de Hollander M, Garcia AA, Ramírez CA, Mendes R, Raaijmakers JM (2017) Linking rhizospheremicrobiome composition of wild and domesticated Phaseolus vulgaris to genotypic and root phenotypic traits. ISME J 11(10):2244

Perrig D, Boiero ML, Masciarelli OA, Penna C, Ruiz OA, Cassan FD, Luna MV (2007) Plant-growth promoting compounds produced by two agronomically important strains of Azospirillum brasilense, and implications forinoculant formulation. Appl Microbiol Biotechnol 75:1143–1150

Pierson LS III, Pierson EA (1996) Phenazine antibiotic production in Pseudomonas aureofaciens: role in rhizosphere ecology and pathogen suppression. FEMS Microbiol Lett 136:101–108

Prasanna L, Eijsink VGH, Meadow R, Gåseidnes S (2013) A novel strain of Brevibacillus laterosporus produces chitinases that contribute to its biocontrol potential. Appl Microbiol Biot 97(4):1601–1611

Purkayastha GD, Mangar P, Saha A, Saha D (2018) Evaluation of the biocontrol efficacy of a Serratia marcescens strain indigenous to tea rhizosphere for the management of root rot disease in tea. PLoS ONE 13(2):0191761

Raaijmakers JM, Paulitz TC, Steinberg C, Alabouvette C, Moenne-Loccoz Y (2009) The rhizosphere: a playground and battlefield for soil borne pathogens and beneficial microorganisms. Pl Soil 321:341–361

Rajkumar M, Ae N, Prasad MNV, Freitas H (2010) Potential of siderophore-producing bacteria for improving heavy metal phytoextraction. Trends Biotechnol 28:142–149

Ramachandran VK, East AK, Karunakaran R, Downie JA, Poole SP (2011) Adaptationof Rhizobium leguminosarum to pea, alfalfa and sugar beet rhizosphere investigated by comparative transcriptomics. Genome Biol 12:106–109

Rana KL, Kour D, Sheikh I, Dhiman A, Yadav N, Yadav AN, Rastegari AA, Singh K, Saxena AK (2019a) Endophytic fungi: biodiversity, ecological significance and potential industrial applications. In: Yadav AN, Mishra S, Singh S, Gupta A (eds) Recent Advancement in White Biotechnology through Fungi, vol 1. Diversity and Enzymes Perspectives. Springer, Switzerland, pp 1–62

Rana KL, Kour D, Sheikh I, Yadav N, Yadav AN, Kumar V, Singh BP, Dhaliwal HS, Saxena AK (2019b) Biodiversity of endophytic fungi from diverse niches and their biotechnological applications. In: Singh BP (ed) Advances in endophytic fungal research: present status and future challenges. Springer International Publishing, Cham, pp 105–144. https://doi.org/10.1007/978-3-030-03589-1_6

Rana KL, Kour D, Yadav AN (2018) Endophytic microbiomes: biodiversity, ecological significance and biotechnological applications. Res J Biotechnol 14:1–30

Rashid MH, Schafer H, Gonzalez J, Wink M (2012) Genetic diversity of rhizobia nodulating lentil (Lens culinaris) in Bangladesh. Syst Appl Microbiol 35:98–109

Rastegari AA, Yadav AN, Gupta A (2019) Prospects of renewable bioprocessing in future energy systems. Springer International Publishing, Cham

Raza A, Faisal M (2013) Growth promotion of maize by desiccation tolerant Micrococcus luteus-chp37 isolated from Cholistan desert, Pakistan. Aust J Crop Sci 7:1693–1698

Reddy KRN, Choudary KA, Reddy MS (2007) Antifungal metabolites of Pseudomonas fluorescens isolated from rhizosphere of rice crop. J Mycol Plant Pathol 37(2):280–284

Revillas JJ, Rodelas B, Pozo C, Martinez-Toledo MV, Gonzalez LJ (2000) Productionof B-Group vitamins by two Azotobacter strains with phenolic compoundsas sole carbon source under diazotrophic and adiazotrophic conditions. J Appl Microbiol 89:486–493

Roesch LF, Fulthorpe RR, Riva A, Casella G, Hadwin AK, Kent AD, Daroub SH, Camargo FAO, Farmerie WG, Triplett EW (2007a) Pyrosequencing enumerates and contrasts soil microbial diversity. ISME J 1(4):283

Roesch LF, Fulthorpe RR, Riva A, Casella G, Hadwin AK, Kent AD, Daroub SH, Camargo FA, Farmerie WG, Triplett EW (2007b) Pyrosequencing enumerates and contrasts soil microbial diversity. ISME J 1:283–290

Ryu C, Farag M, Hu C, Reddy M, Wei H, Paré PW, Kloepper JW (2003) Bacterial volatiles promote growth in Arabidopsis. Proc Natl Acad Sci 100:4927–4932

Sandilya SP, Bhuyan PM, Gogoi DK, Kardong D (2016) Phosphorus solubilization and plant growth promotion ability of rhizobacteria of R. communis L. growing in Assam, India. Proc Natl Acad Sci India Sect B Biol Sci 88(3):959–966

Sandilya SP, Bhuyan PM, Vijay N, Gogoi DK, Kardong D (2017) Impact of Pseudomonas aeruginosa strain MAJ PIA03 affecting the growth and phytonutrient production of castor, a primary host-plant of Samia ricini. J Soil Sci Plant Nutr 17(2):499–515

Saleh SS, Glick BR (2001) Involvement of gacS and rpoS in enhancement of the plant growth-promoting capabilities of Enterobacter cloacae CAL2 and UW4. Can J Microbiol 47(8):698–705

Saraf M, Pandya U, Thakkar A (2014) Role of allelochemicals in plant growth promoting rhizobacteria for biocontrol of phytopathogens. Microbiol Res 169:18–29

Selim S, Negrel J, Govaerts C, Gianinazzi S, Van Tuinen D (2005) Isolation and partial characterization of antagonistic peptides produced by Paenibacillus sp. strain B2 isolated from the sorghum mycorrhizosphere. Appl Environ Microbiol 71(11):6501–6507
Shafi J, Tian H, Ji M (2017) Bacillus species as versatile weapons for plant pathogens: a review. Biotechnol Biotechnol Equipment. https://doi.org/10.1080/13102818.2017.1286950
Shaharoona B, Arshad M, Zahir ZA (2006) Effect of plant growth promoting rhizobacteria containing ACC-deaminase on maize (Zea mays L.) growth under axenic conditions and on nodulation in mung bean (Vigna radiata L.). Lett Appl Microbiol 42(2):155–159
Shanmugam PM, Veeraputhran R (2000) Effect oforganic manure, biofertilizers, inorganic nitrogen and zincon growth and yield of rabi rice. Madras Agric J 2:87–90
Sharma A, Kumar A, Dhaka TS (2012a) Impact on sugar factory effluent on chlorophyll and protein contents of Cicer arietinum and Tigonella foenum-gracecum. Curr Adv Agri Sci 4(1):62–63
Sharma A, Kumar A, Dhaka TS (2012b) Impact of sugar factory effluent on seed germination, seedling growth of Cicer arietinum and Trigonella foenum-graecum. Bioinfolet 9(2):220–221
Sharma P, Sardana V, Kandola SS (2011) Response of groundnut (Arachis hypogaea L.) to Rhizobium Inoculation. Libyan Agric Res Centre J Int 2:101–104
Sheng XF, Xia JJ (2006) Improvement of rape (Brassica napus) plant growth and cadmium uptake by cadmium-resistant bacteria. Chemosphere 64:1036–1042
Shih HD, Liu YC, Hsu FL, Mulabagal V, Dodda R, Huang JW (2003) Fungichromin: a substance from Streptomyces padanus with inhibitory effects on Rhizoctoniasolani. J Agric Food Chem 51(1):95–99
Shridhar BS (2012) Review: nitrogen fixing microorganisms. Int J Microbiol Res 3(1):46–52
Singh PP, Shin YC, Park CS, Chung YR (1999) Biological control of Fusarium wilt of cucumber by chitinolytic bacteria. Phytopathology 89:92–99
Singh R, Pandey DK, Kumar A et al (2017) PGPR isolates from the rhizosphere of vegetable crop Momordica charantia: characterization and application as biofertilizer. Int J Curr Microbiol App Sci 6(3):1789–1802
Singh RP, Jha PN (2016) A halotolerant bacterium Bacillus licheniformis HSW-16 augments induced systemic tolerance to salt stress in wheat plant (Triticumaestivum). Front Plant Sci 7:1890
Sirohi MH, Jackson J, Edwards M, Ollerton J (2015) Diversity and abundance of solitary and primitively eusocial bees in an urban centre: a case study from Northampton (England). J Insect Consev 123–136
Sneha S, Anitha B, Sahair RA, Raghu N, Gopenath TS, Chandrashekrappa GK, Basalingappa KM (2018) Biofertilizer forcrop production and soil fertility. Acad J Agric Res 6(8):299–306
Sofia IA, Paula P, Castro ML (2014) Phosphate solubilizing rhizobacteria enhance Zea mays growth in agricultural P-deficient soils. Ecol Eng 73:526–535
Spaepen S, Vanderleyden J (2011) Auxin and plant-microbe interactions cold spring harbor. Perspect Biol 3(4):a001438
Stutz EG, Défago G, Kern H (1986) Naturally occurring fluorescent pseudomonads involved in suppression of black root rot of tobacco. Phytopathol 76:181–185
Subrahmanyam G, Archana G (2011) Plant growth promoting activity of Enterobacter sp. C1D in heavy metal contaminated soils. In: Plant growth-promoting rhizobacteria (PGPR) for sustainable agriculture, pp 440
Subrahmanyam G, Sharma RK, Kumar GN, Archana G (2018). Vigna radiata var. GM4 plant growth enhancement and root colonization by a multi-metal-resistant plant growth-promoting bacterium Enterobacter sp. C1D in Cr (VI)-amended soils. Pedosphere 28(1):144–156
Suman A, Yadav AN, Verma P (2016) Endophytic microbes in crops: diversity and beneficial impact for sustainable agriculture. In: Singh D, Abhilash P, Prabha R (eds) Microbial inoculants in sustainable agricultural productivity, research perspectives. Springer, India, pp 117–143. https://doi.org/10.1007/978-81-322-2647-5_7

Tian BY, Cao Y, Zhang KQ (2015) Metagenomic insights into communities, functions of endophytes, and their associates with infection by root-knot nematode, Meloidogyne incognita, in tomato roots. Sci Rep 5:17087

Timmusk S, Nicander B, Granhall U, Tillberg E (1999) Cytokinin production by Paenibacillus polymyxa. Soil Biol Biochem 31(13):1847–1852

Vaishnav A, Kumari S, Jain S, Varma A, Tuteja N, Choudhary DK (2016) PGPR mediated expression of salt tolerance gene in soybean through volatiles under sodium nitroprusside. J Basic Microbiol 56:1274–1288

Van Loon LC (2007) Plant responses to plant growth-promoting rhizobacteria. Eur J Plant Pathol 119:243–254

Venkataraman GS, Neelakantan S (1967) Effect of cellular constituents of the nitrogen fixing blue-green algae. Cylindrospermum musciola on the rootgrowth of rice seedlings. J General Appl Microbiol 13:53–61

Verma P, Yadav AN, Khannam KS, Kumar S, Saxena AK, Suman A (2016a) Molecular diversity and multifarious plant growth promoting attributes of Bacilli associated with wheat (Triticum aestivum L.) rhizosphere from six diverse agro-ecological zones of India. J Basic Microbiol 56:44–58

Verma P, Yadav AN, Khannam KS, Mishra S, Kumar S, Saxena AK, Suman A (2016b) Appraisal of diversity and functional attributes of thermotolerant wheat associated bacteria from the peninsular zone of India. Saudi J Biol Sci. https://doi.org/10.1016/j.sjbs.2016.01.042

Verma P, Yadav AN, Khannam KS, Panjiar N, Kumar S, Saxena AK, Suman A (2015a) Assessment of genetic diversity and plant growth promoting attributes of psychrotolerant bacteria allied with wheat (Triticum aestivum) from the northern hills zone of India. Ann Microbiol 65:1885–1899

Verma P, Yadav AN, Kumar V, Singh DP, Saxena AK (2017) Beneficial plant-microbes interactions: biodiversity of microbes from diverse extreme environments and its impact for crop improvement. In: Singh DP, Singh HB, Prabha R (eds) Plant-microbe interactions in agro-ecological perspectives: volume 2: microbial interactions and agro-ecological impacts. Springer Singapore, Singapore, pp 543–580. https://doi.org/10.1007/978-981-10-6593-4_22

Verma P, Yadav AN, Shukla L, Saxena AK, Suman A (2015b) Alleviation of cold stress in wheat seedlings by Bacillus amyloliquefaciens IARI-HHS2-30, an endophytic psychrotolerant K-solubilizing bacterium from NW Indian Himalayas. Natl J Life Sci 12:105–110

Voisard C, Bull CT, Keel C, Laville J, Maurhofer M, Schnider U, Défago G, Haas D (1994) Biocontrol of root diseases by Pseudomonas fluorescens CHA0: current concepts and experimental approaches. In: O'Gara F, Dowling DN, Boesten B (eds) Molecular ecology of rhizosphere microorganisms. VCH, Weinheim, Germany, pp 67–89

Wagg C, Bender SF, Widmer F, van der Heijden MGA (2014) Soil biodiversity and soil community composition determine ecosystem multifunctionality. Proc Natl Acad Sci 111:5266–5270

Walia A, Mehta P, Chauhan A, Shirkot CK (2013) Antagonistic activity of plant growth promoting rhizobacteria isolated from tomato rhizosphere against soil borne fungal plant pathogens. Inte J Agri Environ Biotechnol 6(4):571–580

Wani PA, Khan MS (2010) Bacillus species enhance growth parameters of chickpea (Cicer arietinum L.) in chromium stressed soils. Food Chem Toxicol 48:3262–3267

Wani SA, Chand S, Ali T (2013) Potential use of Azotobacter chroococcum incrop production: an overview. Curr Agric Res J 1:35–38

Waqas M, Khan AL, Kamran M, Hamayun M, Kang SM, Kim YH, Lee IJ (2012) Endophytic fungi produce gibberellins and indoleacetic acid and promotes host-plant growth during stress. Mole 17:10754–10773

Weller DM (2007) Pseudomonas biocontrol agents of soilborne pathogens: looking back over 30 years. Phytopathology 97(2):250–256

Widawati S (2011) Diversity and phosphate solubilization by bacteria isolated from laki island coastal ecosystem. Biodiversitas 12(1):17–21

Xie X, Zhang H, Paré P (2009) Sustained growth promotion in Arabidopsis with longterm exposure to the beneficial soil bacterium Bacillus subtilis (GB03). Plant Signal Behav 4:948–953

Xu J, Zhang Y, Zhang P, Trivedi P, Riera N, Wang Y Liu X, Fan G, Tang J, Coletta-Filho HD, Cubero J (2018) The structure and function of the global citrus rhizosphere microbiome. Nat Commun 9(1):4894

Xue L, Xue Q, Chen Q, Lin C, Shen G, Zhao J (2013) Isolation and evaluation of rhizosphere actinomycetes with potential application for biocontrol of Verticillium wilt of cotton. Crop Prot 43:231–240

Yadav AN (2017a) Agriculturally important microbiomes: biodiversity and multifarious PGP Attributes for Amelioration of Diverse Abiotic Stresses in Crops for Sustainable Agriculture. Biomed J Sci Tech Res 1:1–4

Yadav AN (2017b) Beneficial role of extremophilic microbes for plant health and soil fertility. J Agric Sci 1:1–4

Yadav AN (2019) Microbiomes of wheat (Triticum aestivum L.) endowed with multifunctional plant growth promoting attributes. EC Microbiol 15:1–6

Yadav AN, Gulati S, Sharma D, Singh RN, Rajawat MVS, Kumar R, Dey R, Pal KK, Kaushik R, Saxena AK (2019a) Seasonal variations in culturable archaea and their plant growth promoting attributes to predict their role in establishment of vegetation in Rann of Kutch. Biologia 74:1031–1043. https://doi.org/10.2478/s11756-019-00259-2

Yadav AN, Kour D, Sharma S, Sachan SG, Singh B, Chauhan VS, Sayyed RZ, Kaushik R, Saxena AK (2019b) Psychrotrophic microbes: biodiversity, mechanisms of adaptation, and biotechnological implications in alleviation of cold stress in plants. In: Sayyed RZ, Arora NK, Reddy MS (eds) Plant growth promoting rhizobacteria for sustainable stress management: volume 1: rhizobacteria in abiotic stress management. Springer Singapore, Singapore, pp 219–253. https://doi.org/10.1007/978-981-13-6536-2_12

Yadav AN, Kumar R, Kumar S, Kumar V, Sugitha T, Singh B, Chauhan VS, Dhaliwal HS, Saxena AK (2017a) Beneficial microbiomes: Biodiversity and potential biotechnological applications for sustainable agriculture and human health. J Appl Biol Biotechnol 5:1–13

Yadav AN, Kumar V, Prasad R, Saxena AK, Dhaliwal HS (2018a) Microbiome in crops: diversity, distribution and potential role in crops improvements. In: Prasad R, Gill SS, Tuteja N (eds) Crop improvement through microbial biotechnology. Elsevier, USA, pp 305–332

Yadav AN, Mishra S, Singh S, Gupta A (2019c) Recent advancement in white biotechnology through fungi: volume 1: diversity and enzymes perspectives. Springer International Publishing, Cham

Yadav AN, Sachan SG, Verma P, Kaushik R, Saxena AK (2016a) Cold active hydrolytic enzymes production by psychrotrophic Bacilli isolated from three sub-glacial lakes of NW Indian Himalayas. J Basic Microbiol 56:294–307

Yadav AN, Sachan SG, Verma P, Saxena AK (2016b) Bioprospecting of plant growth promoting psychrotrophic Bacilli from cold desert of north western Indian Himalayas. Indian J Exp Biol 54:142–150

Yadav AN, Singh S, Mishra S, Gupta A (2019d) Recent advancement in white biotechnology through fungi: volume 2: perspective for value-added products and environments. Springer International Publishing, Cham

Yadav AN, Singh S, Mishra S, Gupta A (2019e) Recent advancement in white biotechnology through fungi: volume 3: perspective for sustainable environments. Springer International Publishing, Cham

Yadav AN, Verma P, Kaushik R, Dhaliwal HS, Saxena AK (2017b) Archaea endowed with plant growth promoting attributes. EC Microbiol 8:294–298

Yadav AN, Verma P, Kour D, Rana KL, Kumar V, Singh B, Chauahan VS, Sugitha T, Saxena AK, Dhaliwal HS (2017c) Plant microbiomes and its beneficial multifunctional plant growth promoting attributes. Int J Environ Sci Nat Resour 3:1–8 https://doi.org/10.19080/ijesnr.2017.03.555601

Yadav AN, Verma P, Kumar M, Pal KK, Dey R, Gupta A, Padaria JC, Gujar GT, Kumar S, Suman A, Prasanna R, Saxena AK (2015) Diversity and phylogenetic profiling of niche-specific Bacilli from extreme environments of India. Ann Microbiol 65:611–629

Yadav AN, Verma P, Kumar S, Kumar V, Kumar M, Singh BP, Saxena AK, Dhaliwal HS (2018b) Actinobacteria from rhizosphere: molecular diversity, distributions and potential biotechnological applications. In: Singh B, Gupta V, Passari A (eds) New and future developments in microbial biotechnology and bioengineering. USA, pp 13–41. https://doi.org/10.1016/b978-0-444-63994-3.00002-3

Yadav AN, Verma P, Kumar S, Kumar V, Kumar M, Sugitha TCK, Singh BP, Saxena AK, Dhaliwal HS (2018c) Actinobacteria from rhizosphere: molecular diversity, distributions, and potential biotechnological applications. In: New and future developments in microbial biotechnology and bioengineering, pp. 13–41

Yadav AN, Verma P, Singh B, Chauhan VS, Suman A, Saxena AK (2017c) Plant growth promoting bacteria: biodiversity and multifunctional attributes for sustainable agriculture. Adv Biotechnol Microbiol 5:1–16

Yadav AN, Yadav N (2018) Stress-adaptive microbes for plant growth promotion and alleviation of drought stress in plants. Acta Sci Agr 2:85–88

Yadav AN, Yadav N, Kour D, Kumar A, Yadav K, Kumar A, Rastegari AA, Sachan SG, Singh B, Chauhan V, Saxena AK (2019). Bacterial community composition in lakes. In: Freshwater microbiology. Academic Press, pp 1–71

Yadav AN, Yadav N, Sachan SG, Saxena AK (2019f) Biodiversity of psychrotrophic microbes and their biotechnological applications. J Appl Biol Biotechnol 7:99-108

Yi HS, Ahn YR, Song GC, Ghim SY, Lee S, Lee G, Ryu CM (2016) Impact of a bacterial volatile 2, 3-butanediol on Bacillus subtilis rhizosphere robustness. Front Microbiol 7:993

Zeller SL, Brand H, Schmid B (2007) Host-plant selectivity of rhizobacteria in a crop/weed model system. PLoS ONE 2(9):846

Zhang H, Sun Y, Xie X, Kim M, Dowd S, Paré P (2009) A soil bacterium regulates plant acquisition of iron via deficiency-inducible mechanisms. Plant J 58:568–577

Zou C, Li Z, Yu D (2010) Bacillus megaterium strain XTBG34 promotes plant growth by producing 2-pentylfuran. J Microbiol 48:460–466

Chapter 2
Culturable Endophytic Fungal Communities Associated with Cereal Crops and Their Role in Plant Growth Promotion

Hira Saleem, Hareem Mohsin, Rabia Tanvir and Yasir Rehman

Abstract Many microorganisms are known to live in association with plants. Endophytes are the microorganisms that live in the internal tissues of plants. Endophytic fungi hold great importance for the roles that they play in association with the host plants. Endophytes are known to promote the growth of the host plants by various activities such as detoxification of toxic compounds, protection against pathogens, and production of plant growth promoting hormones. Many biotechnologically important metabolites are also produced by endophytes such as anticancer and antimicrobial compounds. There is a rich diversity of endophytes that needs to be explored for biotechnological purposes. This chapter focuses on the endophytic fungi of cereal crops and the roles they play.

Keywords Cereal crops · Endophytic · Fungal communities · Plant growth promotion · Sustainable agriculture

2.1 Introduction

Plants play a vital role in the ecosystem. They are the producers that interact with different microbial communities and help in maintaining the biodiversity and stability of the ecosystem. There can be two types of such microbial communities with respect to the location, i.e., epiphytic fungi and endophytic fungi (Lindow and Brandl 2003). Epiphytic Fungi (epi; *upon*, phytic; plant) refers to the fungal communities propagating or adhering to the plant surface. Common examples include *Pestalotia* and

H. Saleem · H. Mohsin
Department of Microbiology and Molecular Genetics, University of the Punjab, 54590 Lahore, Pakistan

R. Tanvir
University Diagnostic Lab (UDL), Department of Microbiology,
University of Veterinary and Animal Sciences(UVAS), 54000 Lahore, Punjab, Pakistan

Y. Rehman (✉)
Department of Life Sciences, School of Science, University of Management and Technology, 54770 Lahore, Pakistan
e-mail: Yasir.rehman@umt.edu.pk

A. N. Yadav et al. (eds.), *Plant Microbiomes for Sustainable Agriculture*, Sustainable Development and Biodiversity 25, https://doi.org/10.1007/978-3-030-38453-1_2

Botryosphaeria (Santamaría and Bayman 2005). Endophytic Fungi (endo; *within*, phytic; plant) refers to the fungal communities which are present within the plant tissues asymptomatically. Common examples include *Colletotrichum*, *Xylaria,* and *Guignardia* (Schulz et al. 2002). Majority of these microbial communities comprise of endophytic fungal species.

Plants majorly depend on different kinds of associations with microbial species, especially the endophytic fungi, which play essential roles in plant growth, resistance against diseases, and tolerance against environmental stresses (Yang et al. 2018). The preliminary definition for the term endophytes was given by Hallmann et al. (1997) who stated that these are the microbial communities present within the plant tissue that cause no visible symptoms of any disease in the plant. However, this definition cannot be taken as the sole source to describe the endophytes (Hallmann et al. 1997; Lugtenberg et al. 2016). Mainly because there has been not much discussion about the unculturable endophytic species in the plant microbiome. Secondly, the latent plant pathogens are difficult to be recognized by culturable techniques and thus cause problems if they are in association with the unculturable endophytic communities. Due to these two major reasons, the endophytes, especially the endophytic fungal communities, are a challenge to study (Card et al. 2015; Hardoim et al. 2015; Lugtenberg et al. 2016).

The endophytes are most commonly associated with either roots or shoots of the plant. Such a relationship is termed as a symbiotic relationship in which two organisms are associated with each other. This can either be mutualism, commensalism, or parasitism. Furthermore, one fungal association which is in a mutualistic relation with one type of plant can be pathogenic to other types of plants (Bokati et al. 2016; Johnson et al. 1997). The root associated fungal endophytes usually have a mutualistic relation. They play a vital role in the better uptake of water and minerals and in return utilize carbohydrates from the host plant (Jumpponen 2001). In the case of shoots, most of such association has been found in the foliar regions of the plants at the point of contact of the leaves and the soil (Clement et al. 1997; Saikkonen et al. 2013). They play an important role in alleviating the stress, which can be abiotic or biotic, experienced by the host plant, thus making the host fit for survival (Bokati et al. 2016).

A research conducted by Bokati et al. (2016) demonstrated that not just the presence of fungal communities as endophytes is important, but the timing of their colonization also plays a critical role (McGonigle et al. 1999). The earlier the colonization and propagation of the endophytes, the better the absorption of nutrients and minerals from the soil with reduced effects of environmental stresses (Bokati et al. 2016). Endophytic fungal species are also reported to be promising candidates for the control of diseases (Kusari et al. 2012). Reports also state that the rice blast disease can be effectively controlled by the application of endophytic fungi (Atugala and Deshappriya 2015). Endophyte application, therefore, can limit the long term use of fertilizers and fungicides which are harmful to environment and human health (Priyadarshani et al. 2018; Tian et al. 2004).

Cereal crops, also known as grain crops, are the most commonly consumed group of plants all over the world. Approximately 50% of the total intake of food constitutes cereals and thus is considered as the staple food (Awika 2011). The word cereal is derived from the Latin word "cerealis" which means "grain". These crops belong to the grass family Poaceae (Gramineae) characterized by thin and long stalks. The grain usually comprises of three parts, i.e., endosperm (the part present around the embryo which provides nutrition for growth), germ (the part which becomes a plant), and the bran (hard outer covering of the grain) (Sarwar et al. 2013).

In the form of whole grain, it is a rich source of a number of nutrients such as starch, fats, oils, vitamins, and proteins. But when processed and converted into a refined form, bran and germ are removed and the endosperm is left which serves as a rich source of carbohydrates in the human diet (Sarwar et al. 2013). To fulfill the need of carbohydrates in our body, cereals are included in the daily diet. In wheat, proteins comprise of about 7–22% and in rice, it is the second major nutrient. Monosaccharides and disaccharides are also present but in a lesser quantity. Different kinds of oils and triglycerides along with some other nutrients are also present.

A diversity of endophytes has been isolated from cereal crops as well and the most extensively studied crops are wheat and maize (Larran et al. 2007; Sapkota et al. 2015; Yadav 2017). These endophytic species are reported to enhance the growth of the aforementioned cereal crops in nutrient depleted areas (Bokati et al. 2016; Yadav et al. 2019a). Research has been performed to assess the effect of abiotic factors on the growth of the plants in association with fungal strains. Different factors such as the intensity of light, availability of nutrients, and pH have been studied. Variations in plant growth were observed in the presence of fungal endophytes; however, the changes were specific for specific plant–fungal association. This indicated that the influence of abiotic factors usually depends on the plant–symbiont association and interaction (Kia et al. 2017).

This chapter describes the importance of endophytic fungal association with cereal crops. These symbiotic fungal species play a significant role in enhancing the plant growth along with providing resistance and tolerance against diseases and stress factors. They have also proven themselves as promising candidates in a number of biotechnological applications. Due to the production of a plethora of substances, secondary metabolites, and enzymes, they have the potential to be used in various industries such as medicine and agriculture.

2.2 Biodiversity of Endophytic Fungal Communities

Fungal endophytes, in association with the plants, are an indispensable element of the plant microbiome. These endophytes are ubiquitously present in nature (Herrera et al. 2013) and variation is found among these endophytes in different geographical locations (Weiss et al. 2011) such as Arctic tundra, mangroves, grasslands and savannahs, tropical and temperate forests as well as hot deserts (Arnold and Lutzoni 2007). It has also been reported that there is a limited number of fungal endophytes in places at higher latitudes but the number increases in the tropical regions. Plants that are reported to possess these fungal communities include mosses, non-vascular plants, ferns, flowering plants, and conifers (Arnold 2007; Suman et al. 2016; Yadav et al. 2018b).

The diversity of these species is also marked by the host specificity. Studies show that there can be only one host plant or a particular range of host plants with which the association can be made. Through this "host specificity", fungal species are restricted to a particular plant or a group of plants. Another important term is the "host recurrence" which is defined by the frequency of occurrence of an endophyte within a particular plant host, or a range of plant hosts, in a particular habitat. The phenomenon of "host selectivity" is also there according to which the endophyte has the capability to be in relationship with a number of plants but still the endophytic species will show a preference toward one particular plant. Host selectivity is most commonly termed as "host preference" by the mycologists (Cohen 2004; Dequn 2001; Huang et al. 2008; Rana et al. 2019a, b).

It has been reported by Dreyfuss and Chapela (1994) that approximately one million endophytic species inhabit the plants. These endophytic species are said to shape the diversity and structure of the plant microbial communities (Sanders 2004). It has been reported that out of these one million species, only 80000–100000 have been studied (Ainsworth 2008). Kumar and Hyde (2004) reported that in the tropical and temperate forests, the plants contain a high number of fungal endophytes. Prior studies have demonstrated that around as many as 17 endophytic fungal species can be isolated from a single leaf of a plant. Such observations lead to the conclusion that endophytes can be present in a large number in a single part of a plant (Gamboa and Bayman 2001). Arnold et al. (2000) also reported a large number of unculturable endophytes residing in plants of tropical forests.

2.3 Importance of Cereal Crops

Cereal crops are the staple foods which hold great importance in an individual's diet. The significance of these crops are listed below (Papageorgiou and Skendi 2018):

- These crops provide energy of 10000–15000 kJ/Kg. This is almost 10–20 times more as compared to energy provided by citrus fruits.
- They provide almost 30% of the total calories in the daily diet. In Asia, the cereals fulfill around 70–80% of the daily energy requirement of the people.
- It is a rich source of minerals which include magnesium, calcium, and potassium in a combined form of sulfate and phosphates. In low quantities, zinc, manganese, and copper are also present.
- Wheat products, which are rich in phytosterols, play an important role in minimizing the chances of breast cancer as they stimulate estrogen production.
- There are fibers present in the cereals, both soluble and insoluble, such as cellulose, pectin, and hemicellulose. They are efficient in making the peristaltic movements more effective which ultimately prevents constipation.

- There is extremely low sugar content which helps in lowering the blood sugar level.

2.4 Relation of Endophytic Fungal Communities and Cereal Crops

With the rising demands of the crops, their production needs to be increased as well. However, there are certain limitations and challenges with the traditional methods of crop production. The challenges include the application of herbicides, pesticides, and fertilizers and the impact of these chemical formulations on the environment and soil. Furthermore, the crop rotation system and increased cultivation of the crops have led to a decrease in the nutrient supply from the soil, reduction of the soil fertility and to some extent reduction in the crop yield (Kour et al. 2019b; Rees et al. 2013). Due to these problems, the challenge remains on how to meet the rising demand of cereal crops while maintaining the quality of the crops and avoiding any harm to the environment.

To address these problems, scientists have been studying the symbiotic relation of fungal communities which can enhance the productivity and sustainability of the crops. These endophytic symbionts reside in the healthy tissues within the plants and propagate within asymptomatically. They have been reported to act as "biological trigger" which will switch on the response systems of the host plants more effectively and rapidly as compared to the plants without any such symbiotic associations (Bandara et al. 2006; Redman et al. 2002). Such symbiotic relations not only help in enhancing the crop growth and yield but can also be used as biofertilizers thus causing a reduction in the harmful effects on the environment caused by the use of chemical fertilizers. These fungal communities also play a vital role in recycling nutrients in the soil and in making the quality of soil better (Karthik et al. 2016; Khan et al. 2013; Kumar et al. 2019; Ripa et al. 2019; Yadav 2019b).

2.5 Role and Importance of Endophytic Fungal Communities

In recent years, the role of endophytic fungi has been extensively studied related to plants and crops from all over the world (Vandenkoornhuyse et al. 2015). A number of important roles have been found related to this symbiotic relation:

- Increased supply and recycling of nutrients
- Protection of cereal crops and plants from abiotic and biotic stresses
- Enhancement in crop growth and yield
- Improving the biodiversity of plants in the ecosystem
- Plants' protection against different predators and other plant pathogens.

2.6 Isolation of Endophytic Fungi

2.6.1 Prerequisites for Isolation of Endophytic Fungi

For the isolation of these endophytic fungal communities, some prerequisites are to be taken into consideration. These are as follows:

- The plants at the sampling site should be healthy and free of any kind of disease. But if the research is being conducted with respect to any kind of biotic stress, then the site should be having plants that are infected and facing high malady pressure.
- There should be no previous study or research work done with the respective soil.
- Minimum exposure to the contaminants whether they are from the land, air, or water pollution.
- The piece of land under study should be inhabiting plant species of the same kind.
- There should be the active growth stage of the plants.
- Sampling is to be done under normal conditions, not after a heavy rain or a cold spell or any other infrequent weather condition, without any kind of disturbance.

During sampling, all other factors such as pH, salinity, type of soil, temperature, humidity, moisture content of soil, and nutrients in the soil should be recorded (Murphy et al. 2018).

2.6.2 Method for Isolation

The protocol for isolation of endophytic fungi is followed as given by Strobel et al. (1996). Modifications can be done according to the plant species or the type of fungal community being targeted (Radji et al. 2011). Following are the common steps which are carried out for the isolation purpose:

- Collection of plant samples which include leaves, roots, stem, or branches.
- Washing of plant samples under tap water for 10 min.
- Air-drying the samples.
- Cutting of sample into small pieces of approximately 1.0 cm in length with the help of a sterile surgical blade.
- Surface sterilization by immersing the sample in 70% ethanol for 1 min.

- Immersion of samples in 5.25% sodium hypochlorite solution for 5 min.
- Immersion of samples again in 70% ethanol for 30 s.
- Lastly, the immersion of samples in autoclaved distilled water for 3–5 s.
- Placement of the sterilized sample pieces on a plate of potato dextrose agar (PDA) containing antibiotics such as chloramphenicol (50ug/ml) and streptomycin (250ug/ml). The presence of antibiotics will not allow bacterial growth.
- Incubation of plates at 28°C for almost 10–14 days. The plates should be checked within that time for the growth of fungal isolates.
- Transfer of pure cultures into agar plates of PDA and again incubation at 28°C for almost 10–14 days.

Another method for isolation of endophytic fungi has been described by Huang et al. (2001). Following steps are included:

- Removing the outer bark of the sample.
- Cutting of the outer bark into small pieces with a sterile surgical blade.
- Grinding the pieces in the form of a paste.
- Adding the paste in autoclaved PDA medium just when the medium reaches the pouring temperature, i.e., 40–45 °C.
- Pouring the media in autoclave plates under aseptic conditions.
- Incubating the poured plates at 25 °C and keep checking the plates till the growth of hyphal tips is observed.
- Removing the hyphal tips and sub-culturing in fresh PDA medium followed by incubation at 25 °C for at least 14 days.
- Purification of cultures by transferring them into a new medium plate by the hyphal tip method.
  - The hyphal tip method involves the observation of the hyphae under a dissecting microscope at high magnification. The hyphae of interest are cut (a segment of about 1 mm thick) with the help of a sterile surgical blade and then transferred to the agar plate.

2.6.3 Characterization of Endophytes

After isolation of endophytic microorganisms, identification and characterization come as the most important step. A number of methods have been adopted by researchers for this purpose. Traditional techniques such as morphological characterization, biochemical profiling, as well as sequencing of the PCR products have been extensively used for the identification of molecular markers. Advanced technologies are now being used which include terminal restriction fragment length polymorphism (T-RFLP), denaturing gradient gel electrophoresis (DGGE), and metagenomics. These methods have paved the way toward better and precise characterization of the endophytic communities (Rodriguez et al. 2009; Yang et al. 2018). Tao et al. (2008) characterized *Bletillaochracea* for endophytic diversity and phylogenetic relationship through the DGGE technique. Berg et al. (2005) characterized the microbial communities associated with potato plant using T-RFLP technique. Sessitsch et al. (2012) characterized the endophytic microbial communities associated with the roots of rice plants through shotgun clone library construction followed by Sanger sequencing.

2.6.4 Maintenance and Preservation of Endophytic Fungi

For keeping microorganisms and endophytes in a viable state for a longer time, different preservation strategies are used. The metabolic functioning and physiological properties are maintained via these techniques. The simple techniques include the sub-culturing method, storing the cultures under oil (Smith and Onions 1994), water (Burdsall and Dorworth 1994), silica gel, or soil (Smith and Onions 1994). Other methods include Lyophilization (Onions 1971) and cryopreservation (Smith 1998) of the strains.

Following are the five major strategies used for preservation (Freire et al. 2016)

i. **Continuous Sub-culturing**: This technique was devised by Lacaz et al. (1991). This technique involves continuous transferring of the cultures to new vials having fresh sterilized PDA media followed by incubation at 25 °C.
ii. **Use of Mineral Oil for Preservation**: This method of preservation was introduced by Braz et al. (2009). 20 ml glass tubes are filled with 2 ml PDA medium and microbial strain is inoculated in the tube and incubated for 7 days. Following the incubation, a thick layer of 10 ml mineral oil (autoclaved for 2 days consecutively) is used for covering the culture tubes. The tubes are then covered with stopper, sealed with aluminum foil, and stored at 25 °C.
iii. **Use of Autoclaved Water for Preservation**: This technique was introduced by Diogo et al. (2005). 10 ml autoclaved distilled water (autoclaved for 2 days consecutively) is filled in glass tubes (20 ml). Five small segments of the fungal hyphae are taken from a 7-day old culture and transferred to the tubes. The tubes are then covered with the stopper, sealed with aluminum foil, and stored at 28 °C at room temperature.
iv. **Preservation at –20 °C**: This technique was introduced by Girão et al. (2004). Glass tubes of 20 ml are prepared having 8 ml of autoclaved distilled water, 0.5 ml dimethyl sulfoxide (DMSO), and 1 ml of glycerol (DMSO and glycerol act as cryo-protectants). All the components are subjected to autoclaving for 2 days consecutively. Five small segments of the fungal hyphae are taken from a 7-day old culture and transferred to the tubes. The tubes are then covered with stopper, sealed with aluminum foil, and stored at –20 °C.
v. **Preservation at –70 °C**: For this strategy, 0.4 ml of autoclaved distilled water, 0.025 ml of DMSO, 0.05 ml of glycerol, and 10 mg of polypropylene spheres are put in a 1.5 ml eppendorf. Small segments of the fungal hyphae are taken from a 7-day old culture and transferred to the tubes. The tubes are then covered with stopper, sealed with aluminum foil, and stored at –70 °C (Freire et al. 2016).

2.7 Specific Cereal Crops and Their Associated Endophytic Fungal Communities

2.7.1 Triticum aestivum *(Wheat) and Its Associated Fungal Communities*

Wheat is the major cereal crop being consumed globally. It has diverse endophytic fungal communities associated with it via symbiosis. The fungal isolates *Aspergillus flavus, Cladosporium cladosporioides, Trichoderma harzianum,* and *Fusarium proliferatum* have been majorly reported with the wheat crop. It was reported by Ripa et al. (2019) that these isolates were involved in producing plant growth promoting hormones, majorly indole acetic acid and siderophores. Majority of these isolates were able to resist the high salinity pressure of up to 7.5% and showed good growth in PDA medium. Metal resistance against copper, nickel, and cadmium was also observed which makes them promising candidates for the better growth of the crops even at areas where there is heavy metal pollution (Ripa et al. 2019).

2.7.2 Oryza sativa *(Rice) and Its Associated Fungal Communities*

Rice is the second major cereal crop cultivated all over the world. A number of fungal communities are associated with rice as well. Naik et al. (2009) conducted a research in which it was concluded that the major endophytes associated with rice crops were *Penicillium chrysogenum, Fusarium oxysporum,* and *Cladosporium cladosporioides.* These species were reported to act antagonistically against pathogenic compounds by producing different bioactive compounds. These facts can be exploited for better and enhanced crop production.

There are also reports that fungal species are also residing in the roots of the plant. Most noteworthy is the presence of *Ascomycota* phylum in which the most prominent members are *Aspergillus, Penicillium, Fusarium,* and *Trichoderma* (Santos-Medellín et al. 2017; Sharma et al. 2019). The species of *Aspergillus* and *Penicillium* are reported for the production of different organic acids (Ding et al. 2019; Khan et al. 2014).

2.7.3 Zea mays *(Maize) and Its Associated Fungal Communities*

Maize being an important cereal crop has a high content of vital nutrients (Ngachan et al. 2011). Maize crop is reported to be in symbiotic relation with the endophytic fungi mainly in the root area (Orole and Adejumo 2009; Potshangbam et al. 2017).

It has been reported by Potshangbam et al. (2017) that major endophytic fungal communities associated with the leaves of maize plants were *Acremonium, Fusarium,* and *Penicillium.* The root part of the maize plant inhabited a large number of fungal communities including *Fusarium, Trichoderma, Aspergillus, Alternaria alternate,* and *Botryodiplodia.* The study reported that *Fusarium* sp. is present in different parts of the maize plant. These endophytes are also involved in the production of different biologically important molecules that help the plant to grow better. Such beneficial associations can be exploited to synthesize biofertilizers thus minimizing the use of chemical fertilizers and the associated risks (Potshangbam et al. 2017; Shweta et al. 2010).

2.7.4 Hordeum vulgare *(Barley) and Its Associated Fungal Communities*

Barley being an important nutritional crop has a high content of minerals including magnesium, manganese, and selenium. Furthermore, it is considered as a storehouse for dietary fibers which play a role in keeping the gut healthy (Pourkheirandish and Komatsuda 2007). Barley plant is vulnerable to attack by several pests such as *Diuraphis noxia*, *Metopolophium dirhodum* and *Mayetiola destructor.* The endophytic fungi *Neotyphodium* is reported to be anti-herbivore and proves to be a biologically controlling agent against pest attack. This can also have many biotechnological interventions to stop the accumulation of harmful substances in plants by eliminating the use of pesticides (Clement et al. 2005).

2.8 Role of Endophytic Fungi in Plant Growth Promotion

2.8.1 Growth Promotion Mechanisms

2.8.1.1 Phosphate Solubilization

Phosphate is one of the essential macronutrients needed for plant growth. Many microbial species including fungi have the ability to solubilize phosphate leading to its mineralization (Yadav et al. 2015a, b, c). Among total population of fungi, 0.1–0.5% constitute the phosphate solubilizing fungi in which the endophytic fungi contribute majorly. Important endophytic P-solubilizing fungi belong to genera *Aspergillus, Penicillium, Curvularia, Fusarium,* and *Candida* (Mehta et al. 2019). Phosphate solubilization is highly dependent on the tendency of the microorganisms to produce organic acids in the surrounding environment. These organic acids form complexes with calcium, iron, and aluminum present in the soil to convert low soluble phosphate into soluble one through chelation and exchange reactions. Low pH is required to make insoluble phosphate available to the plants in the form of soluble phosphates. Organic acids such as citric acids, gluconic, and ketogluconic acid lower the pH of the environment, thus providing the medium in which insoluble metal part of the phosphate is exchanged with sodium or magnesium resulting in soluble phosphate salt (Behera et al. 2017; Yadav et al. 2016). Organic acids also increase the accessibility of phosphate for the plants by blocking phosphate adsorption sites on soil particles or by forming complexes with cations on mineral surfaces of soil (Rodríguez and Fraga 1999). On the other hand, insoluble part of phosphate can be released with the help of enzymes such as phytase and phosphatases which are released by the microorganisms including endophytic fungi. Acid phosphatases (AcPase) are unique sets of enzymes mostly located in the cell wall of many microorganisms. They hydrolyze phosphomonoesters at acidic pH by transferring a phosphoryl group to alcohol in the presence of certain phosphate acceptors. Acid phosphatases production by *Serratia* sp. is well documented and directly related to plant growth promotion (Behera et al. 2017). With the help of this mechanism, soil fertility and plant growth are enhanced by the "phosphate solubilizing microorganisms" (Adhikari and Pandey 2019). Spagnoletti et al. (2017) described the solubilization of phosphate in the presence of iron, calcium, and aluminum by dark septate endophytic fungi and reported that phosphate solubilization was maximum in calcium phosphate ranging from 42.87 ± 5.37 to 51.33 ± 1.87 μgml^{-1}. Microbial phosphate solubilization is one of the major events that contribute to promotion of the plant growth (Tarafdar and Gharu 2006).

2.8.1.2 Synthesis of Siderophores

Siderophores are iron binding compounds of low molecular weight (500–1000 Dalton). These extracellular agents are synthesized by the many microorganisms including bacteria and fungi present in iron deficient soil (Crowley 2006). These compounds are helpful in chelating Fe(III) and its subsequent transport into cells for growth. Microbial siderophores, therefore, help the plants to scavenge iron required for their growth. Endophytic fungi *Acremonium sclerotigenum* from host tree *Terminalia bellerica* produce siderophores that not only help in the uptake of iron by the plant but also act as a biocontrol agent against many pathogens (Prathyusha et al. 2015; Wang et al. 1993). *Epichloe festucae*, a foliar endophyte of perennial ryegrass, has the ability to produce extracellular fusarinine and intracellular ferricrocin that promote the growth of the host plant by mobilizing iron (Kajula et al. 2010). *Phialocephala fortinii*, a dark septate fungi found in endophytic relationship with *Pinus sylvestris*, *Abies alba*, *Picea abies*, and *Carex curvula*, have the ability to produce three different types of siderophores such as ferricrocin, ferrirubin, and ferrichrome C in low iron conditions (Bartholdy et al. 2001). As these microorganisms contribute to the availability of iron to the host plant, these can also have the potential to serve as biofertilizers (O'Sullivan and O'Gara 1992).

Moreover, siderophores can also serve as a type of biocontrol. In this process, siderophores complement the lytic activity, antibiosis, and hormonal effect of endophytes to decrease the growth of other competing microorganisms and plant pathogens in their vicinity (Neilands and Leong 1986).

2.8.1.3 Phytohormone Production

The roots of plants which are colonized by endophytic fungi such as *Phoma glomerata*, *Porostereum spadiceum*, *Penicillium*, and *Aspergillus sp.* produce many phytohormones as compared to the plants which are not colonized (Joshi et al. 2018). Thus, one of the major contributions of the endophytic fungi is the production of plant growth promoting stimulators known as phytohormones which include abscisic acids, ethylene, auxins, cytokinins, and gibberellins. Their classes, roles, and functions are discussed in Table 2.1.

Table 2.1 PGR produced by endophytic fungi and their attributes towards plant growth promotion

PGRs	Attributes toward plant growth	Endophytic fungi producing PGPR	Reported studies
Auxins (IAA)	Positive effect on root growth, Cell division	*Phoma glomerata, Penicillium* sp.	Vessey (2003), Gravel et al. (2007), Waqas et al. (2012)
Cytokinins	Cell division, Inhibits senescence	*Fusarium* sp.	Li et al. (2012), Waqas et al. (2012), Shah et al. (2019)
Gibberellins	Cell elongation, Promote Flowering	*Porostereum spadiceum, Phoma glomerata, Penicillium* sp.	Waqas et al. (2012), Tanimoto (2005), Hamayun et al. (2017)
Abscisic acids	Abscission of leaves and fruits, Dormancy induction of buds and seeds	*Aspergillus nidulans, Glomus intraradice.*	Forchetti et al. (2007), Herrera-Medina et al. (2007), Brader et al. (2014), Xu et al. (2018)
Ethylene	Promotes senescence, epinasty, Major role in fruit ripening	*Piriformospora indica, Arabidopsis thalian, Fusarium solani.*	Camehl et al. (2010), Kavroulakis et al. (2007)

Among different types of auxin, indole-3-acetic acid (IAA) is a natural and most common auxin and its beneficial effects on the growth of plant roots enable them to absorb nutrients efficiently from the soil (Vessey 2003). It promotes the cell division and root growth leading to the differentiation of root nodules and increase in the surface area of roots. However, root tissues are sensitive to IAA concentration and when its concentration exceeds beyond the required limit, the root elongation is stopped (Gravel et al. 2007; Tanimoto 2005).

Auxin is responsible for cell elongation by increasing the turgor pressure due to the presence of dissolved solute. Here is the acid-growth hypothesis which postulates this mode of action of auxin.

- Production of IAA stimulates the H^+ pumps present in the cell membrane.
- Once H^+ pumps are activated, H^+ ions are pumped into the cell wall decreasing the pH.
- When the environment of the cell wall is acidified, it stimulates the pH dependent enzymes which cause bond breakage between the microfibrils of cellulose present in the cell wall.
- The cell wall expands due to nutrients and solutes which elongate the cell wall and ultimately enlarge the size of the cell.

Endophytic Fungi can also tolerate stress of drought and salinity and can still support the plant cells by the production of plant growth stimulators. Such stress conditions can cause plants to wilt and start senescence and can even cause the death of the plants (Iqbal and Ashraf 2013). Waqas et al. (2012) reported that endophytic fungi such as *Penicillium* sp. and *Phoma glomerata* association result in tremendous shoot growth, plant biomass, and chlorophyll ratio. These endophytes are involved in the production of phytohormones such as gibberellins with class GA_3. GA_3 helps in flowering of the plants as well as in cell elongation. Another study by Khan et al. (2012) reported endophyte *Paecilomyces variotii* LHL 10 present in cucumber roots that could produce a high amount of IAA and gibberellins.

2.8.1.4 Biological Nitrogen Fixation

Biological nitrogen fixation is one of the most important biological processes exhibited by microorganisms. The association between the host plant and the nitrogen fixing microorganisms is either symbiotic or asymbiotic. Many endophytes are also involved in this process. Plant internal environment is high in carbon and low in oxygen which is a favorable environment for the fixation of nitrogen by endophytes. This again is beneficial for the host plant (Ladha and Reddy 2003; Patle et al. 2018).

2.8.2 Endophytic Fungi as "Biocontrol" Agents

One of the major roles of endophytic fungal communities is as the biocontrolling agents which play an essential role in protecting the host plant against its predators and insects (Rana et al. 2018; Yadav et al. 2019d). It has been reported by Mirlohi et al. (2004) that the plant roots in absence of any endophytic fungi get infected with insects and pests while those which are in association with endophytic fungi are not harmed by the predators. The fungal endophytes act as biocontrol agents by the production of different fungal metabolites which include indole di-terpenes, ergot alkaloids, peramine, and the lolines which are described as follows (Malinowski and Belesky 2000):

- The class indole di-terpenes include lolitriol and paxilline. This class chiefly acts as neurotoxins. Moreover, they are tremorgenic toward small insects and mice, therefore, known for the anti-insect activity.
- Among the ergot alkaloids, the compound ergovaline is the most effective in keeping the insects away especially the stem weevil.
- The remaining two groups of metabolites, i.e., loline and peramine are less essential but are known for their activities against the insects and pests (Mirlohi et al. 2004).

2.9 Biotechnological Significance of Endophytic Fungi

2.9.1 Production of Secondary Metabolites

Secondary metabolites are the compounds that are not supporting plant growth directly but are involved in its protection. Approximately 80% of fungal endophytes produce bioactive compounds with antimicrobial and herbicidal properties. They are also a rich source for many other secondary metabolites such as antifungal, anticancer, and antiparasitic compounds (Joshi et al. 2018; Krohn et al. 2002). Many endophytic fungi have the ability to protect host plant against soil borne pathogens such as *Aspergillus fumigatus, Botrytis cinerea, Blumeria graminis, Fusarium culmorum, Globisporangium ultimum, Monilinia laxa, Moniliophthora perniciosa, Penicillium expansum, Phytophthora* sp, *Plasmopara viticola, Puccinia polygoni-amphibii,* and *Sclerotinia sclerotiorum* and thus help plant to survive (Yadav 2018; Yadav et al. 2019a, b, c). Reported mechanisms involve the mycoparasitism, antibiosis, cell wall degradation, and induction of defense response (Zhang et al. 2014). *Piriformospora indica* is a plant root endophytic fungi showing resistance against phytopathogen *Fusarium culmorum* (Waller et al. 2005; Zheng et al. 2016).

Apart from bioactivity against plant pathogens, the fungal endophytes, from the plant *Aquilaria sinensis*(agarwood),have exhibited anticancer property against different human cancer cell lines such as HL-60, 293-T, and HepG2 (Cui et al. 2011). Antimicrobial activity has been displayed by the endophytes from *Dendrobium devonianum* and *Dendrobium thyrsiflorum* which have the ability to produce inhibitory molecules against *Candida albicans, Cryptococcus neoformans, Escherichia coli, Staphylococcus aureus, Aspergillus fumigates,* and *Bacillus subtilis.* Medicinal plant *Stryphnodendron adstringens* harbors endophytic fungi that produce antimicrobial agents active against *Candida albicans* and *Cladosporium sphaerospermum.* They are also reported to inhibit the growth of the cancer cells MCF-7 and TK-10(Carvalho et al. 2012).

2.9.2 Resistance Against Heavy Metals

Heavy metals are harmful to plants as they produce oxidative stress. Plants have many mechanisms to tolerate them (Idris et al. 2004). Some plants are metal tolerant while some are hyperaccumulators. Hyperaccumulators such as *Alyssum bertolonii, Alnusfirma, Brassica napus, Nicotiana tabacum, Thlaspi caerulescens, T. goesingense,* and *Solanum nigrum* have certain endophytes associated with them that help them tolerate metal stress. Endophytes such as *Microsphaeropsis, Mucor, Phoma, Alternaria, Peyronellaea, Steganosporium, and Aspergillus* are known to help the plant in metal resistance and detoxification (Bai et al. 2012; Yadav et al. 2018a).

2.9.3 Role in Phytoremediation

Endophytic fungi are also known to facilitate the host plant in phytoremediation by protecting the plant from different contaminants present in the polluted contaminated soil. Endophytic fungi *Phomopsis liquidambari* have ability to produce the enzyme which can degrade phenolic acid allelochemicals such as 4-hydroxybenzoic acid released by the decomposing foliage. These allelochemicals have negative impacts on the plants as these slow down the growth of the plants and the bacterial population present in the soil (Chen et al. 2011; Kour et al. 2019b; Rana et al. 2019b).

2.9.4 Production of Extracellular Enzymes

Many endophytic fungi species have the ability to produce enzymes that have biotechnological and industrial importance. Many extracellular enzymes such as cellulases, chitinases, laccase, pectinases, xylanases, proteases, amylases, β-galactosidase, and other catabolic enzymes are produced by endophytic fungi. Chitinases produced by the endophytic fungi *Neotyphodium* sp. has the ability to break down the chitin which is present in the phytopathogen cell wall. Another fungal endophyte *Sarocladium zeae* isolated from maize produces hemicellulase which is an enzyme involved in bioconversion of lignocellulosic biomass into sugars which are then easily fermented (Bischoff et al. 2009; Kour et al. 2019a; Zheng et al. 2016).

2.9.5 Biotransformation and Nutrient Recycling

Biotransformation is the conversion of one chemical compound to another one with the help of a biological agent. Many endophytic fungi are also known to perform this function. Different types of transformation reactions may activate or inactivate a certain compound. Zikmundova et al. (2002) reported biotransformation of phytoanticipins 2-benzoxazolinone (BOA) to N-(2-hydroxyphenyl) malonamic acid by endophytic fungi isolated from *Aphelandra arborea* (Malyan et al. 2019).

Nutrient recycling is one of the important mechanisms exhibited by endophytic fungi. In this process, essential nutrients are balanced so that they become available for the natural ecosystem. *Phomopsis liquidambari* is one such endophytic fungi that have the ability to stimulate mineralization and facilitate the ammonium release in the environment which provokes the nitrification process by bacteria present in the soil (Chen et al. 2013; Yadav and Yadav 2018).

2.10 Ecological Significance of Endophytic Fungi in Plant Protection

Endophytic fungi have a major role in the degradation of dead host plants and thus help in the recycling of the nutrients in the environment (Boberg et al. 2011). Fungal endophytes enable the plant to tolerate stress conditions such as drought, salinity, and pH by Induced Systemic Resistance (ISR), bioremediation, and protection (Yadav 2019a). Endophytic fungi produce metabolites against insects. One of the anti-insect metabolites is anthraquinone metabolite called rugulosin discovered from the endophyte *Abies balsamea* and later from *Picea scopiformis* and *Picea glauca*. The compound was found to reduce the growth rate of *Choristoneura fumiferana*, *Lambdina fiscellaria*, and *Zeiraphera canadensis* (Joshi et al. 2018; Sumarah et al. 2008).

2.11 Conclusion and Future Prospects

It can be concluded that plant–fungal associations thrive under the principle of mutualism where both the partners are getting benefit from each other. Such endophytes play an important role in plant growth promotion as these provide resistance to plant against different environmental stresses and toxic compounds, protect host plant against several pathogens, and produce many plant growth promoting hormones. Endophytic fungi are also significantly important as biotransformers of different chemicals and help in the recycling of nutrients. These are also known to produce many metabolites that have medicinal importance such as anticancer and antimicrobial compounds. The endophytes also find many industrial usages as they are known for the production of many important enzymes and metabolites. Advance studies in genetic engineering, metagenomics, metatranscriptomics, and proteomics could be employed for better understanding of the molecular mechanisms behind these abilities, and to exploit them further for different biotechnological processes.

References

Adhikari P, Pandey A (2019) Phosphate solubilization potential of endophytic fungi isolated from *Taxus wallichiana* Zucc. roots. Rhizosphere 9:2–9

Ainsworth GC (2008) Ainsworth & Bisby's dictionary of the fungi. Cabi, Kew

Arnold AE (2007) Understanding the diversity of foliar endophytic fungi: progress, challenges, and frontiers. Fungal Biol Rev 21:51–66

Arnold AE, Lutzoni F (2007) Diversity and host range of foliar fungal endophytes: are tropical leaves biodiversity hotspots? Ecology 88:541–549

Arnold AE, Maynard Z, Gilbert GS, Coley PD, Kursar TA (2000) Are tropical fungal endophytes hyperdiverse? Ecol Lett 3:267–274

Atugala D, Deshappriya N (2015) Effect of endophytic fungi on plant growth and blast disease incidence of two traditional rice varieties. J Natl Sci Found Sri Lanka 43:173–187

Bai LY, Weng JR, Lo WJ, Yeh SP, Wu CY, Wang CY et al (2012) Inhibition of Hedgehog signaling induces monocytic differentiation of HL-60 cells. Leuk Lymphoma 53:1196–1202

Bandara W, Seneviratne G, Kulasooriya SA (2006) Interactions among endophytic bacteria and fungi: effects and potentials. J Biosci 31:645–650

Bartholdy B, Berreck M, Haselwandter K (2001) Hydroxamate siderophore synthesis by *Phialocephala fortinii*, a typical dark septate fungal root endophyte. Biometals 14:33–42

Behera B, Yadav H, Singh S, Mishra R, Sethi B, Dutta S et al (2017) Phosphate solubilization and acid phosphatase activity of *Serratia* sp. isolated from mangrove soil of Mahanadi river delta, Odisha. India. J Genet Eng Biotechnol 15:169–178

Berg G, Krechel A, Ditz M, Sikora RA, Ulrich A, Hallmann J (2005) Endophytic and ectophytic potato-associated bacterial communities differ in structure and antagonistic function against plant pathogenic fungi. FEMS Microbiol Ecol 51:215–229

Bischoff KM, Wicklow DT, Jordan DB, de Rezende ST, Liu S, Hughes SR et al (2009) Extracellular hemicellulolytic enzymes from the maize endophyte *Acremonium zeae*. Curr Microbiol 58:499–503

Boberg JB, Ihrmark K, Lindahl BD (2011) Decomposing capacity of fungi commonly detected in Pinus sylvestris needle litter. Fungal Ecol 4:110–114

Bokati D, Herrera J, Poudel R (2016) Soil influences colonization of root-associated fungal endophyte communities of maize, wheat, and their progenitors. J Mycol 2016:9

Brader G, Compant S, Mitter B, Trognitz F, Sessitsch A (2014) Metabolic potential of endophytic bacteria. Curr Opin Biotechnol 27:30–37

Braz SCdM, Motta CMdS, Massa DMdL, Neves RP, Magalhães OMC (2009) Viabilidade, confirmação taxonômica e detecção enzimática de espécies de Acremonium preservadas sob óleo mineral na Coleção de Culturas University Recife Mycology. Rev Soc Bras Med Trop 42:63–66

Burdsall HH Jr, Dorworth EB (1994) Preserving cultures of wood-decaying Basidiomycotina using sterile distilled water in cryovials. Mycologia 86:275–280

Camehl I, Sherameti I, Venus Y, Bethke G, Varma A, Lee J et al (2010) Ethylene signalling and ethylene-targeted transcription factors are required to balance beneficial and nonbeneficial traits in the symbiosis between the endophytic fungus Piriformospora indica and Arabidopsis thaliana. New Phytol 185:1062–1073

Card SD, Hume DE, Roodi D, McGill CR, Millner JP, Johnson RD (2015) Beneficial endophytic microorganisms of *Brassica*–a review. Biol Control 90:102–112

Carvalho CR, Gonçalves VN, Pereira CB, Johann S, Galliza IV, Alves TM et al (2012) The diversity, antimicrobial and anticancer activity of endophytic fungi associated with the medicinal plant *Stryphnodendron adstringens* (Mart.) *Coville* (Fabaceae) from the Brazilian savannah. Symbiosis 57:95–107

Chen Y, Peng Y, Dai C-C, Ju Q (2011) Biodegradation of 4-hydroxybenzoic acid by *Phomopsis liquidambari*. Appl Soil Ecol 51:102–110

Chen Y, Ren C-G, Yang B, Peng Y, Dai C-C (2013) Priming effects of the endophytic fungus *Phomopsis liquidambari* on soil mineral N transformations. Microb Ecol 65:161–170

Clement S, Elberson L, Bosque-Pérez N, Schotzko D (2005) Detrimental and neutral effects of wild barley–*Neotyphodium* fungal endophyte associations on insect survival. Entomol Exp Appl 114:119–125

Clement S, Wilson AD, Lester D, Davitt C (1997) Fungal endophytes of wild barley and their effects on *Diuraphis noxia* population development. Entomol Exp Appl 82:275–281

Cohen SD (2004) Endophytic-host selectivity of *Discula umbrinella* on *Quercus alba* and *Quercus rubra* characterized by infection, pathogenicity and mycelial compatibility. Eur J Plant Pathol 110:713–721

Crowley DE (2006) Microbial siderophores in the plant rhizosphere. In: Barton LL, Abadia J (eds) Iron nutrition in plants and rhizospheric microorganisms. Springer, Netherlands, Dordrecht, pp 169–198. https://doi.org/10.1007/1-4020-4743-6_8

Cui J-l, Guo S-x, Xiao P-g (2011) Antitumor and antimicrobial activities of endophytic fungi from medicinal parts of *Aquilaria sinensis*. J Zhejiang Univ Sci B 12:385–392

Dequn Z (2001) Host-specificity, host-exclusivity, and host-recurrence in saprobic fungi. Mycol Res 105:1449–1457

Ding L-J, Cui H-L, Nie S-A, Long X-E, Duan G-L, Zhu Y-G (2019) Microbiomes inhabiting rice roots and rhizosphere. FEMS Microbiol Ecol 95:fiz040

Diogo HC, Sarpieri A, Pires MC (2005) Fungi preservation in distilled water. An Bras Dermatol 80:591–594

Dreyfuss M, Chapela IH (1994) Potential of fungi in the discovery of novel, low-molecular weight pharmaceuticals. In: Gullo VP (ed) Discovery of novel natural products with therapeutic potential. Elsevier, pp 49–80

Forchetti G, Masciarelli O, Alemano S, Alvarez D, Abdala G (2007) Endophytic bacteria in sunflower (*Helianthus annuus* L.): isolation, characterization, and production of jasmonates and abscisic acid in culture medium. Appl Microbiol Biotechnol 76:1145–1152

Freire AKL, dos Santos Bentes A, de Lima Sampaio I, de Lima AM, Botineli LF, da Rocha LC et al (2016) Availability and morphological characteristics of endophytic fungi held in different methods of preservation. Sci Res Essays 11:76–79

Gamboa MA, Bayman P (2001) Communities of endophytic fungi in leaves of a tropical timber tree (*Guarea guidonia*: Meliaceae) 1. Biotropica 33:352–360

Girão MD, Prado MRd, Brilhante RSN, Cordeiro RA, Monteiro AJ, Sidrim JJC et al (2004) Viabilidade de cepas de Malassezia pachydermatis mantidas em diferentes métodos de conservação. Rev Soc Bras Med Trop 37:229–233

Gravel V, Antoun H, Tweddell RJ (2007) Growth stimulation and fruit yield improvement of greenhouse tomato plants by inoculation with *Pseudomonas putida* or *Trichoderma atroviride*: possible role of indole acetic acid (IAA). Soil Biol Biochem 39:1968–1977

Hallmann J, Quadt-Hallmann A, Mahaffee W, Kloepper J (1997) Bacterial endophytes in agricultural crops. Can J Microbiol 43:895–914

Hamayun M, Hussain A, Khan SA, Kim H-Y, Khan AL, Waqas M et al (2017) Gibberellins producing endophytic fungus *Porostereum spadiceum* AGH786 rescues growth of salt affected soybean. Front Microbiol 8:686

Hardoim PR, Van Overbeek LS, Berg G, Pirttilä AM, Compant S, Campisano A et al (2015) The hidden world within plants: ecological and evolutionary considerations for defining functioning of microbial endophytes. Microbiol Mol Biol Rev 79:293–320

Herrera-Medina MJ, Steinkellner S, Vierheilig H, Ocampo Bote JA, García Garrido JM (2007) Abscisic acid determines arbuscule development and functionality in the tomato arbuscular mycorrhiza. New Phytol 175:554–564

Herrera J, Poudel R, Bokati D (2013) Assessment of root-associated fungal communities colonizing two species of tropical grasses reveals incongruence to fungal communities of North American native grasses. Fungal Ecol 6:65–69

Huang W, Cai Y, Hyde K, Corke H, Sun M (2008) Biodiversity of endophytic fungi associated with 29 traditional Chinese medicinal plants. Fungal Divers 33:61–75

Huang Y, Wang J, Li G, Zheng Z, Su W (2001) Antitumor and antifungal activities in endophytic fungi isolated from pharmaceutical plants *Taxus mairei*, *Cephalataxus fortunei* and *Torreya grandis*. FEMS Immunol Med Microbiol 31:163–167

Idris R, Trifonova R, Puschenreiter M, Wenzel WW, Sessitsch A (2004) Bacterial communities associated with flowering plants of the Ni hyperaccumulator *Thlaspi goesingense*. Appl Environ Microbiol 70:2667–2677

Iqbal M, Ashraf M (2013) Gibberellic acid mediated induction of salt tolerance in wheat plants: growth, ionic partitioning, photosynthesis, yield and hormonal homeostasis. Environ Exp Bot 86:76–85

Johnson NC, Graham J, Smith F (1997) Functioning of mycorrhizal associations along the mutualism–parasitism continuum. New Phytol 135:575–585

Joshi S, Sahgal M, Sahu S, Prakash A (2018) Fungal endophytes and their secondary metabolites: role in sustainable agriculture. In: Gehlot P, Singh J (eds) Fungi and their role in sustainable development: current perspectives. Springer, Singapore, Singapore, pp 121–146. https://doi.org/10.1007/978-981-13-0393-7_8

Jumpponen A (2001) Dark septate endophytes–are they mycorrhizal? Mycorrhiza 11:207–211

Kajula M, Tejesvi MV, Kolehmainen S, Mäkinen A, Hokkanen J, Mattila S et al (2010) The siderophore ferricrocin produced by specific foliar endophytic fungi in vitro. Fungal Biol 114:248–254

Karthik C, Oves M, Thangabalu R, Sharma R, Santhosh S, Arulselvi PI (2016) Cellulosimicrobium funkei-like enhances the growth of *Phaseolus vulgaris* by modulating oxidative damage under Chromium (VI) toxicity. J Adv Res 7:839–850

Kavroulakis N, Ntougias S, Zervakis GI, Ehaliotis C, Haralampidis K, Papadopoulou KK (2007) Role of ethylene in the protection of tomato plants against soil-borne fungal pathogens conferred by an endophytic *Fusarium solani* strain. J Exp Bot 58:3853–3864

Khan AL, Hamayun M, Kang S-M, Kim Y-H, Jung H-Y, Lee J-H et al (2012) Endophytic fungal association via gibberellins and indole acetic acid can improve plant growth under abiotic stress: an example of *Paecilomyces formosus* LHL10. BMC Microbiol 12:3

Khan AL, Waqas M, Khan AR, Hussain J, Kang S-M, Gilani SA et al (2013) Fungal endophyte *Penicillium janthinellum* LK5 improves growth of ABA-deficient tomato under salinity. World J Microbiol Biotechnol 29:2133–2144

Khan MS, Zaidi A, Ahmad E (2014) Mechanism of phosphate solubilization and physiological functions of phosphate-solubilizing microorganisms. In: Khan MS, Zaidi A, Musarrat J (eds) Phosphate solubilizing microorganisms. Springer International Publishing, Cham, pp 31–62. https://doi.org/10.1007/978-3-319-08216-5_2

Kia SH, Jurkechova M, Glynou K, Piepenbring M, Maciá-Vicente JG (2017) The effects of fungal root endophytes on plant growth are stable along gradients of abiotic habitat conditions. FEMS Microbiol Ecol 94

Kour D, Rana KL, Yadav N, Yadav AN, Kumar A, Meena VS et al (2019a) Rhizospheric Microbiomes: Biodiversity, Mechanisms of Plant Growth Promotion, and Biotechnological Applications for Sustainable Agriculture. In: Kumar A, Meena VS (eds) Plant growth promoting rhizobacteria for agricultural sustainability: from theory to practices. Springer, Singapore, Singapore, pp 19–65. https://doi.org/10.1007/978-981-13-7553-8_2

Kour D, Rana KL, Yadav N, Yadav AN, Singh J, Rastegari AA et al (2019b) Agriculturally and industrially important fungi: current developments and potential biotechnological applications. In: Yadav AN, Singh S, Mishra S, Gupta A (eds) Recent advancement in white biotechnology through fungi: volume 2: perspective for value-added products and environments. Springer International Publishing, Cham, pp 1–64. https://doi.org/10.1007/978-3-030-14846-1_1

Krohn K, Biele C, Drogies K, Steingrover K (2002) Endophytic fungi: a source of novel biologically active secondary metabolites. Eur J Org Chem 14:2331–2336

Kumar A, Chaturvedi AK, Yadav K, Arunkumar KP, Malyan SK, Raja P et al (2019) Fungal phytoremediation of heavy metal-contaminated resources: current scenario and future prospects. In: Yadav AN, Singh S, Mishra S, Gupta A (eds) Recent advancement in white biotechnology through fungi: volume 3: perspective for sustainable environments. Springer International Publishing, Cham, pp 437–461. https://doi.org/10.1007/978-3-030-25506-0_18

Kumar DSS, Hyde KD (2004) Biodiversity and tissue-recurrence of endophytic fungi in *Tripterygium wilfordii*. Fungal Diversity

Kusari S, Hertweck C, Spiteller M (2012) Chemical ecology of endophytic fungi: origins of secondary metabolites. Chem Biol 19:792–798

Lacaz CdS, Porto E, Martins JEC (1991) Micologia médica: fungos, actinomicetos e algas de interesse médico. Rev Inst Med Trop São Paulo 33:332–332

Ladha J, Reddy P (2003) Nitrogen fixation in rice systems: state of knowledge and future prospects. Plant Soil 252:151–167

Larran S, Perelló A, Simón MR, Moreno V (2007) The endophytic fungi from wheat (*Triticum aestivum* L.). World J Microbiol Biotechnol 23:565–572

Li H-Y, Wei D-Q, Shen M, Zhou Z-P (2012) Endophytes and their role in phytoremediation. Fungal Divers 54:11–18

Lindow SE, Brandl MT (2003) Microbiology of the phyllosphere. Appl Environ Microbiol 69:1875–1883

Lugtenberg BJ, Caradus JR, Johnson LJ (2016) Fungal endophytes for sustainable crop production. FEMS Microbiol Ecol 92

Awika JM (2011) Major cereal grains production and use around the world. Advances in cereal science: implications to food processing and health promotion, vol 1089, pp 1–13. https://doi.org/10.1021/bk-2011-1089.ch001

Malinowski DP, Belesky DP (2000) Adaptations of endophyte-infected cool-season grasses to environmental stresses: mechanisms of drought and mineral stress tolerance. Crop Sci 40:923–940

Malyan SK, Kumar A, Baram S, Kumar J, Singh S, Kumar SS et al (2019) Role of fungi in climate change abatement through carbon sequestration. In: Yadav AN, Singh S, Mishra S, Gupta A (eds) Recent advancement in white biotechnology through fungi: volume 3: perspective for sustainable environments. Springer International Publishing, Cham, pp 283–295. https://doi.org/10.1007/978-3-030-25506-0_11

McGonigle TP, Miller MH, Young D (1999) Mycorrhizae, crop growth, and crop phosphorus nutrition in maize-soybean rotations given various tillage treatments. Plant Soil 210:33–42

Mehta P, Sharma R, Putatunda C, Walia A (2019) Endophytic fungi: role in phosphate solubilization. In: Singh BP (ed) Advances in endophytic fungal research. Springer International Publishing, Cham, pp 183–209. https://doi.org/10.1007/978-3-030-03589-1_9

Mirlohi A, Sabzalian MR, Khayyam Nekouei M (2004) Endophytic fungi, characteristics and their potential for genetic manipulation. Iran J Biotechnol 2:75–83

Murphy B, Doohan F, Hodkinson T (2018) From concept to commerce: developing a successful fungal endophyte inoculant for agricultural crops. J Fungi 4:24

Naik BS, Shashikala J, Krishnamurthy Y (2009) Study on the diversity of endophytic communities from rice (*Oryza sativa* L.) and their antagonistic activities in vitro. Microbiol Res 164:290–296

Neilands J, Leong SA (1986) Siderophores in relation to plant growth and disease. Annu Rev Plant Physiol 37:187–208

Ngachan S, Mohanty A, Pattanayak A (2011) Status paper on rice in North East India. ICAR Research Complex for NEH Region, Rice Knowledge Management Portal (RKMP) Directorate of Rice Research

O'sullivan DJ, O'Gara F (1992) Traits of fluorescent *Pseudomonas* spp. involved in suppression of plant root pathogens. Microbiol Mol Biol Rev 56:662–676

Onions AH (1971) Chapter IV preservation of fungi. In: C Booth (ed) Methods in microbiology, vol 4. Elsevier, pp 113–151

Orole O, Adejumo T (2009) Activity of fungal endophytes against four maize wilt pathogens. Afr J Microbiol Res 3:969–973

Papageorgiou M, Skendi A (2018) 1-Introduction to cereal processing and by-products. In: Galanakis CM (ed) Sustainable recovery and reutilization of cereal processing by-products. Woodhead Publishing, pp 1–25. https://doi.org/10.1016/B978-0-08-102162-0.00001-0

Patle P, Navnage N, Ramteke P (2018) Endophytes in plant system: roles in growth promotion, mechanism and their potentiality in achieving agriculture sustainability. Int J Chem Stud 6:270–274

Potshangbam M, Devi SI, Sahoo D, Strobel GA (2017) Functional characterization of endophytic fungal community associated with *Oryza sativa* L. and *Zea mays* L. Front Microbiol 8:325

Pourkheirandish M, Komatsuda T (2007) The importance of Barley genetics and domestication in a global perspective. Ann Bot 100:999–1008

Prathyusha P, Rajitha Sri A, Ashokvardhan T, Satya Prasad K (2015) Antimicrobial and siderophore activity of the endophytic fungus *Acremonium sclerotigenum* inhabiting *Terminalia bellerica* Roxb. Int J Pharm Sci Rev Res 30:84–87

Priyadarshani C, Deshappriya N, Sandamali T (2018) Effect of fungal endophytes of rice variety Ld 368 on growth and brown spot disease incidence of rice. Trop Plant Res 5(3):292–302

Radji M, Sumiati A, Rachmayani R, Elya B (2011) Isolation of fungal endophytes from *Garcinia mangostana* and their antibacterial activity. Afr J Biotechnol 10:103–107

Rana KL, Kour D, Sheikh I, Dhiman A, Yadav N, Yadav AN et al (2019a) Endophytic fungi: biodiversity, ecological significance and potential industrial applications. In: Yadav AN, Mishra S, Singh S, Gupta A (eds) Recent advancement in white biotechnology through fungi, vol 1. Diversity and enzymes perspectives. Springer, Switzerland, pp 1–62

Rana KL, Kour D, Sheikh I, Yadav N, Yadav AN, Kumar V et al (2019b) Biodiversity of endophytic fungi from diverse niches and their biotechnological applications. In: Singh BP (ed) Advances in endophytic fungal research: present status and future challenges. Springer International Publishing, Cham, pp 105–144. https://doi.org/10.1007/978-3-030-03589-1_6

Rana KL, Kour D, Yadav AN (2018) Endophytic microbiomes: biodiversity, ecological significance and biotechnological applications. Res J Biotechnol 14:1–30

Redman RS, Sheehan KB, Stout RG, Rodriguez RJ, Henson JM (2002) Thermotolerance generated by plant/fungal symbiosis. Science 298:1581–1581

Rees RM, Baddeley JA, Bhogal A, Ball BC, Chadwick DR, Macleod M et al (2013) Nitrous oxide mitigation in UK agriculture. Soil Sci Plant Nutr 59:3–15

Ripa FA, Cao WD, Tong S, Sun JG (2019) Assessment of plant growth promoting and abiotic stress tolerance properties of wheat endophytic fungi. BioMed Res Int

Rodríguez H, Fraga R (1999) Phosphate solubilizing bacteria and their role in plant growth promotion. Biotechnol Adv 17:319–339

Rodriguez R, White J Jr, Arnold AE, Redman RS (2009) Fungal endophytes: diversity and functional roles. New Phytol 182:314–330

Saikkonen K, Ruokolainen K, Huitu O, Gundel PE, Piltti T, Hamilton CE et al (2013) Fungal endophytes help prevent weed invasions. Agric Ecosyst Environ 165:1–5

Sanders IR (2004) Plant and arbuscular mycorrhizal fungal diversity–are we looking at the relevant levels of diversity and are we using the right techniques? New Phytol 164:415–418

Santamaría J, Bayman P (2005) Fungal epiphytes and endophytes of coffee leaves (*Coffea arabica*). Microb Ecol 50:1–8

Santos-Medellín C, Edwards J, Liechty Z, Nguyen B, Sundaresan V (2017) Drought stress results in a compartment-specific restructuring of the rice root-associated microbiomes. MBio 8:e00764-00717

Sapkota R, Knorr K, Jørgensen LN, O'Hanlon KA, Nicolaisen M (2015) Host genotype is an important determinant of the cereal phyllosphere mycobiome. New Phytol 207:1134–1144

Sarwar MH, Sarwar MF, Sarwar M, Qadri NA, Moghal S (2013) The importance of cereals (Poaceae: Gramineae) nutrition in human health: a review. J Cereals Oilseeds 4:32–35

Schulz B, Boyle C, Draeger S, Römmert A-K, Krohn K (2002) Endophytic fungi: a source of novel biologically active secondary metabolites. Mycol Res 106:996–1004

Sessitsch A, Hardoim P, Döring J, Weilharter A, Krause A, Woyke T et al (2012) Functional characteristics of an endophyte community colonizing rice roots as revealed by metagenomic analysis. Mol Plant-Microbe Interact 25:28–36

Shah S, Shrestha R, Maharjan S, Selosse M-A, Pant B (2019) Isolation and characterization of plant growth-promoting endophytic fungi from the roots of Dendrobium moniliforme. Plants 8:5

Sharma S, Kour D, Rana KL, Dhiman A, Thakur S, Thakur P et al (2019) Trichoderma: biodiversity, ecological significances, and industrial applications. In: Yadav AN, Mishra S, Singh S, Gupta A (eds) Recent advancement in white biotechnology through fungi: volume 1: diversity and enzymes perspectives. Springer International Publishing, Cham, pp 85–120. https://doi.org/10.1007/978-3-030-10480-1_3

Shweta S, Zuehlke S, Ramesha B, Priti V, Kumar PM, Ravikanth G et al (2010) Endophytic fungal strains of *Fusarium solani*, from *Apodytes dimidiata* E. Mey. ex Arn (Icacinaceae) produce camptothecin, 10-hydroxycamptothecin and 9-methoxycamptothecin. Phytochemistry 71:117–122

Smith D (1998) The use of cryopreservation in the ex-situ conservation of fungi. Cryo-letters 19:79–90

Smith D, Onions AH (1994) The preservation and maintenance of living fungi, vol Ed. 2. CAB international

Spagnoletti FN, Tobar N, Di Pardo AF, Chiocchio VM, Lavado RS (2017) Dark septate endophytes present different potential to solubilize calcium, iron and aluminum phosphates. Appl Soil Ecol 111:25–32

Strobel G, Yang X, Sears J, Kramer R, Sidhu RS, Hess W (1996) Taxol from *Pestalotiopsis microspora*, an endophytic fungus of Taxus wallachiana. Microbiology 142:435–440

Suman A, Yadav AN, Verma P (2016) Endophytic microbes in crops: diversity and beneficial impact for sustainable agriculture. In: Singh D, Abhilash P, Prabha R (eds) Microbial inoculants in sustainable agricultural productivity, research perspectives. Springer, India, pp 117–143. https://doi.org/10.1007/978-81-322-2647-5_7

Sumarah MW, Puniani E, Blackwell BA, Miller JD (2008) Characterization of polyketide metabolites from foliar endophytes of *Picea glauca*. J Nat Prod 71:1393–1398

Tanimoto E (2005) Regulation of root growth by plant hormones—roles for auxin and gibberellin. Crit Rev Plant Sci 24:249–265

Tao G, Liu Z, Hyde K, Lui X, Yu Z (2008) Whole rDNA analysis reveals novel and endophytic fungi in *Bletilla ochracea* (Orchidaceae). Fungal Divers 33:101–112

Tarafdar J, Gharu A (2006) Mobilization of organic and poorly soluble phosphates by *Chaetomium globosum*. Appl Soil Ecol 32:273–283

Tian X, Cao L, Tan H, Zeng Q, Jia Y, Han W et al (2004) Study on the communities of endophytic fungi and endophytic actinomycetes from rice and their antipathogenic activities in vitro. World J Microbiol Biotechnol 20:303–309

Vandenkoornhuyse P, Quaiser A, Duhamel M, Le Van A, Dufresne A (2015) The importance of the microbiome of the plant holobiont. New Phytol 206:1196–1206

Vessey JK (2003) Plant growth promoting rhizobacteria as biofertilizers. Plant Soil 255:571–586

Waller F, Achatz B, Baltruschat H, Fodor J, Becker K, Fischer M et al (2005) The endophytic fungus *Piriformospora indica* reprograms barley to salt-stress tolerance, disease resistance, and higher yield. Proc Natl Acad Sci 102:13386–13391

Wang Y, Brown H, Crowley D, Szaniszlo P (1993) Evidence for direct utilization of a siderophore, ferrioxamine B, in axenically grown cucumber. Plant, Cell Environ 16:579–585

Waqas M, Khan AL, Kamran M, Hamayun M, Kang S-M, Kim Y-H et al (2012) Endophytic fungi produce gibberellins and indoleacetic acid and promotes host-plant growth during stress. Molecules 17:10754–10773

Weiss M, Sýkorová Z, Garnica S, Riess K, Martos F, Krause C et al (2011) Sebacinales everywhere: previously overlooked ubiquitous fungal endophytes. PLoS ONE 6:e16793

Xu G, Yang S, Meng L, Wang B-G (2018) The plant hormone abscisic acid regulates the growth and metabolism of endophytic fungus *Aspergillus nidulans*. Sci Rep 8:6504. https://doi.org/10.1038/s41598-018-24770-9

Yadav AN (2017) Agriculturally important microbiomes: biodiversity and multifarious PGP attributes for amelioration of diverse abiotic stresses in crops for sustainable agriculture. Biomed J Sci Tech Res 1:1–4

Yadav AN (2018) Biodiversity and biotechnological applications of host-specific endophytic fungi for sustainable agriculture and allied sectors. Acta Sci Microbiol 1:01–05

Yadav AN (2019a) Endophytic fungi for plant growth promotion and adaptation under abiotic stress conditions. Acta Sci Agric 3:91–93

Yadav AN (2019b) Fungal white biotechnology: conclusion and future prospects. In: Yadav AN, Singh S, Mishra S, Gupta A (eds) Recent advancement in white biotechnology through fungi: volume 3: perspective for sustainable environments. Springer International Publishing, Cham, pp 491–498. https://doi.org/10.1007/978-3-030-25506-0_20

Yadav AN, Kumar V, Prasad R, Saxena AK, Dhaliwal HS (2018a) Microbiome in crops: diversity, distribution and potential role in crops improvements. In: Prasad R, Gill SS, Tuteja N (eds) Crop improvement through microbial biotechnology. Elsevier, USA, pp 305–332

Yadav AN, Mishra S, Singh S, Gupta A (2019a) Recent advancement in white biotechnology through fungi: volume 1: diversity and enzymes perspectives. Springer International Publishing, Cham

Yadav AN, Sachan SG, Verma P, Saxena AK (2015a) Prospecting cold deserts of north western Himalayas for microbial diversity and plant growth promoting attributes. J Biosci Bioeng 119:683–693

Yadav AN, Sachan SG, Verma P, Saxena AK (2016) Bioprospecting of plant growth promoting psychrotrophic *Bacilli* from cold desert of north western Indian Himalayas. Indian J Exp Biol 54:142–150

Yadav AN, Sachan SG, Verma P, Tyagi SP, Kaushik R, Saxena AK (2015b) Culturable diversity and functional annotation of psychrotrophic bacteria from cold desert of Leh Ladakh (India). World J Microbiol Biotechnol 31:95–108

Yadav AN, Sharma D, Gulati S, Singh S, Kaushik R, Dey R et al (2015c) Haloarchaea endowed with phosphorus solubilization attribute implicated in phosphorus cycle. Sci Rep 5:12293

Yadav AN, Singh S, Mishra S, Gupta A (2019b) Recent advancement in white biotechnology through fungi: volume 2: perspective for value-added products and environments. Springer International Publishing, Cham

Yadav AN, Singh S, Mishra S, Gupta A (2019c) Recent advancement in white biotechnology through fungi: volume 3: perspective for sustainable environments. Springer International Publishing, Cham

Yadav AN, Verma P, Kumar V, Sangwan P, Mishra S, Panjiar N et al (2018b) Biodiversity of the genus *Penicillium* in different habitats. In: Gupta VK, Rodriguez-Couto S (eds) New and future developments in microbial biotechnology and bioengineering, *Penicillium* system properties and applications. Elsevier, Amsterdam, pp 3–18. https://doi.org/10.1016/b978-0-444-63501-3.00001-6

Yadav AN, Yadav N (2018) Stress-adaptive microbes for plant growth promotion and alleviation of drought stress in plants. Acta Sci Agric 2:85–88

Yadav AN, Yadav N, Sachan SG, Saxena AK (2019b) Biodiversity of psychrotrophic microbes and their biotechnological applications. J Appl Biol Biotechnol 7:99–108

Yang H, Ye W, Ma J, Zeng D, Rong Z, Xu M et al (2018) Endophytic fungal communities associated with field-grown soybean roots and seeds in the Huang-Huai region of China. PeerJ 6:e4713

Zhang W, Xu L, Yang L, Huang Y, Li S, Shen Y (2014) Phomopsidone A, a novel depsidone metabolite from the mangrove endophytic fungus *Phomopsis* sp. A123. Fitoterapia 96:146–151

Zheng Y-K, Qiao X-G, Miao C-P, Liu K, Chen Y-W, Xu L-H et al (2016) Diversity, distribution and biotechnological potential of endophytic fungi. Ann Microbiol 66:529–542

Zikmundova M, Drandarov K, Bigler L, Hesse M, Werner C (2002) Biotransformation of 2-benzoxazolinone and 2-hydroxy-1, 4-benzoxazin-3-one by endophytic fungi isolated from *Aphelandra tetragona*. Appl Environ Microbiol 68:4863–4870

Chapter 3
Current Perspectives on Phosphate-Solubilizing Endophytic Fungi: Ecological Significances and Biotechnological Applications

Edla Sujatha, Kuraganti Gunaswetha and Pallaval Veera Bramhachari

Abstract Phosphorus is one of the essential nutrients for optimum plant growth after nitrogen. Their structural and chemical complexity greatly reduces their availability to the plants and is one of the major limiting macroelements to plant growth. Phosphorus is present in both organic and inorganic forms. Though abundant amount of phosphorus is present in the soil, its availability is reduced by various environmental factors that influence bio-geo-cycling of phosphorus. Current research is mainly focused on the exploitation of endophytic fungi for solubilization of phosphorus in an efficient way. Endophytic fungi including the genera *Aspergillus, Penicillium, Piriformospora, Trichoderma, Curvularia,* and other class of endophytic symbionts such as AM fungi are identified as potent Phosphate solubilizers. Endophytic fungi promote plant growth by a variety of mechanisms such as solubilization of "P"-like macronutrients by different reactions, able to produce bio-control agents, i.e., antibiotics and siderophores and plant protecting agents against pathogens, synthesis of growth hormones such as gibberellins, cytokines, and auxins. Phosphate-solubilizing endophytic fungi are promising and efficient organisms capable of increasing "P" availability and the best alternative approach to chemical fertilizers.

Keywords Endophytic fungi · Biotechnological applications · Ecological significance · Inorganic and organic phosphates solubilization

3.1 Introduction

Soil microorganisms greatly influence the nature of the soil and its health through beneficial and harmful activities. Microorganisms present in the rhizosphere mediate certain functions, for instance, decomposition, nutrient immobilization, mineralization, nitrogen fixation, and release of nutrients. In addition to these, microorganisms

E. Sujatha (✉) · K. Gunaswetha
Department of Microbiology, Kakatiya University, Warangal, Telangana, India
e-mail: sujathaedla_1973@kakatiya.ac.in

P. V. Bramhachari
Department of Biotechnology, Krishna University, Machilipatnam 521001, Andhra Pradesh, India

A. N. Yadav et al. (eds.), *Plant Microbiomes for Sustainable Agriculture*, Sustainable Development and Biodiversity 25, https://doi.org/10.1007/978-3-030-38453-1_3

also possess phosphate-solubilizing ability by converting insoluble phosphates to soluble phosphorus (Pradhan and Sukla 2005) in soil and make them readily accessible to plants.

The father of plant pathology de Bary proposed the term endophyte to refer any organism that resides within plant tissue exclusive of any disease manifestations in host plant (Nisa et al. 2015). All types of plants harbor a wide variety of microorganisms, for instance, fungi, bacteria, and cyanobacteria that posses endophytic properties and play an imperative role in plant metabolism and physiology (Hardoim et al. 2015). During symbiotic association established between endophytes and host plants, both the organisms get mutually benefited, the plant provides nutrients to colonizing endophytes, while the endophytes accelerate biosynthetic pathways for metabolite synthesis that have many applications in agronomy. For example, management of plant growth and novel disease-resistant mechanisms against pathogens.

Endophytes can be isolated from exterior or interior part of sterilized plant tissues. Significant biochemical molecules such as Terpenoids, isoflavonoids, flavonoids, and phenolics are released from plant roots. They may attract the fungi from root region to colonize within the plant as an endophyte. However, the endophytes are depicted to be colonized in different plant tissues associated with the different ecosystems. Fungi can be classified into diverse groups derived from their role and survival, such as epiphytic, endophytic, pathogenic, and mycorrhizal fungi (Porras-Alfaro and Bayman 2011). Some endophytic fungi can find their way to either vertical or horizontal root region and penetrate to the deeper regions of plants.

Endophytic fungi colonize the tissues of host plant by particular route of transmission and this can be either vertical or horizontal method. Endophytic fungi transmit from the mother plant to offspring via seeds (true endophytes). A study carried out by Hodgson et al. (2014) in forbe species, common poppy, knapweed, cornflower, sheep's sorrel, groundsel, and ribwort plantain, and two endophyte species, *Cladosporium sphaerospermum* and *Alternaria alternata* primarily investigated the vertical transmission of species. Horizontal transmission occurs by airborne spores or through soil. Endophytes colonize forbes via leaves of the host through horizontal transmission.

According to earlier fossil records, evolutionary tendency reveals an association between a diverse group of plants and endophytic fungi. Plant endophyte communications resulted in plant growth promotion, uptake of micronutrients, and synthesis of different types of secondary metabolites and bioactive compounds with potential applications in industry, medicine, and agriculture. Endophytic fungi provide protection to plants against plant pathogens, reduce biotic and abiotic stresses, and for the reason that these organisms are considered as eco-friendly bioresources. Endophytic fungi may enhance plant growth by solubilization of potassium, phosphorus and zinc, produces phytohormones, viz., cytokines, gibberellic acids, indole acetic acids, hydrolytic enzymes and Fe-chelating compounds, ammonia and hydrogen cyanide (Rai et al. 2014). Different classes of fungi, for instance, Mucoromycota, Basidiomycota, Oomycota, and Ascomycota were depicted as plant growth promoters and protect the plants under anomalous and abiotic stress conditions. Natural products produced by endophytes were previously reported with potential anti-bacterial, anti-fungal, and anti-protozoal properties. Nonetheless, the secondary metabolites produced by endophytes are proved to have remarkable anti-cancer, antimicrobial, insecticidal properties, pharmaceutical sciences, and in other biotechnological applications (Kusari et al. 2011; Aly et al. 2010; Uzma et al. 2018; Mishra et al. 2017).

3.2 Diversity of Phosphate-Solubilizing Microorganisms

A great extent of microbial species exhibit phosphate-solubilizing ability, these include archaebacteria, bacteria, actinomycetes, and fungi. These microbes reside in the plant tissues exclusive of causing any harmful effects to the host. Generally, they prevail in tissues of the host plant with the symbiotic association. These microbes were previously isolated from different types of plants, including *Triticum* (Yadav et al. 2018a; Verma et al. 2015, 2016a, b), *Oryza sativa* (Piromyou et al. 2015), *Zea mays, Capsicum annuum L., Saccharum officinarum* (Montanez et al. 2012; Thanh and Diep 2014), mustard, citrus (Kasotia and Choudhary 2014), *Solanum tuberosum* (Rado et al. 2015; Manter et al. 2010;), *Glycine max* (Mingma et al. 2014), *Pisum sativum* (Narula et al. 2013; Tariq et al. 2014), *Phaseolus vulgaris* (Suyal et al. 2015), *Helianthus* (Forchetti et al. 2010; Ambrosini et al. 2012), and *Cicer arietinum* (Saini et al. 2015). Fungal endophytes pertaining to diverse genera including *Acremonium, Aspergillus, Paecilomyces, Cryptococcus, Fusarium, Curvularia, Rhodotorula, Cladosporium, Alternaria, Phaeomoniella, Chaetomium, Colletotrichum, Berkleasmium, Rhizoctonia, Geomyces, Leptospora, Phyllosticta, Microdochium, Neotyphodium, Ophiognomonia, Glomus, Penicillium, Rhizopus, Trichoderma, Xylaria,* and *Wallemia* have been isolated from various host plants (Suman et al. 2016; Verma et al. 2017; Yadav et al. 2018a, b). Recently, a nematode fungus *Arthrobotrys oligospora* was identified to solubilize rock phosphate Togo, Tilemi rock phosphate, Kodjari phosphate rock. The fungi solubilized all three types of rock phosphates. Given these, Duponnogs and group (2006) demonstrated the phosphate solubilization ability in vivo conditions.

3.3 Biotechnological Applications of Natural Products from Endophytic Fungi

Biotechnology has opened up numerous avenues for exploitation of endophytic microorganisms in medicine, agriculture, and industry from diverse ecological niches and their applications in agriculture are aptly essential for plant growth, plant protection, and yield (Yadav et al. 2018a; Rana et al. 2019c). Because of their ability to promote plant growth and adapt under extreme abiotic stresses, the endophytic microorganisms have, in fact, captured the attention of the scientific community

(Soni et al. 2018; Yadav et al. 2019b, c, d). With the exceptional capacity to produce secondary metabolites, the endophytic fungi may perhaps unearth novel applications in pharmaceuticals, industrial processes, and horticulture (Joseph and Priya 2011). Endophytic bacterial and fungal organisms possess extensive applications as bio-control agents, bio-inoculants, and bio-fortification of micronutrients (Yadav 2019; Yadav et al. 2019a; Yadav and Yadav 2019).

The excessive usage of chemical phosphorus (P) fertilizers to increase agricultural yield sequentially to meet the requirements of escalating global food demand potentially causes soil and water pollution, eutrophication, depletion of soil fertility, and deposition of toxic heavy metals such as arsenic (As), lead (Pb), and selenium (Se) in the soil. Abundant soil microbes together with bacteria, fungi, actinomycetes, and algae are efficient in solubilizing insoluble soil phosphate to convert into the soluble P and making it available to plants. Strikingly, these microbes promote the growth and yield of a wide variety of crops. Thus, it is essential to inoculate phosphate-solubilizing microorganisms (PSM) via the seeds, to the crop and soil, which is a potential strategy to improve the crop yield. Despite their immense significance in the improvement of soil fertility, phosphorus-solubilizing microbes are yet to replace conventional chemical fertilizers in commercial agriculture. Extensive studies are obligatory to comprehend recent approaches in a diversity of phosphate-solubilizing endophytes and their colonizing ability and application to enhance agronomic yield.

3.4 Endophytic Fungi as P-Solubilizers and Growth Promoters

Phosphate-solubilizing fungi enhance plant growth by different mechanisms and those are (Fig. 3.1): provide nutrients to plant by solubilization process, produce biological control substances, i.e., antibiotics and siderophores, provide protection against the plant pathogens, and stimulate growth hormones production (auxins, gibberellins, and cytokines). In agriculture, phosphate-solubilizing fungi play a significant role as bio-inoculants for improvement of plant growth (Khan et al. 2010; Kour et al. 2019b, c; Rana et al. 2019a, b). The competent phosphate-solubilizing fungi were tested under in vitro conditions and selected for large-scale production and eventually distributed to farmers. In addition to that, suitable carrier selection is also paramount for the development of fungal inoculants such as peat, farmyard manure, soil, cow dung, and cafe powder which are being used as suitable carriers. However, a perfect carrier is designed to possess some unique qualities like good absorption ability, sufficient level of moisture pH, aeration stability, pH buffering capacity, and porousness. In addition to these characteristics, the carrier must be eco-friendly, non-hazardous to microbes, plants, animal, and humans. In addition to this, it should be easy to handle, mix, sterilizable, and store. Keeping in view of cost–benefit ratio, the carriers ought to be cheaper and easily available. The carrier enhances the persistence of phosphorus-solubilizing activity; fungal spores after mixing with a carrier can be

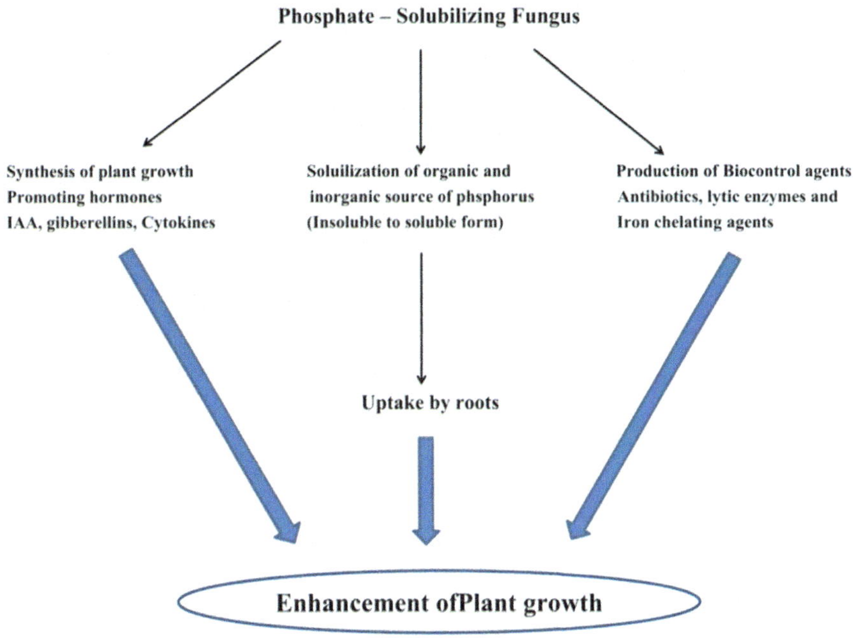

Fig. 3.1 Mechanisms of plant growth acceleration by endophytic P-solubilizing fungi

stored for about 3 months at 30 ± 2 °C. Plethora of examples of commercially available inoculants, i.e., *Penicillium radicum* and *Penicillium bilaiae* were demonstrated in large-scale production and shown to possess “P” absorption ability. Various stages involved in bulk scale production and application of P-solubilizing fungi are clearly depicted (Fig. 3.2).

Various studies indicating the effect of a single culture and/or mixed culture of phosphate-solubilizing fungus on different plant growth parameters observed the effect of mixed inoculation of P-solubilizing fungal strains (two strains of *A. awamori* and four of *P. citrinum*) on growth and seed production of chickpea in pot experiment. Notably, all the isolates were shown synergistic effect and resulted in noteworthy stimulation of root and shoot lengths of legume, height of the plant, seed weight, and number compared to the un-inoculated control.

Phosphorus is a vital nutrient and a part of structural compounds and mediates catalytic reactions in plant metabolism. Phosphorus plays a major role in capturing solar energy and is converted into useful plant compound. Phosphorus is a key component of DNA and RNA. Two phosphate-solubilizing fungi, i.e., *Penicillium oxallicum* P4 and *Aspergillus niger* P85 were isolated by Yin et al. (2015) from calcium-rich soils of China. A remarkable increase in plant fresh weight was observed in strain p24 when rock phosphate was supplemented externally. A study carried out on *Aspergillus aculeatus* P93 has also shown a significant increase in the availability of soluble phosphorus of maize grown in non-amended soil (Yin et al. 2017).

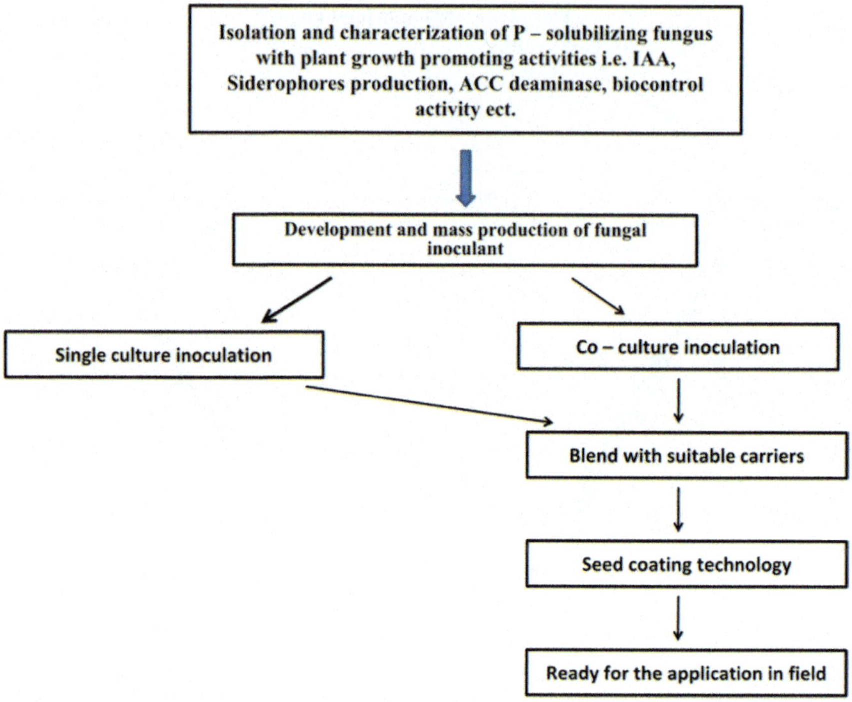

Fig. 3.2 Production and application of phosphate-solubilizing endophytic fungal inoculants

In a recent study, phosphate-solubilizing microorganisms depicted a synergistic effect on plant growth parameters and leaf chlorophyll content (Senthil kumar et al. 2018). In addition to solubilizing phosphates, some may produce potential bio-control agents against plant pathogens. PSM can produce anti-fungal compounds such as flavonoids and phenolics, siderophores, antibiotics and hydrolytic compounds. All of which inhibits growth of plant pathogens.

3.5 Phosphate Solubilization Mechanism by Endophytic Fungi

Based on the availability of type of phosphates (organic or inorganic), endophytic fungi employ suitable mechanism for solubilization of phosphates and endophytic fungi are capable to synthesize organic acids, proteins, OH^- ions, Ca^{+2} exopolysaccharides, CO, siderophores, and enzymes, those may play a significant role in phosphate solubilization (Fig. 3.3).

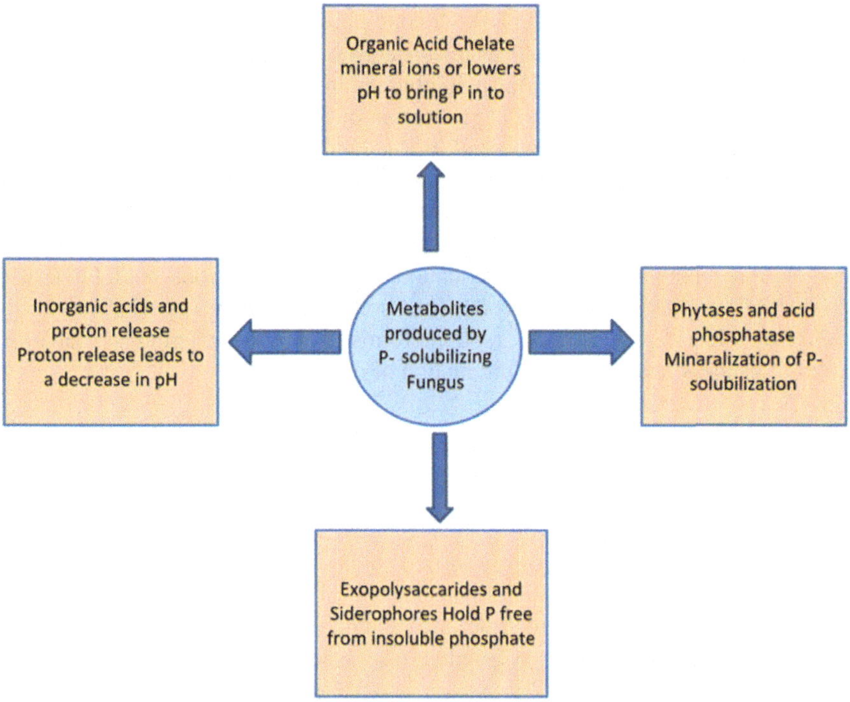

Fig. 3.3 Production of metabolites by P-solubilizing fungi

3.5.1 Inorganic P-Solubilization

Various theories explained the key mechanism underlying the inorganic phosphate solubilization. The principal mechanism is dissolving compounds such as hydroxyl ions, siderophores, organic acids, carbon monoxides, proteins, extracellular enzymes, and exopolysaccharides (Sharma et al. 2013). Secretion of organic acids by endophytic fungi into soil seems to be a most important mechanism for inorganic phosphate solubilization primarily organic acids such as glycolic, maleic, formic, lactic acid, gluconic acid, oxalic, tartaric propionic, and succinic acids. The quantity of organic acid varies with the endophytic fungal strain and also the type of organic acid produced is greatly determined by source of insoluble phosphorus. According to Mendes et al. (2013), *Aspergillus niger* FS1 primarily secrete oxalic acid in higher quantity in treatments with $FePO_4$ and $AlPO_4$, whereas gluconic acid was produced in meager quantity in $AlPO_4$. In contrast to this, *Penicillium canescens* FS23 produced citric and gluconic acids after treatment with $Ca_3(PO_4)_2$, $AlPO_4$, and rock phosphate. Organic matter present in soil is an excellent source of organic phosphorus. The total quantity of organic phosphorous present in soil is as high as 30–50% of total phosphorus. Organic phosphorus in the soil is principally in the form of inositol phosphate (or) soil phytates. Organic phosphorus can be mineralized with the involvement of enzymes such as phosphatases, phytases.

Many studies explained about p-solubilization based on an organic acid concept. Several genes are involved in production of organic acids. Among all organic acids, gluconic acid is more essential in P-solubilization produced by endophytic fungi. The genetic basis of P-solubilization was studied by Kusari et al. (2012). He observed upregulation of pyrroloquinoline quinine and glucose dehydrogenase genes in solubilization of phosphorous. PQQ-dependent glucose dehydrogenase present on cytoplasmic membrane stimulates oxidation of glucose to gluconic acid. Because of production of gluconic acid, the pH of the soil further decreases which make the following ions HO_4^{-2} and HPO_4^{-3} (soluble forms of phosphorus) more available. An array of genes is involved in the production of organic acids.

3.5.2 Organic P-Solubilization

Mineralization of organic phosphate carried out by involvement of various enzymes, i.e., phytases, phosphonatases, and phosphatases. Phosphatases dephosphorylation or hydrolyze and phosphoanhydride and organic phospho-ester bonds of organic matter. Among all phosphatases, predominant types of enzymes are phosphomonoesterases (Nannipieri et al. 2011). Based on pH optima they are classified as acid and alkaline phosphatase (Behera et al. 2014). Several genes encoding for alkaline and acid phosphatases with broad substrate specificity were cloned and characterized. However, a considerable amount of phosphatases was secreted by plant roots, and it has been reported that microbial phosphates possess a stronger affinity for substrate when compared to derived by plant phosphatases.

A large quantity of phosphorus is found in fruits and seeds for the reason that it is important for the development of seeds. Phytin is a significant form of "P" in seeds. Phytin is naturally degraded by phytases. This is the main source of inositol phosphate and constitutes for more than fifty percent of organic phosphorus in the soil. Phytases act upon phytate and make available free form of phosphorus. Noteworthy that the phosphonatases and carbon-phosphorus lyases hydrolyze carbon-phosphorus bond of organophosphates and release free phosphate (Rodriguez et al. 2006). Because of scarcity, organo-phosphatases do not add much to the soluble form of phosphate in the soil solution. A plethora of studies revealed that organic acids released by P-solubilizing fungi are much superior to bacteria; therefore, endophytic fungi exhibit greater P-solubilization activity. Motsara et al. (1995) revealed that solubilization of rock phosphate was much higher under in vitro conditions by *Rhizoctonia solani*, Penicillium, *Fusarium oxysporum*, and *Aspergillus niger*.

3.6 Development of Phosphate-Solubilizing Endophytic Fungal Inoculants

Phosphates-solubilizing endophytic fungal inoculants are used as major bio-fertilizers. These bio-inoculants are more eco-friendly when compared to chemical fertilizers. Fungi are depicted as predominant P-solubilizing inoculants when compared to other bacterial inoculants; hence, these strains hold much importance in agriculture. Nonetheless, several fungal bio-fertilizers were already developed by IARI as bio-inoculants which comprises *Aspergillus awamori, A. niger,* and *P. digitatum, P.bilaii* strains which were commercialized by Novozymes Biologicals Limited (Canada). Interestingly, the strain *P.radicum* was recently developed by Bio-Care Technology (Australia) by Gupta and Rodriguez Couto (2018). Similarly, in India, P-bio-fertilizers were produced by Ambika Biotech and Agro Services (Madhya Pradesh) (Pal et al. 2015).

For production of P-solubilizing fungal inoculums, huge amount of endophytic fungal strains are required. There are broadly three phases in development of bio-fertilizers. In the first phase, there is a selection and screening of potential phosphate solubilizers followed by a selection of fungal inoculants. The screening process can be carried but by the cultivation of fungi in modified Pikovskaya's medium. In second phase, proper endophytic fungal bio-fertilizers can be developed and the third phase includes checking the quality and persistence of P-solubilizing microorganisms and distribution to farmers (Khan et al. 2010; Kumar et al. 2017). Notably, few potent microorganisms are selected, screened, and cultivated in large scale for production of bio-fertilizers under optimized conditions in a suitable fermentation broth. For cultivation of fungi, lower pH (acidic condition) is more suitable at the same time inhibits the bacterial contamination (Nelofer et al. 2016). Once an adequate amount of growth is obtained, the biomass can be extracted and mixed with suitable pre-sterilized carrier material, purified, packed under aseptic conditions, and stored under appropriate conditions before commercialization. At every stage of bio-fertilizer production, it is mandatory to assess the level of contamination as well as for the amount of desired microorganisms.

Definite problems may also be associated with the commercial-scale synthesis of bio-fertilizers, among those sometimes microorganisms unable to survive under in vivo conditions. This may be due to the fact that the bio-inoculants are either difficult to survive under unfavorable environmental conditions or outcompeted by presented microflora (Walia et al. 2017). One of the important strategies to surmount this problem is amalgamation of bio-fertilizers with suitable carriers.

Carriers being used in production of bio-fertilizers should possess definite characteristics like it should be easily mixed with microbes to enhance the sustainability and survival of microorganisms by maintaining optimum pH, an adequate level of moisture and aeration, etc. Thus, the carrier material should possess an excellent moisture absorption ability and pH buffering capacity, nonetheless, it should also be non-toxic to microorganisms, and eco-friendly to plants, animals, and humans. In addition to there, it is easy to sterilize, and easy to mix, handle, and store. In view of the cost–benefit ratio, the carrier should be effortlessly available and cheap. Nowadays, different types of carriers are being used for the production of bio-fertilizers. Smith (1995) has classified the carriers into different categories, first category comprises different types of soils, peat and coal, waste cake powder farmyard manure, plant debris and second group comprises barnyard compost, soya bean oil, shelled nut oil, barnyard compost. The last group consists of inert materials like perlite, rock phosphate, and calcium sulfate vermiculite. These carriers have also been used in combinations. Wang et al. (2015) reported the utilization of different carriers for developing a bio-fertilizer of *A. niger* and reported a mixture of wheat husk and perlite to enhance the availability of "P" content.

Among soil microorganisms, AM fungi have been found to be a noteworthy component of soil–plant systems (Schreiner and Bethlenfalvay 2003). An AM fungus plays a major role in nutrient and water uptake by plants and provides other benefits to host, such as tolerance under adverse environmental conditions and disease resistance (Pal et al. 2014). Due to their obligatory symbiosis, it is highly difficult to produce AM-based bio-fertilizers in *in vitro* conditions.

Mass scale production AMF is highly difficult because of its specific nutritional requirements (Pal et al. 2015). The different strategies were reported by Berruti et al. (2016) for the utilization of AMF as bio-fertilizers. According to first strategy AMF-harboring rhizosphere, soil can be used as bio-inoculants; however, this method may not be reliable and may perhaps result in colonization of weeds and pathogens. In other strategies, AM spores that were isolated from root region can be used for this AM fungal organism inoculated on a host trap plant in an inert medium. The trap plant is highly susceptible for Arbuscular mycorrhizal fungal growth, and therefore it is used for production of AM fungi for bulk scale. This is frequently used inoculums for inculcation to crop plants in large scale. This consists of a set of some kind of AM spores present in soil inoculants.

In this method, trap plant should be highly amicable for inoculation of desired AM fungi and also should be ideal for large-scale production of propagates. In addition to these, the trap plant should show intense root development within a short period of time and resist to harsh environmental conditions, suitable for synthesis of fungal propagules (Sadhana 2014). An important observation came from a study of Selvakumar et al. (2016) that maize could be the suitable host trap plant when compared to the Sudan grass for the propagation of *Claroideoglomus etunicatum*. Other trap plants including *Chloris gayana, Sorghum vulgare, Zea mays, Sorghum bicolor var. sudanense,* and *Ipomea batatas* are most common trap plants used for mass scale culturing of Arbuscular mycorrhizal fungi (Sadhana 2014).

By wet sieving and decantation, the AM fungi is regularly isolated from soil (Singh et al. 2010) followed by microscopic observation of AM fungi. Mass multiplication is carried out by collecting a large number of spores by pot culture method. Host trap plant and AM fungi were cultured in natural solid medium containing clay, peat, sand, perlite soil, and different types of composted plant debris. Tamil Nadu Agricultural University designed a method, and according to this, a trench lined with polythene sheet is being used as plant growth pot or tub. Fifty kg of vermiculite and 5 kg of sterilized soil are filled in trench up to 20 cm height. To this 1 kg of AM spores inoculated 2–5 cm below the surface of vermiculite. Sterilized seeds of trap plant are sown in a trench along with an appropriate dosage of nitrogen source urea and superphosphate. After the period of 60 days roots of trap, plants are cut, spores, a mixture of vermiculite, hyphal fragments, and infected root pieces obtained were used as AM fungal inoculants. Without using soil also some researchers developed hydroponics and aeroponics for the cultivation of AM fungi in the presence of trap plants. The major advantage of these methods is that there is a feasibility to produce pure and clean AM spores (Ijdo et al. 2011).

3.7 Application of Phosphate-Solubilizing Endophytic Fungal Bio-Inoculants

Treatments of seed surface with suitable bio-inoculants are the most common choice of inoculation prior to seeding and reported to be the popularly used method (Walia et al. 2013b). However, there are few techniques that are widely used for the inoculation of endophytic microorganisms, viz., soil application, seed treatment, and foliar spraying. In seed treatment process, carrier-coated fungal inoculums are immersed with seeds in a liquid culture medium. In this method, a fungus adhered firmly to the seed surface. Conversely, there are some constraints in this method. Amount of viable fungi adhered on to the seed surface may not be adequate. The plant species are under cultivation at commercial scale by vegetative propagation, and the endophytic P-solubilizing bio-inoculants are usually applied to plant parts before planting in the field (Panhwar et al. 2013; Kour et al. 2019a; Kumar et al. 2019; Yadav et al. 2019e). The shoots developed from such plants are deemed to be more amenable for bacterization by endophytic microorganisms. Application of endophytes to the soil is another method of bio-inoculants application (supplementation of soil with endophytic bio-inoculants). These methods have many advantages which include the following:

- A high number of P-solubilizing fungi may disseminate per unit area.
- Less number of interactions may occur between bio-inoculants and chemically treated seeds.
- This method is more rapid in comparison with seed inoculation technique.
- These bio-inoculants are more tolerant to dry and desiccated conditions.

In view of above aspects, phosphate-solubilizing endophytic fungal inoculants can be applied by two approaches.

1. Single culture of phosphate-solubilizing fungi can be used as inoculants as single culture approach (SCA).
2. Two cultures can be used as bio-inoculants are called mixed culture approach (MCA).

3.8 Effect of Phosphate-Solubilizing Fungal Bio-Inoculants

Soil is a natural habit for wide variety of microbial communities. The interactions occurring between microbial communities basically impact a physico-chemical property of the soil, and soil fungi performs several imperative roles in the maintenance of soil biochemistry directly or indirectly. Direct endophytic fungi accelerate plant growth promotion through the production of phytohormones, mineralization of salts and ions (Guleria et al. 2014) and in the indirect mechanism; it plays an important role in bio-control agents against phytopathogenic microbe (Walia et al. 2013a). Mehta et al. (2011) studied the growth enhancement of groundnut in association with endophytic fungi. They studied two fungi, i.e., *Aspergillus niger and Penicillium notatum* supplement of soils tri-calcium phosphate (TCP) under pot culture conditions and reported that there is a remarkable improvement in dry weight and height of the plant. When a mixed culture of fungal strains was employed as inoculants, a substantial improvement was observed in plant height as 81% and plant dry weight as 105% compared to controls (Prasanna et al. 2011; Mehta et al. 2011).

It is pertinent that a number of plants and weight of seeds enhanced remarkably with single or multiple inoculations of fungal strains. Other studies carried out by Priyadharsini and Muthukumar (2017) on pigeon pea revealed that when inoculated with the fungi *Curvularia geniculata* has shown a significant impact on growth parameters. *C. geniculata* inoculated seedlings of pigeon pea were taller (26.53%) and showed increased shoot and root dry weight (16.67–33.33%) as compared to uninoculated control seedlings. In addition to P-solubilization, endophytic fungi also play an imperative role in phytohormone production that can remarkably enhance plant growth. Thus, the exploitation of phosphate-solubilizing fungi is considered eco-friendly, profitable, and sustainable approach for enhancement of crop yield.

3.9 Application of RDNA Technology in Developing Phosphate-Solubilizing Endophytic Fungi

Interaction of endophytic fungi with host plant is relatively an intricate process. Set of genes are involved in such interaction including nitrogen, phosphorus, and other nutrient exchanges between endophytic fungi and host plant tissues were studied; nonetheless, widespread research is desirable to better comprehend the genetic aspects of such interactions. Comprehensive and enhanced knowledge is essential for the involvement of genes and their regulation to undertake genetic manifestation of fungi, which consecutively can be employed for better phosphorus uptake and improved plant growth. The molecular approaches signify a vital role in understanding the genetic aspects of host fungal interactions. Among several molecular approaches, the cloning and gene sequencing methods are most promising and consent to determine which techniques are time-consuming. Nucleic acid hybridizations and probing techniques are required to possess sufficient knowledge of microbial community. However, other molecular methods such as amplified ribosomal DNA restriction analysis or ribosomal intergenic spacer analysis can be employed for the endophytic fungal colonizations.

3.10 Future Applications of Endophytic Fungal Phosphate Solubilizers

In recent times, usage of phosphate fertilizers is highly expensive that cannot be afforded by farmers, particularly in developing countries. Scientists thus have a great responsibility toward society to find some innovative ways from biological sources to make available "P" to crop plants, by an economically efficient alternative for chemical fertilizers. Most of the soils are deficient in available phosphorus to plants and chemical fertilizers are expensive. Due to this interest has been developed in the application of rhizosphere microbes and endophytic fungi with phosphate-solubilizing capacity as bio-inoculants to solubilize phosphate from poorly available sources in soil. Although the potentiality increased for developing such inoculants, their vast applications remain incomplete by intricacy in an understanding of microbial inoculants. These endophytic fungi not only enhance the phosphate availability to the plants but also provide protection to the plants against plant pathogens and stimulate plant growth. The major challenge associated with endophytic fungi is a commercial-scale application, in fact, managing microbial communities to favor plant colonization by beneficial endophytic microorganisms. The contribution of endophytic fungal research may have environmental and economic impacts. Molecular-level research in this aspect is necessary for a better understanding of host endophytic interaction.

Acknowledgements The authors sincerely acknowledge the support extended from Kakatiya University, Warangal and Krishna University, Machilipatnam.

Conflict of Interest We declare no Conflict of Interest.

References

Aly AH, Debbab A, Kjer J, Proksch P (2010) Fungal endophytes from higher plants: a prolific source of phytochemicals and other bioactive natural products. Fungal Divers 41:1–16

Ambrosini A, Beneduzi A, Stefanski T, Pinheiro F, Vargas L, Passaglia LP (2012) Screening of plant growth promoting rhizobacteria isolated from sunflower (*Helianthus annuus* L.). Plant Soil 356:245–264. https://doi.org/10.1007/s11104-011-1079-1

Behera BC, Singdevsachan SK, Mishra RR, Dutta SK, Thatoi HN (2014) Diversity, mechanism and biotechnology of phosphate solubilizing microorganisms in mangrove—a review. Biocatal Agric Biotechnol 3:97–110

Berruti A, Lumini E, Balestrini R, Bianciotto V (2016) Arbuscular mycorrhizal fungi as natural bio-fertilizers: let's benefit from past successes. Front Microbiol 6:1559

Duponnois R, Kisa M, Plenchette C (2006) Phosphate solubilizing potential of the nematofungus *Arthrobotrys oligospora*. J Plant Nutr Soil Sci 169:280–282

Forchetti G, Masciarelli O, Izaguirre MJ, Alemano S, Alvarez D, Abdala G (2010) Endophytic bacteria improve seedling growth of sunflower under water stress, produce salicylic acid, and inhibit growth of pathogenic fungi. Curr Microbiol 61:485–493

Guleria S, Sharma K, Walia A, Chauhan A, Shirkot CK (2014) Population and functional diversity of phosphate solubilizing bacteria from apricot (*Prunus Armeniaca*) of mid and high regions of Himachal Pradesh. Bioscan 9(2):1435–1443

Gupta VG, Rodriguez-Couto S (eds) (2018) New and future developments in microbial biotechnology and bioengineering: penicillium system properties and applications. Elsevier

Hardoim PR, van Overbeek LS, Berg G, Pirttilä AM, Compant S, Campisano A et al (2015) The hidden world within plants: ecological and evolutionary considerations for defining functioning of microbial endophytes. Microbiol Mol Biol Rev 79:293–320

Hodgson S, Cates C, Hodgson J, Morley NJ, Sutton BC, Gange AC (2014) Vertical transmission of fungal endophytes is widespread in forbs. Ecol Evol 4:1199–1208

IJdo M, Cranenbrouck S, Declerck S (2011) Methods for large-scale production of AM fungi: past, present, and future. Mycorrhiza 21:1–16

Joseph B, Priya RM (2011) Bioactive compounds from endophytes and their potential in pharmaceutical effect: a review. Am J Biochem Mol Biol 1(3):291–309

Kasotia A, Choudhary DK (2014) Role of endophytic microbes in mitigation of abiotic stress in plants. In: Ahmad P, Rasool S (eds) Emerging technologies and management of crop stress tolerance. Elsevier, New York, pp 97–108

Khan MS, Zaidi A, Ahemad M, Oves M, Wani PA (2010) Plant growth promotion by phosphate solubilizing fungi-current perspective. Arch Agron Soil Sci 56:73–98

Kour D, Rana KL, Yadav AN, Yadav N, Kumar V, Kumar A, Sayyed RZ, Hesham AE-L, Dhaliwal HS, Saxena AK (2019a) Drought-tolerant phosphorus-solubilizing microbes: biodiversity and biotechnological applications for alleviation of drought stress in plants. In: Sayyed RZ, Arora NK, Reddy MS (eds) Plant growth promoting rhizobacteria for sustainable stress management, Volume 1: Rhizobacteria in abiotic stress management. Springer, Singapore, pp 255–308. https://doi.org/10.1007/978-981-13-6536-2_13

Kour D, Rana KL, Yadav N, Yadav AN, Kumar A, Meena VS, Singh B, Chauhan VS, Dhaliwal HS, Saxena AK (2019b) Rhizospheric microbiomes: biodiversity, mechanisms of plant growth promotion, and biotechnological applications for sustainable agriculture. In: Kumar A, Meena VS (eds) Plant growth promoting rhizobacteria for agricultural sustainability: from theory to practices. Springer, Singapore, pp 19–65. https://doi.org/10.1007/978-981-13-7553-8_2

Kour D, Rana KL, Yadav N, Yadav AN, Singh J, Rastegari AA, Saxena AK (2019c) Agriculturally and industrially important fungi: current developments and potential biotechnological applications. In: Yadav AN, Singh S, Mishra S, Gupta A (eds) Recent advancement in white biotechnology through fungi, Volume 2: Perspective for value-added products and environments. Springer, Cham, pp 1–64. https://doi.org/10.1007/978-3-030-14846-1_1

Kumar V, Yadav AN, Verema P, Sangwan P, Abhishake S, Singh B (2017) β-Propeller phytases: diversity, catalytic attributes, current developments and potential biotechnological applications. Int J Biol Macromolec 98:595–609

Kumar M, Kour D, Yadav AN, Saxena R, Rai PK, Jyoti A, Tomar RS (2019) Biodiversity of methylotrophic microbial communities and their potential role in mitigation of abiotic stresses in plants. Biologia 74:287–308. https://doi.org/10.2478/s11756-019-00190-6

Kusari S, Zuhlke S, Spiteller M (2011) Effect of artificial reconstitution of the interaction between the plant *Camptotheca acuminata* and the fungal endophyte *Fusarium solani* on camptothecin biosynthesis. J Nat Prod 74(4):764–775

Kusari S, Hertweck C, Spiteller M (2012) Chemical ecology of endophytic fungi: origins of secondary metabolites. Chem Biol 19(7):792–798

Manter DK, Delgado JA, Holm DG, Stong RA (2010) Pyrosequencing reveals a highly diverse and cultivar-specific bacterial endophyte community in potato roots. Microb Ecol 60:157–166

Mehta P, Walia A, Chauhan A, Shirkot CK (2011) Accelerated solubilization of inorganic phosphate and production of antifungal activity in soil by plant growth promoting rhizobacteria isolated from apple rhizosphere. J Mycol Plant Pathol 41(3):342–349

Mendes GO, Moreira de Freitas AL, Pereira OL, da Silva IR, Vassilev NB, Costa MD (2013) Mechanisms of phosphate solubilization by fungal isolates when exposed to different P sources. Ann Microbiol 64:239. https://doi.org/10.1007/s13213-013-0656-3

Mingma R, Pathom-aree W, Trakulnaleamsai S, Thamchaipenet A, Duangmal K (2014) Isolation of rhizospheric and roots endophytic actinomycetes from Leguminosae plant and their activities to inhibit soybean pathogen, *Xanthomonas campestris* pv. *glycine*. World J Microbiol Biotechnol 30:271–280

Mishra VK, Passari AK, Chandra P, Leo VV, Kumar B, Gupta VK, Singh BP (2017) Determination and production of antimicrobial compounds by *Aspergillus clavatonanicus* strain MJ31, an endophytic fungus from Mirabilis jalapa L. using UPLC-ESI-MS/MS and TD GC-MS. PLoS One 12(10):1–24. https://doi.org/10.1371/journal.pone.0186234

Montanez A, Blanco AR, Barlocco C, Beracochea M, Sicardi M (2012) Characterization of cultivable putative endophytic plant growth promoting bacteria associated with maize cultivars (*Zea mays* L.) and their inoculation effects *in vitro*. Appl Soil Ecol 58:21–28

Motsara MR, Bhattacharyya PB, Srivastava B (1995) Bio-fertilizers their description and characteristics. In: Biofertilizer technology, marketing and usage, a sourcebook-cum-glossary, Fertilizer development and consultation organization 204–204. A Bhanot Corner, 1–2 Pamposh Enclave, New Delhi, 110048, India, pp 9–18

Nannipieri P, Giagnoni L, Landi L, Renella G (2011) Role of phosphatase enzymes in soil. In: Bunemann E, Oberson A, Frossard E (eds) Phosphorus in action: biological processes in soil phosphorus cycling, soil biology, vol 26. Springer, Heidelberg, pp 251–244

Narula S, Anand R, Dudeja S, Pathak D (2013) Molecular diversity of root and nodule endophytic bacteria from field pea (*Pisum sativum* L.). Legum Res 36:344–350

Nelofer R, Syed Q, Nadeem M, Bashir F, Mazhar S, Hassan A (2016) Isolation of phosphorus-solubilizing fungus from soil to supplement biofertilizer. Arab J Sci Eng 41:2131–2138

Nisa H, Kamili AN, Nawchoo IA, Shafi S, Shameem N, Bandh SA (2015) Fungal endophytes as prolific source of phytochemicals and other bioactive natural products: a review. Mic Pathog 82:50–59

Pal S, Singh HB, Rakshit A (2014) The arbuscular mycorrhizal symbiosis: an underground world wide web. In: Singh DP, Singh HB (eds) Microbial communities for sustainable soil health and ecosystem productivity. Studium Press LLC, Houston, pp 219–253

Pal S, Singh HB, Farooqui A, Rakshit A (2015) Fungal biofertilizers in Indian agriculture: perception, demand and promotion. J Eco-friendly Agric 10:101–113

Panhwar QA, Jusop S, Naher UA, Othman R, Razi MI (2013) Application of potential phosphate-solubilizing bacteria and organic acids on phosphate solubilization from phosphate rock in aerobic rice. Sci World J 2013:272409

Piromyou P, Greetatorn T, Teamtisong K, Okubo T, Shinoda R, Nuntakij A, Tittabutr P, Boonkerd N, Minamisawa K, Teaumroong N (2015) Preferential association of endophytic *Bradyrhizobia* with different rice cultivars and its implications for rice endophyte evolution. Appl Environ Microbiol 81:3049–3061

Porras-Alfaro A, Bayman P (2011) Hidden fungi, emergent properties: endophytes and microbiomes. Annu Rev Phytopathol 49:291–315

Pradhan S, Sukla LB (2005) Solubilization of inorganic phosphates by fungi isolated from agriculture soil. Afr J Biotechnol 5:850–854

Prasanna A, Deepa V, Murthy PB, Deecaraman M, Sridhar R, Dhandapani P (2011) Insoluble phosphate solubilization by bacterial strains isolated from rice rhizosphere soils from southern India. Int J Soil Sci 6(2):134–141

Priyadharsini P, Muthukumar T (2017) The root endophytic fungus *Curvularia geniculata* from *Parthenium hysterophorus* roots improves plant growth through phosphate solubilization and phytohormone production. Fungal Ecol 27:69–77

Rado R, Andrianarisoa B, Ravelomanantsoa S, Rakotoarimanga N, Rahetlah V, Fienena F, Andriambeloson O (2015) Biocontrol of potato wilt by selective rhizospheric and endophytic bacteria associated with potato plant. Afr J Food Agric Nutr Dev 15:9762–9776

Rai M, Rathod D, Agarkar G, Dar M, Brestic M, Pastore GM, Junior MRM (2014) Fungal growth promotor endophytes: a pragmatic approach towards sustainable food and agriculture. Symbiosis 62:63–79

Rana KL, Kour D, Sheikh I, Dhiman A, Yadav N, Yadav AN, Rastegari AA, Singh K, Saxena AK (2019a) Endophytic fungi: biodiversity, ecological significance and potential industrial applications. In: Yadav AN, Mishra S, Singh S, Gupta A (eds) Recent advancement in white biotechnology through fungi, vol 1. Diversity and enzymes perspectives. Springer, Switzerland, pp 1–62

Rana KL, Kour D, Sheikh I, Yadav N, Yadav AN, Kumar V, Singh BP, Dhaliwal HS, Saxena AK (2019b) Biodiversity of endophytic fungi from diverse niches and their biotechnological applications. In: Singh BP (ed) Advances in endophytic fungal research: present status and future challenges. Springer, Cham, pp 105–144. https://doi.org/10.1007/978-3-030-03589-1_6

Rana KL, Kour D, Yadav AN (2019c) Endophytic microbiomes: biodiversity, ecological significance and biotechnological applications. Res J Biotechnol 14:142–162

Rodriguez H, Fraga R, Gonzalez T, Bashan Y (2006) Genetics of phosphate solubilization and its potential applications for improving plant growth-promoting bacteria. Plant and soil 287:15–21

Sadhana B (2014) Arbuscular mycorrhizal fungi (AMF) as a bio-fertilizer-a review. Int J Curr Microbiol App Sci 3:384–400

Saini R, Dudeja SS, Giri R, Kumar V (2015) Isolation, characterization, and evaluation of bacterial root and nodule endophytes from chickpea cultivated in Northern India. J Basic Microbiol 55:74–81

Schreiner RP, Bethlenfalvay GJ (2003) Crop residue and Collembola interact to determine the growth of mycorrhizal pea plants. Biol Fertil Soils 39(1):1–8

Selvakumar G, Kim K, Walitang D, Chanratana M, Kang Y, Chung B, Sa T (2016) Trap culture technique for propagation of arbuscular mycorrhizal fungi using different host plants. Korean J Soil Sci Fertil 49:608–613

Senthil Kumar CM, Jacob TK, Devasahayam S, Thomas S, Geethu C (2018) Multifarious plant growth promotion by an entomopathogenic fungus *Lecanicillium psalliotae*. Microbiol Res 207:153–160

Sharma SB, Sayyed RZ, Trivedi MH, Gobi TA (2013) Phosphate solubilizing microbes: sustainable approach for managing phosphorus deficiency in agricultural soils. Springerplus 2:587

Singh SR, Singh U, Chaubey AK, Bhat MI (2010) Mycorrhizal fungi for sustainable agriculture—a review. Agric Rev 31:93–104

Smith SR (1995) Agricultural recycling of sewage sludge and the environment. CAB international

Soni R, Yadav SK, Rajput AS (2018) ACC-deaminase producing rhizobacteria: prospects and application as stress busters for stressed agriculture. In: Panpatte DG, Jhala YK, Shelat HN, Vyas RV (eds) Microorganisms for green revolution. Springer, New Delhi, pp 161–175

Suman A, Yadav AN, Verma P (2016) Endophytic microbes in crops: diversity and beneficial impact for sustainable agriculture. In: Singh DP, Abhilash P, Prabha R (eds) Microbial inoculants in sustainable agricultural productivity, research perspectives. Springer, New Delhi, pp 117–143. https://doi.org/10.1007/978-81-322-2647-5_7

Suyal DC, Yadav A, Shouche Y, Goel R (2015) Bacterial diversity and community structure of Western Indian Himalayan red kidney bean (*Phaseolus vulgaris*) rhizosphere as revealed by 16S rRNA gene sequences. Biologia 70:305–313

Tariq M, Hameed S, Yasmeen T, Zahid M, Zafar M (2014) Molecular characterization and identification of plant growth promoting endophytic bacteria isolated from the root nodules of pea (*Pisum sativum* L.). World J Microbiol Biotechnol 30:719–725

Thanh DTN, Diep CN (2014) Isolation, characterization and identification of endophytic bacteria in maize (*Zea mays* L.) cultivated on Acrisols of the Southeast of Vietnam. Am J Life Sci 2:224–233

Uzma F, Hashem A, Murthy N, Mohan HD, Kamath PV, Singh BP, Venkataramana M, Gupta VK, Siddaiah CN, Chowdappa S, Alqaeawi AA, Abd Allah EF (2018) Endophytic fungi alternative sources of cytotoxic compounds: a review. Front Pharmacol 9(309):1–37. https://doi.org/10.3389/fphar.2018.00309

Verma P, Yadav AN, Shukla L, Saxena AK, Suman A (2015) Hydrolytic enzymes production by thermotolerant *Bacillus altitudinis* IARI-MB-9 and *Gulbenkiania mobilis* IARI-MB-18 isolated from Manikaran hot springs. Int J Adv Res 3:1241–1250

Verma P, Yadav AN, Khannam KS, Kumar S, Saxena AK, Suman A (2016a) Molecular diversity and multifarious plant growth promoting attributes of bacilli associated with wheat (*Triticum aestivum* L.) rhizosphere from six diverse agro-ecological zones of India. J Basic Microbiol 56:44–58

Verma P, Yadav AN, Khannam KS, Mishra S, Kumar S, Saxena AK, Suman A (2016b) Appraisal of diversity and functional attributes of thermotolerant wheat associated bacteria from the peninsular zone of India. Saudi J Biol Sci. https://doi.org/10.1016/j.sjbs.2016.01.042

Verma P, Yadav AN, Kumar V, Singh DP, Saxena AK (2017) Beneficial plant-microbes interactions: biodiversity of microbes from diverse extreme environments and its impact for crops improvement. In: Singh DP, Singh HB, Prabha R (eds) Plant-microbe interactions in agro-ecological perspectives. Springer Nature, Singapore, pp 543–580. https://doi.org/10.1007/978-981-10-6593-4_22

Walia A, Mehta P, Chauhan A, Shirkot CK (2013a) Antagonistic activity of plant growth promoting rhizobacteria isolated from tomato rhizosphere against soil borne fungal plant pathogens. Int J Agri Environ Biotechnol 6(4):587–595

Walia A, Mehta P, Chauhan A, Shirkot CK (2013b) Effect of *Bacillus* sp. strain CKT1 as inoculums on growth of tomato seedlings under net house conditions. Proc Natl Acad Sci India Sect B Biol Sci 84(1):144–155

Walia A, Guleria S, Chauhan A, Mehta P (2017) Endophytic bacteria: role in phosphate solubilization. In: Maheshwari DK, Annapurna K (eds) Endophytes: crop productivity and protection, sustainable development and biodiversity. Springer, Berlin, pp 1–33

Wang H, Liu S, Zhai L, Zhang J, Ren T, Fan B, Liu H (2015) Preparation and utilization of phosphate bio-fertilizers using agricultural waste. J Int Agric Adv 14:158–167

Yadav AN (2019) Endophytic fungi for plant growth promotion and adaptation under abiotic stress conditions. Acta Sci Agric 3:91–93

Yadav N, Yadav AN (2019) Actinobacteria for sustainable agriculture. J Appl Biotechnol Bioeng 6:38–41

Yadav AN, Kumar V, Prasad R, Saxena AK, Dhaliwal HS (2018a) Microbiome in crops: diversity, distribution and potential role in crops improvements. In: Prasad R, Gill SS, Tuteja N (eds) Crop improvement through microbial biotechnology. Elsevier, New York, pp 305–332

Yadav AN, Verma P, Kumar V, Sangwan P, Mishra S, Panjiar N, Gupta VK, Saxena AK (2018b) Biodiversity of the genus penicillium in different habitats. In: Gupta VK, Rodriguez-Couto S (eds) New and future developments in microbial biotechnology and bioengineering, penicillium system properties and applications. Elsevier, Amsterdam, pp 3–18. https://doi.org/10.1016/b978-0-444-63501-3.00001-6

Yadav AN, Gulati S, Sharma D, Singh RN, Rajawat MVS, Kumar R et al (2019a) Seasonal variations in culturable archaea and their plant growth promoting attributes to predict their role in establishment of vegetation in Rann of Kutch. Biologia 74:1031–1043. https://doi.org/10.2478/s11756-019-00259-2

Yadav AN, Yadav N, Sachan SG, Saxena AK (2019b) Biodiversity of psychrotrophic microbes and their biotechnological applications. J Appl Biol Biotechnol 7:99–108

Yadav AN, Mishra S, Singh S, Gupta A (2019c) Recent advancement in white biotechnology through fungi. Volume 1: Diversity and enzymes perspectives. Springer, Cham

Yadav AN, Singh S, Mishra S, Gupta A (2019d) Recent advancement in white biotechnology through fungi. Volume 2: Perspective for value-added products and environments. Springer, Cham

Yadav AN, Singh S, Mishra S, Gupta A (2019e) Recent advancement in white biotechnology through fungi. Volume 3: Perspective for sustainable environments. Springer, Cham

Yin Z, Shi F, Jiang H, Roberts DP, Chen S, Fan B (2015) Phosphate solubilization and promotion of maize growth by *Penicillium oxalicum* P4 and *Aspergillus niger* P85 in a calcareous soil. Can J Microbiol 61(12):913–923

Yin Z, Fan B, Roberts DP, Chen S, Shi F, Buyer JS, Jiang H (2017) Enhancement of maize growth and alteration of the rhizosphere microbial community by phosphate-solubilizing fungus *Aspergillus aculeatus* P93. J Agric Biotechnol 2(2):1–10

Chapter 4
Endophytic Microbes from Medicinal Plants and Their Secondary Metabolites for Agricultural Significances

Chanda V. Parulekar Berde, Prachiti P. Rawool, Pallaval Veera Bramhachari and Vikrant B. Berde

Abstract Endophytes constitute an important component of microbial diversity since 20 years, remarkable progress in the field revealed the significance of endophytic microorganisms. Endophytic fungi are an unexplored group of organisms that has huge potential for innovative pharmaceutical substances; they are established as anticancer, antioxidants, antifungal, and anti-inflammatory. Likewise in recent years, incredible progress was made in developing them as therapeutic molecules against diverse ailments. In recent years, more studies are warranted in bioprospecting new endophytic microorganisms and their applications. Bacterial and fungal endophytes ubiquitously reside in internal tissue of living plants. Endophytic fungi distributed out from tropical region to arctic region, possess vast potential in terms of secondary metabolite production. It is pertinent to know that the various bioactive indispensable compounds evaluated by these endophytic fungi are host-specific. They are very significant in augmenting the adaptability of the endophyte and its host plants for instance biotic and abiotic stress tolerance. The ensuing effect is to produce metabolites either primary or secondary that are obliging for fungi themselves, the host plant in addition to the human race thereof. This chapter primarily emphasizes on the ecology, colonization, biodiversity, secondary metabolites from endophytic fungal cultures.

Keywords Endophytic microbial diversity · Medicinal plants · Bioactive compounds · Secondary metabolites

---

C. V. Parulekar Berde (✉) · P. P. Rawool
Department of Microbiology, Gogate Jogalekar College, Ratnagiri 415612, Maharashtra, India
e-mail: cvberde27@gmail.com

P. V. Bramhachari
Department of Biotechnology, Krishna University, Machilipatnam 521001, Andhra Pradesh, India

V. B. Berde
Department of Zoology, Arts, Commerce and Science College, Lanja 416701, Maharastra, India

A. N. Yadav et al. (eds.), *Plant Microbiomes for Sustainable Agriculture*, Sustainable Development and Biodiversity 25, https://doi.org/10.1007/978-3-030-38453-1_4

4.1 Introduction

With the human population rising, a simultaneous boost in diseases along with an increase in the incidence of reemerging diseases is noticed. The key origin of these medical emergencies being drug resistance in pathogenic bacteria, while the solution lies in the discovery of newer drugs to combat resistant bacteria. Numerous ecosystems were previously explored for the bioprospection of antibiotic producing microorganisms counting with terrestrial, marine, freshwater ecosystems. Notably, with the advent of endophytic microorganisms, a new niche is open for drug discovery (Berde 2015).

Endophyte was commonly associated with fungi initially, but now also includes bacteria as well. These microbes may perhaps exert beneficial or detrimental effects on plants and their metabolism. The endophytic microbes dwell in different tissue types within numerous plant species. The endophytic bacteria are ubiquitous in nature. An endophyte completes its life cycle colonizing inter and intracellularly contained by the healthy tissues of the host plant, with no visible symptoms of disease (Wilson 1995). And thus some endophytic microorganisms live in plant tissue without causing considerable harm to the plant.

Plant-associated microbes have been discovered in the fossilized tissues of stems and leaves (Taylor and Taylor 2000). The endophytic microorganisms are believed to have devised genetic systems and acquired as well as donated characteristics (Stierle et al. 1993). The growth, survival, and transmission of endophytic microorganisms to other plants occur vertically or via vectors and are firmly dependant on the host plants. When at least one stage of the life cycle of the endophyte is outside the host plant, it is facultative endophytes. Many substances of medicinal importance found in plants have also been extracted from their endophytes (Yadav 2017; Yadav et al. 2017; Yadav and Yadav 2018). Therefore, the attention is now on studies that are focused on the isolation and application of endophytes from medicinal plants.

Endophytic microorganisms survive inside the host plant tissues and produce novel metabolic compounds, having activity against various pathogens. Researchers have discovered new therapeutic alternatives in the form of bioactive secondary metabolites in endophytes such as antiviral, antibacterial, anticancer, and antidiabetic compounds (Kumar et al. 2015). Recently endophytes are considered as an essential source of secondary metabolites and bioactive antimicrobial natural products. The endophytic bacterial natural products, for example, munumbicins, ecomycins, pseudomycins, and xiamycins are antibacterial, antimycotic, and antiplasmodial in addition to antiviral in nature, respectively (Berde 2015).

Recent studies evidenced counts of natural products including compounds made up of terpenoids, flavonoids, alkaloids, steroids, etc. Metabolites of endophytes have been reported to hinder the growth of a number of microorganisms (Rana et al. 2019c; Yadav 2018). Microbial metabolites are considered as antifungal and antibacterial chemotherapeutic. In 2008, Moricca and Ragazzi reported that genes regulating the communication between an endophyte and a plant are strictly modulated by the environment. Hostile environment induces the production of defense chemicals.

Plants growing in extreme habitats ought to be screened for isolation of endophytes and their metabolites. Plants present in various environmental conditions including tropic, temperate, xerophytic, and aquatic, harbor the endophytic microorganisms. The endophyte associated plants produce various metabolites that induce resistance. It is notable that symbiotic plant triggers defense system more promptly than non symbiotic plants following a pathogen encounter (Jalgaonwala et al. 2011).

In the plant–microbe relationship, the endophytes supply nutrients to the plant, protect the plant from the invasion of pathogenic microorganisms, and in return, get space to colonize. The various activities of endophytes such as nitrogen fixation, solubilizing iron, and production of metabolites for plant protection enable the endophyte to help in the associate plant growth (Marx 2004; Porras-Soriano et al. 2009; Ryan et al. 2008; Rana et al. 2019a, b). The endophytes produce excess substances of impending use to modern medicine, agriculture, and industry. Endophytes were documented to produce a variety of biological activities such as antibiotic, anti-inflammatory, antiviral, anticancer, and antioxidant (Kado 1992; Kobayashi and Palumboo 2000; Yadav et al. 2019a, b, c).

4.2 Ecology of Endophytic Bacteria

The endophytic bacteria were perhaps isolated from monocotyledonous as well as dicotyledonous plants, including woody tree species, for example, oak (Brooks et al. 1994) and pear (Whitesides and Spotts 1991), as well as herbaceous crop plants, such as sugar beets (Jacobs et al. 1985) and maize (Gutierrez-Zamora and Martinez-Romero 2001). Diversity linked with bacterial endophytes occurs in plant species and also in colonizing bacterial taxa.

Plants can be colonized concurrently by a variety of endophytic bacteria. Plant endophytic bacteria have been classified into 82 genera within Alphaproteobacteria, Betaproteobacteria, Gammaproteobacteria, Actinobacteria Firmicutes, and *Bacteroidetes* and most of them fit into Alphaproteobacteria, Betaproteobacteria, and Gammaproteobacteria (Lodewyckx et al. 2002; Rosenblueth and Martínez-Romero 2006). While new endophytes are being continuously reported from different plant species. Within a particular plant, in the different tissues, different bacterial and fungal species can colonize. Thus, the plant itself forms a complex microecosystem, providing niche/habitats for the endophytes as per their preferences (Kour et al. 2019b, c). These habitats are not only exemplified by plant external surfaces, where epiphytic bacteria predominate, but also by internal tissues especially in xylem and phloem, where many microorganisms penetrate and survive.

The distribution of endophytic bacteria in different parts of the plant was first observed by Gardner et al. (1982). The endophytic bacteria present in the xylem fluid of Florida citrus tree roots was identified by the authors. Among the 13 genera found, the most dominant species were *Pseudomonas* amounting to 40% while *Enterobacter* comprised of 18%. This stable biodiversity is considered to be the most important condition in the establishment of any ecosystem.

4.3 Colonization of Plant by Endophytes

In some cases, endophytic bacteria may be transferred through the seed. These endophytic bacteria can stimulate plant growth and their progress in the host tissue. Endophytic microorganisms having the ability to fix atmospheric nitrogen and solubilize phosphate, as well as the ability to eliminate soil contaminants, play a crucial role in increasing the soil fertility.

Even though bacteria are prokaryotes and fungi are eukaryotes, they share many qualities of their relationship with plant hosts. In both the cases, colonization of root tissues is internal as well as external. Mostly it is systemical. The mode of colonization, however, differs in the two. Bacteria primarily colonize intercellularly and are mostly found in the vascular tissues of host plants. This helps the endophyte in its distribution. Asymptomatic colonization of the roots by fungi may be intercellular or intracellular in nature.

4.4 Natural Products from Endophytic Bacteria as Secondary Metabolites

Knowing the importance of endophytic microbial community, the endophytic microbial composition of the medicinal plants should be studied, identity of the microorganisms should be carried out, and the endophytic microbial preservation should be given preference in research. Secondly, there is a need to bioprospect the endophytes for other industrial applications also apart from antibiotics. The relationship between the geographical distribution of plants and their endophytic composition, as well as the plant–microbe relation needs to be established.

The endophytes colonize a particular niche, i.e., the plant tissue, which helps them in their role as biocontrol agents similar to that of phytopathogens. There are numerous reports about the role of the endophytic microorganisms in controlling plant pathogens, insects, and nematodes, and also in accelerating seedling emergence, enhancing growth of plants, and helping in plant establishment under adverse conditions (Kour et al. 2019a; Suman et al. 2016). Disease development is prevented due to the de novo production of varied new compounds and antifungal metabolites.

According to Lodewyckx et al. (2002), endophytes include bacterial genera that was found in soil, such as *Pseudomonas, Burkholderia,* and *Bacillus*. A wide range of diverse range of secondary metabolic products including antibiotics, anticancer compounds, volatile organic compounds, antifungal, antiviral, insecticidal, and immunosuppressant agents are obtained from these genera. Extensive number of biologically active compounds have been isolated from endophytic microorganisms; however, there still remains a largely untapped source of novel natural products.

Guo et al. 2000 reported cytonic acids that act on viruses namely cytomegaloviruses. There are very few reports of antiviral from endophyte bacteria. Sun et al. (2006) have worked on the endophyte *B. amyloliquefaciens* (ES-2) isolated from a herb *Scutellaria baicalensis* Georgi. The endophyte produces fengycins and surfactins which have antibacterial and antifungal properties. Zhou et al. (2015) has reported an endophytic fungi *Aspergillus versicolor* producing antiviral butyrolactones. Settu et al. (2010) have worked on the endophytes of *Andrographis paniculola*. Endophytes of these plants possess activity against Gram-positive and Gram-negative bacteria. Khaled et al. (2018) have worked on endophytic fungi of Egyptian medicinal plants and have found these isolates to possess antiviral and antioxidant activities.

4.4.1 Endophytes as a Source of Antibiotics

Medicinal plants have been used in the treatment of numerous infections and diseases, with their medicinal applications described in the Ayurveda. The compounds responsible for these medical applications are present in the plants and are alo obtained from the endopytic bacteria and fungi, endophytic in these plants. These natural compounds offer a great diversity of chemical structures that can be researched and applied for betterment of mankind (Berde 2015). Research on secondary metabolites with antimicrobial activity is essential with the development of antibiotic resistance in pathogens and the problem of emerging and reemerging diseases. The potential endophytic microorganisms can be utilized to address these problems.

Numerous synthetic drugs have been developed based on the lead compounds isolated from natural products. The classical example cited is that of the prototypical taxane isolated by Wani et al. in 1971, from the bark of a yew tree *Taxus brevifolia*. In 1996, Strobel et al. reported an endophytic fungus (*Pestalotiopsis microspora*) found in Yew tree with ability to produce Taxol. Like fungal endophytic cultures, endophytic bacteria also have potential of synthesizing novel natural products. Work is being focussed in order to explore endophytic bacteria for new and unique natural products of commercial importance. Endophytes thus are a source for antibacterial, antifungal, antidiabetic, antioxidant, and immunosuppressive products. Ecomycins, Pseudomycins, Munumbicins are some examples of the unique antibiotics obtained from endophytes.

4.4.1.1 Diterpenes

A large number of compounds with cytotoxic activity have been found to be produced by endophytic fungi over the years. Paclitaxel or Taxol is an antitumor compound reported from endophytes of number of plants. *Pestalotiopsis microspora* endophytic fungal species isolated from *Taxodium distichum* (Li et al. 1996) and *Taxus wallichiana* (Strobel et al. 1996) have been reported to produce paclitaxel. The endophytic fungal isolate *Penicillium raistrickii*endophytic in *Taxus brevifolia* was found to produce paclitaxel as well as baccatin III (Stierle and Stierle 2000). Earlier, from the same plant species, an endophytic fungi *Taxomyces andreanae* was reported, producing paclitaxel and baccatin (Strobel et al. 1993). There are two more reports of Paclitaxel production by *Pestalotiopsis* species. *Pestalotiopsis pausiceta* associated with *Cardiospermum helicacabum* (Gangadevi et al. 2008) and *Pestalotiopsis terminaliae* endophytic in the plant *Terminalia arjuna* (Gangadevi and Muthamary 2009a). Scientists have reported another endophytic fungus isolated from *Terminalia arjuna*, identified as *Chaetomella raphigera*, also to produce paclitaxel (Gangadevi and Muthamary 2009b). Production of paclitaxel was reported from an endophyte, *Bartalinia robillardoides*, of the medicinal plant *Aegle marmelos* or Indian bael (Gangadevi and Muthamary 2008).

4.4.1.2 Polyketides

Curvularia geniculate, an endophytic fungus, isolated from *Catunaregam tomentosa*, is reported to produce 5 hybrid peptide–polyketides, curvularides A–E. Curvularide B showed antifungal activity against *C. albicans*in in addition to synergistic activity with a fluconazole drug. A number of polyketides have been reported from endophytic fungal strains, shown in Table 4.1.

4.4.1.3 Lignans

Podophyllotoxin is a lignin with antimitotic and tubulin polymerase inhibition activity. Its derivatives namely, etoposide, teniposide, and etoposide phosphate, are preferably used in the treatment of cancer. A number of endophytes have been reported for podophyllotoxin production. Endophytic fungi *Tramates hirsute* isolated from dried rhizomes of *Podophyllum hexandrum* produces podophyllotoxin and its derivatives (Puri et al. 2006). Another report of fungal endophyte producing podophyllotoxin from *Podophyllum peltatum* has been cited. The endophyte was identified as *Phialocephala fortinii* (Eyberger et al. 2006). *Fusarium oxysporum*, an endophyte isolated from *Juniperus recurve*, a medicinal plant found in the Himalayas, was found to produce Podophyllotoxin (Kour et al. 2008). Podophyllotoxin has also been isolated from an *Alternaria* species, an endophyte of the plant *Juniperus vulgaris*(Lu et al. 2006) and *Aspergillus fumigatus*, endophyte of *Juniperus communis* L. Horstmann (Kusari et al. 2009).

4.4.1.4 Terpenoids

Stierle and Stierle (2000) have reported the isolation of endophytic *Penicillium* species from *Taxus brevifolia* (Yew). Of these endophytes, *Penicillium brevicompactum* is reported to produce a terpenoid, mycophenolic acid. This compound is antifungal, an immune suppressant drug and used in the treatment of Dengue.

Table 4.1 Polyketides produced by endophytic fungi

Compound	Uses	Host plant	Endophytic fungi	References
Griseofulvin	Antifungal	*Abies holophylla (Manchurian fir)*	*Xylaria* sp.	Park et al. (2005)
		Pinus strobus (White pine)	*Xylaria* sp.	Richardson et al. (2014)
		Vaccinium augustifolium (Blue berry shrub)		
		Taxus brevifolia	*Penicillium* sp.	Stierle et al. (1995), Stierle and Stierle (2000)
Brefeldin A	Antifungal, antiviral, and anticancer, protein transport inhibitor	*Texus mairei* *Torruia grandis*	*Paecilomycs* sp.	Wang et al. (2002)
		Torruia grandis	*Aspergillus clavatus*	
Sequoiatones	Antitumor (Breast Cancer)	*Sequoia sempervirens*	*Aspergillus paraciticus*	Stierle et al. (2001)
Sequoiamonascins	Antitumor (Breast, lung and CNS)	*Sequoia sempervirens*	*Aspergillus paraciticus*	Stierle et al. (2003)
Torreyanic acid	Anticancer	*Torrueya taxifolia*	*Pestalotiopsis microspore*	Lee et al. (1996)
Nodulisporins A, B, C	Antifungal	*Juniperus cedrus*	*Nodulis porium* sp	Dai et al. (2006)
Nodulisporins D, E, F	Antibacterial, antifungal, antialgal	*Arica arborea*	*Nodulis porium* sp	Dai et al. (2009)
Phomopsolides	Insecticidal (antibeetle), antibacterial	*Taxus brevifolia*	*Phomopsis oblonga*	Grove (1985)
		Taxus brevifolia	*Penicillium* sp	Steirle et al. (1997)
Rugulosin	Antilarval (Sprous budworm	*Balsam fir*	*Hormonema dematioids*	Calhoun et al. (1992)
Methyl(2Z,4E)-6(acetyloxy)-5-formyl-7-oxoocta-2,4-dienoate (Macrolide pyrinopherol)	Antifungal	*Pinus strobus*	*Lophodermium* sp.	Sumarah et al. (2011)

4.4.1.5 Terpenes

Many compounds belonging to this group have been isolated from endophytic fungal cultures and have been tabulated below in Table 4.2.

4.4.2 Heterospirocyclic Compounds

Two heterospirocyclic compounds useful in the treatment of cancer have been found to be produced by endophytic fungal cultures. Pseurotin A is antibacterial in addition to being anticancer. It is produced by *Penicillium raistrickii*, an endophyte of the plant *Taxus brevifolia* (Stierle and Stierle 2000) and *Penicillium janczewskii* KM Zalessky, associated with the *Prumnopitys andina* (Schmeda-Hieschmann et al. 2008). Tauramin is an anticancer compound produced by the endophytic fungi *Phyllosticta spinarum* isolated from *Platycladus orientilis* (Wijeratne et al. 2008).

4.4.2.1 Antimicrobial Peptides

Antimicrobial Peptides (AMPs) are the new generation of native peptide molecules. These are found in all living beings. They are being referred to as natural antibiotics. The AMPs are reported to have a very wide activity against a large spectrum of pathogenic microorganisms as well as protozoan and metazoan parasites (Liu et al. 2000; Vizioli and Salzet 2002). All of these components are main elements involved directly in the innate immune response of their hosts. This activity comprises of the expression of fluid phase proteins that recognize pathogen-associated molecular patterns. The response of the antibiotic peptides is quick, highly efficient, and with broad host activity range (Hoffmann and Reichhart 2002).

4.4.2.2 Antimicrobial Peptides from Endophytes

Endophytic bacteria compete with the pathogenic organisms and prevent them from colonizing the plant tissues. Secondary metabolites produced by the endophytes prevents the growth of pathogens microorganisms, thus playing a role in the plant defense mechanisms. AMPS are molecules of choice for drug development due to specificity for their targets with higher degree of interactions. Antibacterial cyclo-(Pro-Thr) and cyclo-(Pro-Tyr) are produced by endophytic fungus *Penicillium* sp, endophytic fungi found in mangrove plant *Acrostichum aureurm*. Both peptides demonstrated activity against *Staphylococcus aureus* and *Candida albicans*.

Epichlicin, a novel cyclic peptide was reported from the endophytic fungus *Epichloe typhina*, found in plant *Phleum pretense* L. The peptide was antagonistic at low concentrations, against the *Cladosporium phlei* spores, the fungal pathogen of the timothy plant (Seto et al. 2007).

Table 4.2 Terpenes isolated from fungal endophytic isolates

Compound	Uses	Host plant	Endophytic fungi	References
5-(Hydroxymethyl)-2-(20-trimethyltetrahydro-2H-Pyran-2-yl)-Phenol	Antifungal	*Pinus strobus*	*Lophodermium* sp.	Sumarah et al. (2011)
Phomadecalin	Activity against Pseudomonas aeruginosa	*Pinus* sp.	*Microdiplodia* sp. *KS 75-1*	Hatakeyama et al. (2010)
Xylarenones	Antitumour antimicrobial	*Torreya jackii CHUN*	*Xylaria* sp. *NCYZ*	Hu et al. (2008)
Tuberculariols	Anticancer	*Taxux mairei*	*Tubercularia* sp. *TF5*	Xu et al. (2009)
Enfumafungin	Antifungal	*Juniperus communis*	*Hormonema* sp.	Pelaez et al. (2000)
Periconicins (DitStrobelerpene)	Antibacterial, antifugal against Tricophyton rubrum	*Taxus cuspidate*	*Periconia* sp.	Kim et al. (2004)
Heptelidic acid (Konigic acid)	Anticancer, antimalarial, antilarval (Spruce budworm)	*Abies balsamea*	*Phyllosticta* sp.	Calhoun et al. (1992), Kim and Choong (2009), Tanak et al. (1998)
Isopimaratriene-dione and trion analogs (diterpenoids)	Antilarval (Spruc budworm) and insect toxin	*Balsom fir*	Unidentified endophyte	Findlay et al. (1995a)
Remulosin and Mellein analogs	Antilarval (Spruce budworm)	*Picea muriana BSP*	*Canoplea elegantula (Cooke) M.B.Ellis*	Findlay et al. (1995b)

Leucinostatin A, an antitumor and antifungal peptide was isolated from extracts of *Acremonium* sp. associated with *Taxus baccata* (Strobel et al. 1997). The endophytic fungi, *Penicillium raistrickii* endophytic in *Taxus brevifolia*, produced the peptide Cycloaspeptide A, and also other nitrogen containing compounds such as Benzomalvin C (analgesic anti-inflammatory), Fiscalin B (mycotoxins), Oxaline (anticancer), and Roquefortine C (anticancer) (Stierle and Stierle 2000). Noble et al. (1991) reported the isolation of compound Echinocandin from endophytic fungal cultures *Cryptosporiosis* sp and *Pezicula* sp. endophytic in the plants *Pinus sylvestris* and *Fagus sylvatica.*

4.4.2.3 Ecomycins

The endophytic bacterium, *Pseudomonas viridiflava* was reported to produce compounds called as Ecomycins. The ecomycins, lipopeptides in nature, contain unusual amino acids such as homoserine and β-hydroxy aspartic acid. Three lipopeptides produced by *P. viridiflava* strain EB273 were identified and characterized (Harrison et al. 1991).

4.4.2.4 Pseudomycins

Pseudomycins are antifungal produced by *Pseudomonas syringae*, a plant-associated bacterium (Harrison et al. 1991). These antifungal peptides are lipopeptides containing amino acids like L-chlorothreonine, D- and L-diaminobutyric acid, and L-hydroxyl aspartic acid. Pseudomycin A shows activity against *Candida albicans,* an oppurtunistic pathogen. Pseudomycins A–C contain hydroxyaspartic acid, arginine, lysine, serine, and diaminobutyric acid. They are active against fungal plant pathogens including *C. albicans* and *C. neoformans.*

4.4.2.5 Munumbicins

The munumbicins are made up of 4 bioactive substances having a broad activity spectrum against fungal and bacterial plant pathogens as well as *Plasmodium* species. Castillo et al. (2002) have reported the production of munumbicins by *Streptomyce*s NRRL 30562, an endophytic bacterium of *Kennedia nigriscans*, a medicinal plant native to Australia. The activity of these compounds was against Gram-positive bacteria including the methicillin-resistant strain of *S.aureus* (MRSA, ATCC 33591) and a vancomycin-resistant strain of *E.faecalis* (VREF, ATCC 51299). Munumbin B is effective against multiple-drug-resistant (MDR) *Mycobacteriumtuberculosis,* an acid-fast bacterium, while munumbicins C and D are effective against the malarial parasite *Plasmodium falciparum.*

4.4.2.6 Polyhydroxy Butyrate

Polyhydroxy butyrate (PHB) and poly-3-hydroxyalkanoate (PHA) are the most widely produced microbial bioplastics. These are gaining attention due to their commercial value. Many naturally occurring species of bacteria have the ability to produce bioplastics, as proven by their genomic analysis studies (Kalia et al. 2003). *Herbaspirillum seropedicae*, a diazotrophic endophyte, is found in a variety of higher plants. Catalán et al. (2007) have shown that *H. seropedicae* produces significant levels of PHB, when grown on a range of carbon sources. Degradation of polymer in the host system takes place over the time. Hence there is a possibility of the use of these polymers in drug delivery in cases where slow release of compound is needed.

4.5 Conclusion and Future Perspectives

This chapter highlights the need for novel pharmaceutical solutions to fight emerging and reemerging infections. Endophytic microorganisms are a promising source, as these fungi and bacteria are constantly at war with pathogenic microbes to create an ecological niche for themselves. They produce secondary metabolites as a source of communication and defense. These need to be bioprospected in order to tackle the medical problems being faced presently and will appear in the future too.

Acknowledgements The authors are grateful to their respective academic institutions for the support extended. The authors declare that they have no competing interests.

References

Berde CV (2015) Bioprospecting of endophytes of medicinal plants. J Pharma Biol Sci 3:210–211

Brooks DS, Gonzales CF, Appel DN, Filer TH (1994) Evaluation of endophytic bacteria as potential biological control agents for oak wilt. Biol Cont 4:373–381

Calhoun LA, Findlay JA, Miller DJ, Whitney NJ (1992) Metabolites toxic to spruce budworm from balsam fir needle endophytes. Mycological Res 96:281–286

Castillo UF, Strobel GA, Ford EJ, Hess WM, Porter H, Jensen JB, Albert H, Robinson R, Condron MA, Teplow DB, Stevens D, Yaver D (2002) Mumumbicins, wide spectrum antibiotics produced by *Streptomyces* NRRL 30562, endophytic on *Kennedia nigriscans*. Microbiology 148:2675–2685

Catalán AI, Ferreira F, Gill PR, Batista S (2007) Production of polyhydroxyalkanoates by *Herbaspirillum seropedicae* grown with different sole carbon sources and on lactose when engineered to express the lacZ, lacY genes. Enzyme Microbial Technol 40:1352–1367

Dai J, Krohn K, Flörke U, Draeger S, Schulz B, Kiss-Szikszai A (2006) Metabolites from the endophytic fungus *Nodulisporium* sp. from *Juniperus cedrus*. European J Org Chem 2006:3498–3506

Dai J, Krohn K, Draeger S, Schulz B (2009) New naphthalenechroman coupling products from the endophytic fungus, *Nodulisporium* sp. from *Erica arborea*. European J Org Chem 2009:1564–1569

Eyberger AL, Dondapati R, Porter JR (2006) Endophyte fungal isolates from *Podophyllum peltatum* produce podophyllotoxin. J Nat Prods 69:1121–1124

Findlay JA, Buthelezi S, Lavoie R, Rodriguez L (1995a) Bioactive isocoumarins and related metabolites from conifer endophytes. J Nat Prods 58:1759–1766

Findlay JA, Li G, Penner PE (1995b) Novel diterpenoid insect toxins from a conifer endophyte. J Nat Prods 58:197–200

Gangadevi V, Muthumary J (2008) Taxol, an anticancer drug produced by an endophytic fungus *Bartalinia robillardoides Tassi*, isolated from a medicinal plant, *Aegle marmelos Correa ex Roxb.* World J Microbiol Biotechnol 24:717–724

Gangadevi V, Muthumary J (2009a) Taxol production by *Pestalotiopsis terminaliae*, an endophytic fungus of *Terminalia arjuna* (arjun tree). Biotechnol Appl Biochem 158:675–684

Gangadevi V, Muthumary J (2009b) A novel endophytic taxol-producing fungus *Chaetomella raphigera* isolated from a medicinal plant, *Terminalia arjuna*. Biotechnol Appl Biochem 158:675–684

Gangadevi V, Murugan M, Muthumary J (2008) Taxol determination from *Pestalotiopsis pauciseta*, a fungal endophyte of a medicinal plant. Chin J Biotechnol 24:1433–1438

Gardner JM, Feldman AW, Zablotowicz RM (1982) Identity and behavior of xylem-residing bacteria in rough lemon roots of Florida citrus trees. Appl Environ Microbiol 43:1335–1342

Grove JFJ (1985) Metabolic products of *Phomopsis oblonga*. Part 2. Phomopsolides A and B, tiglic esters of two 6-substituted 5,6-dihydo-5-hydroxypyran-2-ones. Chem Soc Perkin Trans 1:865–869

Guo B, Dai JR, Ng S, Huang Y, Leong C, Ong W, Carte BK (2000) Cytonic acids A and B: novel tridepside inhibiyors of hCMV protease from the endophytic fungus *Cytonaema* species. J Nat Prod 63:602–604

Gutiérrez-Zamora ML, Martínez-Romero E (2001) Natural endophytic association between Rhizobium etli and maize (Zea mays L.). J Biotechnol 91:117–126

Harrison LH, Teplow DB, Rinaldi M, Strobel G (1991) Pseudomycins, a family of novel peptides from *Pseudomonas syringae* possessing broad spectrum antifungal activities. J Gen Microbiol 137:2857–2865

Hatakeyama T, Koseki T, Murayama T, Shiono Y (2010) Eremophilane sesquiterpenes from the endophyte *Microdiplodia* sp. KS 75-1 and revision of the stereochemistries of phomadecalins C and D. Phytochem Letts 3:148–151

Hoffmann JA, Reichhart JM (2002) *Drosophila* innate immunity: an evolutionary perspective. Nat Immunol 3:121–126

Hu ZY, Li YY, Huang YJ, Su WJ, Shen YM (2008) Three new sesquiterpenoids from *Xylaria* sp. NCY2. Helv Chim Acta 91:46–52

Jacobs MJ, Williams MB, David AG (1985) Innumeration, location, characterisation of endophytic bacteria within sugar beet roots. Can J Bot 63:1262–1265

Jalgaonwala RE, Mohite BV, Mahajan RT (2011) A review: natural products from plant associated endophytic fungi. J Microbiol Biotechnol Res 1(2):21–32 Scholar Research Library

Kado CI (1992) Plant pathogenic bacteria. In Balous A, Truper HG, Dworkin M, Harder W, Schlerifer KH (eds) The prokaryotes, vol I. Springer, New York, pp 659–674

Kalia VC, Chauhan A, Bhattacharyya G (2003) Genomic databases yield novel bioplastic producers. Nat Biotechnol 21:845–846

Khaled AS, Elkhateeb WA, Ahmed MT, El-Beih AA, Tahany MA, El-Diwany AI, Ahmed EF (2018) Antiviral and antioxidant potential of fungal endophytes of Egyptian medicinal plants. Fermentation 4:49–60

Kim JH, Choong HL (2009) Heptelidic acid, a sesquiterpene lactone, inhibits etoposide-induced apoptosis in human leukemia U937 cells. J Microbiol Biotechnol 19:787–791

Kim S, Shin DS, Lee T, Oh KB (2004) Periconicins, two new fusicoccane diterpenes produced by an endophytic fungus *Periconia* sp. with antibacterial activity. J Nat Products 67:448–450

Kobayashi DY, Palumboo JD (2000) Bacterial endophytes and their effects on plants and uses in agriculture. In: Bacon CW, White JF (eds) Microbial endophytes. Marcel Dekker Inc, New York, pp 199–233

Kour A, Shawl AS, Rehman S, Sultan P, Qazi PH, Suden P, Khajuria RK, Verma V (2008) Isolation and identification of an endophytic strain of *Fusarium oxysporum* producing podophyllotoxin from Juniperus recurva. World J Microbiol Biotechnol 24:1115–1121

Kour D, Rana KL, Kumar A, Rastegari AA, Yadav N, Yadav AN, Gupta VK (2019a) Extremophiles for hydrolytic enzymes productions: biodiversity and potential biotechnological applications. In: Molina G, Gupta VK, Singh BN, Gathergood N (eds) Bioprocessing for biomolecules production. Wiley, USA, pp 321–372

Kour D, Rana KL, Yadav N, Yadav AN, Kumar A, Meena VS, Singh B, Chauhan VS, Dhaliwal HS, Saxena AK (2019b) Rhizospheric Microbiomes: biodiversity, mechanisms of plant growth promotion, and biotechnological applications for sustainable agriculture. In: Kumar A, Meena VS (eds) Plant growth promoting rhizobacteria for agricultural sustainability: from theory to practices. Springer, Singapore, pp 19–65. https://doi.org/10.1007/978-981-13-7553-8_2

Kour D, Rana KL, Yadav N, Yadav AN, Singh J, Rastegari AA, Saxena AK (2019c) Agriculturally and industrially important fungi: current developments and potential biotechnological applications. In: Yadav AN, Singh S, Mishra S, Gupta A (eds) Recent advancement in white biotechnology through fungi. Volume 2: Perspective for value-added products and environments. Springer, Cham, pp 1–64. https://doi.org/10.1007/978-3-030-14846-1_1

Kumar AG, Antony RA, Kannan VR (2015) Exploration of endophytic microorganisms from selected medicinal plants and their control potential to multi drug resistant pathogen. J Med Plant Studies 3:49–57

Kusari S, Lamshöft M, Spiteller M (2009) *Aspergillus fumigatus Fresenius*, an endophytic fungus from *Juniperus communis L. Horstmann* as a novel source of the anticancer pro-drug deoxypodophyllotoxin. J Appl Microbiol 107:1019–1030

Lee JC, Strobel GA, Lobkovsky E, Clardy J (1996) Torreyanic acid: a selectively cytotoxic quinone dimer from the endophytic fungus. Pestalotiopsis microspora J Org Chem 61:3232–3233

Li JY, Strobel G, Sidhu R, Hess WM, Ford EJ (1996) Endophytic taxol-producing fungi from bald cypress, *Taxodium distichum*. Microbiol 142:2223–2226

Liu Y, Luo J, Xu C, Ren F, Peng C, Wu G, Zhao J (2000) Purification, characterization, and molecular cloning of the gene of a seed-specific antimicrobial protein from pokeweed. Plant Physiol 122:1015–1024

Lodewyckx C, Vangronsveld J, Porteous F, Moore ERB, Taghavi S, Mezgeay M, van der Lelie D (2002) Endophytic bacteria and their potential applications. Crit Rev Plant Sci 21(6):583–606

Lu L, He J, Yu X, Li G, Zhang X (2006) Studies on isolation and identification of endophytic fungi strain SC13 from pharmaceutical plant *Sabina vulgaris* Ant. and metabolites. Acta Botany Boreal-Occident Sinica 15:85–89

Marx J (2004) The roots of plant–microbe collaborations. Science 304:234–236

Moricca S, Ragazzi A (2008) Fungal endophytes in Mediterranean oak forest: a lesson from Discula Quercina. Phytopathol 98:380–386

Noble HM, Langley D, Sidebottom PJ, Lane SJ, Fisher PJ (1991) An echinocandin from an endophytic *Cryptosporiopsis* sp. and *Pezicula* sp. in *Pinus sylvestris* and *Fagus sylvatica*. Mycol Res 95:1439–1440

Park JH, Choi GJ, Lee SW, Lee HB, Kim KM, Jung HS, Jang KS, Cho KY, Kim JC (2005) *Griseofulvin* from *Xylaria* sp. strain F0010, an endophytic fungus of *Abies holophylla* and its antifungal activity against plant pathogenic fungi. J Microbiol Biotechnol 15:112–117

Pelaez F, Cabello A, Platas G, Diez MT, Gonzalez del Val A, Basilio A, Martan I, Vicente F, Bills GF, Giacobbe RA, Schwartz RE, Onishi JC, Meinz MS, Abruzzo GK, Flattery AM, Kong L, Kurtz MB (2000) The discovery of enfumafungin, a novel antifungal compound produced by an endophytic *Hormonema* species biological activity and taxonomy of the producing organisms. Syst Appl Microbiol 23:333–343

Porras-Soriano A, Soriano-Martín ML, Porras-Piedra A, Azcón R (2009) Arbuscular mycorrhizal fungi increased growth, nutrient uptake and tolerance to salinity in olive trees under nursery conditions. J Plant Physiol 166:1350–1359

Puri SC, Nazir A, Chawla R, Arora R, Riyaz-ul-Hasan S, Amna T, Ahmed B, Verma V, Singh S, Sagar R, Sharma A, Kumarc R, Sharma RK, Qazi GN (2006) The endophytic fungus *Trametes hirsuta* as a novel alternative source of podophyllotoxin and related aryl tetralin lignans. J Biotechnol 122:494–510

Rana KL, Kour D, Sheikh I, Dhiman A, Yadav N, Yadav AN, Rastegari AA, Singh K, Saxena AK (2019a) Endophytic fungi: biodiversity, ecological significance and potential industrial applications. In: Yadav AN, Mishra S, Singh S, Gupta A (eds) Recent advancement in white biotechnology through fungi, vol 1. Diversity and Enzymes Perspectives. Springer, Switzerland, pp 1–62

Rana KL, Kour D, Sheikh I, Yadav N, Yadav AN, Kumar V, Singh BP, Dhaliwal HS, Saxena AK (2019b) Biodiversity of endophytic fungi from diverse niches and their biotechnological applications. In: Singh BP (ed) Advances in endophytic fungal research: present status and future challenges. Springer, Cham, pp 105–144. https://doi.org/10.1007/978-3-030-03589-1_6

Rana KL, Kour D, Yadav AN (2019c) Endophytic microbiomes: biodiversity, ecological significance and biotechnological applications. Res J Biotechnol 14:142–162

Richardson SN, Walker AK, Nsiama TK, McFarlane J, Sumarah MW, Ibrahim A, Miller JD (2014) Griseofulvin-producing *Xylaria* endophytes of *Pinus strobus* and *Vaccinium angustifolium*: evidence for a conifer-understory species endophyte ecology. Fungal Ecol 11:107–113

Rosenblueth M, Martínez-Romero E (2006) Bacterial endophytes and their interactions with hosts. Mol Plant-Microbe Interact 19:827–837

Ryan RP, Germaine K, Franks A, Ryan DJ, Dowling DN (2008) Bacterial endophytes: recent developments and applications. FEMS Microbiol Lett 278:1–9

Schmeda-Hirschmann G, Hormazabal E, Rodriguez JA, Theoduloz C (2008) Cycloaspeptide A and pseurotin A from the endophytic fungus *Penicillium janczewskii*. Zeitschrift für Naturforschung C 63:383–388

Seto Y, Takahashi K, Matsura H, Kogami Y, Yada H, Yoshihara T, Nabita K (2007) Novel cyclic peptide Epichlicin from the endophytic fungus *Epichloe typhin*. Biosci Biotech Biochem 71:1470–1475

Settu S, Arunachalam S, Gayatri S (2010) Endophytic fungi: a review on pharmaceulogical activities and its industrial applications. Inter J Pharma Sci 53:82–89

Stierle AA, Stierle DB (2000) Bioactive compounds from four endophytic *Penicillium* sp. isolated from the Northwest Pacific yew tree. In: Atta-Ur-Rahman (ed) Bioactive natural products, vol. 24. Elsevier Science, Amsterdam, pp 933–978

Stierle A, Strobel G, Stierle D (1993) Taxol and taxane production by *Taxomyces andreanae* an endophytic fungus of Pacific yew. Science 260:214–216

Stierle A, Stierle D, Strobel G, Bignami G, Grothaus P (1995) Bioactive metabolites of the endophytic fungi of Pacific yew, *Taxus brevifolia*: paclitaxel, taxanes, and other bioactive compounds. In: Georg GI, Chen TT, Ojima I, Vyas DM (eds) Taxane anticancer agents: basic science and current status. Washington, DC, p. 81–97. ACS Symposium Series 583

Stierle DB, Stierle AA, Ganser B (1997) New phomopsolides from a *Penicillium* sp. J Nat Prod 60:1207–1209

Stierle A, Stierle DB, Bugni T (2001) Sequoiatones C-F, constituents of redwood endophyte *Aspergillus parasiticus*. J Nat Prod 64:1350–1353

Stierle DB, Stierle AA, Bugni T (2003) Sequoiamonascins A-D: novel anticancer metabolites isolated from a redwood endophyte. J Org Chem 68:4966–49699

Strobel GA, Stierle A, Stierle D (1993) *Taxomyces andreanae*, a proposed new taxon for a bulbilliferous hyphomycete associated with Pacific yew. Mycotaxon 47:71–78

Strobel G, Yang X, Sears J, Kramer R, Sidhu RS, Hess WM (1996) Taxol from *Pestalotiopsis microspora*, an endophytic fungus of *Taxus wallachiana*. Microbiology 142:435–440

Strobel GA, Torczynski R, Bollon A (1997) *Acremonium* sp.—aleucinostatin A producing endophyte of European yew (*Taxus baccata*). Plant Sci 128:97–108

Suman A, Yadav AN, Verma P (2016) Endophytic microbes in crops: diversity and beneficial impact for sustainable agriculture. In: Singh D, Abhilash P, Prabha R (eds) Microbial inoculants in sustainable agricultural productivity, research perspectives. Springer, India, pp 117–143. https://doi.org/10.1007/978-81-322-2647-5_7

Sumarah MW, Kesting JR, Sorensen D, Miller JD (2011) Antifungal metabolites from fungal endophytes of *Pinus strobus*. Phytochem 72:1833–1837

Sun L, Lu Z, Bie X, Lu F, Yang S (2006) Isolation and characterization of a coproducer of fengycins and surfactins, endophytic *Bacillus amyloliquefaciens* ES-2, from *Scutellaria baicalensis* Georgi. World J Microbiol Biotechnol 22:1259–1266

Tanak Y, Shiomi K, Kamaei K, Sugoh-Hagino M, Enomoto Y, Fang F, Yamaguchi Y, Masuma R, Zhang CG, Zhang XW, Omura S (1998) Antimalarial activity of radicicol, heptelidic acid and other fungal metabolites. J Antibiotics 51:153–160

Taylor TN, Taylor EL (2000) The rhynie chert ecosystem, a model for understanding fungal interactions. In: Bacon CW, White JF (eds) Microbial endophytes. Marcel Decker Inc., New York, pp 31–48

Vizioli J, Salzet M (2002) Antimicrobial peptides versus parasitic infections. Trends Parasitol 18:475–476

Wang J, Huang Y, Fang M, Zhang Y, Zheng Z, Zhao Y, Su W (2002) Brefeldin A a cytotoxin produced by *Paecilomyces* sp. and *Aspergillus clavatus* isolated from *Taxus mairei* and *Torreya grandis*. FEMS Immunol Med Microbiol 34:51–57

Wani MC, Taylor HL, Wall ME, Coggon P, McPhail AT (1971) Plant Antitumor Agents VI. The isolation and structure of taxol, a novel antileukemic and antitumor agent from *Taxus brevifolia*. J Amer Chem Soc 93:2325–2327

Whitesides S, Spotts RA (1991) Susceptibility of pear cultivars to blossom blast caused by *Pseudomonas syringae*. Hortscience 26

Wijeratne EMK, Paranagama PA, Marron MT, Gunatilaka MK, Arnold AE, Gunatilaka AAL (2008) Sesquiterpene quinones and related metabolites from *Phyllosticta spinarum*, a fungal strain endophytic in *Platycladus orientalis*, of the sonoran desert. J Nat Prod 71:218–222

Wilson D (1995) Endophyte: the evolution of a term, and clarification of its use and definition. Oikos 73:274–276

Xu R, Wang MZ, Lu CH, Zheng ZH, Shen YM (2009) Tuberculariols A-C, new sesquiterpenes from the mutant strain M-741 of Tubercularia sp. TF 5. Helv Chim Acta 92:1514–1519

Yadav AN (2017) Agriculturally important microbiomes: biodiversity and multifarious PGP attributes for amelioration of diverse abiotic stresses in crops for sustainable agriculture. Biomed J Sci Tech Res 1:1–4

Yadav AN (2018) Biodiversity and biotechnological applications of host-specific endophytic fungi for sustainable agriculture and allied sectors. Acta Sci Microbiol 1:01–05

Yadav AN, Yadav N (2018) Stress-adaptive microbes for plant growth promotion and alleviation of drought stress in plants. Acta Sci Agric 2:85–88

Yadav AN, Kumar R, Kumar S, Kumar V, Sugitha T, Singh B, Chauhan VS, Dhaliwal HS, Saxena AK (2017) Beneficial microbiomes: biodiversity and potential biotechnological applications for sustainable agriculture and human health. J Appl Biol Biotechnol 5:1–13

Yadav AN, Mishra S, Singh S, Gupta A (2019a) Recent advancement in white biotechnology through fungi. Volume 1: Diversity and enzymes perspectives. Springer, Cham

Yadav AN, Singh S, Mishra S, Gupta A (2019b) Recent advancement in white biotechnology through fungi. Volume 2: Perspective for value-added products and environments. Springer, Cham

Yadav AN, Singh S, Mishra S, Gupta A (2019c) Recent advancement in white biotechnology through fungi. Volume 3: Perspective for sustainable environments. Springer, Cham

Zhou K, Qiao K, Edgar S, Stephanopoulos GPT (2015) Distributing a metabolic pathway among a microbial consortium enhances production of natural products. Nat Biotechnol 33:377–383

Chapter 5
Phyllospheric Microbiomes: Diversity, Ecological Significance, and Biotechnological Applications

Natesan Sivakumar, Ramamoorthy Sathishkumar, Gopal Selvakumar, Rajaram Shyamkumar and Kalimuthu Arjunekumar

Abstract The phyllosphere referred to the total aerial plant surfaces (above-ground portions), as habitat for microorganisms. Microorganisms establish compositionally complex communities on the leaf surface. The microbiome of phyllosphere is rich in diversity of bacteria, fungi, actinomycetes, cyanobacteria, and viruses. The diversity, dispersal, and community development on the leaf surface are based on the physiochemistry, environment, and also the immunity of the host plant. A colonization process is an important event where both the microbe and the host plant have been benefited. Microbes commonly established either epiphytic or endophytic mode of life cycle on phyllosphere environment, which helps the host plant and functional communication with the surrounding environment. To the scientific advancement, several molecular techniques like metagenomics and metaproteomics have been used to study and understand the physiology and functional relationship of microbes to the host and its environment. Based on the available information, this chapter describes the basic understanding of microbiome in leaf structure and physiology, microbial interactions, especially bacteria, fungi, and actinomycetes, and their adaptation in the phyllosphere environment. Further, the detailed information related to the importance of the microbiome in phyllosphere to the host plant and their environment has been analyzed. Besides, biopotentials of the phyllosphere microbiome have been reviewed.

Keywords Biotechnological applications · Diversity · Ecological significance · Phyllospheric microbiomes · Plant growth promotion

N. Sivakumar (✉) · R. Sathishkumar · K. Arjunekumar
Department of Molecular Microbiology, School of Biotechnology, Madurai Kamaraj University, Madurai 625 021, Tamilnadu, India
e-mail: microshivaak@gmail.com

G. Selvakumar
Department of Microbiology, Allagappa University, Karaikudi-03, Tamilnadu, India

R. Shyamkumar
Department of Biotechnology, Kamaraj College of Engineering and Technology, Vellakovil, Kallikudi, Madurai, Tamilnadu, India

A. N. Yadav et al. (eds.), *Plant Microbiomes for Sustainable Agriculture*, Sustainable Development and Biodiversity 25, https://doi.org/10.1007/978-3-030-38453-1_5

5.1 Introduction

The term phyllosphere is referred to as "the aerial part of the plant or the parts of a plant above the ground usually surface of leaves, considered as a habitat for microorganisms." This is a place where normally a variety of microorganism (bacteria, yeasts, and fungi) colonizes. The global leaf area corresponds to both upper and lower surfaces, has approximately twice as great as the land surface area (Vorholt 2012). The phyllosphere is the ambient region for microbes to colonize and establish its association with plants usually epiphytes. Microbial communities in the phyllosphere are highly complex and consist of many cultured and uncultured microorganisms (Müller and Ruppel 2014). It has a heterogeneous group of the microbial association at the micrometer scale area due to its diverse microenvironments (habitats). The phyllospheric microbes are adapted to the insensitive environmental conditions, specifically microbial epiphytes are highly exposed to atmospheric temperature, light, UV radiation, less water, and nutrient availability. These external factors affect the composition and diversity of phyllospheric microbial communities (Vorholt 2012). However, the type of plant and invading microbial populations (pathogens) are also influencing the commensals and/or mutualistic relationship with their host plant (Lindow and Brandl 2003). Less number of studies are available for the microbiology of phyllosphere rather than plant root. Moreover, with increasing anthropogenic stresses, the diversity and community structure of phyllosphere microflora have been continually modified. In this chapter, we focused on the phyllospheric microbiome, structure and diversity, epiphytic mechanism, molecular interactions, ecological significance, and the microbial importance in biotechnology.

5.2 Basic Understanding of Leaf Structure

The leaf is a highly organized and multi-layered plant organ (Fig. 5.1), which consists of the epidermis (upper and lower) covered by a waxy cuticle that provides a physical barrier against abiotic and biotic stresses. The epidermis involves many regulatory processes of leaf physiology including gas exchange, temperature regulation, primary production, secretion of secondary metabolites, and water mobilization. Also, a specialized epidermal cell such as stomata, hydathodes (modified stomata), and trichomes (outgrowth) are there in the epidermis. The stomata are surrounded by two cupped hand cells called guard cells, which may open or close due to internal water pressure. Inside the leaf, a layer of cells called the mesophyll, is present, usually two layers, namely, palisade layer and the spongy layer. They contain chlorophyll and photosynthesis occurs in these cells. The palisade cells are more column cells and the spongy cells are more loosely packed between the palisade layer and the lower epidermis, and it allows for gas exchange. The veins of the leaf contain the vascular tissue, xylem and phloem are found in it. Veins run from tips of the roots and are extended up to the edges of the leaves. The outer layer cells are called bundle sheath

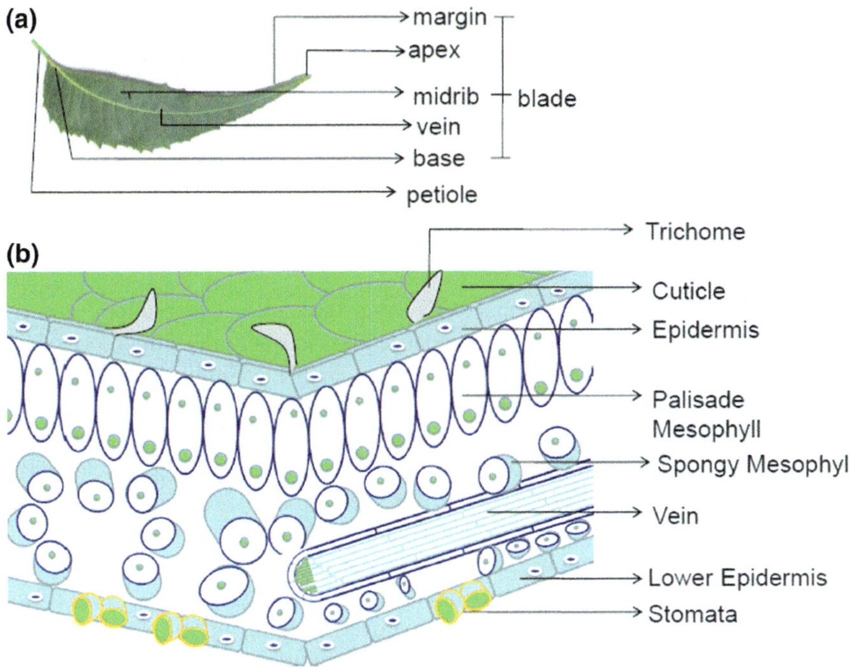

Fig. 5.1 Structural organization of a leaf

cells which circle the xylem and the phloem. The xylem transports water and phloem transports sugar (food).

Glandular trichome of the epidermis releases a wide spectrum of leaf exudates, such as polysaccharide salts, lipids, volatile compounds, and proteins, and its function is associated with plant–microbe and plant–insect interactions (Hirano and Upper 1983). But, non-glandular trichome involves regulation of water tension, light absorption, and protect the leaf from UV radiation and heat as well as drought tolerance (Hirano and Upper 1983).

5.3 Phyllosphere Habitat

The phyllosphere is a unique and dynamic habitat which constitutes irregular, and sometimes relatively large microbial community inhabitant in the ecosystem (Whipps et al. 2008). The total terrestrial phyllosphere area estimated is around 6.4×10^8 km^2 (Morris and Kinkel 2002), and it exhibits numerous microhabitat which represents a major source of microorganism. Variety of bacteria, filamentous fungi, and yeasts are naturally colonized on the phyllosphere region and less frequently, protozoa and nematodes. These microorganisms exhibit commensalism and/or mutualism (symbionts) or antagonism type of relationship on their host plants. The microbial association in phyllosphere has several advantages and importance to global processes including biogeochemical cycles (carbon and nitrogen) and environmental impact.

5.3.1 Microbial Assembly on Leaf

The arrangement of leaf epidermal cells describes the leaf physiology and the microenvironment which allow the abundance and distributions of microorganisms on the leaf surface (Shiraishi et al. 2015; Esser et al. 2015). Simply, epiphytes make biofilm-like growth, most preferably larger bacterial aggregates are on the trichomes, veins, and epidermal cell groves (Brewer et al. 1991; Morris et al. 1997), where the leaf exudates containing nutrient-rich region. The presence of outer cuticle and its physiology help the microbes to colonize this site. Presence of aliphatic compounds in the cuticle layer determines the physicochemical properties of the leaf surface and renders the permeability and wettability, which facilitate the adherence of microorganisms (Sadler et al. 2016). Water permeability of this site may play a vital role in the survival and growth of the epiphytes. Moreover, leaching the nutrients along with water makes the epiphytes to utilize and develop colonies on the phyllosphere (Burch et al. 2014). The leaf surface with higher water and nutrient penetration is heavily colonized by bacterial communities (Beattie 2011). In general, bacteria maintain the cuticular permeability by secretion of biosurfactants, for example, *Pseudomonas syringae* release syringafactin on the cuticle layer of the leaf which facilitate the availability of sugar for persistent epiphytic growth (Van der Wal and Leveau 2011). Similarly, fructose availability by *Pantoea eucalypti* 299R and *Pantoea agglomerans* (Leveau and Lindow 2001). Figure 5.2 represents the phyllosphere microbial assemblage, wherein the epiphytic microorganism exploits this microenvironment for special distribution of microbes, survival as well as blooming (colonization). At the same time, surface microorganisms change the phyllosphere chemistry, and they render the heterogeneous oligotrophic mode of epiphytic life. Besides, microorganism establishes special niches on the leaf surface with the interactive mode of life (Agler et al. 2016) in this microhabitat microbial population can be constantly maintained.

5.4 Microbial Diversity in the Phyllosphere

The phyllosphere consists of diverse numerous microbial communities including bacteria, filamentous fungi, yeasts, algae, and protozoans (Whipps et al. 2008; Verma et al. 2013, 2015, 2016a, b). The nature of various microorganisms (epiphytic and endophytic) associated with phyllosphere is given in Fig. 5.3. Among the diverse community of microbes, bacteria are the predominant community on leaves and its

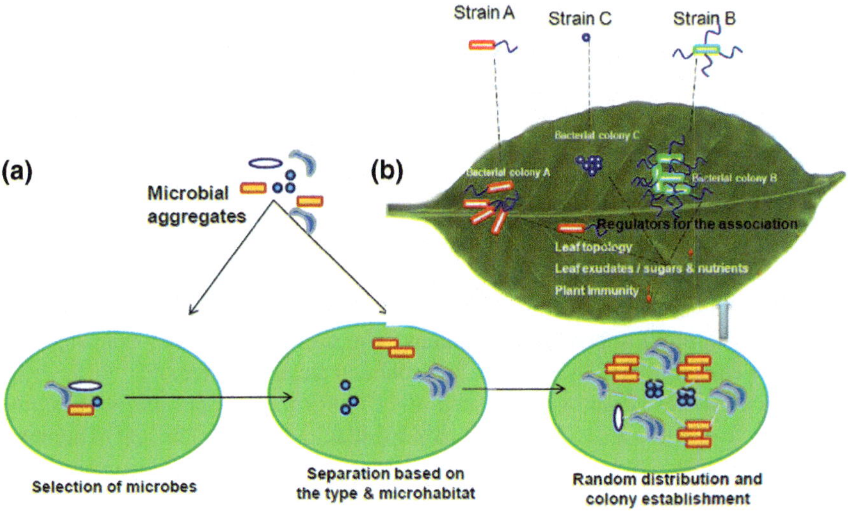

Fig. 5.2 Structure of phyllosphere microbial assemblage. **a** stages for microbial community structure development, **b** regulations for the microbial community structure in phyllosphere

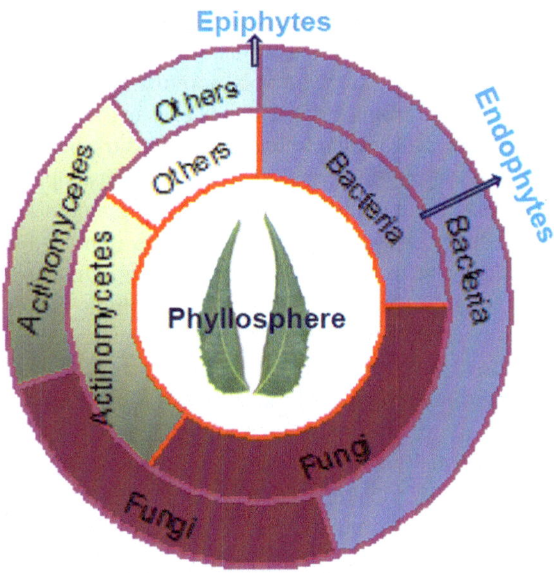

Fig. 5.3 Epiphytic and endophytic microbes in phyllosphere

range is between 10^2 and 10^{12} g^{-1} of the leaf (Inacio et al. 2002). The conventional culture-based method has been used for the identification of different microbial communities of the leaves. Thompson et al. (1993) identified 78 bacterial species from the sugar beet, and Legard et al. (1994) screened 88 bacterial species from 37 genera.

However, the culture-dependent method based profiling of phyllosphere communities is likely to be incorrect and miscalculates diversity (Rasche et al. 2006). The culture-independent approaches like 16S rDNA sequences of the whole microbial mass of phyllosphere could give the complete and complex microbial community structure of the environment. Molecular studies suggested that alpha-, beta- and gammaproteobacteria and firmicutes are the dominant bacterial inhabitants of the phyllosphere. Frequently, acidobacteria, actinobacteria, and cyanobacteria are also occurring in the phyllosphere environment (Kadivar and Stapleton 2006). Lambais et al. (2006) identified that 97% of the bacterial sequences of the phyllosphere have been new and unidentified. Yang et al. (2001) reported large numbers of novel bacteria from the phyllosphere of crop plants. The number of studies confirmed the diversity of yeast in the phyllosphere environment as an epiphyte.

The cultivable yeasts genera such as *Cryptococcus, Sporobolomyces*, and *Rhodotorula* and its species have been largely inhabitant in the plant leaf (Thompson et al. 1993; Glushakova and Chernov 2004). Moreover, the culture-dependent methods have been used to study the abundance of filamentous fungi, ranging from 10^2 to 10^8 CFU g^{-1}. Genera such as *Cladosporium, Alternaria, Penicillium, Acremonium, Mucor*, and *Aspergillus* are the frequent filamentous fungi colonizing as epiphytes and endophytes (Arnold et al. 2000; Inacio et al. 2002; Rana et al. 2019a, b, c).

However, the culture-independent strategy is the best to investigate the diversity and distribution of specific bacterial groups of interest (Miyamoto et al. 2004; Sessitsch et al. 2006). Other than the 16S/18S rDNA sequences, multiplex terminal restriction fragment length polymorphism (TRFLP) has been used to analyze several phylogenetic groups or functional genes in the microenvironment (Singh et al. 2006). Soils, water, air, tree buds, and plant debris from the previous crops are the sources for microbes in phyllosphere (Manceau and Kasempour 2002). Those microorganisms may be habited in phyllosphere either transient or residual epiphytes (Suslow 2002; Zak 2002). The atmospheric microflora, rainfall, humidity, wind, etc. can directly influence the transients of microorganisms to the phyllosphere (Lighthart 1997). During the plant growth period, the epiphytic bacterial population will increase in quantity (Inacio et al. 2002). The microorganisms on the seed or roots may be established as epiphytes or endophytes (Wulff et al. 2003). Some epiphytes may be injected into the internal space of the leaf and colonize as endophytes. The distribution pattern of the phyllosphere microorganisms is not even, mostly bacteria colonize at the epidermal wall junctions, specifically in the grooves and the veins or stomata or at the base of trichomes (Melotto et al. 2008), also found in the cuticle layer, near hydathodes and stomatal pits (Aung et al. 2018). The microbial load is higher at the lower leaf surface perhaps the lower leaf surface contains thin cuticle, stomata, and/or trichomes (Beattie and Lindow 1999). Mostly, all microorganisms that appear in the phyllosphere are capable to colonize and grow (Whipps et al. 2008), and it disperses throughout the surface by rain splash, bounce-off, wash-off, water movement, or removal by insects or pest (Kinkel 1997; Yang et al. 2001; Lambais et al. 2006).

5.4.1 Bacterial Diversity in the Phyllosphere

Phyllosphere is a heterogeneous environment (Koskella 2013), bacteria are considered the most abundant inhabitants of the leaves, and its average number is being around 10^6–10^8 cells cm^{-2} (Andrews and Harris 2000; Hirano and Upper 2000). But the population of epiphytic bacteria differs depending on the plant species and its surrounding environment. The variation is mainly due to the physical and nutritional conditions of the phyllosphere. Commonly, the broad-leaf plants have the highest number of bacteria than the grasses or waxy broad-leaf plants (Kinkel et al. 2000).

Generally, the phyllosphere contains four major phyla of bacteria such as the Proteobacteria, Firmicutes, Bacteroides, and Actinobacteria (Kembel et al. 2014; Durand et al. 2018). Methylotrophic bacteria are predominant in phyllosphere which includes genera such as *Methylobacterium, Methylophilus, Methylibium, Hyphomicrobium, Methylocella, Methylocapsa*, and *Methylocystis* (Mizuno et al. 2013; Iguchi et al. 2013; Kwak et al. 2014; Krishnamoorthy et al. 2018). *Methylobacterium* and *Sphingomonas* are the predominant genera belonging to the class alphaproteobacteria reported in several plant phyllospheres (Delmotte et al. 2009; Kumar et al. 2019a). The bacterial community organization on phyllosphere is controlled by specific assemblage regulations (Buee et al. 2009; Reinhold-Hurek et al. 2015). Normally, soil type, plant genotype and species, immune system of the plant, age, climatic condition, and the geographic region are the factors forcing the bacterial community assembly (Leff et al. 2015; Zarraonaindia et al. 2015; Copeland et al. 2015). Extensive studies are available for the soil and rhizosphere bacterial community on phyllosphere bacterial colonization in Arabidopsis *thaliana* (Bodenhausen et al. 2013; Maignien et al. 2014; Bai et al. 2015; Muller et al. 2015) and maize (Peiffer et al. 2013). *Proteobacteria, Actinobacteria*, and *Bacteroidetes* are the most abundant phyla colonizing the leaf and root of *A.* thaliana (Delmotte et al. 2009; Redford et al. 2010; Bodenhausen et al. 2013). *Massilia, Flavobacterium, Pseudomonas,* and *Rathayibacter* are a prevalent bacterial genus in *A. thaliana* (Bodenhausen et al. 2013), *Deinococcus thermus* on tree phyllosphere (Redford et al. 2010), and *Bacillus* and *Pantoea dominate* the lettuce (Rastogi et al. 2012).

Kembel et al. (2014) studied the bacterial communities on tropical tree leaves, around 400 bacterial taxa the phyllosphere has been dominated with Actinobacteria, Alpha-, Beta-, and Gammaproteobacteria, and Sphingobacteria. However, Archaea is the profuse members of the plant-associated microbe, commonly Thaumarchaeota, Crenarchaeota, and Euryarchaeota make the endophytic mode of life in plants (Müller et al. 2015). Durand et al. (2018) characterized the bacterial genera such as *Methylobacterium, Kineococcus, Sphingomonas*, and *Hymenobacter* of the phylum Firmicutes from the leaf surface. The phyllosphere of the grapevine contains Acinetobacter, Bacillus, Citrobacter, Curtobacterium, Enterobacter, Erwinia, Frigoribacterium, Methylobacterium, Pantoea, Pseudomonas, and Sphingomonas as dominant genera (Kecskeméti et al. 2016). Steven et al. (2018) characterized Pseudomonas and Enterobacteriaceae as predominant taxa from apple. Several studies revealed Pseudomonas as the most abundant genus of phyllosphere region (Aleklett et al. 2014;

Kecskeméti et al. 2016; Steven et al. 2018). Seed coat associated bacteria that have been reported in phyllosphere are mainly Actinobacteria, Bacteroidetes, Firmicutes, and Proteobacteria (Johnston-Monje and Raizada 2011; Rodríguez-Escobar et al. 2018).

The most notable bacterial pathogen is *Pseudomonas syringae*, it causes diseases in a wide range of economically important plant species (Mansfield et al. 2012; Morris et al. 2013; Burch et al. 2014). Hamd Elmagzob et al. (2019) identified taxa such as Rhizobiales, Clostridiales, Pseudomonadales, Burkholderiales, Bacteroidales, Enterobacteriales, Rhodocyclales, Sphingomonadales, Lactobacillales, and Bacillales from the leaves of *Cinnamomum camphora* (L.) Presl. Several studies reported diazotrophic bacteria on phyllosphere (Fürnkranz et al. 2008; Rico et al. 2014). Diazotrophic bacteria can use atmospheric dinitrogen (N_2) as nitrogen source for its metabolic activities. Bacterial diazotrophic include *Beijerinckia, Azotobacter, Klebsiella*, and Cyanobacteria (e.g., *Nostoc*, *Scytonema*, and *Stigonema*). Diazotrophic nitrogen fixation has been reported in many species which contains an enzyme nitrogenase (encoded by *nif* genes) (Rico et al. 2014). Recently, 16 s rRNA gene-based high-throughput sequencing technology has been used for the diversity analysis of phyllosphere, for example, the distribution of endophytic bacteria of *C. camphora* (L.) Presl leaves has been analyzed by 16S rRNA gene metagenomics, revealing Proteobacteria, Firmicutes, Bacteroidetes, Actinobacteria, Gemmatimonadetes, Acidobacteria, Planctomycetes, Chloroflexi, and Fusobacteria are the major phyla of the polymicrobial community (Hamd Elmagzob et al. 2019).

5.4.2 Fungal Microbiota of Phyllosphere

Fungi are saprophytic and they may be associated with plants either epiphytic or endophytic, and mostly they are known for their pathogenesis on plant system (Voříšková and Baldrian 2013; Yadav et al. 2019b, c, d). There are several reports revealed that phyllosphere fungi have a profound role in the residing host. Both epiphytic and endophytic fungi inhabiting the leaf are of high species diversity with diverse metabolic functions (Yao et al. 2019), such as leaf litter decomposition and recycling the carbon and nitrogen (Kannadan and Rudgers 2008; Guerreiro et al. 2018). In general, endophytic fungi can help plant growth and also provide resistance to biotic (pathogens) and abiotic (drought and salinity) stresses, (Arnold et al. 2007; Purahong and Hyde 2011; Guerreiro et al. 2018; Yadav et al. 2018c). In culture-dependent approaches, several fungal species have been isolated from small herbs to larger woody plants. Inácio et al. (2010) reported that the density of yeast-like fungi may vary from plant to plant and approximately 5×10^4 cells cm^{-2}. *Aureobasidium pullulans* are yeast-like fungi abundant in phyllosphere (Cordier et al. 2012; Setati et al. 2012). Apart from yeast-like fungi, many filamentous fungi have been reported from health as well as infected plant leaves. Through the culture-dependent method, Ripa et al. (2019) isolated *Aspergillus niger, Fusarium oxysporum, Penicillium aurantiogriseum, Fusarium incarnatum, Alternaria alternata, Alternaria tenuissima, Cladosporium cladosporioides, Talaromyces funiculosus, Aspergillus flavus, Trichoderma aureoviride, Trichoderma harzianum, Penicillium janthinellum, Fusarium proliferatum, Fusarium equiseti,* and *Aspergillus stellatus* from wheat plant.

Dhayanithy et al. (2019) isolated twenty endophytic fungi from the leaves and stem of *Catharanthus roseus,* among them Colletotrichum, Alternaria, and Chaetomium were the dominant genera. Many of them make endophytic association begin with epiphytic initiation (Rodriguez et al. 2009; Porras-Alfaro and Bayman 2011), and some endophytes later turned to pathogens. The olive tree phyllosphere is found to be highly diverse having more than 149 genera and 68 families of fungi (Martins et al. 2016) in a Mediterranean ecosystem (Portugal), but Abdelfattah et al. (2015) reported only 13 endophytic fungal taxa in the leaves and twigs of olive trees. There has been a discrepancy to understand the phyllosphere fungi as endophytic or epiphytic, occasionally it is uncertain, for the reason that some can reside both epiphytic and endophytic modes of association. In general, phyllosphere endophytic fungi are the epiphytic habitats and are penetrated into the plant tissues to form an endophytic association (Kharwar et al. 2010; Porras-Alfaro and Bayman 2011). Though they are phyllospheric, the soil has acted as a reserve for these potential endophytic inoculums of the above-ground organs (Zarraonaindia et al. 2015). For example, *Ascochyta* sp. and *Fagus crenata* B1 (Osono 2006), *Colletotrichum gloeosporioides* and *Phomopsis* sp. (Rivera-Vargas et al. 2006; Twizeyimana et al. 2013), and Table 5.1 listed some examples of phyllosphere fungal endophytes.

Osono (2008) reported that endophytic *Colletotrichum gloeosporioides* and *C. acutatum*, and epiphytes *Pestalotiopsis* sp., *Aureobasidium pullulans*, *Phoma* sp., and *Ramichloridium* sp. are the phyllosphere fungi in the plant *Camellia japonica*. However, the abundance and diversity of the fungi differ in plant species as well as in different eco-climatic conditions. Moreover, seasonal and leaf age-dependent variations also occur in the epiphytic and endophytic phyllosphere fungal assembly, for example, *Geniculosporium* sp. is varied in leaf age, and *Cladosporium cladosporioides* has been varied in both season and leaf age of the plant Camellia japonica (Osona 2008). Phyllosphere fungi play an important function in mineral absorption and mineral recycling process, specifically carbon, nitrogen, and phosphorus recycling in the forest ecosystem. Therefore, the study about the phyllosphere fungi and its physiology with host plant is important.

5.4.3 Actinomycetes Diversity in Phyllosphere

In addition to bacterial diversity, actinobacteria share a considerable interest in epiphytic and endophytic life forms in the phyllosphere. They are soil-inhabiting saprophytic microbes and have been extensively studied for their therapeutic secondary metabolites. This versatile group of gram-positive bacteria has adapted to diverse environments including the phyllosphere of the plant (Singh et al. 2018). Some actinobacteria form symbiotic association residing in plant tissues have generated

Table 5.1 Phyllosphere fungal endophytes

Endophytic fungi	Host plant	Type	References
Aspergillus, Phomopsis, Wardomyces, Penicillium	*Euterpe oleracea* (palm)	Palm	Rodrigues (1994)
Ramularia spp.	*Vitis riparia* (grapevine)	Wild	Kernaghan et al. (2007)
Absidia sp., *Aspergillus* sp., *Cladosporium* sp., *Cunninghamella* sp., *Fusarium* sp., *Nigrospora* sp., *Paecilomyces* sp., *Penicillium* sp., *Rhizopus* sp.,	*Meyna spinosa Roxb.*	Medicinal plant	Bhattacharyya et al. (2017)
Penicillium chrysogenum, and *Penicillium crustosum*	*Teucrium polium*	Medicinal plant	El-Din Hassan (2017)
Alternaria alternata, *Setosphaeria* sp., *Cochliobolus* sp., *Alternaria* sp. *Phoma herbarum, Davidiella tassiana, Botryosphaeria dothidea, Ulocladium alternariae, Phoma macrostoma* var. incolorata, *Phoma exigua* var. exigua, *Cladosporium cladosporioides* strain, *Botryosphaeria* sp., *Guignardia mangiferae, Pyrenophora tritici-repentis, Guignardia alliacea, Rhizopus oryzae*	*Catharanthus roseus*	Medicinal plant	Sreekanth et al. (2017)

(continued)

enormous significance to the host and its environment through their novel metabolites. Diversity and distribution of endophytic actinobacteria have been largely documented, from medicinal plants, crop plants, and some other terrestrial plants (Qin et al. 2011; Masand et al. 2015; Dinesh et al. 2017; Nalini and Prakash 2017). Several species of actinobacteria have been reported from plants such as *Triticum aestivum, Lupinus termis, Lobelia clavatum, Acacia auriculiformis, Aquilaria crassna, Oryza*

Table 5.1 (continued)

Endophytic fungi	Host plant	Type	References
Ascomycetes: *Trichoderma, Penicillium, Fusarium,* and *Aspergillus*. Non-ascomycetes: *Mucor* (Mucoromycota) and *Schizophyllum* (Basidiomycota)	*Stanhopea tigrina*	Orchid	Salazar-Cerezo et al. (2018)
Trichothecium sp., *Epicoccum nigrum, Alternaria alternaria, Alternaria arborescens, Nigrospora sphaerica, Epicoccum* sp., *Alternaria* sp. *Nigrospora* sp., *Colletotrichum gloeosporioides, Fusarium oxysporum, Trichothecium roseum*	*Vitis vinifera* (Grape fruit cells)	Fruit plant	Huang et al. (2018)
Aspergillus japonicus	*Euphorbia indica* L.	Wild plant	Ismail et al. (2018)
Alternaria spp., *Trichophyton* spp., *Geotrichum* spp., *Candida* spp., *Aspergillus* spp., *Aureobasidium* spp., *Fusarium* spp., *Exserohilum* spp., *Curvularia* spp., *Coccidioides* spp., *Bipolaris* spp.	*Epipremnum aureum, Azadirachta indica, Piper betle, Catharanthus roseus, Ficus religiosa, Musa acuminate, Ficus Benghalensis, Ficus racemosa, Calotropis procera, Ocimum tenuiflorum*	Medicinal plant	Jariwala and Desai (2018)
Nigrospora sphaerica, Acremonium falciforme, Allomyces arbuscula, Penicillium chrysogenum, Acrophialophora sp, Mycelia sterilia	*Litsea cubeba*	Medicinal plant	Deka and Jha (2018)

(continued)

Table 5.1 (continued)

Endophytic fungi	Host plant	Type	References
Colletotrichum gloeosporioides f. sp. camelliae and Pleosporales sp.	*Camellia sinensis*	Tea	Win et al. (2019)
Tremellales, Davidiellaceae, Basidiomycota, Rhodotorula, Tremellales, Meria, Cryptococcus, Cladosporium, Acaromycetes, Erythrobasidium, etc.	*Aegiceras corniculatum* (*Myrsinaceae*), *Avicennia marina* (*Verbenaceae*), *Bruguiera gymnorrhiza*, *Kandelia candel* and *Rhizophora stylosa* (*Rhizophoraceae*), and *Excoecaria agallocha* (*Euphorbiaceae*)	Mangrove	Yao et al. (2019)

sativa, Xylocarpus granatum, and Elaeagnus angustifolia from various environments like arid, semiarid, and mangrove are *Actinoplane missouriensis, Actinoallomurus acacia, Actinoallomurus coprocola, Actinomadura glauciflava, Amycolatopsis tolypomycina, Actinoallomurus oryzae, Jishengella endophytica, Kribbella* sp., *Microbispora mesophila, Microbispora* sp., *Micromonospora* sp., *Nocardioides* sp., *Nocardia alba, Nonomuraea rubra, Micromonospora* sp. *Nonomuraea* sp., *Pseudonocardia* sp., *Planotetraspora* sp., *Pseudonocardia endophytica, Pseudonocardia halophobica, Streptomyces* sp., *and Streptomyces javensis* (Coombs and Franco 2003; Thamchaipenet et al. 2010; Chen et al. 2011; Xie et al. 2011; Yadav 2017; Yadav and Yadav 2018). Reports revealed that the actinomycetes diversity in phyllosphere is high in the tropical and temperate ecosystem (Strobel and Daisy 2003; Yadav et al. 2018b; Yadav and Yadav 2019). Moreover, the physiology of the plant and the environment determines the actinobacterial association in plants and allows them to establish endophytic life (Du et al. 2013). Some important actinobacterial diversity in various plant sources is discussed in the following (Table 5.2).

5.5 Mechanism of Microbial Interaction with the Phyllosphere

The leaf physiology determines the microbial diversity and abundance on the phyllosphere. It establishes the microhabitat where the microorganisms adapt to their physiology to survive in this habitat (Staley et al. 2014; Shiraishi et al. 2015). The epiphytic microbes formed as colonial form, which gives protection to the microorganisms from this harsh microhabitat (Lindow and Brandl 2003; Remus-Emsermann

Table 5.2 Diversity of endophytic actinobacteria

Endophytic actinobacteria	Host plant	Habitat	References
Microbispora sp., *Micromonospora* sp., *Nocardioides* sp., *Streptomyces* sp.,	*Triticum aestivum*	Arid	Coombs and Franco (2003)
Actinoplane missouriensis	*Lupinus termis*	Arid	El-Tarabily (2003)
Pseudonocardia endophytica	*Lobelia clavatum*	Arid	Chen et al. (2009)
Actinoallomurus acaciae, *Streptomyces* sp., *Actinoallomurus coprocola*, *Amycolatopsis tolypomycina*, *Kribbella* sp., *Microbispora mesophila*	*Acacia auriculiformis*	Arid	Thamchaipenet et al. (2010)
Actinomadura glauciflava, *Pseudonocardia halophobica*, *Nocardia alba*, *Nonomuraea rubra*, *Streptomyces javensis*	*Aquilaria crassna*	Mangrove	Nimnoi et al. (2010)
Actinoallomurus oryzae	*Oryza sativa*	Aquatic	Indananda et al. (2011)
Jishengella endophytica	*Xylocarpus granatum*	Mangrove	Xie et al. (2011)
Micromonospora sp. *Nonomuraea* sp., *Pseudonocardia* sp., *Planotetraspora* sp.	*Elaeagnus angustifolia*	Arid	Chen et al. (2011)
Streptomyces phytohabitans	*Curcuma phaeocaulis*	Arid	Bian et al. (2012)
Nonomuraea solani	*Solanum melongena*	Arid	Wang et al. (2013b)
Actinoplanes hulinensis, *Streptomyces harbinensi*, *Wangella harbinensis*	*Glycine max*	Arid	Jia et al. (2013), Liu et al. (2013), Shen et al. (2013)
Micromonospora sonneratiae	*Sonneratia apetala*	Mangrove	Li et al. (2013)
Modestobacter roseus	*Salicornia europaea*	Saline	Qin et al. (2013)
Promicromonospora endophytica	*Eucalyptus microcarpa*	Arid	Kaewkla and Franco (2013)
Blastococcus endophyticus, *Plantactinospora endophytica*	*Camptotheca acuminate*	Arid	Zhu et al. (2013)

(continued)

Table 5.2 (continued)

Endophytic actinobacteria	Host plant	Habitat	References
Actinoplanes brasiliensis, Couchioplanes caeruleus, Gordonia otitidis, Micrococcus aloeverae, Streptomyces zhaozhouensis	*Aloe arborescens*	Arid	He et al. (2014)
Micromonospora schwarzwaldensis Streptomyces sp., *Wenchangensis*	*Centella asiatica*	Mangrove	Ernawati et al. (2016)
Glutamicibacter halophytocola, Kineococcus endophytica, Streptomyces sp.,	*Limonium sinense*	Saline	Feng et al. (2017)
Marmoricola endophyticus	*Thespesia populnea*	Mangrove	Jiang et al. (2017)
Allostreptomyces psammosilenae	*Psammosilene tunicoides*	Arid	Huang et al. (2017)
Micromonospora terminaliae	*Terminalia mucronata*	Mangrove	Kaewkla et al. (2017)
Nocardiopsis sp., *Pseudonocardia* sp. *Streptomyces* sp.,	*Dracaena cochinchinensis*	Semiarid	Salam et al. (2017)
Mangrovihabitans endophyticus	*Bruguiera sexangula*	Mangrove	Liu et al. (2017)
Actinoplanes sp., *Agrococcus* sp., *Amnibacterium* sp., *Brachybacterium* sp., *Brevibacterium* sp., *Citricoccus* sp., *Curtobacterium* sp., *Dermacoccus* sp., *Glutamicibacter* sp., *Gordonia* sp., *Isoptericola* sp., *Janibacter* sp., *Kocuria* sp., *Leucobacter* sp., *Mycobacterium* sp., *Micrococcus* sp., *Nocardioides* sp., *Kineococcus* sp., *Kytococcus* sp., *Marmoricola* sp., *Microbacterium* sp. *Micromonospora,* sp., *Nocardia* sp., *Nocardiopsis* sp., *Pseudokineococcus,* sp., *Sanguibacter* sp., *Streptomyces* sp., *Verrucosispora* sp.,	*Avicennia marina, Aegiceras corniculatum, Kandelia obovota, Bruguiera gymnorrhiza,* and *Thespesia populnea*	Mangrove	Jiang et al. (2018)
Glycomyces anabasis	*Anabasis aphylla*	Arid	Zhang et al. (2018)

et al. 2012). Commonly, bacteria develop larger sized colonial association on the leaf surface, especially at veins as well as the groves of epidermal cells (Morris et al. 1997; Hirano and Upper 2000). The epidermal grooves are rich in nutrients specifically sugar and water. This region is less waxy cuticle, usually the leaf surface is fully coved with waxy cuticle which prevents the permeability and wettability of the leaf surface and regulates the colonization of the microbes on phyllosphere (Lindow and Brandl 2003; Burch et al. 2014).

The leaf surface water droplets diffuse the waxy cuticle and improve the permeability by which the compounds are diffused from the apoplast to phyllosphere surface (Schreiber 2005). These leached compounds and water on the phyllosphere are making the availability of nutrients to the microorganisms. Most commonly, the flow of water from the stomata (transpiration) is increasing the permeability and wettability of guard cells and its surface cuticles (Schönherr 2006). Hence, higher permeation of the cuticle layer permits the microbes to colonize densely (Krimm et al. 2005). Moreover, the surface bacteria are able to produce certain compounds like biosurfactants (syringafactin produced by *Pseudomonas syringae*) (Krimm et al. 2005; Burch et al. 2014) which can modify the cuticle surfaces of the leaf and establish its association. This can facilitate water availability and alter sugar availability that can improve living conditions for epiphytic bacterial growth (Lindow and Brandl 2003; Van der Wal and Leveau 2011). Epiphytes such as *Pseudomonas* sp., *Stenotrophomonas* sp., and *Achromobacter* increase the water permeability of the lipophilic cuticle present in *Hedera* and *Prunus*, which increases the availability of the compounds at the phyllosphere which will improve the epiphytic fitness on the leaf surface (Schreiber et al. 2005).

It has been experimentally proved in the bean phyllosphere containing fructose facilitates the growth of *Erwinia herbicola* and *Pantoea agglomerans* (Remus-Emsermann et al. 2013; Tecon and Leveau 2016). However, irregular distribution of fructose differentially promotes the *P. eucalypti* population on bean leaves (Mercier and Lindow 2000; Leveau and Lindow 2001; Remus-Emsermann et al. 2011). These studies suggested that the permeated carbon sources on the leaf surface are merrily exploited by the epiphytic microorganisms for their growth and multiplication. At the same time, the phyllosphere microbial population can influence the modulation of the physicochemical properties of the leaf with the help of both biotic and abiotic surroundings (Bringel and Couée 2015; Ohshiro et al. 2016; Quan and Liang 2017). Soil microbial community may also influence the determination of phyllosphere microbial diversity. However, the microbes can construct the niches in the phyllosphere microhabitat wherein it can sustain and establish its population steadily (Agler et al. 2016; San Roman and Wagner 2018). Recent studies revealed the special relationships between the bacterial species in the phyllosphere community. Presence of sugars and nutrients in this environment significantly change the individual bacterial cells within the microbial aggregates (Fig. 5.2) would spatially be established with cell-to-cell interactions along with direct physical interactions (Levy et al. 2018; Tecon et al. 2018). The community structure is organized based on the driven factors such as dispersal, selection of microbes, diversification, and ecological drift. The fitness of the community is due to internal (strain types) and external determinants (environment) of the phyllosphere (Schlechter et al. 2019).

In general, the internal factors of the community are based on the microbial relationship within the aggregates. The microbes usually have either commensal or antagonistic or mutualistic or cooperative association by which the community structure can be established. Both cooperative and mutualistic microbial interactions shape the community structure as well as to develop larger colonial association containing the maximum microbial population. While commensals have weak interactive partners in the community, they are randomly distributed in the habitat. The commensals should not influence the interactive association within the structured community (Stubbendieck et al. 2016). Besides, antagonistic microbes have a negative interaction within the community, one can outcompete the other and the sensitive microbes have been eliminated from the environment. The effect of cooperative microbial interactions on the phyllosphere community structure establishment is not demonstrated (Schlechter et al. 2019).

Bacteria can ascertain the cell-to-cell communication system and establish a larger community structure with heterogeneous populations, usually with mutualistic and cooperative partners. However, some kind of mutualistic relationship may occur between rapid growing bacteria and pathogenic fungi, which leads to cause superficial infection on the host plant which increases the nutrient accessibility of the bacteria to rich its population (Suda et al. 2009; Zeilinger et al. 2016; Amine Hassani et al. 2018). Inversely, fungal–fungal interactions seem to decrease the bacterial population, for example, oomycete species *Dioszegia* sp. and *Albugo* sp. outcompete the bacterial microbiota on *A. thaliana* leaf (Chou et al. 2000; Agler et al. 2016). Moreover, competitive interactions of microbes involve negative effects on at least one species of the habitat. Some competitive microbes produce certain toxic chemical substances (antibiotics and siderophores) as secondary metabolites which pose a negative effect on its competitor microbes. The best example of such interaction is a gram-negative *Pantoea agglomerans* bacteria which inhibit the growth of *Erwinia amylovora*, a phytopathogen of apple by antibiotic activity (Wright et al. 2001; Pusey et al. 2011).

Generally, the competition of microbes is mainly for their nutrition and space. The phyllosphere is a nutrient-limited environment, wherein the competitive partner has compromised their growth by either coexisting or excluded from the site (Saleem et al. 2017). Besides, the phyllosphere is greatly colonized by both oligotrophic and competitive microbes which play an important role in community structure formation (Schlechter et al. 2019). However, the key factors of the phyllosphere community assemblage are currently vague. Hence, more studies required to find the key factors determining the phyllosphere community structure assemblage.

5.6 Factors Controlling Phyllosphere Microbiomes

Once microbes arrived at the phyllosphere, a variety of factors resolve whether microbial cells are competent to colonize the leaf and become confined. Colony establishment depends on the leaf–atmosphere environmental interaction with the residing microorganisms in the phyllosphere. At the beginning, the microbe reaches the cuticle layer, a waxy surface that protects the leaf from the pathogens. In general, cuticle restricts the microbial association due to the functions such as barrier, reducing water and solute lass, aqueous pollution, reflectance to minimize the temperature, conferring water repellent, etc. (Beattie 2002; Whipps et al. 2008). The whole-cell biosensor-based study revealed the available nutrients on the leaf surface facilitate the growth of residing microbes at a limited level (Miller et al. 2001). This was confirmed by the microscopic observation of leaf surfaces, at the low nutrient region contains less dense microbial colonization than the nutrient-rich surface (Monier and Lindow 2005). Naturally, nutrient enrichment may happen by pollen deposits and honeydew at the phyllosphere surface (Lindow and Brandl 2003), besides plant leaves release a large array of volatile organic substances into the margin layer around leaves (Jackson et al. 2006). Nutrients that include CO_2, acetone, terpenoids, aldehydes, alcohols, long-chain hydrocarbons, sesquiterpenoids, and nitrogen-containing compounds (Whipps et al. 2008) are available nutrients for microbial growth. Some of the compounds may act as growth inhibitor or toxic to microbial growth (Dingman 2000; Shepherd et al. 2005). Hence, microbes establish several adaptive mechanisms for maintaining their growth in adverse conditions.

5.6.1 Microbial Adaptations in Phyllosphere Environment

Microbes like bacteria establishing colonies at the phyllosphere are limited by various factors including both biotic and abiotic. Abiotic factors such as the available nutrient (Delmotte et al. 2009), seasonal variation, rainfall, temperature, plant immunity, and competitor microbes (Rastogi et al. 2013) are influencing surveillance of microbes in the phyllosphere. Metaproteomic studies on the leaf surface communities have been identified as microbes producing vitamins and siderophores which give adaptation to the microbes at the environment. For example, phyllosphere of soybean, clover, and Arabidopsis plants largely colonized by Sphingomonas and Methylobacterium provides vitamins and siderophores to the plant (Green 2006; Delmotte et al. 2009) and it competes for other microbes. *Methylobacterium* spp. are involved in the assimilation of methanol at the phyllosphere, a by-product of demethylated pectin during the cell wall metabolism of the plant (Galbally and Kirstine 2002; Delmotte et al. 2009), and it gives epiphytic fitness to the microbes. Proteome studies revealed that some unique properties of rhizosphere bacteria have been found in the phyllosphere microbiota. For example, genes of methanol dehydrogenase and formaldehyde-activating enzyme (of Rhizosphere *Methylobacterium* spp.) and nitrogen fixation (Rhizobium sp.) are also reported in both phyllosphere and rhizosphere samples of rice (Knief et al. 2012). Gourion et al. (2006) observed upregulation of methylotrophic proteins such as MxaF and Fae and stress-related protein PhaA during epiphytic growth of *Methylobacterium extorquens*.

Phyllosphere colonization may occur in two different habitats, (1) the surface (epiphytic) and (2) the apoplast or leaf interior (endophytic). During the epiphytic life, many of the environmental factors regulate the growth such as solar radiation, temperature, water availability, nutrient, humidity, etc., whereas the endophytes are challenged with a plant defense mechanism. A bacteria colonizing at both habitats may differentially express their genes, for example, *P. syringae* pv. syringae B728a at epiphytic growth express the genes involved in motility, chemosensing, phosphate mobilization, and utilization of tryptophan which is higher than in endophytic growth (Yu et al. 2013). However, the secondary metabolite (syringomycin, syringopeptin) production was higher in the endophytic stage. One such adaptation is the production of pigments, bacteria such as Pseudomonas, Sphingomonas, and Methylobacterium produce pigmentation by which they give protection against UV light (Lindow and Brandl 2003). Presence of extracellular polysaccharide is another protective measure of plant-bacteria against desiccation and osmotic stress (Monier and Lindow 2004). Delmotte et al. (2009) found several stress-resistant proteins (PhyR and EcfG) from the phyllosphere of soybean, clover, and Arabidopsis through metaproteogenomic survey. Flagellin-like protein is high in pseudomonas at the epiphytic growth which enables the bacteria to access the nutrition by the chemostatic model (Yu et al. 2013).

5.6.2 Plant Immunity/Responses to Control Microbial Colonization

The plant has its immune system which plays an important role in determining microbial assembly (Jacoby et al. 2017). Plants contain two layers of defense, the primary immunity is named pattern-triggered immunity (PTI), it has a conserved molecule named microbe/pathogen-associated molecular patterns (MAMPs/PAMPs). The PTI is a localized immunity mediated at the plasma membrane containing pattern recognition receptors (Monaghan and Zipfel 2012; Wang et al. 2019). The MAMP/PAMP limits the growth of bacterial pathogens. For example, the flagellin-sensitive receptor 2 (FLS2) is a pattern receptor which recognizes the *P. syringae* pathovar (pv.) bacterial flagellin (flg22) (Chinchilla et al. 2006; Newman et al. 2013; Trdá et al. 2015). However, the plant response to limits its defense against non-pathogenic bacteria is still unknown. The effector's protein-mediated destabilization of plant immunity and immune escape is also reported (Jones and Dangl 2006; Cui et al. 2009). Plant immunity is targeted with specific proteins, which involves the self-protection against the microbial association has been deactivated by the interaction of microbial effector proteins and it makes protein–protein networks (Bogdanove 2002; Snelders et al. 2018). Besides, plants have evolved with intracellular receptor molecules called nucleotide-binding leucine-rich repeat proteins (NLRs), which either openly or ultimately recognize effector proteins to give the second layer of plant immunity named effector-triggered immunity (ETI) (Jacob et al. 2013; Wu et al.

2014). Both PTI and ETI generate more specific and diverse immunity against phyllosphere microflora. Beneficial or the synergistic microbes interact with signaling pathways (MAMPs) of the plant to elevate the production of its immune response. However, if pathogen could interact by using MAMPs, the immune output will be higher and will restrict the colonial establishment of pathogens. Pathogens that live in host tissues use hemibiotrophs and necrotrophs mode of life (Table 5.3). Some chemicals of the plant tissues inhibit the microbial association either biotrophs (salicylic acid) or necrotrophs (jasmonic acid) type and also the reactive O_2 species may have an inhibitory effect on the pathogens (Lehmann et al. 2015). Plants use jasmonic acid, methyl jasmonate, ethylene, flavonoid, 12-oxo-phytodienoic acid, and salicylic acid-mediated signals for quenching pathogens on its surface (Table 5.4). Recently, pathogens with biotrophy-necrotrophy switch have been identified in fungi such as *Colletotrichum sp*, *Phytophthora capsici*, *Moniliophthora roreri,* and *Macrophomina phaseolina* in which pathogen evokes a differential response of growth in host tissues (Chowdhury et al. 2015). Some important research in the mode of immune evoke by the pathogen has been listed in Tables 5.3 and 5.4.

Phyllosphere region is usually colonized by a variety of microorganisms. Naturally, leaf epidermises are always contacted to external and internal environments and are enriched with a diverse group of bacteria, yeast, fungi, and viruses. The cuticle layer of the leaf surface plays a significant role during the contact with leaf

Table 5.3 Mode of life cycle of pathogen established against plant immunity

Organism	Life cycle	Host	References
Fusarium graminearum	Hemibiotrophic	Wheat	Ding et al. (2011)
Colletotrichum graminicola	Biotrophic and necrotrophic	Maize	Vargas et al. (2012)
Septoria tritici	Hemibiotrophic	Wheat	Yang et al. (2013)
Phytophthora capsici	Hemibiotrophic	Tomato	Jupe et al. (2013)
Colletotrichum sp.	Hemibiotrophic	Plants	Gan et al (2013)
Moniliophthora roreri	Hemibiotrophic and necrotrophic	Cacao	Meinhardt et al. (2014)
Fusarium verticillioides	Biotrophic	Maize	Lanubile et al. (2014)
Botrytis sp	Necrotrophic	Plants	Van Kan et al. (2014)
Botrytis fabae	Necrotrophic	Faba bean	El-Komy (2014)
Sclerotinia sclerotiorum	Biotrophic, hemibiotrophic, and necrotrophic	Plants	Kabbage et al. (2015)
Zymoseptoria tritici	Hemibiotrophic	Wheat	Rudd et al. (2015)
Phytophthora infestans	Hemibiotrophic	Tomato	Zuluaga et al. (2016)
Rhizoctonia solani	Necrotrophic	Wheat	Foley et al. (2016)

Note "Hemibiotrophs" - an organism that is parasitic in living tissue for some time and then continues to live in dead tissue. "Necrotrophs" - can kill the host cells and feed on the contents

Table 5.4 Signaling pathway inhibits pathogenic microbes in phyllosphere

Molecules/signals	Pathogen	Host	References
Flavonoid pathway	Bacterial pathogens	*Melampsora medusae*	Miranda et al. (2007)
Jasmonic acid, ethylene, and the flavonoid	*Phymatotrichopsis omnivora*	Medicago truncatula	Uppalapati et al. (2009)
Methyl jasmonate and ethylene	*Macrophomina phaseolina*	*Medicago truncatula*	Gaige et al. (2010)
Jasmonic acid and ethylene	*Fusarium graminearum*	Wheat	Sun et al. (2010)
Jasmonate and ethylene	*Fusarium sp*	Wheat	Gottwald et al. (2012)
Ethylene and jasmonate	*Pythium ultimum*	Apple	Shin et al. (2014)
Methyl jasmonate, 12-oxo-phytodienoic acid, salicylic acid, and flavonol	*Fusarium oxysporum* f.sp. *lycopersici*	Tomato	Krol et al. (2015)

microbiota (Vacher et al. 2016). Though some group of microorganism may not multiply after it reaches on the surface, many continue to survive and multiply, until they can attain maximum number (Schönherr 2006; Innerebner et al. 2011; Pusey et al. 2011). To multiply, microorganisms require carbon, nitrogen, inorganic, and organic energy sources. However, in the absence of such nutrients, phyllosphere is still usually colonized by a large number of bacteria (105–107 CFU/g of the leaf) in the presence of high relative humidity and free water at suitable environmental conditions (Schönherr 2006; Baldwin et al. 2017). This is due to the release of nutrients or leaf exudates which adequately supported the microbial growth. There are varieties of molecules leached from the plant leaves such as sugar, amino acids, organic acids, minerals, etc. (Beattie 2011; Remus-Emsermann et al. 2011; Meiners et al. 2017). These leaching materials may differ with plant species and the environmental condition (Beattie 2011; Remus-Emsermann et al. 2011; Mendes et al. 2013).

Nutrients such as sugar photosynthates from the leaf interior may be diffused through the cuticle reached the outer surface (Schreiber 2005), and are chiefly used by phyllosphere bacteria. Moreover, water droplets on a leaf surface facilitate the outward diffusion of these sugars (Van der Wal et al. 2013). Both non-pathogenic and pathogenic microorganisms establish colonization on the leaf surface. To survive and thrive, epiphytic microbes have several adaptive properties such as the production of antibiotics, extracellular polymeric substances (EPS), biosurfactant for increasing cuticle permeability, and availability of nutrients volatile organic compounds (VOCs) to the leaf surface. However, in order to avoid the entry of pathogens, plants develop defense reactions. The preliminary defense is activated by recognition of the chemical compounds released during the contact with microbes (Boller and Felix

2009). Pathogen-induced molecular patterns (PAMPs)-triggered immunity (PTI) is a broad spectrum of defenses against the pathogen invades. However, effectors produced by the pathogens often interfere with PTI activation and are recognized by specific proteins, which stimulate effector-triggered immunity (ETI) that induces a hypersensitive response (Craig et al. 2009).

5.7 How to Study Phyllosphere Microbiome?

The diversity and community structure of phyllosphere microbes have been intensely studied by culture-independent methods. However, this approach failed to isolate and identify the complete microbiome of the environment. Therefore, scientist used the culture-independent mass sequencing methods which have been carried out by high-throughput molecular methods, especially PCR-amplified DNA-level conserved taxonomic markers such as 16S rRNA, 18S rRNA, and internal transcribed spacer (ITS) sequences-based metagenome of phyllosphere total microbiome (Mao et al. 2012; Santhanam et al. 2014; Williams and Marco 2014; Jo et al. 2015; Copeland et al. 2015) (Fig. 5.4). The first-generation molecular techniques such as Sanger sequencing, denaturing gradient gel electrophoresis (DGGE), and terminal restriction fragment length polymorphism have been used to describe the community structure variation in plant phenotype, and geographical location (Hunter et al. 2010; Vokou et al. 2012; Izhaki et al. 2013). Those techniques are low throughput and highly expensive that can be used to detect the superficial microbial community of the environment (Rastogi and Sani 2011).

Advancements in molecular techniques, next-generation DNA sequencing is the potent method that significantly reduces the costs and allows to perform hundreds of samples in a single attempt. These techniques open up new windows of omics, specifically "environmental omics." The 454 pyrosequencing is the first to be widely executed to study in microbial community analysis. This method comprises rRNA or ITS amplicon sequencing, whole-genome sequencing, shotgun metagenomics, and transcriptional profiling (Delmotte et al. 2009; Rastogi et al. 2012). Recently, Illumina platform has been performed better and allows ultra-high-throughput sequencing of microbial communities with high-quality reads (Degnan and Ochman 2012). Proteogenomic is another method used for the microbial community structure analysis (Delmotte et al. 2009), a combination of genomics and proteomics to a great extent makes easy the structural and functional differences of microbiota in the phyllosphere environments. Through those methods, microbial diversity of several host plants such as Arabidopsis, Apple tree, Beech, grapevine, oak, poplar, Prunus, rice, soybean, spinach, tomato wheat, etc. was documented. The metadata of the metagenomic studies helps to understand the growth behavior, colonization ability, genus-level community structure formation (or) association, low and high index of diversity, and the host genotype effects on the self-defense as well as the cell wall integrity have been reported.

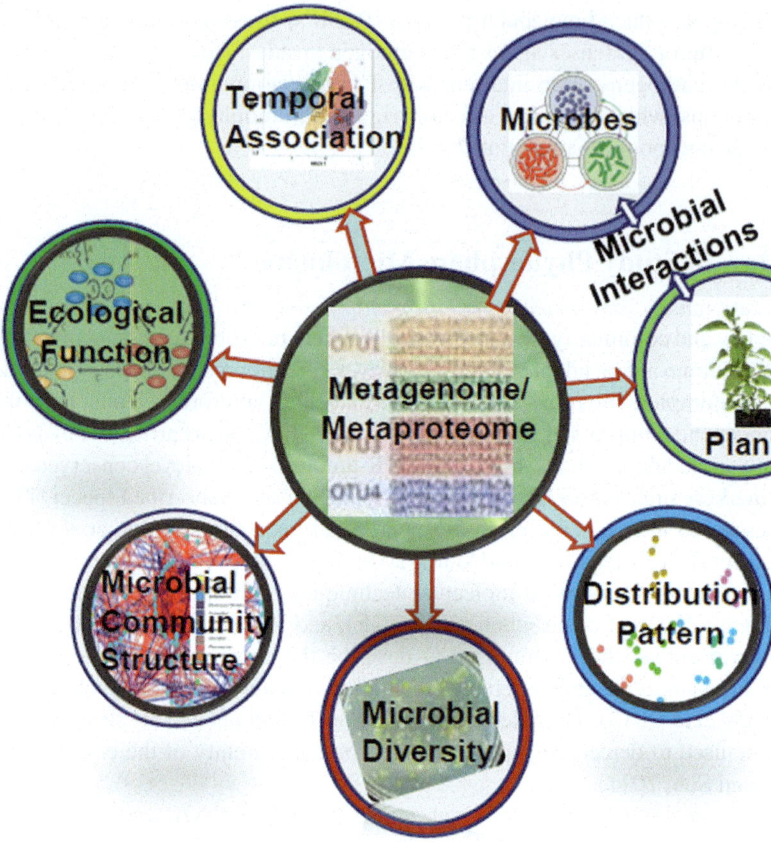

Fig. 5.4 Scope of metagenome in phyllosphere microbes and their functions

Whole microbiome analysis by environmental sequencing is popular today to explain the plant's phyllosphere containing complex microbial communities. There are many methods for mapping the diversity of microbiome which could associate with any of the living and nonliving objects. Also, the environmental sequencing approach determines the whole microbiome of the plant and it illustrates the significant association of microbes on its host under controlled conditions. Recently, studies revealed that genome-wide association (GWA) is the best method which shows potential merits for identifying the microbial communities associated with different kinds of host–microbe interactions. The high-throughput environmental sequencing approach has guided to the discoverer to find the complex microbial ecosystem of leaves. Using this strategy, many studies revealed the microbial association in the phyllosphere of different plants such as mountain shrubs (Ruiz-Pérez et al. 2016), seagrass (Fahimipour et al. 2017), subarctic grass (Uroz et al. 2016), and equatorial forest canopies (Lambais et al. 2006). The studies revealed that plant leaves are colonized by a huge and diverse group of microorganisms, including bacteria, fungi, and viruses (Rastogi et al. 2013; Morella et al. 2018; Sapp et al. 2018; Beilsmith et al. 2019). High-throughput molecular methods or culture-independent molecular techniques have interpreted the phyllosphere microbial community today (Table 5.5). Through this technique, Actinobacteria, Bacteroidetes, Firmicutes, and Proteobacteria are common microbiome of plant leaves (Bulgarelli et al. 2013), and it suggests that *Pseudomonas, Sphingomonas, Methylobacterium, Bacillus, Massilia, Arthrobacter,* and *Pantoea* are predominant genera consistently firm in the phyllosphere. Findings of the studies disclose the variation of microbial community structure mainly based on the genotypic nature of the plant species and also its geographical location. For example, Finkel et al. (2011) observed similar bacterial communities from the different species of Tamarix (*T. aphylla, T. nilotica,* and *T. tetragina*) grown in the same geographical location; however, differences in community structure of microbiota have been strongly related to its geographical distances (Rastogi et al. 2012).

Moreover, the high-throughput studies revealed the special functions/metabolism of the microbes associated with leaf surfaces, specifically carbohydrate transport, leaf litter decomposition, light-driven ATP pumps, methanol metabolism, C1 metabolism (Ottesen et al. 2013; Shade et al. 2013), and the effect of ecological factors such as climate change, temperature, seasonal variation, sporadic contact to soil, and/or anthropogenic activities such as the use of agricultural chemicals and pesticides (Ikeda et al. 2011; Shade et al. 2013; Karlsson et al. 2014; Copeland et al. 2015; Glenn et al. 2015). To attain better perceptive of the phyllosphere ecosystem and understand the functional relationship among plants, microbiota, and environment, metaproteome and metagenomics have been used (Rastogi et al. 2012; Bálint et al. 2013; Dees et al. 2015).

5.8 Impact of Phyllosphere Microbiome on Ecosystem

Phyllosphere microflora significantly influences the ecological relationship of the plants. The phyllosphere usually has bacteria, fungi, lichens, algae, and viruses that have actively participated in the adaptation, growth, resistance, and infection of the plant host (Walker et al. 2017; Verma et al. 2017; Yadav et al. 2018a). The phyllosphere microbiota has not been completely studied with their ecological significance, specifically plant and ecosystem level (Remus-Emsermann and Schlechter 2018). From seed germination to plant reproduction, studies have revealed how the phyllosphere microbiome affects the leaf functions and longevity, seed mass, apical growth, flowering, and fruit development (Jones and Dangl 2006; Sawinski et al. 2013; Kembel et al. 2014); however, the net interplay of the phyllosphere ecosystem in and around the plant is scanty. Recent scientific advancements that simplify the phyllosphere microbial life become understandable. The high-throughput genomics, such as environmental genomics and metagenomics, have greatly expanded our perceptive

Table 5.5 Studies on high-throughput molecular approaches to phyllosphere communities

Method	Plant	Study	References
16S rRNA gene pyrosequencing	Soybean, clover, Arabidopsis	Epiphytic fitness of Sphingomonas and Methylobacterium	Delmotte et al. (2009)
	Pine and other trees	Phyllosphere bacteria community composition	Redford et al. (2010)
	Spinach	Genus-level communities of Proteobacteria and Firmicutes-associated spinach leaves	Lopez-Velasco et al. (2011)
	Grape	Bacterial communities on the surface of leaves and berries from grapevine	Leveau and Tech (2011)
	Lettuce	A "core" community composed of Pseudomonas, Bacillus, Massilia, Arthrobacter, and Pantoea found in lettuce foliage	Rastogi et al. (2012)
	Lettuce	Variation in phyllosphere microbiota composition. Effect of *E. coli* O157:H7 inoculation on microbiota composition	Williams et al. (2013)
	Rice	Metagenomic analysis of rice phyllospheric bacterial communities in relation to blast disease	Prasad Sahu and Kumar (2015)
	Common bean, soybean, and canola	Seasonal community succession of the phyllosphere microbiome	Copeland et al. (2015)
	Espeletia species	Microbial and functional diversity within the phyllosphere.	Ruiz-Pérez et al. (2016)
16/18S rRNA gene pyrosequencing	Oak	Fungal communities in the oak phyllosphere	Jumpponen and Jones (2009)
	Tamarix aphylla, T nilotica, T. tetragina	Geographical location is a major determinant of phyllosphere bacterial communities	Finkel et al. (2011)
	Beech	Plant genotype-based fungal communities on leaf surfaces	Cordier et al. (2012)
	Balsam poplar	Plant species-based fungal community composition	Balint et al. (2013)

(continued)

Table 5.5 (continued)

Method	Plant	Study	References
	Pine	Rapid microbial community changes during initial stages of pine litter decomposition	Gołębiewski et al. (2019)
Metaproteogenomics	Soybean, clover, Arabidopsis	Metabolic adaptations contribute to the epiphytic fitness of Sphingomonas and Methylobacterium	Delmotte et al. (2009)
	Rice	Several methylotrophic enzymes and their role in the carbon cycle by Methylobacterium	Knief et al. (2012)
	Maize	Functional genes that distinguish maize phyllosphere metagenomes in drought and well-watered conditions	Methé et al. (2017)

and understanding on the functional life of phyllosphere microbial communities in plant–environment and the impact on the ecosystem.

Environmental factors are drastically influencing the microbiome changes on phyllosphere. This is common to epiphytic microorganisms, exposed with heavy stress during the season cycle, the day/night cycle, and the growth, age, and anatomical dynamics of the plant. For instance, at drought condition, the epiphytic microbial community was notably increased on Holm oak (Rico et al. 2014). Similarly, at hot condition, bacterial endophytic communities are altered in lower leaves of paddy, but not in the epiphytes (Ren et al. 2014). However, the epiphytic fungal community responded well in worming seasons (Coince et al. 2014; Bálint et al. 2015). Besides, an increase of CO_2 at the phyllosphere region never affects the bacterial abundance (Ren et al. 2014; Vacher et al. 2016), except a few fungal genera.

Microbes have flexible metabolic adaptations, which helps them to survive in the phyllosphere microenvironment. During the metabolic functions, the plant releases carbohydrates, polyols, amino acids, amines, isoprenoids, halogenated compounds, or alcohols, as well as water and salts, which are the available nutrients for epiphytic microorganisms (Trouvelot et al. 2014). However, leaf surface commonly exhibits desolate properties such as saline or alkaline pH which generates stress in phyllosphere microbes (Finkel et al. 2012). Several alphaproteobacteria express PhyR-based stress regulation and colonization on leaf surface (Iguchi et al. 2013). Additionally, they develop multiple mode adaptation to survive in phyllosphere such as tolerance, antimicrobial, and immunity compounds against a microbial competitor (Trouvelot et al. 2014), synthesis of extracellular polysaccharides, and also synthesize phytohormonal compounds.

Besides, biotic and abiotic factors induce molecular level regulations in plants to synthesize a diverse range of phytohormones. Generally, the gaseous ethylene, jasmonate, methyl jasmonate, salicylate, and methylsalicylate are induced by bacterial pathogens (Bodenhausen et al. 2014; Horton et al. 2014). For example, many plant defense mechanisms are induced by the interaction of the biotic component of the ecosystem through signals like volatile and nonvolatile chemicals, and microbes can degrade such chemicals resulting in reduced activity (Mason et al. 2014).

The phyllosphere microbiome acts as a vital role for leaf surface environment and their surrounding ecosystem functions (Ortega et al. 2016). Phyllosphere microbes have interacted with their environment through their metabolic functions (Fig. 5.5). In general, plants release a variety of volatile organic compounds (VOCs) and its precursors on the surface of leaves (Schäfer et al. 2010), and it could regulate the microorganisms in response with the environment. Plants are the major VOCs emitter of the biosphere (>1000 Tg/year) and can release compounds such as terpenes, monoterpenes, flavones, methanol, methane, and halogenated methane (C1 compounds). The epiphytic microbes on the surface of the plant, as well as the airborne bacteria, effectively consumed the emitted VOCs through bacterial metabolism (Junker and Tholl 2013), and this effects of climate change would impact the diversity, species richness, and abundance in the phyllosphere community, and its capability on filtering of plant-emitted volatile substances.

Methane (CH_4) is the most important greenhouse gas (~1.8 ppm), and it has been detected from the leaves, roots, and stems and is released to the atmosphere (Keppler et al. 2006). Phyllosphere microbes especially methanogens use the plant-emitted methane along with leaf exudates (Lenhart et al. 2015; Bringel and Couée 2015). Phyllospheric microbes are often rich in methylotrophic bacteria and can utilize the plant-emitted C1 compounds such as methanol, formaldehyde, and chloromethane

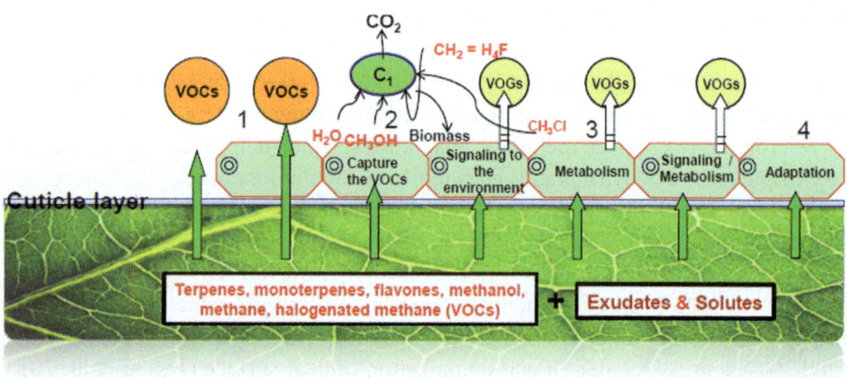

Fig. 5.5 Environmental impact of phyllosphere microbes. Utilization of plant emitting volatile organic compound (VOCs) and C1 compounds by phyllosphere microbes. (1) Free diffusion of VOCs to the atmosphere; (2) Capturing the VOCs by the surface microbes, act as filters; (3) Through specialized metabolic activities microbes metabolize the VOCs; (4) Adaptive response of microbes in the specialized environment. VOGs—Volatile organic gases

(Knief et al. 2012; Jo et al. 2015). Studies proved that the C1 metabolic epiphytic bacteria such as *Methylobacterium extorquens, Methylobacterium radiotolerans, and* Methylocystis use methanol and acetate as their carbon and energy source at the phyllosphere (Belova et al. 2011; Verginer et al. 2010; Iguchi et al. 2013; Jo et al. 2015; Iguchi et al. 2015; Krishnamoorthy et al. 2018). The *Methylobacterium extorquens* contains the methanol-dehydrogenase-like protein XoxF which is expressed during the colonization on *Arabidopsis thaliana* (Schmidt et al. 2010). Besides, chloromethane metabolism (cmu pathway) in methylotrophs has been identified from the surface leaves of *A. thaliana* harbor (Nadalig et al. 2011; Krishnamoorthy et al. 2018). Table 5.6 shows the various phyllosphere methanogenic bacteria and its metabolism.

Chloromethane (CH_3Cl) is one of the abundant chlorinated organic compounds in the atmosphere (currently ~550 ppt) and is to be responsible for the depletion of stratospheric ozone over 16% (World Meteorological Organization 2014). The fluorescence-based bacterial bioreporter study reported that phyllosphere microbes, *M. extorquens* CM4 (Roselli et al. 2013) and *Hyphomicrobium* sp. (Nadalig et al. 2011), having the genes for chloromethane utilization (*cmu*), and also volatile dimethylsulphide (DMS) and dimethylsulfoniopropionate (DMSP), considered as global climate regulator (Schäfer et al. 2010; Nevitt 2011). In the biosphere, a small number of plants like salt marsh grasses *Spartina* and sugarcanes (*Saccharum* sp.) are reported as producers of DMSP. Microbes that are associated with these plants have adaptive metabolism by which it transforms or metabolizes the DMS and DMSP (Ansede et al. 2001). Hence, the phyllosphere microbes are the major source of carbon and sulfur biogeochemical cycles, in the ecosystem and climate regulation through their active filtration or utilization of plant-related volatile compounds (DeLeon-Rodriguez et al. 2013; Šantl-Temkiv et al. 2013).

Microbial populations reside at phyllosphere as epiphytes or as endophytes, and have close contact with the rhizosphere. A microbe can be established as an epiphytic and endophytic association has the metabolic plasticity required for them to thrive. Many experimental evidences suggested that microorganisms commonly associated with plants maybe vital for nutrient accessibility and decomposition of biomass (Bernal et al. 2006; Ramírez Gómez 2011; Lizarazo-Medina and Gómez-Vásquez 2015). The functional ecology of the plant influences the composition and interaction of the phyllosphere microbes (Bodenhausen et al. 2013; Ruiz-Pérez et al. 2016). Many of the phyllosphere microbial communities share the common metabolic properties of the soil microbes. For example, the major phyllosphere bacterial communities such as Bacillus, Burkholderia, Methylobacterium, Pseudomonas, Sphingomonas, and Xanthomonas are the soil inhabitant, which have carbohydrate metabolizing genes involved in utilization of starch, hemicellulose, pectin, and cellulose, rich in humus materials (Rawat et al. 2012; Bodenhausen et al. 2013; Bulgarelli et al. 2013). The nitrogen metabolism such as ammonification, denitrification, and anammox, and the degradation of aromatic compounds are also reported in foliar microbes (Usubillaga et al. 2001; Rawat et al. 2012; Ruiz-Pérez et al. 2016).

Tropospheric microbes (aerosols) play a vital function in global carbon cycles and also metabolize the organic compounds. Some airborne Gammaproteobacteria have

Table 5.6 Studies on Phyllospheric methylotrophic metabolism

Epiphyte	Host plant	Function	References
Methanotrophs and Methylobacteria	Linden, pine and blue spruce lilac, maple, and apple	Diversity of Methanotrophs in woody plant tissues within the winter period	Doronina et al. (2004)
Methylobacterium extorquens	*Medicago truncatula*	Methylotrophic metabolism is advantageous for colonization under competitive conditions	Sy et al. (2005)
Methylobacterium extorquens	–	A proteomic study of *Methylobacterium extorquens* reveals a response regulator essential for epiphytic growth	Gourion et al. (2006)
Methylobacterium extorquens AM1	–	PhyR is involved in the general stress response	Gourion et al. (2008)
Methylocystis heyeri H2(T) and *M. echinoides* IMET10491(T)	Peat	Acetate utilization metabolism as a survival strategy	Belova et al. (2011)
Methylobacterium extorquens DSM 21961	Strawberry	Monitoring the plant epiphyte *Methylobacterium extorquens* DSM 21961	Verginer et al. (2010)
Methylobacterium extorquens	Arabidopsis thaliana or Medicago truncatula	The influence of the factor site, host plant species, time and the presence of other phyllosphere bacteria on Methylobacterium community composition and population size	Knief et al. (2010)
Candida boidinii	*Arabidopsis thaliana*	Yeast methylotrophy and autophagy in a methanol-oscillating environment on growing leaves	Kawaguchi et al. (2011)

(continued)

Table 5.6 (continued)

Epiphyte	Host plant	Function	References
Methylobacterium sp. (NC4), (NC28)	Sugarcane, pigeon pea, mustard, potato, and radish	Epiphytic pink-pigmented methylotrophic bacteria enhance germination and seedling growth of wheat (*Triticum aestivum*) by producing phytohormone	Meena et al. (2012)
Methylobacterium sp. strain OR01	Perilla plants	Dominant colonization and inheritance of Methylobacterium sp.	Mizuno et al. (2013)
Methylosinus sp. B4S	–	Stress resistance and C1 metabolism involved in plant colonization of a methanotroph Arch	Iguchi et al. (2013)
Methylobacterium oryzae	Rice	plant-probiotic methylotroph in the phyllosphere	Kwak et al. (2014)
Methylobacterium radiotolerans VRI8-A4	Groundnut	Diversity of culturable methylotrophic bacteria in different genotypes of groundnut and their potential for plant growth promotion	Krishnamoorthy et al. (2018)

ice nucleation-active (INA) property and contains specific gene (*ina*) via deposition of cloud droplets (Hill et al. 2014) on the leaf surface and mineralize the carbon compounds (Vaïtilingom et al. 2013). Reports confirmed the relationship of INA bacteria and phyllosphere microbiota, combined activities of both phyllosphere microbiota and cloud microbiota actively participating carbon cycle, and strong support for climate regulation (Bringel and Couée 2015). The above information suggested that the phyllosphere microbiome not only supports the health of its host but is also beneficial to the environment, specifically it regulates plant-derived greenhouse and other gaseous pollutants.

5.9 Biotechnological Potential of Phyllosphere Microbiota

The plant beneficial microbes are agriculturally important bioresources, and it can stimulate the plant growth and enhance plant nutrient uptake through solubilization and mobilization (of P, K, and Zn), nitrogen fixation, and siderophore production (microbes-mediated bio-fortification of Fe in different crops). Beneficial microbes can play an important role in increasing yields of the crop, remove contaminants, inhibit pathogens, and produce novel substances. The growth stimulation by beneficial microbes can be a consequence of biological nitrogen fixation, production of plant growth regulators such as IAA, gibberellic acids, and cytokines, and biocontrol of phytopathogens through the production of antibiotic, antifungal, or antibacterial, Fe-chelating compounds, induction of acquired host resistance, enhancing the bioavailability of minerals (Kour et al. 2019; Kumar et al. 2019b; Yadav et al. 2019a).

In this contest, the phyllosphere microbes may positively influence the growth of host plant and produce some antagonistic compound against pathogens. Phyllosphere endophytes with properties such as nitrogen fixation (Jones 1970; Freiberg 1998; Furnkranz et al. 2008), bioremediation of harmful chemicals/pollutants, and biocontrol agents against important foliar plant pathogens (Beattie and Lindow 1995; Balint-Kurti et al. 2010; De Marco et al. 2004) have been documented. Further, the microbiome of phyllosphere is a reflection of environmental conditions; they can contribute significantly to global food webs and nutrient linkages. Many beneficial microbes such as *Achromobacter, Bacillus, Beijerinckia, Burkholderia, Flexibacterium, Methylobacterium, Micrococcus, Micromonospora, Nocardioides, Pantoea, Penicillium, Planomonospora, Pseudomonas, Streptomyces,* and *Xanthomonas* have been reported from the phyllosphere environment of different crop plants (Verma et al. 2013a, b; Mukhtar et al. 2010; Meena et al. 2012; Dobrovol'skaya et al. 2017). However, compared with most other microbial habitats, the investigation of phyllosphere microbes is quite limited. Some of its important biotechnological potentials are listed below.

5.9.1 Biocontrol Agents

Biocontrol is the measure to control pathogens and disease-causing pest including nematodes weeds, insects, and mites by other beneficial microbes or harmless living materials. In nature, plant diseases are caused by bacterial pathogens which provide a substantial decline in the development of agricultural products. For sustainable agriculture, scientific approaches use the antagonistic properties of beneficial microbes against the harmful pathogens instead of using toxic harmful chemicals as biological control (Erwin and Ribeiro 1996; Sharma et al. 2012). Biological treatment is a desirable strategy for controlling plant diseases (You et al. 2015) and there are an increasing number of biocontrol agents (BCAs), such as Bacillus spp., Pseudomonas spp., Trichoderma spp., etc. being commercialized for various crops (Trabelsi and

Mhamdi 2013; Cha et al. 2016). Most of them habitat either on phyllosphere or soil and can play a significant role in killing the number of plant pathogens on the surface of the leaves by competitive principle.

Pathogenic microbial interactions in phyllosphere decrease the fitness of plants, the productivity of crops, and question the safety of horticultural products for human consumption. Phyllosphere actinomycetes have been reported to inhibit the growth and colonization of plant pathogens (Lindow and Brandl 2003). For example, the endophytic isolate *Gordonia* sp. has been reported to produce imidazole-2-yl amino acids that have antifungal properties (Mikolasch et al. 2003) and an acidic polysaccharide called Gordon as the main component in biofilms, which is considered essential for pathogenicity against plant disease (Kondo et al. 2000). Various Streptomyces sp. including S. griseus have been reported as producing various antifungal compounds such as 1-H-pyrrole-2-carboxylic acid (PCA), cycloheximide, and streptomycin which were successfully used to control fungal and bacterial diseases in plants (Leben and Keitt 1954; Nguyen et al. 2015). Wiwiek et al. (2017) studied the rice phyllosphere actinomycetes could be used as potential biocontrol agents against fungal leaf blast disease. Wang and Ma (2011) reported that exogenous actinomycete XN-1 has the potential to act as an antagonistic agent in controlling the occurrence and development of cucumber leaf spot in the greenhouse. This also confirms that phyllosphere microorganisms play an important role in combating the infection of pathogens and have a promising future in developing biocontrol products. Table 5.7 shows the plant-associated bacteria and its biological activities.

Microbes with the production of compounds like indole acetic acid and N-acyl homoserine lactone (AHL) assist the bacteria to colonize on plant surface (Lindow and Brandl 2003). Sartori et al. (2015) studied the biocontrol potential of phyllosphere microorganisms from maize against Exserohi lumturcicum, the causal agent of leaf blight. Shrestha et al. (2016) investigated the prospects of biological control of rice-associated Bacillus against sheath blight and panicle blight of rice caused by Rhizoctonia solani and Burkholderia glumae, respectively. A variety of Bacillus isolates were observed to inhibit the sclerotial germination of the fungus, which could be attributed to the various antimicrobial secondary metabolites produced by the bacteria. Various gram-negative bacteria also show plant protection activity. For example, *Pseudomonas graminis* isolated from the apple phyllosphere showed control against fire blight caused by Erwinia amylovora (Mikiciński et al. 2016), *Pseudomonas protegens* CS1 from the lemon phyllosphere are used as a biocontrol against citrus canker (Michavila et al. 2017).

Further, microbial production of siderophores quenches the phytopathogens and protects the host plant from their infection (Scavino and Pedraza 2013; Ahmed and Holmström 2014; Harsonowati et al. 2017; Sabaté et al. 2018) as a biocontrol agent. For example, the siderophore produced by *Pseudomonas syringae* pv. *syringae* 22d/93 shows biocontrol activity against *Pseudomonas syringae* pv. *glycinea* 1a/96, a plant pathogen (Wensing et al. 2010). The siderophore pyochelin produced by the endophyte control rice blast is caused by Pyricularia oryzae (Harsonowati et al. 2017). Plant-associated *Pseudomonas* spp. has been employed efficiently as commercial biocontrol agents (Loper and Lindow 1987; Walsh et al. 2001). Cyanogenic

Table 5.7 Plant-associated microbial compounds and bioactivity

Compound	Source	Bioactivity	References
Blasticidin-S (VIII):	*Streptomyces griseochromogenes I*	Control the rice blast caused by *Pyricularia oryzae*	Fukunaga (1955)
Kasugamycin (IX)	*Streptomyces kasugaensis*	Rice blast caused by *Pyricularia oryzae*, leaf spot in sugar beet and celery by *Cercospora* spp., and scab in pears and apples caused by *Venturia* spp.	Umezawa et al. (1965)
Methoxyphenone	*Streptomyces griseolus*	Herbicides	Ito et al. (1974)
AM-toxin	*Alternaria mali*	phytotoxin	Park et al. (1977)
Milbemycin (XI):	*S. hygroscopicus* subsp. *aureolacrimosus*	Insecticidal and acaricidal	Mishima et al. (1983)
Diabroticin A	*Bacillus subtilis* and *Bacillus cereus*	Polar insecticide	Stierle et al. (1990)
Spinosad (X): spinosyn A and spinosyn D	*Saccharopolyspora spinosa*	Controls the caterpillar (*Helicoverpa zea* Boddie, *Pieris rapae* (L.), *Keiferia lycopersicella* (Walsingham), thrips (*Ceratitis capitata* (L.), *Thrips palmi* (Karny)) and beetles (*Leptinotarsa decemlineata* (Say))	Mertz and Yao (1990)
AF-toxins	*Alternaria fragariae*	Maculosin is a cyclic dipeptide—phytotoxin	Stierle et al. (1990), Uneo (1990)
Maculosin (XVI)	*Phoma lingam*	Phytotoxin	Stierle et al. (1990)
Efrapeptins	*Tolypocladium* spp.	Pesticide and insecticide	Krasnoff and Gupta (1991), Krasnoff et al. (1991)
Abamectin	*Streptomyces avermitilis*	Insecticide and acaricide	Jansson and Dybas (1996)
Nodulisporic acid	*Nodulisporium* sp.	Insecticidal activity	Ondeyka, et al. (1997)
Pyrizadocidin (VII)	*Streptomyces* #620061	Herbicides	Gerwick et al. (1997)
Syringomycin E:	*Pseudomonas syringae* ESC 10/11	Fungicide-citrus green mold *Penicillium digitatum*	Bull et al. (1998)

(continued)

Table 5.7 (continued)

Compound	Source	Bioactivity	References
Destruxin A and B	*M. anisopliae*	Insecticide	Strasser et al. (2000)
Oosporein	*Beauveria brongniartii*	Insecticide	Strasser et al. (2000)
Beauvericin A and B	*Beauveria bassiana* and *Paecilomyces* spp	Hexadepsipeptide—insecticide	Lane et al. (2000)
Borrelidin	*Streptomyces* species- neau-D50	Antifungal activity against *Phytophthora sojae*	Worapong et al. (2001)
Bialaphos (V)	*Streptomyces hygroscopicus* and *Streptomyces viridochromogenes*	Herbicide—weed control	Tachibana (2003)
Tartrolone C	*Streptomyces* sp. CP1130	Insecticidal macrodiolide	Lewer et al. (2003)
Coronatine	*Pseudomonas coronafacience*	Insecticide—herbicide	Block et al. (2005)
Macrolactin A:	*Bacillus* sp. *sunhua*	*Fungicide—Fusarium oxysporum and Streptomyces scabies*	Han et al. (2005)
Bt-Toxins	*Bacillus thuringiensis*	Bioinsecticides endotoxins	Collier et al. (2005)
Tabtoxin	*Pseudomonas syringae* var. *tabaci*	Phytotoxic—Herbicide	Hoagland et al. (2007)
Phyllostictine A	*Phyllosticta cirsii*	Mycoherbicide	Zonno et al. (2008)
Cinnacidin (XXII):	*Nectria* sp. DA60047	Phytotoxic	Irvine et al. (2008)
Beauvericin A and B	*Beauveria bassiana* and *Paecilomyces* spp.	Hexadepsipeptide—insecticide	Miller et al. (2008)
Herbimycin (VI)	*Streptomyces hygroscopicus* AM3672	Benzaquinoid ansamycin antibiotic with potential herbicidal a	Hahn et al. (2009)
Albucidin	*Streptomyces albus* subsp. *chlorinus* NRRL B-24108	Herbicides	Hahn et al. (2009)
Zinniol	*Alternaria cirsinoxia*	Phytotoxic to *Cirsium arvense* L.	Berestetskii et al. (2010)
Ascaulitoxin aglycone	*Ascochyta caulina*	Phytotoxin	Duke et al. (2011)
Antibiotic 1233A (XXIV)	*Cephalosporium* sp., *Fusarium* sp.,	Phytotoxin	Duke and Dayan (2011)

(continued)

Table 5.7 (continued)

Compound	Source	Bioactivity	References
AK-toxin (XV):	*Alternaria kikuchiana*	Phytotoxin	Saxena (2014)
Bipolaroxin (XVIII)	*Bipolaris cynodontis*	Phytotoxic to *Cynodon dactylon*	Saxena (2014)
Bt-Toxins	*Bacillus thuringiensis*	Bioinsecticides endotoxins	Radhakrishnan et al. (2017)
Phaseolotoxin (III)	*Pseudomonas* sp.	Phytotoxins—herbicide	Aguilera et al. (2017)

fluorescent *Pseudomonas* produces siderophores in the presence of a strong chelator 8-Hydroxyquinoline which inhibits pathogens such as *Rhizoctonia solani* and *Sclerotium rolfsii* (Kotasthane et al. 2017). Table 5.8 listed some important findings as endophytes as biocontrol agents. Mostly, the biocontrol agents use either nonribosomal peptide synthetase (NRPS) gene and/or type 1 polyketide synthase gene for respective compound production.

5.9.2 Plant Growth-Promoting Compounds

Plant growth is regulated by the growth hormones, available nutrient, good environmental condition, and beneficial microbial interaction. Many of the microbes are the prime producers of plant growth hormones, specifically plant-associated or phyllosphere microbial communities produce IAA, gibberellic acids, and cytokines and could fix nitrogen and mobilize nutrients (Dourado et al. 2015). There are many bacteria and fungi which produce IAA, similar to those of plants (Sun et al. 2014; Venkatachalam et al. 2016; Thapa et al. 2018. Microbes use plant tryptophan to produce IAA, which can effectively improve plant growth and enhance overall health (Hayat et al. 2010; Yadav et al. 2015a, b). The genus *Methylobacterium* is among the most commonly observed leaf epiphytes and represents an abundant and stable member of the phyllosphere microbial community of a wide range of crop plants such as sugarcane (*S. officinarum* L.), pigeon pea (*Cajanus cajan* L.), mustard (*Brassica campestris* L.), potato (*Solanum tuberosum* L.), and radish (*Raphanus sativus* L.) (Meena et al. 2012), and has produced variety of growth-promoting phytohormones. The association of plant growth-promoting bacteria (PGPB), especially *Methylobacterium* sp., with plant hosts greatly benefits plant growth by production of phytohormones like auxins and cytokinins, and increased activity of enzymes such as urease and 1-aminocyclopropane-1-carboxylate deaminase (ACCD), which promotes growth and enhances the production of siderophores, thereby enhancing the uptake of essential nutrients.

The benefits associated with plant–microbe interactions are also dependent on the variety of inoculation methods such as soil, foliar, and combination of both soil

Table 5.8 Phyllosphere endophytic fungi as biocontrol agent

Endophyte	Host	Pathogen	References
Fusarium proliferatum	Grape	*Plasmopara viticola*	Falk et al. (1996)
Acremonium zeae	Maize	*Aspergillus flavus* and *Fusarium verticillioides*	Wicklow et al (2005)
Trichoderma sp.	Apple Tobacco, bean, iris Radish, strawberry, cucumber, potato, and tomato	*Nectria galligena, Botrytis cinerea* *Sclerotium rolfsii* *Rhizoctonia solani* *Chondrostereum purpureum*	Reino et al. (2008)
Colletotrichum gloeosporioides, Clonostachys rosea, and Botryosphaeria ribis	*Theobroma cacao*	*Moniliophthora roreri (frosty pod rot), Phytophthora palmivora (black pod rot),* and *Moniliophthora perniciosa (witches broom)*	Mejía et al. (2008)
Trichoderma martiale	Cacao (*Theobroma* cacao)	*Phytophthora palmivora*	Hanada et al. (2009)
Clonostachys rosea	wheat	*Gibberella zeae*	Hue et al. (2009)
Cladosporium, Colletotrichum, Gibberella, Hypocrea, and Trichoderma	*Smallanthus sonchifolius* (Poepp.) H. Rob.	*Lecythophora* sp. *and Fusarium oxysporum*	Rosa et al. (2012)

(continued)

Table 5.8 (continued)

Endophyte	Host	Pathogen	References
Penicillium spp.	*Vitis vinifera* L.	*Botrytis cinerea*	Núñez-Trujillo et al. (2012)
Colletotrichum gloeosporioides, Flavodon flavus, Diaporthe helianthi, Diaporthe phaseolorum, Aporospora terricola	*Vitis labrusca* L.	*Fusarium oxysporum*	Brum et al. (2012)
Cladosporium cladosporioides	*Huperzia serrata*	*Colletotrichum acutatum, Colletotrichum fragariae, Colletotrichum gloeosporioides,* and *P. viticola*	Wang et al. (2013a)
Bionectria ochroleuca, Aureobasidium pullulans, Chaetomium spirochaete	*Vitis vinifera*	*Botrytis cinerea*	Cosoveanu et al. (2014)
Ramularia endophylla	Plant	*Mycosphaerella labyrinth*	Videira et al. (2015)

and foliar inoculations (Lee et al. 2011). A study has been conducted to investigate the inoculation of *Erwinia herbicola* on plant growth by IAA production. The test results showed that about 65% of the *E. herbicola* strain recovered from the leaves showed higher expression of the ipdC gene than in culture. The study indicated that physical or chemical microclimates directly influence the differential expression of ipdC (Brandl et al. 2001). Similarly, endophytic bacteria such as *Bacillus pumilus* E2S2 (Luo et al. 2012), *B. amyloliquefaciens* NBRI-SN13 (Nautiyal et al. 2013), *B. atrophaeus* EY6 and *B. sphaericus* B EY30, *B. subtilis* EY2, *S. kloosii* EY37, and *K. erythromyxa* EY43 (Karlidag et al. 2011) also produce PGPs.

Endophytic *Bacillus* produces phytohormones such as abscisic acid, auxins, brassinosteroids, cytokinins, ethylene, gibberellins, jasmonates, and strigolactones, and increases nutrient (nitrogen and phosphorous) accessibility to the host (Reinhold-Hurek and Hurek 2011; Brader et al. 2014; Santoyo et al. 2016; Shahzad et al. 2016; Ek-Ramos et al. 2019). Zeiller et al. (2015) reported that *C. botulinum* 2301 significantly produce PGPs in a field experiment of clover. A cold-tolerant bacterial strain *Exiguobacterium acetylicum* 1P promotes wheat seedlings growth (Selvakumar et al. 2010), *Brevibacillus* brevis improve the growth of cotton crop (Nehra et al. 2016) *and Bacillus* spp. induce phosphate solubilization more efficiently when present as endophytes in citrus (Giassi et al. 2016). The diazotrophic bacteria associated with phyllosphere gives benefits to the plant by fixing atmospheric nitrogen, solubilization of phosphorus (P), and utilization of available nutrients through its organic end product-mediated solubilization of rock phosphates (Mohammadi 2012; Kembel et al. 2014; Mwajita et al. 2013; Batool et al. 2016; Lambais et al. 2017).

5.9.3 Biopharmaceutical Importance

Biological activity of medicinal plants and their applications in various healing properties have been documented well. In recent years, microbes associated with plants themselves proved with high therapeutic values particularly endophytes. Endophytic microbes are known for their beneficial effects to the host, specifically phytohormones, enzymes, and stress-resistant physiology, and its biotechnological potentials (Parthasarathi et al. 2012; Singh and Dubey 2015; Gouda et al. 2016). Endophytes are known to produce bioactive metabolites, which served as a potent drug for medical and cosmetic industries (Shukla et al. 2014; Gouda et al. 2016). Secondary metabolites produced by the endophytic bacteria, actinomycetes, and fungi have economically valuable compounds such as alkaloids, flavonoids, phenolic acids, quinones, steroids, saponins, terpenoids, tetralones, xanthones, etc. (Strobel and Daisy 2003; Joseph and Priya 2011; Godstime et al. 2014; Shukla et al. 2014; Gouda et al. 2016). For example, endophytic microbes are well-known producers of taxol, a diterpene alkaloid, and lignin such as cathartics, emetics, and cholagogue used for cancer treatment (Konuklugil 1995; Zhang et al. 2009; Nair and Padmavathy 2014; Soliman

Table 5.9 Pharmaceutical valuable products from phyllosphere microbes

Producer	Compound	Activity against	References
Hypericum perforatum, Diaporthe helianthi	Hypericin, emodin, tyrosol	*Salmonella sp.*	Joseph and Priya (2011), Specian et al. (2012)
Ganoderma boninense	Rapamycin, cyclododecane, petalostemumol	*Bacillus subtilis*	Parthasarathi et al. (2012), Ismail et al. (2014)
Fusarium sp. *Cryptosporiopsis quercina*	Xularosides, munumbicins, Saadamycin, cryptocandin	*Candida albicans*	Jalgaonwala et al. (2011), Dutta et al. (2014)
Streptomyces sp., *Kennedia nigricans*	Munumbicins	*Vibrio cholerae*	Kumar et al. (2014)
Cryptosporiopsis quercina	Saadamycin	*Campylobacter jejuni*	Dutta et al. (2014)
Streptomyces sp.	Kakadumycin A, hypericin	*Shigella* sp.	Golinska et al. (2015), Joseph and Priya (2011)
Streptomyces tsusimaensis	Valinomycin	*Corona virus*	Alvin et al. (2014)
Fusarium proliferatum	Kakadumycin, beauvericin	*Listeria monocytogenes*	Golinska et al. (2015)
Boesenbergia rotunda Streptomyces coelicolor	Munumbicins	*Escherichia coli*	Golinska et al. (2015), Singh and Dubey (2015)
Grammothele lineata	Paclitaxel	*Anticancer*	Das et al. (2017), Kasaei et al. (2017), Soliman and Raizada (2018)

and Raizada 2018). There are many novel metabolites with antibacterial, antifungal, antiviral, anticancer, and antihelminthic activity isolated from plant-associated microbes (Gouda et al. 2016; Kasaei et al. 2017) (Table 5.9).

5.9.4 *Other Applications*

Besides the use of phyllosphere microbes for enhanced growth as well as biocontrol agent, some plant-associated bacteria helps the plant to improve phytoremediation of toxins. For example, hydroxamate siderophores producing bacteria compact heavy metal toxicity and improve the phytoremediation property in *A. thaliana* (Grobelak and Hiller 2017). Some endophytes provide additional functions to the host plant like drought tolerance, for example, endophytic *B. subtilis* strain B26 induces drought resistance to *Brachypodium distachyon* grass. The drought resistance mechanism is due to a specific carbohydrate metabolism, the endophytic bacteria increases stress-responsive raffinose-related family carbohydrates in the host (Gagné-Bourque et al. 2015). In another example, the endophytic association increases osmotic responses of the host plant. Endophytic strains such as Arthrobacter sp. and Bacillus spp. in pepper plant increase the proline accumulation, which gives osmotic tolerance (Sziderics et al. 2007).

Further, endophytic bacterial inoculants provide abiotic stress tolerance mechanism to the host by its extracellular enzymes. For example, the endophytic association of various *Bacillus* spp. increases the superoxide dismutase, phenylalanine lyase, catalase, and peroxidase enzymes activity in gladiolus plants under sodium high concentration conditions (Damodaran et al. 2014). Little studies reported that isolation of endophytic bacteria and their enzyme production potential vary when it colonizes in the plant tissues. Moreover, Jalgaonwala et al. (2011) observed maximum proteolytic activity in *Lactobacillus fermentum* isolated from leaves of *Vinca rosea*, which is considered greater to nonendophytic isolates. Similarly, endophytic fungi isolated from *Ocimum sanctum* and *Aloe vera* has better enzymatic activity (Yadav et al. 2015a, b). Besides these mechanisms, plant-associated microorganisms improve nutrient acquisition by supplying minerals and other micro/macronutrients from the soil (Singh et al. 2017; Singh and Singh 2017). Above all merits provide new insights in the field of phyllosphere microbiome and its essentiality of interactions to host plant growth and protection and also its significant role in the ecosystem.

5.9.5 Conclusion and Future Prospects

The phyllosphere is a unique environment colonized by a wide variety of microorganisms including epiphytes and endophytes, beneficial and pathogenic, bacteria, fungus, viruses, etc. Understanding the phyllosphere community structure, networking, and physiology is a great challenge. However, extensive research on phyllosphere microbiota gives great potential for the applications in economic plant productivity, specifically agriculture and forestry, ecosystem cleaning, and health. Hitherto, both in vitro and in vivo experiments are required to improve the understanding of microbial aggregations in the phyllosphere and dynamic play in the ecosystem. Based on the literature understanding, further and future studies should aim to (1) study the community interplay within the closely related and distanced microbial interactions and its stimulatory response on host plant and ecosystem, (2) to know the potentials of beneficial microbes and their commercial value, (3) impact on climate change on phyllosphere microbiome, and their contribution to climate change, (4) moreover, documentation of host-specific, geographic-specific, and seasonal-specific microbial interactions—guiding host–parasite and beneficial–pathogen interactions. Besides, phyllosphere microbiome research assures to understand the current challenges highlighting the terrestrial ecosystem change and the impact of global warming, especially the dominance of pathogenesis.

Acknowledgements Authors thank DST-PURSE, Madurai Kamaraj University for providing Internet and computer facility and Department of Biotechnology, India for providing the infrastructure facilities.

References

Abdelfattah A, Nicosia MGLD, Cacciola SO, Droby S, Schena L (2015) Metabarcoding analysis of fungal diversity in the phyllosphere and carposphere of olive (*Olea europaea*). PLoS ONE 10:e0131069

Agler MT, Ruhe J, Kroll S, Morhenn C, Kim ST, Weigel D et al (2016) Microbial hub taxa link host and abiotic factors to plant microbiome variation. PLoS Biol 14:e1002352

Aguilera S, Alvarez-Morales A, Murillo J, Hernández-Flores JL, Bravo J, De la Torre-Zavala S (2017) Temperature-mediated biosynthesis of the phytotoxin phaseolotoxin by *Pseudomonas syringae* pv. phaseolicola depends on the autoregulated expression of the *phtABC* genes. PLoS ONE 12(6): e0178441. https://doi.org/10.1371/journal.pone.0178441

Ahmed E, Holmström SJM (2014) Siderophores in environmental research: roles and applications: siderophores in environmental research. Microbial Biotechnol 7:196–208

Aleklett K, Hart M, Shade A (2014) The microbial ecology of flowers: an emerging frontier frontier in phyllosphere research. Botany 92:253–266

Alvin A, Kristin I, Miller B, Neilan A (2014) Exploring the potential of endophytes from medicinal plants as sources of antimycobacterial compounds. Microbiol Res 169:483–495

Amine Hassani M, Durán P, Hacquard S (2018) Microbial interactions within the plant holobiont. Microbiome 6:58

Andrews JH, Harris RF (2000) The ecology and biogeography of microorganisms on plant surfaces. Ann Rev Phytopathol 38:145–180

Ansede JH, Frieman R, Yoch DC (2001) Phylogenetic analysis of culturable dimethyl sulfide-producing bacteria from a spartina-dominated salt marsh and estuarine water. Appl Environ Microbiol 67:1210–1217

Arnold AE, Maynard Z, Gilbert GS, Coley PD, Kursar TA (2000) Are tropical fungal endophytes hyperdiverse? Ecol Lett 3:267–274

Arnold AE, Henk DA, Eells RL, Lutzoni F, Vilgalys R (2007) Diversity and phylogenetic affinities of foliar fungal endophytes in loblolly pine inferred by culturing and environmental PCR. Mycologia 99:185–206

Aung K, Jiang Y, Yang He S (2018) The role of water in plant–microbe interactions. The Plant Journal 93:771–780

Bai Y, Muller DB, Srinivas G, Garrido-Oter R, Potthoff E, Rott M, Dombrowski N, Münch PC, Spaepen S, Remus-Emsermann M, Hüttel B, McHardy AC, Vorholt JA, Schulze-Lefert P (2015) Functional overlap of the Arabidopsis leaf and root microbiota. Nature 528:364

Baldwin HE, Bhatia ND, Friedman A, Eng RM, Seite S (2017) The role of cutaneous microbiota harmony in maintaining a functional skin barrier. J Drugs Dermatol 16:12–18

Balint M, Tiffin P, Hallstrom B, O'Hara RB, Olson MS, Fankhauser JD, Piepenbring M, Schmitt I (2013) Host genotype shapes the foliar fungal microbiome of balsam poplar (Populus balsamifera). PLoS ONE 8:e53987

Balint-Kurti P, Simmons SJ, Blum JE, Ballare CL, Stapleton A (2010) Maize leaf epiphytic bacteria diversity patterns are genetically correlated with resistance to fungal pathogen infection. Mol Plant-Microbe Interact 23:473–484

Beattie GA (2002) Leaf surface waxes and the process of leaf colonization by microorganisms. In: Lindow SE, Hecht-Poinar EI, Elliott VJ (eds) Phyllosphere microbiology. APS Press, St. Paul, USA, pp 3–26

Beattie GA (2011) Water relations in the interaction of foliar bacterial pathogens with plants. Ann Rev Phytopathol 49:533–555

Beattie GA, Lindow SE (1999) Bacterial colonization of leaves: a spectrum of strategies. Phytopathology 89:353–359

Beilsmith K, Thoen MPM, Brachi B, Gloss AD, Khan MH, Bergelson J (2019) Genome-wide association studies on the phyllosphere microbiome: embracing complexity in host–microbe interactions. Plant J 97:164–181

Belova SE, Baani M, Suzina NE, Bodelier PLE, Liesack W, Dedysh SN (2011) Acetate utilization as a survival strategy of peat-inhabiting *Methylocystis* spp. Environ Microbiol Rep 3(1):36–46. https://doi.org/10.1111/j.1758-2229.2010.00180.x

Berestetskii AO, Yuzikhin OS, Katkova AS, Dobrodumov AV, Sivogrivov DE, Kolombet LV (2010) Isolation, identification, and characteristics of the phytotoxin produced by the fungus *Alternaria cirsinoxia*. Appl Microbiol Biot 46:75–79

Bernal E, Celis S, Galíndez X, Moratto C, Sánchez J, García D (2006) Microflora cultivable endomicorrizas obtenidas en hojarasca de bosque (Páramo Guerrero finca Puente de Tierra) Zipaquirá, Colombia. Acta Biol Colomb 11:125–130

Bhattacharyya LH, Borah G, Parkash V, Bhattacharyya PN (2017) Fungal endophytes associated with the ethnomedicinal plant *Meyna spinosa* Roxb. Current Life Sci 3(1):1–5

Bian GK, Qin S, Yuan B, Zhang YJ, Xing K, Ju XJ et al (2012) *Streptomyces phytohabitans* sp. nov., a novel endophytic actinomycete isolated from medicinal plant *Curcuma phaeocaulis*. Antonie Van Leeuwenhoek 102:289–296

Block A, Schmetz E, Jones JB, Klee HJ (2005) Coronatine and salicyclic acid: the battle between Arabidopsis and *Pseusomonas* for phytohormone control. Mol Plant Pathol 6:79–83

Bodenhausen N, Horton MW, Bergelson J (2013) Bacterial communities associated with the leaves and the roots of Arabidopsis thaliana. PLoS ONE 8:e56329

Bodenhausen N, Bortfeld-Miller M, Ackermann M, Vorholt JA (2014) A synthetic community approach reveals plant genotypes affecting the phyllosphere microbiota. PLoS Genet 10:e1004283

Bogdanove AJ (2002) Protein-protein interactions in pathogen recognition by plants. Plant Mol Biol 50(6):981–989

Boller T, Felix G (2009) A renaissance of elicitors: perception of microbe-associated molecular patterns and danger signals by pattern-recognition receptors. Annu Rev Plant Biol 60:379–406

Brader G, Compant S, Mitter B, Trognitz F, Sessitsch A (2014) Metabolic potential of endophytic bacteria. Curr Opin Biotechnol 27:30–37

Brandl MT, Quinones B, Lindow SE (2001) Heterogeneous transcription of an indoleacetic acid biosynthetic gene in Erwinia herbicola on plant surfaces. Proc Natl Acad Sci 98 (6):3454–3459

Brewer CA, Smith WK, Vogelmann TC (1991) Functional interaction between leaf trichomes, leaf wettability and the optical properties of water droplets. Plant Cell Environ 14:955–962

Bringel F, Couee I (2015) Pivotal roles of phyllosphere microorganisms at the interface between plant functioning and atmospheric trace gas dynamics. Front Microbiol 6:486

Brum MCP, Araújo WL, Maki CS, Azevedo JL (2012) Endophytic fungi from *Vitis labrusca* L. ('Niagara Rosada') and its potential for the biological control of *Fusarium oxysporum*. Genet Mol Res 11:4187–4197

Buee M, De Boer W, Martin F, Van Overbeek L, Jurkevitch E (2009) The rhizosphere zoo: an overview of plant-associated communities of microorganisms, including phages, bacteria, archaea, and fungi, and of some of their structuring structuring factors. Plant Soil 321:189–212

Bulgarelli D, Schlaeppi K, Spaepen S, Loren V, van Themaat E, Schulze- Lefert P (2013) Structure and functions of the bacterial microbiota of plants. Ann Rev Plant Biol 64:807–838

Bull C, Wadsworth M, Sorensen K, Takemoto J, Austin R, Smilanick J (1998) Syringomycin E produced by the biological control agents control green mold on lemons. Biol Control 12:89–95

Burch AY, Zeisler V, Yokota K, Schreiber L, Lindow SE (2014) The hygroscopic biosurfactant syringafactin produced by *Pseudomonas syringae* enhances fitness on leaf surfaces during fluctuating humidity. Environ Microbiol 16:2086–2098

Carlos A, Ruiz-Pérez SR, Zambranoa MM (2016) Microbial and functional diversity within the phyllosphere of Espeletia Species in an Andean high-mountain ecosystem. Appl Environ Microbiol 82:6

Cha JY, Han S, Hong HJ, Cho H, Kim D, Kwon Y, Kwon SK, Crüsemann M, Yong BL, Kim JF (2016) Microbial and biochemical basis of a Fusarium wilt-suppressive soil

Chen HH, Qin S, Li J, Zhang YQ, Xu LH, Jiang CL et al (2009) *Pseudonocardia endophytica* sp. nov., isolated from the pharmaceutical plant *Lobelia clavata*. Int J Syst Evol Microbiol 59:559–563

Chen MH, Zhang L, Zhang X (2011) Isolation and inoculation of endophytic actinomycetes in root nodules of *Elaeagnus angustifolia*. Mod Appl Sci 5:264–267

Chinchilla D, Bauer Z, Regenass M, Boller T, Felix G (2006) The Arabidopsis receptor kinase FLS2 binds flg22 and determines the specificity of flagellin perception. Plant Cell 18:465–476

Chou HM, Bundock N, Rolfe SA, Scholes JD (2000) Infection of Arabidopsis thaliana leaves with Albugo candida (white blister rust) causes a reprogramming of host metabolism. Mol Plant Pathol Mar 1:99–113

Chowdhury S, Basu A, Kundu S (2015) Cloning, characterization and bacterial overexpression of an osmotin-like protein gene from *Solanum nigrum* L. with antifungal activity against three necrotrophic fungi. Mol Biotechnol 57:371–381

Coince A, Cordier T, Lengelle J, Defossez E, Vacher C, Robin C, Buée M, Marçais B (2014) Leaf, root-associated fungal assemblages do not follow similar elevational diversity patterns. PLoS ONE 9:e100668

Collier FA, Elliot SL, Ellis RJ (2005) Spatial variation in Bacillus thuringiensis/cereus populations within the phyllosphere of broad-leaved dock (Rumex obtusifolius) and surrounding habitats. FEMS Microbiol Ecol 54(2005):417–425

Coombs JT, Franco CMM (2003) Isolation and identification of actinobacteria from surface-sterilized wheat roots. Appl Environ Microbiol 69:5603–5608

Copeland JK, Yuan L, Layeghifard M, Wang PW, Guttman DS (2015) Seasonal community succession of the phyllosphere microbiome. Mol Plant Microbe Interact 28:274–285

Cordier T, Robin C, Capdevielle X, Fabreguettes O, Desprez-Loustau ML, Vacher C (2012) The composition of phyllosphere fungal assemblages of European beech (*Fagus sylvatica*) varies significantly along an elevation gradient. New Phytol 196:510–519

Cosoveanu A, Gimenez-Mariño C, Cabrera Y, Hernandez G, Cabrera R (2014) Endophytic fungi from grapevine cultivars in Canary Islands and their activity against phytopatogenic fungi. Int J Agric Crop Sci 7(15):1497–1503

Craig A, Ewan R, Mesmar J, Gudipati V, Sadanandom A (2009) E3 ubiquitin ligases and plant innate immunity. J Exp Bot 60:1123–1132

Cui H, Xiang T, Zhou JM (2009) Plant immunity: a lesson from pathogenic bacterial effector proteins. Cell Microbiol 11:1453–1446

Damodaran T, Rai RB, Jha SK, Kannan R, Pandey BK, Sah V et al (2014) Rhizosphere and endophytic bacteria for induction of salt tolerance in gladiolus grown in sodic soils. J Plant Interact 9:577–584

Das A, Rahman MI, Ferdous AS, Amin A, Rahman MM, Nahar N et al (2017) An endophytic Basidiomycete, *Grammothele lineata*, isolated from *Corchoru solitorius*, produces paclitaxel that shows cytotoxicity. PLoSONE 12(6):e0178612

Dees MW, Lysøe E, Nordskog B, Brurberg MB (2015) Bacterial communities associated with surfaces of leafy greens: shift in composition and decrease in richness over time. Appl Environ Microbiol 81:1530–1539

Degnan PH, Ochman H (2012) Illumina-based analysis of microbial community diversity. ISME J 6:183–194

Deka D, Jha DK (2018) Antimicrobial activity of Endophytic Fungi from leaves and barks of *Litsea cubeba* Pers., A Traditionally important medicinal plant of north east India. Jordan J Biol Sci 11(1):73–79

DeLeon-Rodriguez N, Lathem TL, Rodriguez-R LM, Barazesh JM, Anderson BE, Beyersdorf AJ et al (2013) Microbiome of the upper troposphere: species composition and prevalence, effects of tropical storms, and atmospheric implications. Proc Natl Acad Sci USA 110:2575–2580

Delmotte N, Knief C, Chaffron S, Innerebner G, Roschitzki B, Schlapbach R et al (2009) Community proteogenomics reveals insights into the physiology of phyllosphere bacteria. Proc Natl Acad Sci USA 106:16428–16433

Dhayanithy G, Subban K, Chelliah J (2019) Diversity and biological activities of endophytic fungi associated with Catharanthus roseus. BMC Microbiol 19(1):22

Dinesh R, Srinivasan VTES, Anandaraj M, Srambikkal H (2017) Endophytic actinobacteria: diversity, secondary metabolism and mechanisms to unsilence biosynthetic gene clusters. Crit Rev Microbiol 43:546–566

Ding L et al (2011) Resistance to hemi-biotrophic *Fusarium graminearum* infection is associated with coordin

Dingman DW (2000) Growth of *Escherichia coli* O157:H7 in bruised apple (*Malus domestica*) tissue as influenced by cultivar, date of harvest, and source. Appl Environ Microbiol 66:1077–1083

Dobrovol'skaya T, Khusnetdinova K, Manucharova N, Golovchenko A (2017) Structure of epiphytic bacterial communities of weeds. Microbiology 86:257–263

Doronina NV, Ivanova EG, Suzina NE, Trotsenko YA (2004) Methanotrophs and methylobacteria are found in woody plant tissues within the winter period. Microbiology 73(6):702–709

Dourado MN, Neves AAC, Santos DS, Araujo WL (2015) Biotechnological and agronomic potential of endophytic pink-pigmented methylotrophic trophic Methylobacterium spp. Biomed Res Int 2015:909016

Du HJ, Zhang YQ, Liu HY, Su J, Wei YZ, Ma BP et al (2013) *Allonocardiopsis opalescens* gen. nov., sp. nov., a new member of the suborder Streptosporangineae, from the surface- sterilized fruit of a medicinal plant. Int J Syst Evol Microbiol 63:900–904

Duke SO, Dayan FE (2011) Modes of action of microbial produced phytotoxins. Toxins 3:1038–1064

Duke SO, Evidente A, Fiore M, Rimando AM, Dayan FE, Vurro M, Christiansen N, Looser R, Hutzler J, Grossman K (2011) Effect of aglycone of ascaulitoxin in amino acid metabolism of *Lemna paucicostata*. Pestic Biochem Physiol 100:41–50

Durand A, Maillard F, Alvarez-Lopez V, Guinchard S, Bertheau C, Valot B et al (2018) Bacterial diversity associated with poplar trees grown on a Hg-contaminated site: community characterization and isolation of Hg-resistant plant growth promoting bacteria. Sci Total Environ 622:1165–1177

Dutta D, Puzari KC, Gogoi R, Dutta P (2014) Endophytes: exploitation as a tool in plant protection. Braz Arch Biol Technol 57:621–629

Ek-Ramos MJ, Gomez-Flores R, Orozco-Flores AA, Rodríguez-Padill C, González-Ochoa G, Tamez-Guerra P (2019) Bioactive products from plant-endophytic gram-positive bacteria. Front Microbiol 10:463

El-Din Hassan S (2017) Plant growth-promoting activities for bacterial and fungal endophytes isolated from medicinal plant of Teucrium polium L J Adv Res 8(6):687–695

El-Komy MH (2014) Comparative analysis of defense responses in chocolate spot-resistant and susceptible faba bean (*Vicia faba*) cultivars following infection by the necrotrophic fungus *Botrytis fabae*. Plant Pathol J 30:355–366

El-Tarabily KA (2003) An endophytic chitinase-producing isolate of *Actinoplanes missouriensis*, with potential for biological control of root rot of lupin caused by *Plectosporium tabacinum*. Aust J Bot 51:257–266

Ernawati M, Solihin DD, Lestari Y (2016) Community structures of endophytic actinobacteria from medicinal plant *Centella asiatica* L. Urbanbased on metagenomic approach. Int J Pharm Pharm Sci 8:292–297

Erwin DC, Ribeiro OK (1996) Phytophthora diseases worldwide. APS press, New York

Esser DS, Leveau JHJ, Meyer KM, Wiegand K (2015) Spatial scales of interactions among bacteria and between bacteria and the leaf surface. FEMS Microbiol Ecol. https://doi.org/10.1093/femsec/fiu034

Fahimipour AK, Kardish MR, Lang JM, Green JL, Eisen JA, Stachowicz JJ (2017) Global-scale structure of the Eelgrass microbiome. Appl Environ Microbiol 83:e03391–16

Falk SP, Pearson RC, Gadoury DM, Seem RC, Sztejnberg A (1996) *Fusarium proliferatum* as a biocontrol agent against grape downy mildew. Phytopathology 86:1010–1017

Feng WW, Wang T, Bai JL, Ding P, Xing K, Jiang JH et al (2017) *Glutamicibacter halophytocola* sp. nov., an endophytic actinomycete isolated from the roots of a coastal halophyte, *Limonium sinense*. Int J Syst Evol Microbiol 67:1120–1125

Finkel OM, Burch AY, Lindow SE, Post AF, Belkin S (2011) Geographical location determines the population structure in phyllosphere microbial communities of a salt-excreting desert tree. Appl Environ Microbiol 77:7647–7655

Finlay AC, Hobby GL, Pan SY, Regna PP, Routein JB, Seeley DB, Shull GM, Sobin BA, Solomans IA, Vinson JW, Kane JH (1950) Terramycin, a new antibiotic. Science 111:85

Foley RC, Kidd BN, Hane JK, Anderson JP, Singh K (2016) Reactive oxygen species play a role in the infection of the necrotrophic fungi Rhizoctoniasolani in wheat. PLoS ONE 11:e0152548

Freiberg E (1998) Microclimatic parameters influencing nitrogen fixation in the phyllosphere in a Costa Rican premontane rain forest. Oecologia 117(1):9–18

Fukunaga K (1955) Blasticidin, a new antiphytopathogenic fungal substance. Part I. Bull Agric Chem Soc Jpn 19:181–188

Furnkranz M, Wolfgang W, Andreas R, Guy A, Frank R, Angela S (2008) Nitrogen fixation by phyllosphere bacteria associated with higher plants plants and their colonizing epiphytes of a tropical lowland rainforest of Costa Rica. ISME J 2(5):561–570

Gagné-Bourque F, Mayer BBF, Charron JBB, Vali H, Bertrand A, Jabaji S (2015) Accelerated growth rate and increased drought stress resilience of the model grass Brachypodium distachyon colonized by Bacillus subtilis B26. PLoS ONE 10:e130456

Gaige AR, Ayella A, Shuai B (2010) Methyl jasmonate and ethylene induce partial resistance in *Medicago truncatula* against the charcoal rot pathogen *Macrophomina phaseolina*. Physiol Mol Plant Pathol 74:412–418

Galbally IE, Kirstine W (2002) The production of methanol by flowering plants and the global cycle of methanol. J Atmosph Chem 43:195–229

Gan P, Ikeda K, Irieda H, Narusaka M, O'Connell RJ, Narusaka Y, Takano Y, Kubo Y, Shirasu K (2013) Comparative genomic and transcriptomic analyses reveal the hemibiotrophic stage shift of *Colletotrichum* fungi. New Phytol 197:1236–1249

Gerwick BC, Fields SS, Chapin EL, Cleveland JA, Heim DR (1997) Pyrizadocidin, a new microbial phytotoxin with activities in a Mehler's reaction. Weed Sci 45:654–657

Giassi V, Kiritani C, Kupper KC (2016) Bacteria as growth-promoting agents for citrus rootstocks. Microbiol Res 190:46–54

Glenn DM, Bassett C, Dowd SE (2015) Effect ofpest management system on 'Empire' apple leaf phyllosphere populations. Sci Horticult 183:58–65

Glushakova AM, Chernov IY (2004) Seasonal dynamics in a yeast population on leaves of the common wood sorrel *Oxalis acetosella* L. Microbiology 73:184–188

Godstime OC, Enwa FO, Augustina JO, Christopher EO (2014) Mechanisms of antimicrobial actions of phytochemicals against enteric pathogens—a review. J Pharm Chem Biol Sci 2:77–85

Gołębiewski M, Tarasek A, Sikora M, Deja-Sikora E, Tretyn A, Niklińska M (2019) Rapid microbial community changes during initial stages of pine litter decomposition. Microb Ecol 77:56–75

Golinska P, Wypij M, Agarkar G, Rathod D, Dahm H, Rai M (2015) Endophytic actinobacteria of medicinal plants: diversity and bioactivity. Antonie Van Leeuwenhoek 108:267–289

Gottwald S, Samans B, Luck S, Friedt W (2012) Jasmonate and ethylene dependent defense gene expression and suppression of fungal virulence factors: two essential mechanisms of *Fusarium* head blight resistance in wheat? BMC Genom 13:369

Gouda S, Das G, Sen SK, Shin HS, Patra JK (2016) Endophytes: a treasure house of bioactive compounds of medicinal importance. Front Microbiol 7:1538

Gourion B, Rossignol M, Vorholt JA (2006) A proteomic study of *Methylobacterium extorquens* reveals a response regulator essential for epiphytic growth. Proc Natl Acad Sci USA 103:13186–13191

Gourion B, Francez-Charlot A, Vorholt JA (2008) PhyR is involved in the general stress response of Methylobacterium extorquens AM1. Journal of Bacteriol 190(3):1027–1035. https://doi.org/10.1128/JB.01483-07

Green PN (2006) Methylobacterium. The Prokaryotes, vol 5 (Falkow S, Rosenberg E, Schleifer K-H, Stackebrandt E, Dworkin M, eds). Springer, New York, NY, pp 257–265

Grobelak A, Hiller J (2017) Bacterial siderophores promote plant growth: Screening of catechol and hydroxamate siderophores. Int J Phytorem 19(9):825–833

Guerreiro MA, Brachmann A, Begerow D, Peršoh D (2018) Transient leaf endophytes are the most active fungi in 1-year-old beech leaf litter. Fungal Divers 89:237–251

Hahn DR, Graupner RC, Chapin E, Gray J, Heim D, Gilbert JR, Gerwick BC (2009) Albucidin: a novel bleaching herbicide from *Streptomyces albus* subsp. *chlorinus* NRRL B- 24108. J Antibiot 62:191–194

Hamd Elmagzob AA, Ibrahim MM, Zhang G-F (2019) Seasonal diversity of endophytic bacteria associated with Cinnamomum camphora (L.) Presl. Diversity 11:112

Han JS, Cheng JH, Yoon TM, Song J, Rajkarnikar A, Kim WG, Yoo ID, Yang YY, Suh JW (2005) Biological control of common scab diseases by antagonistic strain *Bacillus* sp. sunhua. J Appl Microbiol 99:213–221

Hanada RE, Pomella AWV, Soberanis W, Loguercio LL, Pereira JO (2009) Biocontrol potential of *Trichoderma martiale* against the black-pod disease (*Phytophthora palmivora*) of cacao. Biol Control 50:143–149

Harsonowati W, Astuti RI, Wahyudi AT (2017) Leaf blast disease reduction by rice-phyllosphere actinomycetes producing bioactive compounds. J Gen Plant Pathol 83:98–108

Hayat R, Ali S, Amara U, Khalid R, Ahmed I (2010) Soil beneficial bacteria and their role in plant growth promotion: a review. Ann Microbiol 60:579–598

He H, Liu C, Zhao J, Li W, Pan T, Yang L et al (2014) *Streptomyces zhaozhouensis* sp. nov., an actinomycete isolated from Candelabra aloe (Aloe arborescens Mill). Int J Syst Evol Microbiol 64:1096–1101

Hill TC, Moffett BF, Demott PJ, Georgakopoulos DG, Stump WL, Franc GD (2014) Measurement of ice nucleation-active bacteria on plants and in precipitation by quantitative PCR. Appl Environ Microbiol 80:1256–1267

Hirano SS, Upper CD (1983) Ecology and epidemiology of foliar bacterialplant pathogens. Ann Rev Phytopathol 21:243–269

Hirano SS, Upper CD (2000) Bacteria in the leaf ecosystem with emphasis on *Pseudomonas syringae*—a pathogen, ice nucleus, and epiphyte. Microbiol Mol Biol Rev 64:624–653

Hoagland RE, Douglas Boyette C, Weaver MA, Abbas HK (2007) Bioherbicides: research and risks. Toxin Rev 26:313–342

Holland MA, Davis R, Moffitt S, O'Laughlin K, Peach D, Sussan S et al (2000) Using "leaf prints" to investigate a common bacterium. Am Biol Teach 62:128–131

Horton MW, Bodenhausen N, Beilsmith K, Meng DZ, Muegge BD, Subramanian S, Vetter MM, Vilhjálmsson BJ, Nordborg M, Gordon JI, Bergelson J (2014) Genome-wide association study of Arabidopsis thaliana leaf microbial community. Nat Commun 5:5320

Huang MJ, Rao MP, Salam N, Xiao M, Huang HQ, Li WJ (2017) *Allostreptomyces psammosilenae* gen. nov., sp. nov., an endophytic actinobacterium isolated from the roots of *Psammosilene tunicoides* and emended description of the family Streptomycetaceae. Int J Syst Evol Microbiol 67:288–293

Huang LH, Yuan MQ, Ao XJ, Ren AY, Zhang HB, Yang MZ (2018) Endophytic fungi specifically introduce novel metabolites into grape flesh cells in vitro. PLoS ONE 13(5):e0196996

Hue AG, Voldeng HD, Savard ME, Fedak G, Tian X, Hsiang T (2009) Biological control of fusarium head blight of wheat with *Clonostachys rosea* strain ACM941. Can J Plant Pathol 31:169–179

Hunter PJ, Hand P, Pink D et al (2010) Both leaf properties and microbe–microbe interactions influence within-species variation in bacterial population diversity and structure in the lettuce (Lactuca species) phyllosphere. Appl Environ Microbiol 76:8117–8125

Ichihara AK, Shiraishi K, Sato H, Sakamura S, Nishiyama K, Sasaki R, Furusaki A, Matsumoto T (1977) The structure of coronatine. J Am Chem Soc 99:636–637

Iguchi H, Sato I, Yurimoto H, Sakai Y (2013) Stress resistance and C1 metabolism involved in plant colonization of a methanotroph Methylosinus sp. B4S. Arch Microbiol 195:717–726

Iguchi H, Yurimoto H, Sakai Y (2015) Interactions of methylotrophs with plants and other heterotrophic bacteria. Microorganisms 3(2):137–151

Ikeda S, Anda M, Inaba S et al (2011) Autoregulation of nodulation interferes with impacts of nitrogen fertilization levels on the leaf-associated bacterial community in soybeans. Appl Environ Microbiol 77:1973–1980

Inacio J, Pereira P, de Carvalho M, Fonseca A, Amaral- Collaco MT, Spencer-Martins I (2002) Estimation and diversity of phylloplane mycobiota on selected plants in a Mediterranean-type ecosystem in Portugal. Microbial Ecol 44:344–353

Indananda C, Thamchaipenet A, Matsumoto A, Inahashi Y, Duangmal K, Takahashi Y (2011) *Actinoallomurus oryzae* sp. nov., an endophytic actinomycete isolated from roots of a Thai jasmine rice plant. Int J Syst Evol Microbiol 61:737–741

Innerebner G, Knief C, Vorholt JA (2011) Protection of Arabidopsis thaliana against leaf-pathogenic Pseudomonas syringae by Sphingomonas strains in a controlled model system. Appl Environ Microbiol 77:3202–3210

Irvine NM, Yerkes CN, Graupner PR, Roberts RE, Hahn DR, Pearce C, Gerwick BC (2008) Synthesis and characterization of synthetic analogs of cinnacidin, a novel phytotoxin from *Nectria* sp. Pest Manag Sci 64:891–899

Ismail K, Abdullah S, Chong K (2014) Screening for potential antimicrobial compounds from Ganoderma boninense against selected food borne and skin disease pathogens. Int J Pharm Pharm Sci 6:771–774

Ismail MH, Hussain A, Iqbal A, Khan SA, Lee I-J (2018) Endophytic fungus Aspergillus japonicus mediates host plant growth under normal and heat stress conditions. BioMed Res Int. https://doi.org/10.1155/2018/7696831

Ito K, Futatsuya F, Hibi K, Ishida S, Yamada O, Munakata K (1974) Herbicidal activity of 3, 3', 3'-dimethyl-4-methoxybenzophenone (NK-049) in paddy fields. I. Herbicidal characteristics of NK-049 on weeds. Weed Sci 18:10–15

Izhaki I, Fridman S, Gerchman Y, Halpern M (2013) Variability of bacterial community composition on leaves between and within plant species. Curr Microbiol 66:227–235

Jackson EF, Echlin HL, Jackson CR (2006) Changes in the phyllosphere community of the resurrection fern, *Polypodium polypodioides*, associated with rainfall and wetting. FEMS Microbiol Ecol 58:236–246

Jacob F, Vernaldi S, Maekawa T (2013) Evolution and conservation of plant NLR functions. Front Immunol 4:297

Jacoby R, Peukert M, Succurro A, Koprivova A, Kopriva S (2017) The role of soil microorganisms in plant mineral nutrition—current knowledge and future directions. Front Plant Sci 8:1617

Jalgaonwala RE, Mohite BV, Mahajan RT (2011) Natural products from plant associated endophytic fungi. J Microbiol Biotechnol Res 1:21–32

Jansson RK, Dybas RA (1996) Avermectins: biochemical mode of action, biological activity, and agricultural importance. In: Ishaaya I (ed) Insecticides with novel modes of action: mechanisms and applications. Springer Verlag, Berlin

Jariwala B, Desai B (2018) Isolation and identification of endophytic fungi from various medicinal plants. BMR Microbiology 4(1):1–7

Jia F, Liu C, Wang X, Zhao J, Liu Q, Zhang J et al (2013) Wangella harbinensis gen. nov., sp. nov., a new member of the family Micromonosporaceae. Antonie Van Leeuwenhoek 103:399–408

Jiang ZK, Pan Z, Li FN, Li XJ, Liu SW, Tuo L et al (2017) *Marmoricola endophyticus* sp. nov., an endophytic actinobacterium isolated from the *Spesia populnea*. Int J Syst Evol Microbiol 67:4379–4384

Jiang ZK, Tuo L, Huang DL, Osterman IA, Tyurin AP, Liu SW et al (2018) Diversity, novelty, and antimicrobial activity of endophytic actinobacteria from mangrove plants in Beilun Estuary National Nature Reserve of Guangxi China. Front Microbiol 9:868

Jo Y, Cho JK, Choi H, Chu H, Lian S, Cho WK (2015) Bacterial communities in the phylloplane of Prunus species. J Basic Microbiol 55:504–508

Johnston-Monje D, Raizada MN (2011) Conservation and diversity of seed associated endophytes in Zea across boundaries of evolution, ethnography and ecology. PLoS ONE 6:e20396

Jones K (1970) Nitrogen fixation in the phyllosphere of the Douglas Fir, Pseudotsuga Douglasii. Ann Bot 34:239–244

Jones JDG, Dangl JL (2006) The plant immune system. Nature 444:323–329

Joseph B, Priya RM (2011) Bioactive compounds from endophytes and their potential in pharmaceutical effect: a review. Am J Biochem Mol Bio 1:291–309

Jumpponen A, Jones KL (2009) Massively parallel 454 sequencing indicates hyperdiverse fungal communities in temperate Quercus macrocarpa phyllosphere. New Phytol 184:438–448

Junker RR, Tholl D (2013) Volatile organic compound mediated interactions at the plant-microbe interface. J Chem Ecol 39:810–825

Jupe J et al (2013) *Phytophthora capsici*-tomato interaction features dramatic shifts in gene expression associated with a hemi-biotrophic lifestyle. Genome Biol 14:R63

Kabbage M, Yarden O, Dickman MB (2015) Pathogenic attributes of *Sclerotinia sclerotiorum*: Switching from a biotrophic to necrotrophic lifestyle. Plant Sci 233:53–60

Kadivar H, Stapleton AE (2006) Ultraviolet radiation alters maize phyllosphere bacterial diversity. Microbial Ecol 45:353–361

Kaewkla O, Franco CM (2013) Rational approaches to improving the isolation of endophytic actinobacteria from Australian native trees. Microb Ecol 65:384–393

Kaewkla O, Thamchaipenet A, Franco CM (2017) *Micromonospora terminaliae* sp. nov., an endophytic actinobacterium isolated from the surface sterilized stem of the medicinal plant Terminalia mucronata. Int J Syst Evol Microbiol 225–230

Kannadan S, Rudgers JA (2008) Endophyte symbiosis benefits a rare grass under low water availability. Funct Ecol 22:706–713

Karlidag H, Esitken A, Yildirim E, Donmez MF, Turan M (2011) Effects of plant growth promoting bacteria on yield, growth, leaf water content, membrane permeability, and ionic composition of strawberry under saline conditions. J Plant Nutr 34:34–45

Karlsson I, Friberg H, Steinberg C, Persson P (2014) Fungicide effects on fungal community composition in the wheat phyllosphere. PLoS ONE 9:e111786

Kasaei A, Mobini-Dehkordi M, Mahjoubi F, Saffar B (2017) Isolation of taxol-producing endophytic fungi from iranian yew through novel molecular approach and their effects on human breast cancer cell line. Curr Microbiol 74(6):702–709

Kawaguchi K, Yurimoto H, Oku M, Sakai Y (2011) Yeast methylotrophy and autophagy in a methanol-oscillating environment on growing Arabidopsis thaliana leaves. PLoS ONE 6(9):e25257. https://doi.org/10.1371/journal.pone.0025257

Kecskeméti E, Berkelmann-Löhnertz B, Reineke A, Cantu D (2016) Are epiphytic microbial communities in the carposphere of ripening grape clusters (Vitis vinifera L.) different between conventional, organic, and biodynamic grapes? PLoS ONE 11:e0160852

Kembel SW, O'Connor TK, Arnold HK, Hubbell SP, Wright SJ, Green JL (2014) Rela- tionships between phyllosphere bacterial communities and plant functional traits in a neotropical forest. Proc Natl Acad Sci USA 111:13715–13720

Keppler F, Hamilton JT, Brass M, Röckmann T (2006) Methane emissions from terrestrial plants under aerobic conditions. Nature 439:187–191

Kernaghan G, Mayerhofer M, Griffin A (2017) Fungal endophytes of wild and hybrid *Vitis* leaves and their potential for vineyard biocontrol. Can J Microbiol 63(7):583–595

Kharwar RN, Gond SK, Kumar A, Mishra AA (2010) A comparative study of endophytic and epiphytic fungal association with leaf of *Eucalyptus citriodora* Hook., and their antimicrobial activity. World J Microbiol Biotechnol 26:1941–1948

Kinkel LL (1997) Microbial population dynamics on leaves. Ann Rev Phytopathol 35:327–347

Kinkel LL, Wilson M, Lindow SE (2000) Plant species and plant incubation conditions influence variability in epiphytic bacterial population size. Microb Ecol 39:1–11

Knief C, Frances L, Vorholt JA (2010) Competitiveness of diverse Methylobacterium strains in the phyllosphere of Arabidopsis thaliana and identification of representative models, including M. extorquens PA1. Microb Ecol 60(2):440–52. https://doi.org/10.1007/s00248-010-9725-3

Knief C, Delmotte N, Chaffron S, Stark M, Innerebner G, Wassmann R, von Mering C, Vorholt JA (2012) Metaproteogenomic analysis of microbial communities in the phyllosphere and rhizosphere of rice. ISME J 6:1378–1390

Kondo T, Yamamoto D, Yokota A, Suzuki A, Nagasawa H, Sakuda S (2000) Gordonan, an acidic polysaccharide with cell aggregation-inducing activity in insect BM-N4 cells, produced by *Gordonia* sp. Biosci Biotechnol Biochem 64:2388–2394

Konuklugil B (1995) The importance of Aryltetralin (Podophyllum) lignans and their distribution in the plant kingdom. Ankara Univ Eczacilik Fak Derg 24:109–125

Koskella B (2013) Phage-mediated selection on microbiota of a long-lived host. Curr Biol 23:1256–1260

Kotasthane AS, Agrawal T, Waris Zaidi N, Singh US (2017) Identification of siderophore producing and cynogenic fluorescent *Pseudomonas* and a simple confrontation assay to identify potential bio-control agent for collar rot of chickpea. 3 Biotech 7(2):137

Kour D, Rana KL, Yadav N, Yadav AN, Singh J, Rastegari AA, Saxena AK (2019) Agriculturally and industrially important fungi: current developments and potential biotechnological applications. In: Yadav AN, Singh S, Mishra S, Gupta A (eds) Recent advancement in white biotechnology through fungi, Volume 2: Perspective for value-added products and environments. Springer International Publishing, Cham, pp 1–64. https://doi.org/10.1007/978-3-030-14846-1_1

Krasnoff SB, Gupta S (1991) Identification and biosynthesis of efrapeptins in the fungus *Tolypocladium geodes* Gams (Deuteromycotina, Hyphomycetes). J Chem Ecol 17:1953–1960

Krasnoff SB, Gupta S, St Leger RJ, Renwick JA, Roberts DW (1991) Antifungal and insecticidal properties of efrapeptin metabolites of the fungus *Tolypocladium niveum*. J Invertebr Pathol 58:180–188

Krimm U, Abanda-Nkpwatt D, Schwab W, Schreiber L (2005) Epiphytic microorganisms on strawberry plants (*Fragaria ananassa* cv. Elsanta): identification of bacterial isolates and analysis of their interaction with leaf surfaces. FEMS Microbiol Ecol 53:483–492

Krishnamoorthy R, Kwon SW, Kumutha K, Senthilkumar M, Ahmed S, Sa T, Anandham R (2018) Diversity of culturable methylotrophic bacteria in diffrent genotypes of groundnut and their potential for plant growth promotion. 3 Biotech 8:275

Krol P, Iqielski R, Pollmann S, Kepczynska E (2015) Priming of seeds with methyl jasmonate induced resistance to hemi-biotroph *Fusarium oxysporum* f.sp. *lycopersici* in tomato via 12-oxo-phytodienoic acid, salicylic acid and flavonol accumulation. J Plant Physiol 179:122–132

Kumar S, Aharwal RP, Shukla H, Rajak RC, Sandhu SS (2014) Endophytic fungi: as a source of antimicrobials bioactive compounds. World J Pharm Pharm Sci 3:1179–1197

Kumar M, Kour D, Yadav AN, Saxena R, Rai PK, Jyoti A, Tomar RS (2019a) Biodiversity of methylotrophic microbial communities and their potential role in mitigation of abiotic stresses in plants. Biologia 74:287–308. https://doi.org/10.2478/s11756-019-00190-6

Kumar M, Saxena R, Rai PK, Tomar RS, Yadav N, Rana KL, Kour D, Yadav AN (2019b) Genetic diversity of methylotrophic yeast and their impact on environments. In: Yadav AN, Singh S, Mishra S, Gupta A (eds) Recent advancement in white biotechnology through fungi, Volume 3: Perspective for sustainable environments. Springer International Publishing, Cham, pp 53–71. https://doi.org/10.1007/978-3-030-25506-0_3

Kwak MJ, Jeong H, Madhaiyan M, Lee Y, Sa TM, Oh TK, Kim JF (2014) Genome information of Methylobacterium oryzae, a plant-probiotic methylotroph methylotroph in the phyllosphere. PLoS ONE 9(9):e106704

Lambais MR, Crowley DE, Cury JC, Bull RC, Rodrigues RR (2006) Bacterial diversity in tree canopies of the Atlantic forest. Science 312:1917

Lane GA, Christensen MJ, Miles CO (2000) Coevolution of fungal endophytes with grasses: the significance of secondary metabolites. In: Bacon CW, White JFJ (eds) Microbial endophytes. Marcel Dekker, New York, pp 341–388

Lanubile A et al (2014) Functional genomic analysis of constitutive and inducible defense responses to *Fusarium verticillioides* infection in maize genotypes with contrasting ear rot resistance. BMC Genom 15:710

Leben C, Keitt GW (1954) Antibiotics and plant disease: effects of antibiotics in control of plant diseases. J Agric Food Chem 2:234–239

Lee MK, Chauhan PS, Yim WJ, Lee GJ, Kim YS, Park K, Sa TM (2011) Foliar colonization and growth promotion of red pepper (*Capsicum annuum* L.) by *Methylobacterium oryzae* CBMB20. J Appl Biol Chem 54 (2):120–125

Leff JW, Del Tredici P, Friedman WE, Fierer N (2015) Spatial structuring of bacterial communities within individual Ginkgo biloba trees. Environ Microbiol 17:2352–2356

Legard DE, McQuilken MP, Whipps JM, Fenlon JS, Fermor TR, Thompson IP, Bailey MJ, Lynch JM (1994) Studies of seasonal changes in the microbial populations on the phyllosphere of spring wheat as a prelude to the release of a genetically modified microorganism. Agricult Ecosyst Environ 50:87–101

Lehmann S, Serrano ML, Haridon F, Tjamos SE, Metraux JP (2015) Reactive oxygen species and plant resistance to fungal pathogens. Phytochemistry 112:54–62

Lenhart K, Althoff F, Greule M, Keppler F (2015) Technical note: methionine, a precursor of methane in living plants. Biogeosciences 12:1907–1914

Leveau JH, Lindow SE (2001) Appetite of an epiphyte: quantitative monitoring of bacterial sugar consumption in the phyllosphere. Proc Natl Acad Sci USA 98:3446–3453

Leveau JHJ, Tech JJ (2011) Grapevine microbiomics: bacterial diversity on grape leaves and berries revealed by high-throughput sequence analysis of 16S rRNA amplicons. Acta Hort (ISHS) 905:31–42

Levy A, Salas Gonzalez I, Mittelviefhaus M, Clingenpeel S, Herrera Paredes S, Miao J et al (2018) Genomic features of bacterial adaptation to plants. Nat Genet 50:138–150

Lewer P, Chapin E, Graupner PR, Gilbert JR, Peacock C (2003) Tartrolone C: a novel insecticidal macrodiolide produced by *Streptomyces* sp. CP 1130. J Nat Prod 66:143–145

Lighthart B (1997) The ecology of bacteria in the alfresco atmosphere. FEMS Microbiol Ecol 23:263–274

Lindow SE, Brandl MT (2003) Microbiology of the phyllosphere. Appl Environ Microbiol 69:1875–1883

Liu C, Wang X, Zhao J, Liu Q, Wang W, Guan X et al (2013) *Streptomyces harbinensis* sp. nov., an endophytic, ikarugamycin-producing actinomycete isolated from soybean root [Glycine max (L.) Merr.]. Int J Syst Evol Microbiol 63:3579–3584

Liu Y, Jianwei Guo J, Li L, Asem MD, Zhang Y, Mohamad OA et al (2017) Endophytic bacteria associated with endangered plant Ferula sinkiangensis K. M. Shen in an arid land: diversity and plant growth-promoting traits. J Arid Land 9:432–445

Lizarazo-Medina PX, Gómez-Vásquez D (2015) Microbiota rizosférica de Espeletia spp. de los Páramos de Santa Inés y de Frontino-Urrao en Antioquia. Colombia. Acta Biol Colombiana 20:175–182

Loper JE, Lindow SE (1987) Lack of evidence for the in situ fluorescent pigment production by *Pseudomonas syringae* pv. *syringae* on bean leaf surfaces. Phytopathology 77:1449–1454

Lopez-Velasco G, Welbaum GE, Boyer RR, Mane SP, Ponder MA (2011) Changes in spinach phylloepiphytic bacteria communities following minimal processing and refrigerated storage described using pyrosequencing of 16S rRNA amplicons. J Appl Microbiol 110:1203–1214

Luo S, Xu T, Chen L, Chen J, Rao C, Xiao X et al (2012) Endophyte-assisted promotion of biomass production and metal-uptake of energy crop sweet sorghum by plant-growth-promoting endophyte Bacillus sp. SLS18. Appl Microbiol Biotechnol 93:1745–1753

Lurdes Inácio M, Henriques J, Sousa E (2010) Mycobiota associated with Platypus cylindrus Fab. (Coleoptera: Platypodidae) on cork oak in Portugal. Integrated Protection in Oak Forests. IOBC/wprs Bull 57:87–95

Maignien L, DeForce EA, Chafee ME, Eren AM, Simmons SL (2014) Ecological succession and stochastic variation in the assembly of Arabidopsis thaliana phyllosphere communities. MBio 5:e00682–13

Manceau CR, Kasempour MN (2002) In Endophytic versus epiphytic colonization of plants: what comes first? In: Lindow SE, Hecht-Poinar EI, Elliott VJ (eds) Phyllosphere microbiology. APS Press, St Paul, USA, pp 115–123

Mansfield J, Genin S, Magori S, Citovsky V, Sriariyanum M, Ronald P et al (2012) Top 10 plant pathogenic bacteria in molecular plant pathology. Mol Plant Pathol 13:614–629

Mao DP, Zhou Q, Chen CY, Quan ZX (2012) Coverage evaluation of universal bacterial primers using the metagenomic datasets. BMC Microbiol 12:66

Martins F, Pereira JA, Bota P, Bento A, Baptista P (2016) Fungal endophyte communities in above- and belowground olive tree organs and the effect of season and geographic location on their structures. Fungal Ecol 20:193–201

Masand M, Jose PA, Menghani E, Jebakumar SRD (2015) Continuing hunt for endophytic actinomycetes as a source of novel biologically active metabolites. World J Microbiol Biotechnol 31:1863–1875

Mase S (1984) Meiji Herbiace (MW-801, SF-1293) (common name: bialaphos) A new herbicide. Jpn Pestic Inf 45:27–30

Mason CJ, Couture JJ, Raffa KF (2014) Plant-associated bacteria degrade defense chemicals and reduce their adverse effects on an insect defoliator. Oecologia 175:901–910

Meena KK, Kumar M, Kalyuzhnaya MG, Yandigeri MS, Singh DP, Saxena AK et al (2012) Epiphytic pink-pigmented methylotrophic bacteria enhance germination and seedling growth of wheat (*Triticum aestivum*) by producing phytohormone. Antonie Van Leeuwenhoek 101:777–786

Meiners SJ, Phipps KK, Pendergast TH 4th, Canam T, Carson WP (2017) Soil microbial communities alter leaf chemistry and influence allelopathic potential among coexisting plant species. Oecologia 183:1155–1165

Meinhardt LW et al (2014) Genome and secretome analysis of the hemibiotrophic fungal pathogen, *Moniliophthora roreri*, which causes frosty pod rot disease of cacao: mechanisms of the biotrophic and necrotrophic phases. BMC Genom 15:164

Mejía LC, Rojas EI, Maynard Z, Bael SV, Arnold AE, Hebbar P et al (2008) Endophytic fungi as biocontrol agents of *Theobroma cacao* pathogens. Biol Control 46:4–14

Melotto M, Underwood W, He SY (2008) Role of stomata in plant innate immunity and foliar bacterial diseases. Ann Rev Phytopathol 46:101–122

Mendes R, Garbeva P, Raaijmakers JM (2013) The rhizosphere microbiome: significance of plant beneficial, plant pathogenic, and human pathogenic microorganisms. FEMS Microbiol Rev 37:634–663

Mercier J, Lindow SE (2000) Role of leaf surface sugars in coloni- zation of plants by bacterial epiphytes. Appl Environ Microbiol 66:369–374

Mertz FP, Yao RC (1990) *Saccharopolyspora spinosa* sp. nov. isolated from soil collected in a sugar mill rum still. Int J Syst Bacteriol 40:34–39

Methé BA, Li K, Talley SP, Gupta N, Frank B, Xu W, Gordon SG, Goodner B, Stapleton AE (2017) Functional genes that distinguish Maize phyllosphere metagenomes in drought and well-watered conditions. BioRxiv preprint online http://dx.doi.org/10.1101/104331

Michavila G, Adler C, De Gregorio PR, Lami MJ, Caram Di Santo MC, Zenoff AM, Cristobal RE, Vincent PA (2017) *Pseudomonas protegens* CS1 from the lemon phyllosphere as a candidate for citrus canker biocontrol agent. Plant Biol 19:608–617

Mikiciński A, Sobiczewski P, Puławska J, Maciorowski R (2016) Control of fire blight (Erwini a amylovora) by a novel strain 49 M of *Pseudomonas graminis* from the phyllosphere of apple (Malus spp.) Eur J Plant Pathol 145:265–276

Mikolasch A, Hammer E, Schauer F (2003) Synthesis of imidazol- 2-yl amino acids by using cells from alkane-oxidizing bacteria. Appl Environ Microbiol 69:1670–1679

Miller WG, Brandl MT, Quinones B, Lindow SE (2001) Biological sensor for sucrose availability: relative sensitivities of various reporter genes. Appl Environ Microbiol 67:1308–1317

Miller JD, Sumarah M, Adams G (2008) Effect of rugulosin producing endophyte is *Picea galuca* on *Choristoneura fumifera*. J Chem Ecol 34:362–368

Miranda M, Ralph SG, Mellway R, White R, Heath MC, Bohlmann J, Constabel CP (2007) The transcriptional response of hybrid poplar (Populas trichocarpa × P. deltoides) to infection by Melamspora medusa leaf rust involves induction of flavonoid pathway genes leading to accumulation of Proanthocyanidins. Mol Plant Microbe Interact 20:816–831

Mishima H, Ide J, Muramatsu S, Ono M (1983) Milbemycins, a new family of macrolide antibiotics. Structure determination of milbemycins D, E, F, G, H, J and K. J Antibiot 36:980–990

Miyamoto T, Kawahara M, Minamisawa K (2004) Novel endophytic nitrogen-fixing clostridia from the grass *Miscanthus sinensis* as revealed by terminal restriction fragment length polymorphism analysis. Appl Environ Microbiol 70:6580–6586

Mizuno M, Yurimoto H, Iguchi H, Tani A, Sakai Y (2013) Dominant Colonization and Inheritance of *Methylobacterium* sp. Strain OR01 on Perilla Plants. Biosci Biotechnol Biochem 77(7): 1533–1538

Mohammadi K (2012) Phosphorus solubilizing bacteria: occurrence, mechanisms mechanisms and their role in crop production. Resour Environ 2:80–85

Monaghan J, Zipfel C (2012) Plant pattern recognition receptor complexes at the plasma membrane. Curr Opin Plant Biol 15:349–357

Monier JM, Lindow SE (2004) Frequency, size, and localization of bacterial aggregates on bean leaf surfaces. Appl Environ Microbiol 70:346–355

Monier JM, Lindow SE (2005) Aggregates of resident bacteria facilitates survival of immigrant bacteria on leaf surfaces. Microbial Ecol 49:343–352

Morella NM, Gomez AL, Wang G, Leung MS, Koskella B (2018) The impact of bacteriophages on phyllosphere bacterial abundance and composition. Mol Ecol 27:2025–2038

Morris CE, Kinkel LL (2002) Fifty years of phylosphere microbiology: significant contributions to research in related fields. In: Lindow SE, Hecht-Poinar EI, Elliott V (eds) Phyllosphere microbiology. APS Press, St. Paul, Minn, pp 365–375

Morris CE, Monier J, Jacques M (1997) Methods for observing microbial biofilms directly on leaf surfaces and recovering them for isolation of culturable microorganisms. Appl Environ Microbiol 63:1570–1576

Morris CE, Monteil CL, Berge O (2013) The life history of *Pseudomonas syringae*: linking agriculture to earth system processes. Ann Rev Phytopathol 51:85–104

Mukhtar I, Khokhar I, Sobia M, Ali A (2010) Diversity of epiphytic and endophytic microorganisms in some dominant weeds. Pak J Weed Sci Res 16:287–297

Müller T, Ruppel S (2014) Progress in cultivation-independent phyllosphere microbiology. FEMS Microbiol Ecol 87:2–17

Müller H, Berg C, Landa BB, Auerbach A, Moissl-Eichinger C, Berg G (2015) Plant genotype-specific archaeal and bacterial endophytes but similar *Bacillus antagonists* colonize Mediterranean olive trees. Front Microbiol 6:138

Mwajita M, Murage H, Tani A, Kahangi EM (2013) Evaluation of rhizosphere, rhizoplane and phyllosphere bacteria and fungi isolated from rice in Kenya for plant growth promoters. Springerplus 2:606

Nadalig T, Ul F, Haque M, Roselli S, Schaller H, Bringel F, Vuilleumier S (2011) Detection and isolation of chloromethane-degrading bacteria from the Arabidopsis thaliana phyllosphere, and characterization of chloromethane utilization genes. FEMS Microbiol Ecol 77:438–448

Nair DN, Padmavathy S (2014) Impact of endophytic microorganisms on plants, environment and humans. Sci World J. (https://doi.org/10.1155/2014/250693

Nalini MS, Prakash HS (2017) Diversity and bioprospecting of actinomycete endophytes from the medicinal plants. Lett Appl Microbiol 64:261–270

Nautiyal CS, Srivastava S, Chauhan PS, Seem K, Mishra A, Sopory SK (2013) Plant growth-promoting bacteria *Bacillus amyloliquefaciens* NBRISN13 modulates gene expression profile of leaf and rhizosphere community in rice during salt stress. Plant Physiol Biochem 66:1–9

Nehra V, Saharan BS, Choudhary M (2016) Evaluation of *Brevibacillus brevis* as a potential plant growth promoting rhizobacteria for cotton (*Gossypium hirsutum*) crop. SpringerPlus 5:948

Nevitt GA (2011) The neuroecology of dimethyl sulfide: a global-climate regulator turned marine infochemical. Integr Comp Biol 51:819–825

Newman MA, Sundelin T, Nielsen JT, Erbs G (2013) MAMP (microbe-associated molecular pattern) triggered immunity in plants. Front Plant Sci 4:139

Nguyen XH, Naing KW, Lee YS, Kim YH, Moon JH, Kim KY (2015) Antagonism of antifungal metabolites from *Streptomyces griseus* H7602 against Phytophthora capsici. J Basic Microbiol 55:45–53

Nimnoi P, Pongsilp N, Lumyong S (2010) Endophytic actinomycetes isolated from *Aquilaria crassna* Pierre ex. Lec and screening of plant growth-promoters production. World J Microbiol Biotechnol 26:193–203

Núñez-Trujillo G, Cabrera R, Burgos-Reyes RL, Silva ED, Giménez C, Cosoveanu A et al. (2012) Endophytic fungi from *Vitis vinifera* L. isolated in Canary Islands and Azores as potential biocontrol agents of *Botrytis cinerea* Pers.: Fr J Hortic For Biotechnol 16:1–6

Ohshiro M, Hossain MA, Nakamura I, Akamine H, Tamaki M, Bhowmik PC et al (2016) Effects of soil types and fertilizers on growth, yield, and quality of edible Amaranthus tricolor lines in Okinawa, Japan. Plant Prod Sci 19:61–72

Ondeyka JG, Helms GL, Hensens OD, Goetz MA, Zink DL, Tsipouras A, Shoop WL, Slayton L, Dombrowskii AW, Polishook JD, Ostlind DA, Tsou NN (1997) Nodulisporic acid A, a novel and potent insecticide from *Nodulisporium* sp. Isolation, structure determination and chemical transformation. J Am Chem Soc 119:8809–8816

Osono T (2006) Role of phyllosphere fungi of forest trees in the development of decomposer fungal communities and decomposition processes of leaf litter. Can J Microbiol 52:701–716

Osono T (2008) Endophytic and epiphytic phyllosphere fungi of *Camellia japonica*: seasonal and leaf age-dependent variations. Mycologia 100:387–391

Ottesen AR, González Peña A, White JR, Pettengill JB, Li C, Allard S et al (2013) Baseline survey of the anatomical microbial ecology of an important food plant: Solanum lycopersicum (tomato). BMC Microbiol 13:114

Park P, Tsuda H, Hayashi Y, Uneo T (1977) Effect of host specific toxin (AM toxin-I) produced by *Alternaria mali*, an apple pathogen, on ultrastructure of plasma membrane of cells in apple and japanese pear leaves. Can J Bot 55:2383–2393

Parthasarathi S, Sathya S, Bupesh G, Samy DR, Mohan MR, Selva GK et al (2012) Isolation and characterization of antimicrobial compound from marine *Streptomyces hygroscopicus* BDUS 49. World J Fish Mar Sci 4:268–277

Peiffer JA, Spor A, Koren O, Jin Z, Tringe SG, Dangl JL, Buckler ES, Ley RE (2013) Diversity and heritability of the maize rhizosphere microbiome under field conditions. Proc Natl Acad Sci 110:6548–6553

Porras-Alfaro A, Bayman P (2011) Hidden fungi, emergent properties: endophytes and microbiomes. Annu Rev Phytopathol 49:291–315

Prasad Sahu K, Kumar A (2015) Metagenomic analysis of rice phyllospheric bacterial communities in relation to blast disease. M.Sc., Thesis, Krishikosh Indian Agricultural Research Institute, New Delhi

Purahong W, Hyde KD (2011) Effects of fungal endophytes on grass and non-grass litter decomposition rates. Fungal Divers 47:1–7

Pusey PL, Stockwell VO, Reardon CL, Smits THM, Duffy B (2011) Antibiosis activity of *Pantoea agglomerans* biocontrol strain E325 against *Erwinia amylovora* on apple flower stigmas. Phytopathology 101:1234–1241

Qin S, Xing K, Jiang JH, Xu LH, Li WJ (2011) Biodiversity, bioactive natural products and biotechnological potential of plant-associated endophytic actinobacteria. Appl Microbiol Biotechnol 89:457–473

Qin S, Bian GK, Zhang YJ, Xing K, Cao CL, Liu CH et al (2013) *Modestobacter roseus* sp. nov., an endophytic actinomycete isolated from the coastal halophyte *Salicornia europaea* Linn, and emended description of the genus Modestobacter. Int J Syst Evol Microbiol 63:2197–2202

Quan M, Liang J (2017) The influences of four types of soil on the growth, physiological and biochemical characteristics of *Lycoris aurea* (L' Her.) Herb Sci Rep 7. (https://doi.org/10.1038/srep43284)

Radhakrishnan R, Hashem A, Abd_Allah EF (2017) *Bacillus*: a biological tool for crop improvement through bio-molecular changes in adverse environments. Front Physiol 8:667. https://doi.org/10.3389/fphys.2017.00667

Ramírez Gómez M (2011) Importancia de los microorganismos y la edafofauna en los páramos. Colombia tiene Páramos 2011(1):42–57

Rana KL, Kour D, Sheikh I, Dhiman A, Yadav N, Yadav AN, Rastegari AA, Singh K, Saxena AK (2019a) Endophytic fungi: biodiversity, ecological significance and potential industrial applications. In: Yadav AN, Mishra S, Singh S, Gupta A (eds) Recent advancement in white biotechnology through fungi, vol 1. Diversity and Enzymes Perspectives. Springer, Switzerland, pp 1–62

Rana KL, Kour D, Sheikh I, Yadav N, Yadav AN, Kumar V, Singh BP, Dhaliwal HS, Saxena AK (2019b) Biodiversity of endophytic fungi from diverse niches and their biotechnological applications. In: Singh BP (ed) Advances in endophytic fungal research: present status and future challenges. Springer International Publishing, Cham, pp 105–144. https://doi.org/10.1007/978-3-030-03589-1_6

Rana KL, Kour D, Yadav AN (2019c) Endophytic microbiomes: biodiversity, ecological significance and biotechnological applications. Res J Biotechnol 14:142–162

Rasche F, Trondl R, Naglreiter C, Reichenauer TG, Sessitsch A (2006) Chilling and cultivar type affect the diversity of bacterial endophytes colonizing sweet pepper (*Capsicum anuum* L.). Can J Microbiol 52:1036–1045

Rastogi G, Sani RK (2011) Molecular techniques to assess microbial community structure, function, and dynamics in the environment. In: Ahmad I, Ahmad F, Pichtel J (eds) Microbes and Microbial Technology. Springer, New York, pp. 29–57

Rastogi G, Sbodio A, Tech JJ, Suslow TV, Coaker GL, Leveau JHJ (2012) Leaf microbiota in an agroecosystem: spatiotemporal variation in bacterial community composition on field-grown lettuce. ISME J 6:1812–1822

Rastogi G, Coaker GL, Leveau JHJ (2013) New insights into the structure and function of phyllosphere microbiota through high-throughput molecular approaches. FEMS Microbiol Lett 348:1–10. https://doi.org/10.1111/1574-6968.12225

Rawat SR, Mannisto MK, Bromberg Y, Haggblom MM (2012) Comparative genomic and physiological analysis provides insights into the role of Acidobacteria in organic carbon utilization in Arctic tundra soils. FEMS Microbiol Ecol 82:341–355

Redford AJ, Bowers RM, Knight R, Linhart Y, Fierer N (2010) The ecology of the phyllosphere: geographic and phylogenetic variability in the distribution of bacteria on tree leaves. Environ Microbiol 12:2885–2893

Reinhold-Hurek B, Hurek T (2011) Living inside plants: bacterial endophytes. Curr Opin Plant Biol 14:435–443

Reinhold-Hurek B, Bunger W, Burbano CS, Sabale M, Hurek T (2015) Roots shaping their microbiome: global hotspots for microbial activity. Ann Rev Phytopathol 53:403

Reino JL, Guerrero RF, Hernández-Galán R, Collado IG (2008) Secondary metabolites from species of the biocontrol agent *Trichoderma*. Phytochem Rev 7:89–123

Remus-Emsermann MNP, Schlechter RO (2018) Phyllosphere microbiology: at the interface between microbial individuals and the plant host. New Phytol 218:1327–1333

Remus-Emsermann MNP, de Oliveira S, Schreiber L, Leveau JHJ (2011) Quantification of lateral heterogeneity in carbohydrate permeability of isolated plant leaf cuticles. Front Microbiol 2:197

Remus-Emsermann MNP, Tecon R, Kowalchuk GA, Leveau JHJ (2012) Variation in local carrying capacity and the individual fate of bacterial colonizers in the phyllosphere. ISME Journal 6:756–765

Remus-Emsermann MNP, Kim EB, Marco ML, Tecon R, Draft Leveau JHJ (2013) Genome sequence of the phyllosphere model bacterium *Pantoea agglomerans* 299R. Genome Announc. https://doi.org/10.1128/genomea.00036-13

Ren G, Zhang H, Lin X, Zhu J, Jia Z (2014) Response of phyllosphere bacterial communities to elevated CO_2 during rice growing season. Appl Microbiol Biotechnol 98:9459–9471

Rico L, Ogaya R, Terradas J, Peñuelas J (2014) Community structures of N_2-fixing bacteria associated with the phyllosphere of a Holm oak forest and their response to drought. Plant Biol 16:586–593

Ripa FA, Cao WD, Tong S, Sun JG (2019) Assessment of plant growth promoting and abiotic stress tolerance properties of wheat endophytic fungi. BioMed Res Int 6105865:12. (https://doi.org/10.1155/2019/6105865

Rivera-Vargas LI, Lugo-Noel Y, McGovern RJ, Seijo T, Davis MJ (2006) Occurrence and distribution of *Colletotrichum* spp. on Mango (*Mangifera indica* L.) in Puerto Rico and Florida, USA. Plant Pathol J 5:191–198

Rodrigues KF (1994) The foliar fungal endophytes of the Amazonian palm *Euterpe oleracea*. Mycologia 86(3):376–385

Rodriguez RJ, White JF, Arnold AE, Redman RS (2009) Fungal endophytes: diversity diversity and functional roles. New Phytol 182:314–330

Rodríguez-Escobar C, Mitter B, Barret M, Sessitsch A, Compant S (2018) Commentary: seed bacterial inhabitants and their routes of colonization. Plant Plant Soil 422:129–134

Rosa LH, Tabanca N, Techen N, Pan Z, Wedge DE, Moraes RM (2012) Antifungal activity of extracts from endophytic fungi associated with *Smallanthus* maintained in vitro as autotrophic cultures and as pot plants in the greenhouse. Can J Microbiol 58(10):1202–1221

Roselli S, Nadalig T, Vuilleumier S, Bringel F (2013) The 380 kb pCMU01 plasmid encodes chloromethane utilization genes and redundant genes for vitamin B12- and tetrahydrofolate-dependent chloromethane metabolism in Methylobacterium extorquens CM4: a proteomic and bioinformatics study. PLoS ONE 8:e56598

Rudd JJ et al (2015) Transcriptome and metabolite profiling the infection cycle of *Zymoseptoria tritici* on wheat (*Triticum aestivum*) reveals a biphasic interaction with plant immunity involving differential pathogen chromosomal contributions, and a variation on the hemibiotrophic lifestyle definition. Plant Physiol 167:1158–1185

Ruiz-Pérez CA, Restrepo S, Zambrano MM (2016) Microbial and functional diversity within the phyllosphere of *Espeletia* species in an Andean high-mountain ecosystem. Appl Environ Microbiol 82:1807–1817

Sadler C, Schroll B, Zeisler V, Waßmann F, Franke R, Schreiber L (2016) Wax and cutin mutants of Arabidopsis: quantitative characterization of the cuticular transport barrier in relation to chemical composition. Biochim Biophys Acta 1861:1336–1344

Salam N, Khieu TN, Liu MJ, Vu TT, Ky SC, Quach NT et al (2017) Endophytic actinobacteria associated with *Dracaena cochinchinensis* Lour: isolation, diversity, and their cytotoxic activities. Biomed Res Int 2017:1–11

Salazar-Cerezo S, Martinez-Montiel N, del Carmen C-LM, Martinez-Contreras RD (2018) Fungal diversity and community composition of culturable fungi in *Stanhopea trigrina* Cast Gibberellin producers. Front Microbiol 9:612

Saleem M, Meckes N, Pervaiz ZH, Traw MB (2017) Microbial interactions in the phyllosphere increase plant performance under Herbivore biotic stress. Front Microbiol 8:41

San Roman M, Wagner A (2018) An enormous potential for niche construction through bacterial cross-feeding in a homogeneous environment. PLoS Comput Biol 14(7):e1006340
Santhanam R, Groten K, Meldau DG, Baldwin IT (2014) Analysis of plant-bacteria interactions in their native habitat: bacterial communities associated with wild tobacco are independent of endogenous jasmonic acid levels and developmental stages. PLoS ONE 9:e94710
Šantl-Temkiv T, Finster K, Dittmar T, Hansen BM, Thyrhaug R, Nielsen NW et al (2013) Hailstones: a window into the microbial and chemical inventory of a storm cloud. PLoS ONE 8:e53550
Santoyo G, Moreno-Hagelsieb G, del Orozco-Mosqueda MC, Glick BR (2016) Plant growth-promoting bacterial endophytes. Microbiol Res 183:92–99
Sapp M, Ploch S, Fiore-Donno AM, Bonkowski M, Ros LE (2018) Protists are an integral part of the Arabidopsis thaliana microbiome. Environ Microbiol 20:30–43
Sawinski K, Mersmann S, Robatzek S, Böhmer M (2013) Guarding the green: pathways to stomatal immunity. Mol Plant Microbe Interact 26:626–632
Saxena S (2014) Microbial metabolites for development of ecofriendly agrochemicals. Allelopath J 33(1):1–24
Scavino AF, Pedraza RO (2013) The role of siderophores in plant growth-promoting bacteria. In: Maheshwari DK, Saraf M, Aeron A (eds) *Bacteria in agrobiology:* crop productivity. Springer, Berlin, pp 265–285. https://doi.org/10.1007/978-3-642-37241-4-11
Schäfer H, Myronova N, Boden R (2010) Microbial degradation of dimethylsulphide and related C_1-sulphur compounds: organisms and pathways controlling fluxes of sulphur in the biosphere. J Exp Bot 61:315–334
Schlechter RO, Mitja MM, Remus-Emsermann NP (2019) Driving factors of epiphytic bacterial communities: A review. Journal of Advanced Research 19:57–65
Schönherr J (2006) Characterization of aqueous pores in plant cuticles and permeation of ionic solutes. J Exp Bot 57:2471–2491
Schreiber L (2005) Polar paths of diffusion across plant cuticles: new evidence for an old hypothesis. Ann Bot 95:1069–1073
Selvakumar G, Kundu S, Joshi P, Nazim S, Gupta AD, Gupta HS (2010) Growth promotion of wheat seedlings by *Exiguobacterium acetylicum* 1P (MTCC 8707) a cold tolerant bacterial strain from the Uttarakhand Himalayas. Indian J Microbiol 50:50–56
Sessitsch A, Hackl E, Wenzl P, Kilian A, Kostic T, Stralis-Pavese N, Sandjong BT et al (2006) Diagnostic microbial microarrays in soil ecology. New Phytol 171:719–736
Setati ME, Jacobson D, Andong UC, Bauer FF (2012) The vineyard yeast microbiome, a mixed model microbial map. PLoS ONE 7:e52609
Shade A, McManus PS, Handelsman J (2013) Unexpected diversity during community succession in the apple flower microbiome. MBio 4:e00602–12
Shahzad R, Waqas M, Khan AL, Asaf S, Khan MA, Kang SM et al (2016) Seed-borne endophytic *Bacillus amyloliquefaciens* RWL-1 produces gibberellins and regulates endogenous phytohormones of *Oryza sativa*. Plant Physiol Biochem 106:236–243
Sharma PS, D'Souza F, Kutner W (2012) Molecular imprinting for selective chemical sensing of hazardous compounds and drugs of abuse. Trac Trend Anal Chem 34:59–77
Shen Y, Liu C, Wang X, Zhao J, Jia F, Zhang Y (2013) *Actinoplanes hulinensis* sp nov., a novel actinomycete isolated from soybean root (Glycine max (L.) Merr). Antonie Van Leeuwenhoek 103:293–298
Shepherd RW, Bass WT, Houtz RL, Wagner GJ (2005) Phylloplanins of tobacco are defensive proteins deployed on aerial surfaces by short glandular trichomes. Plant Cell 17:1851–1886
Shin S, Lv J, Fazio G, Mazzola M, Zhu Y (2014) Transcriptional regulation of ethylene and jasmonate mediated defense response in apple (*Malus domestica*) root during *Pythium ultimum* infection. Hort Res 1:14053
Shiraishi K, Oku M, Kawaguchi K, Uchida D, Yurimoto H, Sakai Y (2015) Yeast nitrogen utilization in the phyllosphere during plant lifespan under regulation of autophagy. Sci Rep 5:09719

Shrestha BK, Karki HS, Groth DE, Jungkhun N, Ham JH (2016) Biological control activities of rice-associated Bacillus sp. Strains against sheath blight and bacterial panicle blight of rice. PLoS One 11:e0146764

Shukla ST, Habbu PV, Kulkarni VH, Jagadish KS, Pandey AR, Sutariya VN (2014) Endophytic microbes: a novel source for biologically/pharmacologically active secondary metabolites. Asian J Pharmacol Toxicol 2:1–16

Singh R, Dubey AK (2015) Endophytic actinomycetes as emerging source for therapeutic compounds. Indo Global J Pharm Sci 5:106–116

Singh BK, Nazaries L, Munro S, Anderson IC, Campbell CD (2006) Use of multiplex terminal restriction fragment length polymorphism for rapid and simultaneous analysis of different components of the soil microbial community. Appl Environ Microbiol 72:7278–7285

Singh R, Ashok K, Dubey (2018) Diversity and applications of endophytic actinobacteria of plants in special and other ecological niches 9:1767

Snelders NC, Kettles GJ, Rudd JJ, Thomma BPHJ (2018) Plant pathogen effector proteins as manipulators of host microbiomes? Mol Plant Pathol 19:257–259

Soliman SSM, Raizada MN (2018) Darkness: a crucial factor in fungal taxol production. Front Microbiol 9:353

Specian V, Sarragiotto MH, Pamphile JA, Clemente E (2012) Chemical characterization of bioactive compounds from the endophytic fungus *Diaporthe helianthi* isolated from Luehea divaricata. Braz J Microbiol 43:1174–1182

Sreekanth D, Kristin IM, Brett AN (2017) Endophytic fungi from *Cathranthus roseus*: a potential resource for the discovery of antimicrobial polyketides. Nat Prod Chem Res 5:256

Staley C, Dunny GM, Sadowsky MJ (2014) Environmental and animal-associated enterococci. Adv Appl Microbiol 87:147–186

Steven B, Huntley RB, Zeng Q (2018) The influence of flower anatomy and apple cultivar cultivar on the apple flower phytobiome. Phytobiomes J 2:171–179

Stierle A, Cardellina H, Strobel G (1990) Maculosin, a host-specific phytotoxin from *Alternaria alternata* on spotted knapweed. Am Chem Soc Symp Ser 439:53–62

Strasser H, Vey A, Butt TM (2000) Are there any risks in using entomopathogenic fungi for pest control with particular reference to bioactive metabolites of *Metarrhizium*, *Tolypocladium* and *Beauveria* sp.? Biocontrol Sci Tech 10:717–735

Strobel GA, Daisy B (2003) Bioprospecting for microbial endophytes and their natural products. Microbiol Mol Biol Rev 67:491–502

Stubbendieck RM, Vargas-Bautista C, Straight PD (2016) Bacterial communities: interactions to scale. Front Microbiol 7:1234

Suda W, Nagasaki A, Shishido M (2009) Powdery mildew-infection changes bacterial community composition in the phyllosphere. Microbes Environ 24:217–223

Sun Y et al (2010) The role of wheat Jasmonic acid and ethylene pathways in response to *Fusarium graminearum* infection. Plant Growth Regul 80:69–77

Suslow TV (2002) Production practices affecting the potential for persistent contamination of plants by microbial foodborne pathogens. In: Lindow SE, Hecht-Poinar EI, Elliott VJ (eds) Phyllosphere microbiology. APS Press, St Paul, USA, pp 241–256

Sy A, Timmers AC, Knief C, Vorholt JA (2005) Methylotrophic metabolism is advantageous for Methylobacterium extorquens during colonization of Medicago truncatula under competitive conditions. Appl Environ Microbiol 71:7245–7252

Sziderics AH, Rasche F, Trognitz F, Sessitsch A, Wilhelm E (2007) Bacterial endophytes contribute to abiotic stress adaptation in pepper plants (*Capsicum annuum* L.). Can J Microbiol 53:1195–1202

Tachibana K (2003) Bialaphos, a natural herbicide. Meiji Seika Kenkyo Nenpo 42:44–57

Tecon R, Leveau JHJ (2016) Symplasmata are a clonal, conditional, and reversible type of bacterial multicellularity. Sci Rep 6:31914

Tecon R, Ebrahimi A, Kleyer H, Erev Levi S, Or D (2018) Cell-to-cell bacterial interactions promoted by drier conditions on soil surfaces. PNAS 115(39):9791–9796

Thamchaipenet A, Indananda C, Bunyoo C, Duangmal K, Matsumoto A, Takahashi Y (2010) Actinoallomurus acaciae sp. nov., an endophytic actinomycete isolated from *Acacia auriculiformis*, A Cunn. ex Benth. Int J Syst Evol Microbiol 60:554–559

Thapa S, Ranjan K, Ramakrishnan B, Velmourougane K, Prasanna R (2018) Influence of fertilizers and rice cultivation methods on the abundance and diversity of phyllosphere microbiome. J Basic Microbiol 58:172–186

Thompson IP, Bailey MJ, Fenlon JS, Fermor TR, Lilley AK, Lynch JM, McCormack PJ, McQuilken MP et al (1993) Quantitative and qualitative seasonal changes in the microbial community from the phyllosphere of sugar beet (Beta vulgaris). Plant Soil 150:177–191

Trabelsi D, Mhamdi R (2013) Microbial inoculants and their impact on soil microbial communities: a review. Biomed Res Int

Trdá L, Boutrot F, Claverie J, Brulé D, Dorey S, Poinssot B (2015) Perception of pathogenic or beneficial bacteria and their evasion of host immunity: pattern recognition receptors in the frontline. Front Plant Sci 6:219

Trouvelot S, Héloir MC, Poinssot B, Gauthier A, Paris F, Guillier C et al (2014) Carbohydrates in plant immunity and plant protection: roles and potential application as foliar sprays. Front Plant Sci 5:592

Twizeyimana M, Förster H, McDonald V, Wang DH, Adaskaveg JE, Eskalen A (2013) Identification and pathogenicity of fungal pathogens associated with stem-end rot of avocado in California. Plant Dis 97(12):1580–1584

Umezawa H, Okami Y, Hashimoto T, Suhara Y, Otake N (1965) A new antibiotic kasugamycin. J Antibiot Ser A 18:101–103

Uneo T (1990) Secondary metabolites related to host selection by plant pathogenic fungi. Pure Appl Chem 62:1347–1352

Uppalapati SR et al (2009) Global gene expression profiling during Medicago truncatula-*Phymatotrichopsis omnivora* interaction reveals a role for Jasmonic acid, Ethylene and the Flavonoid pathway in disease development. Mol Plant Microbe Interac 22:7–17

Uroz S, Buee M, Deveau A, Mieszkin S, Martin F (2016) Ecology of the forest microbiome: highlights of temperate and boreal ecosystems. Soil Biol Biochem 103:471–488

Vacher C, Hampe A, Porte AJ, Sauer U, Compant S, Morris CE (2016) The phyllosphere: microbial jungle at the plant-climate interface. Ann Rev Ecol Evol Syst 47:1–24

Vaïtilingom M, Deguillaume L, Vinatier V, Sancelme M, Amato P, Chaumerliac N et al (2013) Potential impact of microbial activity on the oxidant capacity and organic carbon budget in clouds. Proc Natl Acad Sci USA 110:559–564

Van der Wal A, Leveau JHJ (2011) Modelling sugar diffusion across plant leaf cuticles: the effect of free water on substrate availability to phyllosphere bacteria. Environ Microbiol 13:792–797

Van der Wal A, Tecon R, Kreft J-U, Mooij WM, Leveau JHJ (2013) Explaining bacterial dispersion on leaf surfaces with an individual-based model (PHYLLOSIM). PLoS ONE 8(10):e75633

Van Kan JAL, Shaw MW, Grant-Downton RT (2014) *Botrytis* species: relentless necrotrophic thugs or endophytes gone rogue? Mol Plant Pathol 15:957–961

Vargas WA et al (2012) Plant defense mechanisms are activated during biotrophic and necrotrophic development of *Colletotricum graminicola* in maize. Plant Physiol 158:1342–1358

Venkatachalam S, Ranjan K, Prasanna R, Ramakrishnan B, Thapa S, Kanchan A (2016) Diversity and functional traits of culturable microbiome members, including cyanobactera in the rice phyllosphere. Plant Biol 18:627–637

Verginer M, Leitner E, Berg G (2010) Production of volatile metabolites by grape associated microorganisms. J Agric Food Chem 58:8344–8350

Verma P, Yadav AN, Kazy SK, Saxena AK, Suman A (2013) Elucidating the diversity and plant growth promoting attributes of wheat (*Triticum aestivum*) associated acidotolerant bacteria from southern hills zone of India. Natl J Life Sci 10:219–227

Verma P, Yadav AN, Khannam KS, Panjiar N, Kumar S, Saxena AK, Suman A (2015) Assessment of genetic diversity and plant growth promoting attributes of psychrotolerant bacteria allied with wheat (*Triticum aestivum*) from the northern hills zone of India. Ann Microbiol 65:1885–1899

Verma P, Yadav AN, Khannam KS, Kumar S, Saxena AK, Suman A (2016a) Molecular diversity and multifarious plant growth promoting attributes of Bacilli associated with wheat (*Triticum aestivum* L.) rhizosphere from six diverse agro-ecological zones of India. J Basic Microbiol 56:44–58
Verma P, Yadav AN, Khannam KS, Mishra S, Kumar S, Saxena AK, Suman A (2016b) Appraisal of diversity and functional attributes of thermotolerant wheat associated bacteria from the peninsular zone of India. Saudi J Biol Sci. https://doi.org/10.1016/j.sjbs.2016.01.042
Verma P, Yadav AN, Kumar V, Singh DP, Saxena AK (2017) Beneficial plant-microbes interactions: biodiversity of microbes from diverse extreme environments and its impact for crop improvement. In: Singh DP, Singh HB, Prabha R (eds) Plant-microbe interactions in agro-ecological perspectives, vol 2. Microbial Interactions and Agro-Ecological Impacts. Springer, Singapore, pp 543–580. https://doi.org/10.1007/978-981-10-6593-4_22
Videira SIR, Groenewald JZ, Verkley GJM, Braun U, Crous PW (2015) The rise of *Ramularia* from the *Mycosphaerella* labyrinth. Fungal Biol 119:823–843
Vokou D, Vareli K, Zarali E, Karamanoli K, Constantinidou HI, Monokrousos N, Halley JM, Sainis I (2012) Exploring biodiversity in the bacterial community of the Mediterranean phyllosphere and its relationship with airborne bacteria. Microb Ecol 64:714–724
Vorholt JA (2012) Microbial life in the phyllosphere. Nat Rev 10:828–840
Voříšková J, Baldrian P (2013) Fungal community on decomposing leaf litter undergoes rapid successional changes. ISME J 7:477–486
Walker AP, McCormack ML, Messier J, Myers-Smith IH, Wullschleger SD (2017) Trait covariance: the functional warp of plant diversity? New Phytol 216:976–980
Walsh UF, Morrissey JP, O'Gara F (2001) Pseudomonas for biocontrol of phytopathogens: from functional genomics to commercial exploitation. Curr Opin Biotechnol 12:289–295
Wang M, Ma Q (2011) Antagonistic Actinomycete XN-1 from phyllosphere microorganisms of cucumber to control corynespora cassiicola. Cucurbit Genet Coop Rep 33–34:17–21
Wang X, Radwan MM, Taráwneh AH, Gao J, Wedge DE, Rosa LH et al (2013a) Antifungal activity against plant pathogens of metabolites from the endophytic fungus *Cladosporium cladosporioides*. J Agric Food Chem 61:4551–4555
Wang X, Zhao J, Liu C, Wang J, Shen Y, Jia F et al (2013b) *Nonomuraea solani* sp. nov., an actinomycete isolated from eggplant root (*Solanum melongena* L.). Int J Syst Evol Microbiol 63:2418–2423
Wang J, Hu M, Wang J, Qi J, Han Z, Wang G, Qi Y, Wang HW, Zhou JM, Chai J (2019) Reconstitution and structure of a plant NLR resistosome conferring immunity. Science 364(6435)
Wensing A, Braun SD, Buttner P, Expert D, Volksch B, Ullrich MS et al (2010) Impact of siderophore production by *Pseudomonas syringae* pv. *syringae* 22d/93 on epiphytic fitness and biocontrol activity against *Pseudomonas syringae* pv. *glycinea* 1a/96. Appl Environ Microbiol 76:2704–2711
Whipps JM, Hand P, Pink D, Bending GD (2008) Phyllosphere microbiology with special reference to diversity and plant genotype. J Appl Microbiol 105:1744–1755
Wicklow DT, Roth S, Deyrup ST, Gloer JB (2005) A protective endophyte of maize: *Acremonium zeae* antibiotics inhibitory to *Aspergillus flavus* and *Fusarium verticillioides*. Mycol Res 109:610–618
Williams TR, Moyne A-L, Harris LJ, Marco ML (2013) Season, irrigation, leaf age, and Escherichia coli inoculation influence the bacterial diversity in the lettuce phyllosphere. PLoS ONE 8:e68642
Win PM, Matsumura E, Fukuda K (2019) Diversity of tea endophytic fungi: cultivar- and tissue preferences. Appl Ecol Environ Res 16(1):677–695
Worapong J, Strobel GA, Ford EJ, Li JY, Baird G, Hess WM (2001) *Muscodor albus* anam. nov., an endophyte from *Cinnamomum zeylanicum*. Mycotaxon 79:67–79
Wright SAI, Zumoff CH, Schneider L, Beer SV (2001) *Pantoea agglomerans* Strain EH318 produces two antibiotics that inhibit Erwinia amylovora in vitro. Appl Environ Microbiol 67:284–292
Wu L, Chen H, Curtis C, Fu ZQ (2014) Go in for the kill: How plants deploy effector-triggered immunity to combat pathogens. Virulence 5(7):710–721

Wulff EG, Van Vuurde JWL, Hockenhull J (2003) The ability of the biological control agent *Bacillus subtilis*, strain BB, to colonise vegetable brassicas endophytically following seed inoculation. Plant Soil 255:463–474

Xie QY, Wang C, Wang R, Qu Z, Lin HP, Goodfellow M et al (2011) Jishengella endophytica gen. nov., sp. nov., a new member of the family Micromonosporaceae. Int J Syst Evol Microbiol 61:1153–1159

Yadav AN (2017) Agriculturally important microbiomes: biodiversity and multifarious PGP attributes for amelioration of diverse abiotic stresses in crops for sustainable agriculture. Biomed J Sci Tech Res 1:1–4

Yadav AN, Yadav N (2018) Stress-Adaptive Microbes for Plant Growth Promotion and Alleviation of Drought Stress in Plants. Acta Sci Agric 2:85–88

Yadav N, Yadav AN (2019) Actinobacteria for sustainable agriculture. Journal of Applied Biotechnology and Bioengineering 6:38–41

Yadav AN, Sachan SG, Verma P, Saxena AK (2015a) Prospecting cold deserts of north western Himalayas for microbial diversity and plant growth promoting attributes. J Biosci Bioeng 119:683–693

Yadav AN, Sachan SG, Verma P, Tyagi SP, Kaushik R, Saxena AK (2015b) Culturable diversity and functional annotation of psychrotrophic bacteria from cold desert of Leh Ladakh (India). World J Microbiol Biotechnol 31:95–108

Yadav AN, Kumar V, Prasad R, Saxena AK, Dhaliwal HS (2018a) Microbiome in Crops: Diversity, distribution and potential role in crops improvements. In: Prasad R, Gill SS, Tuteja N (eds) Crop Improvement through Microbial Biotechnology. Elsevier, USA, pp 305–332

Yadav AN, Verma P, Kumar S, Kumar V, Kumar M, Singh BP, Saxena AK, Dhaliwal HS (2018b) Actinobacteria from Rhizosphere: Molecular Diversity, Distributions and Potential Biotechnological Applications. In: Singh B, Gupta V, Passari A (eds) New and Future Developments in Microbial Biotechnology and Bioengineering. USA, pp 13–41. https://doi.org/10.1016/b978-0-444-63994-3.00002-3

Yadav AN, Verma P, Kumar V, Sangwan P, Mishra S, Panjiar N, Gupta VK, Saxena AK (2018c) Biodiversity of the Genus *Penicillium* in Different Habitats. In: Gupta VK, Rodriguez-Couto S (eds) New and Future Developments in Microbial Biotechnology and Bioengineering, *Penicillium* System Properties and Applications. Elsevier, Amsterdam, pp 3–18. https://doi.org/10.1016/b978-0-444-63501-3.00001-6

Yadav AN, Kour D, Sharma S, Sachan SG, Singh B, Chauhan VS, Sayyed RZ, Kaushik R, Saxena AK (2019a) Psychrotrophic Microbes: Biodiversity, Mechanisms of Adaptation, and Biotechnological Implications in Alleviation of Cold Stress in Plants. In: Sayyed RZ, Arora NK, Reddy MS (eds) Plant Growth Promoting Rhizobacteria for Sustainable Stress Management: Volume 1: Rhizobacteria in Abiotic Stress Management. Springer Singapore, Singapore, pp 219–253. https://doi.org/10.1007/978-981-13-6536-2_12

Yadav AN, Mishra S, Singh S, Gupta A (2019b) Recent Advancement in White Biotechnology Through Fungi Volume 1: Diversity and Enzymes Perspectives. Springer International Publishing, Cham

Yadav AN, Singh S, Mishra S, Gupta A (2019c) Recent advancement in white biotechnology through fungi. Volume 2: Perspective for Value-Added Products and Environments. Springer International Publishing, Cham

Yadav AN, Singh S, Mishra S, Gupta A (2019d) Recent Advancement in White Biotechnology Through Fungi. Volume 3: Perspective for Sustainable Environments. Springer International Publishing, Cham

Yang CH, Crowley DE, Borneman J, Keen NT (2001) Microbial phyllosphere populations are more complex than previously realized. PNAS 98:3889–3894

Yang F, Li W, Jorgensen HJL (2013) Transcriptional reprogramming of Wheat and the hemibiotrophic pathogen Septoria tritici during two phases of the compatible interaction. PLoS ONE 8(11):e81606

Yao H, Sun X, He C, Maitra P, Li XC, Guo LD (2019) Phyllosphere epiphytic and endophytic fungal community and network structures differ in a tropical mangrove ecosystem. Microbiome 7:57

Yu X, Lund SP, Scott RA, Greenwald JW, Records AH, Nettleton D, Lindow SE, Gross DC, Beattie GA (2013) Transcriptional responses of *Pseudomonas syringae* to growth in epiphytic versus apoplastic leaf sites. P Natl Acad Sci USA 29:E425–E434

Zak JC (2002) Implications of a leaf surface habitat for fungal community structure and function. In: Lindow SE, Hecht-Poinar EI, Elliott VJ (eds) Phyllosphere Microbiology. APS Press, St Paul, USA, pp 299–315

Zarraonaindia I, Owens SM, Weisenhorn P, West K, Hampton-Marcell J, Lax S, Bokulich NA, Mills DA, Martin G, Taghavi S et al (2015) The soil microbiome influences grapevine-associated microbiota. MBio 6:e02527–e02514

Zeilinger S, Gupta VK, Dahms TES, Silva RN, Singh HB, Upadhyay RS, Vieira Gomes E, Kin-Ming Tsui C, Chandra Nayak S (2016) Friends or foes? Emerging insights from fungal interactions with plants. FEMS Microbiol Rev. Mar 40(2): 182–207

Zeiller M, Rothballer M, Iwobi AN, Böhnel H, Gessler F, Hartmann A et al (2015) Systemic colonization of clover (Trifolium repens) by clostridium botulinum strain 2301. Front. Microbiol. 6:1207. https://doi.org/10.3389/fmicb.2015.01207

Zhang P, Zhou PP, Yu LJ (2009) An endophytic taxol-producing fungus from Taxus media, *Cladosporium cladosporioides* MD2. Curr Microbiol 59:227–232

Zhang YG, Wang HF, Alkhalifah DHM, Xiao M, Zhou XK, Liu YH et al (2018) Glycomyces anabasis sp. nov., a novel endophytic actinobacterium isolated from roots of Anabasis aphylla L. Int J Syst Evol Microbiol 68:1285–1290

Zhu WY, Zhang JL, Qin YL, Xiong ZJ, Zhang DF, Klenk HP et al (2013) Blastococcus endophyticus sp. nov., an actinobacterium isolated from Camptotheca acuminata. Int J Syst Evol Microbiol 63:3269–3273

Zonno MC, Vurro M, Luceretti S, Andolfi A, Perrone C, Evidente A (2008) Phyllostictine A, potential herbicide produced by *Phyllosticta cirsii*: *in vitro* production and toxicity. Plant Sci 175:818–828

Zuluaga AP, Vega-Arreguín JC, Fei Z, Ponnala L, Lee SJ, Matas AJ, Patev S, Fry WE, Rose JK (2016) Transcriptional dynamics of *Phytophthora infestans* during sequential stages of hemibiotrophic infection of tomato. Mol Plant Pathol 17:42–54

Chapter 6
Biofilms Forming Microbes: Diversity and Potential Application in Plant–Microbe Interaction and Plant Growth

Ajay Kumar and Joginder Singh

Abstract Global climatic change and increasing worldwide population pose challenges for crop production. The promising sustainable solution is the integration of beneficial plant–microbes integration with microbiome to improve agriculture production. Microbial biofilms have a significant role in agriculture because they increase soil fertility and promote plant growth. Bacterial quorum sensing (QS) regulated process is biofilm formation. The plant growth promoting bacteria (PGPB) or Rhizobacteria (PGPR) has the ability to increase the crop yield. PGPR-based formulations have been commercialized to enhance agricultural productivity.

Keywords Biofilm · Quorum sensing · Phytohormones · Biocontrol · Biofertilizers · Models

6.1 Introduction

Plant and its organs in natural conditions are surrounded by microbes. Several rhizobacteria, mycorrhizal fungi, protozoa, actinomycetes, and algae are colonized around the root of plants in rhizosphere. Colonizing microorganisms are dominated by bacterial population. The role of plant growth promoting bacteria (PGPB) or Rhizobacteria (PGPR), either by a direct or indirect mechanism in plant development has been cited by many researchers (Timmusk et al. 2017). Plant growth promoting bacteria (PGPB) or Rhizobacteria (PGPR) can stimulate the production of phytohormones,1-aminocyclopropane-1-carboxylate (ACC) deaminase enzyme, and secondary metabolites for the growth and development of the plant and root system. An arbuscular mycorrhizal fungus (AMF) also triggers the root hair growth in mycorrhizal plants (Yadav et al. 2015a, b, 2016; Zhang et al. 2019). Thus, the colonization of bacteria and fungus promotes the lateral roots and hairs and reduces

A. Kumar (✉)
School of Bioengineering and Biosciences, Lovely Professional University, Phagwara 144411, Punjab, India
e-mail: kumarajaybiotech@gmail.com

J. Singh
Department of Microbiology, Lovely Professional University, Phagwara, Punjab, India

A. N. Yadav et al. (eds.), *Plant Microbiomes for Sustainable Agriculture*, Sustainable Development and Biodiversity 25, https://doi.org/10.1007/978-3-030-38453-1_6

the growth of primary roots. Plant growth promoting bacteria (PGPB) or Rhizobacteria (PGPR) influences the growth of plant via nitrogen fixation to increase the crop yield, siderophores production, phosphate solubilization, osmotic stress alleviation, by effecting the auxin level in plant roots, and modify the root physiology by affecting the gene transcription (Vacheron et al. 2013; Verma et al. 2016; Yadav 2017, 2019).

PGPR such as *Bacillus amyloliquefaciens, Bacillus* megaterium, and *Bacillus subtilis* are able to produce phytohormones such as auxin (IAA) or promote growth under phosphate limitation by excreting phytase (*EC* 3.1.3.8) in the presence of phytate. The main hypothesis to explain colonization efficiency by bacterial and fungal strains is influenced by the chemotaxis effect. The proposed mechanism of root colonization by bacterial and fungal strains depends on the abilities (i) to move toward the place for root colonization, (ii) to use carbon and nitrogen sources present in the soil, (ii) provided by root exudates to shape the root microbiome, (iii) to withstand plant response reaction including reactive oxygen species (ROS), and (iv) to form a biofilm or microcolonies formation at the root surface. Bacillus sp. Strains (*Bacillus subtilis, Bacillus licheniformis, Bacillus amyloliquefaciens, Bacillus cereus*, etc.) are able to produce a set of secondary metabolites such as surfactin, fengycin, and iturin A, which belong to cyclic lipopeptides (CLPs) members and exhibit antifungal activities and inhibit the growth of several plant pathogens (Al-Ali et al. 2018). The impact of phyto stimulating PGPR on root architecture is shown in Fig. 6.1.

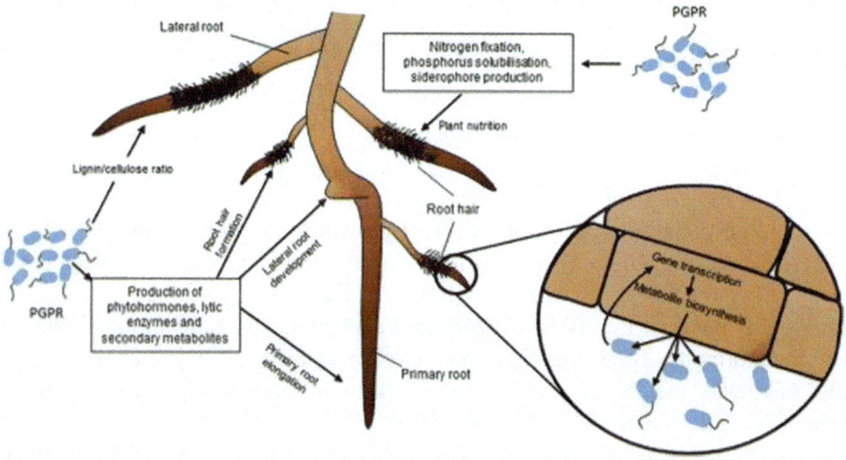

Fig. 6.1 Impact of phyto stimulating PGPR on RSA (root system architecture), nutrient acquisition, and root functioning. *Sources* Vacheron et al. (2013)

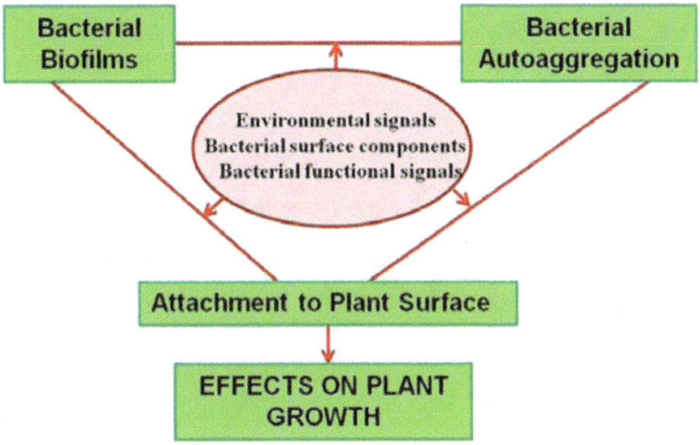

Fig. 6.2 Bacterial auto aggregation and biofilm development, and their relationship with plant colonization. *Sources* Bogino et al. (2013)

6.2 Biofilm Formation by PGPR

It has been observed that most of the bacteria develop bacterial biofilm attached to the surface. Biofilm is made up of protein, lipid, polysaccharide, and DNA in a self-produced extracellular matrix and is found on various surfaces such as soil, potable water system, and living tissues (Singh et al. 2017). It is a secreted extracellular matrix that holds cells together which is common to all bacterial biofilms and provides robustness to biofilm architecture (Molina-Santiago et al. 2019). Several factors such as environmental signals (Cyclic di-AMP act as extracellular signal), nutrient limitation of growth (Iron limitation), quorum sensing (QS) signal molecules such as acylated homoserine lactones(AHLs), exopolysaccharides (EPSs),bacterial surface components particularly flagella, lipopolysaccharides (LPSs), and other factors regulate the cell aggregation and biofilm formation in plant–bacterial associations (Primo et al. 2019; Bogino et al. 2013). Biofilm formation by PGPR and colonization with plant are demonstrated in Fig. 6. 2.

6.3 Biofilm Formation Steps by Bacteria

Biofilm formation is a complex and dynamic process (D'Acunto et al. 2017), formed by several steps (i) adhesion of planktonic cell to support surface (ii) formation of monolayer and cell proliferation (iii) microcolonies formation (iv) macrocolonies formation (v) development of matured biofilm (vi) detachment or dispersal of bacterial cells. Figure 6.3, represents the stages of the biofilm process and Table 6.1, shows some important microorganisms forming biofilms.

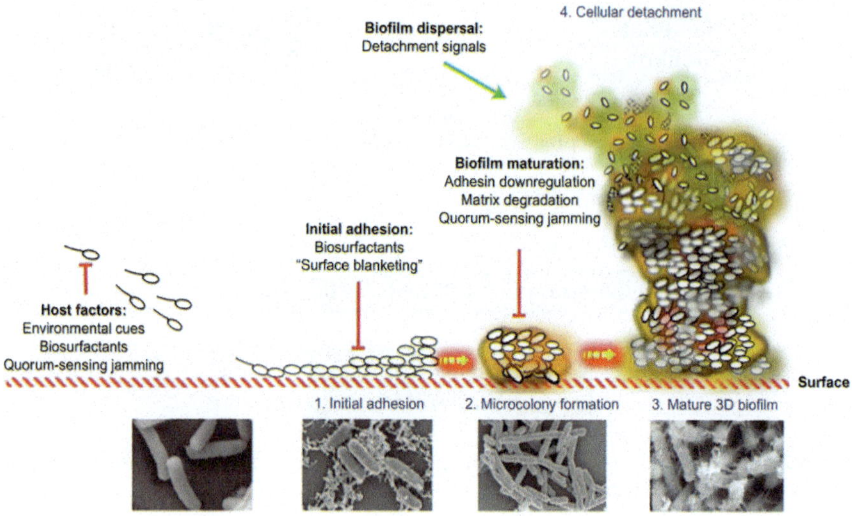

Fig. 6.3 Stages in the biofilm formation process, including scanning electron microscopy imaging of each stage. *Source* Hadla and Halabi (2018)

6.4 Factors Affecting Biofilm Formation

Several factors have been investigated which affects the biofilm formation such as temperature of the surroundings, pH of the soil, nutrient availability, minerals, flow of the fluid, plant defense compounds such as phenazine, surfactants such as cyclic lipopeptides (CLPs), bacterial strains, EPS, flagella, gene expression, Quorum sensing, surface characteristics (hydrophobicity, roughness, wettability), organic fouling, and presence of other microbes (Velmourougane et al. 2017).

Mass transport characteristics have also influenced the biofilm's density (Fysun et al. 2019). The rate of nutrient transport to the biofilm is determined by the rates of convective external mass transfer $[k_l(C_b - C_s)]$ and internal mass transfer $(D_f \frac{dC}{dZ})$ be equal at biofilm surfaces (Beyenal and Lewandowski 2002) as shown in Eq. (1).

$$N_s = k_l(C_b - C_s) = D_f \frac{dC}{dZ} \quad (1)$$

where N_s, the nutrient flux.

k_l, the external mass transfer coefficient
C_b, the substrate concentration at bulk phase
C_s, the substrate concentration at surface of the biofilm
D_f, the effective diffusivity in the biofilm.

Factors affecting biofilm formation are shown in Fig. 6.4.

Table 6.1 Biofilm formation by agriculturally important microorganisms

Microbes	Attributes	References
Trichoderma–Azotobacter	Improves soil nutrient availability and plant growth	Velmourougane et al. (2019a,b)
B. amyloliquifaciens, B. licheniformis, B. megaterium, B. pumilus, B. subtilis,	Promotes plant growth and development	Tiwari et al. (2019)
Anabaena–Azotobacter	Enhanced the availability of nitrogen, phosphorus, and micronutrients in the soil	Kanchan et al. (2019)
Azotobacter chroococcum (Az)—Trichoderma viride (Tv)	Plant growth promotion and biocontrol	Velmourougane et al. (2019a, b)
Agrobacteriumtumefaciens	Tumor formation	Choi et al. (2018)
Azospirillum brasilense	Fix atmospheric nitrogen and synthesize phytohormones such as indole-3-acetic acid (IAA)	Jijón-Moreno et al. (2019)
Rhizobium leguminosarum	Nitrogen fixation	Shahid et al. (2019)
Pseudomonas sp	Plant growth promotion activity, such as synthesis of phytohormones, phosphorus solubilization, production of ammonia, HCN, and siderophore	Singh et al. (2019)
Pseudomonas aeruginosa-Bradyrhizobium sp.	Improving plant growth, nutrient acquisition, and soil health	Kumawat et al. (2019)
Gluconacetobacter diazotrophicus	Plant growth and protect the plants against abiotic stresses	Filgueiras et al. (2019)
Herbaspirillum huttiense, Enterobacter asburiae, and *Staphylococcus sp.*	Nitrogen fixation	Andreozzi et al. 2019
Sinorhizobium meliloti	Fix atmospheric nitrogen	Primo et al. (2019)

Source Besset-Manzoni et al. (2018)

6.5 Mathematical Model of Biofilm Formation

A mathematical model of biofilm formation developed by Eberl et al. (2001) which consists of a set of nonlinear density dependent reaction–diffusion equation which is applicable for single species QS-regulated biofilm system.

$$\frac{\partial C(t)}{\partial t} = D_c \Delta C - \frac{k_1 C M}{k_2 + C} \tag{2}$$

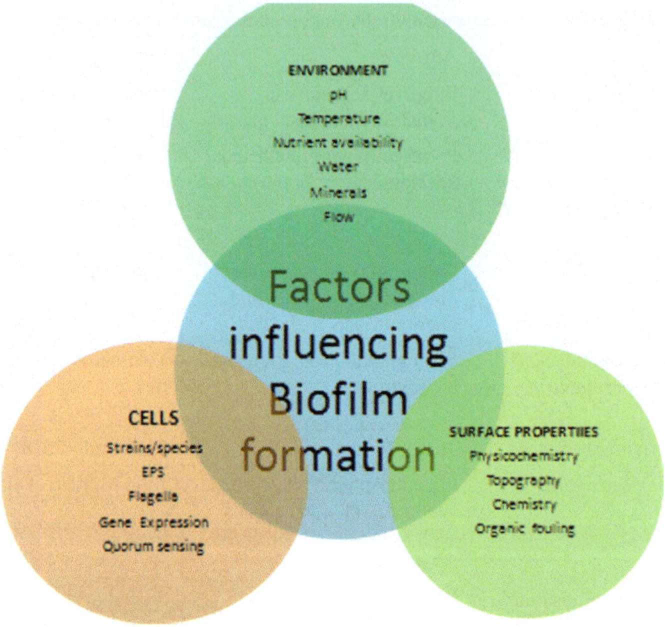

Fig. 6.4 A complex interplay of factors results in biofilm formation, architecture, and hence functionality. *Source* Whitehead and Verran (2015)

$$\frac{\partial M(t)}{\partial t} = \nabla.(D_M(M)\nabla M) + M\left(\frac{k_3 C}{k_2 + C} - k_4\right) \tag{3}$$

$$D_M(M) = \delta \frac{M^a}{(1-M)^b} \tag{4}$$

where D_C, D_M, k_1, k_2, k_3, k_4 are positive constant parameters and $a > 1$, $b > 1$. M denotes biomass density, δ is biomass mobility coefficient, and C is the growth-limiting substrate.

Sometimes the phenomenon of floating biofilm (Fig. 6.5) is observed when a microbial aggregate does not attach to the surface. Fickian diffusion equation is used to describe the phenomenon of dispersed cell diffusion within the biofilm and bulk liquid. The growth is modeled by a hyperbolic partial differential equation and diffusion of dispersed cells by a parabolic partial differential equation. The transport of substrates from the bulk liquid to the biofilm is achieved by diffusion. The role of signaling molecule such as Nitric oxide (NO) in the regulation of dispersal in biofilm has been demonstrated by many researchers (D'Acunto et al. 2019; Zhu et al. 2019). Biofilm formation

$$\frac{\partial X_i}{\partial t} + u\frac{\partial X_i}{\partial z} = \rho_i R_{Mi}(z, t, X, S, \psi) \tag{5}$$

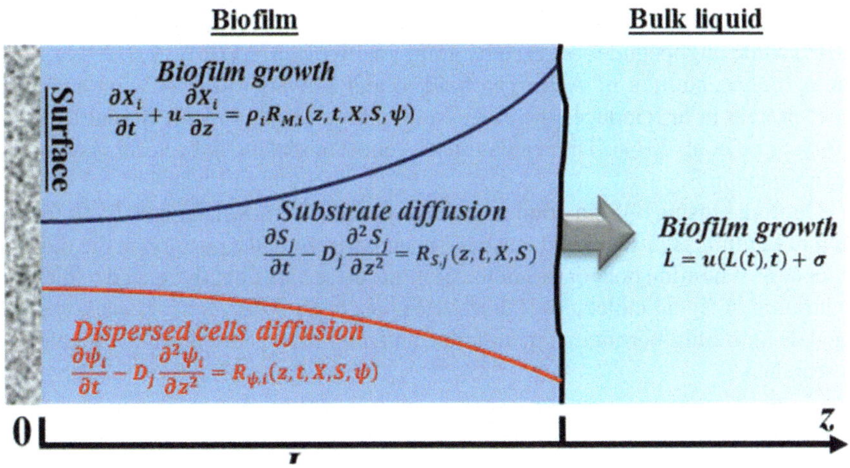

Fig. 6.5 Phenomenon of biofilm growth. *Source* D'Acunto et al. (2019)

Substrate diffusion

$$\frac{\partial S_j}{\partial t} - D_j \frac{\partial^2 S_j}{\partial z^2} = R_{S,j}(z, t, X, S) \tag{6}$$

Dispersed cell diffusion

$$\frac{\partial \psi_i}{\partial t} - D_j \frac{\partial^2 \psi_i}{\partial z^2} = R_{\psi,i}(z, t, X, S, \psi) \tag{7}$$

Biofilm Growth

$$L = u(L(t).t) + \sigma \tag{8}$$

where concentration of dispersed cells is defined by the variables $\psi_{,i}(z, t)$, X_i is biomass concentration, S_j is substrate concentration, L is biofilm thickness, ρ_i is biofilm density, D is diffusion coefficient, where $D_{M,i}$ denotes the diffusivity coefficient of planktonic species i. $R\psi$, $i(z, t, \mathbf{X}, \mathbf{S})$ is the species i conversion rate into the planktonic state, u is superficial velocity, and σ is biomass flux between biofilm and bulk solution.

6.6 Mechanism of Quorum Sensing

Quorum sensing (QS) is a bacterial communication mechanism which is achieved through diffusible signal molecules that enables biofilm formation. Small amounts of chemical signaling molecules (autoinducers) such as N-Acyl Homoserine Lactones

(AHL) are produced and released by gram-negative bacterial cells (Emerenini et al. 2015) while oligopeptides are used by gram-positive bacteria (Roy et al. 2011). High enough concentration of AHL, can bind to and activate a transcription activator, or R protein in turn induces the expression of target gene (De Kievit and Iglewski 2000; Saraf et al. 2014). The regulation of a quorum sensing in bacteria systems are represented in Fig. 6.6.

Quorum sensing (QS) in gram-negative organisms is shown in Fig. 6.7. As formation of biofilm is QS-regulated mechanism and therefore, the position in the biofilm, thickness of biofilm, boundaries of the biofilm surface are directly related to the concentration of signal molecules (Pérez-Velázquez et al. 2016). Several mathematical models for biofilm formation, maturation, and dissolution have been investigated by researchers.

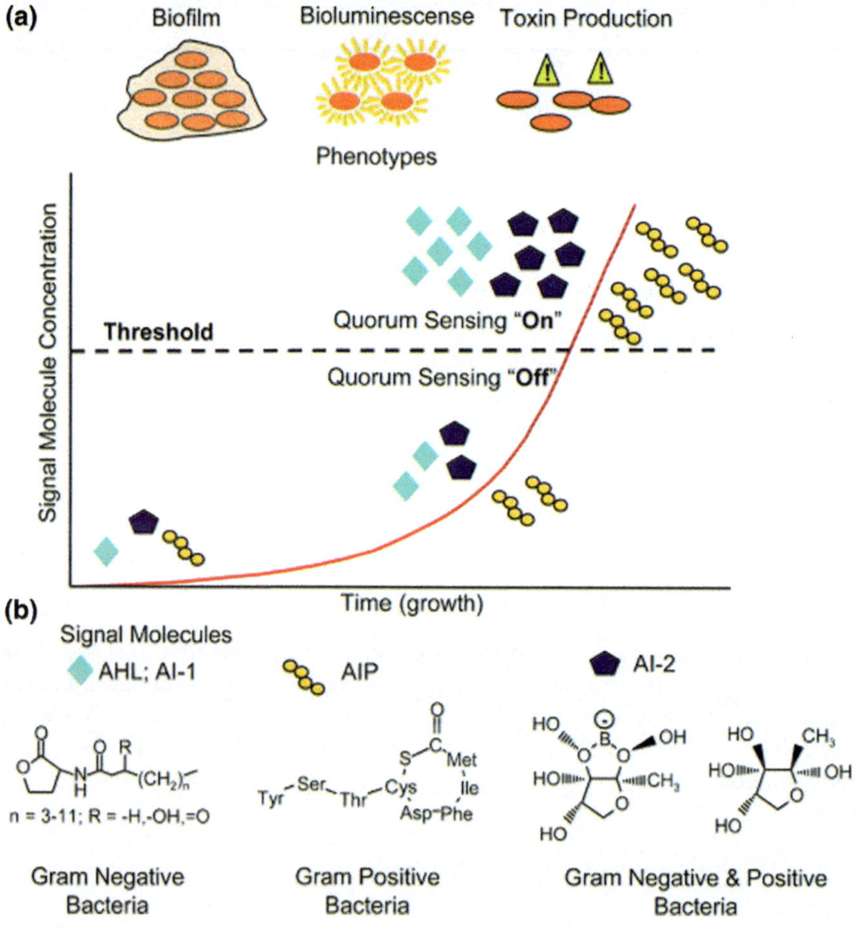

Fig. 6.6 Schematic representation of the triggering of a quorum sensing system (**a**) and the main signaling molecules involved in Gram-positive and Gram-negative bacteria (**b**). *Source* Roy et al. (2011)

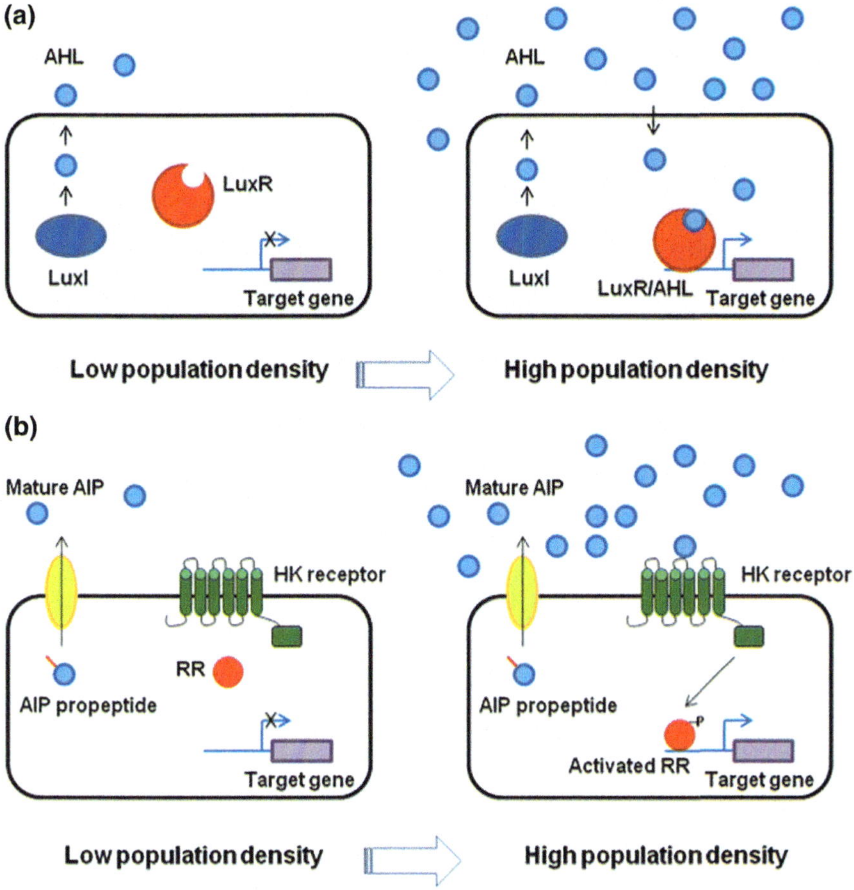

Fig. 6.7 Quorum sensing in gram-negative (**a**) and gram-positive organisms (**b**). *Source* Choudhary and Schmidt-Dannert (2010)

6.7 Applications of Biofilm

Different applications of biofilms have been investigated by investigators in plant–microbe interaction and plant growth as represented in Fig. 6.8 (Velmourougane et al. 2017).

6.7.1 Biofertilizers

Some of the plant growth promoting bacteria (PGPB) or Rhizobacteria (PGPR) are classified as biofertilizers and biocontrol agents or biopesticides. Biofertilizer is a

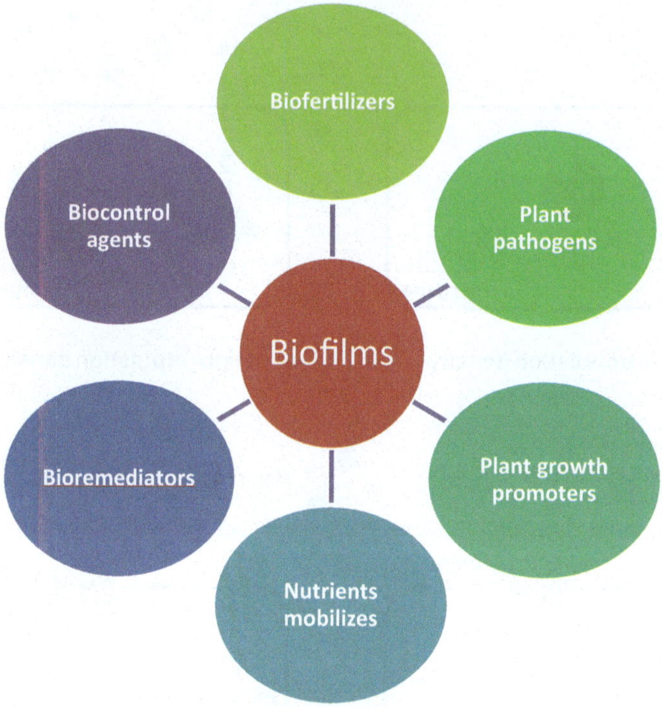

Fig. 6.8 Overview of the significance of biofilms in agriculture. *Source* Velmourougane et al. (2017)

group of beneficial bacteria such as Rhizobacteria enhancing the productivity of the soil (Olanrewaju et al. 2019; Yadav et al. 2017a, 2019a; Yadav and Yadav 2019). Microbial inoculants used in agricultural crops are shown in Fig. 6.9.

Microorganisms sre involved in the production of biofertilizer.

The following are the types of biofertilizer (Barman et al. 2019)

- Nitrogen-fixing biofertilizers (*Rhizobium, Brady rhizobium, Azospirillum, Beijerinckia, Clostridium*, and *Azotobacter*).
- Phosphorus-solubilizing biofertilizers (*Bacillus, Penicillium, Pseudomonas, Xanthomonas Fusarium*, and *Aspergillus*).
- Phosphate-mobilizing biofertilizer (Arbuscular Mycorrhiza, Ectomycorrhiza, Orchid Mycorrhiza).
- Plant growth promoting biofertilizer *Rhizobium, Pseudomonas*, *Azotobacter, Azospirillum*).

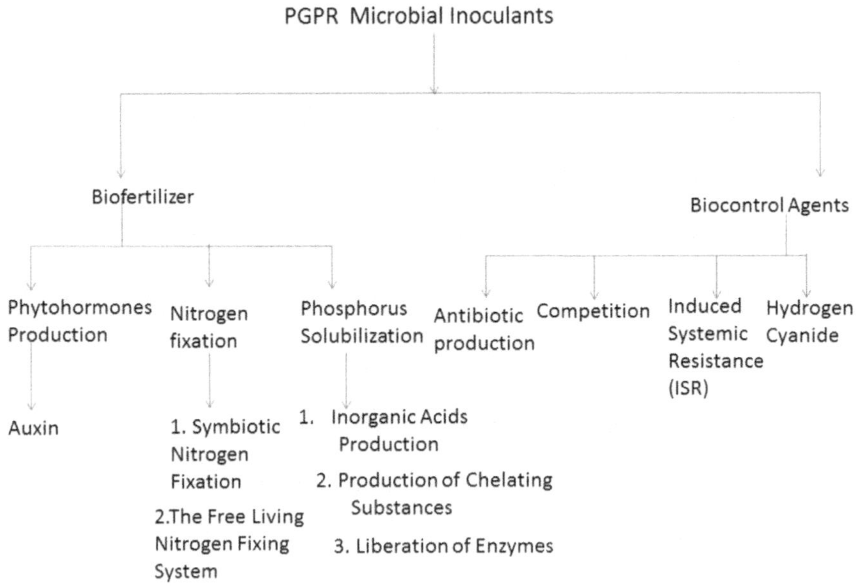

Fig. 6.9 Schematic representation of some importance of microbial inoculant in agriculture and the mechanism of actions. *Source* Alori and Babalola (2018)

6.7.2 Biocontrol Agents

The main function of biocontrol agents (BCA) or biopesticides such as *B. thuringiensis, B. popilliae, P. Fluorescens* is suppressing or controlling plant disease by acting as antifungal or antimicrobial agents (Alori and Babalola 2018; Glick 2012). *Pseudomonas* spp. produces metabolites that include 2,4-diacetylphloroglucinol (2,4-DAPG) [$C_{10}H_{10}O_5$, Molecular Weight: 210.18 g/mol], phenazine (PHZ) [$C_{12}H_8N_2$, Molecular Weight: 180.2 g/mol], pyrrolnitrin (PRN) [$C_{10}H_6Cl_2N_2O_2$, Molecular Weight: 257.07 g/mol], pyoluteorin (PLT) [$C_{11}H_7Cl_2NO_3$ Molecular Weight: 272.08 g/mol], hydrogen cyanide [HCN, Molecular Weight: 27.025 g/mol), and 2-hexyl-5-propyl resorcinol (HPR) [$C_{15}H_{24}O_2$, Molecular Weight: 236.35 g/mol]. It has been observed that 2-hexyl-5-propyl resorcinol (HPR) pay a key role in biofilm formation (Calderon et al. 2019). Structures of microbial metabolites and their interaction with plants are shown in Fig. 6.10.

6.7.3 Plant Growth Promoters

Several factors such as abiotic factors and biotic factors influence the growth and yield of the crop plants (Fig. 6.11). Abiotic factors of soil include moisture, pH, salt concentration, nutrient, temperature, humidity, heavy metals, antibiotics, etc. Biotic

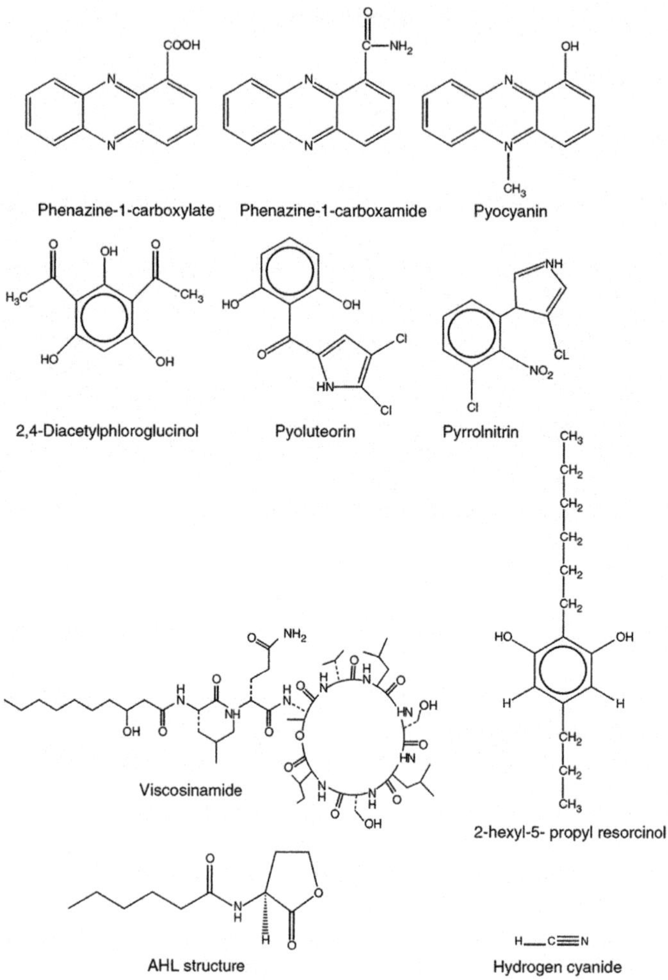

Fig. 6.10 Structures of important metabolites which play a role in the interaction between plant beneficial bacteria and plants. *Source* Pliego et al. (2011)

factors such as plant genotype and soil macro and microflora including soil bacteria, viruses, and protozoans are contributed for the regulation of bacterial populations in the rhizosphere (Ansari et al. 2017; Kumar et al. 2019a, b; Yadav 2018; Yadav et al. 2019b).

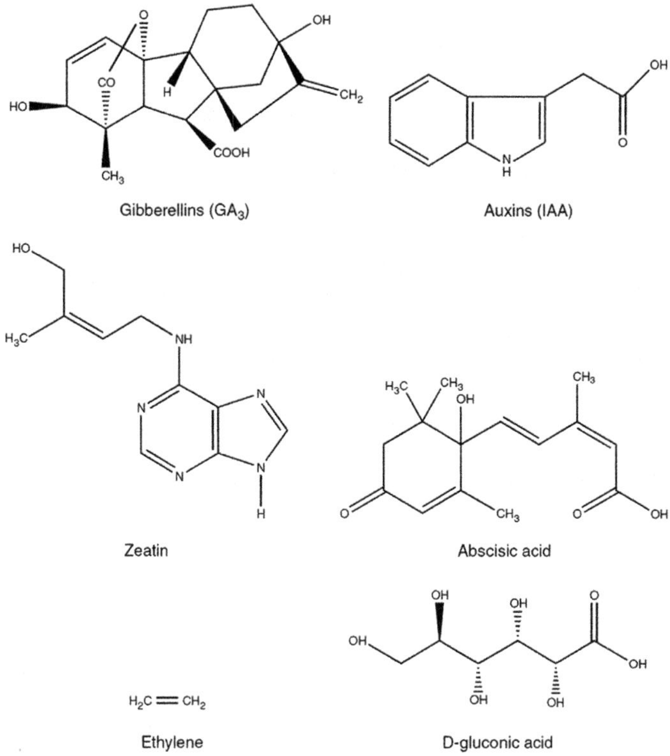

Fig. 6.10 (continued)

6.7.3.1 Direct Mechanism of Plant Growth

Nitrogen Fixation

Biofilm formation enables free living nitrogen-fixing rhizobacteria such as *Rhizobium*, *Gluconacetobacter*, *Pseudomonas*, and *Azospirillum* to fix nitrogen under aerobic conditions with the help of *nif* gene (Yadav et al. 2017a, b, c; Wang et al. 2017). The rhizospheric nitrogen-fixing bacteria have the potential to increase the productivity of the agricultural crops such as rice, wheat, and corn (Suman et al. 2016; Verma et al. 2017a, b).

The biological nitrogen fixation processes are shown in Fig. 6.12.

Phosphorus Solubilisation

Biofilm forming phosphate-solubilizing bacteria (PSB) such as *Bacillus, Erwinia, Pseudomonas, Burkholderia,* and *Rhizobium* are solubilizing the inorganic phosphorus (Pi) released by root exudates (Taktek et al. 2017). Phosphorus solubilization by bacteria is a complex mechanism (Fig. 6.13). Production of organic acid such as gluconic acid (GA) is considered as the primary mechanism of phosphate solubilization

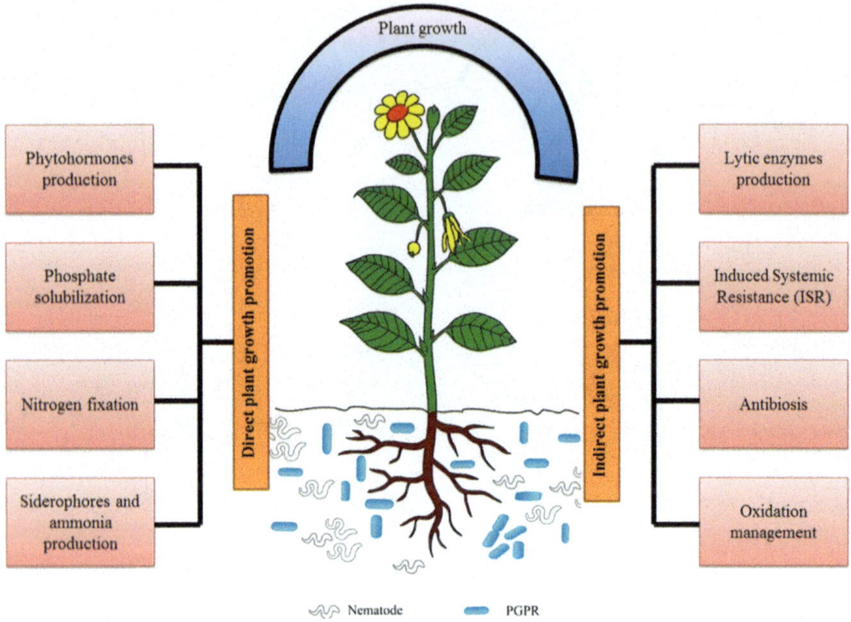

Fig. 6.11 Schematic diagram represents the plant growth promoting and nematicidal activity of PGPR. *Source* Mhatre et al. (2018)

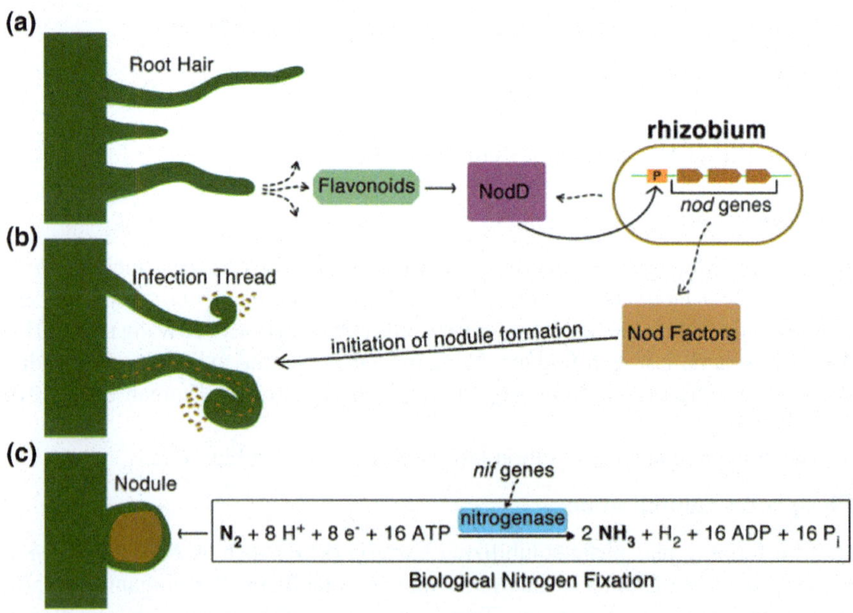

Fig. 6.12 Schematic overview of the nodulation process and biological nitrogen fixation. *Source* Laranjo et al. (2014)

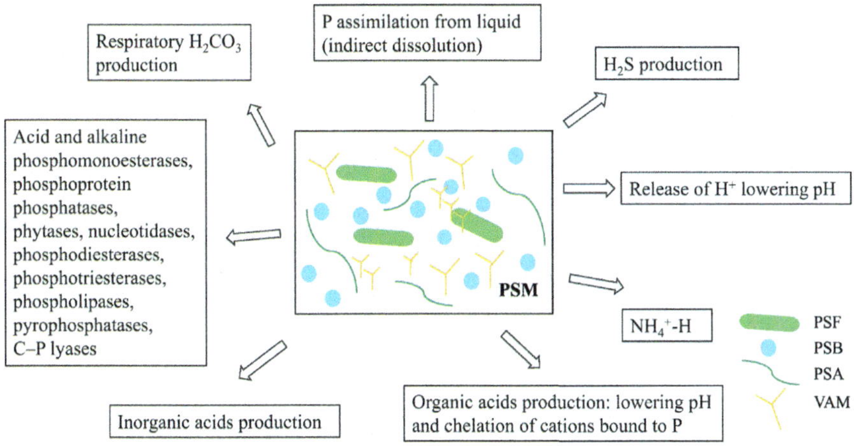

Fig. 6.13 Potential mechanisms for the solubilization of insoluble Phosphate by phosphate-solubilizing microorganisms (PSMs). PSF: phosphate-solubilizing fungi; PSB: phosphate-solubilizing bacteria; PSA: phosphate-solubilizing actinomycete; VAM: vesicular-arbuscular mycorrhizae. *Source* Zhu et al. (2018)

by lowing the pH of surroundings (Liu et al. 2019; Prabhu et al. 2019; Yadav et al. 2015c).

Siderophores Production

Siderophores are ferric ion-specific chelating organic compounds that areproduced under low iron stress conditions. Several microorganisms such as *Pseudomonas, Bacillus,* and *Beauveria* spp, are reported to secrete siderophores (Verma et al. 2016; Yadav et al. 2015a, b; Yadav et al. 2016). The primary function of siderophores is to chelate the ferric iron [Fe(III)]. Siderophore iron complex is adsorbed by plants to meet their iron demands. (Ahmed and Holmström 2014; Pedraza 2015; Rajkumar et al. 2010). Figure 6.14 shows the role of siderophores producing bacteria (SPB) in chelating the heavy metal.

Biocontrol Activity

Due to the microbial diseases and their harmful effects, about one third of the crop yields are lost every year. Harmful activity of one organism is limited by the application of other microorganisms during biocontrol (Kumar et al. 2019a, b; Rana et al. 2019; Yadav et al. 2019b). *P. fluorescens* contains toxic producing genes from *B. thuringiensis* (Bt) and is used as a biopesticide to limit the activity of black cutworms on maize. Biocontrol activity are mediated by the synthesis of bacterial allelochemicals (phenazines, pyoluteorin, 2, 4-diacetylphloroglucinol, etc.) including iron-chelating siderophores, antibiotics (Streptomycin, Cephamycin C, Tetracyclines,

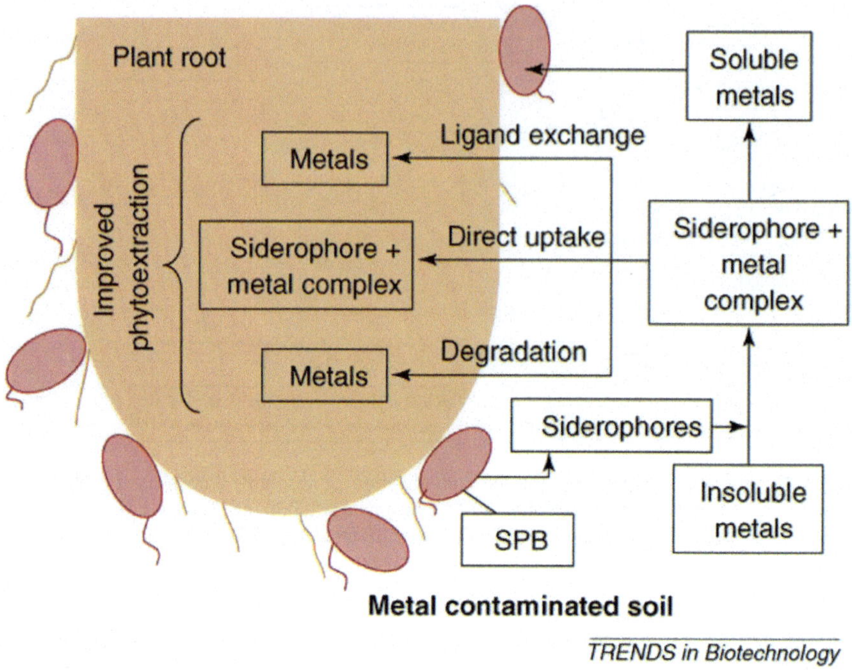

Fig. 6.14 Role of SPB in phytoextraction of heavy metal contaminated soils. *Source* Rajkumar et al. (2010)

Chloramphenicol, etc.),biocidal volatiles (Dimethyldisulfide, Furfural, Benzaldehyde, etc.) lytic enzymes (Glucanases, chitinases, pectinase, etc.), and detoxification enzymes (Pandin et al. 2017; Compant et al. 2005). Plant growth promoting bacteria (PGPB) produces hydrogen cyanide and antibiotics to control pathogens. Microbial compounds used in biocontrol activities are tabulated in Table 6.2 and mechanisms of interactions by plant growth promoting bacteria (PGPB) in the rhizosphere are demonstrated in Fig. 6.15.

6.8 Commercialization of PGPR

Gram-positive bacteria such as *Bacillus* spp. are preferred as commercially available PGPR because of inoculant stability and ease of storage of inoculant product. Other potential PGPR stains such as *Azospirillum, Pseudomonas* do not produce spores and hence difficult to formulate as a biocontrol agent. In the global market, United States, China, India, Russia, and Australia are the main players for biocontrol product as the prospects of PGPR for agricultural crop yield enhancement and disease production

Table 6.2 Example of microbial compounds implicated in biocontrol and microbial cooperation

	Compounds	Signalling activity	Biocontrol activity	References
QS MEDIATOR	<u>Acyl homoserine lactones (AHLs)</u> 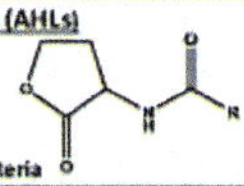 Producer: Gram-negative bacteria	· Cell division · Mobility · Biofilm production · Virulence · Luminescence · Metabolite production	**Regulation of antibiotic production** (e.g phenazine) **Promote ISR**	Venturi and Fuqua 2013
	<u>Diffusible fatty acid (DSF)</u> 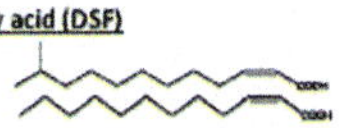 **Producers:** ***Xanthomonas campestris; Burkolderia cenocepacia; Stenotrophomonas maltophilia***	· Cell division · Mobility · Biofilm production · Virulence · Luminescence · Metabolite production	***S. maltophilia*** **biocontrol properties**	Venturi and Keel 2016
QQ MICROBIAL MEDIATOR	<u>Enzyme:</u> · Lactonase · Reductase · Amidase · Oxidase <u>QQ enzyme mechanism:</u> 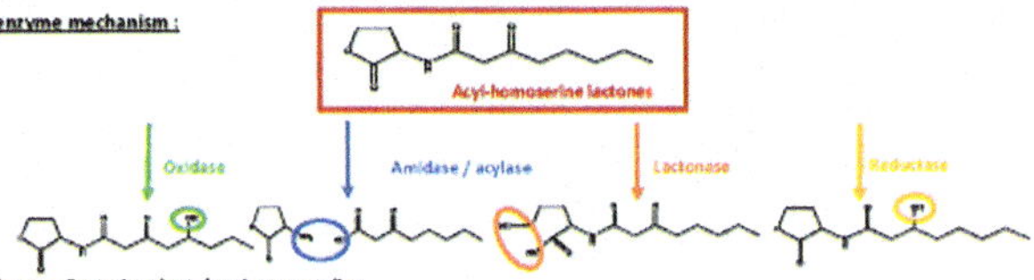 **Producers: Bacteria, plant, fungi, mammalian**	<u>**AHL degradation**</u>: modification of bacterial behaviour of both pathogen and PGPR strains. ⇒ Involved in biocontrol and cell-cell signalling		Grandclément et al. 2016
ANTIBIOTICS	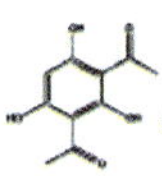 DAPG: 2,4-diacetylphloroglucinol **Producer: fluorescent *Pseudomonas***	· Modification of metabolite production · Biofilm formation · Plant growth	**Antimicrobial activity against:** · Bacteria · Fungi · Viruses · Helminths **Promote ISR**	Jousset et al. 2010 Combes-Meynet et al. 2011
	 Phenazine nucleus *<u>Example of phenazine compounds</u>:* 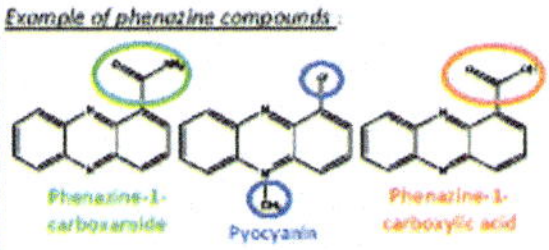 **Producers: *Pseudomonas sp*; *Streptomyces sp***	· Control of bacterial colony size · Biofilm formation · Iron chelation	**Antimicrobial activity against:** · Bacteria · Fungi *Example:* · *Gaeumannomyces graminis* · *Fusarium oxysporum* **Promote ISR**	Price-Whelan et al. 2006
LIPOPEPTIDES	**Cyclolipopeptide** 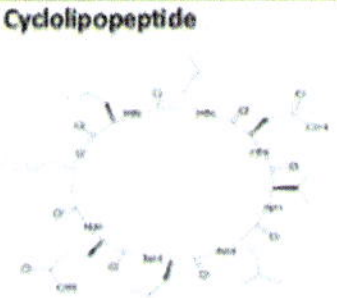 <u>Example of lipopeptides from soil bacteria:</u> · <u>***Pseudomonas***</u>: massetolide A, putisolvin I and II, amphasin, · <u>***Bacillus***</u>: surfactin	· Mobility, swarming · Biofilm formation and maturation	**Antimicrobial activity against:** · Bacteria · Fungi · Oomycete · Viruses · Protozoa *Example:* · *P. syringue* · *P. infestans*	Raaijmakers et al. 2010 Ongena and Jacques 2008 Tran et al. 2007 De Bruijn et al. 2008
VOCs	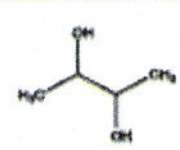 **2,3 butanediol** 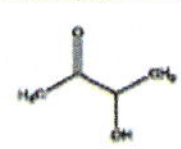 **Acetoin**	· Biofilm production · Motility · Plant growth stimulation	**Promote ISR** *Example: protect plants against P. carotovorum* May increase phenazine production in the rhizosphere *Pseudomonas*.	Ryu 2004 Audrain et al. 2015
	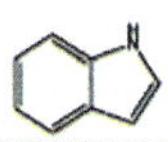 **Indole and its derivatives**	· Biofilm production · Quorum quenching · Motility	**Promote ISR** Modify metabolite biosynthesis in host plants	Lee et al. 2015
	NO **Nitric oxide**	· Biofilm dispersion · Biofilm formation · Plant growth stimulation	**Antimicrobial property** Promote lateral root formation by *A. brasilense*	Audrain et al. 2015

Source Besset-Manzoni et al. (2018)

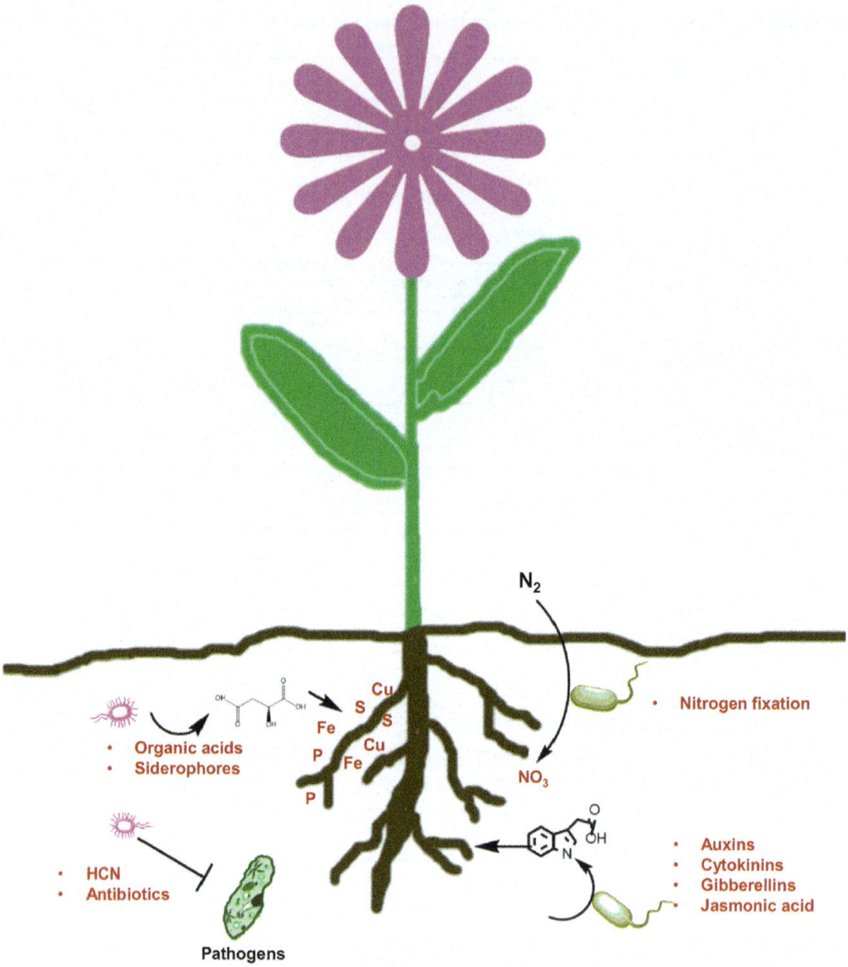

Fig. 6.15 Mechanisms used by plant growth promoting bacteria (PGPB) to enhance interactions in the rhizosphere for higher grain yield and nutrient content. PGPB produce plant hormones, siderophores, and organic acids and solubilize phosphate. They produce hydrogen cyanide and antibiotics to control pathogens. *Source* Ramakrishna et al. (2019)

can never be ignored (Tabassum et al. 2017; Kour et al. 2019a, b, c). The biopesticide formulation technology by industry is shown in Fig. 6.16 and PGPR-based commercialized bioformulation in Table 6.3.

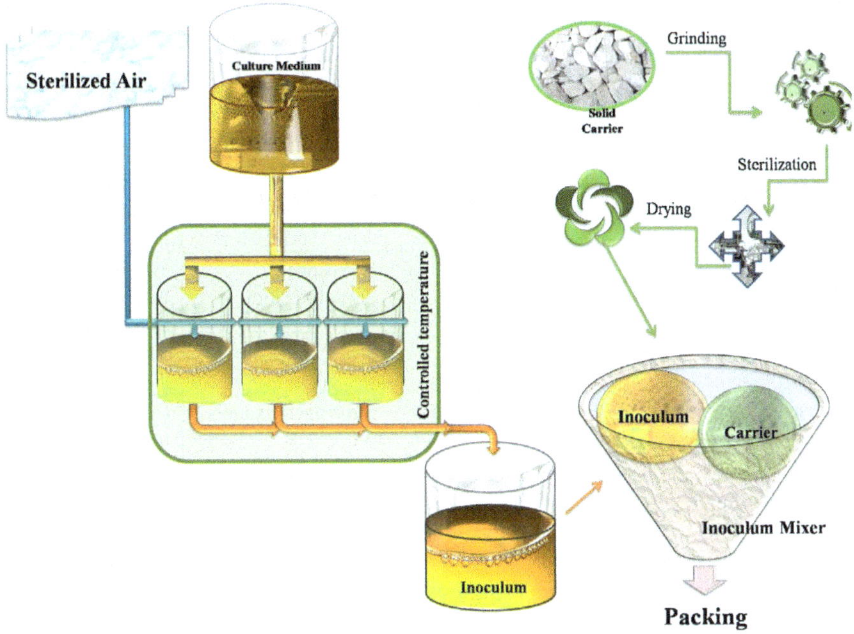

Fig. 6.16 Generalized scheme of the biopesticide formulation by industry where PGPRs are preserved in an appropriate carrier molecule and packaged for commercial application at farmer's end. Several PGPR formulations are commercially available in the market by different commercial producers and different Government institutions. *Source* Tabassum et al. (2017)

6.9 Conclusions and Future Prospects

Biofilm forming bacteria are growing on and around the plants and show the significant plant–microbes interaction to enhance the yield of the agricultural crops. Cell to cell communication and quorum sensing (QS) play a critical role in the formation of biofilm. Biofilms can provide protection from plant responses and thus promoting colonization. The Phyto microbiome offers enormous potentials for agricultural benefits. For the enhancement of the crop yield, researchers are looking for the development of mathematical model based customized inocula of PGPB/PGRP. Thus, the sustainable use of microbial inoculants is a viable alternative for enhancing crop production rather than the use of agrochemicals such as insecticides, pesticides, and inorganic fertilizers. In the coming future, the microbial inoculant technology will ensure sustainable crop production.

Table 6.3 PGPR-based commercialized bioformulation

Product Name	Composition	Country	Benefit/target pathogen/disease	Applications
Bio-phospho	*Bacillus Subtilis*	India	Phosphate solubilization	Wheat, jowar, rice, sorghum, maize, sugarcane, cotton
Rhizobium bio promotor	*Rhizobium sp.*	India	Nitrogen fixation	Leguminous plants
Symbion-P	*Bacillus megaterium var. phosphaticum*	India	Phosphate solubilization	Wheat, rice, teff, barley, maize, corn (sweet and baby corn), sorghum, pearl millets
Azospirillum Inoculant	*Azospirillum*	India	Nitrogen fixation	Crops, cereals, millets
Paddy Azospirillum	*Azospirillum*	India	Nitrogen Fixation	Paddy crops
Actinovate AG	*Streptomyces lydicus* strain WYEC 108	USA	Soil borne diseases like powdery mildew, downy	Fruit and vegetable crops
Symbion-K	*Frateuria aurantia*	India	k-solubilizing	Wheat, rice, teff, Barley, maize, corn (sweet and baby corn), sorghum, pearl millets
Symbion-N	*Rhizobium, Azospirillum, Acetobactor,* and *Azotobactor*	India	Nitrogen fixation	Legumes and pulses
TerraMax's MicroAZ-ST dry	*Azospirillum brasilense* and *lipoferum*	USA	Nitrogen fixation	Corn
NPK liquid	*Azotobacter chroocomccum, Pseudomonas*	India	Nitrogen-fixing, P-solubilizing, k-solubilizing	All crops

References

Ahmed E, Holmström SJ (2014) Siderophores in environmental research: roles and applications. Microb Biotech 7:196–208

Alori ET, Babalola OO (2018) Microbial inoculants for improve crop quality and human health. Front Microbiol 9:2213

Al-Ali A, Deravel J, Krier F, Béchet M, Ongena M, Jacques P (2018) Biofilm formation is determinant in tomato rhizosphere colonization by *Bacillus velezensis* FZB42. Environ Sci Poll Res 25:29910–29920

Andreozzi A, Prieto P, Mercado-Blanco J, Monaco S, Zampieri E, Romano S, Valè G, Defez R, Bianco C (2019) Efficient colonization of the endophytes Herbaspirillum huttiense RCA24 and enterobacter cloacae RCA25 influences the physiological parameters of Oryza sativa L. cv. Baldo rice. Environ Microbiol. https://doi.org/10.1111/1462-2920.14688

Ansari FA, Jafri H, Ahmad I, Abulreesh HH (2017) Factors Affecting Biofilm Formation in in vitro and in the Rhizosphere. In: Ahmad I, Husain FM (eds) Biofilms in plant and soil health. Wiley, Hoboken, USA, p 275

Barman S, Das S, Bhattacharya SS (2019) The prospects of bio-fertilizer technology for productive and sustainable agricultural growth. In: Singh JS, Singh DP (eds) New and future developments in microbial biotechnology and bioengineering. Elsevier Radarweg, Amsterdam, the Netherlands, pp 233–253

Besset-Manzoni Y, Rieusset L, Joly P, Comte G, Prigent-Combaret C (2018) Exploiting rhizosphere microbial cooperation for developing sustainable agriculture strategies. Environ Sci Poll Res 25:29953–29970

Beyenal H, Lewandowski Z (2002) Internal and external mass transfer in biofilms grown at various flow velocities. Biotech Progress 18:55–61

Bogino PC, Oliva MD, Sorroche FG, Giordano W (2013) The role of bacterial biofilms and surface components in plant-bacterial associations. Int J Mol Sci 14:15838–15859

Calderon CE, Tienda S, Heredia Z, Diez EM, Cárcamo-Oyarce G, Eberl L, Cazorla FM (2019) The compound 2-hexyl, 5-propyl resorcinol has a key role in biofilm formation by the biocontrol rhizobacterium Pseudomonas chlororaphis PCL1606. Front Microbiol 10:396

Choi O, Kang DW, Cho SK, Lee Y, Kang B, Bae J, Kim S, Lee JH, Lee SE, Kim J (2018) Anti-quorum sensing and anti-biofilm formation activities of plant extracts from South Korea. Asian Pacific J Trop Biomed 8:411

Choudhary S, Schmidt-Dannert C (2010) Applications of quorum sensing in biotechnology. Appl Microbiol Biotech 86:1267–1279

Compant S, Duffy B, Nowak J, Clément C, Barka EA (2005) Use of plant growth-promoting bacteria for biocontrol of plant diseases: principles, mechanisms of action, and future prospects. Appl Environ Microbiol 71:4951–4959

D'Acunto B, Frunzo L, Mattei MR (2017) Continuum approach to mathematical modelling of multispecies biofilms. Ricerche mat 66:153–169

D'Acunto B, Frunzo L, Klapper I, Mattei MR, Stoodley P (2019) Mathematical modeling of dispersal phenomenon in biofilms. Math Biosci 307:70–87

De Kievit TR, Iglewski BH (2000) Bacterial quorum sensing in pathogenic relationships. Infect Immun 68:4839–4849

Eberl HJ, Parker DF, Van Loosdrecht M (2001) A new deterministic spatio-temporal continuum model for biofilm development. Comput Math Methods Med 3:161–175

Emerenini BO, Hense BA, Kuttler C, Eberl HJ (2015) A mathematical model of quorum sensing induced biofilm detachment. PLoS ONE 10:e0132385

Filgueiras L, Silva R, Almeida I, Vidal M, Baldani JI, Meneses CH (2019) *Gluconacetobacter diazotrophicus* mitigates drought stress in *Oryza sativa* L. Plant Soil 1–7

Fysun O, Kern H, Wilke B, Langowski HC (2019) Evaluation of factors influencing dairy biofilm formation in filling hoses of food-processing equipment. Food Bioprod Process 113:39–48

Glick BR (2012) Plant growth-promoting bacteria: mechanisms and applications. Scientifica

Hadla M, Halabi MA (2018) Effect of quorum sensing. In: Chormey DS, Bakırdere S, Turan NB, Engin GÖ (eds) Comprehensive analytical chemistry. Elsevier, Radarweg, Amsterdam, the Netherlands 81:95–116

Jijón-Moreno S, Baca BE, Castro-Fernández DC, Ramírez-Mata A (2019) TyrR is involved in the transcriptional regulation of biofilm formation and D-alanine catabolism in *Azospirillum brasilense* Sp7. PLoS One 14:e0211904

Kanchan A, Simranjit K, Ranjan K, Prasanna R, Ramakrishnan B, Singh MC, Hasan M, Shivay YS (2019) Microbial biofilm inoculants benefit growth and yield of chrysanthemum varieties under protected cultivation through enhanced nutrient availability. Plant Biosyst Int J Deal Aspect Plant Biol 153:306–316

Kour D, Rana KL, Kumar A, Rastegari AA, Yadav N, Yadav AN, Gupta VK (2019a) Extremophiles for hydrolytic enzymes productions: biodiversity and potential biotechnological applications. In: Molina G, Gupta VK, Singh BN, Gathergood N (eds) Bioprocessing for biomolecules production. Wiley, USA, pp 321–372

Kour D, Rana KL, Yadav AN, Yadav N, Kumar V, Kumar A, Sayyed RZ, Hesham AE-L, Dhaliwal HS, Saxena AK (2019b) Drought-tolerant phosphorus-solubilizing microbes: biodiversity and biotechnological applications for alleviation of drought stress in plants. In: Sayyed RZ, Arora NK, Reddy MS (eds) Plant growth promoting rhizobacteria for sustainable stress management: volume 1: rhizobacteria in abiotic stress management. Springer, Singapore, pp 255–308. https://doi.org/10.1007/978-981-13-6536-2_13

Kour D, Rana KL, Yadav N, Yadav AN, Kumar A, Meena VS, Singh B, Chauhan VS, Dhaliwal HS, Saxena AK (2019c) Rhizospheric microbiomes: biodiversity, mechanisms of plant growth promotion, and biotechnological applications for sustainable agriculture. In: Kumar A, Meena VS (eds) Plant growth promoting rhizobacteria for agricultural sustainability: from theory to practices. Springer, Singapore, pp 19–65. https://doi.org/10.1007/978-981-13-7553-8_2

Kumar A, Patel JS, Meena VS, Srivastava R (2019a) Recent advances of PGPR based approaches for stress tolerance in plants for sustainable agriculture. Biocat Agricul Biotechnol 20:101271

Kumar M, Kour D, Yadav AN, Saxena R, Rai PK, Jyoti A, Tomar RS (2019b) Biodiversity of methylotrophic microbial communities and their potential role in mitigation of abiotic stresses in plants. Biologia 74:287–308. https://doi.org/10.2478/s11756-019-00190-6

Kumawat KC, Sharma P, Sirari A, Singh I, Gill BS, Singh U, Saharan K (2019) Synergism of Pseudomonas aeruginosa (LSE-2) nodule endophyte with Bradyrhizobium sp. (LSBR-3) for improving plant growth, nutrient acquisition and soil health in soybean. World J Microbiol Biotech 35:47

Laranjo M, Alexandre A, Oliveira S (2014) Legume growth-promoting rhizobia: an overview on the Mesorhizobium genus. Microbiol Res 169(1):2–17

Liu C, Mou L, Yi J, Wang J, Liu A, Yu J (2019) The Eno Gene of Burkholderia cenocepacia Strain 71-2 is involved in Phosphate Solubilization. Current Microbial 76:495–502

Mhatre PH, Karthik C, Kadirvelu K, Divya KL, Venkatasalam EP, Srinivasan S, Ramkumar G, Saranya C, Shanmuganathan R (2018) Plant growth promoting rhizobacteria (PGPR): a potential alternative tool for nematodes bio-control. Biocat Agricul Biotechnol 17:119–128

Molina-Santiago C, Pearson JR, Navarro Y, Berlanga-Clavero MV, Caraballo-Rodriguez AM, Petras D, García-Martín ML, Lamon G, Haberstein B, Cazorla FM, de Vicente A (2019) The extracellular matrix protects Bacillus subtilis colonies from Pseudomonas invasion and modulates plant co-colonization. Nat Commun 10:1919

Olanrewaju OS, Ayangbenro AS, Glick BR, Babalola OO (2019) Plant health: feedback effect of root exudates-rhizobiome interactions. Appl Microbiol Biotechnol 103:1155–1166

Pandin C, Le Coq D, Canette A, Aymerich S, Briandet R (2017) Should the biofilm mode of life be taken into consideration for microbial biocontrol agents? Microbial Biotech 10:719–734

Pedraza RO (2015) Siderophores production by *Azospirillum*: biological importance, assessing methods and biocontrol activity. In: Cassán FD, Okon Y, Creus CM (eds) Handbook for Azospirillum. Springer, Switzerland, pp 251–262

Pérez-Velázquez J, Gölgeli M, García-Contreras R (2016) Mathematical modelling of bacterial quorum sensing: a review. Bull Math Biol 78:1585–1639

Pliego C, Kamilova F, Lugtenberg B (2011) Plant growth-promoting bacteria: fundamentals and exploitation. In: Maheshwari D (ed) Bacteria in agrobiology: crop ecosystems. Springer, Berlin, Heidelberg, pp 295–343

Prabhu N, Borkar S, Garg S (2019) Phosphate solubilization by microorganisms: overview, mechanisms, applications and advances. In: Meena SN (ed) Advances in biological science research, a practical approach. Academic Press, Elsevier, London, pp 161–176

Primo ED, Cossovich S, Giordano W (2019) A simple method to evaluate biofilm formation in exopolysaccharide-producing strains of *Sinorhizobium meliloti*. J Biol Edu 7:1–8
Rajkumar M, Ae N, Prasad MN, Freitas H (2010) Potential of siderophore-producing bacteria for improving heavy metal phytoextraction. Trends Biotech 28:142–149
Ramakrishna W, Yadav R, Li K (2019) Plant growth promoting bacteria in agriculture: two sides of a coin. Appl Soil Ecol 138:10–18
Rana KL, Kour D, Yadav AN (2019) Endophytic microbiomes: biodiversity, ecological significance and biotechnological applications. Res J Biotechnol 14:142–162
Roy V, Adams BL, Bentley WE (2011) Developing next generation antimicrobials by intercepting AI-2 mediated quorum sensing. Enzyme Microbial Tech 49:113–123
Saraf M, Pandya U, Thakkar A (2014) Role of allelochemicals in plant growth promoting rhizobacteria for biocontrol of phytopathogens. Microbiol Res 169:18–29
Shahid M, Khan MS, Kumar M (2019) Kitazin-pea interaction: understanding the fungicide induced nodule alteration, cytotoxicity, oxidative damage and toxicity alleviation by Rhizobium leguminosarum. RSC Adv 9:16929–16947
Singh S, Singh SK, Chowdhury I, Singh R (2017) Understanding the mechanism of bacterial biofilms resistance to antimicrobial agents. Open Microbiol J 11:53
Singh SK, Singh PP, Gupta A, Singh AK, Keshri J (2019) Tolerance of heavy metal toxicity using PGPR strains of *Pseudomonas* species. PGPR amelioration in sustainable agriculture, food security and environmental management. Woodhead Publishing, Elsevier, Duxford, pp 239–252
Suman A, Yadav AN, Verma P (2016) Endophytic microbes in crops: diversity and beneficial impact for sustainable agriculture. In: Singh D, Abhilash P, Prabha R (eds) Microbial inoculants in sustainable agricultural productivity, research perspectives. Springer, India, pp 117–143. https://doi.org/10.1007/978-81-322-2647-5_7
Tabassum B, Khan A, Tariq M, Ramzan M, Khan MS, Shahid N, Aaliya K (2017) Bottlenecks in commercialisation and future prospects of PGPR. Appl Soil Ecol 121:102–117
Taktek S, St-Arnaud M, Piché Y, Fortin JA, Antoun H (2017) Igneous phosphate rock solubilization by biofilm-forming mycorrhizobacteria and hyphobacteria associated with Rhizoglomus irregulare DAOM 197198. Mycorrhiza 27:13–22
Timmusk S, Behers L, Muthoni J, Muraya A, Aronsson AC (2017) Perspectives and challenges of microbial application for crop improvement. Front Plant Sci 9(8):49
Tiwari S, Prasad V, Lata C (2019) Bacillus: plant growth promoting bacteria for sustainable agriculture and environment. New and future developments in microbial biotechnology and bioengineering—microbial biotechnology in agro-environmental sustainability. Elsevier, Radarweg, Amsterdam, the Netherlands, pp 43–55
Vacheron J, Desbrosses G, Bouffaud ML, Touraine B, Moënne-Loccoz Y, Muller D, Legendre L, Wisniewski-Dyé F, Prigent-Combaret C (2013) Plant growth-promoting rhizobacteria and root system functioning. Front Plant Sci 4:356
Velmourougane K, Prasanna R, Saxena AK (2017) Agriculturally important microbial biofilms: present status and future prospects. J Basic Microbiol 57:548–573
Velmourougane K, Prasanna R, Chawla G, Nain L, Kumar A, Saxena AK (2019a) Trichoderma-Azotobacter biofilm inoculation improves soil nutrient availability and plant growth in wheat and cotton. J Basic Microbiol. https://doi.org/10.1002/jobm.201900009
Velmourougane K, Prasanna R, Supriya P, Ramakrishnan B, Thapa S, Saxena AK (2019b) Transcriptome profiling provides insights into regulatory factors involved in *Trichoderma viride-Azotobacter chroococcum* biofilm formation. Microbiol Res 227:126292
Verma P, Yadav AN, Khannam KS, Kumar S, Saxena AK, Suman A (2016) Molecular diversity and multifarious plant growth promoting attributes of Bacilli associated with wheat (*Triticum aestivum* L.) rhizosphere from six diverse agro-ecological zones of India. J Basic Microbiol 56:44–58
Verma P, Yadav AN, Khannam KS, Saxena AK, Suman A (2017a) Potassium-solubilizing microbes: diversity, distribution, and role in plant growth promotion. In: Panpatte DG, Jhala YK, Vyas RV,

Shelat HN (eds) Microorganisms for green revolution: volume 1: microbes for sustainable crop production. Springer, Singapore, pp 125–149. https://doi.org/10.1007/978-981-10-6241-4_7

Verma P, Yadav AN, Kumar V, Singh DP, Saxena AK (2017b) Beneficial plant-microbes interactions: biodiversity of microbes from diverse extreme environments and its impact for crop improvement. In: Singh DP, Singh HB, Prabha R (eds) Plant-microbe interactions in agro-ecological perspectives: volume 2: microbial interactions and agro-ecological impacts. Springer, Singapore, pp 543–580. https://doi.org/10.1007/978-981-10-6593-4_22

Wang D, Xu A, Elmerich C, Ma LZ (2017) Biofilm formation enables free-living nitrogen-fixing rhizobacteria to fix nitrogen under aerobic conditions. ISME J 11(7):1602

Whitehead KA, Verran J (2015) Formation, architecture and functionality of microbial biofilms in the food industry. Curr Opin Food Sci 2:84–91

Yadav AN (2017) Agriculturally important microbiomes: biodiversity and multifarious PGP attributes for amelioration of diverse abiotic stresses in crops for sustainable agriculture. Biomed J Sci Tech Res 1:1–4

Yadav AN (2018) Biodiversity and biotechnological applications of host-specific endophytic fungi for sustainable agriculture and allied sectors. Acta Sci Microbiol 1:01–05

Yadav AN (2019) Microbiomes of wheat (*Triticum aestivum* L.) endowed with multifunctional plant growth promoting attributes. EC Microbiol 15:1–6

Yadav N, Yadav AN (2019) Actinobacteria for sustainable agriculture. J Appl Biotechnol Bioeng 6:38–41

Yadav AN, Sachan SG, Verma P, Saxena AK (2015a) Prospecting cold deserts of north western Himalayas for microbial diversity and plant growth promoting attributes. J Biosci Bioeng 119:683–693

Yadav AN, Sachan SG, Verma P, Tyagi SP, Kaushik R, Saxena AK (2015b) Culturable diversity and functional annotation of psychrotrophic bacteria from cold desert of Leh Ladakh (India). World J Microbiol Biotechnol 31:95–108

Yadav AN, Sharma D, Gulati S, Singh S, Kaushik R, Dey R, Pal KK, Saxena AK (2015c) Haloarchaea endowed with phosphorus solubilization attribute implicated in phosphorus cycle. Sci Rep 5:12293

Yadav AN, Sachan SG, Verma P, Saxena AK (2016) Bioprospecting of plant growth promoting psychrotrophic Bacilli from cold desert of north western Indian Himalayas. Indian J Exp Biol 54:142–150

Yadav AN, Kumar R, Kumar S, Kumar V, Sugitha T, Singh B, Chauhan VS, Dhaliwal HS, Saxena AK (2017a) Beneficial microbiomes: biodiversity and potential biotechnological applications for sustainable agriculture and human health. J Appl Biol Biotechnol 5:1–13

Yadav AN, Verma P, Kour D, Rana KL, Kumar V, Singh B, Chauahan VS, Sugitha T, Saxena AK, Dhaliwal HS (2017b) Plant microbiomes and its beneficial multifunctional plant growth promoting attributes. Int J Environ Sci Nat Resour 3:1–8. https://doi.org/10.19080/IJESNR.2017.03.555601

Yadav AN, Verma P, Singh B, Chauhan VS, Suman A, Saxena AK (2017c) Plant growth promoting bacteria: biodiversity and multifunctional attributes for sustainable agriculture. Adv Biotechnol Microbiol 5:1–16

Yadav AN, Gulati S, Sharma D, Singh RN, Rajawat MVS, Kumar R, Dey R, Pal KK, Kaushik R, Saxena AK (2019a) Seasonal variations in culturable archaea and their plant growth promoting attributes to predict their role in establishment of vegetation in Rann of Kutch. Biologia 74:1031–1043. https://doi.org/10.2478/s11756-019-00259-2

Yadav AN, Yadav N, Sachan SG, Saxena AK (2019b) Biodiversity of psychrotrophic microbes and their biotechnological applications. J Appl Biol Biotechnol 7:99–108

Zhang F, Wang P, Zou YN, Wu QS, Kuča K (2019) Effects of mycorrhizal fungi on root-hair growth and hormone levels of taproot and lateral roots in trifoliate orange under drought stress. Arch Agron Soil Sci 65(9):1316–1330

Zhu J, Li M, Whelan M (2018) Phosphorus activators contribute to legacy phosphorus availability in agricultural soils: a review. Sci Total Environ 612:522–537

Zhu X, Rice SA, Barraud N (2019) Nitric oxide and iron signalling cues have opposing effects on biofilm development in pseudomonas aeruginosa. Appl Environ Microbiol 85(3):e02175–18

Chapter 7
Actinobacteria: Diversity, Plant Interactions and Biotechnology Applications

Monnanda Somaiah Nalini and Harischandra Sripathy Prakash

Abstract Actinobacteria are Gram-positive members of the novel phylum and are present in diverse ecosystems on the earth. Basically, they are saprophytes, thriving in all soil types, litter and are one of the successful rhizocolonizers. Actinobacteria are associated with plants as litter degrading, symbiotic, endophytic or as pathogenic microorganisms. These associations often have benefited microbiologists, biotechnologists and chemists to introspect the potentials of strains and their secondary metabolites for manifold applications. Mainly, soil actinobacteria are potential producers of life-saving antibiotics or antimicrobial metabolites with myriad applications in medicine and agriculture. Actinobacteria of the soil and rhizosphere and as endophytes often have excellent plant growth-enhancing traits and function as antagonists in several important plant diseases of agriculturally prominent crops. In biotechnology, they are preferred for the production of antibiotics, drug analogues, engineered drugs, as sources of industrially important enzymes, and in the biodegradation of harmful xenobiotics.

Keywords Actinobacteria · Diversity · Bacterial associations · Biodegradation · Antibiotics · *Streptomyces*

7.1 Introduction

The actinobacteria are a group of microorganisms with high Guanine to cytosine (G + C, >55%) nucleotide content and constitute a major portion of the rhizosphere and its soil. To date, their ubiquitous nature of occurrence is well supported by their isolations from diverse habitats ranging from terrestrial to aquatic and hostile environments such as hyperthermal, hyperacidic, hypersaline, caves, coal mine, volcanic and unusual arid regions. The actinobacteria are placed in the novel phylum

M. S. Nalini (✉)
Department of Studies in Botany, University of Mysore, Manasagangotri, Mysuru 570 006, Karnataka, India

H. S. Prakash
Department of Studies in Biotechnology, University of Mysore, Manasagangotri, Mysuru 570 006, Karnataka, India

A. N. Yadav et al. (eds.), *Plant Microbiomes for Sustainable Agriculture*, Sustainable Development and Biodiversity 25, https://doi.org/10.1007/978-3-030-38453-1_7

(*Actinobacteria Phyl. nov.*) comprising 16 orders, with many of them elevated to the novel orders (*Ord. nov.*) (Ludwig et al. 2012) and are as follows:

Actinomycetales	Actinopolysporales *Ord. nov.*
Bifidobacteriales	Catenulisporales *Ord. nov.*
Corynebacteriales *Ord. nov.*	Frankiales *Ord. nov.*
Glycomycetales *Ord. nov.*	Jiangellales *Ord. nov.*
Kineosporales *Ord. nov.*	Micrococcales *Ord. nov.*
Micromonosporales *Ord. nov.*	Propionibacteriales *Ord. nov.*
Pseudonocardiales *Ord. nov.*	Streptomycetales *Ord. nov.*
Streptosporangiales *Ord. nov.*	*Incertae sedis Ord. nov.*

Soil is a complex substrate varying in composition, pH, and is known to harbour microorganisms. The presence of actinobacteria in soil was documented in 1903. Since then, the use of cultivation media with antibiotics for the selective isolation and enumeration of actinomycete colonies was devised (Williams and Davies 1965). Actinobacteria are broadly classified as commonly occurring and cosmopolitan generic group, the Streptomycetes and non-streptomycetes, also known as rare actinomycetes. *Streptomyces* are primarily soil bacteria comprising 10% of the total soil microbiome (Hayakawa et al. 1996a), are identified by profuse aerial mycelia and spore chains and are recovered easily from a number of substrates and plant litter. The chalky and powdery colonies are readily visible on humic acid–vitamin agar (HV) plates (Hayakawa and Nonomura 1987), which otherwise restricts the growth of filamentous bacteria. The rare actinobacteria grow on HV agar, but require selective isolation techniques and enrichment methods such as the use of chemoattractant, i.e., vanillin for the isolation of *Catenuloplanes* and *Virgosporangium*; pollen baiting for *Actinoplanes*; and rehydration/centrifugation for the isolation of motile actinomycetes (Hayakawa 2008).

Actinobacteria exhibit diversity in plant microbiomes, with a reduction of populations documented in the spermosphere and phyllosphere microbiomes (Lopez-Velasco et al. 2013; Kembel et al. 2014; Yadav and Yadav 2019a). Nevertheless, in the rhizospheric soil and in plants, they occur widely in all tissues. As endophytes, their diversity has been documented from crop plants (Coombs and Franco 2003; Cao et al. 2005; Tian et al. 2007), native plants (Janso and Carter 2010; Kim et al. 2012; Kaewkla and Franco 2013) and medicinal plants (Qin et al. 2009; Zhao et al. 2011). Due to high species diversity, the identification of actinobacteria is challenging as well as cumbersome. In the absence of accurate methods based on morphology, the 16S rRNA as a phylogenetic marker has been reliable in the identification of species. Recently, the multilocus sequence analysis (MLSA) was applied as a molecular tool, for resolving the identification in members of Streptomycetaceae (Labeda et al. 2017), whereas for studies on the actinobacterial communities of phyllosphere and rhizosphere microbiomes, a 16S rRNA microarray technology known as PhyloChip (Mendes et al. 2011) has been designed.

Interest in actinomycete research was generated over the past decades due to the identification and production of antibiotic compounds from soil *Streptomyces* species. ~8,000 antibiotic compounds are described from streptomycetes alone (Berdy 2005) and the number of compounds described for non-streptomycetes are lesser but with anti-pathogenic and antagonistic potentials (Lazzarini et al. 2000; El-Tarabily and Sivasithamparam 2006).

Actinobacteria are useful in improving crop yields through the enhancement of plant growth traits. Rhizoactinobacterial and endophytic strains of *Streptomyces*, *Nocardia* and *Nonomuraea* function as plant growth enhancers by producing the growth regulator, Indole acetic acid (IAA), siderophores, enzymes, by phosphate solubilization or by increasing the nutrient levels (Nimnoi et al. 2014; Vurukonda et al. 2018; Verma et al. 2015; Yadav et al. 2015). Actinobacteria are useful in the degradation of polyhydrocarbons and xenobiotic compounds. Strains of *Rhodococcus* are extensively used in the biodegradation of harmful polyhydrocarbons, aromatic and plasticizer compounds due to their ability to thrive in harsh environments (Yadav et al. 2019f). The physiology and genomic diversity among *Rhodococcus* strains make them amenable to genetic manipulation and greater understanding of the pathways for functional gene (s) involved in the biodegradation processes (Zampolli et al. 2019). Actinobacteria are known sources of industrially important enzymes such as cellulases, pectinases, chitinases, xylanases and proteases (Kour et al. 2019a; Yadav et al. 2016). These enzymes are used in detergent, leather, textiles, paper and pulp and in food industries. High stability, extreme pH and temperature tolerance are some of the criteria to be employed for industrial processes (Yadav et al. 2019c, d, e).

This chapter highlights the actinobacterial diversity associated with soil and rhizosphere microbiomes, and in endophytic plant species, their interactions with plant microbiomes, association with plants and few applications in biotechnology.

7.2 Actinobacterial Interactions with Plants

Microbial associations with plants can be described as by far the most reliable and beneficial in providing increased plant growth, in terms of seedling vigour, nutrition and productivity (Mendes et al. 2013). The 'microbiome concept', which was initially coined to describe the microbial associations in the human gut (Qin et al. 2010), has also been extended to microbial communities associated with plant species. They are dependent on plants for specific functional traits to sequester the physiologically fixed carbon into direct surroundings. Currently, five types of microbiomes are described, namely, spermosphere (Verona 1958), rhizosphere (Berendsen et al. 2012), phyllosphere (Vorholt 2012), endosphere (Hardoim et al. 2015) and mycorrhizosphere (Frey-Klett et al. 2007). Actinobacterial communities are known to exist in all the microbiomes and their interactions are documented.

7.2.1 Plant Microbiome and Actinobacteria

'Plant microbiome' is a term defined as 'the collective genomes of microorganisms living in association with plants', which has in turn led to new inputs on the evolution of plants (Rosenberg et al. 2009). For greater than 100 years, plant microbiome has been considered as significant contributors in maintaining plant health and productivity. The development in research methodologies with inputs from molecular and next-generation sequencing techniques and analytical tools (Jansson et al. 2012; Berg et al. 2013) has led to achieve important goals in understanding the plant-associated microbial communities in several plant species (Mendes et al. 2011; Bulgarelli et al. 2012) with high functional diversity.

'Spermosphere' is defined as the zone surrounding the seeds which interacts with soil, microbial communities and the germinating seed and strictly is applicable to the short timed germination process occurring in seeds (Schiltz et al. 2015). The term was first coined by Verona (1958) to describe the interplay between the soil, seed-borne microbiota and the germinating seeds. During the process, seeds exude substances that either stimulate or inhibit microbial growth and have a direct influence on plant health and growth parameters. Since the timing related to the unfolding of events is of a short duration, the spermosphere remains the less studied zone among the microbiomes. Spinach spermospheric microbial communities were analysed by Pyrosequencing at the germination stage and the actinobacterial abundance was poorly represented by the genera *Corynebacterium* (0.08%), *Sanguibacter* (0.08%) and *Micrococcus* (0.16%) (Lopez-Velasco et al. 2013).

'Rhizosphere' refers to the narrow contact zone between the roots and the soil particles. It is also the first plant environment encountered by the soil microorganisms (Dessaux et al. 2016), which in turn influences plant growth. Three zones are contained in the rhizosphere: the endorhizosphere (root cortex and endodermis; the rhizoplane (mid zone) and the ectorhizosphere (outer zone) extending from rhizoplane to the bulk soil (McNear Jr. 2013). Root exudates, mainly photosynthate secretions mainly serve as source of energy, influencing the soil microbial communities. Actinobacterial communities of rhizosphere primarily comprise of the soil dweller, *Streptomyces*, the species of which composition tends to vary among crop plants (Petrolini et al. 1996; Suarez-Moreno et al. 2019). The rhizosphere dwelling actinobacteria enhance plant growth promotion (PGP) traits by the production of siderophores, phosphate solubilization, nitrogen fixation (Berendsen et al. 2012) and antifungal compounds (Turpaulta et al. 2007). The composition of spermosphere and rhizosphere microbiomes of wild and modern bean accessions cultivated in an agricultural and a native soil from Colombia was characterized by metagenomics and cultivation-dependent approaches (Perez-Jaramillo 2019). Results indicated an increase in the abundance of actinobacteria in the rhizosphere of plants grown from cultivated bean accessions from native to agricultural fields.

'Phyllosphere' is one of the most diverse habitats on earth and a niche for the interaction of epiphytic and endophytic microorganisms. The phyllosphere spans 10^9 sq. km across the dorsal as well as the ventral leaf surfaces and has an estimate of

10^{26} bacterial cells (Vorholt 2012). It encompasses four regions of the above-ground parts: caulosphere, anthosphere, carposphere and phylloplane. The phyllosphere organisms compete for the availability of nutrients and contribute to the diverse architecture and density of the habitat. The actinobacterial diversity is estimated to be lesser than that of proteobacteria and firmicutes. The actinobacteria of the phyllosphere in spinach plants represent <1% of the total bacterial communities (Lopez-Velasco et al. 2013). The relationship between plant functional traits and bacterial communities of the phyllosphere was studied in the leaves of 57 tree species in a neotropical forest in Panama (Kembel et al. 2014). The host leaves harboured 400 bacterial taxa of which, the actinobacteria comprised 5.5% of the core microbiome taxa among the dominant phyla.

'Endosphere' is a term applied to the internal tissues of plants, forming a niche. It is applicable in various ways depending on the purpose for colonization. The microbes called endophytes as well as pathogens dwell within the endosphere, but perform different functions related to either symbioses, mutualism or pathogenic. The route of entry into the endosphere is facilitated by wounds created due to abrasion, via stomata or by the action of cell wall degrading enzymes. Microorganisms interact in the endosphere of plants parts, viz., the roots, root nodules, stem, leaves, floral parts and the fruits (Suman et al. 2016; Verma et al. 2017b; Yadav et al. 2018a, b). Actinobacterial endophytes are dominant colonizers of roots in many agricultural crops or native and medicinal species (Sardi et al. 1992; Coombs and Franco 2003; Janso and Carter 2010; Kaewkla and Franco 2013) and include *Streptomyces* as well as non-streptomycetes. Preferential colonization of *Streptomyces lydicus* WYEC108 in pea root nodules was demonstrated (Tokala et al. 2002).

Mycorhizosphere represents the zone of soil encasing the plant roots (rhizosphere) and the extraradical fungal hyphae protruding from the mycorrhizae. The mycorrhizosphere is the zone surrounded by both the root and the mycorrhizal fungus (Rambelli 1973). The actinobacteria interact with the microbiome for specific production of secondary metabolites, enzymes, plant hormones and growth inhibitors.

7.2.2 Associations of Actinobacteria with Plants

Plant and microbial interactions have often postulated to be old involving beneficial, symbiotic, harmful, mutualistic or free-living types. Actinobacteria form all these associations with plant species (Fig. 7.1). The events leading from the entry to establishment in plants is documented for many host–microbial systems. Such phenomena are well established for fungal and bacterial symbionts, pathogens and saprobes. The following are the associations exhibited by actinobacteria with plant species.

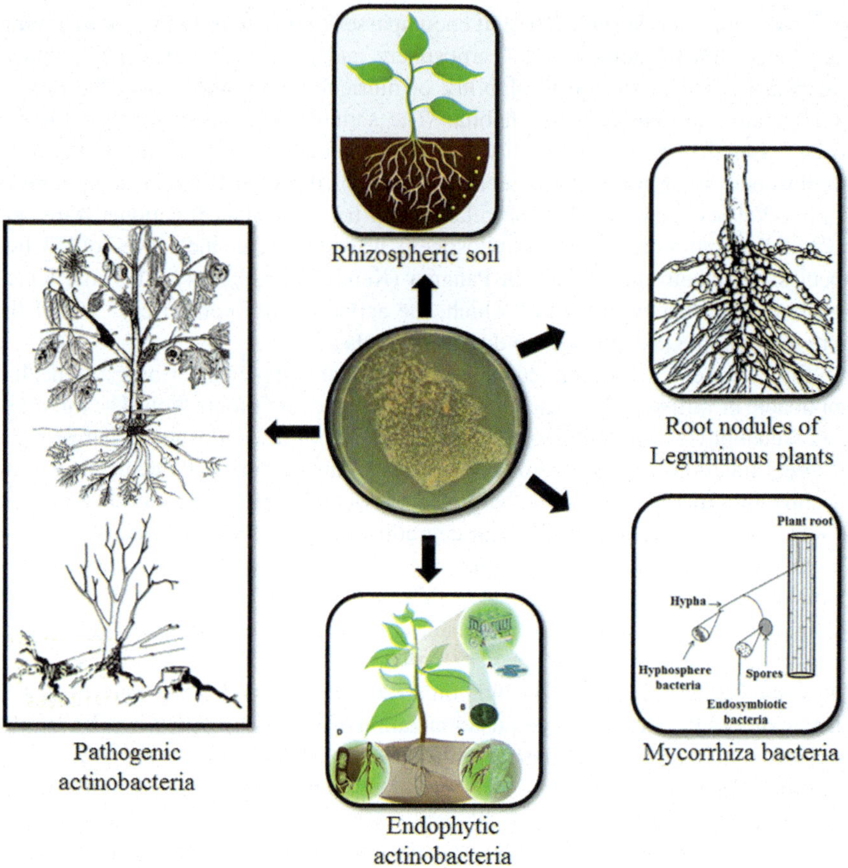

Fig. 7.1 Actinobacterial associations with plants

7.2.2.1 Saprophytic

Actinomycetes reside in the soil types of various habitats and are therefore saprophytic in nature assisting in the decomposition of various substrates into organic matter, which is utilized by other organisms in the food web, thereby balancing the nutrient cycle. One such substrate is the plant litter which serves as principal source of nitrogen (N) and carbon (C) in the soil and has significant development in the terrestrial ecosystem (Sauvadet et al. 2016). In the aquatic habitats such as streams, litter accumulated by leaf fall constitutes one of the chief sources of energy. Microorganisms, both bacteria and fungi are bestowed with the ability to degrade plant remains by secrete cell wall degrading enzymes to breakdown large molecules such as cellulose, lignin and chitin into smaller compounds, which are in turn utilized by saprophytes in the food web (Sinsabaugh and Lenskins 1990). Microbial degradation of litter is the resultant co-effort of fungi and bacteria, with the biochemical and physiological processes of the latter are well established (Das et al. 2007).

The assessment of actinobacterial diversity on the decaying senescent leaves of sugar maple and white oak in a forested stream, along northeastern Ohio (Das et al. 2007) resulted in the lesser diversity of actinobacteria than fungi by Denaturing Gradient Gel electrophoresis technique (DGGE). A study to determine the impact of C cycling on the quality of litter and the bacterial communities of soil in Oak (*Quercus wutaishanica*) secondary forest region in Fuxian Observatory in China was conducted (Zeng et al. 2017). The annual litter accumulation over a two year period due to leaf fall was estimated at 200 g m^{-2} per year. The actinobacteria formed 10–21% of the dominant groups in the primary soil types, which enhanced due to the decomposition of litter. Litter degrading actinobacterial populations are documented from fallen leaves as well as from streams. Actinobacteria such as *Streptomyces, Micromonospora, Actinomadura* and *Pseudonocardia* degrade lignocellulose in plant litter (McCarthy 1987). Fallen leaves from pine litter layers are sources for the isolation of numerous *Streptomyces* and *Microbispora* (Matsukuma et al. 1994). High frequency of *Microbispora* and *Actinokineospora* spp. were isolated from the fallen leaves of bamboo and loosestrife and other plants (Matsumoto et al. 1998; Otoguro et al. 2001).

7.2.2.2 Symbiotic

The term 'Symbiosis' was originally coined by de Bary in 1879 to describe a long-term interaction between two biological species meaning 'living together'. One of the remarkable examples for the plant–microbe symbiotic relationship is that of leguminous plant roots and the bacterium of the genus *Rhizobium* in the fixation of nitrogen. Today, a number of genera, *Azorhizobium, Bradyrhizobium, Mesorhizobium, Sinorhizobium* and *Ensifer*, collectively referred to as 'rhizobia' are documented as nitrogen fixers (Hardoim et al. 2015). In non-leguminous plants, nitrogen fixation is achieved by the Gram-positive actinobacterium, *Frankia*. It is known to establish root nodules in 23 species of non-legumes, called 'actinorhizal plants' of dicotyledonous families. The phylogenetically related groups of actinorhizal plants are Fagales (Betulaceae, Casuarinaceae, Myricaceae), Cucurbitales (Datiscaceae, Coriariaceae) and Rosales (Rosaceae, Elaeagnaceae, Rhamnaceae) (Berry et al. 2011). Actinorhizal plant species are able to colonize in nutrient-poor sites and are ecologically important. Non-*Frankia* actinobacterial strains of Micromonosporaceae and Thermomonosporaceae isolated from the surface-sterilized roots of *Casuarina equisetifolia* was able to fix nitrogen, as analysed by the amplification of *nif*H genes (Valdes et al. 2005).

7.2.2.3 Pathogenic

Plant-associated actinobacteria cause diseases in plant species and are termed pathogenic. Several species of *Streptomyces* infect potato as pathogens and are common agents of potato scab, i.e., *Streptomyces acidiscabies, Streptomyces europaeiscabei, Streptomyces scabies* and *Streptomyces turgidiscabies* (Bignell et al. 2010). *Clavibacter* is a Gram-positive actinobacteria with many subspecies and is the causal agent of bacterial spot of tomato, disease of alfalfa (*C. michiganensis* subsp. *insidiosus*), maize (*C. michiganensis* subsp. *nebraskensis*) and wheat (*C. michiganensis* subsp. *tessellarius*). The ratoon stunting of sugarcane is caused by *Leifsonia xyli* subsp. *xyli* (Young et al. 2006), while *Rathayibacter tritici* incites gumming disease in several grasses (Evtushenko and Dorofeeva 2012). Besides crop plants, garden species, beet, American holly and poinsettia are affected.

7.2.2.4 Endophytic

Endophytes are microorganisms residing inside healthy plant tissues without causing any 'overt, negative effects' on hosts (Bacon and White 2000). In plants, endophytic associations include those of fungi, actinobacteria, bacteria, yeasts, etc. Actinobacteria form endophytic associations in host plants. In 1886, *Frankia* was first isolated as the actinobacterial endophyte from the non-legume root nodules. The first description of an actinobacterial association from the leaves of grass was the new genus *Actinosynnema* (Hasegawa et al. 1978). Endophytic actinobacteria are described from several plants: crop plants (Coombs and Franco 2003; Cao et al. 2004, 2005; Tian et al. 2007), native plants (Janso and Carter 2010; Kim et al. 2012) and medicinal plants (Castillo et al. 2002, 2006; Akshatha et al. 2016). Native tree species do have actinobacterial associations with streptomycetes as major strains (Kaewkla and Franco 2013), while rare actinobacteria were also isolated from fewer tree species (Caruso et al. 2000; Himaman et al. 2016). They are distributed in all plant parts with high diversity in the root tissues than stems and leaves. Root endophytic strains, of *Streptomyces* spp. have excellent antagonistic potentials or as biocontrol agents to several plant pathogens.

7.3 Actinobacterial Diversity: Predictions or Reality?

Of all the microorganisms associated with plant species, the diversity estimate for the total number of fungal species has been documented to be 6,11,000 (Mora et al. 2011), which is much lower than the earlier predicted conservative estimate of 22,70,000 (Hawksworth 2001), solely based on the traditional taxonomical approaches. On a global basis, the actual number of fungal species is underestimated. Conventional techniques employed in the identification of actinobacteria have often resulted in the misidentification of closely related species. Advanced methods using molecular data with next generation technologies have emerged as a reliable tool in the accurate identification of closely related species and in enumerating diversity. Pyrosequencing and Illumina platform technologies have provided reliable methods to enumerate the number of species in microbial communities of several ecosystems (Buee et al. 2009). Recently, a high-density 16S rRNA microarray technology known as PhyloChip (Mendes et al. 2011) was designed to detect the abundance and diversity of bacterial communities in the rhizosphere and phyllosphere microbiome.

7.3.1 Diversity in Soil and Rhizosphere

Soil contains a complex mixture of microorganisms, both pathogenic and beneficial to plants. The diversity of microbes in soil as well as the rhizosphere has been underestimated. The estimate of microbial presence in soil has not been accurately measured and has often resulted in ambiguity. Studies indicate actinobacteria as important components in the rhizosphere and are known to influence plant growth and protect roots against the invasion of plant pathogens. Actinobacterial species diversity estimates predicted ca. 10,000 species per gram of boreal forest soil (Yadav 2019; Torsvik et al. 2002). Using improvised computational estimates, bacterial counts were recorded to be greater than one million per gram of soil (Gans et al. 2005). Different types of soil samples consisting of forest, greenhouse, mountain and waterfalls from the Western Ghats of India were enumerated for the actinobacteria isolated using actinomycete isolation agar (AIA) supplemented with the antibiotics, ketoconozole (30 mg/l) and nalidixic acid (100 mg/l). 12 species of *Streptomyces* were documented (Ganesan et al. 2017).

The most common actinobacteria distributed in soil are the *Streptomycetes*, which account for the total actinomycete populations, but in order to isolate rare genera, selective media and enrichment methods have been designed (Hayakawa 2008). Pre-treatment methods such as physical and chemical, enrichment by rehydration and centrifugation allowed the delineation of streptomycetes and favoured the isolation of rare taxa such as the non-motile as well as the motile actinomycete genera. Actinomycetes are known to utilize humic acid as the sole source of carbon and nitrogen, which facilitated the growth of actinomycete colonies of the genera *Streptomyces, Micromonospora, Microbispora, Streptosporangium, Nocardia, Dactylosporangium, Microtetraspora* and *Thermomonospora* on agar medium, while restricting the development of true bacteria (Kumar et al. 2019b; Yadav et al. 2017a, b, d). The recovery of *Microtetraspora* a rare genus, and four-spored actinomycetes, with LSV-SV agar enriched with Kraft lignin as the source of carbon and nitrogen was reported (Hayakawa et al. 1996b). The distribution of actinobacteria in soil types were documented along a vertical gradient through test borings (Takahashi et al. 1990). Top layers contained actinobacteria at 10^7 g^{-1}, second layer (0.5–5.0 m) and still below (1.0 m) had reduced actinomycete populations.

Fort-five and 93 strains of actinobacteria representing streptomycetes and non-streptomycetes were isolated from the rhizospheres of carrot (El-Tarabilly et al. 1997) and cucumber (El-Tarabilly 2006). Actinobacterial strains were isolated from the rhizospheric soils of *Vitis vinifera* across four Moroccan sites (Loqman et al. 2009). The total count ranged from 16–65 $\times$ 10^6 cfu/g of dry soil. The rhizospheric soil adhered to the roots of seven traditional medicinal plant species from the Panxi plateau, China was analysed for the diversity of actinobacteria (Zhao et al. 2012). 196 strains were grouped into eight suborders and 13 families. In three species, streptomycetes were dominant, while in four species, rare actinobacteria were isolated, each plant species contained unique strains. Soils sampled from ten crop plants at a depth of 5–15 cm contained 156 actinomycete isolates (Kaur et al. 2013). 382 actinobacteria were isolated from 27 rhizospheric soil samples of eucalyptus plantation (Himaman et al. 2016). The total actinomycete counts in two different media of isolation ranged from 9.4×10^4 to 3.4×10^6 and 1.0×10^5 to 2.16×10^6 cfu g^{-1} of dry soil on Starch casein agar and HV agar, respectively. The rhizospheric strains comprised mainly of non-streptomycetes. *Micromonospora* was most frequently isolated from endophytic and rhizospheric samples.

Oak rhizosphere and the surrounding soil had contrasting bacterial diversity, as studied by Pyrosequencing technology (Uroz et al. 2010). Differences in the distribution of bacterial communities of soil and rhizosphere in *Arabidopsis thaliana* were observed (Lunderberg et al. 2012). Using 454 pyrosequencing of the bacterial communities associated with leaves and roots of *A. thaliana*, Bodenhausen et al. (2013) observed abundant distribution of actinobacteria in the samples along with other bacterial communities. Therefore, owing to the contradictions arising in the estimations of actinobacteria from soil and its rhizosphere, the next-generation sequencing techniques have become handy in accurately predicting the actinobacterial populations in several rhizospheric soils of plant species. PhyloChip, a high-density 16S r RNA gene microarray technology, could detect the diversity and abundance of the bacterial communities in the rhizosphere (Mendes et al. 2011), and the results were consistent in estimating the actinobacterial communities in oat rhizosphere (De Angleis et al. 2009). Actinobacteria accounted for 11% of total bacterial populations in two distant potato fields and three cultivars in Netherlands (Weinert et al. 2011), while a study in the rhizosphere of beet seedlings, actinobacteria was recorded as the second largest taxa (Mendes et al. 2011).

Composite soil samples of wheat rhizosphere, from Idar region of Gujarat, resulted in the isolation of *Streptomyces* spp. (Jog et al. 2012) with plant growth-promoting ability. Rhizospheric soils sampled from South African plant species indicated the diversity of *Streptomyces* spp. (Adegboye et al. 2012). Soils from the rhizosphere of legume and Japanese grass species analysed for the actinobacteria contained 84% Streptomycetes, while rare actinomycetes were less (Matsumoto and Takahashi 2017). The rhizosphere is a potential site for the existence of actinobacteria, wherein the filamentous bacteria are found at a count of 10^6 cells per mm^3.

Plantation soils are a niche for microorganisms and do play a major role in the decomposition and degradation of complex organic residues formed from the crop residues and shade trees. Actinobacterial populations are present in the plantation soils and help in degrading the organic matter by producing the extracellular enzymes. Coffee plantation soils are well suited for the isolation of actinomycetes as they are rich in humus and fallen litter. In a study conducted to isolate soil actinomycetes from Coffee plantation of Chikmagalur region, Western Ghats, India, Sameera et al. (2018a) documented the presence of 24 species of actinobacteria across altitudinal gradients and two soil profiles. The physico-chemical parameters of the soil and total actinomycete counts correlated with the soil profiles. The actinobacteria consisted of 54.2% *Streptomyces* spp. and 45.8% of non-streptomycetes, comprising three novel orders.

7.3.2 Diversity in Crop Plants

Actinobacteria are associated with roots of many crop plant species, although they are major components of the rhizosphere. Their presence around the roots is necessary to evade the invasion of root pathogens. 28 healthy plant species cultivated in Northwestern Italy were subjected to actinobacterial isolations from the surface-sterilized roots (Sardi et al. 1992). Of the 499 isolates, 482 strains were *Streptomyces*, two strains were *Streptoverticillium* and four strains were *Nocardia*, while one strain each of *Micromonospora* and *Streptosporangium* were obtained. Streptomycete populations in the roots of 156 plant species were studied along the seven-year sampling period in Italy (Petrolini et al. 1996). 81% of the strains belonged to *Streptomyces*, other rare actinobacteria belonged to *Micromonospora* (215 strains), *Streptosporangium* (12), *Streptoverticillium* (2), *Saccharomonospora* and *Nocardia* (70) species. Healthy wheat plants (*Triticum aestivum* L.), growing in the major regions of southern Australia were analysed for the actinobacteria. The surface-sterilized fragments yielded 88% of *Streptomyces* and 12% of *Microbispora, Micromonospora* and *Nocardioides* (Coombs and Franco 2003). Since the actinobacteria are soil-inhabiting populations, the diversity was assessed in the surface-sterilized roots of banana plants from a plantation in China (Cao et al. 2005). Of the 131 isolates, 99 comprised of *Streptomyces*, followed by morphologically distinct genera *Streptoverticillium* (28) and *Streptosporangium.* The diversity of actinobacterial community in the internal portions of rice (*Oryza sativa* cv. Qilisimiano) stems and root tissues was determined from South China (Tian et al. 2007). 191 strains were morphologically distinguished into 33 groups by the 16S rDNA genes as *Streptomyces* and *Nocardioides.* The gene clones from the library distinguished strains from roots as belonging to nine genera; *Streptomyces* (24%), *Micromonospora* (6.6%), *Actinoplanes* (6.6%), *Frankia* (4.4%), *Dactylosporangium* (4.4%), *Amycolatopsis* (4.4%), *Corynebacterium* and *Rhodococcus* each with 2%. An uncultured actinobacterium was also detected. More diversity of actinobacteria was obtained from roots than stem portions.

619 actinobacteria were isolated from the roots of field-grown tomato cultivars in south China during 2014 (Tan et al. 2006). The isolates were grouped into *Streptomyces aureochromogenes* (30.3%), *Hygroscopicus* and *Viridis* groups in healthy tomato roots, while *Cinereus* and *Flavus* groups were characterized from the roots of *Ralstonia* wilt tomato cultivars. Healthy tomato plant roots collected from southern Australia were analysed for the actinobacteria, and *Streptomyces* spp. were the most

commonly isolated ones, and rare members included *Microbispora* and *Nonomuraea* spp. (Inderiati and Franco 2008). Ten plants comprising the crop and medicinal species were used for the isolation of actinobacteria from plant parts. 50% comprised of the root isolates, while the stem and leaf isolates were 29% and 21%, respectively (Kaur et al. 2013).

The wattle tree (*Acacia auriculiformis* A. Cunn. ex Benth.) grown for its association with rhizobacteria is used to improve the soil fertility and is native to Australia. The diversity of actinobacteria in the leaves and roots was determined from four sites in the University campuses of Thailand (Bunyoo et al. 2009). The four sites differed in the actinobacterial diversity, with *Streptomyces* from two sites, and rare genera such as *Actinoallomurus, Amycolatopsis, Microbispora and Kribbella* with two species each from three sites. 23 root samples were collected from the eucalyptus plantation in Thailand for endophytic actinobacterial analysis (Himaman et al. 2016). Overall, 95 endophytic root isolates were characterized as strains and included genera of rare actinobacteria, namely, *Micromonospora* (28.9%), *Nocardia, Actinomadura, Actinoallomurus, Cryptosporangium* and *Amycolatopsis.*

There is strong evidence that the endophytic populations have originated mainly from the rhizosphere. The rhizospheric populations of actinobacteria consist mainly of *Streptomyces* spp. and are abundant and diverse. Several species of crop plants have developed associations with these groups of microorganisms that have immense applications in agriculture and biotechnology.

7.3.3 Diversity in Medicinal Plants

Actinobacteria occur as abundant microbiota in the plants of medicinal value, as isolated from various plants across biodiverse regions. They are often referred as 'endophytes' and reside in the intercellular spaces of plant parts. Diverse actinobacterial species have yielded array of metabolites with wide applications in agriculture, pharmacy and biotechnology.

Taechowisan et al. (2003) obtained 330 actinomycetes from seven medicinal species from Chiang Mai, Thailand with *Streptomyces* (n = 277) as the dominant isolates followed by *Microbispora* (n = 14), *Nocardia* (n = 8) and *Microbispora* sp. (n = 4). Plant species (300) sampled from the Amazonian rainforest sites yielded 14 endophytes, 12 belonged to *Streptomyces*, one each of *Micromonospora* and *Amycolatopsis* spp. (Bascom-Slack et al. 2009). Three Indian medicinal species, *Aloe vera, Ocimum sanctum* and *Mentha arvensis* were selected for the actinobacterial isolations from the roots, stems and leaves. Of the 40 isolates, *Streptomyces* spp., were dominant genera and the rest were represented by *Saccharopolyspora, Micromonospora* and *Actinopolyspora* spp. (Gangwar et al. 2011).

The Sichuan Province of China has a subtropical climate and medicinal species have been used from 3000 years for human health. 13 native medicinal species were subjected to the isolation and study of actinobacterial diversity in the plant parts (Yuan et al. 2008). The actinobacteria were classified into two genera: *Streptomyces* and *Micromonospora,* of which the former showed high species diversity. The Chinese tropical rainforest is a unique region comprising the transition vegetation and is a home to 3,000 endemic plant species. At Xishuangbanna, medicinal plants were studied for the presence of diverse actinobacterial populations in the plant parts. One plant species, *Maytennus austroyunnanensis,* was selected for the actinobacterial isolations (Qin et al. 2012) due to the importance of maytansinoids. Culture-dependent and -independent methods were applied for the actinobacterial isolations and to study their diversity in plant parts. A total of 312 strains were detected of which, roots comprised of major isolates (40.4%) than the stems (27%) and leaves (32.6%). The actinobacteria were distributed in eight suborders. *Streptomyces* were the most frequently isolated genus with 15 species, and rare ones belonged to 20 genera.

The roots of four medicinal plant species collected from the Chinese Herbal Plant Base, Hebei, were evaluated for the presence of actinobacteria, which comprised of *Streptomyces* spp. and two species of *Glycomyces* (Zhang et al. 2012). Four medicinal plant species from the Western Ghats, namely, *Rauwolfia densiflora, Leucas ciliata, Cajanus lineata* and *Gomphostemma heyneanum,* were evaluated for the presence of actinobacteria from the stem and leaves (Akshatha et al. 2016). 68% of the strains belonged to *Streptomyces* and the rare ones included *Promicromonospora, Arthrobacter, Patulibacter, Rhodococcus* and *Nocardia* spp. The diversity of actinomycetes mentioned in the above paragraph indicates that morphological, physiological and biochemical characteristics were undertaken by Taechowisan et al. (2003) to identify the actinomycetes to the generic level, while in all other studies the identification to the species level was accomplished by the sequencing of 16S rRNA gene. A combination of techniques involving the crumbling of sterilized plant fragments, desiccation with calcium carbonate and enzymatic hydrolysis coupled with differential centrifugation were standardized for the actinobacterial species from 90 tropical rainforest plants in Southwest China (Qin et al. 2009), which facilitated the isolation of rare actinobacteria.

7.3.4 Diversity in Native Tropical Species

The Papua New Guinea along with the adjoining areas of Archipelago is home to a number of tropical plant species and is one of the biodiverse regions on the earth. A range of plant parts and habit were sampled for the isolation and identification of 123 actinobacterial strains from 113 plant species (Janso and Carter 2010). The use of different isolation techniques as well as the enrichment procedures resulted in the isolation of rare genera, viz., *Lentzea, Lechevalieria, Kitasatospora* and *Planotetraspora.* Molecular characterization of the isolates resulted in resolving the diverse actinobacterial taxa into six families: Streptomycetaceae (27%), Streptosporangiaceae (40%), Thermomonosporaceae (16%), Micromonosporaceae (8%), Pseudonocardiaceae (8%) and Actinosynnemaceae (2%). The non-streptomycetes comprised of 74% of the total actinobacterial strains. The diversity of endophytic actinobacteria from the surface-sterilized root samples of eleven native herbaceous plants from Korean Provinces was assessed (Kim et al. 2012). 61 strains identified comprised of *Streptomyces* (45.9%), *Micromonospora* (18.8%), *Rhodococcus* (6.6%), *Microbispora* (4.9%), *Micrococcus* (4.9%), and other strains included *Arthrobacter, Dietzia, Kitasatospora, Herbiconiux, Mycobacterium, Nocardia, Rathayibacter* and *Tsukamurella*.

The tropical neem tree (*Azadirachta indica* A. Juss.) yielded diverse actinobacteria (Verma et al. 2009a, b) from the stems (23.6%), roots (54.5%) and leaves (21.8%). *Streptomyces* was the dominant genus (49.09%), while rare actinobacterial genera identified were *Streptosporangium* (14.5%), *Streptoverticillium* (5.5%), *Microbispora* (10.9%), *Nocardia* (3.6%) *and Saccharomonospora* (5.5%). Kaewkla and Franco (2013) reported diverse *Streptomyces* spp. (72%) from four Australian native trees such as native pine tree (*Callitris preissii*), red gum (*Eucalyptus camaldulensis*), grey box tree (*Eucalyptus microcarpa*) and apricot tree (*Pittosporum phillyraeoides*). The native pine tree and grey box tree contained 33.7% and 33.1% of the isolates, while the apricot and red gum tree isolates comprised of 26.4 and 6.4%, respectively. Rare actinobacteria: *Polymorphospora, Gordonia, Actinomycetospora* and two novel genera *Williamsia* and *Flindersia* were isolated. Of the four native trees, the apricot tree yielded 12 *Streptomyces* spp. and four genera, *Amycolatopsis, Actinopolymorpha, Polymorphospora* and *Nocardiopsis*, which were not recovered from other trees.

7.4 Applications of Actinobacteria in Biotechnology

7.4.1 Antibiotics in Medicine and Agriculture

Actinobacteria are pioneers in their ability to produce wide range of antibiotic compounds, which have immense benefits in medicine and agriculture. Since the discovery of Streptomycin from the soil isolate, *Streptomyces griseus*, soil-derived *Streptomyces* spp. are the most sought microorganisms in the industrial screening programmes. ~8,000 antibiotics are described from *Streptomyces* spp., followed by rare actinomycetes (Berdy 2005).

7.4.1.1 Antibiotics from Soil-Derived Actinobacteria

Soil-derived actinobacteria, especially *Streptomyces*, deserve a special mention as the single largest producer (80%) of life-saving antibiotic drugs. They are typically soil bacteria, with plenty of spores readily recognized for the volatile, earthly odour, geosmin. Though the first antibiotic was discovered in the fermentation product of *Streptomyces griseus*, over the subsequent decades there has been a phenomenal search for the discovery of newer antibiotics owing to the emergence of multidrug-resistant Gram-negative bacterial strains. The mercurial period from 1945 to 1950s is remarkable for the discovery of antibiotics from actinobacterial genera *Actinomyces, Streptomyces, Micromonospora, Nocardia*, etc. *Streptomyces* spp. produce a range of antibiotics with diverse chemical structures and the basic classification deals with two broader classes (Benedict 1953): pigmented antibiotics (Aureomycin, Trichomycin, Griseolutin, Actinorhodine, Rhodocidin, etc.), non-pigmented antibiotics (Antimycin, Nigericin, Cardicine, Nocardamin, Flavomycin, etc.), non-pigmented organic bases (Streptomycin, Streptolin A & B, Neomycin, Flavomycin, Viomycin, Achromycin, etc.) and miscellaneous group (Streptocin, Chromin, Ascosin, Cacaomycetin, etc.).

The selectivity of antibiotic molecules to inhibit growth of the test organism is based on their inhibition against target structures or functions related to cell wall biosynthesis, translation, RNA transcription, DNA synthesis and replication (Lo Grasso et al. 2016). Recent studies on the genes and or gene clusters involved in biosynthesis of antibiotics and their regulation have opened up newer mechanisms to incorporate tailoring steps for operating the genes of interest for a particular antibiotic synthesis. Media composition, alteration of fermentation conditions and other genetic factors are required to produce antibiotics. Since many of the antibiotic-producing actinomycetes resist the genetic manipulation, transferring the antibiotic synthesizing genes to a heterologous expression system represents a successful strategy. Shuttle vectors harbouring the gene(s) of interest expressed in model organisms such as *S. coelicolor, S. avermitilis* and *S. lividans* have been reported (Yadav et al. 2019a; Alduina et al. 2003, 2005; Giardina et al. 2010).

Therapeutic drugs from actinobacteria have wide implications as targets against cancer, immunomodulators. Drug discovery is based on high-throughput platforms to screen thousands of fermentation products of microbes to deliver a new medicine with high market value. Actinobacteria have yielded drugs with potential value in clinical trials (Table 7.1). Rare actinobacteria, namely, *Micromonospora, Actinoplanes, Actinomadura* and *Streptosporangium* produce antibiotic classes such as macrolides, polysaccharides, aminocyclitols (Lancini and Lorenzetti 1993). The years 1966–1998 have provided literature on the antibiotics belonging to macrolides, quinones, diterpenes, anthracyclines and ansa-macrolactams in these rare genera (Lazzarini et al. 2000).

7.4.1.2 Antibiotics from Endophytic Actinobacteria

Antibiotics are important drugs preferred for health care due to their potent therapeutic applications for the clinical use (Farnet and Zazopoulos 2005). Plant-associated endophytic actinomycetes produce wide range of antibiotics (Matsumoto and Takahashi 2017). *Streptomyces* and *Micromonospora* are the potential producers of antibiotics. Munumbicins, the novel peptide antibiotics are produced by the endophytic *Streptomyces* spp., from the ethnomedicinal plants of the Upper Amazon and Northern Territory of Australia and were effective against Gram-positive bacteria *Bacillus anthracis* and *Mycobacterium tuberculosis* (Castillo et al. 2002, 2006). *Streptomyces* sp. 30566 produced Kakadumycins active against *B. anthracis* (MIC 0.2

Table 7.1 Antibiotics as drug targets from actinobacteria and their clinical applications

Drug	Analogue	Producer strain	Target	Mechanism	References
Approved drugs					
Everolimus	Rapamycin	*Streptomyces hygroscopicus*	Interleukin (IL)-2 and IL-15, T-cells and B-cells, Vascular smooth muscles	Arrest of cell cycle at the G1/S phase	Chapman and Perry (2004)
Telithromycin	Erythromycin A	*Saccharopolyspora erythraea*	Peptidyltransferase	Inhibits protein synthesis Antibacterial effect	Zhanel et al. (2002)
Miglustat	Nojirimycin	*Streptomyces lavendulae*	NK	Inhibits glucosylceramide synthase	Pastores et al. (2005)
Daptomycin		*Streptomyces roseosporus*	Bacterial cell membranes	Antibacterial	Fenton et al. (2004)
Amrubicin	Doxorubicin	*Streptomyces peucetius* var. *caesius*	Human tumor xenografts of breast lung and gastric cancer	Antitumour activity	Sugiura et al. (2005)
Biapenem	Thienamycin	*Streptomyces cattaleya*	Gram-positive and -negative bacteria	Antibacterial	Perry and Ibbotson (2002)
Ertapenem	Thienamycin	*Streptomyces cattaleya*	*Escherichia coli*, *Klebsiella* sp., *Citrobacter* sp., *Enterobacter* sp., *Morganella morganii*, *Proteus* sp. and *Serratia marcescens*	Antibacterial	Sader and Gales (2001)

(continued)

Table 7.1 (continued)

Drug	Analogue	Producer strain	Target	Mechanism	References
Pimercolimus	Ascomycin	*Streptomyces hygroscopicus* var. *ascomyceticus*	T-cell	Blocking T-cell activation	Gupta and Chow (2003)
Ozogamycin	Calichemicin	*Micromonospora echinospora Calichensis* sp.	Acute myeloid lymphoma	DNA cleaving antibiotics	Lee et al. (1987)
Drugs in clinical trials					
Elsamitrucin	Chartreusin	*Streptomyces chartreusis*	Topoisomerase II	Antitumour effect	Portugal (2003)
Brostallicin	Distamycin A	*Streptomyces distallicus*	DNA minor groove	Anticancer agent	Broggini et al. (2004)
Geldanamycin		*Streptomyces hygroscopicus*	NK	Inhibition of the protein chaperone heat shock protein (HSP)	Bisht et al. (2003)
Fidaxomicin	Lipiarmycin	*Actinoplanes deccanensis*	Gram-positive aerobes, anaerobes, Gram-negative anaerobes	RNA-polymerase inhibitor	Srivastava et al. (2011)
Lantibiotics	Carbacyclic lantibiotics	*Actinomadura namibiensis*	*Streptococcus aureus, Streptococcus pyogenes* and *Streptococcus faecium*	Antimicrobial activity	Boakes et al. (2011)

(continued)

Table 7.1 (continued)

Drug	Analogue	Producer strain	Target	Mechanism	References
Lantibiotics NAI-107	NK	*Microbispora* sp.	*S. aureus*	NK	Seibert et al. (2009) Meindl et al. (2010) Donadio et al. (2010) Genilloud and Vincente (2013)
Lucensimycins and okilactomycins	NK	*Streptomyces lucensis* MA7349	Lucensimycins A–B bind weakly to the S4 protein	NK	Singh et al. (2006, 2008a, b)
Delaminomycins A–C	NK	*Streptomyces albulus*	*S. aureus, Streptococcus pneumoniae*	Antibiotic activity	Ueno et al. (1993)
Okilactomycin A, B, C and D	NK	*Streptomyces scabrisporus Streptomyces griseoflavus* subsp. *zamamiensis*	*S. pneumoniae Enterococcus faecalis* and *Bacillus subtilis*	Inhibition for RNA synthesis	Imai et al. (1987)
Philipimycin	Thiazolyl peptides	*Actinoplanes philippinensis* MA7347	*S. pneumoniae, E. faecalis, E. faecium, S. aureus*	Inhibitor of protein synthesis	Singh et al. (2013)
NOVO3 and NOVO4	NK	*Streptosporangium* P1532, *Amycolatopsis* Z0363	*S. aureus* and *Enterococci*		Peoples et al. (2011, 2012)
Tetracycline	NK	*Streptomyces aureofaciens*	Inhibits aminoacyl-tRNA binding	Blocks protein synthesis	Van Bambeke (2004)

(continued)

Table 7.1 (continued)

Drug	Analogue	Producer strain	Target	Mechanism	References
Chloramphenicol	NK	NK	*Streptomyces orchidaceus*	Anti peptidoglycan synthesis	Lo Grasso et al. (2016)
Erythromycin	NK	*Saccharopolyspora erythraea*	50S subunit	Blocking the peptidyltransferase activity	
Kanamycin	NK	*Streptomyces kanamyceticus*	Binds to 30S subunit	NK	
Thiostrepton	NK	*Streptomyces laurentii*	NK	Inhibits ribosome-dependent EF-Tu and EF-G GTPase	
Streptomycin	NK	*S. griseus*	NK	Prevents formation of initiation complex	
Rifampicin	NK	*Amycolatopsis mediterranei*	NK	Anti-tuberculosis and inhibits the bacterial RNA polymerase	Schulz and Zillig (1981)
Novobiocin	Albamycin or cathomycin	*Streptomyces niveus*	GyrB subunit of the enzyme	Aminocoumarin antibiotic, inhibitor of bacterial DNA gyrase	Walsh et al. (1993), Maxwell (1999)

NK = not known

to 0.3 μg ml^{-1}) (Castillo et al. 2003). The antimycotic Coronamycin produced by *Streptomyces* NRRL 30562 at 2 μg ml^{-1} (MIC) is effective against pythiaceous fungi and the human pathogen *Cryptococcus neoformans* (MIC 4 μg ml^{-1}) (Ezra et al. 2004). It was tested against agriculturally important plant pathogens along with *S. griseoviridis* formulation (Mycostop) and considered as a potential agricultural agent. Maklamicin, an antibacterial polyketide from *Micromonospora* isolated from Maklam Phueak (*Abrus pulchellus*) has shown activity against Gram-positive bacteria at 0.2–13 μg ml^{-1} (Igarashi et al. 2011). The peptide antibiotic coronamycin from *Streptomyces* sp. (MSU-2110) showed cytotoxic potentials by inhibiting the HMEC and BT20 cell lines (IC_{50} 5–10 μg ml^{-1}) (Ezra et al. 2004). A trehalose-derived antibiotic and a novel inhibitor of metastasis, Brartemicin is produced by *Micromonospora* sp., isolated from *Artemisia vulgaris,* the Brazilian medicinal plant (Igarashi et al. 2009). The compound indicated anti-invasive property in murine colon carcinoma cells (IC_{50} 0.39 μM) without toxicity (Table 7.2).

7.4.2 Anti-pathogenic Potentials of Actinobacteria

Actinobacterial metabolites either produced by soil or endophytic organisms have strong antimicrobial potentials. 70% of these novel metabolites are produced by actinomycetes (Miyadoh 1993). Of all actinobacteria, *Streptomyces* spp., have contributed phenomenally for the production of the life-saving drugs such as the novel antibiotics, enzyme inhibitors, antiviral, antitumor targets and immunomodulators. Rare actinobacteria have contributed to antibiotic database as target drugs. A number of papers have published the anti-pathogenic potentials of soil, rhizospheric and endophytic actinobacteria, which is related to their antagonistic potentials (Table 7.3). Some of the mechanisms of antagonistic nature of these organisms and their metabolites include antibiosis, hyperparasitism and secretion of cell wall degrading enzymes resulting in hyphal lysis (El-Tarabily and Sivasithamparam 2006).

7.4.2.1 Anti-pathogenic Potentials of Soil and Rhizospheric Actinobacteria

The actinomycetes, of the genus *Streptomyces,* are saprophytic bacteria that decompose organic matter, such as lignocellulose, starch and chitin, in soil. Actinomycetes are important in the rhizosphere, where they influence plant growth and protect plant roots against the invasion of root pathogenic fungi (Crawford et al. 1993). A number of diseases of crop plants are caused by the root invading pathogens, panicle and sheath blight and grapevine trunk diseases (Loqman et al. 2009; Harikrishnan et al. 2014; Suarez-Moreno et al. 2019). The potential biocontrol agents from actinobacteria are summarized in Table 7.3. The anti-pathogenic activity against root rot pathogens is by hyperparasitism and well documented in the strains *A. philippinensis, M. carbonacea* against cavity spot disease of carrots (El-Tarabily et al. 1997) and

Table 7.2 Bioactive novel antibiotics produced by endophytic actinobacteria of medicinal plants

Actinobacterial strain	Medicinal species	Antibiotics	Chemical group	Bioactivity	References
Streptomyces NRRL 30562 *Streptomyces* NRRL 3052	*Kennedia nigriscans*	Mumumbicins A–D Munumbicins E-4 & E-5	Peptide antibiotics	Antibacterial; antimalarial	Castillo et al. (2002, 2006)
Streptomyces NRRL 30566	*Grevillea pteridifolia*	Kakadumycin A	Peptide antibiotics	Antibacterial; antimalarial	Castillo et al. (2003)
Streptomyces (MSU-2110)	*Monstera* sp.	Coronamycin	Peptide antibiotics	Antifungal; antimalarial	Ezra et al. (2004)
Micromonospora sp.	*Artemisia vulgaris*	Brartemicin	Trehalose-derived antibiotic	Anti-metastatic/anti-invasive	Igarashi et al. (2009)
Micromonospora sp.	*Abrus pulchellus* subsp. *puchellus*	Maklamicin	Polyketide	Antibacterial	Igarashi et al. (2011)

Table 7.3 Anti-pathogenic activity of actinobacterial strains from rhizosphere and selected plant species

Host plant	Pathogen	Disease	Anti-pathogenic Strain(s)	Mechanism	References
Rhizospheric soil strains					
Taraxacum officinale	*P. ultimum*	Damping off	*Streptomyces* sp.	Antagonism	Crawford et al. (1993)
Raspberry	*Phytophthora fragariae* var. *rubi*	Root rot	Actinomycete strains	Cell wall degradation by β-glucanases	Valois et al. (1996)
Daucus carota L.	*Pythium coloratum*	Cavity spot	*Actinoplanes philippinensis* *Micromonospora carbonacea*	Hyperparasitism	El-Tarabily et al. (1997)
Cucumis sativus L.	*Pythium aphanidermatum*	Root rot	*A. philippinensis* *Microbispora rosea* *Micromonospora chalcea* *Streptomyces griseoflavus* *Actinoplanes campanulatus* *Streptomyces spiralis*	Cell wall degradation by β-glucanases	El-Tarabily (2006), El-Tarabily et al. (2009)
Piper sp.	*Phytophthora capsici*	Root rot	*Streptomyces rochei*	Co-antagonistic with *Trichoderma harzianum*	Ezziyyani et al. (2007)
Traditional medicinal plants	Plant pathogenic Test bacteria	–	*Streptomyces* spp.	Antagonism	Zhao et al. (2012)
O. sativa L.	*R. solani*	Sheath blight of paddy	*Streptomyces aurantiogriseus* VSMGT1014	Antagonism	Harikrishnan et al. (2014)

(continued)

Table 7.3 (continued)

Host plant	Pathogen	Disease	Anti-pathogenic Strain(s)	Mechanism	References
Eucalyptus camaldulensis	*Cylindrocladium* sp.	Leaf and shoot blight	*Streptomyces ramulosus* *Streptomyces himastatinicus*	Antagonism	Himaman et al. (2016)
Vitis vinifera L.		GTD YGD	*Streptomyces* spp.	Antagonism	Alvarez-Perez et al. (2017)
Capsicum annuum L.	*Colletotrichum capsici*	Chilli anthracnose	*Streptomyces violaceoruber*	Antagonism	Thilagam and Hemalatha (2019)
Oryza sativa L.	*Burkholderia glumae*	Panicle blight	*Streptomyces* spp. A20	Antibiotic production	Suarez-Moreno et al. (2019)
Soil strains					
Soil/compost	*Fusarium oxysporum* f. sp. *cucumerinum*	Wilt of cucumber	*Streptomyces* sp. 385	Cell wall degradation by chitinolytic enzymes	Singh et al. (1999)
Culture collection (Malaya)	*Fusarium oxysporum* f. sp. *cubense*	Wilt of Cavendish banana	*Streptomyces* sp. g10	Antagonism	Getha et al. (2005)
Pond soil	*Fusarium oxysporum* f. sp. *lycopersici*	Wilt of tomato	*Streptomyces griseus*	Chitinase production	Anitha and Rabeeth (2009)
Loktak lake area (Manipur)	Rice fungal pathogens	Rice fungal diseases	*Streptomyces vinaceusdrappus*	Antagonism	Ningthoujam et al. (2009)
Soil, Northwest Iran	*Bacterial strains*	–	*Streptomyces* sp.	Antibacterial	Dehnad et al. (2010)

(continued)

Table 7.3 (continued)

Host plant	Pathogen	Disease	Anti-pathogenic Strain(s)	Mechanism	References
Herbal compost	*Fusarium oxysporum* f. sp. *ciceri*	Wilt of chickpea	*Streptomyces tsusimaensis*, *Streptomyces caviscabies*, *Streptomyces setonii*, *Streptomyces africanus*	Antagonism	Gopalakrishnan et al. (2011a)
Herbal vermicompost	*Macrophomina phaseolina*	Charcoal rot of Sorghum	*Streptomyces* sp.	Antagonism	Gopalakrishnan et al. (2011b)
Culture collection (Thailand)	*Fusarium oxysporum* f. sp. *capsici*	Wilt of chilli	*Streptomyces* sp.	Inhibition of conidial germination	Saengnak et al. (2013)
Crop plants	Plant pathogenic fungi	–	*Streptomyces* sp.	Antagonism	Kaur et al. (2013)
Field soil	*Sclerotinia sclerotiorum*	Stem rot of oilseed rape	*Streptomyces felleus* YJ1	Antagonism	Cheng et al. (2014)
Agricultural field	Test bacteria	–	*Streptomyces* sp. SCA 7	Antagonism	Kumar et al. (2014)
Mountain forest soil	*Alternaria alternata*	Brown spot of tobacco	*Streptomyces microflavus*	Pathogen inhibition	Gao et al. (2014)
Soil sample (Korea)	*Colletotrichum gloeosporioides*	Tomato and pepper anthracnose	*Streptomyces* sp. A 1022	Antagonism	Kim et al. (2014)
Vitis vinifera L.	*Botrytis cinerea*	Grey mold Grapevine trunk disease Young grapevine decline	*Micromonospora* sp. *Streptomyces* sp. *Streptomyces* spp.	Hyphal colonization Antagonism	Loqman et al. (2009), Alvarez-Perez et al. (2017)
Soil samples (Malaysia)	*Colletotrichum gloeosporioides*	Chilli anthracnose	*Streptomyces ambofaciens* S2	Antifungal compound	Heng et al. (2015)

(continued)

Table 7.3 (continued)

Host plant	Pathogen	Disease	Anti-pathogenic Strain(s)	Mechanism	References
Soil samples	Test microorganisms	–	*Streptomyces* sp. T-4	Antibiotic production	Atta (2015)
Soils, Western Ghats	Pathogenic fungal/bacterial strains	Human pathogenic	*Streptomyces rimosus*	Growth inhibition	Ganesan et al. (2017)
Endophytic strains					
Native plant species (Italy)	*Micrococcus luteus* *Escherichia coli* *Fusarium oxysporum* f. sp *cyclaminis*	–	*Streptomyces* spp.	Growth inhibition	Petrolini et al. (1996)
Lycopersicon esculentum	*Rhizoctonia solani*	Damping off	*Streptomyces* sp. S30		Cao et al. (2004)
Musa sp. AAA Cavendish	*Fusarium oxysporum* f. sp. *cubense*	*Fusarium* wilt	*Streptomyces griseorubiginous*	Antagonism	Cao et al. (2005)
L. esculentum	*Ralstonia solanacearum*	Bacterial wilt of tomato	*Streptomyces* spp.	Antagonism	Tan et al. 2006
Thottea grandiflora *Polyalthia* sp. *Mapania* sp.	Plant pathogenic fungi Test bacteria	–	*Streptomyces coelicolor* *Streptomyces caelestis* *Streptomyces fulvoviolaceus*	Antagonism	Zin et al. (2007)
L. esculentum	*R. solani* *Alternaria solani* *Phytophthora* sp. *Pythium irregulare*	–	*Streptomyces* spp. *Microbispora* spp.	Antagonism	Inderiati and Franco (2008)

(continued)

Table 7.3 (continued)

Host plant	Pathogen	Disease	Anti-pathogenic Strain(s)	Mechanism	References
Cucumis sativus *Cucumis moschata*	*Colletotrichum orbiculare*	Anthracnose	*Streptomyces* sp.	Hyphal colonization	Shimizu et al. (2009)
A. indica	*Pythium* spp. *Phytophthora* spp.	Damping off; late blight; root rot	*Streptosporangium* sp. *Streptoverticillium* sp. *Nocardia* sp.	Antagonism	Verma et al. (2009a, b)
Traditional medicinal plants	Plant pathogenic Test bacteria	–	*Streptomyces* spp. *Nonomuraea roseola* *Micromonospora chokoriensis*	Antagonism	Zhao et al. (2011)
Ocimum sanctum *Mentha arvensis* *Aloe vera*	Plant pathogenic fungi	–	*Saccharopolyspora* sp.	Antagonism	Gangwar et al. (2011)
Artemisia argyi *Paeonia lactifolia* *Radis platycodi* *Achyranthes bidentata*	*Staphylococcus aureus* MRSA	–	*Streptomyces* spp. *Glycomyces* sp.	Antibacterial inhibition	Zhang et al. (2012)
Crop plants	Plant pathogenic fungi	–	*Streptomyces* sp.	Antagonism	Kaur et al. (2013)
Native Algerian plants	*Rhizoctonia solani*	Damping off	*Streptomyces mutabilis* *Streptomyces cyaneofuscatus*	Antagonism	Goudjal et al. (2014)
Eucalyptus camaldulensis	*Cylindrocladium* sp.	Leaf and shoot blight	*Streptomyces ramulosus* *Streptomyces himastatinicus* *Streptomyces malaysiensis*	Antagonism	Himaman et al. (2016)

by the action of cell wall degrading enzymes in the hyphal lysis of plant pathogens such as glucanases and chitinases (El-Tarabily 2006; El-Tarabily et al. 2009).

In a previous study (El-Tarabily et al. 1996), 352 out of 817 streptomycete and non-streptomycete actinomycete isolates produced inhibitory compounds active against *Pythium coloratum* in vitro using the dual culture agar method. 45 inhibitory isolates were identified as *Streptomyces* spp., *Streptoverticillium* spp., *Actinoplanes* spp., *Micromonospora* spp., *Actinomadura* spp., *Microbispora* spp. and *Streptosporangium* spp. These isolates were chosen for further in vitro and in vivo studies. Non-streptomycete actinobacteria are excellent antagonists of several plant pathogens (El-Tarabily and Sivasithamparam 2006).

7.4.2.2 Anti-pathogenic Potential of Endophytic Actinobacteria

Some actinomycetes form associations with plants and colonize the internal tissues. *Streptomyces scabies* and *Frankia* species can penetrate host plant tissues and form pathogenic or endophytic associations between plants and actinomycetes, respectively (Hasegawa et al. 2006). Until reports of Sardi et al. (1992), work on endophytic actinomycetes other than *Streptomyces* were rare. *Streptomyces* spp., *Microbispora* spp. and *Streptosporangium* spp. were isolated from roots of different plant species in Italy and Brazil that showed antagonistic activities against Gram-positive bacteria and fungi (Sardi et al. 1992; Coombs and Franco 2003) and as biocontrol agents (Verma et al. 2011). The use of actinomycetes as biological control agents of soil-borne root disease of several crop plants is of interest. The endophytic presence of *Streptomyces* spp. may play important roles in plant development and health.

7.4.3 Actinobacteria as Plant Growth Promoters

Sustainable agriculture is a key player for boosting the food productivity. The increasing demand for food production with the use of conventional methods by the application of fungicides, herbicides, insecticides and chemical fertilizers adversely affects human health (Kour et al. 2019c; Rana et al. 2019a, b; Verma et al. 2017a). As an alternative, environment-friendly approach using beneficial microorganisms to improve plant traits for high yields is an efficient strategy (Glick 2012). Soil is a habitat for the multiplication of various microorganisms. The concentration of actinobacteria is greater in the soil attached to the roots, i.e., the rhizosphere and is facilitated by the secretion of amino acids, sugars and micromolecules from the root exudates (Badri et al. 2009). Therefore, actinobacteria are capable of promoting plant growth by a number of ways involving the uptake of nitrogen, phosphorus and iron (Kour et al. 2019b; Kumar et al. 2019b; Yadav et al. 2019b, f; Yadav and Saxena 2018). Plant rhizosphere contains growth-promoting rhizobacteria termed PGPR. The most abundant rhizosphere colonizing actinobacteria are the *Streptomyces* spp. with high species diversity. Rare actinobacteria are documented from these plant species. They are the important components of rhizospheric soils of crop plants, medicinal plants and trees.

Actinobacteria augment plant growth by direct or indirect mechanisms involving the (1) fixation of nitrogen, (2) phosphate solubilization, (3) sequestering of iron, (4) production of growth regulators and (5) ethylene. The genera *Streptomyces, Frankia, Nocardia, Kitasatospora and Thermobifida* are IAA producers (Rana et al. 2019c; Yadav 2018; Yadav et al. 2017c; Yadav and Yadav 2018; Franco-Correa and Chavarro-Anzola 2016). Rhizosphere soil samples collected from three-month-old wheat plants contained 15 morphologically distinct actinobacteria (Jog et al. 2012), while three strains: *S. rochei, S. carpinensis* and *thermolilacinus* with PGP traits were tested positive for the production of Indole acetic acid (IAA, 2.6–19.22 mg l^{-1}), siderophores (1.3–34.17 mg l^{-1}) and high phosphate solubilization (911.6 mg l^{-1}). Rice rhizosphere actinobacterial strain *Streptomyces* A20 showed PGP traits by solubilizing phosphate (79.5 mg l^{-1}), siderophore production, IAA production (4.0 mg l^{-1}) and extracellular enzymes cellulases and proteases (Suarez-Moreno et al. 2019). *Streptomyces violaceolatus,* from the coffee plantation soils, produced 109.24 mg ml^{-1} of IAA and enhanced seedling growth parameters in bean and sorghum as evaluated by the roll towel method (Sameera et al. 2018b). *Streptomyces* spp. isolated from the rhizospheres of plant species including crop plants was able to promote PGP traits.

One of the strategies to enhance plant growth is by adopting co-inoculation of the nodule-inducing species with the actinobacterial strains. Under greenhouse trials, co-inoculation of soybean (*Glycine max*) with *Bradyrhizobium japonicum* and the actinobacterial strains *Nocardia alba*, *Nonomuraea rubra* and *Actinomadura glauciflava* increased acetylene reduction activity ~1.7 to 2.7-fold with increased levels of potassium, N, calcium, iron, magnesium and zinc in the plants (Nimnoi et al. 2014). The association of actinomycetes is known to confer advantages to host plants with the production of IAA, siderophores and nutrient uptake.

Streptomycetes have demonstrated their ability as plant growth enhancers and as biocontrol agents and their products have high commercial value in controlling several plant–pathogenic diseases of agricultural species (Vurukonda et al. 2018). Some of the commercial products are registered as microbial pesticides in European Union, Canada, South Korea, USA in the targeted soil-borne, powder/downy mildews, sheath blight, grey molds, leaf spots of fruits and vegetables, bacterial rots, cankers, basal rots, fire blight etc.

7.4.4 Production of Industrially Important Enzymes

Actinomycetes are abundant taxa in the soil and act as decomposers of complex organic matter helpful for the nutrient cycling in the terrestrial and the aquatic ecosystems. Over the past few decades, considerable interest has been generated in the enzymatic degradation of complex polymers or substrates by microorganisms.

Their abundance, renewable and as inexpensive nature often ensures them as potential candidates in the industrial applications, viz., in food processing, oxychemicals production, and in textiles for the biopolishing of fabrics, and pulp and paper industry (Kasana et al. 2008).

Lignocellulolytic enzymes, one of the potent enzymes produced by actinomycetes, and exploited widely in various lignocelluloses based industries. They are hydrolytic enzymes capable of degrading tough lignocellulose in the plant biomass. Hydrolysis of lignocellulosic biomass by lignocellulolytic enzymes is used in the production of bioethanol and biomethane, textile industry, pulp and paper making, detergents industry, animal feed and food. Hemicellulases are used in biobleaching, deinking of paper waste, clarification of fruit juices, upgradation of feed, fodder and fibres, and saccharification of hemicelluloses to xylose sugars. Actinobacterial cellulases are inducible extracellular enzymes produced on wide variety of substrates. Jeffery et al. (2007) isolated a *Streptomyces griseus* strain from Malaysian soil that produced cellulase with carboxymethyl cellulose property (4.5 mg ml^{-1}). *Streptomyces* spp., *Cellulomonas fimi, Microbispora bispora* and *Thermobifida fusca* are cellulase producing actinomycetes (Saini et al. 2015). Proteases have applications in the bioremediation and degradation of gelatinous wastes generated by food and pharmaceutical industries. *Streptomyces* spp., isolated from Indian soil has protease producing property (Jain et al. 2009). Industrial production of enzymes on large scale is associated mainly with substrate. The use of agriculture residues as low-cost substrates for the production of industrial enzymes is a significant way to reduce production cost.

Xylan is the abundant source of noncellulosic polysaccharide present in hardwoods and annual plants, and comprises 20–35% of the total dry weight in tropical plant biomass (Elegir et al. 1994). Xylanase degrades β-1, 4 xylan, by cleaving the β-1, 4 glycosidic linkages randomly, and the products are xylose and xylo-oligosaccharides like xylobiose (Ninawe et al. 2008). Xylanases are important industrially, which is used in paper manufacturing to bleach paper pulp, increase the brightness of pulp and to improve the digestibility of animal feed and for clarification of fruit juices. Microorganisms are the rich sources of xylanases and are produced by diverse species of actinobacteria. *Streptomyces* sp. strain B-12-2 and *Streptomyces cyaneus* SN32 secrete high amounts of extracellular xylanases (Elegir et al. 1994; Ninawe et al. 2008).

7.4.5 Biodegradation Potentials of Actinobacteria

7.4.5.1 Pesticides

Pesticides are the chemicals employed to kill pests and are known to persist in soils over time. They are pollutants and accumulate in the body parts of plants, animals and humans. Soil microbes are known for their efficiency to degrade pesticidal compounds. Actinobacteria are dominant colonizers of soil types. The process of breakdown of chemical pesticides such as Carbofuran by soil actinobacteria has been documented (Jayabarath et al. 2010). *Streptomyces abnosinicus, Streptomyces atratus, Streptoverticillium album, Nocardia farcinia, Nocardia vaccine, Nocardia amarae* and *Micromonospora chalcea* have shown the ability to resist carbofuran under culture conditions (Kumar et al. 2019a). A soil *Streptomyces* sp. has shown the ability to degrade Chlorpyrifos, the insecticide into a less toxic form by hydrolysis (Briceno et al. 2012).

7.4.5.2 Hydrocarbons

The actinobacterial genus *Rhodococcus* is bestowed with the ability to degrade a wide range of organic and xenobiotic compounds, which poses severe health hazards. Besides, it is able to produce metabolites of biotechnological significance such as carotenoids, wax esters, oils, biosurfactants and bioflocculation agents (Yadav and Yadav 2019b; Jones and Goodfellow 2010). Progress related to the biosynthetic pathway and functional genomics have led to its immense applications in biotechnology. Hydrocarbons are released into the environment by the anthropogenic activities and by natural sources. *Rhodococcus* strains, *Rhodococcus jostii, Rhodococcus opacus, Rhodococcus ruber,* are able to oxidize n-alkanes due to the presence of the enzyme, alkane 1-monooxygenase (Tiancsics et al. 2014), alkane hydroxylases and cytochrome for the oxidation of xenobiotics. Some species are able to degrade highly classified aromatic hydrocarbons exemplified by benzene, toluene, ethylene and *o*-xylene (BTEX). *Rhodococcus* sp. strain DK17 has the ability to degrade Toluene and *o*-Xylene (Kim et al. 2002, 2010), whereas *R. jostii* RHA1 efficiently assimilates ethylbenzene, isopropyl benzene and biphenyl (Seto et al. 1995). Genetic analysis of this strain has identified 203 oxygenases, 86 dioxygenases, 88 flavoprotein monooxygenes and 50 hydroxylases in the conversion of steroids and aromatic compounds (McLeod et al. 2006). Phenols are degraded by *Rhodococcus* strains due to the presence of phenol hydroxylase. *R. erythropolis* UPV-1 efficiently degrades Polyaromatic hydrocarbons (PAH), phenol and a mixture of cresols (Irvine et al. 2000). PAH degradation is by the action of intercellular dioxygenases. Naphthalene is degraded by *R. opacus* R7 via this mechanism followed by the oxidation into salicylate and gentisate (Di Gennaro et al. 2010).

7.4.5.3 Plasticizer Compounds

Plasticizer compounds are released into nature by the industrial processes, which have deleterious effects. *Rhodococcus* strains efficiently degrade toxicants such as phthalates and terephthalates via the protocatechuate ortho-cleavage pathway (Patrauchan et al. 2005). Another noted synthetic polymer generated as plastic wastes is polyethylene and has a complex degradation process involving the oxidation of hydrocarbon chains into aliphatic fragments and subsequently requires specific microbial strains for their mineralization in nature (Koutny et al. 2006). *R. ruber* C208 efficiently degrades polyethylene. *R.* strain AD45 consumes Polyisoprene, the synthetic rubber used in seedlings by oxidation to epoxide and conjugation with glutathione and by dehydrogenation steps (Van Hylckama Vlieg et al. 2000). Gene coding for latex clearing protein (*lcp*) was identified in *Rhodococcus rhodochrous* strain RPK1 (Watcharakul et al. 2016) (Table 7.4).

7.4.6 Bioemulsifiers

Microorganisms produce bioemulsifiers during their growth phase. They contain surface active agents such as proteins and find applications in the form of biofilms as well are used as biomediators (Doshi et al. 2010). Biosurfactants have a hydrophilic moiety, comprising an acid, peptide cations, or anions, mono-, di- or polysaccharides and a hydrophobic moiety of unsaturated or saturated hydrocarbon chains or fatty acids (Lang 2002). These structures confer the ability to lower surface and interfacial tension of liquids and to form micelles and microemulsions between two different phases. These compounds are divided into two classes: low-molecular-weight compounds termed biosurfactants, such as lipopeptides, glycolipids, proteins and high-molecular-weight polymers of polysaccharides, lipopolysaccharide proteins or lipoproteins that are collectively called bioemulsans or bioemulsifiers (Banat et al. 2010). Many microorganisms are producers of bioemulsifiers. The hydrocarbon-degrading actinobacterial strains, *R. ruber and R. erythropolis* were earlier identified to produce bioemulsifiers (Bicca et al. 1999). *Actinopolyspora* sp. A18 isolated from garden soil exhibited emulsification activity and a partially purified glycopeptide (68% protein, 5% lipids, 22.2% non-reducing sugars) bioemulsifier was identified (Doshi et al. 2010). Five strains of *Streptomyces* isolated from the hydrocarbon-contaminated soil of Baghdad, Iraq showed biosurfactant properties, and strain SS20 produced maximum bioemulsifier property (E24% = 100%), and stability of 75% for two weeks (Hayder et al. 2014).

7.4.7 Conclusion and Future Prospects

Actinobacteria are ubiquitous in soil types, plant litter, plant rhizospheres and are associated with plant microbiomes as microbial communities. With plants, they form diverging modes of lifestyles ranging from saprophytic to endophytic associations. Actinobacterial diversity among crops, medicinal and native plant species have focused them as soil or rhizospheric and as endophytic colonizers. Actinobacteria comprising both streptomycetes and non-streptomycetes are producers of antibiotics and antimicrobial metabolites of high therapeutic applications. They are also used often as potential biocontrol agents in several crop diseases and as plant growth promoters. Besides, actinobacteria are known producers of important extracellular

Table 7.4 Biodegradation potentials of hydrocarbons and plasticizer compounds by *Rhodococcus* spp

Hydrocarbon class	Strain	Genes	Mechanism	Enzymes identified	References
Aliphatic hydrocarbons	*R. erythropolis, R. ruber, R. opacus, R. equi* and *R. jostii*	alk B	Oxidation	Alkane 1-monooxygenase	Tiancsics et al. (2014)
	R. jostii RHA 1	Cyp125	–	–	RosÅoniec et al. (2009)
	R. rhodochrous	Cytochrome P450	Hydroxylate octane	–	Cardini and Jurtshuk (1970)
	R. aetherivorans BCP1	prm A, C, B, D genes smo A, B, D, C genes	Degradation of short-chain n-alkanes	(soluble di-iron monooxygenase)	Cappelletti et al. (2015)
Aromatic hydrocarbons					
BTEX	*Rhodococcus* sp. strain DK17,	akb gene	Hydroxylation	Oxygenase, Reductase, a ferredoxin component and a dehydrogenase	Kim et al. (2002, 2010)
	R. jostii RHA1	bph, etb1, and etb2,	–	–	Seto et al. (1995)
	R. opacus R7	Monooxygenases/phenol hydroxylases	o-Xylene degradation pathway	Dioxygenase	Di Canito et al. (2018)
Phenols	*R. erythropolis* CCM2595	pheA1 and pheA2	–	Phenol hydroxylase enzyme	Zídková et al. (2013)
	R. erythropolis UPV-1	–	–	Degrading PAHs, phenol, and a mixture of o-, m-, and p-cresols	Irvine et al. (2000)
	R. opacus 1CP		Ortho-pathway	Catechol−1,2 dioxygenase	Kolomytseva et al. (2007)

(continued)

Table 7.4 (continued)

Hydrocarbon class	Strain	Genes	Mechanism	Enzymes identified	References
Polyaromatic hydrocarbons	*R. opacus* R7	narAa and narAb,	Dioxygenation of the aromatic ring	Naphthalene dioxygenase	Di Gennaro et al. (2010)
Plasticizer compounds					
Phthalate	*R. jostii* RHA1	–	Protocatechuate ortho-cleavage pathway 3-oxoadipate pathway	–	Hara et al. (2007, 2010); Patrauchan et al. (2005)
Polyethylene	*R. ruber* C208	–	Polyethylene degradation	–	Gravouil et al. (2017)
Polyisoprene	*Rhodococcus* sp. AD45	10 genes	Oxidation and dehydrogenation	Glutathione S-transferase and a 1-hydroxy-2-glutathionyl-2-methyl-3-butene dehydrogenase	Van HylckamaVlieg et al. (2000)
	R. rhodochrous RPK1	lcp	Oxidation	–	Watcharakul et al. (2016)
4,4-Dithiodibutyric acid (DTDB)	*R. erythropolis* MI2	MI2	4 MB catabolism pathway	Monooxygenase	Khairy et al. (2016)

enzymes which find applications in the industrial processes. They are also degraders of harmful substances known to be environmental and health hazardous. New insights in the actinobacterial genomics have deciphered the mystery behind antibiotic production, engineered drugs for clinical applications and their efficiency in biodegradation. A thorough understanding of these strategies combined with metagenomic approaches may unravel the true potentials of many more actinobacterial taxa for future applications.

References

Adegboye MF, Babalola OO, Ngoma L, Okoh AI (2012) Analysis of *Streptomyces* spp. native to Mahikeng soils of South Africa. J Pure Appl Microbiol 6(3):1001–1010

Akshatha JV, Prakash HS, Nalini MS (2016) Actinomycete endophytes from the ethnomedicinal plants of southern India: antioxidant activity and characterization studies. J Biol Active Prod Nat 6(2):150–165

Alduina R, De Grazia S, Dolce L, Salerno P, Sosio M, Donadio S, Puglia AM (2003) Artificial chromosome libraries of *Streptomyces coelicolor* A3(2) and *Planobispora rosea*. FEMS Microbiol Lett 218(1):181–186

Alduina R, Giardina A, Gallo G, Renzone G, Ferraro C, Contino A, Scaloni A, Dona-dio S, Puglia AM (2005) Expression in *Streptomyces lividans* of *Nonomuraea* genes cloned in an artificial chromosome. Appl Microbiol Biotechnol 68(5):656–662

Álvarez-Pérez JM, González-García S, Cobos R, Olego MÁ, Ibañez A, Díez-Galán A, Garzón-Jimeno E, Coque JJR (2017) Use of endophytic and rhizosphereactinobacteria from grapevine plants to reduce nursery fungal graft infections that lead to young grapevine decline. Appl Environ Microbiol 83:e01564–17. https://doi.org/10.1128/AEM.01564-17

Anitha A, Rabeeth M (2009) Control of *Fusarium* wilt of tomato by bioformulation of *Streptomyces griseus* in green house condition. Afr J Basic Appl Sci 1:9–14

Atta HM (2015) Biochemical studies on antibiotic production from *Streptomyces* sp.: Taxonomy, fermentation, isolation and biological properties. J Saudi Chem Soc 19:12–22

Bacon CW, White JF (2000) Microbial endophytes. Marcel Dekker, New York, pp 341–388

Badri DV, Weir TL, van der Lelie D, Vivanco JM (2009) Rhizosphere chemical dialogues: plant-microbe interactions. Curr Opin Biotechnol 20(6):642–650

Banat M, Andrea F, Isabella G, Giuseppina B, Maria GM, Letizia F, Smyth TJ, Roger M (2010) Microbial biosurfactants production, applications and future potential. Appl Microbiol Biotechnol 87:427–444

Bascom-Slack CA, Ma C, Moore E, Babbs B, Fenn K, Greene JS, Hann BD, Keehner J, Kelley-Swift EG, Kembaiyan V, Lee SJ, Li P, Light DY, Lin EH, Schorn MA, Vekhter D, Boulanger LA, Hess WM, Vargas PN, Strobel GA, Strobel SA (2009) Multiple, novel biologically active endophytic actinomycetes isolated from upper Amazonian rainforests. Microb Ecol 58(2):374–383

Benedict RG (1953) Antiobiotics produced by actinomycetes. Bot Rev 19:229–320. https://doi.org/10.1007/BF02861819

Berdy J (2005) Bioactive microbial metabolites: a personal view. J Antibiot 58(1):1–26

Berendsen RL, Pieterse CMJ, Bakker PAHM (2012) The rhizosphere microbiome and plant health. Trends Plant Sci 17:478–486. https://doi.org/10.1016/j.tplants.2012.04.001

Berg G, Zachow C, Müller H, Philipps J, Tilcher R (2013) Nextgeneration bio-products sowing the seeds of success for sustainable agriculture. Agronomy 3:648–656

Berry AM, Mendoza-Herrera A, Guo Y-Y, Hayashi J, Persson T, Barabote RD, Demchenko K, Zhang S, Pawlowski K (2011) New perspectives on nodule nitrogen assimilation in actinorhizal symbioses. Funct Plant Biol 38:645–652

Bicca FC, Fleck LC, Ayub AZ (1999) Production of biosurfactant by hydrocarbon degrading *Rhodococcus ruber* and *Rhodococcus erythropolis*. Rev de Microbiol 30:231–236

Bignell DR, Huguet-Tapia JC, Joshi MV, Pettis GS, Loria R (2010) What does it take to be a pathogen: genomic insights from *Streptomyces* species. Anton Van Lee 98:179–194. https://doi.org/10.1007/s10482-010-9429-1

Bisht KS, Bradbury CM, Mattson D, Kaushal A, Sowers A, Markovina S, Ortiz KL, Sieck LK, Isaacs JS, Brechbiel MW, Mitchell JB, Neckers LM, Gius D (2003) Geldanamycin and 17-allylamino-17-demethoxygeldanamycin potentiate the *in vitro* and *in vivo* radiation response of cervical tumor cells via the heat shock protein 90-mediated intracellular signaling and cytotoxicity. Cancer Res 63:8984–8995

Boakes S, Cortes Bargallo J, Dawson MJ (2011) US Patent 7.989.416

Bodenhausen N, Horton MW, Bergelson J (2013) Bacterial communities associated with the leaves and the roots of *Arabidopsis thaliana*. PLoS ONE 8(2):e56329. https://doi.org/10.1371/journal.pone.0056329

Briceno G, Fuentes MS, Palma G, Jorquera MA, Amoroso MA, Deiz MC (2012) Chlorpyrifos biodegradation and 3, 5, 6 trichloropyridinol production by actinobacteria isolated from soil. Int Biodet Biodegrad 73:1–7

Broggini MS, Marchini E, Fontana D, Moneta C, Fowst Geroni C (2004) Brostacillin: A new concept in minor groove DNA binder development. Antican Drugs 15:1–6

Buée M, Reich M, Murat C, Morin E, Nilsson RH, Uroz S, Martin F (2009) 454 pyrosequencing analyses of forest soils reveal an unexpectedly high fungal diversity. New Phytol 184:449–456

Bulgarelli D, Rott M, Schlaeppi K, Loren van Themaat EV, Ahmadinejad N, Assenza F, Rauf P, Huettel B, Reinhardt R, Schmelzer E, Peplies J, Gloeckner FO, Amann R, Eickhorst T, Schulze-Lefert P (2012) Revealing structure and assembly cues for *Arabidopsis* root-inhabiting bacterial microbiota. Nature 488:91–95

Bunyoo C, Duangmal K, Nuntagij A, Thamchaipenet A (2009) Characterisation of endophytic actinomycetes isolated from wattle trees (*Acacia auriculiformis* A. Cunn. ex Benth.) in Thailand. Thai J Genet 2(2):155–163

Cao L, Qiu Z, You J, Tan H, Zhou S (2005) Isolation and characterization of endophytic *streptomycete* antagonists of *Fusarium* wilt pathogen from surface-sterilized banana roots. FEMS Microbiol Lett 247:147–152. https://doi.org/10.1016/j.femsle.2005.05.006

Cao L, Qiu Z, You J, Tan H, Zhou S (2004) Isolation and characterization of endophytic *Streptomyces* strains from surface-sterilized tomato (*Lycopersicon esculentum*) roots. Lett Appl Microbiol 39:425–430. https://doi.org/10.1111/j.1472-765X.2004.01606.x

Cappelletti M, Presentato A, Milazzo G, Turner RJ, Fedi S, Frascari D, Zannoni D (2015) Growth of *Rhodococcus* sp. strain BCP1 on gaseous n-alkanes: new metabolic insights and transcriptional analysis of two soluble di-iron monooxygenase genes. Front Microbiol. https://doi.org/10.3389/fmicb.2015.00393

Cardini G, Jurtshuk P (1970) The enzymatic hydroxylation of n-octane by *Corynebacterium* sp. strain 7E1C. J Biol Chem 245:2789–2796

Caruso M, Colombo AL, Fideli L, Pavesi A, Quaroni S, Saracchi M, Ventrella G (2000) Isolation of endophytic fungi and actinomycetes taxane producers. Ann Microbiol 50:3–13

Castillo U, Harper JK, Strobel GA, Sears J, Alesi K, Ford E, Lin J, Hunter M, Maranta M, Ge H, Yaver D, Jensen JB, Porter H, Robison R, Millar D, Hess WM, Condron M, Teplow D (2003) Kakadumycins, novel antibiotics from *Streptomyces* sp. NRRL 30566, an endophyte of *Grevillea pteridifolia*. FEMS Microbiol Lett 234:183–190

Castillo UF, Strobel GA, Ford EJ, Hess WM, Porter H, Jensen JB, Albert H, Robison R, Condron MA, Teplow DB, Stevens D, Yaver D (2002) Munumbicins, wide-spectrum antibiotics produced by *Streptomyces* NRRL 30562, endophytic on *Kennedia nigriscans*. Microbiol 148:2675–2685

Castillo UF, Strobel GA, Mullenberg K, Condron MM, Teplow DB, Folgiano V, Gallo M, Ferracane R, Mannina L, Viel S, Codde M, Robison R, Porter H, Jensen J (2006) Munumbicins E-4 and E-5: novel broad-spectrum antibiotics from *Streptomyces* NRRL 3052. FEMS Microbiol Lett 255:296–300

Chapman TM, Perry CM (2004) Everolimus. Drugs 64:861–872
Cheng G, Huang Y, Yang Y, Liu F (2014) *Streptomyces felleus* YJ1: potential biocontrol agents against the *Sclerotinia* stem rot (*Sclerotinia sclerotiorum*) of oilseed rape. J Agric Sci 6:91–98
Coombs JC, Franco CMM (2003) Isolation and identification of actinobacteria from surface-sterilized wheat roots. Appl Environ Microbiol 69:5603–5608. https://doi.org/10.1128/AEM.69.9.5603-5608.2003
Crawford DL, Lynch JM, Whipps JM, Ousley MA (1993) Isolation and characterization of actinomycete antagonists of a fungal root pathogen. Appl Environ Microbiol 59:3899–3905
Das M, Royer TV, Leff LG (2007) Diversity of fungi, bacteria, and actinomycetes on leaves decomposing in a stream. Appl Environ Microbiol 73(3):756–767
DeAngelis KM, Brodie EL, DeSantis TZ, Andersen GL, Lindow SE, Firestone MK (2009) Selective progressive response of soil microbial community to wild oat roots. ISME J 3:168–178
Dehnad A, Parsa L, Bakhshi R, Soofiani SA, Mokhtarzadeh A (2010) Investigation antibacterial activity of *Streptomyces* isolates from soil samples, West of Iran. Afr J Microbiol Res 4(14):1542–1549
Dessaux Y, Grandclement C, Faure D (2016) Engineering the rhizosphere. Trends Plant Sci 21:266–278
Di Canito A, Zampolli J, Orro A, D'Ursi P, Milanesi L, Sello G, Steinbüchel A, Di Gennaro P (2018) Genome-based analysis for the identification of genes involved in o-xylene degradation in *Rhodococcus opacus* R7. BMC Genom 19(587):1–17. https://doi.org/10.1186/s12864-018-4965-6
Di Gennaro P, Terreni P, Masi G, Botti S, De Ferra F, Bestetti G (2010) Identification and characterization of genes involved in naphthalene degradation in *Rhodococcus opacus* R7. Appl Microbiol Biotechnol 87:297–308
Donadio S, Maffioli S, Monciardini P, Sosio M, Jabes D (2010) Antibiotic discovery in the twenty-first century: current trends and future perspectives. J Antibiot 63:423–430
Doshi DD, Muniyar JP, Bhuyan SS, Mujumdar SS (2010) Studies on bioemulsifier production by *Actinopolyspora* sp. A18 isolated from garden soil. Ind J Biotechnol 9:391–396
Elegir G, Szakacs G, Jeffries TW (1994) Purification, characterization, and substrate specificities of multiple xylanases from *Streptomyces* sp. strain B-12-2. Appl Environ Microbiol 60(7):2609–2615
El-Tarabily KA (2006) Rhizosphere-competent isolates of streptomycete and non-streptomycete actinomycetes capable of producing cell-wall-degrading enzymes to control *Pythium aphanidermatum* damping-off disease of cucumber. Can J Bot 84(2):211–222. https://doi.org/10.1139/b05-153
El-Tarabily KA, Hardy GEStJ, Sivasithamparam K, Kurtboke DI (1997) The potential for the biological control of cavity spot diesase of carrots, caused by *Pythium coloratum* by streptomycete and non-streptiomycete actinomycetes. New Phytol 137:307–495
El-Tarabily KA, Nassar AH, Hardy GE, Sivasithamparam K (2009) Plant growth promotion and biological control of *Pythium aphanidermatum* a pathogen of cucumber by endophytic actinomycetes. J Appl Microbiol 106:13–26. https://doi.org/10.1111/j.1365-2672.2008.03926.x
El-Tarabily KA, Sivasithamparam K (2006) Non-streptomycete actinomycetes as biocontrol agents of soil-borne fungal plant pathogens and as plant growth promoters. Soil Biol Biochem 38:1505–1520
El-Tarabily KA, Sykes ML, Kurtboke DI, Hardy GESJ, Barbosa AM, Dekker RFH (1996) Synergistic effects of a cellulase-producing *Micromonospora carbonacea* and an antibiotic-producing *Streptomyces violascens* on the suppression of *Phytophthora cinnamomi* root-rot of *Banksia grandis*. Can J Bot 74:618–624
Evtushenko LI, Dorofeeva LV (2012) Genus Rathayibacter. In: M Goodfellow, P. Kämpfer, HJ Busse, ME Trujillo, KI Suzuki,W Ludwig et al (eds) Bergey's manual of systematic bacteriology, vol 5. Springer, NewYork, NY, pp 949–960

Ezra D, Castillo UF, Strobel GA, Hess WM, Porter H, Jensen JB, Condron MA, Teplow DB, Sears J, Maranta M, Hunter M, Weber B, Yaver D (2004) Coronamycins, peptide antibiotics produced by a verticillate *Streptomyces* sp. (MSU-2110) endophytic on *Monstera* sp. Microbiol 150:785–793

Ezziyyani M, Requena ME, Egea-Gilabert C, Candela ME (2007) Biological control of *Phytophthora* root rot of pepper using *Trichoderma harzianum* and *Streptomyces rochei* in combination. J Phytopathol 155:342–349

Farnet CM, Zazopoulos E (2005) Improving the discovery from microorganisms. In: Zhang L, Demain DL (eds) Natural products: drug discovery and therapeutic medicine, Humana, Totowa, p 95

Fenton C, Keating GM, Curran MP (2004) Daptomycin. Drugs 64:445–455

Franco-Correa M, Chavarro-Anzola V (2016) Actinobacteria as plant growth-promoting rhizobacteria. In Actinobacteria-Basics and Biotechnological Applications. IntechOpen, pp 249–270

Frey-Klett P, Garbaye J, Tarkka M (2007) The mycorrhiza helper bacteria revisited. New Phytol 176:22–36

Ganesan P, Reegan AD, David RHA, Gandhi MR, Paulraj MG, Al-Dhabi NA, Ignacimuthu S (2017) Antimicrobial activity of some actinomycetes from Western Ghats of Tamil Nadu, India. Alexan J Med 53:101–110

Gangwar M, Dogra S, Sharma N (2011) Antagonistic bioactivity of endophytic actinomycetes isolated from medicinal plants. J Adv Lab Res Biol 2(4):154–157

Gans J, Wolinsky M, Dunbar J (2005) Computational improvements reveal great bacterial diversity and high metal toxicity in soil. Science 309:1387–1390

Gao F, Wu Y, Wang M (2014) Identification and antifungal activity of actinomycete strain against *Alternaria* sp. Span J Agric Res 12(4):1158–1165

Genilloud O, Vicente F (2013) In: Marinelli F, Genilloud O (eds) Antimicrobials, new and old molecules in the fight against multiresistant bacteria, vol 17. Springer, Germany, pp 327–360

Getha K, Vikineswary S, Wong WH, Seki T, Ward A, Goodfellow M (2005) Evaluation of *Streptomyces* sp. strain g10 for suppression of *Fusarium* wilt and rhizosphere colonization in pot-grown banana plantlets. J Ind Microbiol Biotechnol 32:24–32

Giardina A, Alduina R, Gottardi E, Di Caro V, Süssmuth RD, Puglia AM (2010) Two heterologously expressed *Planobispora rosea* proteins cooperatively induce *Streptomyces lividans* thiostrepton uptake and storage from the extracellular medium. Microb Cell Fact 9:44

Glick BR (2012) Plant growth-promoting bacteria: mechanisms and applcications. Article ID 963401, http://dx.doi.org/10.6064/2012/963401

Gopalakrishnan S, Humayun P, Kiran BK, Kannan IGK, Vidya MS, Deepthi K, Rupela O (2011b) Biocontrol of charcoal-rot of sorghum by actinomycetes isolated from herbal vermicompost. Afr J Biotechnol 10:18142–18152

Gopalakrishnan S, Pandey S, Sharma M, Humayun P, Kiran BK, Sandeep D, Vidya MS, Deepthi K, Rupela O (2011a) Evaluation of actinomycete isolates obtained from herbal vermicompost for the biological control of *Fusarium* wilt of chickpea. Crop Prot 30:1070–1078

Goudjal Y, Toumatia O, Yekkour A, Sabaou N, Mathieu F, Zitouni A (2014) Biocontrol of *Rhizoctonia solani* damping-off and promotion of tomato plant growth by endophytic actinomycetes isolated from native plants of Algerian Sahara. Microbiol Res 169(1):59–66. https://doi.org/10.1016/j.micres.2013.06.014

Gravouil K, Ferru-Clément R, Colas S, Helye R, Kadri L, Bourdeau L, Moumen B, Mercier A, Ferreira T (2017) Transcriptomics and lipidomics of the environmental strain *Rhodococcus ruber* point out consumption pathways and potential metabolic bottlenecks for polyethylene degradation. Environ Sci Technol 51:5172–5181

Gupta AK, Chow M (2003) Pimecrolimus: a review. J Eur Acad Dermatol Venereol 17:493–503

Hara H, Eltis LD, Davies JE, Mohn WW (2007) Transcriptomic analysis reveals a bifurcated terephthalate degradation pathway in *Rhodococcus* sp. strain RHA1. J Bacteriol 189:1641–1647

Hara H, Stewart GR, Mohn WW (2010) Involvement of a novel ABC transporter and monoalkyl phthalate ester hydrolase in phthalate ester catabolism by *Rhodococcus jostii* RHA1. Appl Environ Microbiol 76:1516–1523

Hardoim PR, van Overbeek LS, Berg G, Pirttilä AM, Compant S, Campisano A, Döring M, Sessitsch A (2015) The hidden world within plants: ecological and evolutionary considerations for defining functioning of microbial endophytes. Microbiol Mol Biol Rev 79(3):293–320. https://doi.org/10.1128/MMBR.00050-14

Harikrishnan H, Shanmugaiah V, Balasubramanian N, Sharma MP, Kotchoni SO (2014) Antagonistic potential of native strain *Streptomyces aurantiogriseus* VSMGT1014 against sheath blight of rice disease. World J Microbiol Biotechnol 30:3149–3161

Hasegawa T, Lechevalier MP, Lechevalier HA (1978) A new genus of the Actinomycetales, *Actinosynnema* gen. nov. Int J Syst Bacteriol 28:304–310

Hasegawa S, Meguro A, Shimizu M, Nishimura T, Kunoh H (2006) Endophytic actinomycetes and their interactions with host plants. Actinomycetologica 20(2):72–81

Hawksworth DL (2001) The magnitude of fungal diversity: the 1.5 million species estimate revisited. Mycol Res 108:1422–1432

Hayakawa M (2008) Studies on the isolation and distribution of rare actinomycetes in soil. Actinomycetol 22:12–19

Hayakawa M, Momose Y, Yamazaki T, Nonomura H (1996a) A method for the selective isolation of Microtetraspora glauca and related four-spored actinomycetes from soil. J Appl Microbiol 80(4):375–386

Hayakawa M, Nonomura H (1987) Humic acid-vitamin agar a new medium for the selective isolation of soil actinomycetes. J Ferment Technol 65:501–509

Hayakawa M, Takeuchi T, Yamazaki T (1996b) Combined use of trimethoprim with nalidixic acid for the selective isolation of actinomycetes from soil. Actinomycetol 10:80–90

Hayder NH, Alaa S, Abdulmalik H (2014) Optimized conditions for bioemulsifier production by local *Streptomyces* sp. SS 20 isolated from hydrocarbon contaminated soil. Rom Biotechnol Lett 19(1):8979–8993

Heng JLS, Shah UK, Rahman NAA, Shaari K, Hamzah H (2015) *Streptomyces ambofaciens* S2—a potential biological control agent for *Colletotrichum gleosporioides* the causal agent for anthracnose in red chilli fruits. J Plant Pathol Microbiol S 1:006. https://doi.org/10.4172/2157-7471.S1-006

Himaman W, Hamchaipenet A, Pathom-Aree W, Duangmal K (2016) Actinomycetes from *Eucalyptus* and their biological activities for controlling Eucalyptus leaf and shoot blight. Microbiol Res 188–189:42–52. https://doi.org/10.1016/j.micres.2016.04.011

Igarashi Y, Mogi T, Yanase S, Fujita T, Sakurai H, Saiki I, Ohsaki A (2009) Brartemicin, an inhibitor of tumor cell invasion from the actinomycete *Nonomuraea* sp. J Nat Prod 72:980–982

Igarashi Y, Ogura H, Furihata K, Oku N, Indananda C, Thamchaipenet A (2011) Maklamicin, an antibacterial polyketide from endophytic *Micromonospora* sp. J Nat Prod 74:670–674. https://doi.org/10.1021/np100727h

Imai H, Suzuki KI, Morioka M, Numasaki Y, Kadota S, Nagai K, Sato T, Iwanami M, Saito T (1987) Okilactomycin, a novel antibiotic produced by a *Streptomyces* species. J Antibiot 40:1475–1482

Inderiati S, Franco CMM (2008) Isolation and identification of endophytic actinomycetes and their antifungal activity. J Biotech Res Tropical Region 1:1–6

Irvine VA, Kulakov LA, Larkin MJ (2000) The diversity of extradiol dioxygenase Bedo genes in cresol degrading rhodococci from a creosote-contaminated site that express a wide range of degradative abilities. Anton van Lee 78:341–352

Jain R, Agrawal SC, Jain PC (2009) Proteolytic actinomycetes from Indian habitats. J Cult Coll 6:28–37

Janso JE, Carter GT (2010) Biosynthetic potential of phylogenetically unique endophytic actinomycetes from tropical plants. Appl Environ Microbiol 76:4377–4386

Jansson JK, Neufeld JD, Moran MA, Gilbert JA (2012) Omics for understanding microbial functional dynamics. Environ Microbiol 14:1–3

Jayabarath J, Musfira SA, Giridhar R, Shyam Sundar S, Arulmurugan R (2010) Biodegradation of carbofuran pesticide by saline soil actinomycetes. Int J Biotecnol Biochem 6(2):187–192

Jeffery LSH, Sahilah AM, Son R, Tosiah S (2007) Isolation and screening of actinomycetes from from Malaysian soil for their enzymatic and antimicrobial activities. J Trop Agric and Food Sci 35(1):159–164

Jog R, Naresh kumar G, Rajkumar S (2012) Plant growth promoting potential and soil enzyme production of the most abundant *Streptomyces* spp. from wheat rhizosphere. J Appl Microbiol 113:1154–1164

Jones A, Goodfellow M (2010) Genus II. Rhodococcus (Zopf 1891) emend Goodfellow et al 1998. In: Bergey's manual of systematic bacteriology, vol 4, 2nd edn. Springer, Berlin, pp 1–65

Kaewkla O, Franco CM (2013) Rational approaches to improving the isolation of endophytic actinobacteria from Australian native trees. Microb Ecol 65(2):384–393. https://doi.org/10.1007/s00248-012-0113-z

Kasana RC, Salwan R, Dhar H, Dutt S, Gulati A (2008)A rapid and easy method for the detection of microbial cellulases on Agar plates using gram's iodine. Curr Microbiol 55:503–507

Kaur T, Sharma D, Kaur A, Manhas RK (2013) Antagonistic and plant growth promoting activities of endophytic and soil actinomycetes. Arch Phytopathol Plant Protection 46(14):1756–1768. https://doi.org/10.1080/03235408.2013.777169

Kembel SW, O'connor TK, Arnold HK, Hubbell SP, Wright SJ, Green JL (2014) Relationships between phyllosphere bacterial communities and plant functional traits in a neotropical forest. Proc Natl Acad Sci (USA) 38:13715–13720

Khairy H, Meinert C, Wübbeler JH, Poehlein A, Daniel R, Voigt B, Riedel K, Steinbüchel A (2016) Genome and proteome analysis of *Rhodococcus erythropolis* MI2: elucidation of the 4,4′-Dithiodibutyric acid catabolism. PLoS ONE 11(12):e0167539. https://doi.org/10.1371/journal.pone.0167539

Kim D, Kim YS, Kim SK, Kim SW, Zylstra GJ, Kim YM, Kim E (2002) Monocyclic aromatic hydrocarbon degradation by *Rhodococcus* sp. strain DK17. Appl Environ Microbiol 68:3270–3278

Kim HJ, Lee EJ, Park SH, Lee HS, Chung N (2014) Biological control of anthracnose (*Colletotrichum gloeosporioides*) in pepper and cherry tomato by *Streptomyces* sp. A1022. J Agric Sci 2:54–62

Kim SH, Han HY, Lee YJ, Kim CW, Yang JW (2010) Effect of electrokinetic remediation on indigenous microbial activity and community within diesel contaminated soil. Sci Total Environ 408:3162–3168

Kim TU, Cho SH, Han JH, Shin YM, Lee HB, Kim SB (2012) Diversity and physiological properties of root endophytic actinobacteria in native herbaceous plants of Korea. J Microbiol 50(1):50–57. https://doi.org/10.1007/s12275-012-1417-x

Kolomytseva MP, Baskunov BP, Golovleva LA (2007) Intradiol pathway of para-cresol conversion by *Rhodococcus opacus* 1CP. Biotechnol J 2:886–893

Kour D, Rana KL, Kumar A, Rastegari AA, Yadav N, Yadav AN, Gupta VK (2019a) Extremophiles for hydrolytic enzymes productions: Biodiversity and potential biotechnological applications. In: Molina G, Gupta VK, Singh BN, Gathergood N (eds) Bioprocessing for Biomolecules Production. Wiley, USA, pp 321–372

Kour D, Rana KL, Yadav N, Yadav AN, Kumar A, Meena VS, Singh B, Chauhan VS, Dhaliwal HS, Saxena AK (2019b) Rhizospheric microbiomes: biodiversity, mechanisms of plant growth promotion, and biotechnological applications for sustainable agriculture. In: Kumar A, Meena VS (eds) Plant growth promoting rhizobacteria for agricultural sustainability: from theory to practices. Springer Singapore, Singapore, pp 19–65. https://doi.org/10.1007/978-981-13-7553-8_2

Kour D, Rana KL, Yadav N, Yadav AN, Singh J, Rastegari AA, Saxena AK (2019c) Agriculturally and industrially important fungi: current developments and potential biotechnological applications. In: Yadav AN, Singh S, Mishra S, Gupta A (eds) Recent advancement in white biotechnology through fungi, volume 2: perspective for value-added products and environments. Springer International Publishing, Cham, pp 1–64. doi:https://doi.org/10.1007/978-3-030-14846-1_1

Koutny M, Sancelme M, Dabin C, Pichon N, Delort AM, Lemaire J (2006) Acquired biodegradability of polyethylenes containing prooxidant additives. Polym Degrad Stab 91:1495–1503

Kumar M, Kour D, Yadav AN, Saxena R, Rai PK, Jyoti A, Tomar RS (2019b) Biodiversity of methylotrophic microbial communities and their potential role in mitigation of abiotic stresses in plants. Biologia 74:287–308. https://doi.org/10.2478/s11756-019-00190-6
Kumar PS, Duraipandiyan V, Ignacimuthu S (2014) Isolation, screening and partial purification of antimicrobial antibiotics from soil *Streptomyces* sp. SCA 7. Kaohsiung J Med Sci 30(9):435–46. https://doi.org/10.1016/j.kjms.2014.05.006
Labeda DP, Dunlap CA, Rong X, Huang Y, Doroghazi K-S, Metcalf WW (2017) Phylogenetic relationships in the family Streptomycetaceae using multi-locus sequence analysis. Anton Van Lee 110:563–583
Lancini G, Lorenzetti R (1993) Biotechnology of antibiotics and other bioactive microbial metabolites Plenum Press, New York and London pp 49–57
Lang S (2002) Biological amphiphiles (microbial biosurfactant). Curr Opin Colloid Inter Science 7(1–2):12–20
Lazzarini A, Cavaletti L, Toppo G, Marinelli F (2000) Rare genera of actinomycetes as potential sources of new antibiotics. Anton Van Lee 78:399–405
Lee MD, Dunne TS, Siegel MM, Chang CC, Morton GO, Borders DB (1987) Calichemicins a novel family of antitumor antibiotics. 1. Chemistry and partial structure of calichemicin. J Am Chem Soc 109:3464–3466
Lo Grasso L, Martino DL, Alduina R (2016) Production of antibacterial compounds from actinomycetes. Actinobacteria-Basics Biotechnol Appl 177–198. https://doi.org/10.5772//61525
Lopez-Velasco G, Carder PA, Welbaum GE, Ponder MA (2013) Diversity of the spinach (*Spinacia oleracea*) spermosphere and phyllosphere bacterial communities. FEMS Microbiol Lett 346:146–154
Loqman S, Barka EA, Clément C, Ouhdouch Y (2009) antagonistic actinomycetes from Moroccan soil to control the grapevine gray mold. World J Microbiol Biotechnol 25:81. https://doi.org/10.1007/s11274-008-9864-6
Ludwig W, Euzeby J, Schumann P, Busse H, Trujillo ME, Kaempfer P, Whitman WB (2012) Road map of the phylum actinobacteria In: M Goodfellow, Pr Kämpfer, H-J Busse, ME Trujillo, K Suzuki, W Ludwig, WB Whitman (eds) Bergey's manual® of systematic bacteriology, vol 5. The actinobacteria part A, Springer pp. 1–32
Lundberg DS, Lebeis SL, Paredes SH, Yourstone S, Gehring J, Malfatti S, Tremblay J, Engelbrektson A, Kunin V, Del Rio TG, Edgar RC, Eickhorst T, Ley RE, Hugenholtz P, Tringe SG, Dangl JL (2012) Defining the core *Arabidopsis thaliana* root microbiome. Nature 488:86–90
Matsukuma S, Okuda T, Watanabe J (1994) Isolation of actinomycetes from pine litter layers. Actinomycetol 8:57–65
Matsumoto A, Takahashi Y (2017) Endophytic actinomycetes: promising source of novel bioactive compounds. J Antibiot 70:514–519. https://doi.org/10.1038/ja.2017.20
Matsumoto A, Takahashi Y, Mochizuki M, Seino A, Iwai Y, Omura S (1998) Characterization of actinomycetes isolated from fallen leaves. Actinomycetol 12:46–48
Maxwell A (1999) DNA gyrase as a drug target. Biochem Soc Trans 27(2):48–53
McCarthy AJ (1987) Lignocellulose-degrading actinomycetes. FEMS Microbiol Rev 46:145–163
McLeod MP, Warren RL, Hsiao WWL, Araki N, Myhre M, Fernandes C, Miyazawa D, Wong W, Lillquist AL, Wang D, Dosanjh M, Hara H, Petrescu A, Morin RD, Yang G, Stott JM, Schein JE, Shin H, Smailus D, Siddiqui AS, Marra MA, Jones SJM, Holt R, Brinkman FSL, Miyauchi K, Fukuda M, Davies JE, Mohn WW, Eltis LD (2006) The complete genome of *Rhodococcus* sp. RHA1 provides insights into a catabolic powerhouse. Proc Natl Acad Sci USA 103:15582–15587
McNear DH Jr (2013) The rhizosphere—roots, soil and everything in between. Nat Educ Knowl 4(3):1
Meindl K, Schmiederer T, Schneider K, Reicke A, Butz D, Keller S, Guhring H, Vertesy L, Wink J, Hoffmann JH, Bronstrup M, Sheldrick GM, Sussmuth RD (2010) Labyrinthopeptins: a new class of carbacyclic lantibiotics. Angew Chem Int Ed 49:1151–1154

Mendes R, Garbeva P, Raaijmakers JM (2013) The rhizosphere microbiome: significance of plant beneficial, plant pathogenic, and human pathogenic microorganisms. FEMS Microbiol Rev 37:634–663

Mendes R, Kruijt M, de Bruijn I, Dekkers E, van der Voort M, Schneider JH, Piceno YM, DeSantis TZ, Andersen GL, Bakker PA, Raaijmakers JM (2011) Deciphering the rhizosphere microbiome for disease-suppressive bacteria. Science 332:1097–1100

Miyadoh S (1993) Research on antibiotic producing in Japan over the last decade: a producing microorganism approach. Actinomycetol 9:100–106

Mora C, Tittensor DP, Adl S, Simpson AG, Worm B (2011) How many species are there on Earth and in the ocean? PLoS Biol 9:e1001127

Nimnoi P, Pongsilp N, Lumyong S (2014) Co-inoculation of soybean (G*lycine max*) with actinomycetes and *Bradyrhizobium japonicum* enhances plant growth, nitrogenase activity and plant nutrition. J Plant Nutrit 37(3):432–446

Ninawe S, Kapoor M, Kuhad RC (2008) Purification and characterization of extracellular xylanase from *Streptomyces cyaneus* SN32. Biores Technol 99(5):1252–1258

Ningthoujam S, Sanasam S, Tamreihao K, Nimaieh S (2009) Antagonistic activites of local actinomycete isolates agianst rice fungal pathogens. Afr J Microbiol Res 3:737–742

Otoguro M, Hayakawa M, Yamazaki T, Iimura Y (2001) An integrated method for the selective isolation of *Actinokineospora* spp. in soil and plant litter. J Appl Bacteriol 91:118–130

Pastores GM, Barnett NL, Kolodny EH (2005) An open-label, noncomparative study of miglustat in type I Gaucher disease: efficacy and tolerability over 24 months of treatment. Clin Ther 27:1215–1227

Patrauchan MA, Florizone C, Dosanjh M, Mohn WW, Davies J, Eltis LD (2005) Catabolism of benzoate and phthalate in *Rhodococcus* sp. strain RHA1: redundancies and convergence. J Bacteriol 187:4050–4063

Peoples A, Ling LL, Lewis K, Zhang Z (2011) Novel Antibiotics. Novo Biotic Pharmaceuticals, US Pat 20110136752:A1

Peoples A, Zhang Q, Moore C, Ling L, Rothfeder M, Lewis K (2012) NovoBiotic Pharmaceuticals. US Pat 8097709:B2

Perez Jaramillo JE (2019) Impact of plant domestication on spermosphere and rhizosphere microbiome composition. PhD thesis, Leiden University, URN: ISBN: 9789463324755

Perry CM, Ibbotson T (2002) Biapenem. Drugs 62:2221–2234

Petrolini B, Quaroni S, Saracchi M, Sardi P (1996) Studies on the *Streptomycete* population inhabiting plant roots. Actinomycetes Vol II Int Cent Theor Appl Ecol 7:66–78

Portugal J (2003) 3(6):411–420

Qin J, Li R, Raes J et al (2010) A human gut microbial gene catalogue established by metagenomic sequencing. Nature 464:59–65

Qin S, Li J, Chen HH, Zhao GZ, Zhu WY, Jiang CL, Jiang JH, Li WJ (2009) Isolation diversity and antimicrobial activity of rare actinobacteria from medicinal plants of tropical rain forests in Xishuangbanna China. Appl Environ Microbiol 75:6176–6186. https://doi.org/10.1128/AEM.01034-09

Qin S, Chen HH, Zhao GZ, Li J, Zhu WY, Xu LH, Jiang JH, Li WJ (2012) Abundant and diverse endophytic actinobacteria associated with medicinal plant *Maytenus austroyunnanensis* in Xishuangbanna tropical rainforest revealed by culture-dependent andculture-independent methods. Environ Microbiol Rep 4:522–531

Rambelli A (1973) The Rhizosphere of mycorrhizae. In: Marks GL, Koslowski TT (eds) Ectomycorrhizae. Academic Press, New York, pp 299–343

Rana KL, Kour D, Sheikh I, Dhiman A, Yadav N, Yadav AN, Rastegari AA, Singh K, Saxena AK (2019a) Endophytic fungi: biodiversity, ecological significance and potential industrial applications. In: Yadav AN, Mishra S, Singh S, Gupta A (eds) Recent Advancement in White Biotechnology through Fungi, vol 1. Diversity and Enzymes Perspectives. Springer, Switzerland, pp 1–62

Rana KL, Kour D, Sheikh I, Yadav N, Yadav AN, Kumar V, Singh BP, Dhaliwal HS, Saxena AK (2019b) Biodiversity of endophytic fungi from diverse niches and their biotechnological applications. In: Singh BP (ed) Advances in endophytic fungal research: present status and future challenges. Springer International Publishing, Cham, pp 105–144. https://doi.org/10.1007/978-3-030-03589-1_6

Rana KL, Kour D, Yadav AN (2019c) Endophytic microbiomes: biodiversity, ecological significance and biotechnological applications. Res J Biotechnol 14:142–162

Rosenberg E, Sharon G, Zilber-Rosenberg I (2009) The hologenome theory of evolution contains Lamarckian aspects within a Darwinian framework. Environ Microbiol 11:2959–2962. https://doi.org/10.1111/j.1462-2920.2009.01995.x

RosÅoniec KZ, Wilbrink MH, Capyk JK, Mohn WW, Ostendorf M, van der Geize R, Dijkhuizen L, Eltis LD (2009) Cytochrome P450â 125 (CYP125) catalyses C26-hydroxylation to initiate sterol side-chain degradation in RHA1. Mol Microbiol 74(5):1031–1043

Sader HS, Gales AC (2001) Emerging strategies in infectious diseases: new carbapenem and trinem antibacterial agents. Drugs 61:553–564

Saengnak V, Chaisiri C, Nalumpang S (2013) Antagonistic *Streptomyces* species can protect chilli plants against wilt disease caused by *Fusarium*. J Agric Technol 9:1895–1908

Saini A, Aggarwal NK, Sharma A, Yadav A (2015) Actinomycetes: a aource of lignocellulolytic enzymes. Enzyme Res Article ID 279381, http://dx.doi.org/10.1155/2015/279381

Sameera B, Prakash HS, Nalini MS (2018a) Actinomycetes from the coffee plantation soils of Western Ghats: diversity and enzymatic potentials. Int J Curr Microbiol App Sci 7(8):3599–3611

Sameera B, Prakash HS, Nalini MS (2018b) Indole acetic acid production by the actinomycetes of coffee plantation soils of Western Ghats. Int J Curr Res 10(10):74482–74487

Sardi P, Saracchi M, Quaroni S, Petrolini B, Borgonovi GE, Merli S (1992) Isolation of endophytic *Streptomyces* strains from surface-sterilized roots. Appl Environ Microbiol 58(8):2691–2693

Sauvadet M, Chauvat M, Fanin N, Coulibaly S, Bertrand I (2016) Comparing the effects of litter quantity and quality on soil biota structure and functioning: application to a cultivated soil in Northern France. Appl Soil Ecol 107:261–271

Schiltz S, Gaillard I, Pawlicki-Jullian N, Thiombiano B, Mesnard F, Gontier E (2015) A review: what is the spermosphere and how can it be studied? J Appl Microbiol 119:1467–1481

Schulz B, Boyle C (2006) What are endophytes? In: Schulz B, Boyle C, Sieber TN (eds) Microbial root endophytes. Springer, Heidelberg, pp 1–13

Schulz W, Zillig W (1981) Rifampicin inhibition of RNA synthesis by estabilization of DNA-RNA polymerase-oligonucleotide complexes. Nucleic Acids Res 9:6889–6906

Seibert G, V´ertesy L, Wink J, Winkler I, Broenstrup M, Hoffmann H, Guehring H, Toti L, S¨ussmuth R, Sheldrick G, George M, Meindl K (2009) US Patent 20090298904

Seto M, Kimbara K, Shimura M, Hatta T, Fukuda M, Yano K (1995) A novel transformation of polychlorinated biphenyls by *Rhodococcus* sp. strain RHA1. Appl Environ Microbiol 61:3353–3358

Shimizu M, Yazawa S, Ushijima YA (2009) Promising strain of endophytic *Streptomyces* sp. for biological control of cucumber anthracnose. J Gen Plant Pathol 75:27–36

Singh PP, Shin YC, Park CS, Chung YR (1999) Biological control of *Fusarium* wilt of cucumber by chitinolytic bacteria. Phytopathol 89:92–99

Singh SB, Zhang C, Zink DL, Herath KB, Ondeyka J, Masurekar P, Jayasuriya H, Goetz MA, Tormo JR, Vicente F, Martin J, Gonzalez I, Genilloud O (2013) Occurrence, distribution, dereplication and efficient discovery of thiazolyl peptides by sensitive-resistant pair screening. J Antibiot 66:599–607

Singh SB, Zink DL, Dorso K, Motyl MR, Salazar O, Basilio A, Vicente MF, Byrne KM, Ha SN, Genilloud O (2008a) Isolation, structure, and antibacterial activities of Lucensimycins D–G, discovered from*Streptomyces lucensis* MA7349 Using an antisense strategy. J Nat Prod 72:345–352

Singh SB, Zink DL, Herath KB, Salazar O, Genilloud O (2008b) Discovery and antibacterial activity of lucensimycin C from *Streptomyces lucensis*. Tetrahedron Lett 49:2616–2619

Singh SB, Zink DL, Huber J, Genilloud O, Salazar O, Diez MT, Basilio A, Vicente F, Byrne KM (2006) Discovery of Lucensimycins A and B from *Streptomyces lucensis* MA7349 using an antisense strategy. Org Lett 8:5449–5452

Sinsabaugh RL, Lenkins AE (1990) Enzymatic and chemical analysis of particulate organic matter from a boreal river. Freshwater Biol 23:301–309

Srivastava A, Talaue M, Liu S, Degen D, Sineva RYE, Chakraborty A, Druzhinin SY, Chatterjee S, Mukhopadhyay J, Ebright YW (2011) New target for inhibition of bacterial RNA polymerase: 'switch region'. Curr Opin Microbiol 14:532–543

Suarez-Moreno ZR, Vinchira-Villarraga DM, Vergara-MoralesDI, Castellanos L, RamosFA, GuarnacciaC, Degrassi G, Venturi V, Moreno-Sarmiento N (2019) Plant-growth promotion and biocontrol properties of three *Streptomyces* spp. isolates to control bacterial rice pathogens. Front Mirobiol 10(290):1–17

Sugiura T, Yi Ariyoshi, Negoro S, Nakamura S, Ikegami H, Takada M, Yana T, Fukuoka M (2005) Phase I/II study of amrubicin, a novel 9-aminoanthracycline, in patients with advanced non-small-cell lung cancer. Invest New Drugs 23:331–337

Suman A, Yadav AN, Verma P (2016) Endophytic Microbes in Crops: Diversity and Beneficial impact for Sustainable Agriculture. In: Singh D, Abhilash P, Prabha R (eds) Microbial inoculants in sustainable agricultural productivity, research perspectives. Springer, India, pp 117–143. https://doi.org/10.1007/978-81-322-2647-5_7

Taechowisan T, Peberdy JF, Lumyong S (2003) Isolation of endophytic actinomycetes from selected plants and their antifungal activity. World J Microbiol Biotechnol 19:381–385

Takahashi Y, Seki Y, Tanaka Y, Oiwa R, Iwai Y, Omura S (1990) Vertical distribution of microorganisms in soil. Actinomycetol 4:1–6

Tan HM, Cao LX, He ZF, Su GJ, Lin B, Zhou SN (2006) Isolation of endophytic actinomycetes from different cultivars of tomato and their activities against *Ralstonia solanacearum in vitro*. World J Microbiol Biotechnol 22:1275–1280. https://doi.org/10.1007/s11274-006-9172-y

Thilagam R, Hemalatha N (2019) Plant growth promotion and chilli anthracnose disease suppression ability of rhizosphere soil actinobacteria. J Appl Microbiol 126(6):1835–1849

Tian X, Cao L, Tan H, Han W, Chen M, Liu Y, Zhou S (2007) Diversity of cultivated and uncultivated actinobacterial endophytes in the stems and roots of rice. Microb Ecol 53(4):700–707. https://doi.org/10.1007/s00248-006-9163-4

Tiáncsics A, Benedek T, Farkas M, Máthé I, Márialigeti K, Szoboszlay S, Kukolya J, Kriszt B (2014) Sequence analysis of 16S rRNA, gyrB and catA genes and DNA-DNA hybridization reveal that *Rhodococcus jialingiae* is a later synonym of *Rhodococcus qingshengii*. Int J Syst Evol Microbiol 64:298–301

Tokala RK, Strap JL, Jung CM, Crawford DL, Salove MH, Deobald LA, Bailey JK, Morra MJ (2002) Novel plant-microbe rhizosphere interaction involving *Streptomyces lydicus* WYEC108 and the pea plant (*Pisum sativum*). Appl Environ Microbiol 68:2161–2171

Torsvik V, Ovreas L, Thingstad TF (2002) Prokaryotic diversity—magnitude, dynamics, and controlling factors. Science 296:1064–1066

Turpaulta MP, Gobrang GR, Bonnauda P (2007) Temporal variations of rhizosphere and bulk soil chemistry in a douglas fir stand. Geoderma 137:490–496

Ueno M, Amemiya M, Yamazaki K, Iijima M, Osono M, Someno T, Iimuma H, Hamada M, Ishizuka M, Takeuchi T (1993) Ddelaminomycins, novel extracellular matrix receptor antagonist. J Antibiot 46:1156–1162

Uroz S, Bu_ee M, Murat C, Frey-Klett P, Martin F (2010) Pyrosequencing reveals a contrasted bacterial diversity between oak rhizosphere and surrounding soil. Environ Microbiol Rep 2:281–288

Valanarasu M, Ignacimuthu S, Agastian P (2012) Actinomycetes from Western Ghats of Tamil Nadu with its antimicrobial properties. Asian Pac J Trop Biomed 2:830–837

Valdes M, Peresz N, Satnos PE, Caballero-Mellado J, Pena-Cabriales JJ, Normand P, Hirsch AM (2005) Non-*Frankia* actinomycetes from surface-sterilized roots of *Casuarina equisetifolia* fix nitrogen. Appl Environ Microbiol 71(1):460–466

Valois D, Fayad K, Barasubiye T, Garon M, Dery C, Brzezinski R, Beaulieu C (1996) Glucanolytic actinomycetes antagonistic to *Phytophthora fragariae* var. rubi, the causal agent of raspberry root rot. Appl Environ Microbiol 62:1630–1635

Van Bambeke F (2004) Glycopeptides in clinical development: pharmacological profile and clinical perspectives. Curr Opin Pharmacol 4(5):471–478

Van Hylckama Vlieg JET, Leemhuis H, LutjeSpelberg JH, Janssen DB (2000) Characterization of the gene cluster involved in isoprene metabolism in *Rhodococcus* sp. strain AD45. J Bacteriol 182:1956–1963

Verma P, Yadav AN, Khannam KS, Panjiar N, Kumar S, Saxena AK, Suman A (2015) Assessment of genetic diversity and plant growth promoting attributes of psychrotolerant bacteria allied with wheat (*Triticum aestivum*) from the northern hills zone of India. Ann Microbiol 65:1885–1899

Verma P, Yadav AN, Khannam KS, Saxena AK, Suman A (2017a) Potassium-Solubilizing Microbes: Diversity, Distribution, and Role in Plant Growth Promotion. In: Panpatte DG, Jhala YK, Vyas RV, Shelat HN (eds) Microorganisms for green revolution: volume 1: microbes for sustainable crop production. Springer Singapore, Singapore, pp 125–149. https://doi.org/10.1007/978-981-10-6241-4_7

Verma P, Yadav AN, Kumar V, Singh DP, Saxena AK (2017b) Beneficial plant-microbes interactions: biodiversity of microbes from diverse extreme environments and its impact for crop improvement. In: Singh DP, Singh HB, Prabha R (eds) Plant-Microbe interactions in agro-ecological perspectives, volume 2: microbial interactions and agro-ecological impacts. Springer Singapore, Singapore, pp 543–580. https://doi.org/10.1007/978-981-10-6593-4_22

Verma VC, Gond SK, Kumar A, Mishra A, Kharwar RN, Gange AC (2009a) Endophytic actinomycetes from *Azadirachta indica* A. Juss. Isolation, diversity and anti-microbial activity. Microbiol Ecol 57:749–756. https://doi.org/10.1007/s00248-008-9450-3

Verma VC, Singh SK, Prakash S (2011) Bio-controland plant growth promotion potential of siderophore producing endophytic *Streptomyces* from *Azadirachta indica* A. Juss. J Basic Microbiol 51:550–556. https://doi.org/10.1002/jobm.201000155

Verma VC, Gond SK, Kumar A, Mishra A, Kharwar RN, Gange AC (2009b) Endophytic actinomycetes from *Azadirachta indica* A. Juss.: isolation, diversity, and anti-microbial activity. Microb Ecol 57(4):749–756 https://doi.org/10.1007/s00248-008-9450-3

Verona O (1958) The spermosphere. Ann Inst Pasteur (Paris) 95(6):795–798

Vorholt JA (2012) Microbial life in the phyllosphere. Nat Rev Microbiol 10:828–840

Vurukonda SSKP, Giovanardi D, Stefani E (2018) Plant growth promoting and biocontrol activity of *Streptomyces* spp. as endophytes. Int J Mol Sci 19:952. https://doi.org/10.3390/ijms19040952

Walsh TJ, Standiford HC, Reboli AC, John JF, Mulligan ME, Ribner BS, Montgomerie JZ, Goetz MB, Mayhall CG, Rimland D, Stevens DA, Hansen SL, Gerard GC, Ragual RJ (1993) Randomized double-blinded trial of rifampin with either novobiocin or trimethoprim-sulfamethoxazole against methicillin-resistant *Staphylococcus aureus* colonization: prevention of antimicrobial resistance and effect of host factors on outcome. Antimicrob Agents Chemother 37(6):1334–1342

Watcharakul S, RötherW Birke J, Umsakul K, Hodgson B, Jendrossek D (2016) Biochemical and spectroscopic characterization of purifiedlatex clearing protein (Lcp) from newly isolated rubber degrading *Rhodococcus rhodochrous* strain RPK1 reveals novel properties of Lcp. BMC Microbiol 16:92

Weinert N, Piceno Y, Ding GC, Meincke R, Heuer H, Berg G, Schloter M, Andersen G, Smalla K (2011) PhyloChip hybridization uncovered an enormous bacterial diversity in the rhizosphere of different potato cultivars: many common and few cultivar-dependent taxa. FEMS Microbiol Ecol 75:497–506

Williams ST, Davies FL (1965) Use of antibiotics for selective isolation and enumeration of actinomycetes in soil. J Gen Microbiol 38:251–261

Yadav AN (2018) Biodiversity and biotechnological applications of host-specific endophytic fungi for sustainable agriculture and allied sectors. Acta Sci Microbiol 1:01–05

Yadav AN (2019) Fungal white biotechnology: conclusion and future prospects. In: Yadav AN, Singh S, Mishra S, Gupta A (eds) Recent advancement in white biotechnology through fungi:

volume 3: perspective for sustainable environments. Springer International Publishing, Cham, pp 491–498. https://doi.org/10.1007/978-3-030-25506-0_20

Yadav AN, Kour D, Rana KL, Yadav N, Singh B, Chauhan VS, Rastegari AA, Hesham AE-L, Gupta VK (2019a) Metabolic engineering to synthetic biology of secondary metabolites production. In: Gupta VK, Pandey A (eds) New and future developments in microbial biotechnology and bioengineering. Elsevier, Amsterdam, pp 279–320. https://doi.org/10.1016/B978-0-444-63504-4.00020-7

Yadav AN, Kour D, Sharma S, Sachan SG, Singh B, Chauhan VS, Sayyed RZ, Kaushik R, Saxena AK (2019b) Psychrotrophic microbes: biodiversity, mechanisms of adaptation, and biotechnological implications in alleviation of cold stress in plants. In: Sayyed RZ, Arora NK, Reddy MS (eds) Plant growth promoting rhizobacteria for sustainable stress management, volume 1: rhizobacteria in abiotic stress management. Springer Singapore, Singapore, pp 219–253. https://doi.org/10.1007/978-981-13-6536-2_12

Yadav AN, Kumar R, Kumar S, Kumar V, Sugitha T, Singh B, Chauhan V, Dhaliwal HS, Saxena AK (2017a) Beneficial microbiomes: biodiversity and potential biotechnological applications for sustainable agriculture and human health. J Appl Biol Biotechnol 5:45–57

Yadav AN, Kumar V, Prasad R, Saxena AK, Dhaliwal HS (2018a) Microbiome in crops: diversity, distribution and potential role in crops improvements. In: Prasad R, Gill SS, Tuteja N (eds) Crop Improvement through Microbial Biotechnology. Elsevier, USA, pp 305–332

Yadav AN, Mishra S, Singh S, Gupta A (2019c) Recent advancement in white biotechnology through fungi volume 1: diversity and enzymes perspectives. Springer International Publishing, Cham

Yadav AN, Sachan SG, Verma P, Kaushik R, Saxena AK (2016) Cold active hydrolytic enzymes production by psychrotrophic Bacilli isolated from three sub-glacial lakes of NW Indian Himalayas. J Basic Microbiol 56:294–307

Yadav AN, Sachan SG, Verma P, Saxena AK (2015) Prospecting cold deserts of north western Himalayas for microbial diversity and plant growth promoting attributes. J Biosci Bioeng 119:683–693

Yadav AN, Saxena AK (2018) Biodiversity and biotechnological applications of halophilic microbes for sustainable agriculture. J Appl Biol Biotechnol 6:1–8

Yadav AN, Singh S, Mishra S, Gupta A (2019d) Recent advancement in white biotechnology through fungi. Volume 2: perspective for value-added products and environments. Springer International Publishing, Cham

Yadav AN, Singh S, Mishra S, Gupta A (2019e) Recent advancement in white biotechnology through fungi. Volume 3: perspective for sustainable environments. Springer International Publishing, Cham

Yadav AN, Verma P, Kumar S, Kumar V, Kumar M, Singh BP, Saxena AK, Dhaliwal HS (2018b) Actinobacteria from rhizosphere: molecular diversity, distributions and potential biotechnological applications. In: Singh B, Gupta V, Passari A (eds) New and future developments in microbial biotechnology and bioengineering. USA, pp 13–41. https://doi.org/10.1016/b978-0-444-63994-3.00002-3

Yadav AN, Verma P, Sachan SG, Saxena AK (2017c) Biodiversity and biotechnological applications of psychrotrophic microbes isolated from Indian Himalayan regions. EC Microbiol ECO 01:48–54

Yadav AN, Yadav N, Sachan SG, Saxena AK (2019f) Biodiversity of psychrotrophic microbes and their biotechnological applications. J Appl Biol Biotechnol 7:99–108

Yadav N, Yadav A (2018) Biodiversity and biotechnological applications of novel plant growth promoting methylotrophs. J Appl Biotechnol Bioeng 5:342–344

Yadav N, Yadav AN (2019a) Actinobacteria for sustainable agriculture. J Appl Biotechnol Bioeng 6:38–41

Yadav N, Yadav AN (2019b) Biodegradation of biphenyl compounds by soil microbiomes. Biodivers Int J 3:37–40

Young AJ, Petrasovits LA, Croft BJ, Gillings M, Brumbley SM (2006) Genetic uniformity of international isolates of *Leifsonia xyli* subsp. *xyli*, causal agent of ratoon stunting disease of sugarcane (*Saccharum* interspecific hybrids). Austral Pl Pathol 35:503–511

Yuan HM, Zhang XP, Zhao K, Zhong K, Gu YF, Lindstrom K (2008) Genetic characterisation of endophytic actinobacteria isolated from the medicinal plants in Sichuan. Ann Microbiol 58(4):597–604

Zampolli J, Zeaiter Z, Di Canito A, Di Gennaro P (2019) Genome analysis and–omics approaches provide new insights into the biodegradation potential of *Rhodococcus*. Appl Microbiol Biotechnol 103:1069–1080

Zeng Q, Liu Y, An S (2017) Impact of litter quantity on the soil bacteria community during the decomposition of *Quercus wutaishanica* litter. PeerJ. https://doi.org/10.7717/peerj.3777

Zhanel GG, Walters M, Noreddin A, Vercaigne LM, Wierzbowski A, Embil JM, Gin AS, Douthwaite S, Hoban DJ (2002) The ketolides. Drugs 62(12):1771–1804

Zhang X, Ren K, Zhang L (2012) Screening and preliminary identification of medicinal plants endophytic actinomycetes used for inhibiting Penicillin-resistant *Staphylococcus aureus*. Int J Biol 4(2):119–124

Zhao K, Penttinen P, Chen Q, Guan T, Lindström K, Ao X, Zhang L, Zhang X (2012) The rhizospheres of traditional medicinal plants in Panxi, China, host a diverse selection of actinobacteria with antimicrobial properties. Appl Microbiol Biotechnol 94:1321–1335. https://doi.org/10.1007/s00253-011-3862-6

Zhao K, Penttinen P, Guan T, Xiao J, Chen Q, Xu J, Lindström K, Zhang L, Zhang X, Strobel GA (2011) The diversity and anti-microbial activity of endophytic actinomycetes isolated from medicinal plants in Panxi plateau, China. Curr Microbiol 62:182–190. https://doi.org/10.1007/s00284-010-9685-3

Zídková L, Szoköl J, Rucká L, PátekM Nešvera J (2013) Biodegradation of phenol using recombinant plasmid-carrying *Rhodococcus erythropolis* strains. Int Biodeterior Biodegrad 84:179–184

Chapter 8
Phylogenetic Diversity of Epiphytic Pink-Pigmented Methylotrophic Bacteria and Role in Alleviation of Abiotic Stress in Plants

Ganapathy Ashok, Guruvu Nambirajan, Krishnan Baskaran, Chandran Viswanathan and Xavier Alexander

Abstract Plant and methylotrophic bacterial interactions that improve plant growth and plant fitness are becoming a topic of very important considerable interest. Methylotropic bacteria are distributed in various diverse environments/colonize different habitats and utilize reduced one-carbon compounds as source of energy and play an important role in the biogeochemical cycle. Methylotrophic bacteria colonize in different parts of the plants like endophytes, epiphytes and in roots of plant rhizosphere. Pink-pigmented facultative methylotrophic (PPFM) bacteria present in the phyllosphere enhance plant growth by producing phytohormones such as IAA, Zeatin, Cytokinins, ACC deaminase and diverse secondary metabolites to overcome abiotic stress. Biological interactions of Methylotrophic bacteria enhance plant growth indirectly by increasing the nutrients uptake and beneficial in reduction of greenhouse effects to the environments. Pink-pigmented facultative methylotrophic bacteria colonize in phyllosphere of plants as epiphytes and utilize methanol as a sole carbon source of energy. In plant colonization, the occurrence and distribution of Methylotrophic bacteria may be influenced by various factors like plant genotype, geographical conditions or by interactions with associated microorganisms and phytohormones production which may result and lead to increased plant fitness.

Keywords Methylotrophic bacteria · PPFM · Plant growth promotion · Phyllosphere · Sustainable agriculture

---

G. Ashok (✉) · C. Viswanathan
Department of Biotechnology, Sree Narayana Guru College, Coimbatore 641 105, Tamil Nadu, India
e-mail: ashokmku@gmail.com

G. Nambirajan
Department of Microbiology, Sree Narayana Guru College, Coimbatore 641 105, Tamil Nadu, India

K. Baskaran
Department of Biochemistry, Sree Narayana Guru College, Coimbatore 641 105, Tamil Nadu, India

X. Alexander
National Institute of Pharmaceutical Education and Research, Kolkata 700 054, India

A. N. Yadav et al. (eds.), *Plant Microbiomes for Sustainable Agriculture*, Sustainable Development and Biodiversity 25, https://doi.org/10.1007/978-3-030-38453-1_8

8.1 Introduction

Prolonged biological and chemical research have expanded our agricultural knowledge. Chemical fertilizer contains the most important elements of modern agriculture that provide the required nutrients, which are not present in the soil or other organic sources for crop improvement. The utilization and overexploitation of chemical fertilizers have an 'ecological footprint'. It reduces productivity and disturbs nutrients level in the soil, which further leads to a deterioration in quality of the soil and causes various plant diseases. The excessive use of chemical fertilizers in the field depletes non-renewable resources and dangerous to soil fertility and environments (Dubey et al. 2012). In general, the association of *Methylobacterium* spp. and host plants may be or epiphytic or endophytic in nature (Kumar et al. 2019b; Jourand et al. 2004; Omer et al. 2004b; Lacava et al. 2004). *M. nodulans* and *M. radiotolerans* interact with host plants and fix nitrogen fixation and nodule formation (Sy et al. 2001; Menna et al. 2006), whereas some *Methylobacterium* species are involved in the production of phytohormones (Meena et al. 2006) or interact with plant pathogens (Lacava et al. 2004), promoting plant growth (Madhaiyan et al. 2006b; Tani et al. 2012) and higher rate of photosynthetic activity (Cervantes et al. 2004).

Methylobacterium spp. are in connection with more than 70 plant species that actively colonize in different parts of the plants like branches, roots and leaves. Several studies have reported earlier that *Methylobacterium* spp. are identified as endophytes of various plants, such as citrus fruits, pine, cotton, eucalyptus, strawberries, peanuts, hemp, Catharanthus roseus, mangroves and tobacco.

Methylobacterium spp. are well known to be not phytopathogenic bacteria and reported that few *Methylobacterium* spp. produce enzyme pectinase and cellulose, which may cause systemic resistance during plant colonization of methylotrophs strains. In addition to phytohormone production, *Methylobacterium* spp. are capable of producing valuable biotechnological potential product like bioplastic, which are biodegradable and ecofriendly in nature. Polyhydroxyalkanoate (PHA) and polyhydroxybutyric acid (PHB) are biopolymers that are genetically modified strains like *M. extorquens* to increase higher amount of PHB and PHA production by utilizing methanol as substrate (Hofer et al. 2011).

Methylotropic bacteria colonize in different parts of the host plant as endophytes, epiphytes in the phyllosphere and produce diverse secondary metabolites as biocontrol agents to defense against phytopathogens. This chapter mainly deals with *Methylobacterium* spp. diversity, biotechnological importance of pink-pigmented facultative methylotrophic (PPFM) bacteria and various potential applications in agriculture as biofertilizers, co-inoculants and its role in biogeochemical cycle. This chapter also covers diversity of methylotrophs, genomics, metabolic potential of pink-pigmented facultative methylotrophic bacteria in the plant phyllosphere and role in alleviation of abiotic stress to the host plants.

8.2 Diversity and Metabolism of Methyotrophs

Methylotrophs are classified and subdivided into three subgroups on the basis of their metabolic activity like carbon-substrate utilization: (1) Obligate methylotrophs utilize single carbon compounds as sole source of energy (2) Restricted facultative methylotrophs utilize a limited range of complex carbon compounds apart from C1 compounds and (3) methylotrophs utilize and grow in medium with complex carbon compounds are called less-restricted facultative (Jenkins et al. 1987). Three distinct genera such as *Methylophilus* (Jenkins et al. 1987), *Methylobacillus* (Urakami and Komagata 1986; Yordy and Weaver 1977), and *Methylovorus* (Govorukhina and Trotsenko 1991) of betaproteobacteria are classified and considered as restricted facultative methylotrophs, whereas genus *Methylobacterium* is considered and well known as less-restricted facultative methylotrophs in the Alphaproteobacteria. Recently, Taubert et al. (2016) identified and reported an additional active group of the methylotrophic community. A common one-carbon (C1) substrate for many methylotrophic bacteria is methanol, whereas subgroups of these bacteria have the ability to use methane, methanesulfonate, other methylated sulphur species, methylated amines and the halogenated hydrocarbons chloromethane, bromomethane and dichloromethane, either in addition to methanol or exclusively methane, methanesulfonate, other methylated sulphur species, methylated amines and the halogenated hydrocarbons chloromethane, bromomethane, and dichloromethane as sole source or in addition with methanol as source of energy. The association of Methylobacterium spp. and host plants may be or epiphytic, phlylosphere, rhizosphere or endophytic in nature and produce phtohormones, nitrogen fixation, abiotic stress tolerance and maintain biogeochemical cycles (Kumar et al. 2019b) (Figs. 8.1 and 8.2).

8.3 Methylotrophic Community in the Phyllosphere

The distribution and diversity of phyllosphere microorganisms are influenced by various factors like nutrient availability, stress resistance, motility, growth, bacterial traits and metabolic activity (Bulgarelli et al. 2013; Yadav 2018; Yadav et al. 2017c, 2019). In addition, climate, plant genotype and geography are the major driving forces for methylotrophic bacterial population in the phyllosphere region of plants (Redford et al. 2010; Siefert et al. 2014 and Knief et al. 2010). Knief et al. (2010) reported efficient methylotrophic bacterial colonization, competitiveness and survival are closely linked to bacterial phylogeny and metabolic diversity of microorganisms of *Arabidopsis thaliana* in the phyllosphere. Knief et al. (2010) studied and reported that *Methylobacterium* community composition had strong effects and it varies based on culture-independent metagenome sequencing

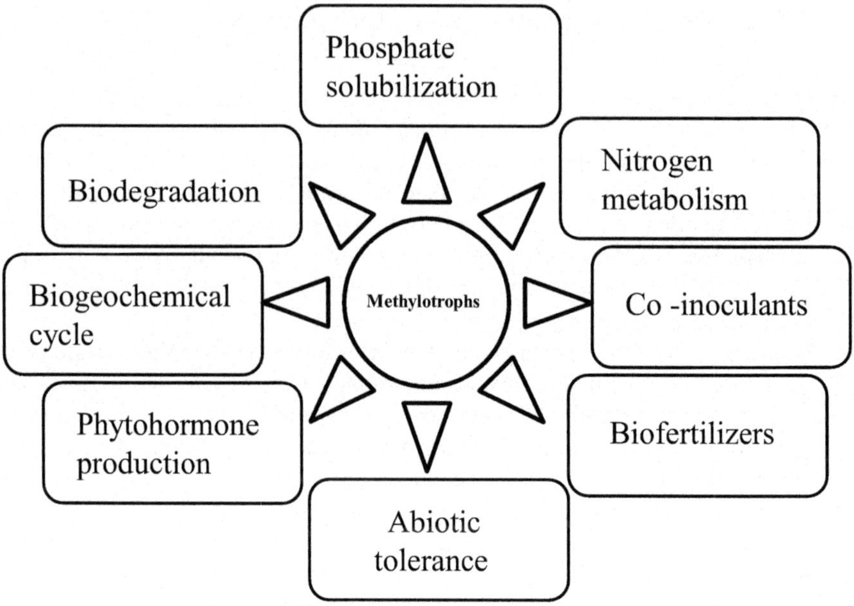

Fig. 8.1 Diverse role of methylotrophic bacteria application

analysis of leaves from *Medicago truncatula, Arabidopsis thaliana* and surrounding plant species at different locations. In *Medicago truncatula*, efficient colonization of phyllosphere Methylotrophs was observed due to the advantage of utilizing methanol as a source of energy and as a solitary carbon substrate (Sy et al. 2005). The association and interactions of different methylotrophic species like *M. mesophilicum, M. radiotolerans* and *M. fujisawaense* reported as strong colonizers with plant species were observed (Mizuno et al. 2013). In phyllosphere, methylotrophic microbes are present in huge numbers and under competitive conditions or during plant colonization, methylotrophic bacteria use plant-derived methanol as a substrate for energy and used for efficient colonization in the phyllosphere region (Abanda-Nkpwatt et al. 2006; Fall and Benson 1996; Sy et al. 2005). Colonization pattern of plant root and leaf surfaces was observed by using of green-fluorescent-marked strain of *Methylobacterium suomiense* (Poonguzhali et al. 2008) (Fig. 8.3).

8.4 Epiphytic PPFM Methylotrophs in the Phyllosphere

Epiphytic Pink-Pigmented Facultative Methylotrophs (PPFMs) are phylogenetically diverse and belong to the genus *Methylobacterium.* PPFMs utilize one-carbon compounds such as methanol, formate, formaldehyde and other multicarbon substrates as a sole source of energy. Pink-Pigmented Facultative Methylotrophs (PPFMs) belong

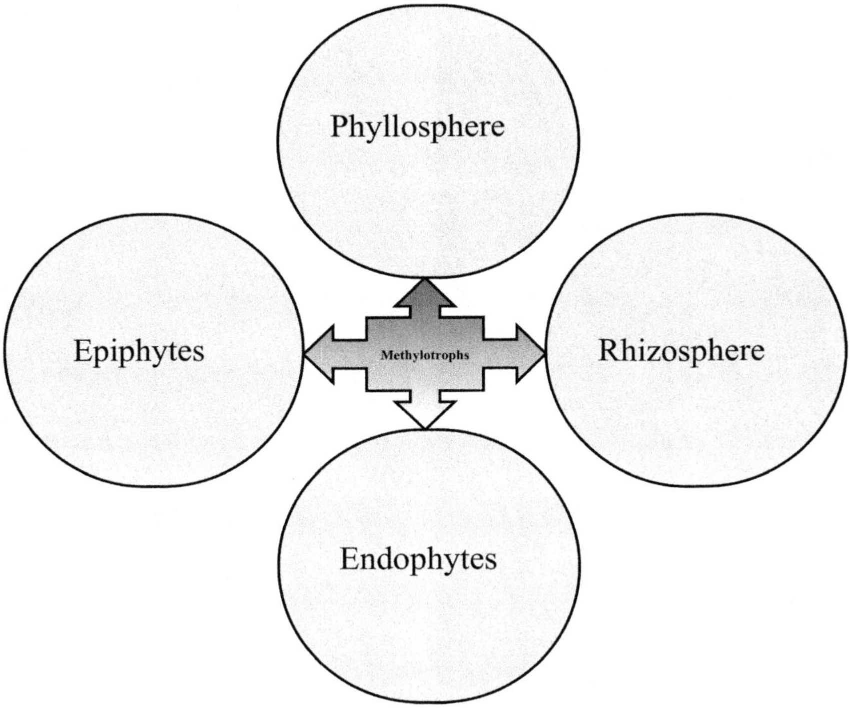

Fig. 8.2 Distribution role of methylotrops associated with different parts of the plants

to *Proteobacteria*, order Rhizobiales and Methylobacteriaceae family (Green and Bousifield 1982). PPFM is found in diverse habitats ubiquitous in nature including phyllosphere, rhizosphere, dust, freshwater, sediments and Lakes (Corpe and Rheem 1989; Green and Bousifield 1982). *Methylobacterium* spp. are generally distributed as epiphytes representing a significant bacterial population on plant leaves and in phyllosphere region of numerous plants (Hirano and Upper 1991; Holland and Polacco 1994). The colonization of *Methylobacterium* in a mucilaginous layer of plant tissues is the first step in colonization of microbes in the plant phyllosphere region (Andreote et al. 2006; Rossetto et al. 2011; Verma et al. 2017; Yadav et al. 2018c). The presence of methanol dehydrogenase (*mxaF*) gene in the genome of Pink-Pigmented Facultative Methylotrophic bacteria oxidizes methanol as an energy source (Anthony et al. 1994). In phyllosphere region of some plants, methane and methanol are emitted in the aerial part and serve as a habitat for distribution of methylotrophic bacterial population were reported earlier (Corpe and Basile 1982). Pink-Pigmented Facultative Methylotrophs were isolated using methanol-based mineral medium using methanol as an exclusive carbon and energy source (Corpe 1985).

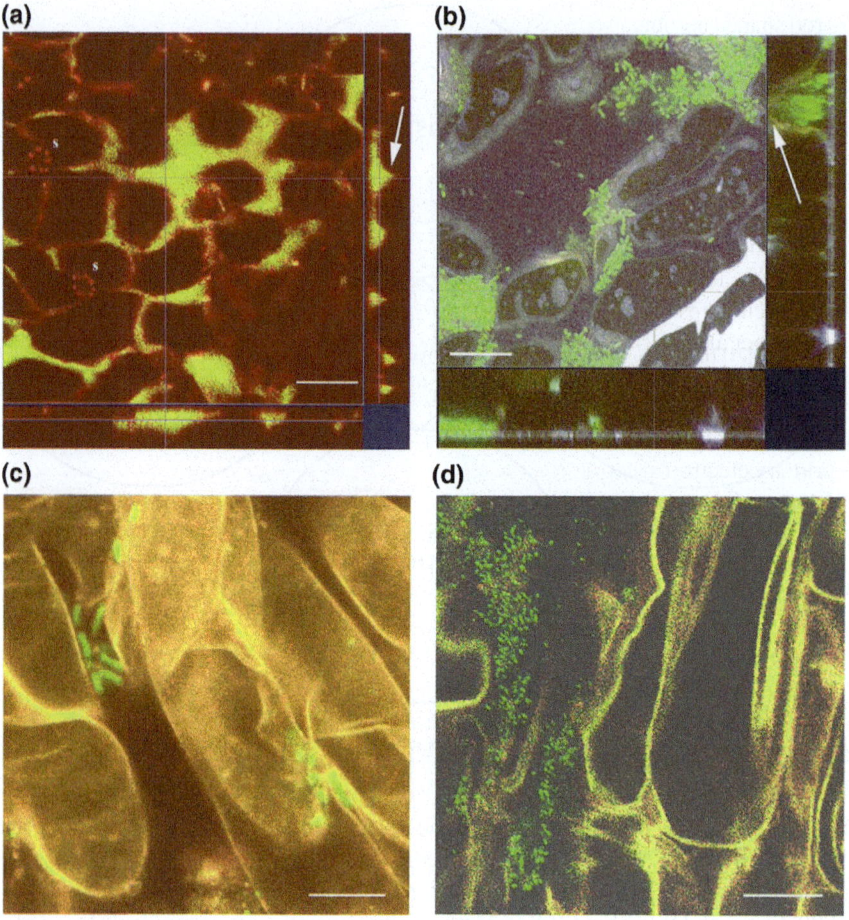

Fig. 8.3 Colonization of methylotrophic bacteria using green-fluorescent-marked strain of *Methylobacterium suomiense*

8.5 Genomics of PPFM Bacteria

The genotype of PPFM bacteria or interactions of associated microorganisms influence bacterial colonization and distribution in the host plant either directly or indirectly (Dourado et al. 2012).

8.6 Genetic Diversity of Methylotrophs

In general, Methylotrophic bacteria appears pink-pigmented in colours due to biosynthetic potential of carotenoids in the bacterium (VanDien et al. 2003). Methylotrophs are rod shaped aerobic in nature and able to grow in medium containing methanol and methylamine as carbon(C1) source for its metabolic activity (Toyama et al. 1998). The most significant characteristic feature of this group is the ability to oxidize and utilize methanol as a substrate by using the enzyme methanol dehydrogenase enzyme (MDH). PPFMs strains were isolated through leaf impression technique from phyllosphere of three different crops, which were further confirmed based on genomic DNA isolation of the isolates and PCR amplification of partial *mxaF* gene (550 bp sized partial *mxaF* gene). In metabolism of methylotrophic bacteria, the enzyme methanol dehydrogenase (MDH), the *mxa*F gene encode for encodes the large subunit, which helps to understand *Methylobacterium* niche-specific plant association (Dourado et al. 2012).

The enzyme methanol dehydrogenase (MDH) oxidizes methanol into formaldehyde metabolism, which starts in the periplasm of methylotrophic bacterium (Zang et a 2003). *The mxa*F and *mxa*I genes encodes for large, small subunits and cytochrome C primary electron acceptor for methanol dehydrogenase are encoded by *mxa*G gene (Mcdonald and Murrell 1997). Methanol dehydrogenase enzyme is mainly composed of two small (8.5 kDa) and two large (66 kDa) subunits. The large subunit (MxaF) is important for the functional activity of methanol dehydrogenase (Skovran et al. 2011). Random amplified polymorphic DNA (RAPD) is a unique molecular fingerprinting technique which was commonly used to distinguish between closely related bacterial strains at species level (Mazurier et al. 1992; Williams et al. 1990).

Van Aken et al. (2004) investigated and reported metabolic and genetic diversity of PPFM bacteria in the phyllosphere region of maize, cotton and sunflower to understand the PPFMs diversity within a particular plant species and different plant species using RAPD molecular fingerprinting and profiling carbon-substrate utilization pattern. Vuilleumier et al. (2009) reported variations in the numbers of insertion elements (IS) and in the organization of the genes have been identified in two different *Methylobacterium* (AM1 and DM4) strains associated with methanol metabolism. *Methylobacterium* bacterial strains have been sequenced and reported *M. extorquens PA1 as an* as a competitive colonizer of the phyllosphere region of *Arabidopsis thaliana* plants (Knief et al. 2010).

8.7 Methylotrophic as Plant Growth Promoters

Methylotrophs promote plant growth through beneficial interactions with plants by producing phytohormones and indirectly by increasing the availability of nutrients (Lidstram and chistordava 2002; Koenig et al. 2002). Methylotrophs colonize in various parts of the plant and produce phytohormones like auxins, cytokinin and zeatin. Plant growth substance promotes growth of both shoot and root system (Verma et al. 2013, 2014, 2015, 2016; Yadav et al. 2016). Doronina et al. (2001) reported aerobic methylotrophic bacteria produce auxins range from 20 mg/ml in the culture medium. In methylotrophic bacteria, biosynthesis of IAA was initiated from tryptophan as precursor and addition of tryphtophan enhances the synthesis of IAA (Schneider and

Wightman 1974). The biosynthesis of IAA through IPA pathway, which involves the transfer of amino group from tryphtophan to IPS, which is catalyzed by aromatic aminotransferases and then to IAA in methylotrophic bacteria. The enzyme aminotransferase activity was observed and identified in several methylotrophic bacteria (Ivanova et al. 2001).

The genes responsible for enzymes such as amine oxidase, aldehyde dehydrogense, N-acyl transferase and amidase were related to auxins biosynthesis and identified in methylotrophic bacteria (Kwak 2014; Madhaiyan et al. 2006c; Tani et al. 2012). Schauer and Kutschera (2011) reported a novel *Methylobacterium funariae* produced phytohormone like auxin and cytokinin were isolated from phyllosphere region of common mosses. In phyllosphere region, inoculation with *Methylobacterium* produced phytohormone IAA, which indirectly alter IAA concentrations in the plant and stimulate the plant growth (Lee et al. 2006). Pink-pigmented facultative bacteria were widely distributed and colonize in the phyllosphere of medicinal, agricultural crops and wild plants in Ukraine region (Romanovskaya et al. 1998). Lee et al. (2004) reported phytohormone IAA from methylotrophic isolates such as *Methylotrophic extorquens* and *Methylotrophic fujisawaense* isolated from the phyllosphere region of rice.

8.7.1 Production of Phytohormones by PPFM

Anitha (2010) reported Pink Pigmented Facultative Methylotrophic bacteria (PPFMs) was isolated from phyllosphere of soybean and groundnut producing phytohormone IAA and enhance plant growth. Keerthi et al. (2015) reported PPFM were used as biofertilizers in green grams isolated from phyllosphere environment. Tani et al. (2015) reported methylotrophic sp. producing both IAA and cytokinin associated with red pepper. Cytokinins are plant growth hormones, which regulate many physiological processes in plants such as to stimulate plant cell division, activate dormant buds, remove apical domination and induce seed germination. Ivanova et al. (2000) reported *M. mosophilicum* isolated from phyllosphere of rye grass lium perenne were able to synthesize cytokinins using biotest with the *Amaranthus candatus* L. seedlings. Holland (1997) reported application of exogenous methanol to the host plant, which stimulates the growth of PPFM bacteria by producing phytohormone cytokinins. In addition to the cytokinin PPFM bacteria isolated from different crops like soybean, barley, maize and Arabidopsis plant contain phytohormone zeatin and zeatin rhiboside (Long et al. 1996). The presence of phytohormone cytokinins and zeatin in the culture liquids of methylotropic bacteria is confirmed through chromotagraphic and enzyme immuno assay analysis (Ivanova et al. 2000). Epiphytic pink-pigmented methylotrophic bacteria produce cytokinin, stimulate germination and growth of wheat (*Triticum aestivum*) seedling was reported Meena et al. (2012). Phytohormone production by methylotrophic bacteria associated with different crops (Table 8.1).

Table 8.1 Phytohormone production by methylotrophic bacteria associated with different crops

Crop plants	Crop associated Methylotrophs	Biofertilizer/Phytohormones production	References
Groundnut	Pink-pigmented facultative methylotroph	IAA production	Anitha (2010)
Green Gram	Pink-pigmented facultative methylotroph	Biofertilizer	Keerthi et al. (2015)
Soybean	Pink-pigmented facultative methylotroph	IAA production	Anitha 2010)
Red Pepper	*Methylobacterium* sp.	IAA and cytokinin production	Tani et al. (2015)
Rice	*Methylobacterium extorquens, Methylobacterium fujisawaense*	IAA production	Lee et al. (2004)
Wheat	*Methylobacterium* sp.	Cytokinin production	Meena et al. (2012)

8.8 PPFM as Biofertilizers

The spraying of PPFM on plants with 20% methanol leads to twofold increase in the PPFM population and increase in soybean plants, when compared to control plants (Nishio et al. 1977; Kumar et al. 2019a; Yadav et al. 2018a, b). Jayajyothi et al. (2014) reported foliar spray of pink-pigmented methylotrophic bacteria and Pseudomonas strains, in addition with biofertilizer enhance the microbial population and increase the nutrient uptake to the plants. Abd El Gawad et al. (2015) studied and reported enhanced growth, antioxidant activities and increased yield in snap bean crops based in field experiments in different seasons using PPFM bacterial isolates. Foliar spray or irrigation of PPFM bacteria along with methanol, ethanol or even both showed improvement in plant growth of cotton, sugarcane and strawberry plants (Madhaiyan et al. 2005; Yavarpanah et al. 2015). Ivanova et al. (2001) reported application of methanol spray on leaf surfaces to promote the growth of plants by producing phytohormones like cytokinin and auxin by PPFM bacteria. Madhaiyan et al. (2006a, b) investigated and reported higher yields of sugarcane (Saccharum officinarum L.), cotton (Gossypium hirsutum L.) were observed through foliar spray of PPFM along with methanol, which increases phytohormone production. Chauhan et al. (2010) also reported that the application of fertilizers with PPFM as foliar spray leads to higher crop yields. ICAR (2013) advocated application of PPFMs as biofertilizers can protect crops from drought stress conditions.

8.9 PPFM in the Nitrogen Metabolism

Nitrogen is considered as one of the essential nutrients required for plant growth, but the availability of nitrogen from the atmosphere was limited for the metabolism of plants (Kour et al. 2019a, b). In nitrogen fixation, the conversion of atmospheric nitrogen into ammonia takes place for the nutrient availability to the plants. The nitrogenase enzyme was involved in the biological reduction of nitrogen to ammonia which was carried out by a few prokaryotic organisms (Menna et al. 2006). PPFM are involved in the nitrogen metabolism of colonized plants indirectly. Soybean plants have several urease isoenzymes: the Eu1 urease located in beans, the Eu4 urease located in all plant tissues and the Eu2 and Eu3 ureases, which are necessary for the normal urease activity of soybean plants. In the soybean plants with the mutant *eu3-e1/eu3-e1* gene, urea was accumulated in the plant tissues because of impaired urease activity. The colonization of such plants by PPFM did not restore their urease activity. At the same time, the colonization of the double *eu1-sun/eu1-sun, eu4/eu4* soybean mutants by PPFM led to the restoration of their urease activity to a level of 20–40% of that of the wildtype plants, due to the PPFM urease (Holland and Polacco 1992).

8.10 PPFM as Bio-inoculants and Co-inoculants

Meena et al. (2012) reported application of methylotrophs as bio-inoculants for seed coating or as seed inoculation enhances seed germination. Methylotrophs are capable of promoting plant growth with different groups of bacteria as co-inoculants, which results in higher yield in pot and crop field conditions Poonguzhali et al. (2008). Meena et al. (2012) suggested development of bio-inoculants and co-inoculation of methylotrophic bacteria results in increased production of cytokinins and higher crop yield. Meenakshi and Savalgi (2009) reported co-inoculation of methylotrophs with *B. japonicum* as foliar spray consequences raise in number of nodules, when compared to seeds with single *B. japonicum* as control. In addition, foliar spray of bio-inoculants with methylotrophs leads to increase in chlorophyll content to the host plants. Nalayani et al. 2014 reported foliar application of different types of microbial consortia strains Pseudomonas, Bacillus and Azospirillum with PPFM results in higher yield of cotton plants.

8.11 PPFM in Abiotic Stress Tolerance

The phyllosphere methylobacteria are highly resistant to UV dehydration, freezing on hygroscopic carriers and ionizing radiation and elevated temperatures. The phyllosphere epiphytic methylotrophic PPFM may remain viable after UV irradiation with higher doses that are lethal to bacterial strains like *Pseudomonas, Enterococci* and *Methanotrophs* (Romanovskaya et al. 1998; Yadav et al. 2017a, b, d; Yadav and Saxena 2018). Plants can regulate phytohormones production during unfavourable conditions and in stressed environments to overcome from biotic or abiotic stresses (Salamone et al. 2005). Ethylene is a plant growth hormone essential for plants, which is produced during various physiological changes in plants and endogenously by plants (Khalid et al. 2006). Saleem et al. (2007) reported earlier ethylene as a plant growth regulator and identified as a stress-related hormone. Saleem et al. (2007) also reported the production of ethylene during unfavourable conditions or stress conditions, the invivo accumulation of ethylene is drastically increased, which negatively alters the overall growth of plant. The overall increased concentration of ethylene may lead to reduced performance of the crop.

Ethylene is a stress associated hormone related to auxin biosynthetic pathway and an increased level of ethylene in plants leads to deleterious effects like plant growth, accelerating abscission, ageing, inhibiting root elongation and senescence. In ethylene biosynthetic pathway, aminocyclopropane-1-carboxylic acid (ACC) is the precursor of the ethylene hormone converted from S-adenosylmethionine (SAM) and to ethylene by ACC synthase (ACS) and ACC oxidase (ACO), enzymes that are transcriptionally regulated separately by both biotic and abiotic factors. ICAR et al. (2013) reported the beneficial application of *Methylobacterium* (PPFMs) as biofertilizer helps the crops to protect and overcome crops drought stress and during high-temperature conditions. PPFMs synthesize phytohormones, 1-aminocyclopropane-1-carboxylate (ACC) to overcome abiotic stress conditions by utilizing methanol produced from plant leaves as a source of carbon and energy (ICAR 2013).

Plant growth-promoting methylotrophic bacteria produce the enzyme 1-aminocyclopropane-1-carboxylate (ACC) deaminase, which indirectly stimulate growth by decreasing ethylene concentrations in plants (Glick 1995). Chinnadurai et al. (2009) revealed that phyllosphere methylobacteria distributed in the rice leaves produce the enzyme ACC deaminase, which control the ethylene concentrations level in the rice plant. In earlier investigations, *Methylobacterium* strains were identified and reported to have ACC deaminase activity and tested for their potential in plant growth-promoting traits in various crops. *Methylobacterium* spp. are not phytopathogenic in nature which help in plant growth promotion by decreasing environmental stress, immobilizing heavy metals, degrading toxic organic compounds and even inhibiting plant pathogens. *Methylobacterium* spp able to synthesize polymer degrading pectinase and cellulase, suggesting that they can indirectly induce systemic resistance during plant colonization.

8.12 Conclusion and Future Prospects

PPFMs isolates and other methylotrophs improve plant growth by controlling or by inhibiting phytopathogens. PPFMs inhibit several phytopathogens including *Fusarium oxysporum, Sclerotium rolfsii, Colletotrichum capsici, Xanthomonas campestris and Cercospora capsici* and serve as biocontrol agents. Methylotrophs are widely used as bio-inoculants as a foliar spray on plants and serve as an alternative to chemical fertilizers to enhance crop yield. The application of methylotrophs as foliar spray regulates plant growth directly or indirectly. Methylotrophs regulate and play a key role in biogeochemical cycle of soil ecosystem, making the soil more suitable for higher crop yield. In addition, several characteristic features of methylotrophs like nitrogen fixation, phytohormone production, nodulation and nutrient acquisition as a promising substitute for synthetic or chemical fertilizers. In conclusion, methylotrophic bacteria serve as an alternative of biological control, plant growth promotion by nitrogen fixation, phosphate solubilization, phytohormone production and ACC deaminase production, along with balanced carbon cycling. Beneficial methylotrophic can be used for effective organic farming in sustainable agriculture in the future.

Acknowledgements The authors wish to thank principal Dr. M. Ilangovan, Sree Narayana Guru College, Coimbatore, Tamil Nadu, India and facility provided by Sree Narayana Guru College Educational Trust, Tamil Nadu, India is greatly acknowledged.

References

Abanda-Nkpwatt D, Musch M, Tschiersch J, Boettner M, Schwab W (2006) Molecular interaction between *Methylobacterium extorquens* and seedlings: growth promotion, methanol consumption, and localization of the methanol emission site. J Exp Bot 57:4025–4032

Abd El-Gawad HG, Ibrahim MFM, Abd El-Hafez AA, Abou El- Yazied A (2015) Contribution of pink pigmented facultative Methylotrophic bacteria in promoting antioxidant enzymes, growth and yield of Snap Bean. Am Eurasian J Agric Environ 15(7):1331–1345

Andreote FD, Lacava PT, Gai CS, Araújo WL, Maccheroni W Jr, Van-Overbeek LS, van-Elsas JD, Azevedo JL (2006) Model plants for studying the interaction between *Methylobacterium mesophilicum* and *Xylella fastidiosa*. Can J Microbiol 52:419–426

Anitha KG (2010) Enhancing seed germination of mono and dicotyledons through IAA production of PPFM. Trends Soil Sci Plant Nutr J 1:14–18

Anthony C, Ghosh M, Blake CC (1994) The structure and function of methanol dehydrogenase and related quinoproteins containing pyrralo-quinoline quinine. Biochem J 304:665–674

Bulgarelli D, Schlaeppi K, Spaepen S, Ver Loren van Themaat E, Schulze-Lefert P (2013) Structure and functions of the bacterial microbiota of plants. Annu Rev Plant Biol 64:807–838

Cervantes-Martınez J, Lopez-Dıaz S, Rodrıguez-Garay B (2004) Detection of the effects of *Methylobacterium* in *Agave tequilana* Weber var. azul by laser-induced fluorescence. Plant Sci 166:889–892

Chauhan PS, Lee GS, Lee MK, Yim WJ, Lee GJ, Kim YS, Chung JB, Sa TM (2010) Effect of *Methylobacterium oryzae* CBMB20 inoculation and methanol spray on growth of red pepper (Capsicum annuum L.) at different fertilizer levels. Korean J Soil Sci Fert 43:514–521

Chinnadurai C, Balachandar D, Sundaram SP (2009) Characterization of 1-aminocyclopropane-1-carboxylate deaminase producing methylobacteria from phyllosphere of rice and their role in ethylene regulation. World J Microbiol Biotechnol 25:1403–1411
Corpe WA (1985) A method for detecting methylotrophic bacteria on solid surfaces. J Microbiol Meth 3:215–221
Corpe WA, Basile DV (1982) Methanol-utilizing bacteria associated with green plants. Dev Ind Microbiol 23:483–493
Corpe WA, Rheem S (1989) Ecology of the methylotrophic bacteria on living leaf surfaces. FEMS Microbiol Ecol 62:243–250
Doronina NV, Kudinova LV, Trotsenko YA (2001) *Methylovorus mays* sp. nov.: a new species of aerobic obligately methylotrophic bacteria associated with plants. Microbiol 69:599–603
Dourado MN, Ferreira A, Araujo WL, Azevedo JL, Lacava PT (2012) The diversity of endophytic methylotrophic bacteria in an oil-contaminated and an oil-free mangrove ecosystem and their tolerance to heavy metals. Biotechnol Res Int 2012:759865
Dubey V, Patel AK, Shukla A, Shukla S, Singh S (2012) Impact of continuous use of chemical fertilizer. Int J Eng Res Dev 3:13–16
Fall R, Benson AA (1996) Leaf methanol—the simplest natural product from plants. Trends Plant Sci 1:296–301
Glick BR (1995) The enhancement of plant growth by free-living bacteria. Can J Microbiol 41:109–117
Govorukhina NI, Trotsenko YA (1991) *Methylovorus*, a new genus of restricted facultatively methylotrophic bacteria. Int J Syst Bacteriol 41:158–162
Green PN, Bousifield IJ (1982) A taxonomic study of some gram negative facultatively methylotrophic bacteria. J Gen Microbiol 128:623–638
Hirano SS, Upper CD (1991) Bacterial community dynamics. In: Andrews JH, Hirano SS (eds) Microbial ecology of leaves. Springer, New York, 271–294
Hofer P, Vermette P, Groleau D (2011) Introducing a new bioengineered bug: *Methylobacterium extorquens* tuned as a microbial bioplastic factory. Bioeng Bugs 2:71–79
Holland MA (1997) Occam's razor applied to hormonology. Are cytokinins produced by plants? Plant Physiol 115:865–868
Holland MA, Polacco JC (1992) Urease-null and hydrogenase-null phenotypes of a phylloplane bacterium reveal altered nickel metabolism in two soybean mutants. Plant Physiol 98:942–948
Holland MA, Polacco JC (1994) PPFMs and other covert contaminants: is there more to plant physiology than just plant? Annu Rev Plant Physiol Plant Mol Biol 45:197–209
ICAR-Rice Knowledge Management Portal (2013) Use of biofertilizer in paddy to withstand drought. http://www.rkmp.co.in/mr/ general-domain/news-and-events/use-of-biofertilizer-in-paddyto- withstand-drought. Accessed 22 April 2016
Ivanova EG, Doronina NV, Shepeliakovskaia AO, Laman AG, Brovko FA, Trotsenko IuA (2000) Facultative and obligate aerobic methylobacteria synthesize cytokinins. Mikrobiologiya 69:764–769
Ivanova EG, Doronina NV, Trotsenko YA (2001) Aerobic methylobacteria are capable of synthesizing auxins. Microbiol 70:392–397
Jenkins O, Byrom D, Jones D (1987) Methylophilus: a new genus of methanol-utilizing bacteria. Int J Syst Bacteriol 37:446–448
Jeyajothi R, Subbalakshmi L, Nalliah D (2014) Effect of PPFM application on microbial population in transplanted tice. Trends Biosci 7:3573–3574
Jourand P, Giraud E, Bena G, Sy A, Willems A, Gillis M, Dreyfus B, De Lajudie P (2004) *Methylobacterium nodulans* sp. nov., for a group of aerobic, facultatively methylotrophic, legume root-nodule-forming and nitrogen-fixing bacteria. Int J Syst Evol Microbiol 54:2269–2273
Keerthi MM, Babu R, Joseph M, Amutha R (2015) Optimizing plant geometry and nutrient management for grain yield and economics in irrigated greengram. Am J Plant Sci 6:1144–1150

Khalid A, Akhtar MJ, Mahmood MH, Arshad M (2006) Effect of substrate-dependent microbial ethylene production on plant growth. Microbiol 75:231–236. https://doi.org/10.1134/S0026261706020196

Knief C, Ramette A, Frances L, Alonso-Blanco C, Vorholt JA (2010) Site and plant species are important determinants of the *Methylobacterium* community composition in the plant phyllosphere. ISME J 4:719–728

Koenig RL, Morris RO, Polacco JC (2002) tRNA is the source of low-level trans-zeatin production in *Methylobacterium* spp. J Bacteriol 184:1832–1842

Kour D, Rana KL, Yadav N, Yadav AN, Kumar A, Meena VS, Singh B, Chauhan VS, Dhaliwal HS, Saxena AK (2019a) Rhizospheric microbiomes: biodiversity, mechanisms of plant growth promotion, and biotechnological applications for sustainable agriculture. In: Kumar A, Meena VS (eds) Plant growth promoting rhizobacteria for agricultural sustainability: from theory to practices. Springer Singapore, Singapore, pp 19–65. https://doi.org/10.1007/978-981-13-7553-8_2

Kour D, Rana KL, Yadav N, Yadav AN, Singh J, Rastegari AA, Saxena AK (2019b) Agriculturally and industrially important fungi: current developments and potential biotechnological applications. In: Yadav AN, Singh S, Mishra S, Gupta A (eds) Recent advancement in white biotechnology through fungi, Volume 2: Perspective for value-added products and environments. Springer International Publishing, Cham, pp 1–64. https://doi.org/10.1007/978-3-030-14846-1_1

Kumar M, Kour D, Yadav AN, Saxena R, Rai PK, Jyoti A, Tomar RS (2019a) Biodiversity of methylotrophic microbial communities and their potential role in mitigation of abiotic stresses in plants. Biologia 74:287–308. https://doi.org/10.2478/s11756-019-00190-6

Kumar M, Saxena R, Rai PK, Tomar RS, Yadav N, Rana KL, Kour D, Yadav AN (2019b) Genetic diversity of methylotrophic yeast and their impact on environments. In: Yadav AN, Singh S, Mishra S, Gupta A (eds) Recent advancement in white biotechnology through fungi: Volume 3: Perspective for sustainable environments. Springer International Publishing, Cham, pp 53–71. https://doi.org/10.1007/978-3-030-25506-0_3

Kwak MJ, Jeong H, Madhaiyan M, Lee Y, Sa TM, Oh TK, Kim JF (2014) Genome information of *Methylobacterium oryzae*, a plant-probiotic Methylotroph in the phyllosphere. PLoS ONE 9:e106704

Lacava PT, Araujo WL, Marcon J, Maccheroni WJ, Azevedo JL (2004) Interaction between endophytic bacteria from citrus plants and the phytopathogenic bacteria *Xylella fastidiosa*, causal agent of citrus-variegated chlorosis. Lett Appl Microbiol 39:55–59

Lee HS, Madhaiyan M, Kim CW, Choi SJ, Chung KY, Sa TM (2006) Physiological enhancement of early growth of rice seedlings (*Oryza sativa* L.) by production of phytohormone of N2-fixing methylotrophic isolates. Biol Fertil Soil 42:402–408

Lee KH, Madhaiyan M, Kim CW, Lee HS, Poonguzhali S, Sa T (2004) Isolation and characterization of the IAA producing Methylotrophic bacteria from phyllosphere of rice cultivars (Oryza sativa L.). Korean J Soil Sci Fertil 37:235–244

Lidstrom ME, Chistoserdova L (2002) Plants in the pink: cytokinin production by *Methylobacterium*. J Bacteriol 184:1818

Long RLG, Holland MA, Stebbins N, Morris RO, Polacco JC (1996) Evidence for cytokinin production by plant-associated Methylotrophs. Plant Physiol 111:316

Madhaiyan M, Poonguzhali S, Lee HS, Hari K, Sundaram SP, Tongmin SA (2005) A Pink-pigmented facultative methylotrophic bacteria accelerate germination growth and yield of sugarcane clone Co86032 (*Saccharum officinarum* L.). Biol Fertil Soils 41:350–358

Madhaiyan M, Poonguzhali S, Ryu JH, Sa TM (2006c) Regulation of ethylene levels in canola (*Brassica campestris*) by 1-aminocyclopropane-1-carboxylate deaminase-containing *Methylobacterium Fujisawaense*. Planta 224:268–278

Madhaiyan M, Poonguzhali S, Senthilkumar M, Sundaram S, Sa T (2009) Nodulation and plant-growth promotion by methylotrophic bacteria isolated from tropical legumes. Microbiol Res 164:114–120

Madhaiyan M, Poonguzhali S, Sundaram SP, Sa T (2006a) A new insight into foliar applied methanol influencing phylloplane methylotrophic dynamics and growth promotion of cotton (*Gossypium hirsutum* L.) and sugarcane (*Saccharum officinarum* L.). Environ Exp Bot 57:168–176

Madhaiyan M, Suresh Reddy BV, Anandham R, Senthikumar M, Poonguzhali S, Sundaram SP, Sa T (2006b) Plant growth promoting *Methylobacterium* induces defense responses in groundnut (*Arachis hypogaea* L.) compared with rot pathogens. Curr Microbiol 53:270–276

Mazurier SI, Audurier A, Marquet-Van der Mee N, Notermans S, Wernars K (1992) A comparative study of randomly amplified polymorphic DNA analysis and conventional phage typing for epidemiological studies of *Listeria monocytogenes* isolates. Res Microbiol 143:507–512

Mcdonald IR, Murrell JC (1997) The methanol dehydrogenase structural gene mxaF and its use as a functional gene probe for methanotrophs and methylotrophs. Appl Environ Microbiol 63:3218–3224

Menna P, Hungria M, Barcellos FG, Bangel EV, Hess PN, Martinez-Romero E (2006) Molecular phylogeny based on the 16S rRNA gene of elite rhizobial strains used in Brazilian commercial inoculants. Syst Appl Microbiol 29:315–332

Meena KK, Kumar M, Kalyuzhnaya MG, Yandigeri MS, Singh DP, Saxena AK, Arora DK (2012) Epiphytic pink-pigmented methylotrophic bacteria enhance germination and seedling growth of wheat (*Triticum aestivum*) by producing phytohormone. Antonie Van Leeuwenhoek 101:777–786

Meenakshi BC, Savalgi VP (2009) The effect of co-inoculation of *Methylobacterium* and *B. japonicum* on plant growth, dry matter content and enzyme activities in soybean. Karnataka J Agric Sci 22:344–348

Mikanova O, Friedlova M, Simon T (2009) The influence of fertilization and crop rotation on soil microbial characteristics in the long-term field experiment. Plant Soil Environ 55:11–16

Mizuno M, Yurimoto H, Iguchi H, Tani A, Sakai Y (2013) Dominant colonization and inheritance of *Methylobacterium* sp. strain OR01 on perilla plants. Biosci Biotechnol Biochem 77:1533–1538

Mizuno M, Yurimoto H, Yoshida N, Iguchi H, Sakai Y (2012) Distribution of pink-pigmented facultative methylotrophs on leaves of vegetables. Biosci Biotechnol Biochem 76:578–580

Nalayani P, Anandham R, Raj SP, Chidambaram P (2014) Pink pigmented facultative methylotrophic bacteria (PPFMB)- a potential bioinoculant for cotton nutrition. Cotton Res J 6:50–53

Nishio N, Tsuchiya Y, Hayashi M, Nagai S (1977) A fed-batch culture of methanol-utilizing bacteria with pH-stat. J Ferment Technol 55:151–155

Omer Z, Tombolini R, Broberg A, Gerhardson B (2004a) Indole-3-acetic acid production by pink-pigmented facultative *Methylotrophic bacteria*. Plant Growth Reg 43:93–96

Omer ZS, Tombolini R, Gerhardson B (2004b) Plant colonization by pink-pigmented facultative methylotrophic bacteria (PPFMs). FEMS Microbiol Ecol 47:319–326

Oyaizu-Masuchi Y, Komagata K (1988) Isolation of free-living nitrogen-fixing bacteria from the rhizosphere of rice. J Gen Appl Microbiol 34:127–164

Pattnaik S, Rajkumari J, Paramanandham P, Busi S (2017) Indole Acetic Acid Production and Growth-Promoting Activity of *Methylobacterium extorquens* MP1 and *Methylobacterium zatmanii* MS4 in Tomato. Int J Veg Sci 23:321–330

Poonguzhali S, Madhaiyan M, Yim WJ, Kim KA, Sa TM (2008) Colonization pattern of plant root and leaf surfaces visualized by use of green-fluorescent-marked strain of *Methylobacterium suomiense* and its persistence in rhizosphere. App Microbiol Biotechnol 78:1033–1043

Poorniammal R, Sundaram SP, Kumutha K (2009) In vitro biocontrol activity of *Methylobacterium extorquens* against fungal pathogens. Int J Plant Prot 2:59–62

Pradhan N, Sukla LB (2006) Solubilization of inorganic phosphate by fungi isolated from agriculture soil. Afr J Biotechnol 5:850–854

Radha TK, Savalgi VP, Alagawadi AR (2009) Effect of methylotrophs on growth and yield of soybean (*Glycine max* (L.) Merrill). Karnataka J Agric Sci 22:118–121

Raja P, Uma S, Sundaram S (2006) Non-nodulating pink pigmented facultative *Methylobacterium* sp. with a functional nifH gene. World J Microbiol Biotechnol 22:1381–1384

Rao L, Dhir KK (1993) Some biochemical aspects of nitrogen fixation under salt stress in mung bean (*Vigna radiate* (L.) Wilczek). In: New trends in plant physiology proceedings, national symposium on growth and differentiation in plants, pp 255–258

Redford AJ, Bowers RM, Knight R, Linhart Y, Fierer N (2010) The ecology of the phyllosphere: geographic and phylogenetic variability in the distribution of bacteria on tree leaves. Environ Microbiol 12:2885–2893

Rekadwad BN (2014) Growth promotion of crop plants by *Methylobacterium organophilum*: efficient bio-inoculant and biofertilizer isolated from mud. Res Biotechnol 5:1–6

Romanovskaya VA, Sokolov IG, Malashenko YR, Rokitko PV (1998) Mutability of epiphytic and soil bacteria of the genus *Methylobacterium* and their resistance to ultraviolet and ionizing radiation. Mikrobiologiya 67:106–115

Rossetto PB, Dourado MN, Quecine MC, Andreote FD, Araújo WL, Azevedo JL, Pizzirani-Kleiner AA (2011) Specific plant induced biofilm formation in *Methylobacterium* species. Braz J Microbiol 42:878–883

Salamone EGD, Hynes RK, Nelson LM (2005) Role of cytokinins in plant growth promotion by rhizosphere bacteria. In: Siddiqui ZA (ed) PGPR: biocontrol and biofertilization. Springer, Amsterdam, pp 173–195

Saleem M, Arshad M, Hussain S, Bhatti S (2007) Perspective of plant growth promoting rhizobacteria (PGPR) containing ACC deaminase in stress agriculture. J Ind Microbiol Biotechnol 34:635–648. https://doi.org/10.1007/s10295-007-0240-6

Savitha P, Sreenivasa MN, Nirmalnath JP (2015) In vitro screening for biocontrol activity of pink pigmented facultative methylotrophs against phytopathogens. Karnataka J Agric Sci 28:286–287

Schauer S, Kutschera U (2011) A novel growth-promoting microbe, *Methylobacterium funariae* sp. nov., isolated from the leaf surface of a common moss. Plant Signal Behav 6:510–515

Schneider EA, Wightman F (1974) Metabolism of auxin in higher plants. Annu Rev Plant Physiol 25:487–513

Siefert A, Fridley JD, Ritchie ME (2014) Community functional responses to soil and climate at multiple spatial scales: when does intraspecific variation matter. PLoS One 9:e111189

Skovran E, Palmer AD, Rountree AM, Good NM, Lidstrom ME (2011) XoxF is required for expression of methano dehydrogenase in *Methylobacterium extorquens* AM1. J Bacteriol 193:6032–6038

Sy A, Giraud E, Jourand P, Garcia N, Willems A, de Lajudie P, Prin Y, Neyra M, Gillis M, Boivin-Masson C, Dreyfus B (2001) Methylotrophic methylobacterium bacteria nodulate and fix nitrogen in symbiosis with legumes. J Bacteriol 183:214–220

Sy A, Timmers AC, Knief C, Vorholt JA (2005) Methylotrophic metabolism is advantageous for *Methylobacterium extorquens* during colonization of *Medicago truncatula* under competitive conditions. Appl Environ Microbiol 71:7245–7252

Tani A, Sahin N, Fujitani Y, Kato A, Sato K, Kimbara K (2015) *Methylobacterium* species promoting rice and barley growth and interaction specificity revealed with whole-cell matrix-assisted laser desorption/ionization-time-of-flight mass spectrometry (MALDI-TOF/MS) analysis. PLoS ONE 10:e0129509

Tani A, Takai Y, Suzukawa I, Akita M, Murase H, Kimbara K (2012) Practical application of methanol-mediated mutualistic symbiosis between *Methylobacterium* species and a roof greening moss. Racomitrium japonicum. PLoS One 7:e33800

Taubert M, Grob C, Howat AM, Burns OJ, Chen Y, Neufeld JD, Murrell JC (2016) Analysis of active methylotrophic communities: when DNA-SIP meets high-throughput technologies. In: Martin F, Uroz S (eds) Microbial environmental genomics (MEG), methods in molecular biology. Humana Press, New York, pp 235–255

Toyama H, Anthony C, Lidstrom ME (1998) Construction of insertion and deletion mxa mutants of *Methylobacterium extorquens* AM1 by electroporation. FEMS Microbiol Lett 166:1–7

Trotsenko YA, Ivanova E, Doronina N (2001) Aerobic methylotrophic bacteria as phytosymbionts. Microbiology 70:623–632

Urakami T, Komagata K (1986) Emendation of *Methylobacillus yordy* and weaver 1977, a genus for methanol-utilizing bacteria. Int J Syst Bacteriol 36:502–511

Van Aken B, Peres CM, Doty SL, Yoon JM, Schnoor JL (2004) *Methylobacterium populi* sp. nov., a novel aerobic, pink pigmented, facultatively methylotrophic, methane-utilizing bacterium isolated from poplar trees (*Populus deltoides* x *nigra* DN34). Intl J Syst Evol Microbiol 54:1191–1196

VanDien SJ, Okubo Y, Hough MT, Korotkova N, Taitano T, Lidstrom ME (2003) Reconstruction of C3 and C4 metabolism in *Methylobacterium extorquens* AM1 using transposon mutagenesis. Microbiology 149:601–609

Verma P, Yadav AN, Kazy SK, Saxena AK, Suman A (2013) Elucidating the diversity and plant growth promoting attributes of wheat (*Triticum aestivum*) associated acidotolerant bacteria from southern hills zone of India. Natl J Life Sci 10:219–227

Verma P, Yadav AN, Kazy SK, Saxena AK, Suman A (2014) Evaluating the diversity and phylogeny of plant growth promoting bacteria associated with wheat (Triticum aestivum) growing in central zone of India. Int J Curr Microbiol Appl Sci 3:432–447

Verma P, Yadav AN, Khannam KS, Mishra S, Kumar S, Saxena AK, Suman A (2016) Appraisal of diversity and functional attributes of thermotolerant wheat associated bacteria from the peninsular zone of India. Saudi J Biol Sci. https://doi.org/10.1016/j.sjbs.2016.01.042

Verma P, Yadav AN, Khannam KS, Panjiar N, Kumar S, Saxena AK, Suman A (2015) Assessment of genetic diversity and plant growth promoting attributes of psychrotolerant bacteria allied with wheat (*Triticum aestivum*) from the northern hills zone of India. Ann Microbiol 65:1885–1899

Verma P, Yadav AN, Kumar V, Singh DP, Saxena AK (2017) Beneficial plant-microbes interactions: Biodiversity of microbes from diverse extreme environments and its impact for crop improvement. In: Singh DP, Singh HB, Prabha R (eds) Plant-Microbe interactions in agro-ecological perspectives: Volume 2: Microbial interactions and agro-ecological impacts. Springer Singapore, Singapore, pp 543–580. https://doi.org/10.1007/978-981-10-6593-4_22

Vuilleumier S, Chistoserdova L, Lee MC, Bringel F, Lajus A, Zhou Y, Gourion B, Barbe V, Chang J, Cruveiller S, Dossat C, Gillett W, Gruffaz C, Haugen E, Hourcade E, Levy R, Mangenot S, Muller E, Nadalig T, Pagni M, Penny C, Peyraud R, Robinson DG, Roche D, Rouy Z, Saenampechek C, Salvignol G, Vallenet D, Wu Z, Marx CJ, Vorholt JA, Olson MV, Kaul R, Weissenbach J, Medigue C, Lidstrom ME (2009) *Methylobacterium* genome sequences: a reference blueprint to investigate microbial metabolism of C1 compounds from natural and industrial sources. PLoS One 4:e5584

Wellner S, Lodders N, Kampfer P (2011) Diversity and biogeography of selected phyllosphere bacteria with special emphasis on *Methylobacterium* spp. Syst Appl Microbiol 34:621–630

Whittenbury R, Davies SL, Wilkinson JF (1970) Enrichment, isolation and some properties of methane-utilizing bacteria. J Gen Microbiol 61:205–218

Wiegel J (1992) The genus *Xanthobacter*. In: Balows A, Triiper HG, Dworkin M, Harder W, Schleifer KH (eds) The prokaryotes, 2nd edn. Springer, New York, pp 2365–2383

Williams JG, Kubelik AR, Livak KJ, Rafalski JA, Tingey SV (1990) DNA polymorphisms amplified by arbitrary primers are useful as genetic markers. Nucleic Acids Res18:6531–6535

Yadav AN (2018) Biodiversity and biotechnological applications of host-specific endophytic fungi for sustainable agriculture and allied sectors. Acta Sci Microbiol 1:01–05

Yadav AN, Kumar R, Kumar S, Kumar V, Sugitha T, Singh B, Chauhan V, Dhaliwal HS, Saxena AK (2017a) Beneficial microbiomes: biodiversity and potential biotechnological applications for sustainable agriculture and human health. J Appl Biol Biotechnol 5:45–57

Yadav AN, Kumar V, Prasad R, Saxena AK, Dhaliwal HS (2018a) Microbiome in crops: diversity, distribution and potential role in crops improvements. In: Prasad R, Gill SS, Tuteja N (eds) Crop improvement through microbial biotechnology. Elsevier, USA, pp 305–332

Yadav AN, Sachan SG, Verma P, Saxena AK (2016) Bioprospecting of plant growth promoting psychrotrophic Bacilli from cold desert of north western Indian Himalayas. Indian J Exp Biol 54:142–150

Yadav AN, Saxena AK (2018) Biodiversity and biotechnological applications of halophilic microbes for sustainable agriculture. J Appl Biol Biotechnol 6:48–55

Yadav AN, Verma P, Kour D, Rana KL, Kumar V, Singh B, Chauahan VS, Sugitha T, Saxena AK, Dhaliwal HS (2017b) Plant microbiomes and its beneficial multifunctional plant growth promoting attributes. Int J Environ Sci Nat Resour 3:1–8. https://doi.org/10.19080/IJESNR.2017.03.555601

Yadav AN, Verma P, Kumar S, Kumar V, Kumar M, Singh BP, Saxena AK, Dhaliwal HS (2018b) Actinobacteria from rhizosphere: molecular diversity, distributions and potential biotechnological applications. In: Singh B, Gupta V, Passari A (eds) new and future developments in microbial biotechnology and bioengineering. USA, pp 13–41. https://doi.org/10.1016/b978-0-444-63994-3.00002-3

Yadav AN, Verma P, Sachan SG, Kaushik R, Saxena AK (2018c) Psychrotrophic microbiomes: molecular diversity and beneficial role in plant growth promotion and soil health. In: Panpatte DG, Jhala YK, Shelat HN, Vyas RV (eds) Microorganisms for green revolution-Volume 2: Microbes for Sustainable agro-ecosystem. Springer, Singapore, pp 197–240. https://doi.org/10.1007/978-981-10-7146-1_11

Yadav AN, Verma P, Sachan SG, Saxena AK (2017c) Biodiversity and biotechnological applications of psychrotrophic microbes isolated from Indian Himalayan regions. EC Microbiol ECO. 01:48–54

Yadav AN, Verma P, Singh B, Chauhan VS, Suman A, Saxena AK (2017d) Plant growth promoting bacteria: biodiversity and multifunctional attributes for sustainable agriculture. Adv Biotechnol Microbiol 5:1–16

Yadav AN, Yadav N, Sachan SG, Saxena AK (2019) Biodiversity of psychrotrophic microbes and their biotechnological applications. J Appl Biol Biotechnol 7:99–108

Yavarpanah Z, Alizadeh M, Seifi E (2015) Effects of foliar and root applications of hydro-alcoholic solutions on physiological and biochemical attributes and fruit yield and weight of strawberry. J Plant Physiol Breed 5:47–54

Yim W, Woo S, Kim K, Sa T (2012) Regulation of ethylene emission in tomato (*Lycopersicon esculentum* Mill.) and red pepper (*Capsicum annuum* L.) inoculated with ACC deaminase producing *Methylobacterium* spp. Korean J Soil Sci Fert 45:37–42

Yordy JR, Weaver TL (1977) *Methylobacillus*: a new genus of obligate methylotrophic bacteria. Int J Syst Bacteriol 27:247–255

Chapter 9
Potassium Solubilizing Microbes: Diversity, Ecological Significances and Biotechnological Applications

Dheeraj Pandey, Harbans Kaur Kehri, Ifra Zoomi, Uma Singh, Kanhaiya L. Chaudhri and Ovaid Akhtar

Abstract Potassium (K) is seventh abundant element on earth and considered as third most significant macronutrient after nitrogen and phosphorus for plants. K plays an important role in metabolism, activation of enzyme, osmoregulation, balancing of charge and preventing unnecessary water loss, and regulating the stomatal movement in plants. The deficiency of K results in poor growth and development which intensely related to crop yield and resistance to disease. In soils, the fixation of K is comparatively fast; however, its release is a slow process. Many potassium solubilizing microbes (KSMs), such as fungi, bacteria and some arbuscular mycorrhizal (AM) fungi assist in mobilization of K from soil/mineral to plants. Microbes secrete various organic acids, which release K from the minerals and make it bioavailable to the plants. Fixation of K in soil, its uptake by the plant and leaching reduce the availability of K content in the soils. Application of microbial inoculants improve the growth and yield by converting the K in available form. KSMs work as biofertilizer in agriculture; improve the productivity, nutrient availability and reduce the use of agrochemicals in an economic, eco-friendly and sustainable manner. Future studies should focus on isolation and inoculation of indigenous K solubilizers to the plants from different environmental conditions such as cold, drought, nutrient-deficient soils, etc., to improve the availability of K for plants.

Keywords Potassium (K) · Potassium fertilizers · Potassium solubilizing microbes (KSMs) · AM fungi · Quality nutrient · Biofertilizer · H^+-ATPase · K cycling

9.1 Introduction

Potassium (K), with atomic number 19, is situated in alkali metal group with density 0.862 grams per cubic centimetre. It is a soft and silvery-white metal which rapidly

D. Pandey · H. K. Kehri · I. Zoomi · U. Singh · K. L. Chaudhri
Sadasivan Mycopathology Laboratory, Department of Botany, University of Allahabad, Allahabad 211002, India

O. Akhtar (✉)
Department of Botany, Kamla Nehru Institute of Physical and Social Sciences, Sultanpur 228118, India

A. N. Yadav et al. (eds.), *Plant Microbiomes for Sustainable Agriculture*, Sustainable Development and Biodiversity 25, https://doi.org/10.1007/978-3-030-38453-1_9

reacts with oxygen and forms dull-white potassium peroxides. K has single valence electron which is easily ionized as K^+. In 1807, Sir Humphry Davy was first isolated K from plant ashes (caustic potash) by electrolysis method (Thomas et al. 2008). K is seventh abundant element on earth which comprises 2.4% of the earth crust (NCBI Pubchem, 2019) and never found as elemental form in nature because of its very reactive nature. K comprises with different compounds to form ores such as langbeinite (K_2Mg_2 $(SO_4)_3$), carnallite ($KCl{\cdot}MgCl_2{\cdot}6H_2O$), polyhalite ($K_2Ca_2Mg$ $(SO_4)_4{\cdot}2H_2O$) and sylvite (KCl). There are many substitutes of K such as Potassium chloride (KCl) utilized as fertilizers, Potassium hydroxide (KOH) utilized in soaps, detergents and drain cleaners, Potassium carbonate ($KHCO_3$), also known as pearl ash and Potassium nitrate (KNO_3) or saltpeter/saltpetre or nitre, utilized in fertilizers etc. K is third most significant macronutrient after nitrogen and phosphorus, which is also considered as the "quality nutrient". K is indispensable macronutrient for plant and is required in large quantity after N and P for proper growth and development. It plays a key role in metabolism and activates important enzymes, osmotic regulation and charge balance in plants. The deficiency of K results in poor growth and development which intensely related to crop yield and resistance to disease (Ahmad et al. 2016; White and Karley 2010; Armengaud et al. 2010).

K exists in soils in different forms of minerals such as mica, orthoclase and also found in seawater. The rock trapped K are nonexchangeable, whereas exchangeable K in ionic form (K^+) held in soil colloids, which is available to the plant. In present scenario where agricultural lands are decreasing rapidly and instance agriculture is in practice that requires important chemical fertilizers including K fertilizers (Zhang et al. 2013). These chemical fertilizers not only increase cost input but also making the soil barren day by day. Because of the limited resources, increasing world population, urbanization and salinization are alarming for challenge of quality feeding in upcoming future.

About 98% of the total K present in soil is unavailable (Mengel and Kirkby 2001). Available K is also slowly solubilizing (Ahmad et al. 2016) by soil microbes such as bacteria, actinomycetes, fungi, AM fungi, etc., and these microbes have developed different modes for energy and nutrients. Some of them inhabit in rhizospheric soil of growing plants. This plant–microbe interaction provides a solution for sustainable agriculture and helps to increase agricultural productivity in sustainable manner. There are many bacteria (K solubilizing bacteria) and fungi (K solubilizing fungi) solubilize K with their enzymatic activities. The microbes make the K available to plants. Bacteria release acids for the solubilization of K minerals, and these microbes become very important in plant rhizosphere. Such bacteria also act as plant growth promoter and in plant pathogens protection (Zoomi et al. 2017). Some soil fungi, e.g. *Aspergillus* sp., *Fomitopsis meliae,* etc. (Anjanadevi et al. 2016; Kasana et al. 2017) are capable to solubilize K and make them available. In the rhizosphere, AM fungi symbiosis with plant roots becomes very beneficial for their growth and development (Pandey et al. 2019). These microbes solubilize different minerals such as phosphorus (P), nitrogen (N), K, etc. There are a number of studies about the role of AM fungi in K uptake (Garcia and Zimmermann 2014; Perner et al. 2007; Veresoglou et al. 2011). AM fungi solubilize K, store and transport to plant

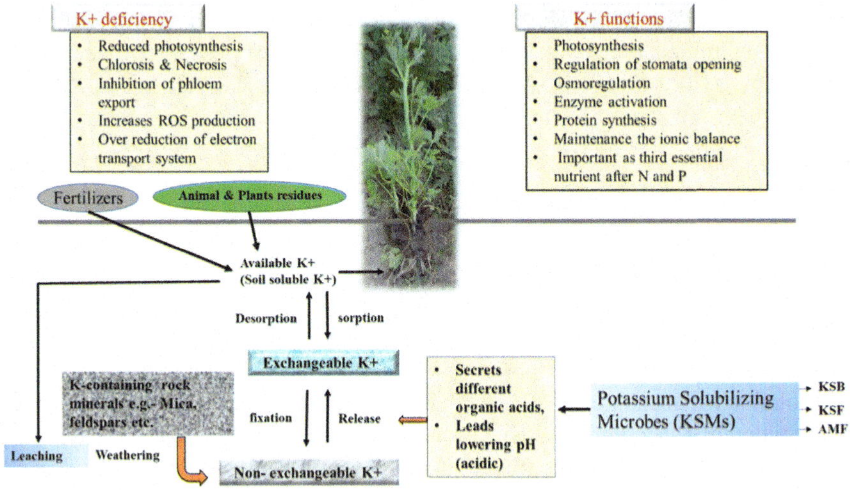

Fig. 9.1 Potassium (K) as a plant nutrient, importance and role of potassium solubilizing microbes (KSMs) in its mineral cycling

(Fig. 9.1). AM fungi accumulate K in different fungal structures like hyphae, vesicles and spores (Johansson and Campbell 1988). Vast applications of chemical fertilizers adversely affect not only the environment and soil but also our economy. There is a need to promote sustainable agriculture and application of beneficial native microbes in modern agriculture.

9.2 Potassium in Soils

9.2.1 Potassium Availability

Insoluble K constitutes about 90–98% in soil. Its concentration varies from 0.04 to 3.0% of total lithospheric K minerals (Mengel and Kirkby 2001; Sparks and Huang 1985). Type of soil and mineral constituents greatly influence the process of release and fixation of K (Braunschweigh 1980). So far to fulfil the requirements of crops, farmers applied synthetic K fertilizers but most of this is fixed with other elements and become inaccessible to crops (Prajapati et al. 2013). Hence, K fertilizer is applied on regular basis, but K unavailability to plants is continued for upcoming crops (Kang et al. 2002). So that unavailable K increases in large amount in soil. This condition interferes with soil physiology and ecology. To overcome this problem, there is a need for enhancement of K cycling soil microbes. K deficiency occurs in agricultural land of 75% of the paddy soils of China, 66% of the wheat belt of Southern Australia and 21% Indian soils (Hasan 2002; Meena et al. 2014a, 2016). In the year 2009–10, Indian farmers applied 42.38 lakh tonnes of muriate of potash fertilizer, and the demand by 2020 would be 9.52 metric tonnes (Kinekar 2011; Pathak et al. 2010).

9.2.2 Fixation of Potassium in Soil

Minerals found in the soil can also fix K and considerably affects the K availability. K fixation in soil comprises the adsorption of K^+ onto weathered sheet of montmorillonite, illite and vermiculite (silicate). Additionally, fixation of K in soils depends on the clay mineral, wetness, opposite ions, pH of the soil and K concentration in soil solution (Schneider et al. 2013). It has also been observed by Oborn et al. (2005) that fixation of K is comparatively fast; however, its release is a slow process. It has been reported by several author that organic acids play a crucial role in the release of fixed K from the clay minerals. Hence, management of the soil pH may be a mean of improving the release of K.

9.3 Role of Potassium

9.3.1 Function of Potassium as a Nutrient

K is very important macronutrient for plants (Hafsi et al. 2014). In plants, it is very important for physiological and biochemical functions (Zhang and Kong 2014). K^+ is absorbed from dissolved soil solution, clay and organic colloids by roots; translocated inside the plants through the xylem and phloem; and then converted into complex compounds (Zandonadi et al. 2010). K is highly mobile macronutrient and has very significant role in regulation of water status in plant cells. K^+ prevents unnecessary water loss and regulates the activity of stomatal cells. It promotes water absorption, maintains osmotic pressor and turgor in cells, and regulates stomatal movement (Meena et al. 2016). K is an important part for the development of chlorophyll and process of photosynthesis such as reduction of CO_2 into carbohydrate (White and Karley 2010). K does not play structural role but help in production of carbohydrate and its translocation in different sites of growth, development and storage and associated with sugar and starch accumulation (Cakmak 2005); however, it plays important roles in cellular organization and membrane permeability, which are important aspects for keeping the appropriate protoplasm hydration. K is transported through K^+ channels, maintains membrane transport and helps in the regulation of 60 different stress enzymes (Meena et al. 2016). K activates necessary enzymes essential for the ATP production as a source of energy, and it is also important in activation of essential enzymes of protein metabolism, amino acid synthesis, progress of cell division and growth-related enzymes of plants, which is also associated with disease resistance against bacteria and fungi.

9.3.2 Potassium Deficiency

Deficiency of K is associated with decrease in crop and food production called hidden hunger. In initial phase, deficiency does not show any specific symptoms. In scarce condition, the K is relocated from older leaves to new growing leaves. Hence, the deficiency symptoms of K appear in older leaves, beginning from tip to base. Leaves start yellowing and eventually develop necrosis at the border of the leaves. K deficiency reduces the strength, slows down the photosynthesis, shortens the internodes, slows down the photosynthate transport and reduces the resistance toward the diseases in plants (Meena et al. 2016) (Fig. 9.1). Blackening of potato tubers and margin of older leaves in cotton, maize, legumes and tobacco is due to K deficiency (Ashley et al. 2005). The accessibility of K in the soil influences root morphology (Thaler and Pages 1998). K-efficient rice had developed better roots as compared to the rice grown under deficient K condition (Yang et al. 2003). *Hordeum maritimum* and *Catapodium rigidum* increase their root length and root surface for uptake of more K (Hafsi et al. 2011). Lateral root number and growth was reduced in barley (Drew 1975) and *Arabidopsis thaliana* (Shin and Schachtman 2004) when raised under low K condition. K deficiency stress affects the dry mass partitioning in root and shoot. This may vary with species and culture conditions (Andrews et al. 1999). There are many studies regarding root shoot dry weight which reported increment in dry weight in *Brassica oleracea* (Singh and Blanke 2000), *H. maritimum* (Hafsi et al. 2011) and *Triticum aestivum* (Andrews et al. 1999).

9.4 Mechanism of K^+ Uptake and Regulation in Plant

In plants, there are various K transport mechanisms, which have important role in K absorption and its relocation (Véry and Sentenac 2003). K absorption is done by H^+-ATPase pump and different K^+ transporter proteins channels of cellular membrane (Hafsi et al. 2014). Absorption of K^+ nutrient is connected with net K^+ influx and low pH (Minjian et al. 2007; Chen and Gabelman 2000). Chen et al. (2008) identified 71 K^+ transporters and channels. K^+ and Na^+ transport system named high-affinity K^+ transporter 1 (HKT1) was reported in *Triticum aestivum* and cloned for the first time by Schachtman and Schroeder (1994). The HKT1 was also identified in *Eucalyptus camaldulensis* by Fairbairn et al. (2000). High-affinity K^+ transporter 5 (AtHAK5) and K^+ channel arabidopsis K^+ transporter 1 (AtAKT1) are two K^+ transporter system identified in *Arabidopsis* (Gierth et al. 2005; Pyo et al. 2010). In plant, K^+ channel is Shaker-type efflux channel, outward-rectifying K^+ channel (SKOR) type which is associated with K^+ transporter from root to xylem. Lack of this SKORK$^+$ channel affects the plant shoot biomass production. SKOR mutant *Arabidopsis thaliana* showed fifty percent decrease in shoot K$^+$; however, root K$^+$ content was not changed (Gaymard et al. 1998). Another Shaker-type efflux channel AKT2 is also found in phloem which is associated with recirculation of K$^+$ (Chérel et al. 2002; Gajdanowicz et al. 2011). KAT1 and KAT2 are inward-directed channels in guard cell that facilitate K$^+$ uptake (Schachtman et al. 1992; Pilot et al. 2001) while GORK is outward-directed K$^+$ channel associated with releasing K$^+$ from guard cell during stomatal movement regulation (Hosy et al. 2003). Many studies at the level of transcriptome and genetic analysis in rice have observed several genes of transporter families, which have significant roles in responses to K$^+$ deficit (Ma et al. 2012). Ma et al. (2012) identified three families of OsHAK genes in an up-regulation to increase K$^+$ uptake by increasing genes expression encoding K$^+$ transporters during K$^+$ deficiency. Buschmann et al. (2000) also identified TaAKT1 gene in *Triticum aestivum* in K deficient condition. Transcription level K regulating factors are also identified, which bind to transport promoter and activate the expression genes such as DDF2, JLO, bHLH121 and TFII-A involved in upregulating with HAK5 promoter (Hong et al. 2013). Du et al. (2019) investigated the role of transcription factor MYB59 in *Arabidopsis*, which regulates NRT1.5/NPF7.3 transporter in K deficit conditions.

9.5 Microbes-Mediated Potassium Solubilization

Different groups of soil microflora are reported which solubilize inaccessible fixed form of K to soluble and make them available to the plants. The potent K solubilizers microbial inoculants enhanced the plant growth and development, which increases the crop yields in eco-friendly way (Meena et al. 2016). Muentz (1890) was first reported the microbial role in rock K solubilization. In present scenario, there are many microbes discovered with K solubilizing potency, and many researches are still going on for their application at large scale. Many potassium solubilizing microbes (KSMs) are discovered till date. Several fungi, bacteria, actinomycetes and some AM fungi (especially rhizospheric microbes) assist in mobilization of K from soil/mineral to plants (Gundala et al. 2013; Verma et al. 2015, 2016a, b). KSM may differ in number and species from soil to soil. Some microbes reported for solubilizing K from silicate mineral (Sheng et al. 2002) such as *B. mucilaginosus* sub spp. *siliceous* can solubilize K from aluminosilicates and feldspar (Aleksandrov et al. 1967; Groudev 1987; Gundala et al. 2013). Various bacterial genera such as *Bacillus*, *Enterobacter*, *Acidithiobacillus*, *Pseudomonas*, *Burkholderia*, *Paenibacillus*, *Cladosporium* and *Arthrobacter* are reported for K solubilizing (Muralikannan 1996; Wu et al. 2005; Sheng 2005; Liu et al. 2012; Singh et al. 2010; Meena et al. 2016). *Bacillus mucilaginosus* strain CS1 is reported as solubilizer of silicate from both rhizospheric and non-rhizospheric soil (Lin et al. 2002; Liu 2001). Mikhailouskaya and Tcherhysh (2005) also isolated the above bacteria from K and silicate-amended soil. AM fungi can increase the solubility of K by the mechanism in which they secrete different organic acids and protons. K solubilization and uptake depends on the soil environment as well as nature of plant and other growth conditions (Clark and Zeto 2000). *Aspergillus terreus* and *Aspergillus niger* were reported for K solubilization from feldspar and potassium aluminium silicate sources (Prajapati and Modi 2012). Sangeeth et al. (2012) reported the role of *Penicillium* sp. in K solubilization from rocks and minerals. KSMs enhanced germination rate, seedling vigour, plant growth and their production (Sangeeth et al. 2012). Similar results were also obtained from some horticultural plants, vegetables and cereals with KSMs (Singh et al. 2010; Prajapati et al. 2013). Under the field test, Xie (1998) concluded that KSMs application in maize, wheat, sudan grass and forage crop significantly reduced the need of chemical fertilizers.

9.6 Diversity of Potassium Solubilizing Microorganisms

The various groups of KSMs including fungi and bacteria have been recovered minerals rocks, soils, industrial area and rhizosphere of a number of plants. These microorganisms have central role in ecosystem functioning and could be utilized as a bioinoculants (Kour et al. 2019a, b, c; Yadav et al. 2019a). These beneficial microorganisms have been cultured from rhizosphere of cotton (Sheng and He 2006), potato–soybean cropping sequence (Biswas 2011), rice (Muralikannan 1996), common bean (Kumar et al. 2012), peanut and sesame (Youssef et al. 2010), black pepper (Sangeeth et al. 2012), sugarcane (Rosa- Magri et al. 2012) rhizospheric soil of different other plants (Zhao et al. 2008; Zhang and Kong 2014; Meena et al. 2014b; Kumar et al. 2015) and from diverse extreme environments (Yadav et al. 2015a, b, c; Yadav et al. 2016). Diep and Hieu (2013) isolated 20 KSB strains from soils/weathered rocks. The microorganisms were also isolated from ceramic industry soil (Prajapati and Modi 2012) and mica core (Gundala et al. 2013). Leaungvutiviroj et al. (2010) isolated these organisms from rhizospheric soil in Thailand and were capable to release K. Based on the above data, it could be concluded that most of the KSMs belong to the bacteria domain.

9.6.1 Bacteria

KSB has been isolated from various habitats including plant microbiomes (Yadav 2017a, b; Yadav et al. 2017b) as well as microbiomes of extreme habitats (Yadav 2019a, b; Yadav and Yadav 2018). Raj (2004) isolated *Bacillus* sp. from the granite crusher yard, Hu et al. (2006) isolated *Bacillus mucilaginosus* from Tianmu mountain, Prajapati and Modi (2012) isolated *Enterobacter hormaechei* from ceramic industry soil, Syed and Patel (2014) isolated *Pseudomonas* spp. and *Bacillus* sp. from loamy sand soil, and Diep and Hieu (2013) isolated *Bacillus megaterium* and *Bacillus coagulans* from soils/weathered rocks. Potassium solubilizing bacteria (KSB) reported to improve the morphogenesis in plants (Lian et al. 2002; Singh et al. 2010). Several bacterial genera such as *Bacillus, Enterobacter, Pseudomonas* and *Azotobacter* were also reported to provide benefit to the plant (Kloepper et al. 1991; Han and Lee 2005; Singh et al. 2010). Sangeeth et al. (2012) reported to enhanced K uptake and biomass in inoculated black pepper. Similarly, Valencia orange trees inoculated with *Bacillus circulans* considerably enhanced K uptake, fruit biomass and yield (Shaaban et al. 2012). Yield of tomato plant was reported to improve when inoculated with *Pseudomonas* sp. (Lynn et al. 2013). Maize plant inoculated with *B. megaterium, B. mucilaginosus* and *A. chroococcum* improved K uptake along with nitrogen and phosphorus and growth of the plant (Wu et al. 2005). Furthermore, KSB also shows resistance towards various stresses such as chilling drought stress (Cakmak 2005; Ramarethinam and Chandra 2006). KSB belongs to different phylum including actinobacteria, firmicutes, proteobacteria and bacteroidetes (Kumar et al. 2019; Rana et al. 2019c; Yadav et al. 2017a, 2019e; Yadav and Saxena 2018).

9.6.2 Fungi

Fungi play paramount role in the release of mineral such as phosphates, silicate (Banfield et al. 1999) and carbonate (Verrecchia and Dumont 1996). Additionally, numerous studies have also been concluded the roles of fungi in releasing the K from minerals (Wallander and Wickman 1999; Yuan et al. 2000, 2004; Glowa et al. 2003; Yadav 2018; Yadav et al. 2019b, c, d). According to Lopes-Assad et al. (2010), filamentous saprophytic fungi of the genera *Aspergillus, Penicillium* and *Fusarium* are well-recognized K solubilizers. Several fungi, e.g. *Aspergillus fumigatus* (Lian et al. 2008), *Aspergillus niger* (Prajapati and Modi 2012), *Penicillium* spp. (Sangeeth et al. 2012), *Aspergillus* spp. and *Aspergillus terreus* (Prajapati et al. 2013), have been reported to mobilize organic and inorganic K and enhance the release of K from minerals and rock. Potassium solubilizing fungi (KSF) enhanced the biomass, yield and nutrient uptake in plants (Badr et al. 2006). Utilization of these beneficial fungi can be an emerging aspect concerning the sustainable development in the preparation of biofertilizer (Priyadharsini and Muthukumar 2016; Raghavendra et al. 2016; Yadav and Sidhu 2016).

Arbuscular mycorrhizal (AM) fungi belonging to the phylum Glomeromycota (Kehri, et al. 2018a) form mutualistic symbiotic association with roots of the plants and are ubiquitous in distribution (natural and agroecosystems) (Kehri et al. 2018b; Akhtar et al. 2019). These fungi provide several benefits to the plant including nutrition and increased the tolerance of plant towards various stresses. AM symbiosis can positively affect the plant growth by solubilizing fixed form of minerals (Verma et al. 2017a, b; Yadav et al. 2018a, b, c, d). AM fungi produce structures such as vesicles, spores, intra- and intercellular hyphae, auxiliary cells and extraradical mycelium (Kehri et al. 2018a). Particle-induced X-ray emission (PIXE) experiments including the *Rhizophagus intraradices* suggested that AM fungi accumulated K in hyphae, vesicles and spores (Olsson et al. 2008, 2011). Moreover, Casieri et al. (2013) identified four sequences in *Rhizophagus irregularis* in which three sequence coding for Shaker-like channels and one for KT/KUP/HAK transporter are reported to involve in K^+ transport from an ETS library.

9.7 Mechanism of Potassium Solubilization

There are various mechanisms used by KSB to release the K from the minerals. These bacteria produced low molecular weight organic acids which decrease the pH (Goldstein 1994) and produce the chelated compounds that form complexes with K, thereby enhancing the availability to the plant (Uroz et al. 2009; Keshavarz et al. 2013; Parmar and Sindhu 2013). The production of different types of organic acids is the foremost mechanism used by the bacteria to release the K from the minerals and make it bioavailable to the plant (Sheng et al. 2003). There are various types of organic acids produced by the bacteria such as propionic acid, lactic acid, 2-ketogluconic acid, citric acid, tartaric acid, acetic acid, oxalic acid, glycolic acid, succinic acid, malonic acid and fumaric acid (Table 9.1) (Wu et al. 2005; Sheng and He 2006). The production of organic acids decreases the pH (acidification) of soil thereby solubilizing K (Keshavarz et al. 2013; Parmar and Sindhu 2013). Römheld and Kirkby (2010) also reported that the bacterial-mediated acidification stimulates the production of chelates that ultimately enhance the crop growth and productivity. In addition to this, KSB also known to solubilize the K by the production of siderophores and lipo-chitooligosaccharides (Calvaruso et al. 2006).

Fungi adopt several mechanisms to solubilize K from minerals, thereby making it bioavailable to the plant. The known and the potential mechanisms used by fungi to solubilize the K comprise the production of organic acids (gluconic, oxalic and citric acid), protons and chelating agents such as siderophores and organic ligands. Lian et al. (2008) have been reported that *Cladosporium, Aspergillus* and *Penicillium* spp. excrete large quantity of citric, gluconic and oxalic acids that solubilize the silicates, mica and feldspar. Lian et al. (2007) also reported that *Aspergillus fumigatus* improved K solubilization when inoculated onto minerals containing K. Besides this, fungi also apply direct biophysical forces that can break the K mineral, thereby reducing the particle sizes (Xiaoxi et al. 2012). Fungi decrease soil pH and enhance the K solubilization in the soil. Lopes-Assad et al. (2006) studied the K solubilizing capacity of two strains of *Aspergillus niger* (CCT4355 and CCT911) and found that the studied fungi significantly decreased the pH of the soil. Similar to this finding, Rosa-Magri et al. (2012) found that the acids released by *Torulaspora globosa* solubilized the alkaline ultramafic rock and solubilized K in the medium within two weeks. Some of the PSF has been listed in Table 9.1.

There are very few reports on the mechanisms of AM fungi for K solubilizing. AM fungi do so by redox redaction, acidification, complexation and metal accumulation (Burgstaller and Schinner 1993; Adeleke et al. 2010). Most of these mechanisms are related to production of various low molecular weight organic acids (oxalate, malate

Table 9.1 Mechanism of potassium solubilizing microorganisms

Microorganisms	Mechanisms	References
Bacteria		
Sphingomonas, Burkholderia	Acidification, complexation	Uroz et al. (2007)
Pseudomonas spp.	Tartaric, citric acids	Krishnamurthy (1989)
Bacillus circulans GY92	Lipo-chitooligosaccharides production	Lian et al. (2001)
Bacillus glathei	Siderophores, organic ligands	Calvaruso et al. (2006)
E. hormaechei	Organic acids	Prajapati et al. (2013)
Pseudomonas spp.	Organic acids	Sheng and Huang (2002)
Pseudomonas aeruginosa	Acetate, citrate, oxalate	Sheng et al. (2003), Badr et al. (2006)
Paenibacillus mucilaginosus	Tartaric, citric, oxalic acids	Liu et al. (2012), Hu et al. (2006)
Bacillus mucilaginosus	Organic acids	Han and Supanjani (2006)
Fungi		
Penicillium frequentans, Cladosporium	Oxalic, citric, gluconic acids	Argelis et al. (1993)
Aspergillus niger, Penicillium sp.	Citric, glycolic, succinic acids	Sperberg (1958)
Aspergillus fumigatus, Aspergillus candidus	Oxalic, tartaric, citric, oxalic acids	Banik and Dey (1982)
Aspergillus fumigatus	Acidification	Lian et al. (2007)
AM fungi		
Arbuscular mycorrhiza	Released protons (H^+) or CO_2 and many different types of organic acids or compounds	Meena et al. (2014b)

and citrate) and ligands (Burgstaller and Schinner 1993; Jain and Sharma 2004). AM fungi can increase the solubility of K from minerals by discharging protons, H^+ or CO_2. Some of the mechanism of AM fungi has been listed in Table 9.1.

9.8 Role of Potassium Solubilizers in Sustainable Agriculture

Fixation of K in soil, its uptake by the plant and leaching reduced the availability of K in the soils (Sheng and Huang 2002). Application of microbial inoculants improved the growth and yield by releasing fix form of K from the soil (Rana et al. 2019a, b). It has been reported that crops inoculated with K solubilizers enhanced the productivity of crops (Yadav et al. 2017c, d; Yadav and Yadav 2019) (Table 9.2)

Table 9.2 Effect of potassium solubilizing microorganisms on the crops

Microbes	Crops	Effect on crops	References
Bacillus sp.	Wheat	Increased yield and disease resistances	Kloepper et al. (1991)
Paenibacillus glucanolyticus	Black pepper	Enhanced potassium uptake and dry weight	Sangeeth et al. (2012)
Bacillus mucilaginosus KCTC 3870	Eggplant	Increased P and K uptake	Han and Lee (2005)
Pseudomonas putida and *P. fluorescens*	Lettuce/tomato	Increased root and shoot growth	Hall et al. (1996), Glick et al. (1997)
Pseudomonas putida	Tea	Theaflavin, thearubigin, highly polymerized substances, total liquor colour, were improved	Bagyalakshmi et al. (2012)
Bacillus circulans	Valencia orange trees	Inoculation significantly improved leaf K content, fruit weight and yield	Shaaban et al. (2012)
Azotobacter and silicate bacteria	Cotton	Yield	Ciobanu (1961)
Bacillus pasteurii	Peanut/sesame	Significant increase in K availability	Youssef et al. (2010)
Aspergillus and *Bacillus*	Sorghum	Increased dry matter yield and uptake	Badr et al. (2006)
Bacillus edaphicus	Cotton rape	Growth, uptake	Sheng (2005)
Azospirillum chroococcum	Wheat	Increased crop productivity and nutrient uptake	Kapulnik et al. (1985, 1987), Kloepper et al. (1989, 1991)
Enterobacter hormaechei, Aspergillus terreus	Okra	Increased root, shoot growth and K uptake in Okra	Prajapati et al. (2013)
AM fungi	Switch grass	Increased plant height, root and shoot weight, root length and P, N contents	Clark et al. (1999)
Bacillus spp.	Chilly	Increased the uptake	Supanjani et al. (2006)
Frateuria aurantia	Brinjal	Improved the growth and yield	Ramarethinam and Chandra (2005)

(continued)

Table 9.2 (continued)

Microbes	Crops	Effect on crops	References
Bacillus spp.	Rice	Improved the yield	Muralikannan (1996)
Bacillus spp.	Rice	Improved the yield	Raj (2004)
Bacillus mucilaginosus, Azotobacter chroococcum	Maize-wheat	Improved growth and yield	Singh et al. (2010)
Bacillus mucilaginosus MCRCp1	Groundnut	Improved the growth and yield	Sugumaran and Janarthanam (2007)
Bacillus mucilaginous	Sorghum	Increased the uptake	Basak and Biswas (2009)
A. brasilense, A. chorococcum, Burkholderia cepacia, Rhizophagus clarus and *R. fasciculatus*	*Lycopersicon esculentum*	Increased the uptake in shoot	Pulido et al. (2003)
Bacillus mucilaginosus, *B. subtilis* and phosphate solubilizer (*Bacillus megaterium*)	Maize (*Zea mays*)	Co-inoculation increased the availability and uptake of P and K in maize. The shoot and root growth enhanced significantly grown on P- and K-limited soils	Abou-el-Seoud and Abdel-Megeed (2012)
Glomus mosseae and *Glomus intraradices*	*Zea mays*	Enhanced K uptake in plant	Wu et al. (2005)
Glomus etunicatum and *Glomus mosseae*	Corn and wheat	Enhanced the uptake of K, Mg and Ca and alleviated the negative effects of soil compaction	Miransari et al. (2009a, b)
Glomus intraradices	*Lycopersicon esculentum*	Enhanced the acquisition of K, Mg and Ca.	Cimen et al. (2010)

Lin et al. (2002) reported considerable increase in the uptake of K and P in *B. mucilaginosus* inoculated. Similar to this, Basak and Biswas (2009) and Meena et al. (2015) reported that the inoculation of bacteria improved the biomass over non-inoculated plants. Ramarethinam and Chandra (2006) reported that *Frateuria aurantia* improved the crop yield. Additionally, these microorganisms also produce vitamins, amino acids, auxins and gibberellic acids that improve the performance of the plant (Ponmurugan and Gopi 2006). Ai-min et al. (2013) reported that the enhancement of K also enhanced the enzymatic activities. In the same year, Prajapati et al. (2013) stated that inoculation of okra with *Enterobacter hormaechei* improved the root and shoot length by K mobilizing.

Pindi and Satyanarayana (2012) concluded from their experiment that these microorganisms could be used with other microorganisms such as *Rhizobium, Azospirillum, Azotobacter,* PSM, etc. The plant growth was also reported to increase by the co-inoculation of beneficial microorganisms. Singh et al. (2010) reported that maize and wheat plants inoculated with *Bacillus mucilaginosus*, *Azotobacter chroococcum* and *Rhizobium* show maximum K mobilization from waste mica and sequentially translocate it to the plant, thereby improving the growth (Singh et al. 2010). A field study was conducted by Han et al. (2006) on cucumber and pepper inoculated with *Bacillus mucilaginosus*, and the authors reported that inoculation of *B. mucilaginosus* mobilized the K from minerals. Thus, application of KSMs as a biofertilizer in agriculture improves the productivity, nutrient availability and reduces the use of agrochemicals in an economic, eco-friendly and sustainable manner (Sheng et al. 2003; Sindhu et al. 2010).

AM fungi play a pivotal role in uptake of mineral nutrient (especially K) and its subsequent translocation to the plants (Perner et al. 2007; Baslam et al. 2013). In the year 1999, Clark et al. reported that, under AM fungi conditions, the acquisition of K^+ was enhanced more in switchgrass grown in acid soil as compared to calcium and magnesium. Similar to this, Kaldorf et al. (1999) reported the acquisition of K^+ in *Zea mays* roots inoculated with the *Glomus* Br1 isolate. Furthermore, *Glomus mosseae* and *Glomus intraradices* enhanced K uptake by *Zea mays* (Wu et al. 2005). Miransari et al. (2009a, b) studied the influence of *Glomus etunicatum* and *Glomus mosseae* on nutrient uptake and concluded that AM fungi enhanced the uptake of K, Mg and Ca and alleviated the negative effects of soil compaction on corn and wheat. Study conducted on tomato (*Lycopersicon esculentum* L.) by Cimen et al. (2010) reported that inoculation of *Glomus intraradices* enhanced the acquisition of K, Mg and Ca. Similar study was conducted and reported that the fungus improved the uptake of K, Ca and Mg in tomato plant (Meena et al. 2015b; Singh et al. 2015). Moreover, Zaefarian et al. (2011) studied the single and combined effects of *Glomus intraradices, Glomus etunicatum, Glomus mosseae, Glomus fasciculatum* and *Gigaspora hartiga* on the uptake of N, K, Fe, Zn, Cu, etc. The role of AM fungi in K nutrition to the plant has been extensively reviewed by Garcia and Zimmermann (2014).

9.9 Conclusion and Future Perspectives

K^+ plays a key role in the metabolism of plant and triggers enzymes, osmotic regulation and charge balance in plants. The deficiency of K results in poor growth and development which intensely related to crop yield. Application of K as a fertilizer and less awareness of the farmers about K aggravates the situation. Therefore, utilization of naturally present K in the soil (fix form or not bioavailable form) is important for sustainable agriculture. Soil microorganisms (bacteria, fungi and AM fungi) are the integral component of soil and play a key role in K solubilization from the minerals. These microorganisms have various mechanism to solubilize the unavailable K and make it available to the plants. Results from various studies specified that soil inoculated with KSMs had positively influence the availability of K to the soil and plants. Additionally, KSMs also increase the availability of other nutrients (Ca, Mg, P) to the plants and also release siderophores, growth hormones and ammonia (NH^{3+}). Application of biological agents such as bacteria, fungi and AM fungi could offer an earlier and never-ending source of K for plant uptake. Therefore, there is a need to design experiments including different types of soil and microbial inoculants to find the best combination. There is also a need to commercialize the KSMs as an inoculum and make them available to the farmers. Future studies should also require focusing on the factors affecting availability of K and isolation of indigenous K solubilizers and inoculation of indigenous K solubilizers to the different environmental conditions such as cold, drought and nutrient-deficient soils.

Acknowledgements The authors are thankful to University Grant Commission (UGC) and CSIR, New Delhi for the financial assistance and also thankful to Head of Botany Department, University of Allahabad for providing all the necessary facilities.

References

Abou-el-Seoud II, Abdel-Megeed A (2012) Impact of rock materials and biofertilizations on P and K availability for maize (*Zea Maize*) under calcareous soil conditions. Saudi J Biol Sci 19:55–63

Adeleke RA, Cloete TE, Bertrand A, Khasa DP (2010) Mobilisation of potassium and phosphorus from iron ore by ectomycorrhizal fungi. World J Microbiol Biotechnol 26(10):1901–1913

Ahmad M, Nadeem SM, Naveed M, Zahir ZA (2016) Potassium-solubilizing bacteria and their application in agriculture. In: Potassium solubilizing microorganisms for sustainable agriculture. Springer, New Delhi, pp 293–313

Ai-min Z, Gang-yong Z, Shuang-feng Z, Rui-ying Z, Bao-cheng Z (2013) Effect of phosphorus and potassium content of plant and soil inoculated with *Paneibacillus kribensis* CX-7 strain antioxidant and antitumor activity of *Phyllanthus emblica* in colon cancer cell lines. Int J Curr Microbiology Appl Sci 2(6):273–279

Akhtar O, Mishra R, Kehri HK (2019) Arbuscular mycorrhizal association contributes to Cr accumulation and tolerance in plants growing on Cr contaminated soils. Proc Natl Acad Sci India Sect B Biol Sci 89(1):63–70. https://doi.org/10.1007/s40011-017-0914-4

Aleksandrov VG, Blagodyr RN, Ilev IP (1967) Liberation of phosphoric acid from apatite by silicate bacteria. Mikrobiol Z 29:111–114

Andrews M, Sprent JI, Raven JA, Eady PE (1999) Relationships between shoot to root ratio, growth and leaf soluble protein concentration of *Pisum sativum, Phaseolus vulgaris* and *Triticum aestivum* under different nutrient deficiencies. Plant Cell Environ 22:949–958

Anjanadevi IP, John NS, John KS, Jeeva ML, Misra RS (2016) Rock inhabiting potassium solubilizing bacteria from Kerala, India: characterization and possibility in chemical K fertilizer substitution. J Basic Microbiol 56:67–77

Argelis DT, Gonzala DA, Vizcaino C, Gartia MT (1993) Biochemical mechanism of stone alteration carried out by filamentous fungi living in monuments. Biogeochemistry 19:129–147

Armengaud P, Breitling R, Amtmann A (2010) Coronatine-insensitive 1 (COI1) mediates transcriptional responses of *Arabidopsis thaliana* to external potassium supply. Mol Plant 3:390–405

Ashley MK, Grant M, Grabov A (2005) Plant responses to potassium deficiencies: a role for potassium transport proteins. J Exp Bot 57:425–436
Badr MA, Shafei AM, Sharaf El-Deen SH (2006) The dissolution of K and P-bearing minerals by silicate dissolving bacteria and their effect on *sorghum* growth. Res J Agric Boil Sci 2:5–11
Bagyalakshmi B, Ponmurugan P, Marimuthu S (2012) Influence of potassium solubilizing bacteria on crop productivity and quality of tea (*Camellia sinensis*). Afr J Agric Res 7:4250–4259
Banfield JF, Barker WW, Welch SA, Taunton A (1999) Biological impact on mineral dissolution: application of the lichen model to understanding mineral weathering in the rhizosphere. Proc Natl Acad Sci 96:3404–3411
Banik S, Dey BK (1982) Available phosphate content of an alluvial soil as influenced by inoculation of some isolated phosphate-solubilizing micro-organisms. Plant Soil 69(3):353–364
Basak BB, Biswas DR (2009) Influence of potassium solubilizing microorganism (*Bacillus mucilaginosus*) and waste mica on potassium uptake dynamics by sudan grass (*Sorghum vulgare* Pers.) grown under two Alfisols. Plant Soil 317:235–255
Baslam M, Garmendia I, Goicoechea N (2013) The arbuscular mycorrhizal symbiosis can overcome reductions in yield and nutritional quality in greenhouse-lettuces cultivated at inappropriate growing seasons. Sci Hort 164:145–154
Biswas DR (2011) Nutrient recycling potential of rock phosphate and waste mica enriched compost on crop productivity and changes in soil fertility under potato–soybean cropping sequence in an Inceptisol of Indo-Gangetic Plains of India. Nutr Cycl Agroecosys 89:15–30
Braunschweigh IC (1980) K^+ availability in relation to clay content. Results of field experiment. Potash Rev 2:1–8
Burgstaller W, Schinner F (1993) Leaching of metals with fungi. Plant Biotechnol J 27:91–116
Buschmann PH, Vaidyanathan R, Gassmann W, Schroeder JI (2000) Enhancement of Na + uptake currents, time-dependent inward-rectifying K^+ channel currents, and K^+ channel transcripts by K^+ starvation in wheat root cells. Plant Physiol 122(4):1387–1398
Cakmak I (2005) The role of potassium in alleviating detrimental effects of abiotic stresses in plants. J Soil Sci Plant Nutr 168:521–530
Calvaruso C, Turpault MP, Frey-Klett P (2006) Root-associated bacteria contribute to mineral weathering and to mineral nutrition in trees: a budgeting analysis. Appl Environ Microbiol 72(2):1258–1266
Casieri L, Lahmidi NA, Doidy J, Veneault-Fourrey C, Migeon A, Bonneau L, Brun A (2013) Biotrophic transportome in mutualistic plant–fungal interactions. Mycorrhiza 23:597–625
Chen J, Gabelman WH (2000) Morphological and physiological characteristics of tomato roots associated with potassium-acquisition efficiency. Sci Hort 83:213–225
Chen YF, Wang Y, Wu WH (2008) Membrane transporters for nitrogen, phosphate and potassium uptake in plants. J Integr Plant Biol 50(7):835–848
Chérel I, Michard E, Platet N, Mouline K, Alcon C, Sentenac H, Thibaud JB (2002) Physical and functional interaction of the *Arabidopsis* K^+ channel AKT2 and phosphatase AtPP2CA. Plant Cell 14:1133–1146
Cimen I, Pirinc V, Doran I, Turgay B (2010) Effect of soil solarization and arbuscular mycorrhizal fungus (*Glomus intraradices*) on yield and blossom-end rot of tomato. Int J Agric Biol 12:551–555
Ciobanu I (1961) Investigation on the efficiency of bacterial fertilizers applied to cotton. Cent Exp Ingras Bact Lucrari Stiint 3:203–214
Clark RB, Zeto SK (2000) Mineral acquisition by arbuscular mycorrhizal plants. J Plant Nutr 23:867–902
Clark RB, Zobel RW, Zeto SK (1999) Effects of mycorrhizal fungus isolates on mineral acquisition by *Panicum virgatum* in acidic soil. Mycorrhiza 9:167–176
Diep CN, Hieu TN (2013) Phosphate and potassium solubilizing bacteria from weathered materials of denatured rock mountain, Ha Tien, Kiên Giang province Vietnam. Am J Life Sci 1:88–92
Drew MC (1975) Comparison of the effects of a localised supply of phosphate, nitrate, ammonium and potassium on the growth of the seminal root system, and the shoot, in barley. New Phytol 75:479–490

Du XQ, WangFL Li H, Jing S, Yu M, Li J, Wu WH, Kudla J, Wang Y (2019) The transcription factor MYB59 regulates $K^+/NO3^-$ translocation in the *Arabidopsis* response to low K^+ stress. Plant Cell 31(3):699–714

Fairbairn DJ, Liu W, Schachtman DP, Gomez-Gallego S, Day SR, Teasdale RD (2000) Characterisation of two distinct HKT1-like potassium transporters from *Eucalyptus camaldulensis*. Plant Mol Biol 43(4):515–525

Gajdanowicz P, Michard E, Sandmann M, Rocha M, Corr^ea LGG, Ramırez-Aguilar SJ, Gomez-Porras JL, Gonzalez W, Thibaud J-B, van Dongen JT et al. (2011) Potassium (K^+) gradients serve as a mobile energy source in plant vascular tissues. Proc Natl Acad Sci USA 108:864–869

Garcia K, Zimmermann SD (2014) The role of mycorrhizal associations in plant potassium nutrition. Front Plant Sci 5:337

Gaymard F, Pilot G, Lacombe B, Bouchez D, Bruneau D, Boucherez J, Sentenac H (1998) Identification and disruption of a plant shaker-like outward channel involved in K^+ release into the xylem sap. Cell 94:647–655

Gierth M, Mäser P, Schroeder JI (2005) The potassium transporter AtHAK5 functions in K^+ deprivation-induced high-affinity K^+ uptake and AKT1 K^+ channel contribution to K^+ uptake kinetics in *Arabidopsis* roots. Plant Physiol 137:1105–1114

Glick BR, Liu C, Ghosh S, Dumbroff EB (1997) Early development of canola seedlings in the presence of the plant growth-promoting rhizobacterium *Pseudomonas putida* GR12-2. Soil Biol Biochem 29:1233–1239

Glowa KR, Arocena JM, Massicotte HB (2003) Extraction of potassium and/or magnesium from selected soil minerals by *Piloderma*. Geomicrobiol J 20:99–111

Goldstein AH (1994) Involvement of the quinoprotein glucose dehydrogenase in the solubilization of exogenous phosphates by gram-negative bacteria. Phosphate in microorganisms: cellular and molecular biology. ASM Press, Washington DC 197–203

Groudev SN (1987) Use of heterotrophic microorganisms in mineral biotechnology. Acta Biotechnol 7:299–306

Gundala PB, Chinthala P, Sreenivasulu B (2013) A new facultative alkaliphilic, potassium solubilizing, Bacillus Sp. SVUNM9 isolated from mica cores of Nellore District, Andhra Pradesh, India. Research and Reviews. J Microbiol Biotechnol 2:1–7

Hafsi C, Debez A, Abdelly C (2014) Potassium deficiency in plants: effects and signaling cascades. Acta Physiol Plant 36:1055–1070

Hafsi C, Romero-Puertas MC, Luis A, Abdelly C, Sandalio LM. (2011). Antioxidative response of *Hordeum maritimum* L to potassium deficiency. Acta Physiol Plant 33(1):193–202

Hall JA, Peirson D, Ghosh S, Glick B (1996) Root elongation in various agronomic crops by the plant growth promoting rhizobacterium *Pseudomonas putida* GR12–2. Isr J Plant Sci 44:37–42

Han HS, Lee KD (2005) Phosphate and potassium solubilizing bacteria effect on mineral uptake, soil availability and growth of eggplant. Res J Agric Biol Sci 1:176–180

Han HS, Supanjani Lee KD (2006) Effects of co-inoculation with phosphate and potassium solubilizing bacteria on mineral uptake and growth of pepper and cucumber. Plant Soil Environ 52:130–136

Hasan R (2002) Potassium status of soils in India. Better Crops Int 16:3–5

Hong JP, Takeshi Y, Kondou Y, Schachtman DP, Matsui M, Shin R (2013) Identification and characterization of transcription factors regulating Arabidopsis HAK5. Plant Cell Physiol 54:1478–1490

Hosy E, Vavasseur A, Mouline K, Dreyer I, Gaymard F, Porée F, Simonneau T (2003) The Arabidopsis outward K^+ channel GORK is involved in regulation of stomatal movements and plant transpiration. Proc Natl Acad Sci 100:5549–5554

Hu X, Chen J, Guo J (2006) Two phosphate-and potassium-solubilizing bacteria isolated from Tianmu Mountain, Zhejiang, China. World J Microb Biot 22:983–990

Jain N, Sharma DK (2004) Biohydrometallurgy for nonsulfidic minerals—a review. Geomicrobiol J 21:135–144

Johansson SA, Campbell JL (1988) PIXE: A novel technique for elemental analysis

Kaldorf M, Kuhn AJ, Schröder WH, Hildebrandt U, Bothe H (1999) Selective element deposits in maize colonized by a heavy metal tolerance conferring arbuscular mycorrhizal fungus. J Plant Physiol 154:718–728

Kang SC, Ha CG, Lee TG, Maheshwari DK (2002) Solubilization of insoluble inorganic phosphates by a soil-inhabiting fungus *Fomitopsis* sp. PS 102. Curr Sci 439–442

Kapulnik Y, Okon Y, Henis Y (1985). Changes in root morphology of wheat caused by *Azospirillum* inoculation. Can J Microbiol 31:881–887

Kapulnik Y, Okon Y, Henis Y (1987). Yield response of spring wheat cultivars (*Triticum aestivum and T. turgidum*) to inoculation with *Azospirillum brasilense* under field conditions. Biol Fertil Soils 4(1–2):27–35

Kasana RC, Panwar NR, Burman U, Pandey CB (2017) Kumar P (2017) Isolation and Identification of Two Potassium Solubilizing Fungi from Arid Soil. Int J Curr Microbiol App Sci 6(3):1752–1762

Kehri HK, Akhtar O, Zoomi I, Pandey D (2018a) Arbuscular mycorrhizal fungi: taxonomy and its systematics. Int J Life Sci Res 6(4):58–71

Kehri HK, Zoomi I, Srivastava P (2018b) Diversity and function of AM fungi in heavy metal contaminated soil. In: Singh S (eds) Biodiversity, monitoring, management and utilization. Astral International (P) Ltd, New Delhi, pp 111–128

Keshavarz ZJ, Aliasgharzad N, Oustan S, Emadi M, Ahmadi A (2013) Isolation and characterization of potassium solubilizing bacteria in some Iranian soils. Arch Agron Soil Sci 59:1713–1723

Kinekar BK (2011) Potassium fertilizer situation in India: current use and perspectives. Karnataka J Agri Sci 24(1)

Kloepper JW, Lifshitz R, Zablotowicz RM (1989) Free-living bacterial inocula for enhancing crop productivity. Trends Biotechnol 7:39–44

Kloepper JW, Zablowicz RM, Tipping EM, Lifshitz R (1991) Plant growth promotion mediated by bacterial rhizosphere colonizers. In: Keister DL, Gregan B (eds) The rhizosphere and plant growth. Kluwer Academic Publishers, Dordrecht, Netherlands, pp 315–326

Kour D, Rana KL, Yadav AN, Yadav N, Kumar V, Kumar A, Sayyed RZ, Hesham AE-L, Dhaliwal HS, Saxena AK (2019a) Drought-Tolerant Phosphorus-Solubilizing Microbes: Biodiversity and Biotechnological Applications for Alleviation of Drought Stress in Plants. In: Sayyed RZ, Arora NK, Reddy MS (eds) Plant Growth Promoting Rhizobacteria for Sustainable Stress Management: Volume 1: Rhizobacteria in Abiotic Stress Management. Springer Singapore, Singapore, pp 255–308. https://doi.org/10.1007/978-981-13-6536-2_13

Kour D, Rana KL, Yadav N, Yadav AN, Kumar A, Meena VS, Singh B, Chauhan VS, Dhaliwal HS, Saxena AK (2019b) Rhizospheric Microbiomes: Biodiversity, Mechanisms of Plant Growth Promotion, and Biotechnological Applications for Sustainable Agriculture. In: Kumar A, Meena VS (eds) Plant Growth Promoting Rhizobacteria for Agricultural Sustainability: From Theory to Practices. Springer Singapore, Singapore, pp 19–65. https://doi.org/10.1007/978-981-13-7553-8_2

Kour D, Rana KL, Yadav N, Yadav AN, Singh J, Rastegari AA, Saxena AK (2019c) Agriculturally and industrially important fungi: current developments and potential biotechnological applications. In: Yadav AN, Singh S, Mishra S, Gupta A (eds) Recent Advancement in White Biotechnology through Fungi, Volume 2: Perspective for Value-Added Products and Environments. Springer International Publishing, Cham, pp 1–64. https://doi.org/10.1007/978-3-030-14846-1_1

Krishnamurthy D, Varamban SV, Thiruvengadasami A, Mathews CK (1989) Solubility of oxygen in sodium-potassium alloys. J Less Common Met. 153:363–372

Kumar A, Bahadur I, Maurya BR, Raghuwanshi R, Meena VS, Singh DK, Dixit J (2015) Does a plant growth-promoting rhizobacteria enhance agricultural sustainability. J Pure Appl Microbiol 9:715–724

Kumar M, Kour D, Yadav AN, Saxena R, Rai PK, Jyoti A, Tomar RS (2019) Biodiversity of methylotrophic microbial communities and their potential role in mitigation of abiotic stresses in plants. Biologia 74:287–308. https://doi.org/10.2478/s11756-019-00190-6

Kumar P, Dubey RC, Maheshwari DK (2012) *Bacillus* strains isolated from rhizosphere showed plant growth promoting and antagonistic activity against phytopathogens. Microbiol Res 167(8):493–499

Leaungvutiviroj C, Ruangphisarn P, Hansanimitkul P, Shinkawa H, Sasaki K (2010) Development of a new biofertilizer with a high capacity for N2 fixation, phosphate and potassium solubilization and auxin production. Biosci Biotech Biochem 74:1098–1101

Lian B, Fu PQ, Mo DM, Liu CQ (2002) A comprehensive review of the mechanism of potassium releasing by silicate bacteria. Acta Mineral Sin 22:179–183

Lian B, Prithiviraj B, Souleimanov Smith DL (2001) Evidence for the production of chemical compounds analogous to nod factor by the silicate bacterium *Bacillus circulans* GY92. Microbiol Res 156(3):289–292

Lian B, Wang B, Pan M, Liu C, Tang HH (2007) Microbial release of potassium from K-bearing minerals by thermophilic fungus *Aspergillus fumigatus*. Geochem Cosmochim Acta 72:87–98

Lian B, Wang B, Pan M, Liu C, Teng HH (2008) Microbial release of potassium from K-bearing minerals by thermophilic fungus *Aspergillus fumigatus*. Geochim Cosmochim Acta 72:87–98

Lin QM, Rao ZH, Sun YX, Yao J, Xing LJ (2002) Identification and practical application of silicate-dissolving bacteria. Agr Sci China 1:81–85

Liu D, Lian B, Dong H (2012) Isolation of *Paenibacillus* sp. and assessment of its potential for enhancing mineral weathering. Geomicrob J 29:413–421

Liu G, Lin Y, Huang Z (2001) Screening of silicate bacteria with potassium-releasing and antagonistical activity. Chinese J Appl Environ Biol 7:66–68

Lopes-Assad ML, Avansini SH, Rosa MM, De Carvalho JR, Ceccato-Antonini SR (2010) The solubilization of potassium-bearing rock powder by *Aspergillus niger* in small-scale batch fermentations. Can J Microbiol 56:598–605

Lopes-Assad ML, Rosa MM, Erler G, Ceccato-Antonini SR (2006) Solubilização de pó-de-rocha por *Aspergillus niger*. Revista Espaço e Geografia 9:1–17

Lynn TM, Win HS, Kyaw EP, Latt ZK, Yu SS (2013) Characterization of phosphate solubilizing and potassium decomposing strains and study on their effects on tomato cultivation. Int J Innov Appl Stud 3:959–966

Ma TL, Wu WH, Wang Y (2012) Transcriptome analysis of rice root responses to potassium deficiency. BMC Plant Biol 12:161

Meena RK, Singh RK, Singh NP, Meena SK, Meena VS (2015a) Isolation of low temperature surviving plant growth–promoting rhizobacteria (PGPR) from pea (*Pisum sativum* L.) and documentation of their plant growth promoting traits. Biocat Agri Biotech 4:806–811

Meena VS, Bahadur I, Maurya BR, Kumar A, Meena RK, Meena SK, Verma JP (2016) Potassium-solubilizing microorganism in evergreen agriculture: an overview. In: Meena VS, Maurya BR, Verma JP, Meena RK (eds) Potassium solubilizing microorganisms for sustainable agriculture. Springer, New Delhi pp 1–20

Meena VS, Maurya BR, Bahadur I (2014a) Potassium solubilization by bacterial strain in waste mica. Bangl J Bot 43:235–237

Meena VS, Maurya BR, Verma JP (2014b) Does a rhizospheric microorganism enhance K^+ availability in agricultural soils? Microbiol Res 169(5–6):337–347

Meena VS, Maurya BR, Verma JP, Aeron A, Kumar A, Kim K, Bajpai VK (2015b) Potassium solubilizing rhizobacteria (KSR): isolation, identification, and K-release dynamics from waste mica. Ecol Eng 81:340–347

Mengel K, Kirkby EA, Kosegarten H, Appel T (2001) Nitrogen. In: Mengel K, Kirkby EA, Kosegarten H, Appel T (eds) Principles of plant nutrition. Springer, Dordrecht pp 397–434

Mikhailouskaya N, Tchernysh A (2005) K-mobilizing bacteria and their effect on wheat yield. J Agron Crop Sci 8:154–157

Minjian C, Haiqiu Y, Hongkui Y, Chunji J (2007) Difference in tolerance to potassium deficiency between two maize inbred lines. Plant Prod Sci 10:42–46

Miransari M, Bahrami HA, Rejali F, Malakouti MJ (2009a) Effects of soil compaction and arbuscular mycorrhiza on corn (*Zea mays* L.) nutrient uptake. Soil Till Res 103:282–290

Miransari M, Bahrami HA, Rejali F, Malakouti MJ (2009b) Effects of arbuscular mycorrhiza, soil sterilization, and soil compaction on wheat (*Triticum aestivum* L.) nutrients uptake. Soil Till Res 104:48–55

Muentz A (1890) Sur la decomposition des roches et la formation de la terre arable. CR Acad Sci 110:1370–1372

Muralikannan M (1996) Biodissolution of silicate, phosphate and potassium by silicate solubilizing bacteria in rice ecosystem. M Sc (Ag) thesis submitted to Tamil Nadu Agricultural University, Coimbatore 125

National Center for Biotechnology Information. PubChem Database. Potassium, AtomicNumber = 19, https://pubchem.ncbi.nlm.nih.gov/element/Potassium

Öborn I, Andrist-Rangel Y, Askekaard M, Grant CA, Watson CA, Edwards AC (2005) Critical aspects of potassium management in agricultural systems. Soil Use and Manag 21:102–112

Olsson PA, Hammer EC, Pallon J, Van Aarle IM, Wallander H (2011) Elemental composition in vesicles of an arbuscular mycorrhizal fungus, as revealed by PIXE analysis. Fungal Boil 115:643–648

Olsson PA, Hammer EC, Wallander H, Pallon J (2008) Phosphorus availability influences elemental uptake in the mycorrhizal fungus *Glomus intraradices,* as revealed by particle-induced X-ray emission analysis. Appl Environ Microbiol 74:4144–4148

Pandey D, Kehri HK, Zoomi I, Akhtar O, Singh AK (2019) Mycorrhizal fungi: biodiversity, ecological significance, and industrial applications. In: Yadav A, Mishra S, Singh S, Gupta A (eds) Recent advancement in white biotechnology through fungi. Fungal biology. Springer, Cham, pp 181–199. https://doi.org/10.1007/978-3-030-10480-1_5

Parmar P, Sindhu SS (2013) Potassium solubilization by rhizosphere bacteria: influence of nutritional and environmental conditions. J Microbiol Res 3:25–31

Pathak H, Mohanty S, Jain N, Bhatia A (2010) Nitrogen, phosphorus, and potassium budgets in Indian agriculture. Nutr Cycl Agroecosys 86:287–299

Perner H, Schwarz D, Bruns C, Mäder P, George E (2007) Effect of arbuscular mycorrhizal colonization and two levels of compost supply on nutrient uptake and flowering of pelargonium plants. Mycorrhiza 17:469–474

Pilot G, Lacombe B, Gaymard F, Chérel I, Boucherez J, Thibaud JB, Sentenac H (2001) Guard Cell Inward K^+ Channel Activity in *Arabidopsis* Involves Expression of the Twin Channel Subunits KAT1 and KAT2. J Biol Chem 276(5):3215–3221

Pindi PK, Satyanarayana SDV (2012) Liquid microbial consortium-a potential tool for sustainable soil health. J Biofertil Biopesticide 3(4):1–9

Ponmurugan P, Gopi C (2006) In vitro production of growth regulators and phosphatase activity by phosphate solubilizing bacteria. Afr J Biotechnol 5(4):348–350

Prajapati K, Sharma MC, Modi HA (2013) Growth promoting effect of potassium solubilizing microorganisms on *Abelmoscus esculantus.* Int J Agric Sci 3:181–188

Prajapati KB, Modi HA (2012) Isolation and characterization of potassium solubilizing bacteria from ceramic industry soil. CIB Tech J Microbiol 1:8–14

Priyadharsini P, Muthukumar T (2016) Interactions between arbuscular mycorrhizal fungi and potassium-solubilizing microorganisms on agricultural productivity. In Potassium solubilizing microorganisms for sustainable agriculture Springer, New Delhi, pp 111–125

Pulido LE, Cabrera A, Medina N (2003) Biofertilization using rhizobacteria and AMF in the production of tomato (*Lycopersicon esculentum* Mill.) and onion (Allium cepa L.) seedlings. II. Root colonization and nutritional status. Cultivos Tropicales 24:5–13

Pyo YJ, Gierth M, Schroeder JI, Cho MH (2010) High-affinity K^+ transport in Arabidopsis: AtHAK5 and AKT1 are vital for seedling establishment and post germination growth under low-potassium conditions. Plant Physiol 153:863–875

Raghavendra MP, Nayaka SC, Nuthan BR (2016) Role of rhizosphere microflora in potassium solubilization. In: Potassium solubilizing microorganisms for sustainable agriculture. Springer, New Delhi, pp 43–59

Raj SA (2004) Solubilization on a silicate and concurrent release of phosphorus and potassium in rice ecosystem. In: Biofertilizer technology for rice based cropping system, India, pp 372–378
Ramarethinam S, Chandra K (2005) Studies on the effect of potash solubilizing/mobilizing bacteria *Frateuria aurantia* on brinjal growth and yield. Pestology 11:35–39
Ramarethinam S, Chandra K (2006) Studies on the effect of potash solubilizing bacteria *Frateuria aurantia* (Symbion-K-liquid formulation) on Brinjal (*Solanum melongena* L.) growth and yield. Pestology 11:35–39
Rana KL, Kour D, Sheikh I, Dhiman A, Yadav N, Yadav AN, Rastegari AA, Singh K, Saxena AK (2019a) Endophytic fungi: biodiversity, ecological significance and potential industrial applications. In: Yadav AN, Mishra S, Singh S, Gupta A (eds) Recent advancement in white biotechnology through fungi, vol 1. Diversity and enzymes perspectives. Springer, Switzerland, pp 1–62
Rana KL, Kour D, Sheikh I, Yadav N, Yadav AN, Kumar V, Singh BP, Dhaliwal HS, Saxena AK (2019b) Biodiversity of endophytic fungi from diverse niches and their biotechnological applications. In: Singh BP (eds) Advances in endophytic fungal research: present status and future challenges. Springer International Publishing, Cham, pp 105–144. https://doi.org/10.1007/978-3-030-03589-1_6
Rana KL, Kour D, Yadav AN (2019c) Endophytic microbiomes: biodiversity, ecological significance and biotechnological applications. Res J Biotechnol 14:142–162
Rosa-Magri MM, Avansini SH, Lopes-Assad ML, Tauk-Tornisielo SM, Ceccato-Antonini SR (2012) Release of potassium from rock powder by the yeast *Torulaspora globosa*. Braz Arch Biol Technol 55:577–582
Sangeeth KP, Bhai RS, Srinivasan V (2012) *Paenibacillus glucanolyticus*, a promising potassium solubilizing bacterium isolated from black pepper (*Piper nigrum* L.) rhizosphere. J Spices Aromat Crops 21:118–124
Schachtman DP, Schroeder JI (1994) Structure and transport mechanism of a high-affinity potassium uptake transporter from higher plants. Nature 370(6491):655–658
Schachtman DP, Schroeder JI, Lucas WJ, Anderson JA, Gaber RF (1992) Expression of an inward-rectifying potassium channel by the Arabidopsis KAT1 cDNA. Science 258:1654–1658
Schneider A, Tesileanu R, Charles R, Sinaj S (2013) Kinetics of soil potassium sorption–desorption and fixation. Commun Soil Sci Plan 44:837–849
Shaaban EA, El-Shamma MS, El-Shazly S, El-Gazzar A, Abdel-Hak RE (2012) Efficiency of rock-feldspar combined with silicate dissolving bacteria on yield and fruit quality of Valencia orange fruits in reclaimed soils. Res J Appl Sci 8:4504–4510
Sheng XF (2005) Growth promotion and increased potassium uptake of cotton and rape by a potassium releasing strain of *Bacillus edaphicus*. Soil Biol Biochem 37:1918–1922
Sheng XF, He LY (2006) Solubilization of potassium-bearing minerals by a wild-type strain of *Bacillus edaphicus* and its mutants and increased potassium uptake by wheat. Can J Microbiol 52:66–72
Sheng XF, He LY, Huang WY (2002) The conditions of releasing potassium by a silicate-dissolving bacterial strain NBT. Agr Sci China 1:662–666
Sheng XF, Xia JJ, Chen J (2003) Mutagenesis of the Bacillus edaphicus strain NBT and its effect on growth of chili and cotton. Agr Sci China 2:409–412
Shin R, Schachtman DP (2004) Hydrogen peroxide mediates plant root cell response to nutrient deprivation. Proc Natl Acad Sci 101:8827–8832
Sindhu SS, Dua S, Verma MK, Khandelwal A (2010) Growth promotion of legumes by inoculation of rhizosphere bacteria. In: Microbes for legume improvement. Springer, Vienna
Singh G, Biswas DR, Marwaha TS (2010) Mobilization of potassium from waste mica by plant growth promoting rhizobacteria and its assimilation by maize (*Zea mays*) and wheat (*Triticum aestivum* L.): a hydroponics study under phytotron growth chamber. J Plant Nutr 33:1236–1251
Singh NP, Singh RK, Meena VS, Meena RK (2015) Can we use maize (*Zea mays*) rhizobacteria as plant growth promoter. Vegetos 28:86–99

Singh P, Blanke MM (2000) Deficiency of potassium but not phosphorus enhances root respiration. Plant Growth Regul 32:77–81

Sparks DL, Huang PM (1985) Physical chemistry of soil potassium. In: Munson RD (ed) Potassium in agriculture ASA. CSSA and SSSA, Madison, pp 201–265

Sperber JI (1958) The incidence of apatite-solubilizing organisms in the rhizosphere and soil. Aust J Agric Res 9:778–781

Sugumaran P, Janarthanam B (2007) Solubilization of potassium containing minerals by bacteria and their effect on plant growth. World J. Agric. Res 3:350–355

Supanjani HH, Jung JS, Lee KD (2006) Rock phosphate potassium and rock solubilizing bacteria as alternative sustainable fertilizers. Agron Sustain Dev 26:233–240

Syed BA, Patel B (2014) Investigation and Correlation of Soil Biotic and Abiotic Factors Affecting Agricultural Productivity in Semi-Arid Regions of North Gujarat, India. Int J Res Stud Biosci 2:18–29

Thaler P, Pagès L (1998) Modelling the influence of assimilate availability on root growth and architecture. Plant Soil 201:307–320

Thomas SJM, Edwards PP, Kuznetsov VL (2008) Sir Humphry Davy: boundless chemist, physicist, poet and man of action. Chem Phys Chem 9:59–66

Uroz S, Calvaruso C, Turpault MP, Frey-Klett P (2009) Mineral weathering by bacteria: ecology, actors and mechanisms. Trends Microbial 17:378–387

Uroz S, Calvaruso C, Turpault MP, Pierrat JC, Mustin C, Frey-Klett P (2007) Effect of the mycorrhizosphere on the genotypic and metabolic diversity of the bacterial communities involved in mineral weathering in a forest soil. Appl Environ Microbiol 73:3019–3027

Veresoglou SD, Shaw LJ, Sen R (2011) *Glomus intraradices* and *Gigaspora margarita* arbuscular mycorrhizal associations differentially affect nitrogen and potassium nutrition of *Plantago lanceolata* in a low fertility dune soil. Plant Soil 340:481–490

Verma P, Yadav AN, Khannam KS, Kumar S, Saxena AK, Suman A (2016a) Molecular diversity and multifarious plant growth promoting attributes of Bacilli associated with wheat (*Triticum aestivum* L.) rhizosphere from six diverse agro-ecological zones of India. J Basic Microbiol 56:44–58

Verma P, Yadav AN, Khannam KS, Mishra S, Kumar S, Saxena AK, Suman A (2016b) Appraisal of diversity and functional attributes of thermotolerant wheat associated bacteria from the peninsular zone of India. Saudi J Biol Sci. https://doi.org/10.1016/j.sjbs.2016.01.042

Verma P, Yadav AN, Khannam KS, Panjiar N, Kumar S, Saxena AK, Suman A (2015) Assessment of genetic diversity and plant growth promoting attributes of psychrotolerant bacteria allied with wheat (*Triticum aestivum*) from the northern hills zone of India. Ann Microbiol 65:1885–1899

Verma P, Yadav AN, Khannam KS, Saxena AK, Suman A (2017a) Potassium-Solubilizing microbes: diversity, distribution, and role in plant growth promotion. In: Panpatte DG, Jhala YK, Vyas RV, Shelat HN (eds) Microorganisms for green revolution: Volume 1: Microbes for sustainable crop production. Springer Singapore, Singapore, pp 125–149. https://doi.org/10.1007/978-981-10-6241-4_7

Verma P, Yadav AN, Kumar V, Singh DP, Saxena AK (2017b) Beneficial plant-microbes interactions: biodiversity of microbes from diverse extreme environments and its impact for crop improvement. In: Singh DP, Singh HB, Prabha R (eds) Plant-Microbe interactions in agro-ecological perspectives: Volume 2: Microbial interactions and agro-ecological impacts. Springer, Singapore, pp 543–580. https://doi.org/10.1007/978-981-10-6593-4_22

Verrecchia EP, Dumont JL (1996) A biogeochemical model for chalk alteration by fungi in semiarid environments. Biogeochemistry 35:447–470

Very AA, Sentenac H (2003) Molecular mechanisms and regulation of K^+ transport in higher plants. Annu Rev Plant Biol 54:575–603

Römheld V, Kirkby EA (2010) Research on potassium in agriculture: needs and prospects. Plant Soil 335(1–2):155–180

Wallander H, Wickman T (1999) Biotite and microcline as potassium sources in ectomycorrhizal and non-mycorrhizal *Pinus sylvestris* seedlings. Mycorrhiza 9:25–32

White PJ, Karley AJ (2010) Potassium. In: Cell biology of metals and nutrients Springer, Berlin, Heidelberg, pp 199–224

Wu SC, Cao ZH, Li ZG, Cheung KC, Wong MH (2005) Effects of biofertilizer containing N-fixer, P and K solubilizers and AM fungi on maize growth: a greenhouse trial. Geoderma 125:155–166

Xiaoxi Z, Liu X, Tang J, Hu S, Jiang P, Li W, Xu L (2012) Characterization and potassium solubilizing ability of *Bacillus circulans* Z_{1-3}. Adv Sci Lett 10:173–176

Xie J, McCobb DP (1998) Control of alternative splicing of potassium channels by stress hormones. Science 280:443–446

Yadav AN (2017a) Agriculturally important microbiomes: biodiversity and multifarious PGP attributes for amelioration of diverse abiotic stresses in crops for sustainable agriculture. Biomed J Sci Tech Res 1:1–4

Yadav AN (2017b) Beneficial role of extremophilic microbes for plant health and soil fertility. J Agric Sci 1:1–4

Yadav AN (2018) Biodiversity and biotechnological applications of host-specific endophytic fungi for sustainable agriculture and allied sectors. Acta Sci Microbiol 1:01–05

Yadav AN (2019a) Endophytic fungi for plant growth promotion and adaptation under abiotic stress conditions. Acta Sci Agric 3:91–93

Yadav AN (2019b) Microbiomes of Wheat (Triticum aestivum L.) Endowed with Multifunctional Plant Growth Promoting Attributes. EC Microbiology 15:1–6

Yadav AN, Kour D, Sharma S, Sachan SG, Singh B, Chauhan VS, Sayyed RZ, Kaushik R, Saxena AK (2019a) Psychrotrophic Microbes: Biodiversity, Mechanisms of Adaptation, and Biotechnological Implications in Alleviation of Cold Stress in Plants. In: Sayyed RZ, Arora NK, Reddy MS (eds) Plant Growth Promoting Rhizobacteria for Sustainable Stress Management: Volume 1: Rhizobacteria in Abiotic Stress Management. Springer Singapore, Singapore, pp 219–253. https://doi.org/10.1007/978-981-13-6536-2_12

Yadav AN, Kumar R, Kumar S, Kumar V, Sugitha T, Singh B, Chauhan V, Dhaliwal HS, Saxena AK (2017a) Beneficial microbiomes: biodiversity and potential biotechnological applications for sustainable agriculture and human health. J Appl Biol Biotechnol 5:45–57

Yadav AN, Kumar V, Prasad R, Saxena AK, Dhaliwal HS (2018a) Microbiome in crops: diversity, distribution and potential role in crops improvements. In: Prasad R, Gill SS, Tuteja N (eds) Crop improvement through microbial biotechnology. Elsevier, USA, pp 305–332

Yadav AN, Mishra S, Singh S, Gupta A (2019b) Recent advancement in white biotechnology through fungi Volume 1: diversity and enzymes perspectives. Springer International Publishing, Cham

Yadav AN, Sachan SG, Verma P, Saxena AK (2015a) Prospecting cold deserts of north western Himalayas for microbial diversity and plant growth promoting attributes. J Biosci Bioeng 119:683–693

Yadav AN, Sachan SG, Verma P, Saxena AK (2016) Bioprospecting of plant growth promoting psychrotrophic Bacilli from cold desert of north western Indian Himalayas. Indian J Exp Biol 54:142–150

Yadav AN, Sachan SG, Verma P, Tyagi SP, Kaushik R, Saxena AK (2015b) Culturable diversity and functional annotation of psychrotrophic bacteria from cold desert of Leh Ladakh (India). World J Microbiol Biotechnol 31:95–108

Yadav AN, Saxena AK (2018) Biodiversity and biotechnological applications of halophilic microbes for sustainable agriculture. J Appl Biol Biotechnol 6:48–55

Yadav AN, Singh S, Mishra S, Gupta A (2019c) Recent advancement in white biotechnology through fungi. Volume 2: Perspective for value-added products and environments. Springer International Publishing, Cham

Yadav AN, Singh S, Mishra S, Gupta A (2019d) Recent advancement in white biotechnology through fungi. Volume 3: Perspective for sustainable environments. Springer International Publishing, Cham

Yadav AN, Verma P, Kaushik R, Dhaliwal HS, Saxena AK (2017b) Archaea endowed with plant growth promoting attributes. EC Microbiol 8:294–298

Yadav AN, Verma P, Kour D, Rana KL, Kumar V, Singh B, Chauahan VS, Sugitha T, Saxena AK, Dhaliwal HS (2017c) Plant microbiomes and its beneficial multifunctional plant growth promoting attributes. Int J Environ Sci Nat Resour 3:1–8. https://doi.org/10.19080/IJESNR.2017.03.555601

Yadav AN, Verma P, Kumar M, Pal KK, Dey R, Gupta A, Padaria JC, Gujar GT, Kumar S, Suman A, Prasanna R, Saxena AK (2015c) Diversity and phylogenetic profiling of niche-specific Bacilli from extreme environments of India. Ann Microbiol 65:611–629

Yadav AN, Verma P, Kumar S, Kumar V, Kumar M, Singh BP, Saxena AK, Dhaliwal HS (2018b) Actinobacteria from rhizosphere: molecular diversity, distributions and potential biotechnological applications. In: Singh B, Gupta V, Passari A (eds) New and future developments in microbial biotechnology and bioengineering. USA, pp 13–41. https://doi.org/10.1016/b978-0-444-63994-3.00002-3

Yadav AN, Verma P, Kumar V, Sangwan P, Mishra S, Panjiar N, Gupta VK, Saxena AK (2018c) Biodiversity of the genus *Penicillium* in different habitats. In: Gupta VK, Rodriguez-Couto S (eds) New and future developments in microbial biotechnology and bioengineering, *Penicillium* system properties and applications. Elsevier, Amsterdam, pp 3–18. https://doi.org/10.1016/b978-0-444-63501-3.00001-6

Yadav AN, Verma P, Sachan SG, Kaushik R, Saxena AK (2018d) Psychrotrophic microbiomes: molecular diversity and beneficial role in plant growth promotion and soil health. In: Panpatte DG, Jhala YK, Shelat HN, Vyas RV (eds) Microorganisms for green revolution-Volume 2: Microbes for sustainable agro-ecosystem. Springer, Singapore, pp 197–240. https://doi.org/10.1007/978-981-10-7146-1_11

Yadav AN, Verma P, Singh B, Chauhan VS, Suman A, Saxena AK (2017d) Plant growth promoting bacteria: Biodiversity and multifunctional attributes for sustainable agriculture. Adv Biotechnol Microbiol 5:1–16

Yadav AN, Yadav N (2018) Stress-adaptive microbes for plant growth promotion and alleviation of drought stress in plants. Acta Sci Agric 2:85–88

Yadav AN, Yadav N, Sachan SG, Saxena AK (2019b) Biodiversity of psychrotrophic microbes and their biotechnological applications. J Appl Biol Biotechnol 7:99–108

Yadav BK, Sidhu AS (2016) Dynamics of potassium and their bioavailability for plant nutrition. In: Potassium solubilizing microorganisms for sustainable agriculture. Springer, New Delhi, pp 187–201

Yadav N, Yadav AN (2019) Actinobacteria for sustainable agriculture. J Appl Biotechnol Bioeng 6:38–41

Yang XE, Liu JX, Wang WM, Li H, Luo AC, Ye ZQ, Yang Y (2003) Genotypic differences and some associated plant traits in potassium internal use efficiency of lowland rice (*Oryza sativa* L.). Nutr Cycl Agroecosys 67:273–282

Youssef GH, Seddik WM, Osman MA (2010) Efficiency of natural minerals in presence of different nitrogen forms and potassium dissolving bacteria on peanut and sesame yields. J Am Sci 6:647–660

Yuan L, Fang DH, Wang ZH, Shun H, Huang JG (2000) Bio-mobilization of potassium from clay minerals: I. By ectomycorrhizas. Pedosphere 10:339–346

Yuan L, Huang J, Li X, Christie P (2004) Biological mobilization of potassium from clay minerals by ectomycorrhizal fungi and eucalypt seedling roots. Plant Soil 262:351–361

Zaefarian F, Rezvani M, Rejali F, Ardakani MR, Noormohammadi G (2011) Effect of heavy metals and arbuscular mycorrhizal fungal on growth and nutrients (N, P, K, Zn, Cu and Fe) accumulation of alfalfa (*Medicago sativa* L.). Am Eurasian J Agric Environ Sci 11(3):346–352

Zandonadi DB, Santos MP, Dobbss LB, Olivares FL, Canellas LP, Binzel ML, Façanha AR (2010) Nitric oxide mediates humic acids-induced root development and plasma membrane H + -ATPase activation. Planta 231:1025–1036

Zhang AM, Zhao GY, Gao TG, Wang W, Li J, Zhang SF, Zhu BC (2013) Solubilization of insoluble potassium and phosphate by *Paenibacillus kribensis* CX-7: a soil microorganism with biological control potential. Afr. J. Microbiol. Res 7:41–47

Zhang C, Kong F (2014) Isolation and identification of potassium-solubilizing bacteria from tobacco rhizospheric soil and their effect on tobacco plants. Appl Soil Ecol 82:18–25

Zhao F, Sheng XF, Huang Z, He L (2008) Isolation of mineral potassium-solubilizing bacterial strains from agricultural soils in Shandong Province. Biodivers Sci 16:593–600

Zoomi I, Narayan RP, Akhtar O, Srivastava P (2017) Role of plant growth promoting rhizobacteria in reclamation of wasteland. In: Microbial biotechnology. Springer, Singapore, pp 61–80. https://doi.org/10.1007/978-981-10-6847-8_3

Chapter 10
Alleviation of Stress-Induced Ethylene-Mediated Negative Impact on Crop Plants by Bacterial ACC Deaminase: Perspectives and Applications in Stressed Agriculture Management

Hassan Etesami, Fatemeh Noori, Ali Ebadi and Narges Reiahi Samani

Abstract The environmental stresses such as heavy metal toxicity, salinity, water deficit, flooding, extreme temperatures, nutrient deficiency, and pathogenicity, which are considered to be the most important limiting factors for agricultural production, are rising all over the world. The occurrence and magnitude of environmental (abiotic and biotic) stresses might augment in the near future because of global climate change. These stresses lead to a significant reduction in yield and growth of stressed plants. It is well known that a sizable portion of the damage that occurs in stress-sensitive crop varieties as a consequence of abiotic and biotic stresses is due to the deleterious action of "stress ethylene" and not necessarily from the direct effects of the stress. In order to reduce the negative effects of "stress ethylene" on growth and yield of stressed plants, application of ecologically compatible and environmentally friendly methods is needed. An alternative approach to reducing stress-induced ethylene production in growing plants in stress-affected soils involves

H. Etesami (✉)
Department of Soil Science, University College of Agriculture and Natural Resources, University of Tehran, Tehran, Iran
e-mail: hassanetesami@ut.ac.ir

F. Noori
Department of Biotechnology and Plant Breeding, Sari Agricultural Sciences and Natural Resources University, Sari, Iran

A. Ebadi
Agriculture Biotechnology Research Institute of Iran (ABRII), AREEO, 31535-1897, Karaj, Iran

N. Reiahi Samani
Department of Agronomy and Plant Breeding, Sari Agricultural Sciences and Natural Resources University, Sari, Iran

A. N. Yadav et al. (eds.), *Plant Microbiomes for Sustainable Agriculture*, Sustainable Development and Biodiversity 25, https://doi.org/10.1007/978-3-030-38453-1_10

employing 1-aminocyclopropane-1-carboxylate (ACC) deaminase-generating bacterial endophytes. These bacteria may promote stressed plant's growth as a consequence of expressing the enzyme 1-aminocyclopropane-1-carboxylate (ACC) deaminase which cleaves 1-aminocyclopropane-1-carboxylate (ACC) (prerequisite of ethylene production) to α-ketobutyrate and ammonia and thereby diminishes ethylene levels in stressed host plants. In this review, the role of bacterial endophytes equipped with the enzyme 1-aminocyclopropane-1-carboxylate (ACC) deaminase in mitigating stress-induced ethylene-mediated negative impact on stressed plants under various environmental stresses (heavy metal toxicity, salinity, water deficit (drought), flooding, extreme temperatures, and nutrient deficiency) is described. In addition, some suggestions that are needed for future research in this context are also presented.

Keywords Bacterial endophytes · Bio-fertilizer · Environmental stress · Drought stress · Heavy metal stress · Salinity stress · Stress ethylene

10.1 Introduction

Food security is one of the most important issues in the world that cannot ever be overlooked by any society. The excessive increase in environmental degradation (due to inappropriate agricultural operations) and the pressure of human population are unpleasant consequences that will soon cause global food production to be inadequate for all people in the world. In this context, the global population, which currently has seven billion people, is projected to augment to 10 billion over the next fifty years (Etesami and Maheshwari 2018; Glick 2014), thus necessitating 70% more food production (FAO 2008). To fulfill the food supply requirements for the projected population by the next 50 years, we must culture a significant expansion of the existing land (increased crop yield through intensification) and significantly increase (an estimated fifty percent) grain yield of major crop plants such as *Oryza sativa* L., *Triticum aestivum* L., and *Zea mays* L. (Godfray et al. 2010). But the excessive use of agricultural arable land and marginal areas (areas where stress factors can be more prevalent) is problematic and may be resulted in degradation of the land (Alexandratos and Bruinsma 2012; Glick 2014; Zahir et al. 2008). On the other hand, non-biotic stresses such as increased carbon dioxide, UV, extreme temperatures, water deficit, flooding, salinity, heavy metals toxicity, nutrient deficiency, severe winds, etc., are the most limiting factors to plant productivity in the world (Etesami 2018; Etesami and Maheshwari 2018) and result in reduced land available for cultivation (Etesami and Maheshwari 2018; Shahbaz and Ashraf 2013). In addition, climate change has affected the frequency and severity of these stresses, mainly drought and high temperatures stresses, and increased the proportion of agricultural land affected from multiple stresses (Ahuja et al. 2010), resulting in a significant decline in yield in main cereals (Carmen and Roberto 2011) and augmenting the productivity and sustainability of the agricultural production systems (Hussain et al. 2015). Across the globe, it is estimated that roughly 70% reduction in crop yields results from direct non-biological stresses (Acquaah 2009). Therefore, due to the urgency of feeding the world's growing human population under different environmental stressful conditions, research on soil and crop productivity (i.e., enhancing tolerance of plants to a wide range of stress factors) to optimize plant growth and secure our future food and feed supply is now in the top priority (Etesami and Maheshwari 2018; Etesami and Beattie 2017).

To attenuate the effects of the biological and non-biological stresses and to augment crop productivity, fitting crop management modifications are necessary. Numerous management system modifications, such as crop rotation, intercropping, row skipping (diminishing planting density by omitting rows), mulching, sheltered cropping (i.e., crops grown under glass, plastic or nets), plant genetic engineering, and bio-fertilization, could be employed to augment crop productivity and decrease the undesirable effects caused by environmental stresses on plant growth (Davies et al. 2011; Wang et al. 2003; Dimkpa et al. 2009). In recent years, sizable attention has been directed toward genetically engineering stress-sensitive crop varieties to be more tolerant to a variety of abiotic stresses and biotic ones (i.e., pathogens including viruses, bacteria, and fungi) (Ahanger et al. 2017). The evolvement of stress-tolerant crop varieties via genetic engineering and plant breeding is absolutely necessary but a long drawn and expensive process. In addition, owing to many hindrances (i.e., proprietary rights and international trade accords on genetically changed crops and restrictions in technology of DNA recombinant in some parts of the world), genetic modification of all plant species is not possible. In addition, it is not sensible to try to engineer crop plants versus all of the biological and non-biological stresses that they might meet within the environment.

There has been a supreme interest in sustainable agriculture with an underscore on the application of useful soil microorganisms. Microbial inoculation to palliate biological and non-biological stresses in plants could be a more economical environmental amiable option which could be attainable in a shorter time frame (Saleem et al. 2007; Etesami and Maheshwari 2018; Etesami and Beattie 2017; Lugtenberg et al. 2002) and helped to achieve sustainable agricultural goals faster (Etesami and Maheshwari 2018). By both extracellular and intracellular microorganisms, plants are colonized in their natural environment (Gray and Smith 2005). Due to the presence of root exudates, a large population of microorganisms is found around plant roots whose populations are more than the population of non-rhizospheric microorganisms. Plant-associated microorganisms can either attach to the outer surface of the plant, such as the roots (rhizosphere microorganisms) or leaves (phyllosphere microorganisms), or they can be located within the plant and form an endophytic relationship (endophytic microorganisms). Among the microorganisms associated with the roots of plants, bacteria have a significant population density. Interactions between plant-associated bacteria and plant roots can be classified as useful, harmful or neutral to the plant (Glick 2014). The bacteria that have mutually beneficial effects with plants can augment plant growth and health either directly through various mechanisms such as N_2 fixation, solubilization of insoluble inorganic phosphates, mineralization of organic phosphates, facilitating the micro- and macronutrient uptake via phytohormone production (e.g., auxin, cytokinin, and gibberellins) or by production of some metabolites (i.e., siderophores, organic acids, etc.), and changing the internal level of plant growth-regulating substances, or indirectly by augmenting the natural resistance of the host against pathogens (Etesami and Maheshwari 2018; Etesami and Beattie 2017; Glick 2012; Hayat et al. 2012; Kloepper et al. 1989; Glick 1995; Patten and Glick 2002; Persello-Cartieaux et al. 2003; Kohler et al. 2006).

The role of these bacteria in plant nutrient management, plant growth promotion, and plant disease control is well known and well established (Etesami and Maheshwari 2018; Compant et al. 2005). In addition to these beneficial effects, the plant growth-promoting bacteria (PGPB) can also impart some degree of tolerance to stress-sensitive crop varieties toward environmental stresses. The term IST (Induced Systemic Tolerance) has been offered for PGPB-induced physical and chemical variations that lead to augmented tolerance to non-biological stress (Yang et al. 2009). In the last ten years, PGPB possessing by distinctive genera including *Methylobacterium*, *Azospirillum*, *Achromobacter*, *Rhizobium*, *Bacillus*, *Variovorax*, *Pseudomonas*, *Pantoea*, *Paenibacillus*, *Burkholderia*, *Klebsiella* sp., *Kosakonia cowanii*, *Microbacterium*, and *Enterobacter* have been reported to provide tolerance to host plants under distinctive biological and non-biological stress environments (Grover et al. 2011; Etesami and Alikhani 2016, 2018; Etesami and Maheshwari 2018; Noori et al. 2018; Yadav et al. 2015a; Yadav et al. 2016; Yadav et al. 2015b). Preceding studies suggest that application of PGPB has become an encouraging alternative to relieve plant stress and the role of these bacteria in managing biological and non-biological stresses is gaining importance. The subject of PGPB elicited tolerance to biotic and non-biological stresses has been reviewed (Etesami 2018; Etesami and Beattie 2017; Etesami and Maheshwari 2018; Glick 2014; Sharma et al. 2013; Kaushal and Wani 2016; Yang et al. 2009; Etesami and Alikhani 2016; Compant et al. 2005; Glick 2012; Vurukonda et al. 2016; Dodd and Pérez-Alfocea 2012; Yao et al. 2010).

To barricade the trauma and ensure survival, crop plants bring out a convoluted and unanimous molecular and cellular responses in response to various biological and non-biological stresses (Fahad et al. 2015). Phytohormones have a vital function in mediating plant's responses to these stresses, by which the plant may try to outlive the stressed conditions (Skirycz and Inzé 2010). One of the phytohormones entangled in stress response is ethylene, which is also considered as a stress hormone in addition to its roles in regulating plant growth and development and in bringing about multifarious physiological alterations in plants at molecular level (Abeles et al. 2012). Under abiotic and biological stress conditions (i.e., nutrient deficiency, salinity, water deficit, waterlogging/flooding, heavy metal toxicity, and pathogenicity), the internal production of ethylene and its direct precursor ACC are quickened considerably which unfavorably affects the root growth, and they finally result in decreasing the growth of the stress-sensitive crop varieties in a general sense (i.e., suppression of root elongation, nodulation and auxin transport, induction of hypertrophies, increase in speed of aging, and promotion of senescence and abscission) (Sun et al. 2006; Abeles 1992; Morgan and Drew 1997).

A number of PGPB are equipped with a pivotal enzyme, ACC (1-Aminocyclopropane-1-Carboxylic Acid) deaminase, which regulates ethylene generation via metabolizing ACC (an instantaneous precursor of ethylene production in higher plants) into alpha-ketobutyrate ($C_4H_6O_3$) and ammonia (NH_3) (Yadav 2017a; Yadav et al. 2017a, b). Inoculation of crop plants with the ACC deaminase-generating PGPB could be beneficial in supporting growth and enlargement of stress-sensitive crop varieties under stressful conditions by diminishing stress-mediated ethylene generation (Glick 2014; Etesami and Maheshwari 2018; Etesami 2018; Etesami and Beattie 2018). The ability to produce this enzyme has been reported in a variety of rhizosphere and endophytic bacterial genera (Glick 2014). One of the leading drawbacks in the large-scale employment of PGPB including ACC deaminase-producing microorganisms is that these bacteria may not always outlive rough environmental conditions such as extremes of pH and temperature, high concentrations of environmental contaminants, salinity, drought, and the presence of other microorganisms that either consume or outcompete these bacteria. A feasible solution to this problem may lie in the application of EPGPB (endophytic plant growth-promoting bacteria) (Sturz and Nowak 2000). Bacteria that are isolated from plant tissues after their surface disinfection and show no symptoms of disease on the plant are known as endophytic bacteria. It seems that the contribution of endophytic bacteria to the production of 1-aminocyclopropane-1-carboxylate deaminase and the abatement of stress ethylene production relative to the contribution of rhizosphere bacteria is much wider (Etesami and Maheshwari 2018). The internal tissues of plants provide a more uniform and safer environment for bacteria than plant surfaces. Considering the important role of 1-aminocyclopropane-1-carboxylate deaminase-producing bacteria in reducing stresses caused by ethylene, aim of this review is to describe the role of bacterial endophytes equipped with the enzyme 1-aminocyclopropane-1-carboxylate (ACC) deaminase in lessening of impact of biological and abiotic stresses onto plants. In addition, some suggestions needed for future research in this field are also presented.

10.2 Abiotic and Biotic Stresses

Apart from physiological and environmental factors such as availability of macro- and micronutrients (i.e., N, P, K, S, Mg, Ca, Fe, Mn, Cu, Zn, etc.), physical and chemical characteristics of soil (EC, soil texture, organic matter, pH, etc.), plant genotype and growth conditions, the growth of stress-sensitive crop varieties in the field may be repressed via a large number of both biological and non-biological stresses. These stresses include temperature extremes (high and low temperatures), high light, waterlogging, global scarcity of water resources (water deficit), the presence of toxic metals (i.e., Cd, Ni, Co, As, etc.), and environmental organic contaminants (i.e., phenols, colorants, alcohols, aldehydes, DDT and metabolites, Dieldrin, PCP, PAHs, RDX, TNT, etc.), radiation, wounding, insect predation, increased salinization of soil and water, mechanical stress, nutritional stress, and damage done to stress-sensitive crop varieties by living organisms, such as bacteria, viruses, fungi, parasites, harmful insects, weeds, and cultivated or native plants (Abeles et al. 2012; Stearns and Glick 2003; Morgan and Drew 1997). These stress factors cause widespread crop losses throughout the world. In the plant grown under the stressed conditions, immense amounts of ROS (reactive oxygen species) are generated that can bring about peroxidation, leading to detriment to cell membranes, protein oxidation, enzyme deterrence, and strand fracture in nucleic acids (Allen 1995).

One of the principal concerns bringing about some of the distinctive ecological and environmental problems is heavy metal pollution of agricultural soils. Due to industrial applications, use of metal-contaminated fertilizers, types of pesticides (i.e., insecticides, herbicides, rodenticides, bactericides, fungicides, and larvicides), and sewage sludge in agricultural land, toxic metals and metalloids accumulate in the environment (Etesami 2018). Some heavy metals such as Fe, Mn, Zn, Cu, etc., are essential or beneficial micronutrients needed by plants for growth and development. But, when present in superfluous, they may act as toxicants and repress the plants growth (Etesami 2018). In addition, high metal amounts in the soil have also been indicated to bring about augmented ethylene synthesis (Rodecap et al. 1981; Safronova et al. 2006) and inhibition of root and shoot development and interfere with numerous biochemical and physiological processes such as CO_2 fixation, respiration, photosynthesis, sugar translocation, N and protein metabolism, and micro- and macronutrients absorption (Zhang et al. 2009; Prasad and Strzalka 2013).

Organic pollutants (i.e., DDT, lindane, polychlorinated biphenyls, dioxins, etc.) in the soil environment, if present above permissible limits, hinder plant growth via many mechanisms such as unusual growth of stress-sensitive crop varieties (Adam and Duncan 1999). This irregular growth of the plant root system might be partly owing to hastened ethylene generation (stress ethylene) in plants grown in soil environment treated with organic contaminants (Coupland and Jackson 1991; De Prado et al. 1999; Jackson 1997).

During periods of flooding/waterlogging, which occur several times in a growing season and may last for periods of from one or two days to several weeks, the root environment (rhizosphere) speedily becomes anaerobic bringing about an inauguration in the expression of 1-aminocyclopropane-1-carboxylate (ACC) synthase, leading to the cumulation of 1-aminocyclopropane-1-carboxylate (ACC) in tissues of root (Cattelan et al. 1999; Olson et al. 1995; Jackson 1997; Else and Jackson 1998; Bradford and Yang 1980; Else et al. 1995). The amassed 1-aminocyclopropane-1-carboxylate (ACC) is transported to the shoots (stem + leaf) where there is an aerobic environment and the ACC may be converted to ethylene via the enzyme 1-aminocyclopropane-1-carboxylate (ACC) oxidase, which has need for oxygen for catalyzing this reaction. Lower diffusion of ethylene in water leads to cumulation of ethylene in waterlogged and/or flooded plants and soil (Steffens et al. 2012). Cumulation of stress ethylene can also bring about epinasty, leaf chlorosis, necrosis, and diminished fruit yield (Grichko and Glick 2001a, b).

Drought affects, in essence, all climatic regions of the world, particularly in arid and semi-arid areas, and more than one-half of the earth is vulnerable to drought every year (Wilhite 2000; Kogan 1997; Grichko et al. 2000; Kour et al. 2019a; Verma et al. 2017b). Drought stress (water deficit), as one of the major non-biological stresses, restricts the growth and productivity of crops (i.e., diminution in germination rates, inhibition of photosynthesis, loss of membrane integrity, decline in leaf water potential and stomatal opening, decrease in leaf size, decrease in water uptake (osmotic stress), suppressed root growth, diminished seed number, size and viability, delayed flowering and fruiting, and increased generation of ROS) (Xu et al. 2016; Bray 1997; Ingram and Bartels 1996; Shinozaki and Yamaguchi-Shinozaki 1998; Kramer and Boyer 1995; Greenberg et al. 2008). Like many other environmental factors, water deficit stress also induces augmented ethylene production in tissues of stress-sensitive crop varieties which inhibiting root growth and development, shoot/leaf expansion, and photosynthesis (Mattoo 2017; Sharp 2002) and causes membrane leakage, eventually brings about senescence in stress-sensitive crop varieties (Hipkins and Hillman 1986).

In fact, plants are sensitive to variations in temperature, and react both to seasonal variations and more so to quotidian alterations in the season. The heat stress with regard to the aspect known as global warming is a critical intimidation to world agriculture (Moeder et al. 2002; Robertson et al. 2000). A vacillation in temperature results in hormonal unbalances in stress-sensitive crop varieties and thus their growth is considerably affected (Cheikh and Jones 1994). Following temperature stress (high and chilling temperatures), the ethylene level is swiftly augmented both in tissues of stress-sensitive crop varieties and microbial species in the rhizosphere of stressed plant (Strzelczyk et al. 1994; Arshad and Frankenberger Jr 2012; Shi et al. 2012).

Microbial pathogens are a leading and serious intimidation to food production and ecosystem stability throughout the world (Etesami and Alikhani 2018). Mostly, like abiotic stresses, plant ethylene production is augmented with severity of pathogenic infection (Bashan 1994; Elad 1988). Among abiotic and biotic stresses, salinity stress is one of the most damaging abiotic stresses (Etesami and Beattie 2017), barricading the attainment of sustainable agriculture (Paul 2013; Greenway and Munns 1980; Zhu 2002; Roy et al. 2014). Soil salinity in arid regions is frequently an important delimiting factor for cultivating agricultural crops because salinity converts arable land into non-arable land and diminishes crop productivity and quality (Shahbaz and Ashraf 2013; Yamaguchi and Blumwald 2005). Soil salinization reduces 1–2% of the area that could be used for agriculture every year. Approximately seven percent of the world's land and 20% of the total arable land throughout the world are negatively influenced by salinity (Munns 2005) with a resultant monetary loss of twelve billion US$ in agricultural production (Shabala 2013). Additionally, the salinized areas are augmenting at a rate of ten percent annually, specifically on agricultural land where growth of stress-sensitive crop varieties is contingent upon irrigation, for various reasons, such as weathering of native rocks, low precipitation, high surface evaporation, irrigation with saline water, and poor cultural practices. It has been calculated approximately that more than fifty percent of the arable land would be salinized by the year 2050 (Jamil et al. 2011).

Salinity stress has been reported to limit growth and development of stress-sensitive crop varieties (i.e., by ion toxicity stress, osmotic stress, oxidative stress, and nutrient deficiency), which brings about variations in physiology, morphology, and biochemical characteristics of stressed plants, probably all occurring simultaneously

(Tao et al. 2015; Chinnusamy et al. 2006; Munns 2002, 2005; Gorham et al. 1985; Ruiz et al. 1997; Arbona et al. 2005; Zhu 2001). Almost all of the most consequential crop plants are salinity-sensitive plants (glycophytes) and are and so susceptible to salt stress (Munns and Tester 2008; Cheeseman 2015). Intermediate levels of environmental salinity (e.g., 100 mM NaCl) are adequate to bring about dramatic decrements in the yield of most crop plants (Frommer et al. 1999; Munns and Tester 2008). It is well known that shortage of nutrients such as N, P, K, Ca, and Fe, which occurs in salt-affected soil, promotes the expressions of many genes connected to ethylene biosynthesis and signaling in the roots and enhances plant sensitivity to ethylene (Blumwald 2000; Cuartero and Fernández-Muñoz 1998; Feng and Barker 1992; O'Donnell et al. 1996; Zheng et al. 2013; Rai and Sharma 2006; He et al. 1992; Borch et al. 1999; Rabhi et al. 2007; Yousfi et al. 2007; García et al. 2015; García et al. 2010; Achard et al. 2006; Dong et al. 2011; Xu et al. 2008). According to the studies mentioned above, during periods of all environmental stresses, stressed plants in response to these stresses produce high levels of endogenous ethylene, called "**stress ethylene**", which increases sensitivity of the plants to various environmental stresses and finally leads to abnormal growth of the plants (inhibitory effect on root growth).

10.3 Ethylene as a Stress Hormone

Stress-sensitive crop varieties employ many strategies in answer to environmental stresses (i.e., variation in developmental and morphological motif and in physiological and biochemical processes) that eventually augment the stressed plant's growth and productivity in stressful agricultural lands (Tuteja 2007; Saud et al. 2014). By accumulating organic solutes such as sugars, polyols (sugar alcohols), betaines and proteinogenic amino (i.e., proline, $C_5H_9NO_2$), protecting cellular machinery, maintaining ionic homeostasis, cleansing free radicals or reactive oxygen species (ROS), expressing certain proteins and upregulating their genes and inducing phytohormones, plant increase your tolerance to the environmental stresses (Parida and Das 2005; Tuteja 2007; Munns and Tester 2008).

It is known that phytohormones are involved in regulating the interaction between stress-sensitive crop varieties and environments (i.e., plant responses to environmental stresses) throughout the life cycles of plants (Bartoli et al. 2013; Baxter et al. 2013; Skirycz and Inzé 2010). Phytohormones are thought to be the most influential internal substances that are crucial in adjusting physiological responses that ultimately result in adjustment to an undesirable environment (Khan and Khan 2013). Ethylene, known as a stress hormone, is an important growth phytohormone synthesized by almost all plants and serves as a key modulator between a wide range of plant responses to environmental stresses and normal growth (Abeles 1992). This phytohormone plays multifarious roles in regulating growth and development of stress-sensitive crop varieties (Abeles 1992). Production of ethylene occurs in all higher plants via a methionine ($C_5H_{11}NO_2S$) relying on pathway. Ethylene in plant is synthesized via 3 enzymatic reaction steps: methionine ($C_5H_{11}NO_2S$) is converted to S-adenosyl-methionine (S-AdoMet) by S-AdoMet synthetase; then the direct substrate of ethylene 1-aminocyclopropane-1-carboxylate (ACC) is generated from S-AdoMet by ACC synthase; and ultimately ethylene (carbon dioxide and hydrogen cyanide) is produced via the oxidation of 1-aminocyclopropane-1-carboxylate (ACC) by 1-aminocyclopropane-1-carboxylate (ACC) oxidase (Lin et al. 2009). Concentrations of ethylene and its direct precursor 1-aminocyclopropane-1-carboxylate (ACC) could be plainly mediated by stresses in numerous stress-sensitive crop varieties (Morgan and Drew 1997; Ellouzi et al. 2014; Ma et al. 2012). Apart from its positive effects (at levels of as low as 10 $\mu g\ L^{-1}$) on plant growth (i.e., role in root initiation, nodule formation, root extension, seed germination, xylem formation, biosynthesis of other phytohormones, and ripening as well as in stress signaling) (Mattoo 2017; Abeles 1992; Arshad and Frankenberger 1990), overproduction of ethylene (at levels of as high as 25 $\mu g\ L^{-1}$) in plant roots in answer to both biological and non-biological processes represses root elongation, nodulation and auxin transport, brings about hypertrophies, speeds aging, advances senescence and abscission, and leads to untypical root growth, which imparts an ostensible dent on plant growth and outreach (Abeles 1992; Pech 2002; Arshad and Frankenberger 1990; Frankenberger Jr and Muhammad 1995; Prayitno et al. 2006; Sun et al. 2006). This implies that no extra ethylene is mandatory for the process of stress adjustment in the early stage of plant self-adjustment (Tao et al. 2015; Peng et al. 2014), and if the ethylene concentration (increase in 1-aminocyclopropane-1-carboxylate content and 1-aminocyclopropane-1-carboxylate oxidase activity) augments above a threshold level, it becomes deleterious for plant growth and hinders growth and development of stress-sensitive crop varieties, which is disadvantageous for plants to outlive under sever stressful conditions (Kukreja et al. 2005; Glick 2014; Tittabutr et al. 2013; Albacete et al. 2009; Ghanem et al. 2008; Tao et al. 2015). Since higher amounts of ethylene ("stress ethylene") have prohibitive impacts on root growth and may lead to abnormal growth of the plants, it is crucial to govern the generation of ethylene in the close environs of stressed plant's roots (rhizosphere) for regular growth and development of stress-sensitive crop varieties (Saleem et al. 2007).

Stress tolerance in plants depends mainly on the capability of root to diminish the endogenous ethylene level. It is well known that bacteria equipped with the enzyme 1-aminocyclopropane-1-carboxylate (ACC) deaminase can diminish the negative effects of this hormone (stress ethylene) via diminishing the level of ACC (precursor of ethylene production). Undoubtedly, the production of 1-aminocyclopropane-1-carboxylate (ACC) deaminase by the bacteria is a major mechanism in promoting plant growth under environmental stresses. In addition to reducing "stress ethylene" level in plants, it is well known that the bacteria equipped with the enzyme 1-aminocyclopropane-1-carboxylate (ACC) deaminase may also provide other many benefits to stress-sensitive crop varieties (Glick 1995; Holguin and Patten 1999; Etesami and Maheshwari 2018). These characteristics make the choice of 1-aminocyclopropane-1-carboxylate (ACC) deaminase-producing PGPB more trustworthy than any other alternative.

10.4 Amelioration of "Stress Ethylene" Using Bacterial ACC Deaminase

As described previously, the surplus production of ethylene in answer to biological and non-biological stresses brings about holding back root growth and accordingly growth of the stress-sensitive crop varieties as a whole. Through certain specific enzymes, the bacteria stimulate some physiological changes in stress-sensitive crop varieties at molecular level. Amid these enzymes, bacterial 1-aminocyclopropane-1-carboxylate (ACC) deaminase plays a well-understood role in regulating the plant hormone ethylene and thus, growth and evolvement of stress-sensitive crop varieties are modulated (Glick 2014). It has been well known that ACC deaminase-generating PGPB can regulate and palliate the stress-induced ethylene-mediated negative effect on plants by metabolizing 1-aminocyclopropane-1-carboxylate (ACC); a precursor of plant produced ethylene. Compared to ethylene synthesis inhibitors such as AVG (aminoethoxyvinylglycine), AOA (aminooxyacetic acid), and 1-MCP (1-methylcyclopropene), which are costly, less workable, or potentially causing damage to the environment, the use of bacteria equipped with the enzyme 1-aminocyclopropane-1-carboxylate (ACC) deaminase is more economical, environmental amicable, and workable in a stressed natural soil and plant system.

The enzyme 1-aminocyclopropane-1-carboxylate (ACC) deaminase (Enzyme Commission number: 4.1.99.4), which catalyzes the breaking of 1-aminocyclopropane-1-carboxylate to NH_3 and alpha-ketobutyrate ($CH_3CH_2CCO_2H$), was first found in forty years ago (in 1978) (Honma and Shimomura 1978). The 1-aminocyclopropane-1-carboxylate (ACC) activity has been known to be associated with a large number of assorted soil microbial community (free-living soil microorganisms), plant-associated bacteria (both Gram-negative bacteria and Gram-positive bacteria), including symbionts like rhizobia (diazotrophic bacteria fixing molecular nitrogen inside the root nodules of legumes), general rhizospheric and endophytic PGPB (*Agrobacterium*, *Burkholderia*, *Enterobacter*, *Methylobacterium*, *Azospirillum*, *Alcaligenes*, *Ralstonia*, *Rhodococcus*, *Rhizobium*, *Pseudomonas*, *Sinorhizobium*, *Variovorax*, and *Bacillus*), some plant pathogens, Eukarya, and fungi (Arshad et al. 2007; Etesami and Maheshwari 2018; Etesami and Beattie 2018; Glick 2014; Nascimento et al. 2014; Babalola et al. 2003; Wang et al. 2000; Belimov et al. 2001; Ghosh et al. 2003; Honma and Shimomura 1978; El-Tarabily 2008; Minami et al. 1998; Viterbo et al. 2010; Singh and Kashyap 2012; Verma et al. 2016a, b; Verma et al. 2015; Yadav et al. 2016).

The 1-aminocyclopropane-1-carboxylate (ACC) deaminase enzyme is encoded by a single gene, denominated *acdS*. It was found that the genes might be come into horizontally (laterally) instead of vertically (Hontzeas et al. 2005). As a matter of fact, there are some proofs that 1-aminocyclopropane-1-carboxylate (ACC) deaminase genes may not constantly be an integral part of the chromosomal DNA of a microorganism, but rather exist on large comparatively steady plasmids (Glick et al.

2007b). On top of that, on the foundation of sequence similarity, the presence of putative 1-aminocyclopropane-1-carboxylate (ACC) deaminase genes in the genomes of several plants (*Arabidopsis*, *Populus*, and *Lycopersicon esculentum* Mill.) was exhibited (McDonnell et al. 2009; Plett et al. 2009; Singh et al. 2015).

The action mechanism of 1-aminocyclopropane-1-carboxylate (ACC) deaminase-producing bacteria to reduce "stress ethylene" level is based on a model previously tendered by Glick et al. (1998). A diagrammatic delineation of this model is shown in Fig. 10.1. It is well known that the population of microorganisms is high in the rhizosphere region due to specific conditions different from the non-rhizosphere region. In this area, because of root exudates, the bacteria including the ACC deaminase-generating bacteria attach to the surface of the seed (spermosphere) or root (rhizoplane bacteria) of the plant and some of these bacteria enter the root (bacterial endophytes). Plant growth-promoting bacteria attached to the root of the plant by various mechanisms lead to increased plant root growth. As the root of plant increases, the amount of root exudates also augments. Root exudates contain easily degradable compounds, including organic acids, amino acids, and other small molecules. L-tryptophan (L-Trp) is one of the amino acids found in root secretions. This amino acid is a precursor to the production of auxin. Some bacteria having auxin-producing genes (i.e., *ipdC* and *iaaM* gene) can convert this amino acid to indole-3-acetic acid (IAA). Part of this bacterial IAA is absorbed by the plant. This bacterial IAA together with the plant IAA (IAA produced inside the plant by the plant itself) can lead to the proliferation and elongation of plant cells. These IAAs can also promote the synthesis of the enzyme 1-aminocyclopropane-1-carboxylate (ACC) synthase. The enzyme 1-aminocyclopropane-1-carboxylate (ACC) synthase catalyzes the formation of ACC.

The synthesized 1-aminocyclopropane-1-carboxylate (ACC) exits the plant along with root exudates, and the bacteria equipped with the enzyme 1-aminocyclopropane-1-carboxylate (ACC) deaminase in the rhizosphere/rhizoplane can use some of the ACC as a nitrogen source by breaking it down into NH_3 and $C_4H_6O_3$ (α-ketobutyrate). As a result of the uptake and consumption of 1-aminocyclopropane-1-carboxylate (ACC) by bacteria equipped with the enzyme ACC deaminase, its content is reduced outside the stress-affected plant. Additionally, in order to establish a balance between the amount of ACC produced inside the plant and the ACC removed from the plant, the plant discharges a greater amount of the stress-induced ACC produced inside the plant into the rhizosphere of host plant. On the other hand, more ACC excretion from the plant provides the carbon and nitrogen requirement of the 1-aminocyclopropane-1-carboxylate (ACC) deaminase-producing microorganisms, which results in a greater activity and population of these microorganisms around the plant root. The end result is ACC depletion inside the plant and consequently ethylene depletion within the plant. Decreased ethylene levels, in turn, leads to increased plant root system and consequently plant growth (Glick et al. 1998). Thus, the net outcome of the interaction of ACC deaminase-generating microorganisms with plant cells is

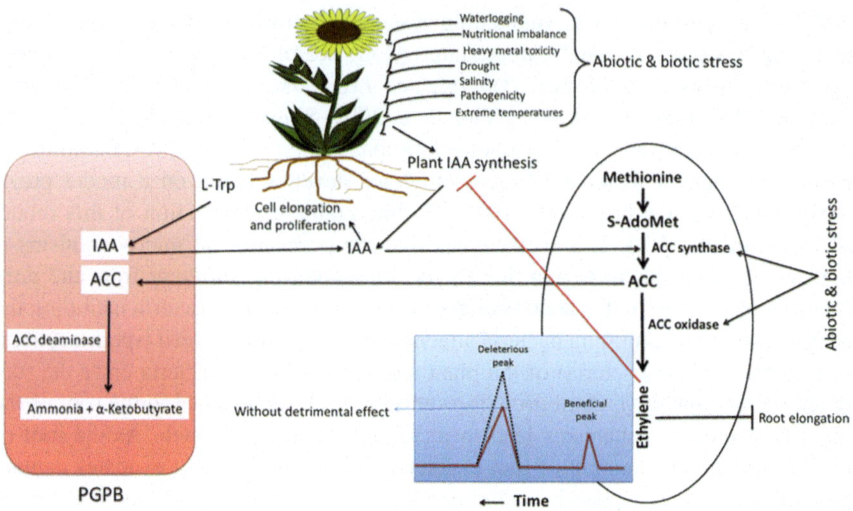

Fig. 10.1 A diagrammatic representation of how plant growth-promoting bacteria (PGPB) having 1-aminocyclopropane-1-carboxylate (ACC) deaminase activity can alleviate the stress-induced ethylene-mediated negative impact on stressed plants. Stressed plants often respond to stressful conditions by producing what is known as "stress ethylene". In the pathway of ethylene biosynthesis, S-AdoMet (S-adenosylmethionine) is converted by ACC synthase to ACC, the immediate precursor of ethylene, and ACC is also converted to ethylene by ACC oxidase. Following the beginning of the stress, a small peak of ethylene is produced after a short time. This small peak of ethylene uses up the existent pool of ACC within plant tissues and likely activates the synthesis of defensive genes within the plant. Afterward, a second much larger peak of ethylene, which is generally detrimental to plant growth, is typically observed following the synthesis of additional ACC within the plant. Due to enhancing transcription of ACC synthase genes, the second peak of ethylene, which mostly is triggered by environmental cues, occurs and acts as a signal to initiate the inhibitory processes to plant growth and survival. The ACC deaminase-producing bacteria can sequester and degrade the stressed plant's ACC (to ammonia and α-ketobutyrate) to supply nitrogen and energy, which reduce the deleterious effect of ethylene and subsequently ameliorate plant stress and promote plant growth. It is expected that ACC deaminase-producing bacteria decrease the magnitude of the second ethylene peak. It has been reported that IAA can also activate the transcription of ACC synthase and ultimately result in the production of relatively high concentrations of ACC and subsequently inhibitory levels of ethylene. However, it was found that with PGPB that both secrete IAA and synthesize ACC deaminase, plant ethylene levels do not become elevated to the same extent as when plants interact with bacteria that secrete IAA but do not synthesize ACC deaminase. In general, when plant ethylene levels increase, the ethylene that is produced feedback inhibits IAA signal transduction thereby limiting the extent that IAA can activate ACC synthase transcription. For more details, see Glick (2014). Abbreviations: L-Trp, Tryptophan; IAA, Indole-3-acetic acid

that the bacteria act as a sink for 1-aminocyclopropane-1-carboxylic acid (Glick et al. 2007a; Glick 2014).

Since the hormone ethylene has been known to be needed for the inspiration in plants of systemic resistance triggered by PGPB (Van Loon et al. 1997), the query arises whether inoculating stress-sensitive crop varieties with ethylene-diminishing bacteria (bacteria equipped with the enzyme 1-aminocyclopropane-1-carboxylate deaminase) might barricade this induction. But, practically, "diminishing ethylene levels by bacterial ACC deaminase does not seem to be irreconcilable with the instigation of systemic resistance. Indeed, some bacterial strains possessing ACC deaminase also incite systemic resistance" (van Loon and Glick 2004). In a general manner, the enzyme 1-aminocyclopropane-1-carboxylate (ACC) deaminase exists in bacteria at a low level until it is incited, and the inspiration of the enzyme 1-aminocyclopropane-1-carboxylate (ACC) deaminase activity is a comparatively slow and intricate process. Shortly following an abiotic or biotic stress, the pool of 1-aminocyclopropane-1-carboxylate (ACC) in the plant is low as is the level of the enzyme 1-aminocyclopropane-1-carboxylate (ACC) deaminase in the collaborated bacterium. Following the proportionately rapid inspiration of a low level of 1-aminocyclopropane-1-carboxylate (ACC) oxidase in the plant, it is likely that there is augmented flux via this enzyme causing the first small peak of ethylene which is of adequate magnitude to trigger a protective/defensive reaction in the plant (Fig. 10.1). With time, bacterial 1-aminocyclopropane-1-carboxylate (ACC) deaminase is motivated (by the augmenting amounts of 1-aminocyclopropane-1-carboxylate (ACC) that ensue from the inspiration of 1-aminocyclopropane-1-carboxylate (ACC) synthase in the plant) so that the dimensions of the second, detrimental, ethylene peak is diminished expressively (Fig. 10.1). The second ethylene peak may be diminished dramatically, but it is not ever entirely abrogated since 1-aminocyclopropane-1-carboxylate (ACC) oxidase has a much higher affinity for 1-aminocyclopropane-1-carboxylate (ACC) than does 1-aminocyclopropane-1-carboxylate (ACC) deaminase. Therefore, when bacteria equipped with the enzyme 1-aminocyclopropane-1-carboxylate (ACC) deaminase are present, ethylene levels are eventually contingent upon the ratio of 1-aminocyclopropane-1-carboxylate (ACC) oxidase to 1-aminocyclopropane-1-carboxylate (ACC) deaminase. For PGPB to be able to diminish plant ethylene levels, level of the 1-aminocyclopropane-1-carboxylate (ACC) deaminase should be at least one hundred- to one thousand-fold greater than level of the 1-aminocyclopropane-1-carboxylate (ACC) oxidase (Glick et al. 1998).

In previous studies (Prayitno et al. 2006; Morgan and Gausman 1966), ethylene stress-induced inhibition of IAA transport and signal transduction has been reported. Whereas bacteria equipped with the enzyme 1-aminocyclopropane-1-carboxylate (ACC) diminish the ethylene amount in roots of stress-sensitive crop varieties, this assuages the ethylene repression of auxin (i.e., indole-3-acetic acid) response factor generation, and indirectly augments growth of development of stress-sensitive crop varieties (Fig. 10.1) (Dharmasiri and Estelle 2004). Therefore, IAA-producing bacteria, despite activating of the transcription of ACC synthase, cannot ultimately bring about the generation of comparatively high concentrations of ACC and afterward repressive levels of ethylene. In general, by diminishing ethylene deterrence of assorted plant processes, and permitting IAA motivation of cell rapid multiplication and extension without the damaging impacts of augmenting 1-aminocyclopropane-1-carboxylate (ACC) synthase and plant ethylene levels, plant growth-promoting bacteria equipped with the enzyme 1-aminocyclopropane-1-carboxylate (ACC) deaminase assist forward the growth and development of stress-sensitive crop varieties (Glick 2014).

The role of bacteria equipped with the enzyme 1-aminocyclopropane-1-carboxylate (ACC) deaminase in diminishing stress-induced ethylene-mediated negative impact on plants is well known and well established (Glick et al. 2007b; Glick 2014). According to antecedent studies (Noori et al. 2018; Ali et al. 2014; Zhao et al. 2015; Zhang et al. 2011; Qin et al. 2014; Nascimento et al. 2012; Palaniyandi et al. 2014; Karthikeyan et al. 2012; Sgroy et al. 2009; Sziderics et al. 2007; Saravanakumar and Samiyappan 2007; Win et al. 2018; Afridi et al. 2019; Yoolong et al. 2019; Barnawal et al. 2016; Gamalero et al. 2017), stress-sensitive crop varieties inoculated with various bacterial strains equipped with the enzyme 1-aminocyclopropane-1-carboxylate (ACC) deaminase could cope with the stressful conditions by lowering ethylene level (Table 10.1). There are some reports that endophytic and rhizosphere ACC deaminase-producing bacteria (i.e., *Burkholderia* sp., and *Pseudomonas fluorescens*) have antagonistic impacts toward pathogenic microorganisms such as *Fusarium oxysporum* (Yuquan et al. 1999; Donate-Correa et al. 2005), *Pythium ultimum* (Wang et al. 2000), *Fusarium proliferatum* (Donate-Correa et al. 2005), *Rhizoctonia solani* (Pandey et al. 2005; Rasche et al. 2006b), *Sclerotinia sclerotiorum* (Pandey et al. 2005), *Erwinia carotovora* sp. *atrospetica* (Eca) (Rasche et al. 2006a), and *Ralstonia solanacearum* (Rasche et al. 2006b). The results of the research revealed that bacteria equipped with the enzyme 1-aminocyclopropane-1-carboxylate (ACC) deaminase were more effective in biocontrol of these pathogens than those without this enzyme. It is also very likely that bacteria equipped with the enzyme 1-aminocyclopropane-1-carboxylate (ACC) deaminase, besides straightly antagonizing pathogenic microorganisms, play a potential role in motivating disease tolerance in stress-sensitive crop varieties (augmented the plant's resistance against pathogen attack) (Saleem et al. 2007; Belimov et al. 2007). But, doing more research in this area is needed for further understanding of this mechanism in the future (Saleem et al. 2007).

It was found that diminished salinity-induced ethylene by bacteria could diminish the detrimental impact of salinity onto growth of stress-sensitive crop varieties (Glick 2014). Previous researches have revealed that plants treated with PGPB equipped with the enzyme 1-aminocyclopropane-1-carboxylate (ACC) deaminase (i.e., *Achromobacter piechaudii*, *Enterobacter* sp., *Pseudomonas* spp., *Mesorhizobium*, *Kocuria rhizophila*, *Cronobacter sakazakii*, *Leclercia adecarboxylata*, *Streptomyces venezuelae*, *Brachybacterium paraconglomeratum*, *Brevibacterium epidermidis*, *Variovorax paradoxus*, *Bacillus mojavensis*, *Serratia* sp., *Klebsiella* sp., *Kosakonia cowanii*, and *P. fluorescens*) were better able to thrive through the salt stress while manifesting a normal pattern of growth (Mayak et al. 2004a, b; Saravanakumar and Samiyappan 2007; Cheng et al. 2007; Nadeem et al. 2006; Noori et al. 2018; Sarkar et al. 2018; Win et al. 2018; Chaudhary and Sindhu 2017; Afridi et al. 2019; Orozco-Mosqueda et al. 2019; Kang et al. 2019; Yoolong et al.

Table 10.1 Alleviation of the stress-induced ethylene-mediated negative impact on plants by 1-aminocyclopropane-1-carboxylate (ACC) deaminase-producing endophytic bacteria

Bacterial strains	Experimental plant	Type of stress	Reference
Pseudomonas fluorescens YsS6 and *P. migulae* 8R6	Tomato	Salinity	Ali et al. (2014)
Bacillus endophyticus	*Salicornia europaea*	Salinity	Zhao et al. (2015)
Bacillus sp. and *Acinetobacter* sp.	*Commelina communis*	Heavy metal	Zhang et al. (2011)
Bacillus flexus and *Streptomyces pactum*	*Limonium sinense*	Salinity	Qin et al. (2014)
Streptomyces sp. PGPA39	Tomato	Salinity	Palaniyandi et al. (2014)
Achromobacter xylosoxidans AUM54	*Catharanthus roseus*	Salinity	Karthikeyan et al. (2012)
Bacillus licheniformis Ps14	*Prosopis strombulifera*	Salinity	Sgroy et al. (2009)
Bacillus sp. TW4	Sweet pepper	Osmotic stress	Sziderics et al. (2007)
P. fluorescens	*Arachis hypogea*	Salinity	Saravanakumar and Samiyappan (2007)
Pseudomonas spp. OFT2	Tomato	Salinity	Win et al. (2018)
Kocuria rhizophila and *Cronobacter sakazakii*	Wheat	Salinity	Afridi et al. (2019)
Streptomyces venezuelae ATCC 10712	Rice	Salinity	Yoolon et al. (2019)
Brachybacterium paraconglomeratum	*Chlorophytum borivilianum*	Salinity	Barnawal et al. (2016)
Pseudomonas migulae 8R6	*Catharanthus roseus*	*Flavescence dorée* Yellows disease	Gamalero et al. (2017)
Paenibacillus and *Bacillus*	Wheat	Pathogenic fungi	Zhao et al. (2015)

2019; Barnawal et al. 2016; Siddikee et al. 2015; Habib et al. 2016; Singh and Jha 2016). These bacteria reduced the production of ethylene by stressed plants.

Under drought stress, PGPB equipped with the enzyme 1-aminocyclopropane-1-carboxylate (ACC) deaminase (i.e., *Variovorax paradoxus* 5C-2, *Achromobacter piechaudii* ARV8, and *Paenibacillus polymyxa*) also diminished the production of ethylene by stressed plant (Mayak et al. 2004c; Dodd et al. 2004; Arshad et al. 2008; Timmusk and Wagner 1999; Glick et al. 2007a; Kumar et al. 2019; Yadav 2017a, b;

Yadav et al. 2019b). According to previous studies, lower 1-aminocyclopropane-1-carboxylate (ACC) content was observed in plants inoculated with ACC deaminase-producing PGPB and in plants genetically engineered to express ACC deaminase than in non-inoculated plants and non-engineered plants under flooded conditions, indicating that the enzyme had a significant role in ACC depletion (Grichko and Glick 2001a, b). Plants inoculated with bacteria equipped with the enzyme 1-aminocyclopropane-1-carboxylate (ACC) deaminase (i.e., *Pseudomonas putida*, *Enterobacter cloacae*, and *P. putida*) showed remarkable tolerance to flooding stress implying that bacterial 1-aminocyclopropane-1-carboxylate (ACC) deaminase diminished the negative impacts of stress-mediated ethylene (Saleem et al. 2007; Grichko and Glick 2001a; Farwell et al. 2006; Rana et al. 2019; Verma et al. 2017a, b).

ACC deaminase-producing bacteria could also mediate heavy metal (i.e., nickel, Ni; lead, Pb; zinc, Zn; copper, Cu; cadmium, Cd; cobalt, Co; and arsenic, As) stress-induced ethylene negative effect in various plants (Arshad et al. 2007; Etesami 2018; Farwell et al. 2006; Safronova et al. 2006; Dell'Amico et al. 2005; Reed and Glick 2005; Belimov et al. 2001). It has studied the significance of PGPB equipped with the enzyme 1-aminocyclopropane-1-carboxylate (ACC) deaminase in augmenting the growth of stress-sensitive crop varieties in the presence of organic contaminants (Saleem et al. 2007; Arshad et al. 2007; Greenberg et al. 2006; Saleh et al. 2004; Huang et al. 2004; Reed and Glick 2005; Yadav et al. 2018a, b, c). The ACC deaminase-producing bacteria (i.e., *Burkholderia phytofirmans* and *P. putida*) helped stressed plants in maintaining normal growth under low and high temperature stress by diminishing "stress ethylene" concentration similar to that under other abiotic and biotic stresses (Bensalim et al. 1998; Barka et al. 2006; Cheng et al. 2007; Kour et al. 2019a, b, c; Yadav et al. 2019a). The results obtained from these few studies indicate that the bacteria equipped with the enzyme 1-aminocyclopropane-1-carboxylate (ACC) deaminase were able to diminish the stress induced in temperature extremes-affected plants. According to studies mentioned above, it can be concluded that use of the bacterial endophytes with 1-aminocyclopropane-1-carboxylate (ACC) deaminase trait can be a useful strategy to recuperate growth and yield of the stressed plant. It seems that these bacteria may be good candidates for bio-fertilizer formulation in stress-based agriculture.

10.5 Conclusions and Future Perspectives

One of the most prominent restrictions to crop production in the world is both biological and non-biological stress conditions dominating in the agricultural land, which are also intensifying with climate change. Under normal conditions, when the plant is not exposed to any stress, the plant produces a small amount of ethylene, which is necessary for the growth and development of the plant, although the amount of ethylene required during fruit ripening is much higher (Abeles 1992). But plants under either biological or non-biological stresses produce augmented levels of ethylene, known as "stress ethylene", which is capable of inhibiting overall plant growth. To mitigate negative effects of high levels of "stress ethylene" on plants and to augment crop productivity, fitting crop management techniques is indispensable. In this context, one of the suggested ways to reduce the negative effects of stress resulting from ethylene production is genetic modification of stress-sensitive crop varieties, but this method, although successful in some cases, is a long and costly process. On the other hand, making a plant resistant to all the stresses that are present in the environment is not an easy task. Bacteria's ability to increase plant resistance to environmental stresses and consequently to plant growth in stress conditions has been well established. In addition, due to their manifold PGP properties, these bacteria are able to reduce most of the stresses imposed on the plant. Because of this bacterial characteristic, it is no longer necessary genetically engineer all stress-sensitive crop varieties to be tolerant to a large number of biological and non-biological stresses. In the present storyline, the use of plant-associated PGPB containing 1-aminocyclopropane-1-carboxylate (ACC) deaminase activity along with other innovations to alleviate ethylene-induced stresses in stress-sensitive crop varieties could be a more cost-efficient environmental amicable option, which could be attainable in a shorter time frame and ensure sustainable agriculture. One of the limitations of using such bacteria (ACC deaminase-producing PGPB) to diminish the effect of stress imposed on plant under natural conditions (i.e., field conditions) is the reduction in their ability and survival under natural conditions (i.e., high concentrations of heavy metals, high salinity and temperature, water deficit, competition with other native microorganisms for water and nutrients, and adverse soil physical and chemical conditions). A possible solution to this problem may lie in the utilization of ACC deaminase-producing bacterial endophytes (Sturz and Nowak 2000).

Since the ability of PGPB including ACC deaminase-producing bacteria to augment plant tolerance to environmental stresses is impacted by environmental stressful conditions (Etesami and Beattie 2018) (i.e., loss in ACC deaminase activity by bacteria with increasing salinity) (Upadhyay et al. 2009), it is suggested testing the production ability of ACC deaminase of isolated bacteria in the presence of various stresses such as salinity, water deficit, heavy metals, etc. In other words, the use of ACC deaminase-producing PGPB that are picked out established upon both high stress tolerance and efficiency in expressing PGP characteristics would outstandingly advance our ability to grow crops in stressed agricultural environments.

In most of the previous studies, the effect of 1-aminocyclopropane-1-carboxylate deaminase-containing bacteria on alleviating a stress has been evaluated. Since most of the environmental stresses, for instance, water deficit, salinity, heavy metals, and imbalanced nutrition, exist in most agricultural land concurrently, it is recommended that the efficacy of ACC deaminase-producing PGPB be investigated concurrently in the presence of these stresses to pick out and introduce the best 1-aminocyclopropane-1-carboxylate deaminase-containing bacteria.

Relative to 1-aminocyclopropane-1-carboxylate (ACC) deaminase-containing rhizobacteria, some beneficial aspects of 1-aminocyclopropane-1-carboxylate (ACC) deaminase-containing bacterial endophytes, i.e., their function in alleviating salinity, water deficit, excess of water, biocontrol, temperature and imbalanced nutrition stresses, nodulation in legumes and field appraisal and use of potential of these bacteria as bio-fertilizers in stressed soil have not been thoroughly exploited. On a commercial scale, application of these bacteria in stressed agriculture might prove effective and could be a sound step in the direction of producing and conserving sustainable crop plants.

Acknowledgements We wish to thank University of Tehran for providing the necessary facilities for doing the study.

References

Abeles FB (1992) Ethylene in plant biology. In: Abeles, FB, Morgan, PW, Saltveit ME (eds) Academic Press, New York

Abeles FB, Morgan PW, Saltveit Jr ME (2012) Ethylene in plant biology. Academic press

Achard P, Cheng H, De Grauwe L, Decat J, Schoutteten H, Moritz T, Van Der Straeten D, Peng J, Harberd NP (2006) Integration of plant responses to environmentally activated phytohormonal signals. Science 311(5757):91–94

Acquaah G (2009) Principles of plant genetics and breeding. Wiley

Adam G, Duncan HJ (1999) Effect of diesel fuel on growth of selected plant species. Environ Geochem Health 21(4):353–357

Afridi MS, Mahmood T, Salam A, Mukhtar T, Mehmood S, Ali J, Khatoon Z, Bibi M, Javed MT, Sultan T (2019) Induction of tolerance to salinity in wheat genotypes by plant growth promoting endophytes: Involvement of ACC deaminase and antioxidant enzymes. Plant Physiol Biochem 139:569–577

Ahanger MA, Akram NA, Ashraf M, Alyemeni MN, Wijaya L, Ahmad P (2017) Plant responses to environmental stresses—from gene to biotechnology. AoB Plants 9(4)

Ahuja I, de Vos RCH, Bones AM, Hall RD (2010) Plant molecular stress responses face climate change. Trends Plant Sci 15(12):664–674

Albacete A, Martinez-Andujar C, Ghanem ME, Acosta M, Sánchez-Bravo J, Asins MJ, Cuartero J, Lutts S, Dodd IC, Pérez-Alfocea F (2009) Rootstock-mediated changes in xylem ionic and hormonal status are correlated with delayed leaf senescence, and increased leaf area and crop productivity in salinized tomato. Plant Cell Environ 32(7):928–938

Alexandratos N, Bruinsma J (2012) World agriculture towards 2030/2050: the 2012 revision

Ali S, Charles TC, Glick BR (2014) Amelioration of high salinity stress damage by plant growth-promoting bacterial endophytes that contain ACC deaminase. Plant Physiol Biochem 80:160–167

Allen RD (1995) Dissection of oxidative stress tolerance using transgenic plants. Plant Physiol 107(4):1049

Arbona V, Marco AJ, Iglesias DJ, López-Climent MF, Talon M, Gómez-Cadenas A (2005) Carbohydrate depletion in roots and leaves of salt-stressed potted *Citrus clementina* L. Plant Growth Regul 46(2):153–160

Arshad M, Frankenberger Jr WT (2012) Ethylene: agricultural sources and applications. Springer Science & Business Media

Arshad M, Frankenberger WT (1990) Ethylene accumulation in soil in response to organic amendments. Soil Sci Soc Am J 54(4):1026–1031

Arshad M, Saleem M, Hussain S (2007) Perspectives of bacterial ACC deaminase in phytoremediation. Trends Biotechnol 25(8):356–362

Arshad M, Shaharoona B, Mahmood T (2008) Inoculation with plant growth promoting rhizobacteria containing ACC-deaminase partially eliminates the effects of water stress on growth, yield and ripening of *Pisum sativum* L. Pedosphere 18(5):611–620

Babalola OO, Osir EO, Sanni AI, Odhiambo GD, Bulimo WD (2003) Amplification of 1-amino-cyclopropane-1-carboxylic (ACC) deaminase from plant growth promoting rhizobacteria in Striga-infested soil. Afr J Biotechnol 2(6):157–160

Barka EA, Nowak J, Clément C (2006) Enhancement of chilling resistance of inoculated grapevine plantlets with a plant growth-promoting rhizobacterium, *Burkholderia phytofirmans* strain PsJN. Appl Environ Microbiol 72(11):7246–7252

Barnawal D, Bharti N, Tripathi A, Pandey SS, Chanotiya CS, Kalra A (2016) ACC-deaminase-producing endophyte *Brachybacterium paraconglomeratum* strain SMR20 ameliorates Chlorophytum salinity stress via altering phytohormone generation. J Plant Growth Regul 35(2):553–564

Bartoli CG, Casalongué CA, Simontacchi M, Marquez-Garcia B, Foyer CH (2013) Interactions between hormone and redox signalling pathways in the control of growth and cross tolerance to stress. Environ Exper Bot 94:73–88

Bashan Y (1994) Symptom expression and ethylene production in leaf blight of cotton caused by *Alternaria macrospora* and *Alternaria alternata* alone and in combination. Can J Bot 72(11):1574–1579

Baxter A, Mittler R, Suzuki N (2013) ROS as key players in plant stress signalling. J Exp Bot 65(5):1229–1240

Belimov AA, Dodd IC, Safronova VI, Hontzeas N, Davies WJ (2007) *Pseudomonas brassicacearum* strain Am3 containing 1-aminocyclopropane-1-carboxylate deaminase can show both pathogenic and growth-promoting properties in its interaction with tomato. J Exp. Bot 58(6):1485–1495

Belimov AA, Safronova VI, Sergeyeva TA, Egorova TN, Matveyeva VA, Tsyganov VE, Borisov AY, Tikhonovich IA, Kluge C, Preisfeld A (2001) Characterization of plant growth promoting rhizobacteria isolated from polluted soils and containing 1-aminocyclopropane-1-carboxylate deaminase. Can J Microbiol 47(7):642–652

Bensalim S, Nowak J, Asiedu SK (1998) A plant growth promoting rhizobacterium and temperature effects on performance of 18 clones of potato. Am J Potato Res 75(3):145–152

Blumwald E (2000) Sodium transport and salt tolerance in plants. Curr Opin Chem Biol 12(4):431–434

Borch K, Bouma TJ, Lynch JP, Brown KM (1999) Ethylene: a regulator of root architectural responses to soil phosphorus availability. Plant, Cell Environ 22(4):425–431

Bradford KJ, Yang SF (1980) Xylem transport of 1-aminocyclopropane-1-carboxylic acid, an ethylene precursor, in waterlogged tomato plants. Plant Physiol 65(2):322–326

Bray EA (1997) Plant responses to water deficit. Trends Plant Sci 2(2):48–54

Carmen B, Roberto D (2011) Soil bacteria support and protect plants against abiotic stresses. Abiotic stress in plants-Mechanisms and daptations, Italy, pp 143–170

Cattelan AJ, Hartel PG, Fuhrmann JJ (1999) Screening for plant growth–promoting rhizobacteria to promote early soybean growth. Soil Sci Soc Am J 63(6):1670–1680

Chaudhary D, Sindhu SS (2017) Amelioration of salt stress in chickpea (Cicer arietinum L.) by coinculation of ACC deaminase-containing rhizospheric bacteria with *Mesorhizobium* strains. Legume Res 40(1):80–86

Cheeseman JM (2015) The evolution of halophytes, glycophytes and crops, and its implications for food security under saline conditions. New Phytol 206(2):557–570

Cheikh N, Jones RJ (1994) Disruption of maize kernel growth and development by heat stress (role of cytokinin/abscisic acid balance). Plant Physiol 106(1):45–51

Cheng Z, Park E, Glick BR (2007) 1-Aminocyclopropane-1-carboxylate deaminase from *Pseudomonas putida* UW4 facilitates the growth of canola in the presence of salt. Can J Microbiol 53(7):912–918

Chinnusamy V, Zhu J, Zhu J-K (2006) Salt stress signaling and mechanisms of plant salt tolerance. In: Genetic engineering. Springer, pp 141–177

Compant S, Duffy B, Nowak J, Clément C, Barka EA (2005) Use of plant growth-promoting bacteria for biocontrol of plant diseases: principles, mechanisms of action, and future prospects. Appl Environ Microbiol 71(9):4951–4959

Coupland D, Jackson MB (1991) Effects of mecoprop (an auxin analogue) on ethylene evolution and epinasty in two biotypes of Stellaria media. Ann Bot 68(2):167–172

Cuartero J, Fernández-Muñoz R (1998) Tomato and salinity. Sci Hortic 78(1–4):83–125

Davies W, Zhang J, Yang J, Dodd I (2011) Novel crop science to improve yield and resource use efficiency in water-limited agriculture. J Agric Sci 149(S1):123–131

De Prado JL, De Prado RA, Shimabukuro RH (1999) The effect of diclofop on membrane potential, ethylene induction, and herbicide phytotoxicity in resistant and susceptible biotypes of grasses. Pestic Biochem Physiol 63(1):1–14

Dell'Amico E, Cavalca L, Andreoni V (2005) Analysis of rhizobacterial communities in perennial Graminaceae from polluted water meadow soil, and screening of metal-resistant, potentially plant growth-promoting bacteria. FEMS Microbiol Ecol 52(2):153–162

Dharmasiri N, Estelle M (2004) Auxin signaling and regulated protein degradation. Trends Plant Sci 9(6):302–308

Dimkpa C, Weinand T, Asch F (2009) Plant–rhizobacteria interactions alleviate abiotic stress conditions. Plant, Cell Environ 32(12):1682–1694

Dodd IC, Belimov AA, Sobeih WY, Safronova VI, Grierson D, Davies WJ (2004) Will modifying plant ethylene status improve plant productivity in water-limited environments, 2005

Dodd IC, Pérez-Alfocea F (2012) Microbial amelioration of crop salinity stress. J Exp Bot 63(9):3415–3428

Donate-Correa J, León-Barrios M, Pérez-Galdona R (2005) Screening for plant growth-promoting rhizobacteria in *Chamaecytisus proliferus* (tagasaste), a forage tree-shrub legume endemic to the Canary Islands. Plant Soil 266(1–2):261–272

Dong H, Zhen Z, Peng J, Chang L, Gong Q, Wang NN (2011) Loss of ACS7 confers abiotic stress tolerance by modulating ABA sensitivity and accumulation in Arabidopsis. J Exp Bot 62(14):4875–4887

Elad Y (1988) Involvement of ethylene in the disease caused by Botrytis cinerea on rose and carnation flowers and the possibility of control. Ann Appl Biol 113(3):589–598

Ellouzi H, Hamed KB, Hernández I, Cela J, Müller M, Magné C, Abdelly C, Munné-Bosch S (2014) A comparative study of the early osmotic, ionic, redox and hormonal signaling response in leaves and roots of two halophytes and a glycophyte to salinity. Planta 240(6):1299–1317

Else MA, Hall KC, Arnold GM, Davies WJ, Jackson MB (1995) Export of abscisic acid, 1-aminocyclopropane-1-carboxylic acid, phosphate, and nitrate from roots to shoots of flooded tomato plants (accounting for effects of xylem sap flow rate on concentration and delivery). Plant Physiol 107(2):377–384

Else MA, Jackson MB (1998) Transport of 1-aminocyclopropane-1-carboxylic acid (ACC) in the transpiration stream of tomato (*Lycopersicon esculentum*) in relation to foliar ethylene production and petiole epinasty. Funct Plant Biol 25(4):453–458

El-Tarabily KA (2008) Promotion of tomato (*Lycopersicon esculentum* Mill) plant growth by rhizosphere competent 1-aminocyclopropane-1-carboxylic acid deaminase-producing streptomycete actinomycetes. Plant Soil 308(1–2):161–174

Etesami H (2018) Bacterial mediated alleviation of heavy metal stress and decreased accumulation of metals in plant tissues: mechanisms and future prospects. Ecot Environ Saf 147:175–191. https://doi.org/10.1016/j.ecoenv.2017.08.032

Etesami H, Alikhani HA (2016) Rhizosphere and endorhiza of oilseed rape (*Brassica napus* L) plant harbor bacteria with multifaceted beneficial effects. Biol Control 94:11–24

Etesami H, Alikhani HA (2018) *Bacillus species* as the most promising bacterial biocontrol agents in rhizosphere and endorhiza of plants grown in rotation with each other. Eur J Plant Pathol 150(2):497–506. https://doi.org/10.1007/s10658-017-1276-8

Etesami H, Beattie GA (2017) Plant-microbe interactions in adaptation of agricultural crops to abiotic stress conditions. In: Probiotics and plant health. Springer, pp 163–200

Etesami H, Beattie GA (2018) Mining halophytes for plant growth-promoting halotolerant bacteria to enhance the salinity tolerance of non-halophytic crops. Front Microbiol 9:148

Etesami H, Maheshwari DK (2018) Use of plant growth promoting rhizobacteria (PGPRs) with multiple plant growth promoting traits in stress agriculture: action mechanisms and future prospects. Ecot Environ Saf 156:225–246. https://doi.org/10.1016/j.ecoenv.2018.03.013

Fahad S, Nie L, Chen Y, Wu C, Xiong D, Saud S, Hongyan L, Cui K, Huang J (2015) Crop plant hormones and environmental stress. In: Sustainable agriculture reviews. Springer, pp 371–400

FAO (2008) FAO land and plant nutrition management service

Farwell AJ, Vesely S, Nero V, Rodriguez H, Shah S, Dixon DG, Glick BR (2006) The use of transgenic canola (*Brassica napus*) and plant growth-promoting bacteria to enhance plant biomass at a nickel-contaminated field site. Plant Soil 288(1–2):309–318

Feng J, Barker AV (1992) Ethylene evolution and ammonium accumulation by tomato plants under water and salinity stresses. Part II. J Plant Nutr 15(11):2471–2490

Frankenberger Jr WT, Muhammad A (1995) Phytohormones in soils: microbial production and function. Marcel Dekker Inc

Frommer WB, Ludewig U, Rentsch D (1999) Taking transgenic plants with a pinch of salt. Science 285(5431):1222–1223

Gamalero E, Marzachì C, Galetto L, Veratti F, Massa N, Bona E, Novello G, Glick B, Ali S, Cantamessa S (2017) An 1-Aminocyclopropane-1-carboxylate (ACC) deaminase-expressing endophyte increases plant resistance to flavescence dorée phytoplasma infection. Plant Biosystems-An Int J Dealing with all Aspects of Plant Biol 151(2):331–340

García MJ, Lucena C, Romera FJ, Alcántara E, Pérez-Vicente R (2010) Ethylene and nitric oxide involvement in the up-regulation of key genes related to iron acquisition and homeostasis in Arabidopsis. J Exp Bot 61(14):3885–3899

García MJ, Romera FJ, Lucena C, Alcántara E, Pérez-Vicente R (2015) Ethylene and the regulation of physiological and morphological responses to nutrient deficiencies. Plant Physiol 169(1):51–60

Ghanem ME, Albacete A, Martínez-Andújar C, Acosta M, Romero-Aranda R, Dodd IC, Lutts S, Pérez-Alfocea F (2008) Hormonal changes during salinity-induced leaf senescence in tomato (*Solanum lycopersicum* L). J Exp Bot 59(11):3039–3050

Ghosh S, Penterman JN, Little RD, Chavez R, Glick BR (2003) Three newly isolated plant growth-promoting bacilli facilitate the seedling growth of canola Brassica campestris. Plant Physiol Biochem 41(3):277–281

Glick BR (1995) The enhancement of plant growth by free-living bacteria. Can Journal Microbiol 41(2):109–117

Glick BR (2012) Plant growth-promoting bacteria: mechanisms and applications. Scientifica

Glick BR (2014) Bacteria with ACC deaminase can promote plant growth and help to feed the world. Microbiol Res 169(1):30–39. https://doi.org/10.1016/j.micres.2013.09.009

Glick BR, Cheng Z, Czarny J, Duan J (2007a) Promotion of plant growth by ACC deaminase-producing soil bacteria. In: New perspectives and approaches in plant growth-promoting rhizobacteria research. Springer, pp 329–339

Glick BR, Penrose DM, Li J (1998) A model for the lowering of plant ethylene concentrations by plant growth-promoting bacteria. J Theor Biol 190(1):63–68

Glick BR, Todorovic B, Czarny J, Cheng Z, Duan J, McConkey B (2007b) Promotion of plant growth by bacterial ACC deaminase. Crit Rev Plant Sci 26(5–6):227–242

Godfray HCJ, Beddington JR, Crute IR, Haddad L, Lawrence D, Muir JF, Pretty J, Robinson S, Thomas SM, Toulmin C (2010) Food security: the challenge of feeding 9 billion people. Science 327(5967):812–818

Gorham J, Jones RGW, McDonnell E (1985) Some mechanisms of salt tolerance in crop plants. In: Biosalinity in action: bioproduction with saline water. Springer, pp 15–40

Gray EJ, Smith DL (2005) Intracellular and extracellular PGPR: commonalities and distinctions in the plant–bacterium signaling processes. Soil Biol Biochem 37(3):395–412

Greenberg BM, Huang XD, Gerwing P, Yu XM, Chang P, Wu SS, Gerhardt K, Nykamp J, Lu X, Glick B (2008) Phytoremediation of salt impacted soils: greenhouse and the field trials of plant growth promoting rhizobacteria (PGPR) to improve plant growth and salt phytoaccumulation, pp 627–637

Greenberg BM, Huang XD, Gurska Y, Gerhardt KE, Wang W, Lampi MA, Zhang C, Khalid A, Isherwood D, Chang P (2006) Successful field tests of a multi-process phytoremediation system for decontamination of persistent petroleum and organic contaminants, pp 389–400

Greenway H, Munns R (1980) Mechanisms of salt tolerance in nonhalophytes. Annu Rev Plant Physiol 31(1):149–190

Grichko VP, Filby B, Glick BR (2000) Increased ability of transgenic plants expressing the bacterial enzyme ACC deaminase to accumulate Cd Co, Cu, Ni, Pb, and Zn. J Biotechnol 81(1):45–53

Grichko VP, Glick BR (2001a) Amelioration of flooding stress by ACC deaminase-containingplant growth-promoting bacteria. Plant Physiol Biochem 39(1):11–17

Grichko VP, Glick BR (2001b) Flooding tolerance of transgenic tomato plants expressing the bacterial enzyme ACC deaminase controlledby the 35S, rolD or PRB-1b promoter. Plant Physiol Biochem 39(1):19–25

Grover M, Ali SZ, Sandhya V, Rasul A, Venkateswarlu B (2011) Role of microorganisms in adaptation of agriculture crops to abiotic stresses. World J Microbiol Biotechnol 27(5):1231–1240

Habib SH, Kausar H, Saud HM (2016) Plant growth-promoting rhizobacteria enhance salinity stress tolerance in okra through ROS-scavenging enzymes. BioMed Res Int

Hayat R, Ahmed I, Sheirdil RA (2012) An overview of plant growth promoting rhizobacteria (PGPR) for sustainable agriculture. In: Crop product agri improve. Springer, pp 557–579

He C-J, Morgan PW, Drew MC (1992) Enhanced sensitivity to ethylene in nitrogen-or phosphate-starved roots of *Zea mays* L during aerenchyma formation. Plant Physiol 98(1):137–142

Hipkins MF, Hillman JR (1986) Plant growth substances and the ionic permeability of membranes. In: Plant growth substances 1985. Springer, pp 151–158

Holguin G, Patten CL (1999) Biochemical and genetic mechanisms used by plant growth promoting bacteria. World Sci

Honma M, Shimomura T (1978) Metabolism of 1-aminocyclopropane-1-carboxylic acid. Agri Bioll Chem 42(10):1825–1831

Hontzeas N, Richardson AO, Belimov A, Safronova V, Abu-Omar MM, Glick BR (2005) Evidence for horizontal transfer of 1-aminocyclopropane-1-carboxylate deaminase genes. Appl Environ Microbiol 71(11):7556–7558

Huang XD, El-Alawi Y, Penrose DM, Glick BR, Greenberg BM (2004) Responses of plants to creosote during phytoremediation and their significance for remediation processes. Environ Pollut 130:453–463

Hussain S, Peng S, Fahad S, Khaliq A, Huang J, Cui K, Nie L (2015) Rice management interventions to mitigate greenhouse gas emissions: a review. Environ Sci Pollut Res 22(5):3342–3360

Ingram J, Bartels D (1996) The molecular basis of dehydration tolerance in plants. Ann Rev Plant Biol 47(1):377–403

Jackson M (1997) Hormones from roots as signals for the shoots of stressed plants. Trends Plant Sci 2(1):22–28

Jamil A, Riaz S, Ashraf M, Foolad M (2011) Gene expression profiling of plants under salt stress. Crit Rev Plant Sci 30(5):435–458

Kang S-M, Shahzad R, Bilal S, Khan AL, Park Y-G, Lee K-E, Asaf S, Khan MA, Lee I-J (2019) Indole-3-acetic-acid and ACC deaminase producing Leclercia adecarboxylata MO1 improves *Solanum lycopersicum* L growth and salinity stress tolerance by endogenous secondary metabolites regulation. BMC Microbiol 19(1):80

Karthikeyan B, Joe MM, Islam MR, Sa T (2012) ACC deaminase containing diazotrophic endophytic bacteria ameliorate salt stress in *Catharanthus roseus* through reduced ethylene levels and induction of antioxidative defense systems. Symbiosis 56(2):77–86

Kaushal M, Wani SP (2016) Plant-growth-promoting rhizobacteria: drought stress alleviators to ameliorate crop production in drylands. Ann Microbiol 66(1):35–42

Khan MIR, Khan NA (2013) Salicylic acid and jasmonates: approaches in abiotic stress tolerance. J Plant Biochem Physiol 1(4)

Kloepper JW, Lifshitz R, Zablotowicz RM (1989) Free-living bacterial inocula for enhancing crop productivity. Trends Biotechnol 7(2):39–44

Kogan FN (1997) Global drought watch from space. Bull Am Meteor Soc 78(4):621–636
Kohler J, Caravaca F, Carrasco L, Roldan A (2006) Contribution of Pseudomonas mendocina and Glomus intraradices to aggregate stabilization and promotion of biological fertility in rhizosphere soil of lettuce plants under field conditions. Soil Use Manag 22(3):298–304
Kour D, Rana KL, Yadav AN, Yadav N, Kumar V, Kumar A, Sayyed RZ, Hesham AE-L, Dhaliwal HS, Saxena AK (2019a) Drought-Tolerant phosphorus-solubilizing microbes: biodiversity and biotechnological applications for alleviation of drought stress in plants. In: Sayyed RZ, Arora NK, Reddy MS (eds) Plant growth promoting rhizobacteria for sustainable stress management: volume 1: rhizobacteria in abiotic stress management. Springer Singapore, Singapore, pp 255–308. https://doi.org/10.1007/978-981-13-6536-2_13
Kour D, Rana KL, Yadav N, Yadav AN, Kumar A, Meena VS, Singh B, Chauhan VS, Dhaliwal HS, Saxena AK (2019b) Rhizospheric microbiomes: biodiversity, mechanisms of plant growth promotion, and biotechnological applications for sustainable agriculture. In: Kumar A, Meena VS (eds) Plant growth promoting rhizobacteria for agricultural sustainability: from theory to practices. Springer Singapore, Singapore, pp 19–65. https://doi.org/10.1007/978-981-13-7553-8_2
Kour D, Rana KL, Yadav N, Yadav AN, Singh J, Rastegari AA, Saxena AK (2019c) Agriculturally and industrially important fungi: current developments and potential biotechnological applications. In: Yadav AN, Singh S, Mishra S, Gupta A (eds) Recent advancement in white biotechnology through fungi, volume 2: perspective for value-added products and environments. Springer International Publishing, Cham, pp 1–64. https://doi.org/10.1007/978-3-030-14846-1_1
Kramer PJ, Boyer JS (1995) Water relations of plants and soils. Academic press
Kukreja S, Nandwal A, Kumar N, Sharma S, Unvi V, Sharma P (2005) Plant water status, H 2 O 2 scavenging enzymes, ethylene evolution and membrane integrity of Cicer arietinum roots as affected by salinity. Biol Plant 49(2):305–308
Kumar M, Kour D, Yadav AN, Saxena R, Rai PK, Jyoti A, Tomar RS (2019) Biodiversity of methylotrophic microbial communities and their potential role in mitigation of abiotic stresses in plants. Biologia 74:287–308. https://doi.org/10.2478/s11756-019-00190-6
Lin Z, Zhong S, Grierson D (2009) Recent advances in ethylene research. J Exp Bot 60(12):3311–3336
Lugtenberg BJJ, Chin-A-Woeng TFC, Bloemberg GV (2002) Microbe–plant interactions: principles and mechanisms. Antonie Van Leeuwenhoek 81(1–4):373–383
Ma H, Song L, Shu Y, Wang S, Niu J, Wang Z, Yu T, Gu W, Ma H (2012) Comparative proteomic analysis of seedling leaves of different salt tolerant soybean genotypes. J Prote 75(5):1529–1546
Mattoo AK (2017) The plant hormone ethylene: 0. CRC press
Mayak S, Tirosh T, Glick BR (2004a) Plant growth-promoting bacteria confer resistance in tomato plants to salt stress. Plant Physiol Biochem: PPB 42(6):565–572. https://doi.org/10.1016/j.plaphy.2004.05.009
Mayak S, Tirosh T, Glick BR (2004b) Plant growth-promoting bacteria confer resistance in tomato plants to salt stress. Plant Physiol Biochem 42(6):565–572
Mayak S, Tirosh T, Glick BR (2004c) Plant growth-promoting bacteria that confer resistance to water stress in tomatoes and peppers. Plant Sci 166(2):525–530
McDonnell L, Plett JM, Andersson-Gunnerås S, Kozela C, Dugardeyn J, Van Der Straeten D, Glick BR, Sundberg B, Regan S (2009) Ethylene levels are regulated by a plant encoded 1-aminocyclopropane-1-carboxylic acid deaminase. Physiol Planta 136(1):94–109
Minami R, Uchiyama K, Murakami T, Kawai J, Mikami K, Yamada T, Yokoi D, Ito H, Matsui H, Honma M (1998) Properties, sequence, and synthesis in *Escherichia coli* of 1-aminocyclopropane-l-carboxylate deaminase from *Hansenula saturnus*. J Biochem 123(6):1112–1118
Moeder W, Barry CS, Tauriainen AA, Betz C, Tuomainen J, Utriainen M, Grierson D, Sandermann H, Langebartels C, Kangasjärvi J (2002) Ethylene synthesis regulated by biphasic induction of 1-aminocyclopropane-1-carboxylic acid synthase and 1-aminocyclopropane-1-carboxylic acid oxidase genes is required for hydrogen peroxide accumulation and cell death in ozone-exposed tomato. Plant Physiol 130(4):1918–1926

Morgan PW, Drew MC (1997) Ethylene and plant responses to stress. Physiol Planta 100(3):620–630
Morgan PW, Gausman HW (1966) Effects of ethylene on auxin transport. Plant Physiol 41(1):45–52
Munns R (2002) Comparative physiology of salt and water stress. Plant, Cell Environ 25(2):239–250
Munns R (2005) Genes and salt tolerance: bringing them together. New Phytol 167(3):645–663
Munns R, Tester M (2008) Mechanisms of salinity tolerance. Annu Rev Plant Biol 59:651–681
Nadeem SM, Zahir ZA, Naveed M, Arshad M, Shahzad SM (2006) Variation in growth and ion uptake of maize due to inoculation with plant growth promoting rhizobacteria under salt stress. Soil Environ 25(2):78–84
Nascimento F, Brígido C, Glick B, Oliveira S, Alho L (2012) Mesorhizobium ciceri LMS-1 expressing an exogenous 1-aminocyclopropane-1-carboxylate (ACC) deaminase increases its nodulation abilities and chickpea plant resistance to soil constraints. Lett Appl Microbiol 55(1):15–21
Nascimento FX, Rossi MJ, Soares CRFS, McConkey BJ, Glick BR (2014) New insights into 1-aminocyclopropane-1-carboxylate (ACC) deaminase phylogeny, evolution and ecological significance. PLoS ONE 9(6):e99168
Noori F, Etesami H, Najafi Zarini H, Khoshkholgh-Sima NA, Hosseini Salekdeh G, Alishahi F (2018) Mining alfalfa (*Medicago sativa* L) nodules for salinity tolerant non-rhizobial bacteria to improve growth of alfalfa under salinity stress. Ecot Environ Saf 162:129–138. https://doi.org/10.1016/j.ecoenv.2018.06.092
O'Donnell PJ, Calvert C, Atzorn R, Wasternack C, Leyser HMO, Bowles DJ (1996) Ethylene as a signal mediating the wound response of tomato plants. Science 274(5294):1914–1917
Olson DC, Oetiker JH, Yang SF (1995) Analysis of LE-ACS3, a 1-aminocyclopropane-1-carboxylic acid synthase gene expressed during flooding in the roots of tomato plants. J Biol Chem 270(23):14056–14061
Orozco-Mosqueda M, Duan J, DiBernardo M, Zetter E, Campos-Garcia J, Glick BR, Santoyo G (2019) The production of ACC deaminase and trehalose by the plant growth promoting bacterium *Pseudomonas* sp. UW4 synergistically protect tomato plants against salt stress. Front Microbiol 10:1392
Palaniyandi S, Damodharan K, Yang S, Suh J (2014) *Streptomyces* sp. strain PGPA39 alleviates salt stress and promotes growth of 'Micro Tom' tomato plants. J Appl Microbiol 117(3):766–773
Pandey P, Kang SC, Maheshwari DK (2005) Isolation of endophytic plant growth promoting *Burkholderia* sp. MSSP from root nodules of Mimosa pudica. Curr Sci 177–180
Parida AK, Das AB (2005) Salt tolerance and salinity effects on plants: a review. Ecot Environ Saf 60(3):324–349
Patten CL, Glick BR (2002) Role of *Pseudomonas putida* indoleacetic acid in development of the host plant root system. Appl Environ Microbiol 68(8):3795–3801
Paul D (2013) Osmotic stress adaptations in rhizobacteria. J Basic Microbiol 53(2):101–110
Pech J-C (2002) Ethylene: agricultural sources and applications
Peng J, Li Z, Wen X, Li W, Shi H, Yang L, Zhu H, Guo H (2014) Salt-induced stabilization of EIN3/EIL1 confers salinity tolerance by deterring ROS accumulation in *Arabidopsis*. PLoS Genet 10(10):e1004664
Persello-Cartieaux F, Nussaume L, Robaglia C (2003) Tales from the underground: molecular plant–rhizobacteria interactions. Plant, Cell Environ 26(2):189–199
Plett JM, McDonnell L, Regan S (2009) Plant encoded 1-aminocyclopropane-1-carboxylic acid deaminase activity implicated in different aspects of plant development. Plant Sign Behav 4(12):1186–1189
Prasad MN, Strzalka K (2013) Physiology and biochemistry of metal toxicity and tolerance in plants. Springer Science & Business Media
Prayitno J, Rolfe BG, Mathesius U (2006) The ethylene-insensitive sickle mutant of *Medicago truncatula* shows altered auxin transport regulation during nodulation. Plant Physiol 142(1):168–180

Qin S, Zhang Y-J, Yuan B, Xu P-Y, Xing K, Wang J, Jiang J-H (2014) Isolation of ACC deaminase-producing habitat-adapted symbiotic bacteria associated with halophyte Limonium sinense (Girard) Kuntze and evaluating their plant growth-promoting activity under salt stress. Plant Soil 374(1–2):753–766

Rabhi M, Barhoumi Z, Ksouri R, Abdelly C, Gharsalli M (2007) Interactive effects of salinity and iron deficiency in Medicago ciliaris. CR Biol 330(11):779–788

Rai AK, Sharma N (2006) Phosphate metabolism in the cyanobacterium *Anabaena doliolum* under salt stress. Current Microbiol 52(1):6–12

Rana KL, Kour D, Yadav AN (2019) Endophytic microbiomes: biodiversity, ecological significance and biotechnological applications. Res J Biotechnol 14:16–142

Rasche F, Marco-Noales E, Velvis H, van Overbeek LS, López MM, van Elsas JD, Sessitsch A (2006a) Structural characteristics and plant-beneficial effects of bacteria colonizing the shoots of field grown conventional and genetically modified T4-lysozyme producing potatoes. Plant Soil 289(1–2):123–140

Rasche F, Velvis H, Zachow C, Berg G, Van Elsas JD, Sessitsch A (2006b) Impact of transgenic potatoes expressing anti-bacterial agents on bacterial endophytes is comparable with the effects of plant genotype, soil type and pathogen infection. J Appl Ecol 43(3):555–566

Reed ML, Glick BR (2005) Growth of canola (*Brassica napus*) in the presence of plant growth-promoting bacteria and either copper or polycyclic aromatic hydrocarbons. Can J Microbiol 51(12):1061–1069

Robertson GP, Paul EA, Harwood RR (2000) Greenhouse gases in intensive agriculture: contributions of individual gases to the radiative forcing of the atmosphere. Science 289(5486):1922–1925

Rodecap KD, Tingey DT, Tibbs JH (1981) Cadmium-induced ethylene production in bean plants. Zeitschrift für Pflanzenphysiologie 105(1):65–74

Roy SJ, Negrão S, Tester M (2014) Salt resistant crop plants. Curr Opin Biotechnol 26:115–124

Ruiz D, Martínez V, Cerdá A (1997) Citrus response to salinity: growth and nutrient uptake. Tree Physiol 17(3):141–150

Safronova VI, Stepanok VV, Engqvist GL, Alekseyev YV, Belimov AA (2006) Root-associated bacteria containing 1-aminocyclopropane-1-carboxylate deaminase improve growth and nutrient uptake by pea genotypes cultivated in cadmium supplemented soil. Biol Fertility Soil 42(3):267–272

Saleem M, Arshad M, Hussain S, Bhatti AS (2007) Perspective of plant growth promoting rhizobacteria (PGPR) containing ACC deaminase in stress agriculture. J Indust Microbiol Biotechnol 34(10):635–648

Saleh S, Huang X-D, Greenberg BM, Glick BR (2004) Phytoremediation of persistent organic contaminants in the environment. In: Applied bioreme phytorem. Springer, pp 115–134

Saravanakumar D, Samiyappan R (2007) ACC deaminase from Pseudomonas fluorescens mediated saline resistance in groundnut (*Arachis hypogea*) plants. J Appl Microbiol 102(5):1283–1292

Sarkar A, Ghosh PK, Pramanik K, Mitra S, Soren T, Pandey S, Mondal MH, Maiti TK (2018) A halotolerant Enterobacter sp. displaying ACC deaminase activity promotes rice seedling growth under salt stress. Res Microbiol 169(1):20–32

Saud S, Li X, Chen Y, Zhang L, Fahad S, Hussain S, Sadiq A, Chen Y (2014) Silicon application increases drought tolerance of Kentucky bluegrass by improving plant water relations and morphophysiological functions. Sci World J

Sgroy V, Cassán F, Masciarelli O, Del Papa MF, Lagares A, Luna V (2009) Isolation and characterization of endophytic plant growth-promoting (PGPB) or stress homeostasis-regulating (PSHB) bacteria associated to the halophyte *Prosopis strombulifera*. Appl Microbiol Biotechnol 85(2):371–381

Shabala S (2013) Learning from halophytes: physiological basis and strategies to improve abiotic stress tolerance in crops. Ann Bot 112(7):1209–1221

Shahbaz M, Ashraf M (2013) Improving salinity tolerance in cereals. Crit Rev Plant Sci 32(4):237–249

Sharma SB, Sayyed RZ, Trivedi MH, Gobi TA (2013) Phosphate solubilizing microbes: sustainable approach for managing phosphorus deficiency in agricultural soils. SpringerPlus 2(1):587

Sharp RE (2002) Interaction with ethylene: changing views on the role of abscisic acid in root and shoot growth responses to water stress. Plant, Cell Environ 25(2):211–222

Shi Y, Tian S, Hou L, Huang X, Zhang X, Guo H, Yang S (2012) Ethylene signaling negatively regulates freezing tolerance by repressing expression of CBF and type-A ARR genes in *Arabidopsis*. Plant Cell 24(6):2578–2595

Shinozaki K, Yamaguchi-Shinozaki K (1998) Molecular responses to drought stress. Elsevier Science BV, Amsterdam

Siddikee MA, Sundaram S, Chandrasekaran M, Kim K, Selvakumar G, Sa T (2015) Halotolerant bacteria with ACC deaminase activity alleviate salt stress effect in canola seed germination. J Korean Soc Appl Biol Chem 58(2):237–241

Singh N, Kashyap S (2012) In silico identification and characterization of 1-aminocyclopropane-1-carboxylate deaminase from *Phytophthora sojae*. J Molec Model 18(9):4101–4111

Singh NP, Singh RK, Meena VS, Meena RK (2015) Can we use maize (*Zea mays*) rhizobacteria as plant growth promoter. Vegetos 28(1):86–99

Singh RP, Jha PN (2016) Mitigation of salt stress in wheat plant (Triticum aestivum) by ACC deaminase bacterium *Enterobacter* sp. SBP-6 isolated from *Sorghum bicolor*. Acta Physiol Planta 38(5):110

Skirycz A, Inzé D (2010) More from less: plant growth under limited water. Curr Opin Biotechnol 21(2):197–203

Stearns JC, Glick BR (2003) Transgenic plants with altered ethylene biosynthesis or perception. Biotechnol Adv 21(3):193–210

Steffens B, Kovalev A, Gorb SN, Sauter M (2012) Emerging roots alter epidermal cell fate through mechanical and reactive oxygen species signaling. Plant Cell 24(8):3296–3306

Strzelczyk E, Kampert M, Pachlewski R (1994) The influence of pH and temperature on ethylene production by mycorrhizal fungi of pine. Mycorrhiza 4(5):193–196

Sturz AV, Nowak J (2000) Endophytic communities of rhizobacteria and the strategies required to create yield enhancing associations with crops. Appl Soil Ecol 15(2):183–190

Sun J, Cardoza V, Mitchell DM, Bright L, Oldroyd G, Harris JM (2006) Crosstalk between jasmonic acid, ethylene and Nod factor signaling allows integration of diverse inputs for regulation of nodulation. Plant J 46(6):961–970

Sziderics A, Rasche F, Trognitz F, Sessitsch A, Wilhelm E (2007) Bacterial endophytes contribute to abiotic stress adaptation in pepper plants (*Capsicum annuum* L). Can J Microbiol 53(11):1195–1202

Tao J-J, Chen H-W, Ma B, Zhang W-K, Chen S-Y, Zhang J-S (2015) The role of ethylene in plants under salinity stress. Front Plant Sci 6:1059

Timmusk S, Wagner EGH (1999) The plant-growth-promoting rhizobacterium Paenibacillus polymyxa induces changes in *Arabidopsis thaliana* gene expression: a possible connection between biotic and abiotic stress responses. Molec Plant-Microbe Interac 12(11):951–959

Tittabutr P, Piromyou P, Longtonglang A, Noisa-Ngiam R, Boonkerd N, Teaumroong N (2013) Alleviation of the effect of environmental stresses using co-inoculation of mungbean by *Bradyrhizobium* and rhizobacteria containing stress-induced ACC deaminase enzyme. Soil Sci Plant Nutr 59(4):559–571

Tuteja N (2007) Mechanisms of high salinity tolerance in plants. In: Methods in enzymology, vol 428. Elsevier, pp 419–438

Upadhyay SK, Singh DP, Saikia R (2009) Genetic diversity of plant growth promoting rhizobacteria isolated from rhizospheric soil of wheat under saline condition. Curr Microbiol 59(5):489–496

Van Loon LC, Bakker P, Pieterse CMJ (1997) Mechanisms of PGPR-induced resistance against pathogens

van Loon LC, Glick BR (2004) Increased plant fitness by rhizobacteria. In: Molecular ecotoxicology of plants. Springer, pp 177–205

Verma P, Yadav AN, Khannam KS, Kumar S, Saxena AK, Suman A (2016a) Molecular diversity and multifarious plant growth promoting attributes of Bacilli associated with wheat (*Triticum aestivum* L.) rhizosphere from six diverse agro-ecological zones of India. J Basic Microbiol 56:44–58

Verma P, Yadav AN, Khannam KS, Mishra S, Kumar S, Saxena AK, Suman A (2016b) Appraisal of diversity and functional attributes of thermotolerant wheat associated bacteria from the peninsular zone of India. Saudi J Biol Sci. https://doi.org/10.1016/j.sjbs.2016.01.042

Verma P, Yadav AN, Khannam KS, Panjiar N, Kumar S, Saxena AK, Suman A (2015) Assessment of genetic diversity and plant growth promoting attributes of psychrotolerant bacteria allied with wheat (*Triticum aestivum*) from the northern hills zone of India. Ann Microbiol 65:1885–1899

Verma P, Yadav AN, Khannam KS, Saxena AK, Suman A (2017a) Potassium-Solubilizing microbes: diversity, distribution, and role in plant growth promotion. In: Panpatte DG, Jhala YK, Vyas RV, Shelat HN (eds) Microorganisms for green revolution: volume 1: microbes for sustainable crop production. Springer Singapore, Singapore, pp 125–149. https://doi.org/10.1007/978-981-10-6241-4_7

Verma P, Yadav AN, Kumar V, Singh DP, Saxena AK (2017b) Beneficial plant-microbes interactions: biodiversity of microbes from diverse extreme environments and its impact for crop improvement. In: Singh DP, Singh HB, Prabha R (eds) Plant-Microbe interactions in agro-ecological perspectives: volume 2: microbial interactions and agro-ecological impacts. Springer Singapore, Singapore, pp 543–580. https://doi.org/10.1007/978-981-10-6593-4_22

Viterbo A, Landau U, Kim S, Chernin L, Chet I (2010) Characterization of ACC deaminase from the biocontrol and plant growth-promoting agent *Trichoderma asperellum* T203. FEMS Microbiol Lett 305(1):42–48

Vurukonda SSKP, Vardharajula S, Shrivastava M, SkZ A (2016) Enhancement of drought stress tolerance in crops by plant growth promoting rhizobacteria. Microbiol Res 184:13–24

Wang C, Knill E, Glick BR, Défago G (2000) Effect of transferring 1-aminocyclopropane-1-carboxylic acid (ACC) deaminase genes into *Pseudomonas fluorescens* strain CHA0 and its gac A derivative CHA96 on their growth-promoting and disease-suppressive capacities. Can J Microbiol 46(10):898–907

Wang W, Vinocur B, Altman A (2003) Plant responses to drought, salinity and extreme temperatures: towards genetic engineering for stress tolerance. Planta 218(1):1–14

Wilhite DA (2000) Drought as a natural hazard: concepts and definitions

Win KT, Tanaka F, Okazaki K, Ohwaki Y (2018) The ACC deaminase expressing endophyte *Pseudomonas* spp. Enhances NaCl stress tolerance by reducing stress-related ethylene production, resulting in improved growth, photosynthetic performance, and ionic balance in tomato plants. Plant Physiol Biochem 127:599–607

Xu J, Li Y, Wang Y, Liu H, Lei L, Yang H, Liu G, Ren D (2008) Activation of MAPK kinase 9 induces ethylene and camalexin biosynthesis and enhances sensitivity to salt stress in Arabidopsis. J Biol Chem 283(40):26996–27006

Xu Z, Jiang Y, Jia B, Zhou G (2016) Elevated-CO2 response of stomata and its dependence on environmental factors. Front Plant Sci 7:657

Yadav AN (2017a) Agriculturally important microbiomes: biodiversity and multifarious PGP attributes for amelioration of diverse abiotic stresses in crops for sustainable agriculture. Biomed J Sci Tech Res 1:1–4

Yadav AN (2017b) Beneficial role of extremophilic microbes for plant health and soil fertility. J Agric Sci 1:1–4

Yadav AN, Kour D, Sharma S, Sachan SG, Singh B, Chauhan VS, Sayyed RZ, Kaushik R, Saxena AK (2019a) Psychrotrophic microbes: biodiversity, mechanisms of adaptation, and biotechnological implications in alleviation of cold stress in plants. In: Sayyed RZ, Arora NK, Reddy MS (eds) Plant growth promoting rhizobacteria for sustainable stress management: volume 1: rhizobacteria in abiotic stress management. Springer Singapore, Singapore, pp 219–253. https://doi.org/10.1007/978-981-13-6536-2_12

Yadav AN, Kumar V, Prasad R, Saxena AK, Dhaliwal HS (2018a) Microbiome in Crops: Diversity, distribution and potential role in crops improvements. In: Prasad R, Gill SS, Tuteja N (eds) Crop improvement through microbial biotechnology. Elsevier, USA, pp 305–332

Yadav AN, Sachan SG, Verma P, Saxena AK (2015a) Prospecting cold deserts of north western Himalayas for microbial diversity and plant growth promoting attributes. J Biosci Bioeng 119:683–693

Yadav AN, Sachan SG, Verma P, Saxena AK (2016) Bioprospecting of plant growth promoting psychrotrophic Bacilli from cold desert of north western Indian Himalayas. Indian J Exp Biol 54:142–150

Yadav AN, Sachan SG, Verma P, Tyagi SP, Kaushik R, Saxena AK (2015b) Culturable diversity and functional annotation of psychrotrophic bacteria from cold desert of Leh Ladakh (India). World J Microbiol Biotechnol 31:95–108

Yadav AN, Verma P, Kour D, Rana KL, Kumar V, Singh B, Chauahan VS, Sugitha T, Saxena AK, Dhaliwal HS (2017a) Plant microbiomes and its beneficial multifunctional plant growth promoting attributes. Int J Environ Sci Nat Resour 3:1–8. https://doi.org/10.19080/IJESNR.2017.03.555601

Yadav AN, Verma P, Kumar S, Kumar V, Kumar M, Singh BP, Saxena AK, Dhaliwal HS (2018b) Actinobacteria from rhizosphere: molecular diversity, distributions and potential biotechnological applications. In: Singh B, Gupta V, Passari A (eds) New and future developments in microbial biotechnology and bioengineering. USA, pp 13–41. https://doi.org/10.1016/b978-0-444-63994-3.00002-3

Yadav AN, Verma P, Sachan SG, Kaushik R, Saxena AK (2018c) Psychrotrophic microbiomes: molecular diversity and beneficial role in plant growth promotion and soil health. In: Panpatte DG, Jhala YK, Shelat HN, Vyas RV (eds) Microorganisms for green revolution-volume 2: microbes for sustainable agro-ecosystem. Springer, Singapore, pp 197–240. https://doi.org/10.1007/978-981-10-7146-1_11

Yadav AN, Verma P, Singh B, Chauhan VS, Suman A, Saxena AK (2017b) Plant growth promoting bacteria: biodiversity and multifunctional attributes for sustainable agriculture. Adv Biotechnol Microbiol 5:1–16

Yadav AN, Yadav N, Sachan SG, Saxena AK (2019b) Biodiversity of psychrotrophic microbes and their biotechnological applications. J Appl Biol Biotechnol 7:99–108

Yamaguchi T, Blumwald E (2005) Developing salt-tolerant crop plants: challenges and opportunities. Trends Plant Sci 10(12):615–620

Yang J, Kloepper JW, Ryu C-M (2009) Rhizosphere bacteria help plants tolerate abiotic stress. Trends Plant Sci 14(1):1–4. https://doi.org/10.1016/j.tplants.2008.10.004

Yao L, Wu Z, Zheng Y, Kaleem I, Li C (2010) Growth promotion and protection against salt stress by *Pseudomonas putida* Rs-198 on cotton. Eur J Soil Biol 46(1):49–54

Yoolong S, Kruasuwan W, Phạm HTT, Jaemsaeng R, Jantasuriyarat C, Thamchaipenet A (2019) Modulation of salt tolerance in Thai jasmine rice (*Oryza sativa* L. cv. KDML105) by Streptomyces venezuelae ATCC 10712 expressing ACC deaminase. Sci Rep 9(1):1275

Yousfi S, Mahmoudi H, Abdelly C, Gharsalli M (2007) Effect of salt on physiological responses of barley to iron deficiency. Plant Physiol Biochem 45(5):309–314

Yuquan X, Rong S, Zhixing L (1999) Quickly screening a strain of Pseudomonas B8 with both ACC deaminase activity and antagonism against *Fusarium oxysporum*

Zahir ZA, Munir A, Asghar HN, Shaharoona B, Arshad M (2008) Effectiveness of rhizobacteria containing ACC deaminase for growth promotion of peas (*Pisum sativum*) under drought conditions. J Microbiol Biotechnol 18(5):958–963

Zhang F, Zhang H, Wang G, Xu L, Shen Z (2009) Cadmium-induced accumulation of hydrogen peroxide in the leaf apoplast of *Phaseolus aureus* and *Vicia sativa* and the roles of different antioxidant enzymes. J Hazard Mater 168(1):76–84

Zhang Y-f, He L-y, Chen Z-j, Zhang W-h, Wang Q-y, Qian M, Sheng X-f (2011) Characterization of lead-resistant and ACC deaminase-producing endophytic bacteria and their potential in promoting lead accumulation of rape. J Hazard Mater 186(2–3):1720–1725

Zhao L, Xu Y, Lai X-H, Shan C, Deng Z, Ji Y (2015) Screening and characterization of endophytic *Bacillus* and *Paenibacillus* strains from medicinal plant Lonicera japonica for use as potential plant growth promoters. Braz J Microbiol 46(4):977–989
Zheng D, Han X, An YI, Guo H, Xia X, Yin W (2013) The nitrate transporter NRT2. 1 functions in the ethylene response to nitrate deficiency in *A rabidopsis*. Plant, Cell Environ 36 (7):1328–1337
Zhu JK (2001) Plant salt stress. e LS
Zhu J-K (2002) Salt and drought stress signal transduction in plants. Ann Rev Plant Biol 53(1):247–273

Chapter 11
Halophilic Microbes from Plant Growing Under the Hypersaline Habitats and Their Application for Plant Growth and Mitigation of Salt Stress

Enespa, Jai Prakash and Prem Chandra

Abstract Salinity of the agriculture soil is the serious issue all over the world, and it is also an important environmental factor for reduction of growth and yield of agricultural crops. The density of more salt available in soil may alter the physiological and metabolic activities in the agricultural crops and reduce the growth and production of crops both qualitative and quantitative ways. For combating against soil salinity, many transgenic salt-tolerant crops have been developed but far too little is success. For solution, in the soils the use of plant growth-promoting rhizobacteria (PGPR) can reduce soil salinity, load of chemical fertilizers, and pesticide in the agricultural field, and improve soil health, seed germination, crop growth, and productivity under saline condition PGPR accepted as potential microbes that can tolerate various atmospheric circumstances like more temperature, pH, and saline soils. In the saline environment, many halophilic/halotolerant bacteria and plants/halophytes are observed/adapted and perform a significant role in saline soil ecosystem. Innumerable microfloral communities and halophytes contain salt-tolerant gene, and they perform as an essential protagonist in subsistence for extreme environmental condition especially salt. It can be concluded that PGPR can be used as a supportable, manageable, sustainable, and economical tool for salinity tolerance and productivity of crops/plants.

Keywords Halophile · Hypersaline habitats · PGPR · Soil salinity · Sustainable agriculture · Salt mitigation

Enespa
Department Plant Pathology, MPDC, University of Lucknow, Lucknow 226007, Uttar Pradesh, India

J. Prakash
Department of Microbiology, School for Environmental Sciences, Babasaheb Bhimrao Ambedkar (a Central) University, Lucknow 226025, Uttar Pradesh, India

P. Chandra (✉)
Department of Environmental Science, School for Environmental Sciences, Babasaheb Bhimrao Ambedkar (a Central) University, Lucknow 226025, Uttar Pradesh, India
e-mail: p.chandrabbau@gmail.com

A. N. Yadav et al. (eds.), *Plant Microbiomes for Sustainable Agriculture*, Sustainable Development and Biodiversity 25, https://doi.org/10.1007/978-3-030-38453-1_11

11.1 Introduction

The population of human will be expected to reach 9.8 billion until 2050 (Magallon and Dinneny 2019). In addition, the demand for food also be increased with the enhancing population, but this demand cannot be fulfilled without soil fertility, beneficial microorganisms, and essential nutrients of the soil (Poeplau et al. 2019; Chandra and Enespa 2017). Currently, many chemical fertilizers and pesticides are used in the soil for production of food; however, these ingredients can be increased for crop growth and productivity (Chandra and Enespa 2017), but simultaneously it increases soil salinity and also reduces soil fertility and beneficial microorganisms present in the soils (Rashid et al. 2016; Yang et al. 2019). The salinity in soil ecosystem is a major agrochemical/abiotic stress problem mainly in the semi-barren and waterless areas (Gu et al. 2016). Approximately, 65% of crop's productivity is adversely affected by saline soil (Machado and Serralheiro 2017).

A significant role is played by microorganisms in the improvement of productive soil and crop production and yield. In addition, some ions (e.g., sodium (Na^+) and potassium (K^+)) also affect the growth of plant and microorganisms and ultimately increase the soil salinity (Yan et al. 2015). Besides these, the climate changes such as drought, shortage of water, low rainfall, and abrupt changes in temperature also increase the soil salinity (Chandra and Enespa 2016). Reactive oxygen species (ROS), hydrogen peroxide (H_2O_2), superoxide (O_2), hydroxyl radicals (OH^-), lipid peroxidation, and the integrity of the membrane are other parameters of soil salinity which are produced by the cellular response (Choudhury et al. 2017; Chakraborty et al. 2018; Singh et al. 2018). In the presence of soil salinity, organic matter, essential nutrients, and beneficial microorganisms are reduced and ultimately it negatively affects the crop's productivity (Egamberdieva et al. 2017). Soil salinity reduces the root and shoots growth and finally decreases the crop's productivity (Glick 2014). For the management of soil salinity, plants used various types of mechanisms (Schmidt et al. 2018). Among all mechanisms, osmolyte is a common mechanism used by the plant. Osmolytes provide protection to the plant cell organelles and also build up compatible solutes (Chakraborty et al. 2018; El-Esawi et al. 2019). Besides, the formation of free radicals stabilizes DNA, stress protein, and prolines during salt stress condition are other factors for survival and growth of the plant (Teh et al. 2016; Chandra and Enespa 2016). Moreover, antioxidant enzymes such as peroxidases (POX), superoxide dismutase (SOD), and catalase (CAT) also protect against salinity and toxicity (Joseph and Jini 2010; Caverzan et al. 2016).

However, these mechanisms are not good for a long time in the reduction of soil salinity; currently, it needs a viable method for reduction of soil salinity and improves the soil fertility and increases microbial population, plant growth, and yield at high saline condition (Ladeiro 2012; Shrivastava and Kumar 2015). Microorganisms play a significant role in the improvement of soil fertility, crop's growth, and yield (Yan et al. 2015; Biswas and Paul 2017). Among all microbial group, plant growth-promoting rhizobacteria (PGPR) is an eco-friendly method for plant growth and sustainable agriculture by various ways such as the production of phytohormones, solubilization of minerals such as potassium, zinc, phosphate, and chelation of iron under saline condition (Verma et al. 2015; Yadav et al. 2015a, b, c; Habib et al. 2016; Ilangumaran and Smith 2017; Numan et al. 2018). This chapter describes the role of PGPR in the improvement of soil fertility and reduction of soil salinity and crop's yield under saline condition. In addition, how halotolerant microbes and plants survive under saline condition has been also explained.

11.2 Halophiles, Classification, and Hypersaline Environments

Those microorganisms can propagate and maintain their spore cycle at more saline concentrations ($\geq$150 g L^{-1}/15%) known as halophile (Ollivier et al. 1994; Oren 2008). The halophile is categorized into three dissimilar groups on the beginning of different salt concentrations: 1) less (1–6% NaCl), temperate (7–15%), and more salt concentrated halophile (15–30%) (de Lourdes Moreno et al. 2013; Chandra and Singh 2014; Yadav et al. 2019a, 2015d). Different concentrations of salt occur in the soil, and these are found at various depths in the soil habitats. According to Or et al. (2007), salt concentration and their variability are found much more than water. In the saline environment, different plants are growing known as halotolerant (halophytes) at different concentrations of salts and recorded well adaptability and perform a key character in the biogeochemical cycles (Nabti et al. 2015; Etesami and Beattie 2018). Microbes play a major character in enhancement of herb adaptation at various saline habitats (Bringel and Couée 2015; Bang et al. 2018; Yadav et al. 2019a). However, a limited microbial diversity is found in the extreme soil habitats/hypersaline environments due to various environmental factors and high salt concentrations (Ulukanli and Digrak 2002; Chandra and Singh 2016; Yadav and Saxena 2018). Besides soil salinity, the saline environment is mainly found in the aquatic water such as lakes, river, pond, and sea (Sánchez-Porro et al. 2003). From saline environments, the food or food-based products, plants, and animals contain salts (Maturrano et al. 2006; Ventosa et al. 2015).

11.3 Halophilic/Halotolerant Microbial Diversity in Soil

Soil salinity affects the structure, composition of microbial species, and also bacteriological populations present in the rhizospheric regions of crops. These communities have different groups, which show modified structural and physiological properties under hypersaline condition (Bever et al. 2012; Mendes et al. 2013). However, bacterial communities are dominant as compared to other microbial communities (e.g., virus, fungi, protozoa, and algae), and it is found in the rhizospheric region of the plant under saline condition (Mukhtar et al. 2017; Yamamoto et al. 2018; Chandra and Enespa 2019b). Besides rhizosphere, bacterial communities are also recorded endophytic region, in salt lakes, river water, and root nodules (Albaggar 2014; Leite et al. 2017). But in the saline soil, bacterial communities do not define a similar group of phylogeny but signify a assemblage which has progressed in altered types of microorganisms that belong to the genera *Actinopolyspora, Bacillus, Halomonas Micrococcus, Marinococcus, Pseudomonas, Salinicoccus,* and *Vibrio* which are mainly found in the hypersaline region (Ventosa et al. 1998; Soto-Padilla et al. 2014; Chandra et al. 2014; Verma et al. 2017b; Yadav et al. 2018a, b, d). These genera belong to both Gram-positive and Gram-negative bacteria showing rod-, comma-, and cocci-shaped cell. However, Gram-negative bacteria appear to be dominant in saline environments (Ventosa et al. 1998; Canfora et al. 2014). In Gram-negative bacteria, root-nodulating bacteria showing root-colonizing property are considered to be a halotolerant group (Zahran 1997). These bacteria have capable of nitrogen fixation and improve soil fertility at high concentration. A halotolerant bacterium *Swaminathania salitolerans* gen. nov., sp. nov. was isolated from the rhizosphere, roots, and stems of mangrove-associated wild rice (Loganathan and Nair 2004). Another bacteria belonging to the genus of *Azospirillum, Bacillus, Enterobacter,* and *Azotobacter* were isolated from the different agricultural under saline soils (Alamri and Mostafa 2009; Fendrihan et al. 2017). The popular nitrogen-fixing bacterium *Rhizobium* is linked with marsh grass *Spartina alterniflora* as a halotolerant plant has also been isolated and identified from hypersaline condition (Bedre et al. 2016).

Besides, another nitrogen-fixing bacterium *Bacillus* was screened from salty soils of Egypt, and it showed acetylene reduction activity at 5% NaCl concentration (Zahran et al. 1995). The genus *Azotobacter* is the free-living nitrogen-fixing bacterium showing a significant role in different environmental conditions such as soil, water, and sediments at the high salt concentration (Akhter et al. 2012; Sahoo et al. 2014). *Azotobacter* strain isolated from agricultural crops showed high nitrogen-fixing ability at 30% NaCl. The nitrogen fixation efficiency of a bacterium *A. vinelandii* was decreased from nonsaline to saline condition as reported by Sahoo et al. (2014). *Azospirillum halopraeferens* was isolated and enhanced the growth of mangrove plant by root colonization irrigated with seawater (Bashan et al. 2000). A little information is available on the halotolerant microbial diversity isolated from saline soils as compared with hypersaline aquatic locales (Oren 2008; Yang et al. 2016).

11.4 Effect of Soil Salinity in the Soil Environment

The salinity soil is considered mainly as a major problem in the ecosystem because these problems increase continuously, disturbing biotic and abiotic soil constituents (Vandegehuchte et al. 2010; Bünemann et al. 2018). It also affects natural circumstances in the barren and semi-barren regions of an ecosystem. Excess of saline soils affects seriously on the micro- and macro-floral structure and on space where it lives

(Getu 2009). Excess salt in the soils known as sodic soils contains sodium and chloride ions in the earthen constituent part (Bianco and Defez 2010). Due to insufficient discharge and drainage of irrigation water, salts accumulated in the soil (Cuevas et al. 2019). However, the chlorides, bicarbonates of calcium, sulfates, carbonates, magnesium, sodium, and potassium salts are present in the irrigation water (Warrence et al. 2002). The soil structure growth and the production of crops adversely are affected by salt concentration (Ondrasek et al. 2011; Shrivastava and Kumar 2015). On the bases of soil and groundwater practices generally, the salinity is of three types: transient, groundwater associated, and irrigation salinities (Greene et al. 2016; Chandra et al. 2020). Salinity affects both soil system and living organisms that are known as most severe abiotic environmental stress (Gupta and Huang 2014). The immediate consequences of soil are found for biological activity or conservation occurs within the pore space or on the surfaces of the particles that forms the pores (Indoria et al. 2017; Totsche et al. 2018). High salinity leads to negative effects on soil structure which is well known.

Soil dispersion and clay platelets to swell and aggregate are caused by elevated sodium concentrations (Warrence et al. 2002). Thus, in the binding of clay particles, the forces involved are dislocated under the stimulus of sodium ions. Clay particles to plug soil pores are caused due to the dispersion of soil (Arora and Dagar 2019). Therefore, the permeability of soil for water and air is reduced and forms apparent crusting (Kooistra and Tovey 1994; Greene and Hairsine 2004).

It is documented that the presence of water in the soil leads to the swelling of the soil particles with high smectite clay content, and the hydration of some minerals as a result of the reduction of the cross-sectional area of soil pores is documented (Mahrous et al. 2018). Under high sodium or low salt concentrations, this process is completed and it causes the mobilization of fine particles and diffusion within the pores (Mahrous et al. 2018; Chandra et al. 2020). The water and air will be obstructed within the soil structure and particles by the particles stored in the small pores (Schjønning et al. 2002).

11.5 Mechanisms for Adaptation of Microorganisms in the Hypersaline Environment

Phylogenetically, the microbial life is very diverse at high concentrations, and the salinity environments are occupied by halophilic and halotolerant microflora of all domains of life, such as archaea, bacteria, and eukarya (Oren 2008; Ma et al. 2010). Using this mechanisms these halophile microorganisms to tolerate the high salt concentrations, and in various cases to acclimatize their structure to alterations in high salinity in their environments, are miscellaneous as well (Oren 2008).

The basic mechanisms for adaptation of microorganisms in the hypersaline environment are given below:

- Biological membranes of the microorganism are absorptive to water containing salt. Consequently, the movement of water inside and outside of microbial cells is possessed by changes in ionic activity between cytoplasm and external medium (Murínová and Dercová 2014; Watson 2015).
- The bacterial cell maintains high osmotic pressure under saline condition; therefore, it is another strategy for adaptation mechanism (Weinisch et al. 2018).
- The high concentrations of inorganic salts inside the microbial cell are accumulated and achieved the osmotic balance. The sodium ions are left out from cells in all three domains of life, and inside the cell the salt strategy is based on KCl rather than NaCl as a main salt of intracellular organism (Oren 2002).
- Di-myoinositol-1, 1-phosphate, cyclic 2,3-diphosphoglycerate, α-diglycerol phosphate, mannosylglycerate, and mannosylglyceramide are compatible solutes which are very strong water structure formers and are excepted from the hydration shell of proteins, thus alleviating the hydration shell and decreasing the water activity coefficients (Gunde-Cimerman et al. 2018).
- In many extremophiles, such low-molecular weight compounds are accumulated to increase the concentrations of salts but also as a reply to other ecological alterations such as temperature stress.
- Di-myoinositol-1, 1-phosphate, cyclic 2, 3-diphosphoglycerate, α-diglycerol phosphate, mannosylglycerate, and mannosylglyceramide are the examples of organic compatible solutes in thermophiles and in psychrophiles (da Costa and Santos 2009).
- Mostly, at low salt concentration, the microorganisms are endured and also accumulate salts inside the cell in the form of solutes from outside medium (Shrivastava and Kumar 2015).

11.5.1 Mechanism of Salt Tolerance

The microbial population in the rhizosphere decreases severely due to increase in pH and salinity (Ibekwe et al. 2010). In hypersaline atmosphere the microbes inhabits using "compatible solute strategy" having capability to strong osmotic pressure to resist the salt stress (Pikuta et al. 2007; Chandra and Singh 2017). Choline, betaine, proline, glutamic acid, and other amino acids are the compatible solutes stored by various halophilic bacteria at high concentrations without interfering with cellular processes (Poolman and Glaasker 1998).

11.5.2 Characteristics and Function of Compatible Solutes

The HPLC and NMR methods are followed for the determination and production of compatible solutes in various archaea and bacteria (Roberts 2005a, b). The compounds in limited numbers comprise the bacteria such as sugars (trehalose), polyols (glycerol and glucosyl glycerol), free amino acids (proline and glutamate), offshoots thereof (proline, betaine, and ectoine), quaternary amines and their sulfonium analogs (glycine betaine, carnitine, and dimethylsulfoniopropionate), sulfate esters (choline-O-sulfate), and N-acetylated diamino acids and small peptides (N-acetylornithine and N-acetylglutaminylglutamine amide) (Kempf and Bremer 1998). Generally, the compatible solutes do not carry a net charge at physiological pH due to their high molecular solubility (Galinski 1993). The vital cellular functions such as DNA replication, DNA–protein interactions, and the cellular metabolic machinery without disturbing the solutes can reach high intracellular concentrations in disparity to mineral salts (Wang and Levin 2009; Long et al. 2018). Compatible solutes such as glycine, betaine, and proline increase the cytoplasmic volume and water content freely of the cells at high osmolality, and their accumulation uninterruptedly permitted proliferation of cells under unfavorable conditions (Kohler et al. 2015).

Various halotolerant nitrogen-fixing bacteria accumulate electrolytes such as K^+ glutamate, as enzymes, ribosomes, and transport proteins of these bacteria require high level of potassium for stability and activity using salt in strategy mechanism (Da Costa et al. 1998a, b). But within the cell physiology, organic solute accumulations are more compatible (Ventosa et al. 1998; Wood et al. 2001). The organic solutes have two mechanisms under saline conditions for their mode of actions: firstly to increase the intracellular osmotic strength and secondly to stabilization; the cellular macromolecules are proposed (Yancey et al. 1982; Csonka 1989; Chandra and Enespa 2019a). After adding these solutes in bacterial culture, the drastic stimulation in growth rate is observed in cells in high osmolality media (Gouffi et al. 1998). Higher internal concentrations of solutes accumulated in the alleviation of osmolality (Patchett et al. 1992). The glucose is oxidized in Entner–Doudoroff pathway modifications by the mostly halotolerant organisms (Fig. 11.1), the synthesis of compatible solutes after formation of pyruvate, and its further oxidation by pyruvate oxidoreductase in tricarboxylic acid cycle (TCA) (Kindzierski et al. 2017).

In salt-tolerant bacteria, the accumulation of organic solutes has been found to require genetic initiation (Roberts 2005a, b). In response to osmotic stress in *Bacillus* sp., intracellular proline to increase rapidly has been observed and the corresponding genes were detected, respectively, proB, proA, and proC encoding γ-glutamyl kinase (γ-GK), γ-glutamyl-phosphate reductase (γ-GPR), and pyrroline-5-carboxylate (P5C) reductase (Pérez-Arellano et al. 2010). L-aspartokinase (Ask), L-2,4-diaminobutyric acid transaminase (EctB), L-2,4-diaminobutyric acid acetyl-transferase (EctA), and L-ectoine synthase (EctC) encoding the structural gene and detected for biosynthesis of major harmonious solute like ectoine in *Halobacillus dabanensis* (Reshetnikov et al. 2006; Czech et al. 2019). Choline or choline-O-sulfate oxidized enzymatically into glycine betaine due to involvement of four genes *betI,*

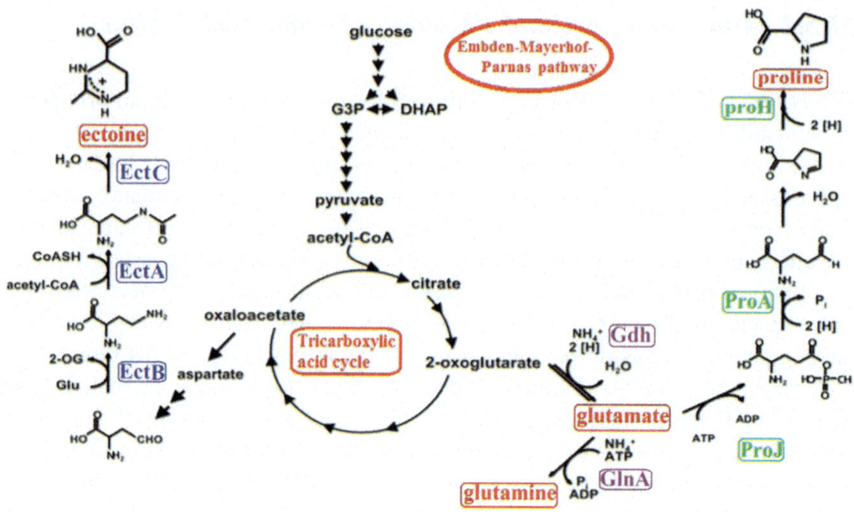

Fig. 11.1 Synthesis of compatible solutes: Proline, ectoine, and glutamine under stress conditions (Figure adopted by Saum and Müller 2008)

betC, betB, and *betA* well characterized at molecular level and organized into one operon (Osteras et al. 1998; Stöveken et al. 2011). Various halotolerant nitrogen-fixing bacteria are also observed in the cell for the maintaining of the balance of Na^+ and K^+ ions (Hanin et al. 2016; Thomas and Apte 1984). A cytoplasmic KCL concentration is maintained by bacteria similar to that of the surrounding medium in order to attain an osmotic equilibrium (Kraegeloh et al. 2005). The Na^+/H^+ antiporter performance is a major character in homeostasis of pH and Na^+ in cells that interchange Na^+ for H^+ (Suárez et al. 2008). The genes that are proved to be involved in halotolerance in nitrogen-fixing bacteria either through knockout studies or through overexpression studies are framed in Table 11.1.

11.5.3 Exchange of Solutes/Ions

Many solutes/ions are present in the soils and perform an important character in the existence of microorganisms in the presence of soil salinity (Shrivastava and Kumar 2015). However, more solutes or ions containing soils can decrease microbial population in the rhizospheric region of plants (Aung et al. 2018). Several microbes reside in hypersaline environment condition proficient passionate osmotic pressure, and thus use compatible solute strategy or salt-in strategy to resist salt stress (Oren 2011). Choline, betaine, proline, glutamic acid, and other amino acids compatible solutes accumulated in most of the bacteria at high salinity without interfering with cellular procedures (Wood et al. 2001).

Table 11.1 Genes conferring salt tolerance response in selected nitrogen-fixing bacteria

Strains	Products	Genes	References
Rahnella aquatilis HX2	1-aminocyclopropane-1-carboxylic acid deaminase	acdS gene	Peng et al. (2019)
Sinorhizobium meliloti B401	Na^+/H^+ antiporter	AtNHX1 gene	Stritzler et al. (2018)
Azospirillum lipoferum FK1	Proline and glycine betaine levels	*PAL, PPO, CHS, CHI, REB2A,* and *IFS*	El-Esawi et al. (2019)
Klebsiella sp. SBP-8	K^+/Na^+ transporters	AcdS gene	Singh et al. (2015)
Pseudomonas aeruginosa	Na^+/H^+ antiporter	nhaP	Inaba et al. (2001)
Sinorhizobium meliloti	(p) pp Gpp synthetase	relA	Wei et al. (2004)
	Glycine betaine/proline	bet genes	Mandon et al. (2003)
	Betaine transporter	betS gene	
	Transcription cleavage factor	greA	
	Potassium-uptake protein	Kup	Nogales et al. (2002)
Rhizobium tropici	Histidine kinase Na^+/H^+ antiporter	ntrY, ndvA and ndvB (synthetic gene), nhaA, nhaB, nhaC	Wai Liew et al. (2007)
Azotobacter vinelandii	Glucosyl glycerol biosynthesis	ggpPS	Klähn et al. (2009)
Enterobacter cloacae	Na^+/H^+ antiporter	nhaA	Lentes et al. (2014)
Synechocystis sp.	Na^+/H^+ antiporter	nhaS1, nhaS2, nhaS3, nhaS4, and nhaS5	Mitschke et al. (2011)
Aphanothece halophytica	Na^+/H^+ antiporter	napA	Laloknam et al. (2006)
Bacillus subtilis	γ-glutamyl kinase	proA, proBproC	Zhao et al. (2011)

11.5.4 Mechanism of Salt-Dependent Lipid Changes

The lipid content present in the microbial plasma membrane shows special character for the survival of stress environmental condition. The phospholipids of *Pseudomonas halosaccharolytica* contain glucosyl phosphatidylglycerol, phosphatidylglycerol, diphosphatidylglycerol, and phosphatidylethanolamine which are responsible for growth under high saline condition (Li et al. 2016), and this result indicates increase of phosphatidylglycerol and reduction in phosphatidylethanolamine (Hiramatsu et al. 1980). Later, Hara and Masui (1985) observed that pulse-chase labeling of lipids with several radioactive originators showed that the rate of synthesis of phosphatidylethanolamine was inhibited by an increase in salt concentration, but the rate of phosphatidylglycerol synthesis was unaffected. The deficiency of motivation of phosphatidylglycerol creation by salt does not settle with compositional data. The radiolabeling experimentations were performed with nongrowing, starved cells, whereas the compositions of lipids were resolute directly on cells collected from culture media (Hara and Masui 1985). The inhibition of phosphatidylethanolamine creation leads to an upsurge in phosphatidylglycerol comfortable in the microbial cell because of the bifurcated phospholipid biosynthetic pathway going inside the cell (Sohlenkamp and Geiger 2016). A similar type of study was performed by Ohno et al. (1979); the little amount of NaCl did not affect the growing bacteria due to the presence of glucosyl phosphatidylglycerol. However, survival mechanisms of halophilic bacteria due to membrane lipid composition cannot judge very easily; this is a very difficult process (Oren 2008). The lots of chemicals, labor, and time may be taken to well understand the interaction between bacterial lipid membrane and salt medium (Pichler and Emmerstorfer-Augustin 2018).

11.5.5 Salt-Tolerant Genes of Bacteria

Many microorganisms contain salt-tolerant gene and perform an important character in survival for extreme environmental condition especially salt (Holmberg and Bülow 1998). The bacterial spores of *Bacillus thuringiensis israelensis, B. sphaericus,* and *B. subtilis* contain osmotolerant protein, i.e., small acid-soluble spore protein (SASP) coded by an ssp gene and this gene can survive at the high salt concentration (Cucchi and Rivas 1995). Cucchi and Rivas (1995) reported a sspE gene from *B. subtilis* and is introduced into another host bacterium *B. thuringiensis israelensis* strain 4Q2 and observed 65–650 times higher level of salt-tolerant property as compared to natural *B. thuringiensis israelensis.* In addition, this bacterium does not cause any side effects in living organisms as well as environments. Some other genes such as ectA (diaminobutyric acid acetyltransferase), ectB (diaminobutyric acid aminotransferase), and ectC (ectoine synthase) genes are reported in *B. halodurans* and showed in the survival of stress tolerance (Reshetnikov et al. 2011).

There are two genes, namely, GspM and EchM have recognized from a metagenomic collection organized from water sample of pond (Kapardar et al. 2010). GspM gene displays comparison with stress proteins, and another gene EchM showed similarity with enoyl-CoA hydratases and both genes were identified to be responsible for halotolerant at high concentration and have latent solicitation in generating halotolerant recombinant bacteria or transgenic crops (Kapardar et al. 2010). The two genes were further isolated from *Rhizobium* sp. BL3 and showed hyper-salt-tolerant ability (Payakapong et al. 2006). Hence, many microbes from rhizosphere can be exploited to isolate novel gene for salt tolerance and their potential application in the plant genetic engineering or plant growth under saline environment condition.

11.5.6 Salt-Tolerant Genes of Yeast

The two genes HAL1 and HAL3 were isolated and showed overexpressed gene from yeast (*Saccharomyces cerevisiae)* and also increased the halotolerant capability by a decreasing intracellular Na^+ and enhanced internal K^+ concentration during salt stress (Ferrando et al. 1995; Locascio et al. 2019). Further, the gene HAL1 has been introduced into tomato crop by *Agrobacterium tumefaciens*-mediated transformation which improves salt tolerance of the transgenic tomato and enhances the growth and productivity (Gisbert et al. 2000). An enzyme mitogen-activated protein kinase (MAPK) coded by a gene HOG1 shows an important role in the osmoregulatory pathway in *S. cerevisiae* (O'Rourke and Herskowitz 1998). This gene is also responsible for salt tolerance in *Torulopsis versatilis* (Wang et al. 2014). A delightful mutant strain *Torulopsis versatilis* T5 showing salt-tolerant ability was fashioned from wild-type *T. versatilis* (T) consuming genome trundling and further isolated two genes T5HOG1 and THOG1, demonstrating upturn of salt tolerance in *T. versatilis* (Cao et al. 2011). Moreover, overexpression of T5HOG1 and THOG1 enhanced the acceptance of salt in *S. cerevisiae* (Cao et al. 2011).

11.5.7 Salt-Tolerant Genes of Plants

A wide range of cruel ecological circumstances such as salinity, heat, cold, drought, and insect attack are normally exposed in plants. Plants have established altered methods being in sessile nature to survive grow and develop under speedily altering environmental conditions (Hayat et al. 2012). For these mechanisms, plants regulate genes for transcription which are known as transcriptomics under stress conditions (Shu et al. 2018). The genes for regulation of transcription play different roles under stressful environmental conditions. However, during the reproductive and seedling stages, plants have more sessile to stress and the stress response studies express novel genes or proteins with imperative roles in plant anxiety reworking during these growth stages (Verma et al. 2016a, b). However, the word salinity acceptance comes from one or more genes that reduce the uptake of the salt content from the soil and the conveyance of salt through the plant (Munns 2005, 1993).

Salinity tolerance is a very complex process that is recycled by plants to regulate (up-regulation or down-regulation) the manufacture of specific gene products in the form of RNA or proteins (Gupta and Huang 2014). This process has been accepted at different stages of central dogma technologies like from initiation of RNA processing, post-transcriptional modification, and initiation translation to post-translational modification of proteins in living organisms especially plants (Zhao et al. 2017). Understanding the transcription or translation of plants delivers thorough knowledge about the gene expression at the mRNA level. The summary of transcriptional or translational level is widely used for isolation and identification of candidate genes involved in stress responses (Xiao et al. 2017).

Transcriptome profiling is the screening processes which down-regulated or up-regulated the transcription processes that are enormous evidence about salt-tolerant genes till now. Further, a genomic method gives an important role in cloning, encoding, screening, and identifying these genes (Lodish et al. 2000). Under salt stress condition, the expression of gene is altered by transcript issues and those up- or down-regulated the expression of the gene in plants or microorganisms by these are most important switches (Lodish et al. 2000).

A gene bZIP was identified and showed up-regulation gene expression in wheat crop under insistent salt stress disorder and gene expression of down-regulation in salt-tolerant variety of wheat crop (Hayano-Kanashiro et al. 2009). The osmotic regulating and ROS-scavenging genes mostly are salt tolerance genes and also up-regulated in salinity toleant species (Amirbakhtiar et al. 2019). According to study, more than 10 genes showed up-regulated genes in halophytes plant species *Spartina alterniflora* under saline condition. Under saline condition, more than 10 genes showed up-regulated genes in *Spartina alterniflora* halophytes plant species, and most of the genes were found to osmotic regulation process among them (Bedre et al. 2016).

11.6 Mechanisms of Plant Growth Promotion of Halophilic Bacteria

11.6.1 Nitrogen Fixation Under Salt Stress Condition

At global level in arid and semi-arid regions, salinity is a serious issue for agriculture. Growth promotion and photosynthesis rate at various stages of plants affected by salinity stress (Magallon and Dinneny 2019). The production of salt-sensitive crops such as legumes is affected by salt stress particularly since these plants depend on nitrogen requirement for symbiotic N_2 fixation (Hussain et al. 2010; Kour et al. 2019b, c, d). The crop productivity mainly depends on the deprived mutual association of nodulation in bacteria and ultimately decreases in nitrogen fixation capacity (Mengel et al. 2001). *Vicia faba, Phaseolus vulgaris,* and *Glycine max* legume plants are more salt-tolerant species than another leguminous plant *Pisum sativum* (Mengel et al. 2001). *V. faba* crop fixed more nitrogen under saline condition due to the presence of rhizobia inside the root nodules and it has been seen (Mengel et al. 2001). *Prosopis, Acacia*, and *Medicago sativa* are the other salt-tolerant leguminous plants but these are less halotolerant than the leguminous plants (Joseph et al. 2015). *Rhizobium* sp. performs a very significant character in symbiosis with plants and nodulation process but, in the presence of salt, inhibits the initial process of rhizobium–legume symbiosis (Maróti and Kondorosi 2014). However, in several reports, the effect of salt stress on nodulation and nitrogen fixation of legumes have been observed (Maróti and Kondorosi 2014). In the presence of salt, the capability of N_2-fixation reduces and is documented to a decrease in the respiration of the nodules and minimize in cytosolic production protein, especially leghaemoglobin by nodulation (Zahran 1999). Saline stress negatively affected on N_2 fixation by legumes is related to the salt-induced decline directly in dry weight and nitrogen content in plant shoot (Delgado et al. 1994).

Glycine betaine is the osmoprotective substances which perform an imperative character in the maintenance of nitrogenase activity in bacteroides under salinity stress (Normand et al. 2015). The halotolerant *Rhizobium* sp. enhanced the growth, nodulation, and fixed N_2 content in *Acacia ampliceps* plant containing 200 mM NaCl concentration in the sand culture medium (Egamberdieva et al. 2013) and one more halotolerant *Rhizobium* sp. designed N_2 fixing symbiosis more effective with soybean than other salt-sensitive strain of bacteria (Egamberdieva et al. 2013). Further, the isolated rhizobial strains from *Acacia nilotica* showed tolerance to 850 mM NaCl concentration formed effective N_2-fixing nodules on *Acacia* trees grown at 150 mM NaCl (Zahran 1999). The salt-tolerant *Rhizobium* strains produce nodulation in legumes and form effective N_2 fixing symbiosis capability in the soil under moderate halophile environment observed in the result (Zahran 1999). Therefore, the booster of salt-tolerant rhizobia strains in the rhizosphere of leguminous crop can enhance the N_2 fixation ability under saline condition. However, host tolerance legume to NaCl is a very key element in influencing the achievement of harmonious *Rhizobium* strains to form symbiosis successfully under the halophilic environment (Egamberdieva et al. 2013).

11.6.2 Phytohormone Production Under Saline Condition

Phytohormones are natural organic compounds which enhance the growth and productivity of cultivars at very less concentrations. These phytohormones support the distinction and improvement of plant growth by the regulation of various progressions. Generally, the phytohormones at plants root locality are the microbial origin recommended for a functional reply in the host crop (Verma et al. 2016a, b; Enespa and Chandra 2019). Indole-3 acetic acid (IAA), gibberellic acid, abscisic acid (ABA), cytokinins, and other plant growth regulators produced by NaCl-tolerant rhizobacteria outwardly maintain the rooting with augmented number of roots, increase root length, shoot length, and number of root tips, and finally lead to increase in the uptake of nutrients and thus progress plant fitness under saline environmental circumstances (Verma et al. 2016a, b). *Bacillus* and *Pseudomonas* strains belong to IAA production that improved the growth of soybean crop at 100 mM NaCl concentration by the increasing antioxidant activity and decreasing the lipid peroxidation (Kumari et al. 2015). Furthermore, an isolated bacterium produced osmotolerant IAA displayed to increase the sprouting of rice seeds in salinity stress are reported (Jha and Subramanian 2013).

11.6.3 ACC Deaminase Production Under Saline Condition

A volatile phytohormone known as ethylene has capacity for growth promotion of plant at very less quantity like nodulations and improvement of various asexual plant parts, rooting, cuttings, and also twisted in the transduction of a signal for the appreciation of saline stress ecosystem (Saravanakumar and Samiyappan 2007). However, a large amount of ethylene is produced under abiotic environmental ecosystem and in the presence of this substrate can inhibit the root growth, shoot growth, and productivity of plants (Morgan and Drew 1997). Some chemical substrates such as aminoethoxyvinylglycine and cobalt ions act as an inhibitor of ethylene synthesis (Arora et al. 2017).

However, these chemical substrates are too much expensive and also can harm plants and environment. Halotolerant rhizobacteria showing plant growth-promoting characters contain aminocyclopropane-1-carboxylate (ACC) deaminase which splits ACC into ammonia and α-ketobutyrate, thereby reducing the near of ethylene in stressed plants (Habib et al. 2016). In the presence of ACC deaminase-producing bacteria, plant 1-aminocyclopropane-1-carboxylate is sequestrated and ruined by the cells of bacteria to fund energy and nitrogen, enhancing the plant growth under saline ecosystem (Tiwari et al. 2018).

The rhizospheric bacteria which belong to Gram-positive and Gram-negative genera such as *Arthrobacter, Bacillus, Brevibacterium, Corynebacterium, Exiguobacterium, Halomonas, Micrococcus, Oceanimonas, Planococcus,* and *Zhihengliuella* have been widely reported for ACC deaminase activity under saline conditions and have recognized as a potential role in enhancement of growth under saline ecosystem through ACC deaminase activity (Siddikee et al. 2015; Yadav et al. 2019c, d, e). *Pseudomonas simiae* strain AU5 is the mutant bacterium overproduced ACC deaminase documented to alleviate salt stress in mung bean plants as compared to wild strain *P. simiae* AU5 and observed decrease the concentration of ethylene and salt-induced membrane (bacteria and plants) damage (Kumari et al. 2016).

11.6.4 Under Salt Condition Phosphate Solubilization

Phosphorus (P) is an indispensable mineral after nitrogen for the growth of plant promotion as it and essential of dissimilar biomolecules such as nucleic acids, nucleotides, phospholipids, and phosphoproteins (Sharma et al. 2013). In the presence of salinity, uptake of P in plants is reduced and deficiency of P is appeared in the form of symptoms such as dark bluish-green in color with leaves and stem becoming purplish, etc. (Sharma et al. 2013). Mostly, insoluble forms of phosphorus in soils, i.e., organic and inorganic phosphate, have less mobilization in the soils (Sharma et al. 2013). Insoluble organic and inorganic phosphate conversion can be possible due to species of rhizobacteria and also helps in the translocation of P from soil to roots. For the solubilization of insoluble phosphates, many rhizobacteria show one of the several mechanisms such as reactions of ion-exchange, chelation, acidification, and the production organic acids of low molecular weight such as gluconic acids (Kalayu 2019; Rana et al. 2019a, b; Verma et al. 2017a). The halotolerant rhizobacteria to be vital for the mobilization of plant nutrients in several types and reduced the acceptability of inorganic fertilizers (Jiang et al. 2019).

However, phosphate solubilization is a common process in the rhizosphere by rhizobacteria that upsurge the mineral accessibility to crop (Jiang et al. 2019). An important role played by the rhizospheric bacteria to the regulation of P from less available forms and are essential for sustaining P is voluntarily available pools. Upadhyay et al. (2011) reported rhizobacterial strains to have well-organized solubilizing ability of phosphate even up to high saline (6% NaCl concentration) condition and enhanced plant growth under similar condition. For example, *Pseudomonas* inoculated in the rhizosphere of *Zea mays* crop showed salt tolerance under 6% NaCl stress condition and increased the crop growth at same salt condition (Bano and Fatima 2009). Additionally, *Herbaspirillum seropedicae* and *Burkholderia* sp. are the phosphate dissolving bacteria; treated plants recorded 1.5–21% dry weight as a compared to control plant under saline condition. Afterward, the better germination of root and shoot growth as compared with control plant after being exposed to NaCl inoculated *Azospirillum* in lettuce seeds (Carrozzi et al. 2012). *P. simiae* solubilizes phosphate by producing acid phosphatase activity along with volatile compounds that enhanced plant storage protein and uptake of P in soybean plants under 100 mM NaCl saline ecosystem (Vaishnav et al. 2015).

11.6.5 Antioxidative Response Under Salt Condition

The compounds inhibit oxidation reaction known as an antioxidant, and this is a chemical/biochemical process that can produce free radicals (Lü et al. 2010). The oxidative stress is caused by the abiotic environmental factor like drought and saline soil and resulted in the formation of reactive oxygen species (ROS) such as singlet oxygen (O_2), hydrogen peroxide (H_2O_2), and hydroxyl radical (^-OH) that damage cellular membranes, proteins, and DNA (Nita and Grzybowski 2016). When the level of ROS increases, this causes oxidative damage to biomolecules such as lipoproteins and at last leads to the death of plants (Sharma et al. 2012). However, some major antioxidative enzymes such as superoxide dismutase (SOD), peroxidase (POX), and catalase (CAT) are produced by rhizospheric bacteria such as *Streptococcus. Proteamaculans*, and *Rhizobium leguminosarum,* and non-antioxidant enzymes/compounds like ascorbic acid, tocopherols, and glutathione contribute in ROS-scavenging mechanism (Sharma et al. 2012). Mycorrhizal-inoculated lettuce plants showed higher superoxide dismutase (SOD) activity and protect the plant in the presence of antioxidant under drought stress condition (Ruiz-Lozano 2003).

Salt resistance plants have been associated to more effective antioxidant schemes, and a salt-tolerant bacterium *P. simiae* strain AU enriched antioxidants (peroxidase and catalase) and gene expression in soybean plants when treated with 100 mM NaCl stress disorder (Vaishnav et al. 2016; Chandra and Enespa 2019c). Drought stress effects in maize plants are alleviated by *Pseudomonas* spp. drought-tolerant rhizobacteria due to decrease in the antioxidant enzyme activity (Afridi et al. 2019). The catalase and peroxidase activity boosted the non-inoculated crops during saline soil, whereas *Azospirillum brasilense* inoculated plants showed lower enzyme activity and expressively ameliorated the deleterious effects of salinity (Omar et al. 2009).

11.6.6 Siderophore Production Under Salt Condition

In the chelation of micronutrients, siderophore plays an imperative character such as iron even under limiting conditions and with the redox activity it serves as a cofactor of many enzymes (Ahmed and Holmström 2014; Chandra and Enespa 2016). Several studies are reported on *Bacillus* to be a good siderophore producer (Kesaulya et al. 2018). Production of siderophores in the rhizosphere by bacteria also helps in dissolving of other ingredients, for example, P, zinc, potassium, and the availability of various ionic ingredients to the plant through chelation of iron from precipitated form (Sharma et al. 2013; Ahmed and Holmström 2014). In the soils, a huge amount of iron is existent, but in an extremely unsolvable ferric hydroxide form, hence the performances of iron as a limiting factor for promotion of plants growth even in ironic soil. However, ferrous (Fe^{++}) iron is oxidized into ferric (Fe^{+++}) form by oxidation process (Kesaulya et al. 2018). Under the biological ecosystem, the ferric ions are inexplicable which forms its achievement by microorganisms, a considerable challenge in the soils (Colombo et al. 2014). Siderophores play important roles in the development of plant growth by rhizospheric microorganisms (Ahmed and Holmström 2014). Plants and bacteria mediate competition using existence of siderophore that results in exclusions of fungal pathogens and other microbial competitors in the rhizosphere by a reduction in the availability of iron for their survival (Ahmed and Holmström 2014).

11.6.7 Halophilic Microbes as Biocontrol Agents

The production of crop yield potentially increased, and its diseases controlled biologically from rhizospheric microflora. Inhibition of phytopathogens using rhizobacteria compromises a more sustainable method to control infection as compared to harmful chemical-based methods (Compant et al. 2010; Etesami and Alikhani 2018). Under the saline condition, a halophilic microbe plays an important role in maintaining morphology, physiology, and reduction in soil salinity and also increases plant susceptibility against phytopathogens (Table 11.2) (Etesami and Beattie 2018).

Halophilic microbes use to hostage the injurious properties of plant pathogens through different mechanisms. Halophilic microbes produce one or more antimicrobial metabolites that act as antifungal, antibacterial, antiviral, antioxidant, cytotoxic,

Table 11.2 Plant growth promotion of plants by halotolerant rhizobacterial inoculation

Microbes	Host Plants	Effect of host plants under saline condition	Salt concentration	References
Bacillus pumilus ST2	*Oryza sativa*	Controlling caspase-like activity, programmed cell death, antioxidative activity	25 mM NaCl	Jha et al. (2011)
B. pumilus STR2, *Exiguobacterium oxidotolerens* STR36	*Bacopa monnieri* L.	High proline content/lipid peroxidation	4 g NaCl/Kg of soil	Bharti et al. (2013)
Burkholderia phytofirmans PsJN, *Enterobacter* sp. FD 17	*Zea Mays*	Decreasing xylem Na^+ concentration/maintain nutrient balance within the plants	25 mM NaCl	Akhtar et al. (2015)
B. pumilus STR2, *Halomonas desiderata* STR8	*Zea mays*	Preventing major shifts indigenous microbial community	50 mM NaCl	Bharti et al. (2015)
P. simiae strain AU-M4	*Glycine Max L.*	Inoculated reduced Na^+ and enhanced K^+ uptake	100 mM NaCl	Vaishnav et al. (2015)
Acinetobacter sp. ACMS25, *Bacillus* sp. PVMX4	*Phyllanthus amarus*	Improved antioxidative defense system	160 mM NaCl	Joe et al. (2016)
P. fluorescens 002	*Zea Mays*	Improved root growth and root formation under salt stress	150 mM NaCl	Zerrouk et al. (2016)
Azotobacter chroococcum AZ6	*Zea mays*	Improved chlorophyll a and total content, reduced proline and amino-acid content	20 mM NaCl	Silini et al. (2016)
Bacillus aquimaris DY-3	*Zea mays* L	Chlorophyll content, leaf relative water content, accumulation of proline, soluble sugar and total phenolic compound, and activities of superoxide dismutase, catalase, peroxidase, and ascorbate peroxidase were enhanced	1% NaCl	Li and Jiang (2017)

(continued)

Table 11.2 (continued)

Microbes	Host Plants	Effect of host plants under saline condition	Salt concentration	References
Bacillus sp., *Actinomycetes* sp., *Rhizobium* sp., *Oceanospirillum* sp.	Paddy crop	Improve rice germination, energy or germination capacity	3–12 g/L NaCl	Shi-Ying et al. (2018)
Bacillus subtilis (BERA 71)	Chickpea crop	Enhanced plant biomass and the synthesis of photosynthetic pigments and reduced the levels of reactive oxygen species (ROS) and lipid peroxidation in plants under conditions of stress.	200 mM NaCl	Abd_Allah et al. (2018)
Pseudomonas PS01	*Arabidopsis thaliana*	Improve the germination rate, transcriptional levels of genes	150 mM NaCl	Chu et al. (2019)
Enterobacter aerogenes LJL-5 and *Pseudomonas aeruginosa* LJL-13	alfalfa plants	Increased the shoot height, fresh and dry weights, yield and crude protein content	150 mM NaCl	Liu et al. (2019)

phytotoxic, and/or antitumor mediators (Olanrewaju et al. 2017). *Bacillus* and *Pseudomonas* bacterial genera secreted this type of metabolites. Halophilic microbes are also able to produce enzymes such as lipase, cellulase, β-1, 3-glucanase, chitinase, and protease which can degrade cell wall and fungal growth (Husson et al. 2017; Vaddepalli et al. 2017). Halophilic microbes compete for nutritive ingredients or for sites binding on roots of plants, and this type of antagonism reduces the growth of phytopathogen or mandatory destroyed proliferation of plant–pathogen (Olanrewaju et al. 2017). Halophilic microbes such as *Alcaligenes, Aeromonas, Bacillus, Rhizobium,* and *Pseudomonas* can produce hydrogen cyanide production, and the presence of this chemical substance may control phytopathogens (El-Rahman et al. 2019; Suman et al. 2016; Verma et al. 2018; Yadav et al. 2018c).

Halophilic microbes activate induced systemic resistance and enhance immunity against phytopathogens (Olanrewaju et al. 2017). Halophilic microbes disrupt signaling pathways of phytopathogens by quorum quenching approach. For interference of signal pathways to minimize pathogen virulence, some specific degrading enzymes, such as lactonase, are responsible (Olanrewaju et al. 2017). Halophilic microbes synthesized siderophore and inhibited the proliferation phytopathogens due to decrease in the iron availability to phytopathogens (Ahmed and Holmström 2014). The halophilic microbes provide biocontrol of phytopathogens by the production of antibiotics and antifungal metabolic substances. *Fusarium sambucinum, F. roseum* var. *sambucinum, F. oxysporum, F. moniliforme, F. graminearum, Penicillium citrinum, Aspergillus flavus*, and *Botrytis cinerea* are phytopathogenic fungi that are controlled by halophilic rhizospheric bacteria *B. subtilis, B. cereus, B. pumilus, B. licheniformis, C. alkalitolerans, Halomonas elongate*, and *Halobacillus halophilus, Halobacillus faecis, Salinicoccus roseus* (Ahmed and Holmström 2014; Olanrewaju et al. 2017; El-Rahman et al. 2019).

11.7 Role of Halophilic Microbes in Sustainable Agriculture

Chemical fertilizers and pesticides are commonly used by the farmers for improvement of soil fertility, growth, and productivity of crops under salt-based and non-salt-based ecosystem (Ju et al. 2018). But their regular use causes an adverse effect on living organism and soils (Bernardes et al. 2015). Apart from these, chemical fertilizers remediate in the crop which feed by the organisms and ultimately reach to top consumers and cause numerous diseases (Gonçalves et al. 2014). However, many transgenic salt-tolerant crops have been developed but far too little is successful (Bharti et al. 2016). An alternative method is available which could replace chemical fertilizers and pesticides and also improve soil health, seed germination, crop growth, and productivity by rhizospheric bacteria (Vejan et al. 2016). These rhizospheric bacteria enhance the growth and improvement of plants either straight or circuitously by colonizing the plant root (Vejan et al. 2016; Kour et al. 2019b; Yadav et al. 2019b).

The uninterrupted character of PGPRs involves the fixation of nitrogen (N_2) secretion of metabolites, for instance, the indole-acetic acid (IAA) production, ammonia, solubilization of phosphate, siderophore, and zinc (Ahemad and Kibret 2014; Chandra and Enespa 2016). Indirect growth promotion can be observed in the prevention and reduction of phytopathogens in plants through biocontrol mechanism. In this mechanism, PGPRs produce some lytic enzymes for fungal pathogens (cellulase, β-1, 3 glucanase, chitinase, and 1-aminocyclopropane-1-carboxylate (ACC) deaminase), reduction of iron (Fe) from the soil/rhizosphere and hydrogen cyanide (HCN), salicylic acid, antibiotics, or antifungal compounds (Odoh 2017; Chandra and Enespa 2019a, b, c). Besides, PGPRs also accepted as capable rhizobacteria that can tolerant environmental stresses such as high salt, high temperature, and pH (Ahemad and Kibret 2014).

The plant growth-promoting rhizobacteria enhance nutrient availability that includes nitrogen fixation and phosphate-solubilizing microorganisms. In indirect means, it reduces the deleterious effect of plant pathogens on crop yield (Ahemad and Kibret 2014). It shows antagonism against phytopathogenic microorganisms by producing siderophore (Vejan et al. 2016). PGPR have been developed and used as biofertilizers. Biofertilizers containing these PGPRs are economical, environment-friendly, and potentially renewable source of necessary enriched plant nutrients that makes it an excellent substitute of harmful fertilizers and chemical (Vejan et al. 2016).

The mechanism-based action can be differentiated into three dissimilar groups, i.e, (1) Biofertilizer, containing PGPR having N_2 fixation and P solubilization capability, (2) biopesticide, containing PGPR that inhibits the growth of phytopathogenic microorganisms, and (3) phytostimulator, containing PGPR that have ability to produce phytohormones (Vejan et al. 2016). Various agronomically imperative PGPR include the species, such as *Alcaligenes* sp., *Caulobacter, Serratia, Erwinia, Bacillus, Enterobacte*r, *Phyllobacterium* sp., and *Bacillus thuringiensis, Hyphomicrobium, Azotobacter, Azospirillum,* and *Acetobacter* (Sharma et al. 2013; Ahemad and Kibret 2014; Vejan et al. 2016; Kour et al. 2019a; Verma et al. 2016a, b). The PGPR used as bio-pesticides and biofertilizers for supportable farming have augmented enormously all over the world. The useful properties of PGPR on the improvement and the production of crops have been studied and reported by worldwide on a wide variety of crops such as pulses, vegetables, cereals, and oilseed crops (Gouda et al. 2018). Numerous PGPRs belonging to genera *Pseudomonas, Bacillus, Azospirillum,* and *Enterobacter* have been screened from the rhizospheric habitat of various economically important crops and were reported for their synergistic effect on plant growth promotion (Egamberdiyeva et al. 2001).

11.8 Conclusions and Future Prospects

Halophilic microbes are isolated from saline soils or rhizosphere of halophytic plants and shows plant growth-promoting characters directly like the production of IAA, solubilization of phosphate, production of siderophore, fixation of N_2, deaminase ACC activity, or indirect ways by controlling phytopathogens under saline condition. However, the habitats of halophilic microbes may be rhizosphere, endophytic, or phyllosphere, and these microbes can augment the biomass and productivity of crops using the halophytic and halotolerant crops. The inoculation of halotolerant microbes in the rhizosphere of crops is a viable strategy for eco-friendly approach and supportable improvement of crop in salt-related farming, which consist of cultivation of crops in dry and semidry regions. Several possibilities of study would move us earlier to accepting these approaches for salt-related cultivation. Knowledge of plant–microbe interactions facilitates policies for the protection of crops and saline soil remediation, and this type of interactions is also observed in the area for ecological appreciative of microbes, which promotes halophyte to adaptability in salinity-rich environment.

References

Abd_Allah EF, Alqarawi AA, Hashem A, Radhakrishnan R, Al-Huqail AA, Al-Otibi FO, Malik JA, Alharbi RI, Egamberdieva D (2018) Endophytic bacterium *Bacillus subtilis* (BERA 71) improves salt tolerance in chickpea plants by regulating the plant defense mechanisms. J Plant Inter 13(1):37–44

Afridi MS, Mahmood T, Salam A, Mukhtar T, Mehmood S, Ali J, Chaudhary HJ (2019) Induction of tolerance to salinity in wheat genotypes by plant growth promoting endophytes: Involvement of ACC deaminase and antioxidant enzymes. Plant Physiol Biochem 139:569–577

Ahemad M, Kibret M (2014) Mechanisms and applications of plant growth promoting rhizobacteria: current perspective. J King Saud Uni Sci 26(1):1–20

Ahmed E, Holmström SJ (2014) Siderophores in environmental research: roles and applications. Microb. Biotech 7(3):196–208

Akhtar SS, Andersen MN, Naveed M, Zahir ZA, Liu F (2015) Interactive effect of biochar and plant growth-promoting bacterial endophytes on ameliorating salinity stress in maize. Funct Plant Biol 42(8):770–781

Akhter MS, Hossain SJ, Amir-Hossain SK, Datta RK (2012) Isolation and characterization of salinity tolerant *Azotobacter* sp. Greener J Biol Sci 2(3):43–51

Alamri SA, Mostafa YS (2009) Effect of nitrogen supply and *Azospirillum brasilense* sp-248 on the response of wheat to seawater irrigation. Saudi J Bio Sci 16(2):101–107

Albaggar A (2014) Investigation of bacterial community composition and abundance in a lowland arable catchment (Doctoral dissertation, University of East Anglia)

Amirbakhtiar N, Ismaili A, Ghaffari MR, Firouzabadi FN, Shobbar ZS (2019) Transcriptome response of roots to salt stress in a salinity-tolerant bread wheat cultivar. PLoS One 14(3):0213305

Arora S, Singh AK, Singh YP (eds) (2017) Bioremediation of salt affected soils: an Indian perspective. Springer

Arora S, Dagar JC (2019) Salinity tolerance indicators. In: Dagar J, Yadav R, Sharma P (eds) Research developments in saline agriculture. Springer, Singapore

Aung K, Jiang Y, He SY (2018) The role of water in plant–microbe interactions. Plant J 93(4):771–780

Bang C, Dagan T, Deines P, Dubilier N, Duschl WJ, Fraune S, Hentschel U, Hirt H, Hülter N, Lachnit T, Picazo D (2018) Metaorganisms in extreme environments: do microbes play a role in organismal adaptation. Zoology 127:1–9

Bano A, Fatima M (2009) Salt tolerance in Zea mays (L). following inoculation with *Rhizobium* and *Pseudomonas*. Biol Fert Soils 45(4):405–413

Bashan Y, Moreno M, Troyo E (2000) Growth promotion of the seawater-irrigated oilseed halophyte Salicornia bigelovii inoculated with mangrove rhizosphere bacteria and halotolerant *Azospirillum* spp. Biol Fert Soils 32(4):265–272

Bedre R, Mangu VR, Srivastava S, Sanchez LE, Baisakh N (2016) Transcriptome analysis of smooth cordgrass (*Spartina alterniflora* Loisel), a monocot halophyte, reveals candidate genes involved in its adaptation to salinity. BMC Genom 17(1):657

Bernardes MF, Pazin M, Pereira LC, Dorta DJ (2015) Impact of pesticides on environmental and human health. Toxicology Studies-Cells, Drugs and Environment (*Andreazza* C y Scola G Eds.). InTech, Croacia 8:195–233

Bever JD, Platt TG, Morton ER (2012) Microbial population and community dynamics on plant roots and their feedbacks on plant communities. Ann Rev Microbiol 66:265–283

Bharti B, Kumar S, Lee HN, Kumar R (2016) Formation of oxygen vacancies and Ti^{3+} state in TiO_2 thin film and enhanced optical properties by air plasma treatment. Sci Rep 30(6):32355

Bharti N, Barnawal D, Maji D, Kalra A (2015) Halotolerant PGPRs prevent major shifts in indigenous microbial community structure under salinity stress. Microb Ecol 70(1):196–208

Bharti N, Yadav D, Barnawal D, Maji D, Kalra A (2013) *Exiguobacterium oxidotolerans*, a halotolerant plant growth promoting rhizobacteria, improves yield and content of secondary metabolites in *Bacopa monnieri* (L.) Pennell under primary and secondary salt stress. Biotech 29(2):379–387

Bianco C, Defez R (2010) Improvement of phosphate solubilization and Medicago plant yield by an indole-3-acetic acid-overproducing strain of *Sinorhizobium meliloti*. Appl Env Microb 76(14):4626–4632

Biswas J, Paul AK (2017) Diversity and production of extracellular polysaccharide by halophilic microorganisms. Biodiversity Int J 1:00006

Bringel F, Couée I (2015) Pivotal roles of phyllosphere microorganisms at the interface between plant functioning and atmospheric trace gas dynamics. Front Microb 22:6–486

Bünemann EK, Bongiorno G, Bai Z, Creamer RE, De Deyn G, de Goede R, Fleskens L, Geissen V, Kuyper TW, Mäder P, Pulleman M (2018) Soil quality–a critical review. Soil Biol Biochem 120:105–125

Canfora L, Bacci G, Pinzari F, Papa GL, Dazzi C, Benedetti A (2014) Salinity and bacterial diversity: to what extent does the concentration of salt affect the bacterial community in a saline soil. PLoS One 9(9):106662

Cao XX, Meng M, Wang YY, Wang CL, Hou LH (2011) Identification of salt-tolerant gene HOG1 in *Torulopsis versatilis*. Biotech Let 33(7):1449

Carrozzi LE, Creus CM, Barassi CA, Monterubbianesi G, Di Benedetto A (2012) Reparation of aged lettuce (*Lactuca sativa*) seeds by osmotic priming and *Azospirillum brasilense* inoculation. Botany 90(12):1093–1102

Caverzan A, Casassola A, Brammer SP (2016) Antioxidant responses of wheat plants under stress. Genet Mol Biol 39(1):1–6

Chakraborty K, Basak N, Bhaduri D, Ray S, Vijayan J, Chattopadhyay K, Sarkar RK (2018) Ionic basis of salt tolerance in plants: nutrient homeostasis and oxidative stress tolerance. In: Plant nutrients and abiotic stress tolerance. Springer, Singapore, pp 325–362

Chandra P, Barsainya M, Singh DP (2014) A fourier transform infrared (FTIR) spectroscopic study on cellular changes in the *Marinococcus Luteus* sslb 1 under different salinity regime. Int J Pharm Bio Sci 5:848–854

Chandra P, Enespa (2016) Applications and mechanisms of plant growth-stimulating rhizobacteria. In: Plant-microbe interaction: an approach to sustainable agriculture. Springer, Singapore, pp 37–62

Chandra P, Enespa (2017) Microbial volatiles as chemical weapons against pathogenic fungi. In: Volatiles and food security. Springer, Singapore, pp 227–254

Chandra P, Enespa (2019a) Mycoremediation of environmental pollutants from contaminated soil. In: Varma A, Choudhary D (eds) Mycorrhizosphere and pedogenesis. Springer, Singapore

Chandra P, Enespa (2019b) Fungal enzymes for bioremediation of contaminated soil. In: Yadav A, Singh S, Mishra S, Gupta A (eds) Recent advancement in white biotechnology through fungi. Fungal Biology. Springer, Cham

Chandra P, Enespa (2019c) Soil–Microbes–Plants: interactions and ecological diversity. In: Varma A, Tripathi S, Prasad R (eds) Plant microbe interface. Springer, Cham

Chandra P, Enespa, Kumar M (2020) Contribution of microbes in the renovation of wetlands. In: Upadhyay A, Singh R, Singh D (eds) Restoration of wetland ecosystem: a trajectory towards a sustainable environment. Springer, Singapore

Chandra P, Singh DP (2014) Removal of Cr (VI) by a halotolerant bacterium *Halomonas* sp. CSB 5 isolated from Sambhar salt Lake Rajasthan (India). Cell Mol Biol 60:64–72

Chandra P, Singh DP (2016) Isolation of alkaliphilic bacterium *Citricoccus alkalitolerans* CSB1: an efficient biosorbent for bioremediation of tannery waste water. Cell Mol Biol 62(3):135

Chandra P, Singh E (2017) Applications and mechanisms of plant growth-stimulating rhizobacteria. In: Choudhary D, Varma A, Tuteja N (eds) Plant-microbe interaction: an approach to sustainable agriculture. Springer, Singapore

Choudhury FK, Rivero RM, Blumwald E, Mittler R (2017) Reactive oxygen species, abiotic stress and stress combination. Plant J 90(5):856–867

Chu TN, Tran BT, Van Bui L, Hoang MT (2019) Plant growth-promoting rhizobacterium *Pseudomonas* PS01 induces salt tolerance in *Arabidopsis thaliana*. BMC Res Notes 12(1):11

Colombo C, Palumbo G, He JZ, Pinton R, Cesco S (2014) Review on iron availability in soil: interaction of Fe minerals, plants, and microbes. J Soils Sediments 14(3):538–548

Compant S, Clément C, Sessitsch A (2010) Plant growth-promoting bacteria in the rhizo-and endosphere of plants: their role, colonization, mechanisms involved and prospects for utilization. Soil Biol Biochem 42(5):669–678

Csonka LN (1989) Physiological and genetic responses of bacteria to osmotic stress. Microbiol Mol Biol Rev 53(1):121–147

Cucchi A, de Rivas CS (1995) ssp genes and spore osmotolerance in *Bacillus thuringiensis israelensis* and *Bacillus sphaericus*. Curr Microbiol 31(4):228–233

Cuevas J, Daliakopoulos IN, del Moral F, Hueso JJ, Tsanis IK (2019) A review of soil-improving cropping systems for soil salinization. Agronomy 9(6):295

Czech L, Höppner A, Kobus S, Seubert A, Riclea R, Dickschat JS, Bremer E (2019) Illuminating the catalytic core of ectoine synthase through structural and biochemical analysis. Sci Rep 9(1):364

da Costa MS, Santos H (2009) Compatible solutes in microorganisms that grow at high temperature. Extremophile 5(3):265

da Costa MS, Santos H, Galinski EA (1998a) An overview of the role and diversity of compatible solutes in Bacteria and Archaea. In: Antranikian G (eds) Biotechnology of extremophiles. Advances in biochemical engineering/biotechnology, vol 61. Springer, Berlin

da Costa MS, Santos H, Galinski EA (1998b) An overview of the role and diversity of compatible solutes in Bacteria and Archaea. In: Biotechnology of extremophiles. Springer, Berlin, pp 117–153

de Lourdes Moreno M, Pérez D, García M, Mellado E (2013) Halophilic bacteria as a source of novel hydrolytic enzymes. Life 3(1):38–51

Delgado MJ, Ligero F, Lluch C (1994) Effects of salt stress on growth and nitrogen fixation by pea, faba-bean, common bean and soybean plants. Soil Biol Biochem 26(3):371–376

Egamberdieva D, Berg G, Lindström K, Räsänen LA (2013) Alleviation of salt stress of symbiotic *Galega officinalis* L. (goat's rue) by co-inoculation of *Rhizobium* with root-colonizing *Pseudomonas*. Plant Soil 369:453–465

Egamberdieva D, Davranov K, Wirth S, Hashem A, Abd_Allah EF (2017) Impact of soil salinity on the plant-growth–promoting and biological control abilities of root associated bacteria. Saudi J Biolog Sci 24(7):1601–1608

Egamberdiyeva D, Davranov KD, Höflich G (2001) Influence of growth-promoting bacteria from Uzbekistan and Germany on the growth and nutrient uptake of cotton and wheat on different soils. Plant nutrition. Springer, Dordrecht, pp 674–675

El-Esawi MA, Al-Ghamdi AA, Ali HM, Alayafi AA (2019) *Azospirillum lipoferum* FK1 confers improved salt tolerance in chickpea (*Cicer arietinum* L.) by modulating osmolytes, antioxidant machinery and stress-related genes expression. Environ Exp Bot 159:55–65

El-Rahman AA, Shaheen HA, El-Aziz RM, Ibrahim DS (2019) Influence of hydrogen cyanide-producing rhizobacteria in controlling the crown gall and root-knot nematode, *Meloidogyne incognita*. Egyptian J Biolog Pest Control 29(1):41

Enespa, Chandra P (2019) Fungal community for novel secondary metabolites. In: Yadav A, Singh S, Mishra S, Gupta A (eds) Recent advancement in white biotechnology through fungi. Fungal biology. Springer, Cham

Etesami H, Alikhani HA (2018) *Bacillus* species as the most promising bacterial biocontrol agents in rhizosphere and endorhiza of plants grown in rotation with each other. Eur J Plant Pathol 150(2):497–506

Etesami H, Beattie GA (2018) Mining halophytes for plant growth-promoting halotolerant bacteria to enhance the salinity tolerance of non-halophytic crops. Front Microbiol 8(9):148

Fendrihan S, Constantinescu F, Sicuia O, Dinu S (2017) *Azospirillum* strains as biofertilizers and biocontrol agents-a practical review. J Adv Agric 7(3):1096–1108

Ferrando A, Kron SJ, Rios G, Fink GR, Serrano R (1995) Regulation of cation transport in *Saccharomyces cerevisiae* by the salt tolerance gene HAL3. Mol Cellular Biol 15(10):5470–5481

Galinski EA (1993) Compatible solutes of halophilic eubacteria: molecular principles, water-solute interaction, stress protection. Experientia 49(6–7):487–496

Getu M (2009) Ethiopian floriculture and its impact on the environment. Mizan Law Rev 3(2):240–270
Gisbert C, Rus AM, Bolarín MC, López-Coronado JM, Arrillaga I, Montesinos C, Caro M, Serrano R, Moreno V (2000) The yeast HAL1 gene improves salt tolerance of transgenic tomato. Plant Physiol 123(1):393–402
Glick BR (2014) Bacteria with ACC deaminase can promote plant growth and help to feed the world. Microb Res 169(1):30–39
Gonçalves AC Jr, Nacke H, Schwantes D, Coelho GF (2014) Heavy metal contamination in brazilian agricultural soils due to application of fertilizers. Environ Risk Assess Soil Contam 26(1):105–135
Gouda S, Kerry RG, Das G, Paramithiotis S, Shin HS, Patra JK (2018) Revitalization of plant growth promoting rhizobacteria for sustainable development in agriculture. Microb Res 206:131–140
Gouffi K, Pichereau V, Rolland JP, Thomas D, Bernard T, Blanco C (1998) Sucrose Is a Nonaccumulated Osmoprotectant inSinorhizobium meliloti. J Bacteriol 180(19):5044–5051
Greene R, Timms W, Rengasamy P, Arshad M, Cresswell R (2016) Soil and aquifer salinization: toward an integrated approach for salinity management of groundwater. Integrated groundwater management. Springer, Cham, pp 377–412
Greene RS, Hairsine PB (2004) Elementary processes of soil–water interaction and thresholds in soil surface dynamics: a review. Earth Surf Process Landf J Br Geomorphol Res Group 29(9):1077–1091
Gu H, Ma C, Gu J, Guo J, Yan X, Huang J, Guo Z (2016) An overview of multifunctional epoxy nanocomposites. J Mat Chem 4(25):5890–5906
Gunde-Cimerman N, Plemenitaš A, Oren A (2018) Strategies of adaptation of microorganisms of the three domains of life to high salt concentrations. FEMS Microbiol Rev 42(3):353–375
Gupta B, Huang B (2014) Mechanism of salinity tolerance in plants: physiological, biochemical, and molecular characterization. Int J Genomic 2014:1–18
Habib SH, Kausar H, Saud HM (2016) Plant growth-promoting rhizobacteria enhance salinity stress tolerance in okra through ROS-scavenging enzymes. Biomed Res Int 2016:1–10
Hanin M, Ebel C, Ngom M, Laplaze L, Masmoudi K (2016) New insights on plant salt tolerance mechanisms and their potential use for breeding. Front Plant Sci 7:1787
Hara H, Masui M (1985) Effect of NaCl concentration on the synthesis of membrane phospholipid in a halophilic bacterium. FEMS Microb Ecol 1(5):279–282
Hayano-Kanashiro C, Calderón-Vázquez C, Ibarra-Laclette E, Herrera-Estrella L, Simpson J (2009) Analysis of gene expression and physiological responses in three Mexican maize landraces under drought stress and recovery irrigation. PLoS One 4(10):e7531
Hayat S, Hayat Q, Alyemeni MN, Wani AS, Pichtel J, Ahmad A (2012) Role of proline under changing environments: a review. Plant Signal Behav 7(11):1456–1466
Hiramatsu T, Yano I, Masui M (1980) Effect of NaCl concentration on the protein species and phospholipid composition of the outer membrane in a moderately halophilic bacterium. FEMS Microbiol Lett 7(4):289–292
Holmberg N, Bülow L (1998) Improving stress tolerance in plants by gene transfer. Trend Plant Sci 3(2):61–66
Hussain N, Sarwar G, Schmeisky H, Al-Rawahy S, Ahmad M (2010) Salinity and drought management in legume crops. Climate change and management of cool season grain legume crops. Springer, Dordrecht, pp 171–191
Husson F, Lê S, Pagès J (2017) Exploratory multivariate analysis by example using R. Chapman and Hall/CRC, p 8
Ibekwe AM, Poss JA, Grattan SR, Grieve CM, Suarez D (2010) Bacterial diversity in cucumber (*Cucumis sativus*) rhizosphere in response to salinity, soil pH, and boron. Soil Biol Biochem 42(4):567–575
Ilangumaran G, Smith DL (2017) Plant growth promoting rhizobacteria in amelioration of salinity stress: a systems biology perspective. Front Plant Sci 23(8):1768
Inaba M, Sakamoto A, Murata N (2001) Functional expression in *Escherichia coli* of low-affinity and high-affinity Na^+ $(Li^+)/H^+$ antiporters of *Synechocystis*. J Bacteriol 183(4):1376–1384

Indoria AK, Rao CS, Sharma KL, Reddy KS (2017) Conservation agriculture–a panacea to improve soil physical health. Curr Sci 112:1–10

Jha Y, Subramanian RB (2013) Paddy plants inoculated with PGPR show better growth physiology and nutrient content under saline condition. Chil J Agric Res 73(3):213–219

Jha Y, Subramanian RB, Patel S (2011) Combination of endophytic and rhizospheric plant growth promoting rhizobacteria in *Oryza sativa* shows higher accumulation of osmoprotectant against saline stress. Acta Physiol Plant 33(3):797–802

Jiang H, Qi P, Wang T, Chi X, Wang M, Chen M, Chen N, Pan L (2019) Role of halotolerant phosphate-solubilizing bacteria on growth promotion of peanut (Arachis hypogaea) under saline soil. Ann App Biol 174(1):20–30

Joe MM, Devaraj S, Benson A, Sa T (2016) Isolation of phosphate solubilizing endophytic bacteria from *Phyllanthus amarus* Schum and Thonn: evaluation of plant growth promotion and antioxidant activity under salt stress. J App Res Med Arom Plant 3(2):71–77

Joseph B, Jini D (2010) Insight into the role of antioxidant enzymes for salt tolerance in plants. Int J Bot 6(4):456–464

Joseph S, Murphy DJ, Bhave M (2015) Identification of salt tolerant *Acacia* species for saline land utilisation. Biologia 70(2):174–182

Ju I, Wj B, Md S, Ia O, Oj E (2018) A review: biofertilizer-a key player in enhancing soil fertility and crop productivity. J Microbiol Biotechnol Rep 7:2

Kalayu G (2019) Phosphate solubilizing microorganisms: promising approach as biofertilizers. Int J Agron 2019:1–7

Kapardar RK, Ranjan R, Grover A, Puri M, Sharma R (2010) Identification and characterization of genes conferring salt tolerance to *Escherichia coli* from pond water metagenome. Bioresour Technol 101(11):3917–3924

Kempf B, Bremer E (1998) Uptake and synthesis of compatible solutes as microbial stress responses to high-osmolality environments. Arch Microbiol 170(5):319–330

Kesaulya H, Hasinu JV, Tuhumury GN (2018) Potential of Bacillus spp produces siderophores insuppressing thewilt disease of banana plants. In: IOP Conference Series: Earth and Environmental Science, vol 102, p 012016

Kindzierski V, Raschke S, Knabe N, Siedler F, Scheffer B, Pflüger-Grau K, Kunte HJ (2017) Osmoregulation in the halophilic bacterium *Halomonas elongata*: a case study for integrative systems biology. PLoS One 12(1):e0168818

Klähn S, Marquardt DM, Rollwitz I, Hagemann M (2009) Expression of the ggpPS gene for glucosylglycerol biosynthesis from *Azotobacter vinelandii* improves the salt tolerance of *Arabidopsis thaliana*. J Exp Bot 60(6):1679–1689

Kohler C, Lourenço RF, Bernhardt J, Albrecht D, Schüler J, Hecker M, Gomes SL (2015) A comprehensive genomic, transcriptomic and proteomic analysis of a hyperosmotic stress sensitive α-proteobacterium. BMC Microbiol 15(1):71

Kooistra MJ, Tovey NK (1994) Effects of compaction on soil microstructure. Dev Agric Eng 11:91–111

Kour D, Rana KL, Kumar A, Rastegari AA, Yadav N, Yadav AN, Gupta VK (2019a) Extremophiles for hydrolytic enzymes productions: biodiversity and potential biotechnological applications. In: Molina G, Gupta VK, Singh BN, Gathergood N (eds) Bioprocessing for biomolecules production. Wiley, USA, pp 321–372

Kour D, Rana KL, Yadav AN, Yadav N, Kumar V, Kumar A, Sayyed RZ, Hesham AE-L, Dhaliwal HS, Saxena AK (2019b) Drought-tolerant phosphorus-solubilizing microbes: biodiversity and biotechnological applications for alleviation of drought stress in plants. In: Sayyed RZ, Arora NK, Reddy MS (eds) Plant growth promoting rhizobacteria for sustainable stress management: volume 1: rhizobacteria in abiotic stress management. Springer Singapore, Singapore, pp 255–308. https://doi.org/10.1007/978-981-13-6536-2_13

Kour D, Rana KL, Yadav N, Yadav AN, Kumar A, Meena VS, Singh B, Chauhan VS, Dhaliwal HS, Saxena AK (2019c) Rhizospheric microbiomes: biodiversity, mechanisms of plant growth promotion, and biotechnological applications for sustainable agriculture. In: Kumar A, Meena VS (eds)

Plant growth promoting rhizobacteria for agricultural sustainability: from theory to practices. Springer Singapore, Singapore, pp 19–65. https://doi.org/10.1007/978-981-13-7553-8_2
Kour D, Rana KL, Yadav N, Yadav AN, Singh J, Rastegari AA, Saxena AK (2019d) Agriculturally and industrially important fungi: current developments and potential biotechnological applications. In: Yadav AN, Singh S, Mishra S, Gupta A (eds) Recent advancement in white biotechnology through fungi, volume 2: perspective for value-added products and environments. Springer International Publishing, Cham, pp 1–64. https://doi.org/10.1007/978-3-030-14846-1_1
Kraegeloh A, Amendt B, Kunte HJ (2005) Potassium transport in a halophilic member of the bacteria domain: identification and characterization of the K^+ uptake systems TrkH and TrkI from *Halomonas elongata* DSM 2581T. J Bacteriol 187(3):1036–1043
Kumari S, Vaishnav A, Jain S, Varma A, Choudhary DK (2015) Bacterial-mediated induction of systemic tolerance to salinity with expression of stress alleviating enzymes in soybean (*Glycine* max L. Merrill). J Plant Growth Regul 34(3):558–573
Kumari S, Vaishnav A, Jain S, Varma A, Choudhary DK (2016) Title. World J Microbiol. World J Microbiol Biotech 32(1):1–10
Ladeiro B (2012) Saline agriculture in the 21st century: using salt contaminated resources to cope food requirements. J Bot 310705–310707
Laloknam S, Tanaka K, Buaboocha T, Waditee R, Incharoensakdi A, Hibino T, Takabe T (2006) Halotolerant cyanobacterium *Aphanothece halophytica* contains a betaine transporter active at alkaline pH and high salinity. Appl. Environ. Microbiol. 72(9):6018–6026
Leite J, Fischer D, Rouws LF, Fernandes-Júnior PI, Hofmann A, Kublik S, Schloter M, Xavier GR, Radl V (2017) Cowpea nodules harbor non-rhizobial bacterial communities that are shaped by soil type rather than plant genotype. Front Plant Sci 20(7):2064
Lentes CJ, Mir SH, Boehm M, Ganea C, Fendler K, Hunte C (2014) Molecular characterization of the Na^+/H^+-antiporter NhaA from *Salmonella typhimurium*. PLoS One 9(7):e101575
Li C, Tan BK, Zhao J, Guan Z (2016) In vivo and in vitro synthesis of phosphatidylglycerol by an *Escherichia coli* cardiolipin synthase. J Biol Chem 291(48):25144–25153
Li HQ, Jiang XW (2017) Inoculation with plant growth-promoting bacteria (PGPB) improves salt tolerance of maize seedling. Russ J Plant Physiol 64(2):235–241
Liu J, Tang L, Gao H, Zhang M, Guo C (2019) Enhancement of alfalfa yield and quality by plant growth-promoting rhizobacteria under saline-alkali conditions. J Sci Food Agric 99(1):281–289
Locascio A, Andrés-Colás N, Mulet JM, Yenush L (2019) *Saccharomyces cerevisiae* as a tool to investigate plant potassium and sodium transporters. Int J Mol Sci 20(9):2133
Lodish H, Berk A, Zipursky SL, Matsudaira P, Baltimore D, Darnell J (2000) Molecular cell biology 4th edition. National Center for Biotechnology Information, Bookshelf
Loganathan P, Nair S (2004) *Swaminathania salitolerans* gen. nov., sp. nov., a salt-tolerant, nitrogen-fixing and phosphate-solubilizing bacterium from wild rice (*Porteresia coarctata* Tateoka). Int J Syst Evol Microbiol 54(4):1185–1190
Long X, Tian J, Liao X, Tian Y (2018) Adaptations of *Bacillus shacheensis* HNA-14 required for long-term survival under osmotic challenge: a multi-omics perspective. RSC Adv 8(48):27525–27536
Lü JM, Lin PH, Yao Q, Chen C (2010) Chemical and molecular mechanisms of antioxidants: experimental approaches and model systems. J Cell Mol Med 14(4):840–860
Ma Y, Galinski EA, Grant WD, Oren A, Ventosa A (2010) Halophiles 2010: life in saline environments. Appl Environ Microbiol 76(21):6971–6981
Machado R, Serralheiro R (2017) Soil salinity: effect on vegetable crop growth. Management practices to prevent and mitigate soil salinization. Horticulturae 3(2):30
Magallon KJ, Dinneny JR (2019) Environmental stress: salinity Ruins a Plant's Day in the Sun. Curr Biol 29(10):360–362
Mahrous M, Šegvić B, Zanoni G, Khadka S, Senadheera S, Jayawickrama P (2018) The role of clay swelling and mineral neoformation in the stabilization of high plasticity soils treated with the fly ash-and metakaolin-based geopolymers. Mineral 8(4):146

Mandon K, Østerås M, Boncompagni E, Trinchant JC, Spennato G, Poggi MC, Le Rudulier D (2003) The *Sinorhizobium meliloti* glycine betaine biosynthetic genes (betICBA) are induced by choline and highly expressed in bacteroids. Mol Plant-Microbe Interact 16(8):709–719
Maróti G, Kondorosi Á (2014) Nitrogen-fixing Rhizobium-legume symbiosis: are polyploidy and host peptide-governed symbiont differentiation general principles of endosymbiosis? Front Microbiol 5. https://doi.org/10.3389/fmicb.2014.00326
Maturrano L, Santos F, Rosselló-Mora R, Antón J (2006) Microbial diversity in Maras salterns, a hypersaline environment in the Peruvian Andes. Appl Env Microbiol 72(6):3887–3895
Mendes R, Garbeva P, Raaijmakers JM (2013) The rhizosphere microbiome: significance of plant beneficial, plant pathogenic, and human pathogenic microorganisms. FEMS Microbiol Rev 37(5):634–663
Mengel K, Kirkby EA, Kosegarten H, Appel T (2001) Nitrogen. Principles of plant nutrition. Springer, Dordrecht, pp 397–434
Mitschke J, Georg J, Scholz I, Sharma CM, Dienst D, Bantscheff J, Hess WR (2011) An experimentally anchored map of transcriptional start sites in the model cyanobacterium *Synechocystis* sp. PCC6803. Proc Natl Acad Sci 108(5):2124–2129
Morgan PW, Drew MC (1997) Ethylene and plant responses to stress. Physiol Plant 100(3):620–630
Mukhtar S, Ishaq A, Hassan S, Mehnaz S, Mirza MS, Malik KA (2017) Comparison of microbial communities associated with halophyte (*Salsola stocksii*) and non-halophyte (*Triticum aestivum*) using culture-independent approaches. Pol J Microbiol 66(3):353–364
Munns R (1993) Physiological processes limiting plant growth in saline soils: some dogmas and hypotheses. Plant Cell Environ 16(1):15–24
Munns R (2005) Genes and salt tolerance: bringing them together. New Phytol 167(3):645–663
Murínová S, Dercová K (2014) Response mechanisms of bacterial degraders to environmental contaminants on the level of cell walls and cytoplasmic membrane. Int J Microbiol 2014:1–16
Nabti E, Schmid M, Hartmann A (2015) Application of halotolerant bacteria to restore plant growth under salt stress. Halophiles. Springer, Cham, pp 235–259
Nita M, Grzybowski A (2016) The role of the reactive oxygen species and oxidative stress in the pathomechanism of the age-related ocular diseases and other pathologies of the anterior and posterior eye segments in adults. Oxidative Med Cell Longev 2016:1–23
Nogales J, Campos R, BenAbdelkhalek H, Olivares J, Lluch C, Sanjuan J (2002) *Rhizobium tropici* genes involved in free-living salt tolerance are required for the establishment of efficient nitrogen-fixing symbiosis with *Phaseolus vulgaris*. Mol Plant-Microbe Interact 15(3):225–232
Normand P, Caumette P, Goulas P, Pujic P, Wisniewski-Dyé F (2015) Adaptations of prokaryotes to their biotopes and to physicochemical conditions in natural or anthropized environments. Environmental microbiology: fundamentals and applications. Springer, Dordrecht, pp 293–351
Numan M, Bashir S, Khan Y, Mumtaz R, Shinwari ZK, Khan AL, Khan A, Ahmed AH (2018) Plant growth promoting bacteria as an alternative strategy for salt tolerance in plants: a review. Microbiol Res 209:21–32
O'Rourke SM, Herskowitz I (1998) The Hog1 MAPK prevents cross talk between the HOG and pheromone response MAPK pathways in *Saccharomyces cerevisiae*. Genes Dev 12(18):2874–2886
Odoh CK (2017) Plant growth promoting rhizobacteria (PGPR): a bio protectant bio inoculant for sustainable agrobiology. A review. Int J Adv Res Biol Sci 4:123–142
Ohno Y, Yano I, Masui M (1979) Effect of NaCl concentration and temperature on the phospholipid and fatty acid compositions of a moderately halophilic bacterium, *Pseudomonas halosaccharolytica*. J Biochem 85(2):413–421
Olanrewaju OS, Glick BR, Babalola OO (2017) Mechanisms of action of plant growth promoting bacteria. World J Microbiol Biotech 33(11):197
Ollivier B, Caumette P, Garcia JL, Mah RA (1994) Anaerobic bacteria from hypersaline environments. Microbiol Mol Biol Rev 58(1):27–38

Omar MN, Osman ME, Kasim WA, El-Daim IA (2009) Improvement of salt tolerance mechanisms of barley cultivated under salt stress using *Azospirillum brasilense*. Salinity and water stress. Springer, Dordrecht, pp 133–147

Ondrasek G, Rengel Z, Veres S (2011) Soil salinisation and salt stress in crop production. Abiotic stress in plants-Mech Adapt. https://doi.org/10.5772/22248

Or D, Smets BF, Wraith JM, Dechesne A, Friedman SP (2007) Physical constraints affecting bacterial habitats and activity in unsaturated porous media-a review. Adv Water Resour 30(6–7):1505–1527

Oren A (2002) Diversity of halophilic microorganisms: environments, phylogeny, physiology, and applications. J Ind Microbiol Biotechl 28(1):56–63

Oren A (2008) Microbial life at high salt concentrations: phylogenetic and metabolic diversity. Sal Syst 4(1):2

Oren A (2011) Thermodynamic limits to microbial life at high salt concentrations. Environ Microbiol 13(8):1908–1923

Osteras M, Boncompagni E, Vincent N, Poggi MC, Le Rudulier D (1998) Presence of a gene encoding choline sulfatase in *Sinorhizobium meliloti* bet operon: choline-O-sulfate is metabolized into glycine betaine. Proc Natl Acad Sci 95(19):11394–11399

Patchett RA, Kelly AF, Kroll RG (1992) Effect of sodium chloride on the intracellular solute pools of *Listeria monocytogenes*. Appl Environ Microbiol 58(12):3959–3963

Payakapong W, Tittabutr P, Teaumroong N, Boonkerd N, Singleton PW, Borthakur D (2006) Identification of two clusters of genes involved in salt tolerance in *Sinorhizobium* sp. strain BL3. Symbiosis 41(1):47–53

Peng J, Wu D, Liang Y, Li L, Guo Y (2019) Disruption of acdS gene reduces plant growth promotion activity and maize saline stress resistance by *Rahnella aquatilis* HX2. J B Microbiol 59(4):402–411

Pérez-Arellano I, Carmona-Álvarez F, Martínez AI, Rodríguez-Díaz J, Cervera J (2010) Pyrroline-5-carboxylate synthase and proline biosynthesis: from osmotolerance to rare metabolic disease. Protein Sci 19(3):372–382

Pichler H, Emmerstorfer-Augustin A (2018) Modification of membrane lipid compositions in single-celled organisms-from basics to applications. Method 147:50–65

Pikuta EV, Hoover RB, Tang J (2007) Microbial extremophiles at the limits of life. Crit Rev Microbiol 33(3):183–209

Poeplau C, Helfrich M, Dechow R, Szoboszlay M, Tebbe CC, Don A, Geerts R (2019) Increased microbial anabolism contributes to soil carbon sequestration by mineral fertilization in temperate grasslands. Soil Biol Biochem 130:167–176

Poolman B, Glaasker E (1998) Regulation of compatible solute accumulation in bacteria. Mol Microbiol 29(2):397–407

Rana KL, Kour D, Sheikh I, Dhiman A, Yadav N, Yadav AN, Rastegari AA, Singh K, Saxena AK (2019a) Endophytic fungi: biodiversity, ecological significance and potential industrial applications. In: Yadav AN, Mishra S, Singh S, Gupta A (eds) Recent advancement in white biotechnology through fungi, vol 1. Diversity and enzymes perspectives. Springer, Switzerland, pp 1–62

Rana KL, Kour D, Sheikh I, Yadav N, Yadav AN, Kumar V, Singh BP, Dhaliwal HS, Saxena AK (2019b) Biodiversity of endophytic fungi from diverse niches and their biotechnological applications. In: Singh BP (ed) Advances in endophytic fungal research: present status and future challenges. Springer International Publishing, Cham, pp 105–144. https://doi.org/10.1007/978-3-030-03589-1_6

Rashid MI, Mujawar LH, Shahzad T, Almeelbi T, Ismail IM, Oves M (2016) Bacteria and fungi can contribute to nutrients bioavailability and aggregate formation in degraded soils. Microbiol Res 183:26–41

Reshetnikov AS, Khmelenina VN, Mustakhimov II, Kalyuzhnaya M, Lidstrom M, Trotsenko YA (2011) Diversity and phylogeny of the ectoine biosynthesis genes in aerobic, moderately halophilic methylotrophic bacteria. Extremophile 15(6):653

Reshetnikov AS, Khmelenina VN, Trotsenko YA (2006) Characterization of the ectoine biosynthesis genes of haloalkalotolerant obligate methanotroph "*Methylomicrobium alcaliphilum* 20Z". Arch Microbiol 184(5):286–297

Roberts MF (2005a) Organic compatible solutes of halotolerant and halophilic microorganisms. Saline Syst 1(1):5–30

Roberts MF (2005b) Organic compatible solutes of halotolerant and halophilic microorganisms. Saline Syst 1(1):5

Ruiz-Lozano JM (2003) Arbuscular mycorrhizal symbiosis and alleviation of osmotic stress. New perspectives for molecular studies. Mycorrhiza 13(6):309–317

Sahoo RK, Ansari MW, Pradhan M, Dangar TK, Mohanty S, Tuteja N (2014) A novel *Azotobacter vinellandii* (SRI Az 3) functions in salinity stress tolerance in rice. Plant Signal Behav 9(7):511–523

Sánchez-Porro C, Martin S, Mellado E, Ventosa A (2003) Diversity of moderately halophilic bacteria producing extracellular hydrolytic enzymes. J Appl Microbiol 94(2):295–300

Saravanakumar D, Samiyappan R (2007) ACC deaminase from Pseudomonas fluorescens mediated saline resistance in groundnut (*Arachis hypogea*) plants. J Appl Microbiol 102(5):1283–1292

Saum SH, Müller V (2008) Regulation of osmoadaptation in the moderate halophile Halobacillus halophilus: chloride, glutamate and switching osmolyte strategies. Sal Syst 4(1):4

Schjønning P, Munkholm LJ, Moldrup P, Jacobsen OH (2002) Modelling soil pore characteristics from measurements of air exchange: the long-term effects of fertilization and crop rotation. Euro J Soil Sci 53(2):331–339

Schmidt RR, Weits DA, Feulner CF, van Dongen JT (2018) Oxygen sensing and integrative stress signaling in plants. Plant Physiol 176(2):1131–1142

Sharma P, Jha AB, Dubey RS, Pessarakli M (2012) Reactive oxygen species, oxidative damage, and antioxidative defense mechanism in plants under stressful conditions. J Bot 2012:1–26

Sharma SB, Sayyed RZ, Trivedi MH, Gobi TA (2013) Phosphate solubilizing microbes: sustainable approach for managing phosphorus deficiency in agricultural soils. SpringerPlus 2(1):587

Shi-Ying Z, Cong F, Yong-xia W, Yun-sheng X, Wei X, Xiao-Long C (2018) Salt-tolerant and plant growth-promoting bacteria isolated from high-yield paddy soil. Canad J Microbiol 64(12):968–978

Shrivastava P, Kumar R (2015) Soil salinity: a serious environmental issue and plant growth promoting bacteria as one of the tools for its alleviation. Saudi J Biol Sci 22(2):123–131

Shu Y, Li W, Zhao J, Liu Y, Guo C (2018) Transcriptome sequencing and expression profiling of genes involved in the response to abiotic stress in *Medicago ruthenica*. Genet Mol Biol 41(3):638–648

Siddikee MA, Sundaram S, Chandrasekaran M, Kim K, Selvakumar G, Sa T (2015) Halotolerant bacteria with ACC deaminase activity alleviates salt stress effect in canola seed germination. J Korean Soc Appl Biol Chem 58(2):237–241

Silini A, Cherif-Silini H, Yahiaoui B (2016) Growing varieties durum wheat (*Triticum durum*) in response to the effect of osmolytes and inoculation by *Azotobacter chroococcum* under salt stress. Afr J Microbiol Res 10(12):387–399

Singh R, Upadhyay AK, Chandra P, Singh DP (2018) Sodium chloride incites reactive oxygen species in green algae *Chlorococcum humicola* and *Chlorella vulgaris*: Implication on lipid synthesis, mineral nutrients and antioxidant system. Bioresour Technol 270:489–497

Singh RP, Jha P, Jha PN (2015) The plant-growth-promoting bacterium *Klebsiella* sp. SBP-8 confers induced systemic tolerance in wheat (*Triticum aestivum*) under salt stress. J Plant Physiol 184:57–67

Sohlenkamp C, Geiger O (2016) Bacterial membrane lipids: diversity in structures and pathways. FEMS Microbiol Rev 40(1):133–159

Soto-Padilla MY, Valenzuela-Encinas C, Dendooven L, Marsch R, Gortáres-Moroyoqui P, Estrada-Alvarado MI (2014) Isolation and phylogenic identification of soil haloalkaliphilic strains in the former Texcoco Lake. Int J Env Health Res 24(1):82–90

Stöveken N, Pittelkow M, Sinner T, Jensen RA, Heider J, Bremer E (2011) A specialized aspartokinase enhances the biosynthesis of the osmoprotectants ectoine and hydroxyectoine in *Pseudomonas stutzeri* A1501. J Bacteriol 193(17):4456–4468

Stritzler M, Elba P, Berini C, Gomez C, Ayub N, Soto G (2018) High-quality forage production under salinity by using a salt-tolerant AtNXH1-expressing transgenic alfalfa combined with a natural stress-resistant nitrogen-fixing bacterium. J Biotech 276:42–45

Suman A, Yadav AN, Verma P (2016) Endophytic microbes in crops: diversity and beneficial impact for sustainable agriculture. In: Singh D, Abhilash P, Prabha R (eds) Microbial inoculants in sustainable agricultural productivity, research perspectives. Springer, India, pp 117–143. https://doi.org/10.1007/978-81-322-2647-5_7

Suárez R, Wong A, Ramírez M, Barraza A, Orozco MDC, Cevallos MA, Iturriaga G (2008) Improvement of drought tolerance and grain yield in common bean by overexpressing trehalose-6-phosphate synthase in rhizobia. Mol Plant-Microbe Interact 21(7):958–966

Teh CY, Shaharuddin NA, Ho CL, Mahmood M (2016) Exogenous proline significantly affects the plant growth and nitrogen assimilation enzymes activities in rice (*Oryza sativa*) under salt stress. Acta Physiol Plant 38(6):151

Thomas J, Apte SK (1984) Sodium requirement and metabolism in nitrogen-fixing cyanobacteria. J Biosci 6(5):771–794

Tiwari G, Duraivadivel P, Sharma SPH (2018) 1-Aminocyclopropane-1-carboxylic acid deaminase producing beneficial rhizobacteria ameliorate the biomass characters of *Panicum maximum* Jacq. by mitigating drought and salt stress. Sci Rep 30(1):17513

Totsche KU, Amelung W, Gerzabek MH, Guggenberger G, Klumpp E, Knief C, Lehndorff E, Mikutta R, Peth S, Prechtel A, Ray N (2018) Microaggregates in soils. J Plant Nutr Soil Sci 181(1):104–136

Ulukanli Z, Digrak M (2002) Alkaliphilic micro-organisms and habitats. Turk J Biol 26(3):181–191

Upadhyay SK, Singh JS, Singh DP (2011) Exopolysaccharide-producing plant growth-promoting rhizobacteria under salinity condition. Pedo 21(2):214–222

Vaddepalli P, Fulton L, Wieland J, Wassmer K, Schaeffer M, Ranf S, Schneitz K (2017) The cell wall-localized atypical β-1, 3 glucanase ZERZAUST controls tissue morphogenesis in *Arabidopsis thaliana*. Development 144(12):2259–2269

Vaishnav A, Kumari S, Jain S, Varma A, Choudhary DK (2015) Putative bacterial volatile-mediated growth in soybean (*Glycine max* L. Merrill) and expression of induced proteins under salt stress. J App Microbiol 119(2):539–551

Vaishnav A, Kumari S, Jain S, Varma A, Tuteja N, Choudhary DK (2016) PGPR-mediated expression of salt tolerance gene in soybean through volatiles under sodium nitroprusside. J Basic Microbiol 56(11):1274–1288

Vandegehuchte ML, de la Peña E, Bonte D (2010) Relative importance of biotic and abiotic soil components to plant growth and insect herbivore population dynamics. PLoS One 5(9):e12937

Verma P, Yadav AN, Khannam KS, Panjiar N, Kumar S, Saxena AK, Suman A (2015) Assessment of genetic diversity and plant growth promoting attributes of psychrotolerant bacteria allied with wheat (*Triticum aestivum*) from the northern hills zone of India. Ann Microbiol 65:1885–1899

Verma P, Yadav AN, Khannam KS, Kumar S, Saxena AK, Suman A (2016a) Molecular diversity and multifarious plant growth promoting attributes of Bacilli associated with wheat (*Triticum aestivum* L.) rhizosphere from six diverse agro-ecological zones of India. J Basic Microbiol 56:44–58

Verma V, Ravindran P, Kumar PP (2016b) Plant hormone-mediated regulation of stress responses. BMC Plant Biol 16(1):86

Verma P, Yadav AN, Khannam KS, Saxena AK, Suman A (2017a) Potassium-solubilizing microbes: diversity, distribution, and role in plant growth promotion. In: Panpatte DG, Jhala YK, Vyas RV, Shelat HN (eds) Microorganisms for green revolution: volume 1: microbes for sustainable crop production. Springer Singapore, Singapore, pp 125–149. https://doi.org/10.1007/978-981-10-6241-4_7

Verma P, Yadav AN, Kumar V, Singh DP, Saxena AK (2017b) Beneficial plant-microbes interactions: biodiversity of microbes from diverse extreme environments and its impact for crop improvement. In: Singh DP, Singh HB, Prabha R (eds) Plant-microbe interactions in agro-ecological perspectives: volume 2: microbial interactions and agro-ecological impacts. Springer Singapore, Singapore, pp 543–580. https://doi.org/10.1007/978-981-10-6593-4_22
Verma P, Yadav AN, Kumar V, Khan MA, Saxena AK (2018) Microbes in termite management: potential role and strategies. In: Khan MA, Ahmad W (eds) Termites and sustainable management: volume 2-economic losses and management. Springer International Publishing, Cham, pp 197–217. https://doi.org/10.1007/978-3-319-68726-1_9
Vejan P, Abdullah R, Khadiran T, Ismail S, Nasrulhaq Boyce A (2016) Role of plant growth promoting rhizobacteria in agricultural sustainability-a review. Mole 29(5):573
Ventosa A, de la Haba RR, Sánchez-Porro C, Papke RT (2015) Microbial diversity of hypersaline environments: a metagenomic approach. Curr Opin Microbiol 25:80–87
Ventosa A, Nieto JJ, Oren A (1998) Biology of moderately halophilic aerobic bacteria. Microbiol Mol Biol Rev 62(2):504–544
Wai Liew C, Illias RM, Muhammad Mahadi N, Najimudin N (2007) Expression of the Na^+/H^+ antiporter gene (g1-nhaC) of alkaliphilic *Bacillus* sp. G1 in *Escherichia coli*. FEMS Microbiol Lett 276(1):114–122
Wang JD, Levin PA (2009) Metabolism, cell growth and the bacterial cell cycle. Nat Rev Microbiol 7(11):822–827
Wang X, Wang C, Wang C, Cao X, Hou L (2014) *Torulopsis versatilis* strains with increased salt tolerance carry mutations in the glycerol transporter gene FPS 1. Int J Food Sci Tech 49(3):673–678
Warrence NJ, Bauder JW, Pearson KE (2002) Basics of salinity and sodicity effects on soil physical properties. Departement of Land Resources and Environmental Sciences, Montana State University-Bozeman, MT, pp 1–29
Watson H (2015) Biological membranes. Essay Biochem 15(59):43–69
Wei W, Jiang J, Yang SS (2004) Mutagenesis and complementation of relA from *Sinorhizobium meliloti* 042BM as a salt tolerance involvement gene. Ann Microbiol 54:317–324
Weinisch L, Kühner S, Roth R, Grimm M, Roth T, Netz DJA, Filker S (2018) Identification of osmoadaptive strategies in the halophile, heterotrophic ciliate *Schmidingerothrix salinarum*. PLoS Biol 16(1):e2003892
Wood JM, Bremer E, Csonka LN, Kraemer R, Poolman B, van der Heide T, Smith LT (2001) Osmosensing and osmoregulatory compatible solute accumulation by bacteria. Comp Biochem Physiol Part A Mol Int Physiol 130(3):437–460
Xiao JP, Zhang LL, Zhang HQ, Miao LX (2017) Identification of genes involved in the responses of tangor (*C. reticulata* × *C. sinensis*) to drought stress. BioMed Res Int 2017:1–15
Yadav AN, Sachan SG, Verma P, Saxena AK (2015a) Prospecting cold deserts of north western Himalayas for microbial diversity and plant growth promoting attributes. J Biosci Bioeng 119:683–693
Yadav AN, Sachan SG, Verma P, Tyagi SP, Kaushik R, Saxena AK (2015b) Culturable diversity and functional annotation of psychrotrophic bacteria from cold desert of Leh Ladakh (India). World J Microbiol Biotechnol 31:95–108
Yadav AN, Sharma D, Gulati S, Singh S, Kaushik R, Dey R, Pal KK, Saxena AK (2015c) Haloarchaea endowed with phosphorus solubilization attribute implicated in phosphorus cycle. Sci Rep 5:12293
Yadav AN, Verma P, Kumar M, Pal KK, Dey R, Gupta A, Padaria JC, Gujar GT, Kumar S, Suman A, Prasanna R, Saxena AK (2015d) Diversity and phylogenetic profiling of niche-specific Bacilli from extreme environments of India. Ann Microbiol 65:611–629
Yadav AN, Saxena AK (2018) Biodiversity and biotechnological applications of halophilic microbes for sustainable agriculture. J Appl Biol Biotechnol 6:48–55
Yadav AN, Gulati S, Sharma D, Singh RN, Rajawat MVS, Kumar R, Dey R, Pal KK, Kaushik R, Saxena AK (2019a) Seasonal variations in culturable archaea and their plant growth promoting

attributes to predict their role in establishment of vegetation in Rann of Kutch. Biologia 74:1031–1043. https://doi.org/10.2478/s11756-019-00259-2

Yadav AN, Kour D, Sharma S, Sachan SG, Singh B, Chauhan VS, Sayyed RZ, Kaushik R, Saxena AK (2019b) Psychrotrophic microbes: biodiversity, mechanisms of adaptation, and biotechnological implications in alleviation of cold stress in plants. In: Sayyed RZ, Arora NK, Reddy MS (eds) Plant growth promoting rhizobacteria for sustainable stress management: volume 1: rhizobacteria in abiotic stress management. Springer Singapore, Singapore, pp 219–253. https://doi.org/10.1007/978-981-13-6536-2_12

Yadav AN, Mishra S, Singh S, Gupta A (2019c) Recent advancement in white biotechnology through fungi volume 1: diversity and enzymes perspectives. Springer International Publishing, Cham

Yadav AN, Singh S, Mishra S, Gupta A (2019d) Recent advancement in white biotechnology through fungi. Volume 2: perspective for value-added products and environments. Springer International Publishing, Cham

Yadav AN, Singh S, Mishra S, Gupta A (2019e) Recent advancement in white biotechnology through fungi. Volume 3: perspective for sustainable environments. Springer International Publishing, Cham

Yadav AN, Kumar V, Prasad R, Saxena AK, Dhaliwal HS (2018a) Microbiome in crops: diversity, distribution and potential role in crops improvements. In: Prasad R, Gill SS, Tuteja N (eds) Crop improvement through microbial biotechnology. Elsevier, USA, pp 305–332

Yadav AN, Verma P, Kumar S, Kumar V, Kumar M, Singh BP, Saxena AK, Dhaliwal HS (2018b) Actinobacteria from rhizosphere: molecular diversity, distributions and potential biotechnological applications. In: Singh B, Gupta V, Passari A (eds) New and future developments in microbial biotechnology and bioengineering. USA, pp 13–41. https://doi.org/10.1016/b978-0-444-63994-3.00002-3

Yadav AN, Verma P, Kumar V, Sangwan P, Mishra S, Panjiar N, Gupta VK, Saxena AK (2018c) Biodiversity of the Genus *Penicillium* in different habitats. In: Gupta VK, Rodriguez-Couto S (eds) New and future developments in microbial biotechnology and bioengineering, *Penicillium* system properties and applications. Elsevier, Amsterdam, pp 3–18. https://doi.org/10.1016/b978-0-444-63501-3.00001-6

Yadav AN, Verma P, Sachan SG, Kaushik R, Saxena AK (2018d) Psychrotrophic microbiomes: molecular diversity and beneficial role in plant growth promotion and soil health. In: Panpatte DG, Jhala YK, Shelat HN, Vyas RV (eds) Microorganisms for green revolution-volume 2: microbes for sustainable agro-ecosystem. Springer, Singapore, pp 197–240. https://doi.org/10.1007/978-981-10-7146-1_11

Yamamoto K, Shiwa Y, Ishige T, Sakamoto H, Tanaka K, Uchino M, Tanaka N, Oguri S, Saitoh H (2018) Tsushima S (2018) Bacterial diversity associated with the rhizosphere and endosphere of two halophytes: *Glaux maritima* and *Salicornia europaea*. Front Microbiol 9:2878

Yan N, Marschner P, Cao W, Zuo C, Qin W (2015) Influence of salinity and water content on soil microorganisms. Int Soil Water Con Res 3(4):316–323

Yancey PH, Clark ME, Hand SC, Bowlus RD, Somero GN (1982) Living with water stress: evolution of osmolyte systems. Science 217(4566):1214–1222

Yang J, Ma LA, Jiang H, Wu G, Dong H (2016) Salinity shapes microbial diversity and community structure in surface sediments of the Qinghai-Tibetan Lakes. Sci Rep 26(6):25078

Yang XD, Ali A, Xu YL, Jiang LM, Lv GH (2019) Soil moisture and salinity as main drivers of soil respiration across natural xeromorphic vegetation and agricultural lands in an arid desert region. Catena 177:126–133

Zahran HH (1997) Diversity, adaptation and activity of the bacterial flora in saline environments. Biol Fert Soil 25(3):211–223

Zahran HH (1999) Rhizobium-legume symbiosis and nitrogen fixation under severe conditions and in an arid climate. Microbiol Mol Biol Rev 63(4):968–989

Zahran HH, Ahmad MS, Afkar EA (1995) Isolation and characterization of nitrogen-fixing moderate halophilic bacteria from saline soils of Egypt. J Basic Microbiol 35(4):269–275

Zerrouk IZ, Benchabane M, Khelifi L, Yokawa K, Ludwig-Müller J, Baluska F (2016) A *Pseudomonas* strain isolated from date-palm rhizospheres improves root growth and promotes root formation in maize exposed to salt and aluminum stress. J Plant Physiol 191:111–119

Zhao BS, Roundtree IA, He C (2017) Post-transcriptional gene regulation by mRNA modifications. Nat Rev Mol Cell Biol 18(1):31

Zhao G, Kong W, Weatherspoon-Griffin N, Clark-Curtiss J, Shi Y (2011) Mg 2^{+} facilitates leader peptide translation to induce riboswitch-mediated transcription termination. EMBO J 30(8):1485–1496

Chapter 12
Microbe-Mediated Drought Tolerance in Plants: Current Developments and Future Challenges

Iti Gontia-Mishra, Swapnil Sapre, Reena Deshmukh, Sumana Sikdar and Sharad Tiwari

Abstract Drought is a conspicuous stress-causing deleterious effect on plant growth and productivity. In order to compensate the yield loss due to drought, efficient and sustainable strategies are required for its management. Drought stress tolerance is complex trait involving clusters of genes; hence, genetic engineering to generate drought-resistant varieties is a challenging task. In this context, the application of plant growth-promoting microbes (PGPM) to mitigate drought stress is gaining attention as an attractive and cost-effective alternative strategy. PGPM have envisaged a plethora of mechanisms to overcome drought stress in plants which encompasses ACC (1-aminocyclopropane-1-carboxylate) deaminase activity, production of exopolysaccharide (EPS) and volatile organic compounds (VOCs), osmolyte and antioxidant production, enhanced uptake of mineral nutrients, phytohormones production, and modulation. These mechanisms either individually or collectively bestow the PGPRs to combat drought stress in plants. The association of arbuscular mycorrhizal fungi (AMF) with the roots of crop plants can significantly promote water and nutrient uptake by host plants and induce tolerance to drought stress. The inoculation of PGPM in crop plants is also capable of modulating host transcriptome for induced drought tolerance. Further, efforts are needed to develop proficient microbial consortia for enhancing plant growth under drought stress. Thus, the application of PGPM/AMF represents a promising approach to increase nutrient availability and expedite the development of sustainable agriculture.

Keywords Drought · Plant growth-promoting rhizobacteria · Arbuscular mycorrhizal fungi · Sustainable agriculture

---

I. Gontia-Mishra (✉) · S. Sapre · R. Deshmukh · S. Sikdar
Biotechnology Centre, Jawaharlal Nehru Agriculture University, Jabalpur, Madhya Pradesh 482004, India

S. Tiwari
Department of Plant Breeding and Genetics, Jawaharlal Nehru Agriculture University, Jabalpur, Madhya Pradesh 482004, India

A. N. Yadav et al. (eds.), *Plant Microbiomes for Sustainable Agriculture*, Sustainable Development and Biodiversity 25, https://doi.org/10.1007/978-3-030-38453-1_12

12.1 Introduction

Plants being sessile are subjected to a wide array of environmental stresses such as salinity, drought, heavy metals, waterlogging, chilling, and high temperature. Furthermore, climate change is influencing the austerity of abiotic stresses, particularly high temperature and drought. Intense drought is mostly associated with the global climate change which is having a major impact of crop productivity (Etesami and Maheshwari 2018). It is being projected that drought will affect more than 50% of the arable lands worldwide by the year 2050 (Vurukonda et al. 2016). Among these stresses, drought is a conspicuous stress-causing deleterious effect on plant growth and productivity. This stress adversely affects many physiological and biochemical processes of plants such as phytosynthesis, respiration, transpiration, carbohydrates metabolism and nutrient uptake, translocation, and assimilation (Khan et al. 2018). In order to compensate the yield loss due to drought, efficient and sustainable strategies are required for its management. From past years, mainly two strategies are mostly focused to combat the drought stress in plants such as traditional breeding methods and genetic engineering of crop for drought-resistant (Naveed et al. 2014). The major constraint for application of these techniques is time taking, tedious, and expensive. Furthermore, the acceptance of a transgenic crop is uncertain in the market regarding the consumer response to genetically modified plant products which varies from country to country (Ullah et al. 2019a).

In addition to several ethical issues, genetic engineering of all crops is not feasible (Etesami and Maheshwari 2018). Besides, drought stress tolerance is complex trait involving clusters of genes; hence, genetic engineering to generate drought-resistant varieties is a challenging task (Nautiyal et al. 2013; Saikia et al. 2018). Another sustainable strategy to limit drought stress is soil resource management by application of mulching, crop residues, crop cover, non-crop mulch material (plastic foil, geotextile), etc. These methods tend to reduce runoff and evaporation from soil surfaces. Besides, use of crop residue as mulch can increase the soil organic matter which in turn increases the soil water storage capacity and its availability to crops during drought conditions (Bodner et al. 2015).

In this context, the use of plant growth-promoting microbes (PGPM) to mitigate drought stress is gaining attention as an attractive and cost-effective alternative strategy. There are various studies which suggest the beneficial role of plant growth-promoting rhizobacteria (PGPR), actinomycetes, and mycorrhiza in boosting plant growth and endurance toward drought stress (Gontia-Mishra et al. 2016; Saikia et al. 2018; Zade et al. 2019). This chapter summarizes the present understanding toward drought stress and their physiological and molecular responses in plants. The later section of the chapter highlights the imperative mechanisms of PGPM for mitigation of drought stress in plants. In the present chapter, we have also attempted to comprehend the perceptive use of PGPM to alleviate the drought stress in crop plant by fine-tuning the metabolic, signaling, and molecular pathways, thereby enhancing crop productivity.

12.2 Impact of Drought on Crop Plants

Similar to other abiotic stresses, drought is also multidimensional in its effects and has various deleterious impacts on plant. The plants respond to drought stress at physiological, biochemical, and molecular levels, from seed germination to maturity and senescence (Tiwari et al. 2017). Nonetheless, adaptability of plants to drought is determined by the severity and time of exposure to stress, plant species as well as the developmental stages (Kaur and Asthir 2017). Under water scarcity, plant exhibits various events to acclimatize such as morphological (reduced biomass and altered root structure), physiological (reduced photosynthesis and altered transpiration and stomatal activity), and biochemical changes (accumulation of osmolytes, increased oxidative enzyme activity) (Conesa et al. 2016).

Excessive loss of water in drought condition leads to closed stomata and controlled gas exchange and desiccation which lead to complete metabolism and cellular structure disruption; this gradually ends in interruption in enzymatic reactions (Jaleel et al. 2007). Aboveground parts of plant are more susceptible to drought stress than their counterparts. During water-deficit conditions, plants respond by shrinking leaf area, spiraling, and in some cases by shedding their leaves. Leaf size reduction is an important strategy to drought stress as it can directly influence the rate of transpiration. However, reduction in leaf size leads to marked decline in the photosynthetic activity of plants. Drought stress in plants is noted by declined leaf water potential, stomatal closure, and marked reduction in cell growth (Farooq et al. 2009). The first plant organs to feel and react to water deficiency are the roots. Under moderate drought stress, the root length is increased which is attributed to the plants need to utilize the groundwater (Forni et al. 2017), but severe drought conditions can retard root growth. Characteristics related to root traits like biomass, root length, root density, and depths of roots have been identified as the major drought avoidance traits under drought environment (Kashiwagi et al. 2006). A decrease in growth is the most obvious plant response to water stress, which results from decrease in water uptake by roots.

Plant growth and morphology are proportional to enlargement, division, and differentiation of cell. Drought is also reported to inhibit mitotic division of cells along with elongation and expansion which results in growth retardation of plants (Hussain et al. 2008). Various physiological changes occur in plants to overcome drought stress as the early response of plant to stress can make the plant to survive. The basic mechanism to show drought response is to decrease the osmotic potential of plant cells; as a result, the turgor potential gets maintained for routine metabolic processes (Levitt 1980). Drought stress leads to an imbalance between antioxidant defenses and the amount of Reactive Oxygen Species (ROS) resulting in oxidative stress. ROS are required to trigger the signaling but eventually at high concentration can cause impairment of plant organelles especially chloroplasts (Smirnoff 1993). Later on, ROS can initiate lipid peroxidation and degradation of vital proteins, lipids, and nucleic acids (Kaushal and Wani 2016).

Plants under drought stress starts synthesis of new metabolites for their proper functioning. Plants can accumulate biomolecules such as dehydrins (DHNs), heat shock proteins (HSPs), late embryogenesis abundant (LEA) proteins (Lipiec et al. 2013), osmolytes like proline, trehalose, and sugars (Ilhan et al. 2015), glycine, and betaine (Chen and Murata 2011). Changes in membrane fluidity, fatty acid, and protein composition of membranes help to maintain the cellular integrity of plants under drought stress (Bohnert et al. 1995). The osmolytes like ectoine, glycine, and betaine interplay in protein solubilization, and the uncharged solutes like mannitol, trehalose, and pinitol play an important role as scavengers of ROS (Ashraf and Foolad 2007). Plants have envisaged several enzymatic mechanisms including superoxide dismutase (SOD), catalase (CAT), ascorbate peroxidase (APX), glutathione reductase (GR) and non-enzymatic components such as cysteine, glutathione, and ascorbic acid, to aid them fight against the oxidative damage caused by ROS (Kaushal and Wani 2016).

Nitric oxide also protects plants under water-deficit condition from oxidative stress. Most of these processes are regulated through a complex network governed by abscisic acid (ABA), ion transport system, and various transcription factors. Water stress in the root tips results in release of a stress-signaling hormone ABA, which is transported from root zone to the aerial parts of plants. It is directly involved in regulation of stomatal aperture closure. Various reports argued that early interaction of drought signals is mediated by transmembrane protein like histidine kinase which functions like osmosensor (Posas et al. 1996; Urao et al. 1999). Other membrane proteins like aquaporins also involved in regulation of cell volume and turgor homeostasis. In addition, some genes are also identified which regulates the synthesis of osmolytes in the cytoplasm in order to maintain the osmotic potential at water stress condition. Other mechanism involved in sensing of drought stress is changed in the fluidity of membrane lipids (Knight and Knight 2001). Phospholipase C and D along with phosphatidyl-4,5-phosphate 5-kinase are found to be involved in this mechanism which leads to control ion and water channels (Mikami et al. 1998).

12.3 PGPM with Special Context to Drought Stress Management

Almost every parts of the plant are colonized by microbes, but the rhizosphere (soil in the proximity of roots) represents the main source of bacteria with plant-beneficial activities. The microbial community residing in the rhizosphere is comparatively different than its surroundings due to the presence of root exudates (Vejan et al. 2016). These bacteria largely utilize root exudates a source of nutrients for their growth and survival, hence termed as PGPR (Kolepper and Schroth 1978; Verma et al. 2015a, b; Yadav et al. 2015). They interact with the plant roots and influence plant growth and yield as well as enhance soil fertility. Besides PGPR, association of arbuscular mycorrhizal fungi (AMF) with the roots of crop plants can significantly promote water and nutrient uptake by host plants and induce tolerance to drought stress (Xu et al. 2018; Hashem et al. 2019; Yadav et al. 2019b, c, d). Many researchers have advocated the positive impact of PGPR and AMF to promote plant growth under drought stress; hence, it is an effective developing technology (Compant et al. 2010; Carmen et al. 2016; Wu et al. 2019; Verma et al. 2017; Yadav et al. 2018a, b). Crop plants in association with PGPRs persuade morphological and biochemical adjustments leading to increased tolerance to drought by eliciting induced systemic tolerance (IST) (Naveed et al. 2014).

It is determined that PGPR can intercede the drought stress in host plant by increasing accumulation of osmolytes (compatible solutes like proline, glycine betaine, polyamines, sugars like trehalose and polyols), improved uptake of nutrients and modulating the activities of antioxidant enzymes (Barnawal et al. 2019; Kour et al. 2019c; d). Phytohormones are known to control signaling of many abiotic and biotic stresses in plants. It is an established fact that phytohormone synthesis and signaling have a great significance in response to extreme environmental conditions (Tiwari et al. 2017; Kour et al. 2019b; Yadav et al. 2019a). Interestingly, this approach is adopted by PGPR to induce drought tolerance in host plant by regulating the level of phytohormones, like ABA, salicylic acid (SA), and ethylene, therefore directly affecting plant signaling networks and altering drought-responsive genes (Lu et al. 2018). Similarly, inoculation of AMF in host plants can accelerate plant growth and yield under drought stress by increased water and nutrient uptake, modulating stress-responsive genes and cell membrane fatty acid composition and degree of unsaturation (Xu et al. 2018; Wu et al. 2019). The use of various PGPR and AMF for alleviation of drought stress and their positive impact on host plants is presented in Tables 12.1 and 12.2.

12.4 Ways Out by Which PGPM Handle Drought Stress

PGPM have envisaged a plethora of mechanisms to overcome drought stress in plants which encompasses ACC (1-aminocyclopropane-1-carboxylate) deaminase activity, production of exopolysaccharide (EPS) and volatile organic compounds (VOCs), osmolyte production, uptake of mineral nutrients (N, P, and K), phytohormones production/modulation [auxin/indole acetic acid (IAA), cytokinins, abscisic acid (ABA), salicylic acid (SA) and jasmonic acid (JA)], and eliciting the activity antioxidant enzymes in host plants. These mechanisms either individually or collectively bestow the PGPRs to combat drought stress in plants.

12.4.1 ACC Deaminase Activity

Ethylene is a crucial plant hormone which controls a number of plant processes ranging from seed germination, fruit ripening, abscission of leaves, and plant senescence

Table 12.1 The role of PGPR in mitigating drought stress in plants

PGPR strain	Source of isolation	Crop	PGPR traits	Effect on plants	References
Bacillus spp.	Rhizosphere of *Pennisetum glaucum*, *Helianthus annuus*, *Zea mays*	*Z. mays*	EPS and IAA, gibberellins, cytokinin production, and P-solubilization	Inoculated maize seedlings showed improved physiological response to drought	Vardharajula et al. (2011)
Bacillus licheniformis	Field soil	*Capsicum annuum*	ACC deaminase activity	PGPR-inoculated pepper plants tolerate the drought stress and showed differentially expressed stress proteins	Lim and Kim (2013)
Burkholderia phytofirmans	Roots of onion	*Triticum aestivum*	IAA production and ACC deaminase activity	PGPR inoculation improved grain yield and nutrients in grains	Naveed et al. (2014)
Pseudomonas aeruginosa	Rhizosphere of *Vigna radiata*	*V. radiata*	IAA, ACC deaminase, P-solubilization, and endogenous proline and glycine betaine accumulation	Increased the levels of antioxidant enzymes, proline, and subsequently modulated the regulation of stress-responsive genes in PGPR-treated plants under water stress conditions	Sarma and Saikia (2014)

(continued)

Table 12.1 (continued)

PGPR strain	Source of isolation	Crop	PGPR traits	Effect on plants	References
Bacillus thuringiensis and *Paenibacillus polymyxa*	*Pinus ponderosa* and *Oryza sativa*	*T. aestivum*	ACC deaminase activity and P-solubilization	Greater plant biomass and fivefold higher survivorship under severe drought; reduced emissions of stress volatiles	Timmusk et al. (2014)
Pseudomonas spp., *Bacillus cereus,* and *Arthrobacter* sp.	Rhizosphere soil	*O. sativa*	IAA production	Improved plant growth and antioxidant defense systems and stability of membranes in plant	Gusain et al. (2015)
Bacillus subtilis	Roots of *Panicum virgatum*	*Brachypodium distachyon*	Phytohormone production and P-solubilization	Amelioration of phenotypic effect of drought and up-regulation of drought-responsive genes	Gagné-Bourque et al. (2015)
Klebsiella sp, *Enterobacter ludwigii,* and *Flavobacterium* sp.	Rhizosphere of *T. aestivum*	*T. aestivum*	IAA and siderophore production, ACC deaminase, P and Zn solubilization	Affected various growth parameters, water status, membrane integrity, osmolyte accumulation, and stress-responsive gene expressions, which were positively altered by PGPR inoculation in wheat under drought	Gontia-Mishra et al. (2016)

(continued)

Table 12.1 (continued)

PGPR strain	Source of isolation	Crop	PGPR traits	Effect on plants	References
Pseudomonas putida and *Bacillus amyloliquefaciens*	–	*Cicer arietinum*	ACC deaminase activity, minerals solubilization, biofilm formation	PGPR inoculation improved the biomass and antioxidant enzymes in plants under drought stress	Kumar et al. (2016)
Pseudomonas putida	Desert regions of Rajasthan	*C. arietinum*	IAA production and P-solubilization	Altered various physiological and biochemical parameters as well as regulation of stress-responsive genes	Tiwari et al. (2016)
Bacillus subtilis	Rhizosphere soil of lemongrass	*T. aestivum*	IAA production and ACC deaminase activity	PGPR inoculation improved the physiological parameters such as net CO_2 assimilation, stomatal conductance, and transpiration rate as well as increased the endogenous IAA and ABA content under drought stress	Barnawal et al. (2017)

(continued)

Table 12.1 (continued)

PGPR strain	Source of isolation	Crop	PGPR traits	Effect on plants	References
Azospirillum spp.	Roots of wheat and maize	*Z. mays*	N_2 fixation, IAA and siderophore production, ACC deaminase activity, P-solubilization	Significant biomass gain and better osmotic balance were noted in PGPR-inoculated plants under drought stress	García et al. (2017)
Enterobacter cloacae and *Citrobacter* sp.	Rhizosphere of *T. aestivum*	*T. aestivum*	IAA and siderophore production, ACC deaminase, P, K, and Zn solubilization	Bio-inoculants showed growth enhancement of wheat seedlings under drought stress	Gontia-Mishra et al. (2017)
Bacillus amyloliquefaciens	Alkaline soil	*O. sativa*	IAA production, ACC deaminase activity, P-solubilization, and proline accumulation	PGPR inoculation positively stimulated membrane integrity and osmolyte accumulation as well as modulated the drought-responsive genes under water-deficit condition	Tiwari et al. (2017)
Pseudomonas fluorescens, Enterobacter hormaechei, and *Pseudomonas migulae*	*Setaria italica*	*S. italica*	ACC deaminase activity and EPS production	PGPR inoculation caused a significant increase in dry biomass	Niu et al. (2018)

(continued)

Table 12.1 (continued)

PGPR strain	Source of isolation	Crop	PGPR traits	Effect on plants	References
Bacillus amyloliquefaciens	–	*Arabidopsis thaliana*	EPS and IAA production	Increased the survival rate of plants, biomass, osmolytes, antioxidant enzyme activities, and modulated stress-responsive genes	Lu et al. (2018)
Ochrobactrum pseudogrignonense, *Pseudomonas* sp, and *Bacillus subtilis*	Rhizosphere soil	*Vigna mungo* and *Pisum sativum*	ACC deaminase activity, IAA production, and P-solubilization	Better plant growth and biomass, increase activity of antioxidant enzymes, down-regulation of ACC-oxidase gene under drought stress	Saikia et al. (2018)
Enterobacter sp. and *Bacillus* sp.	*Mucuna pruriens*	*M. pruriens*	ACC deaminase activity and IAA production	Improved photosynthetic performance and biomass and increased leaf isoprene content	Saleem et al. (2018)
Bacillus licheniformis	Compost	*A. thaliana*	IAA production and P-solubilization	Exposure to PGPR increased the tolerance toward water deficits and positive-modulated stress-responsive genes	Sukkasem et al. (2018)

(continued)

Table 12.1 (continued)

PGPR strain	Source of isolation	Crop	PGPR traits	Effect on plants	References
Bacillus subtilis and *Paenibacillus illinoisensis*	–	*Capsicuum annuum*	Synthesis of nitric oxide and volatile organic compounds	Increase in root length and photosynthetic activity as well as enhanced expression and activity of vacuolar proton pumps	Vigani et al. (2018)
Variovorax paradoxus, *Pseudomonas* spp., *Achromobacter* spp, and *Ochrobactrum anthropi*	*T. aestivum*	*T. aestivum*	ACC deaminase activity	Improved plant growth and foliar nutrient concentrations and significant changes in antioxidant properties in treated with PGPR under drought stress	Chandra et al. (2019)
Pseudomonas spp., *Bacillus* spp.	–	*A. thaliana*	EPS and phytohormone production	Bio-inoculation remarkably increased the biomass and plant water content under drought stress	Ghosh et al. 2019

(continued)

Table 12.1 (continued)

PGPR strain	Source of isolation	Crop	PGPR traits	Effect on plants	References
Paenibacillus beijingensis and *Bacillus* sp.	–	*T. aestivum* and *Cucumis sativus*	–	Bio-inoculation significantly increased seed germination, whereas decreased free proline and soluble sugar in the plants. The modulation of stress-responsive and ROS scavenging genes in PGPR-treated plants under drought stress was noted	Li et al. (2019b)
Streptomyces pactum		*T. aestivum*	Biocontrol agent against antagonists	Enhanced osmotic adjustment and antioxidant capacity of plants via induction of abscisic acid accumulation and up-regulation of drought resistance-related gene expression	Li et al. (2019c)

(continued)

Table 12.1 (continued)

PGPR strain	Source of isolation	Crop	PGPR traits	Effect on plants	References
Pseudomonas simiae	Rhizosphere of *Glycine max*	*G. max*	ACC deaminase activity	PGPR inoculation increased the accumulation of osmolytes and sugars as well as down-regulated the drought stress-responsive genes	Vaishnav and Choudhary (2019)
Bacillus pumilus	Roots of *Glycyrrhiza uralensis*	*G. uralensis*	–	PGPR inoculation increased the total biomass, oxygen species, and antioxidants in plants. It also enhanced the flavonoids, total polysaccharide, and glycyrrhizic acid contents in plants under drought conditions	Xie et al. (2019)
Streptomyces rimosus and *S. monomycini*	–	*Mentha piperita*	Auxin and siderophore production, biocontrol activity against *Phytophthora drechsleri*	PGPR inoculation increased the biomass and growth and essential oils as well as menthol content under drought stress	Zade et al. (2019)

Table 12.2 The role of AMF in mitigating drought stress in plants

Crop/plant	Effect of AMF inoculation on plant	References
Strawberry	AMF inoculation plants had greater water-use efficiency under the regulated deficit irrigation (RDI) regime in comparison to non-mycorrhizal inoculated plants	Boyer et al. (2015)
Phaseolus vulgaris	AMF inoculation and methyl jasmonate prevented inhibition of root hydraulic conductivity under drought conditions, by causing reduction in root salicylic acid contents	Sanchez-Romera et al. (2016)
Zea mays	AMF symbiosis induced an improvement in physiological parameters in drought-sensitive plants including efficiency of photosystem II, membrane stability, accumulation of soluble sugars and plant biomass production. In addition, the drought-responsive genes were down-regulated by the AMF inoculation	Quiroga et al. (2017)
Sorghum bicolor	AMF inoculation alleviated plant growth retardation and prolonged plant lifespan under drought. The improved biomass and the specific leaf area were noted in inoculated plants under drought stress conditions	Sun et al. (2017)
Poncirus trifoliate	AMF inoculation significantly increased leaf sucrose, glucose, and fructose concentration under drought stress, accompanied with a significant increase of leaf sucrose phosphate synthase, neutral invertase, and net activity of sucrose-metabolized enzymes	Wu et al. (2017)
Damask rose	AMF colonization can enhance growth, flower quality, and adaptation of rose plants under drought stress levels, particularly at high level of drought stress via improving their water relations and photosynthetic status	Abdel-Salam et al. (2018)
Poncirus trifoliate	AMF inoculates seedlings showed significantly higher root density, length, and diameter and root IAA level under stress. Mycorrhization caused the up-regulation in IAA biosynthesis. The inoculation also down-regulated the transcript level of root auxin efflux under drought stress	Liu et al. (2018)
Zea mays	AMF association modifies root hydraulic responses to drought. AMF plants showed increased hydrostatic root hydraulic conductivity and osmotic root hydraulic conductivity	Quiroga et al. (2018)
Sorghum bicolor	AMF inoculation improved their transpiration efficiency and increased the nitrogen and phosphorus content of sorghum, especially under water was limiting conditions	Symanczik et al. (2018)
Glycyrrhiza uralensis	The inoculation of AMF to the plants demonstrated improved growth and physiological status such as stomatal conductance, photosynthesis rate, and water-use efficiency compared with non-AM plants. The AMF association also up-regulated the expression of an aquaporin gene PIP and decreased root abscisic acid concentrations	Xie et al. (2018)
Leymus chinensis and *Hemarthria altissima*	AMF inoculation to plants enhanced plant biomass, photosynthetic rate, stomatal conductance, intrinsic water-use efficiency, and SOD activity of the *L.chinensis* and reduced levels of malondialdehyde. The inoculation also demonstrated increased shoot growth in *H.altissima*	Li et al. (2019a)

(Sapre et al. 2019). It is also produced in plant in response to several environmental stresses including drought stress (Gontia-Mishra et al. 2014). Consequently, the ethylene generated under stressful condition is often termed as "stress ethylene" (Glick 2014). The ethylene produced during stress can subsequently induce the defoliation, retarded root, and stem growth along with the expression of genes leading to plant senescence, leading to inferior crop performance (Vejan et al. 2016). Interestingly, it is noted that ACC works as precursor for ethylene biosynthesis (Shaharoona et al. 2006). Among the different suggested strategies for improvement in plant growth under drought stress, the most plausible one is alteration in the endogenous levels of ethylene caused by the PGPR (Kumar et al. 2019a, b).

Furthermore, many PGPR possessing the enzyme ACC deaminase catalyze the conversion of ACC to ammonia and α-ketobutyrate, which indirectly decrease the ethylene concentration in plants under drought stress (Glick et al. 1998). By facilitating the development of longer roots, these PGPR may enhance the survival of seedlings, which help in combating the effect of stress ethylene. The root elongation plants under drought stress can allow a better access to water and uptake of nutrients. Ethylene is also known to compromise the nodule formation and nitrogen fixation in legume (Sapre et al. 2019). *Rhizobium* with ACC deaminase activity can diminish the deleterious effect of ethylene under drought stress by increasing the nodulation and nitrogen fixation in its symbiotic legume partner (Belimov et al. 2009). In this regard, numerous researchers have documented the application of ACC deaminase-producing PGPR in ameliorating drought stress in crop plant such as chickpea (Tiwari et al. 2016), mung bean (Sarma and Saikia 2014), wheat (Gontia-Mishra et al. 2016; Barnawal et al. 2017), rice (Tiwari et al. 2017), foxtail millet (Niu et al. 2018), and other tropical crop plants (Kumar et al. 2019a, b; Kour et al. 2019a; Yadav et al. 2017a; b; Yadav and Yadav 2018).

12.4.2 EPS Production

PGPR have the unique ability to produce exopolysaccharide (EPS)/extracellular polymeric substances. The biofilm formation and EPS production by PGPR are important mechanisms to tolerate drought stress in the environment. The EPS has multifarious function in bacterial cells ranging from quorum-sensing signals, development, survival, and host colonization (Nocelli et al. 2016). The EPS largely constitutes of high-molecular-weight macromolecules like polysaccharide along with smaller proportions of protein, lipids, and uronic acid (Naseem et al. 2018). The EPS-producing PGPR can better clamp and colonize the root surface under adverse conditions (Ali et al. 2014).

It can protect PGPR and its host plant under drought stress by enhancing water retention than the surrounding environment (Hepper 1975; Vurukonda et al. 2016). The EPS secreted by PGPR into the soil can be absorbed by soil particle due to their different anionic functional groups (e.g., sulfhydryl, carboxyl, hydroxyl, sulfonate, amine, and amide) and increase the water-holding capacity of soil and improve physicochemical properties of soil under prolong desiccation conditions (Sandhya et al. 2009). Hence, plants inoculated with EPS-producing PGPR can maintain the higher water potential, boost root-adhering soil/root tissue ratio, and accelerate the uptake of nutrients by plant, thereby enhancing plant growth and yield under drought stress (Selvakumar et al. 2012; Rolli et al. 2014; Kaushal 2019). Many researchers have suggested the use of EPS-producing rhizobacteria in alleviating drought stress in important crop plants such as maize (Vardharajula et al. 2011), sunflower (Sandhya et al. 2009), wheat (Timmusk et al. 2014), and foxtail millet (Niu et al. 2018).

12.4.3 Production of VOCs

Unlike plants, soil bacteria produce a range of volatile compounds, which have specific function in their life cycles as well as interplay with other microbes and plants (Sharifi and Ryu 2018). The bacterial VOCs are chemically characterized as alkenes, ketones, and alcohols. PGPR can stimulate plant growth by synthesizing and releasing volatile compounds, which is now known as an essential mechanism of plant–microorganism interactions (Froni et al. 2017). The role of these VOCs is largely associated as activator against plant pathogens, leading to induced systemic resistance in plants (Ruzzi and Aroca 2015). Remarkably, it is noted that besides their role in biotic stress tolerance, these compounds can actively alleviate several abiotic stresses including drought (Timmusk et al. 2014). The VOCs produced by PGPRs can promote plant growth by increasing photosynthesis, carbon assimilation, enhancing mineral uptake, altering root structure, and intensive phytohormone signaling under abiotic stress conditions (Sharifi and Ryu 2018).

It was demonstrated that a PGPR *Pseudomonas chlororaphis* releases characterized as 2R, 3R-butanediol, can induce drought tolerance in *Arabidopsis thaliana* (Cho et al. 2008). Moreover, few studies suggest the role of VOCs (produced by PGPR) in modulation of the transcript levels in plants, resulting in enhanced biosynthesis of choline and glycine betaine which in turn to shielded *A. thaliana* plants from drought stress (Cho et al. 2008; Zhang et al. 2010). It was noted that VOCs produced by soil bacteria such as acetic acid can stimulate the formation of biofilms/EPS, which can indirectly influence drought stress in plants (Chen et al. 2015). Another study of bacterial inoculation with *Bacillus thuringiensis* in wheat seedlings under drought stress markedly decreased the stress-induced volatile compounds emitted by plants and increased plant biomass and photosynthesis (Timmusk et al. 2014). VOCs producing PGPR are potential for application as bio-stimulants to improve plant health under drought stress. The mechanism of PGPR-induced VOCs in ameliorating drought stress in crop plants is limited and requires to be explored extensively.

12.4.4 Phytohormones Production

The production of phytohormones in plants is essential because of their physiological effects on its growth. The phytohormones such as auxins, cytokinins, gibberellins and ethylene, and abscisic acid (ABA) have a particular function in the regulation of plant growth and development (Vurukonda et al. 2016). PGPR have been widely known to produce these phytohormones which can help in promoting plant growth under stressed conditions by stimulating cell growth and division (Kaushal 2019). One of the important aspects of the bacterial–plant interaction that has received worldwide attention is the bacterial production of IAA/auxins. Production of IAA, a plant hormone that does not any apparent function as a hormone in bacterial cells, may have evolved in bacteria because of its significance in the bacterium–plant relationship (Patten and Glick 2002). A positive correlation is noted between the in vitro production of auxins by PGPR strains and their growth promotion effects (Jha et al. 2012). The auxin-producing PGPR can influence root proliferation and formation of lateral and adventitious roots, which results in an increased mineral and water uptake by the plants rotoscoping them against drought stress (Gontia-Mishra et al. 2016). Besides, some of the PGPR have known to modulate the expression of auxin-responsive genes in host–plant roots (Lakshmanan et al. 2013) resulting in stress tolerance in plants. Several authors have suggested the IAA production by PGPR as a mechanism to drought tolerance enhancement in crop plants such as maize, wheat, mung bean, etc. (Naveed et al. 2014; Sarma and Saikia 2014; García et al. 2017).

Gibberellins have a crucial function in plant growth such as stem elongation, germination, flowering, and senescence (Kaushal 2019). In contrast, cytokinins play a vital role in regulation of cell division and nutrient allocation, and maintain photosynthetic activity under drought stress (Ullah et al. 2019b). The application of cytokinin-producing PGPR, *Bacillus subtilis* in *Platycladus orientalis,* conferred drought stress tolerance by increasing shoot growth (Liu et al. 2013). Similarly, maize and soybean plants inoculated with gibberellin-producing PGPRs (*Pseudomonas putida* and *Azospirillum lipoferum*) registered improved plant growth under drought stress (Cohen et al. 2009; Kang et al. 2014). Consequently, there are evidences from many studies which advocate that PGPM have the ability to positively alter phytohormone levels of plant, leading to drought stress tolerance (Saakre et al. 2017; Ali et al. 2018).

12.4.5 Osmolytes Production and Alteration in Antioxidant Enzyme Activities

Water-deficit conditions result in altered osmotic balance in plants; hence, the water-absorbing capacity of plants get reduced and plant tries to change various physiological and biochemical processes for adaptation under stressed environment. Under these conditions, plants tend to produce a wide range of osmolytes or commonly known as osmoprotectants. Some cellular events like protein and membrane stabilization are supported by osmoprotectants (Nahar et al. 2016). Due to their diverse chemical properties, osmolytes also protect plant cells from oxidative stress by attacking on ROS (Zhu 2002). There are several osmoregulators like proteins, sugars, and free amino acids reported to play a key role in balancing osmotic pressure in plant cells (Hasegawa et al. 2000). It is reported in various studies that microbes residing in soil also produce and secrete osmolytes when encounter to drought stress. These osmolytes function as produced by plants. Maize inoculated with *Pseudomonas fluorescence* growing under drought has potential for accumulating proline and resulted in increased water content and plant biomass (Ansary et al. 2012). Soil microbes also reported to alter the amount of osmoprotectants and antioxidant enzyme in plants (Kaushal 2019).

Plants in response to drought also generate various types of ROS oxidative damage by reacting with importing biomolecules like protein and lipids of cell. To protect from this oxidative damage, cell has developed antioxidant defense system which includes enzymatic as well as non-enzymatic pathway (Miller et al. 2010). Inoculation of plants with beneficial microbes suppresses the activity of antioxidant enzyme activity to alleviate the drought stress. Various species of *Pseudomonas* and *Bacillus* inoculated with maize are reported to reduce the activity of antioxidant enzymes when compared to un-inoculated plants under drought (Sandhya et al. 2010; Vardharajula et al. 2011).

12.4.6 Nutrient Availability

Water deficit can reduce the uptake of nutrients in the soil due to poor soil structure. PGPR and AMF have the capability to increase the accessibility of nutrient in the rhizosphere either by fixing nutrients (N_2) or by solubilizing insoluble minerals such as P, K, and Zn into the soluble form (Vejan et al. 2016). The explanation for enhanced nutrient status in AMF inoculation in plant has increased absorption surface of extraradical hyphae for extensive acquisition of nutrients from soil. The extraradical hyphae of AMF have a deep network into the soil readily absorption of nutrients which in turn is transported to arbuscules in cortical cells and are finally released into the apoplast to ameliorate nutrient deficiency caused by stress (Zhao et al. 2015; Kaushal 2019). Several reports suggest that PGPR inoculation can improve uptake of nutrients especially P under drought stress (Sandhya et al. 2010; Timmusk et al. 2014). In addition, there are evidences of increased mobility of nutrients (P, N, K, Ca, and Zn) in plants inoculated with AMF during drought stress (Gholamhoseini et al. 2013; Zhao et al. 2015; Abdel-Salam et al. 2018).

12.5 Modulation of Host Transcriptome by PGPR Inoculation

The PGPRs have been efficiently utilized in crops to alleviate disease stress. Nevertheless, they can also be used as potential targets for inducing drought tolerance in crop species. Physiological studies in few crop species suggest an efficient application of PGPRs to reduce drought stress (Khan et al. 2018; Niu et al. 2018) (Fig. 12.1). Drought stress is a complex trait which influences various cross-linked signaling between biotic and abiotic stresses. Hence, the beneficial effects of PGPRs in biotic stress can indirectly induce tolerance to drought stress. This mode of genetic improvement comprises understanding the PGPR-induced biochemical differential regulation. These biochemical changes are governed by differential expression of genes involved in induced systemic response (ISR), which are ultimately triggered through salicylic acid (SA)-signaling pathway (Zhang et al. 2002). The PGPR inoculation greatly modifies the transcriptome of the plant species, regulating expression of several genes (Rekha et al. 2018). Microarrays and RNA-seq studies have been employed in certain crop species to understand the molecular mechanism of PGPR-induced drought tolerance.

PGPRs often colonize the root surface and may induce production of several phytochemicals that regulate phytohormone signaling like auxins and ABA (Srivastava

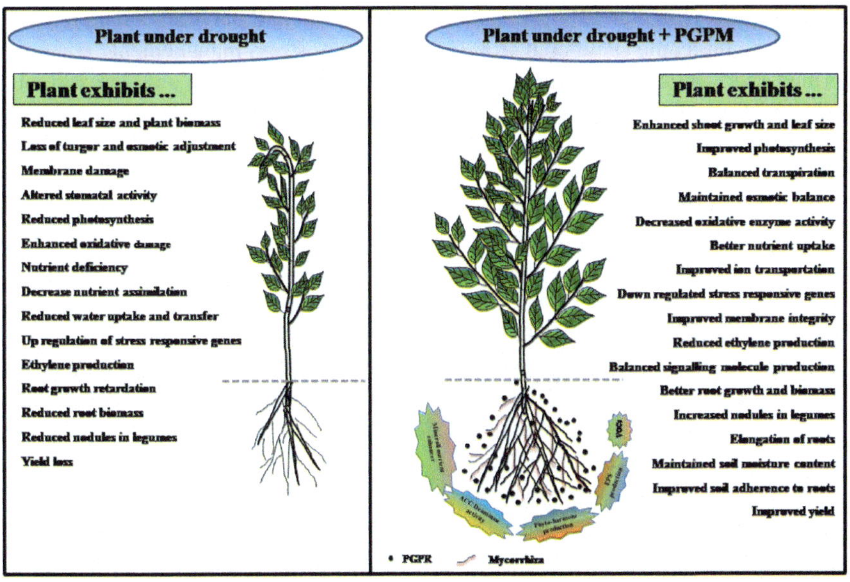

Fig. 12.1 Comparative account of effect of drought stress on plants without PGPM inoculation and with PGPR and AMF inoculation. The PGPR and AMF have envisaged diverse mechanisms for alleviation of drought stress in plants

et al. 2012). Systemic acquired resistance (SAR) is strongly regulated by phytohormones and their *in planta* levels. The inoculation of PGPR *Pseudomonas* sp. in rice plants induced the expression of LEA (late embryogenesis abundant) genes, which encodes IAA amido synthetases resulting in drought stress tolerance (Yasmin et al. 2017). Similarly, the inoculation of *A. thaliana* with PGPR *Paenibacillus polymyxa* induced drought tolerance through enhanced transcription of Early Responsive to Dehydration 15 (ERD15) gene (Timmusk and Wagner 1999). The drought alleviation by PGPR is mediated primarily through differential regulation of ABA-responsive pathway. The ABA-responsive signaling pathway transcription factor genes bZIP1, COC1, and Hsp20 proteins were overexpressed in susceptible rice cultivar upon PGPR *Pseudomonas fluorescens* inoculation, conferring drought tolerance (Saakre et al. 2017).

DHNs are a class of proteins of group 2 LEA proteins and proportionately related to their active accumulation during water stress tolerance. Microarray studies of barley seedlings overexpressing DHNs genes were drought-tolerant (Rodriguez et al. 2005). Plants expressing drought-tolerant CaDHN gene were highly expressed in pepper plant inoculated with PGPR *Bacillus licheniformis* (Lim and Kim 2013). DHN genes are interacting members of the SA-induced SAR response (Jing et al. 2016), commonly observed in PGPR-associated stress alleviation. The inoculation of water-stressed *A. thaliana* roots with *Pseudomonas chlororaphis* O6 induced an overexpression of LEA and dehydrin genes to over 100-folds (Cho et al. 2013). Trehalose is a nonreducing disaccharide, which is actively synthesized in bacteroids of *Rhizobium* sp. (Streeter 1985). It is an osmoprotectant and plays an essential role as a signaling molecule (Paul et al. 2008) during water stress management. The transformed *Rhizobium etli* and *Azospirillum brasilense* mutant with an overexpressing trehalose-6-phosphate synthase gene inoculated to *Phaseolus vulgaris* and maize, respectively, resulted in the enhanced expression of drought tolerance genes (Suárez et al. 2008; Rodríguez-Salazar et al. 2009).

To identify the transcriptional regulation of plants with drought stress in the presence of PGPRs, few transcriptomic studies have been undertaken. The PGPR inoculation improves drought stress tolerance by repressing the enhanced expression of abiotic stress response genes, viz., ABA and ethylene. The transcriptome study of sugarcane plants colonizing *Gluconacetobacter diazotrophicus* identified a reverse regulation of drought stress genes to that of stressed non-inoculated roots. The DREB1A/CBF3, DREB1B/CBF1, and NCED3 homologs were down-regulated in water-stressed plants treated with *G. diazotrophicus*. Although the auxin metabolism and ABA pathway were equally enriched in both the inoculated and non-inoculated plants, cytokinin hormone pathway up-regulation was observed only in the inoculated plants. However, the auxin, ABA, and ethylene-mediated stress-responsive signaling were mainly down-regulated (Vargas et al. 2014). Similar results have been demonstrated through transcriptome analysis of maize plants inoculated with *Pseudomonas putida* strain FBKV2 displayed drought stress alleviation, by Ali et al. (2018).

The expressions of ABA and ethylene signaling pathway genes were down-regulated, including down-regulation of bZIP transcription factor (TFs), and 1-aminocyclopropane-1-carboxylate synthase2 and ethylene-responsive TFs, respectively. The transcriptome of *A. thaliana* treated with PGPR *P. chlororaphis* O6 identified remarkable up-regulation of calmodulin and calcium-binding proteins. These genes play important role in cell-to-cell communication. The stress-responsive down-regulated genes in PGPR inoculation also include class of MYB and AP2 domain transcription factors (Cho et al. 2013). This clearly presents that plant growth-promoting bacteria reduce the drought stress by reducing the expression of stress-induced molecules, keeping optimum environment to plants. It is evident from several studies that PGPRs prime the stress-responsive pathway and following subjection of plants to drought stress reduces the production or biosynthesis of stress molecules.

12.6 Concluding Remarks and Future Prospects

Agricultural productivity is largely dependent on climatic conditions. Climate change is expected to reduce water accessibility for agriculture in coming years. Drought has the noxious effects on growth and development of plant. Hence, it is the need of the hour to search for the effectual solution to overcome the problem of drought stress in plants. Moreover, drought is a complex trait, so developing transgenic plants resistant to drought stress is also a challenging task. Under such stressful conditions, the interaction of plant and beneficial microorganisms is of great importance. The application of drought-tolerant PGPM has gained abundant attention as an alternative and eco-friendly option to mitigate drought stress in crop plants. Another effective option to combat drought stress in plant is the exogenous application of PGPM in combination with either the plant growth regulators (SA, JA, Trinexapacethyl, and ABA), polyamine-like putrescine, biochar (organic carbon), silicon nanoparticles, or seaweed extracts (Ali et al. 2017; Khan et al. 2019; Hashem et al. 2019).

In the current scenario, research must be concentrated to increase the number and diversity of effective and competitive drought-tolerant PGPM from drought-stricken agricultural ecosystems. The drought-tolerant PGPM could be useful to design new bio-inoculants/biofertilizers, especially for arid regions. Additionally, the performance of potential PGPR strains should be essentially assessed under field conditions, as plants usually face cyclic drought conditions rather than continuous drought. Further efforts are needed to develop proficient microbial consortia for enhancing plant growth under drought stress. Regardless of several findings, on the adaptation of plants under drought stress and their association with PGPM for mitigation of stress, substantial efforts are required to explore the underlying molecular mechanisms of interplay between plant and PGPM in soil to hasten the process of stress amelioration in crops. Thus, it could be concluded that the use of PGPM represents a promising approach to increase nutrient availability and expedite the development of sustainable agriculture under drought stress.

Acknowledgements The author I. Gontia-Mishra acknowledges the funding provided by Science and Engineering Research Board, New Delhi, India, grant number PDF/2017/001001.

References

Abdel-Salam E, Alatar A, El-Sheikh MA (2018) Inoculation with arbuscular mycorrhizal fungi alleviates harmful effects of drought stress on damask rose. Saudi J Biol Sci 25:1772–1780

Ali SkZ, Sandhya V, Rao LV (2014) Isolation and characterization of drought-tolerant ACC deaminase and exopolysaccharide-producing fluorescent *Pseudomonas* sp. Ann Microbiol 64:493–502

Ali F, Bano A, Fazal A (2017) Recent methods of drought stress tolerance in plants. Plant Growth Regul 82:363–375

Ali SkZ, Vardharajula S, Vurukonda SS (2018) Transcriptomic profiling of maize (*Zea mays* L.) seedlings in response to *Pseudomonas putida* stain FBKV2 inoculation under drought stress. Ann Microbiol 68:331–349

Ansary MH, Rahmani HA, Ardakani MR, Paknejad F, Habibi D, Mafakheri S (2012) Effect of *Pseudomonas fluorescent* on proline and phytohormonal status of maize (*Zea mays* L.) under water deficit stress. Ann Biol Res 3:1054–1062

Ashraf M, Foolad MR (2007) Roles of glycine betaine and proline in improving plant abiotic stress resistance. Environ Exp Bot 59:206–216

Barnawal D, Bharti N, Pandey SS, Pandey A, Chanotiya CS, Kalra A (2017) Plant growth-promoting rhizobacteria enhance wheat salt and drought stress tolerance by altering endogenous phytohormone levels and TaCTR1/TaDREB2 expression. Physiol Plantarum 161:502–514

Barnawal D, Singh R, Singh RP (2019) Role of plant growth promoting rhizobacteria in drought tolerance: regulating growth hormones and osmolytes. In: Singh AK, Kumar A, Singh PK (eds) PGPR amelioration in sustainable agriculture. Woodhead Publishing, pp 107–128

Belimov AA, Dodd IC, Hontzeas N, Theobald JC, Safronova VI, Davies WJ (2009) Rhizosphere bacteria containing 1-aminocyclopropane-1-carboxylate deaminase increase yield of plants grown in drying soil *via* both local and systemic hormone signaling. New Phytol 181:413–423

Bodner G, Nakhforoosh A, Kaul HP (2015) Management of crop water under drought: a review. Agron Sustain Dev 35:401–442

Bohnert HJ, Nelson DE, Jensen RG (1995) Adaptations to environmental stresses. Plant Cell 7:1099–1111

Boyer LR, Brain P, Xu X, Jeffries P (2015) Inoculation of drought-stressed strawberry with a mixed inoculum of two arbuscular mycorrhizal fungi: effects on population dynamics of fungal species in roots and consequential plant tolerance to water deficiency. Mycorrhiza 25:215–227

Carmen CA, Patricia P, Rubén B, Victoria SM (2016) Plant-rhizobacteria interaction and drought stress tolerance in plants. In: Hossain MA, Wani SH, Bhattacharjee S, Burritt DJ, Tran LSP (eds), Drought stress tolerance in plants, vol 1. Springer, Cham, pp 287–308

Chandra D, Srivastava R, Gupta VV, Franco CM, Sharma AK (2019) Evaluation of ACC-deaminase-producing rhizobacteria to alleviate water-stress impacts in wheat (*Triticumaestivum* L.) plants. Can J Microbiol 65:387–403

Chen TH, Murata N (2011) Glycine betaine protects plants against abiotic stress: mechanisms and biotechnological applications. Plant Cell Environ 34:1–20

Chen Y, Gozzi K, Yan F, Chai Y (2015) Acetic acid acts as a volatile signal to stimulate bacterial biofilm formation. MBio 6:e00392

Cho SM, Kang BR, Han SH, Anderson AJ, Park JY, Lee YH et al (2008) 2R, 3R-butanediol, a bacterial volatile produced by *Pseudomonas chlororaphis* O6, is involved in induction of systemic tolerance to drought in *Arabidopsis thaliana*. Mol Plant Microb Interact 21:1067–1075

Cho SM, Kang BR, Kim YC (2013) Transcriptome analysis of induced systemic drought tolerance elicited by *Pseudomonas chlororaphis* O6 in *Arabidopsis thaliana*. Plant Pathol J 29:209–220
Cohen AC, Travaglia CN, Bottini R, Piccoli PN (2009) Participation of abscisic acid and gibberellins produced by endophytic *Azospirillum* in the alleviation of drought effects in maize. Botanique 87:455–462
Compant E, van der Heijden MGA, Sessitsch A (2010) Climate change effects on beneficial-plant microorganism interactions. FEMS Microbiol Ecol 73:197–214
Conesa MR, Rosa JM, Domingo R, Banon S, Perez-Pastor A (2016) Changes induced by water stress on water relations, stomatal behaviour and morphology of table grapes (cv. Crimson seedless) grown in pots. SciHort 202:9–16
Etesami H, Maheshwari DK (2018) Use of plant growth promoting rhizobacteria (PGPRs) with multiple plant growth promoting traits in stress agriculture: action mechanisms and future prospects. Ecotoxicol Environ Saf 156:225–246
Farooq M, Wahid A, Kobayashi N, Fujita D, Basra SMA (2009) Plant drought stress: effects, mechanisms and management. Agron Sustain Dev 29:185–212
Forni C, Duca D, Glick BR (2017) Mechanisms of plant response to salt and drought stress and their alteration by rhizobacteria. Plant Soil 410:335–356
Gagné-Bourque F, Mayer BF, Charron J-B, Vali H, Bertrand A, Jabaji S (2015) Accelerated growth rate and increased drought stress resilience of the model grass *Brachypodium distachyon* colonized by *Bacillus subtilis* B26. PLoS ONE 10:e0130456
García JE, Maroniche G, Creus C, Suárez-Rodríguez R, Ramirez-Trujillo JA, Groppa MD (2017) *In vitro* PGPR properties and osmotic tolerance of different *Azospirillum* native strains and their effects on growth of maize under drought stress. Microbiol Res 202:21–29
Gholamhoseini M, Ghalavand A, Dolatabadian A, Jamshidi E, Khodaei-Joghan A (2013) Effects of arbuscular mycorrhizal inoculation on growth, yield, nutrient uptake and irrigation water productivity of sunflowers grown under drought stress. Agric Water Manag 117:106–114
Ghosh D, Gupta A, Mohapatra S (2019) A comparative analysis of exopolysaccharide and phytohormone secretions by four drought-tolerant rhizobacterial strains and their impact on osmotic-stress mitigation in *Arabidopsis thaliana*. World J Microbiol Biotechnol 35:90
Glick BR (2014) Bacteria with ACC deaminase can promote plant growth and help to feed the world. Microbiol Res 169:30–39
Glick BR, Penrose DM, Li J (1998) A model for the lowering of plant ethylene concentrations by plant growth promoting bacteria. J Theor Biol 190:63–68
Gontia-Mishra I, Sapre S, Kachare S, Tiwari S (2017) Molecular diversity of 1-aminocyclopropane-1-carboxylate (ACC) deaminase producing PGPR from wheat (*Triticum aestivum* L.) rhizosphere. Plant Soil 414:213–227
Gontia-Mishra I, Sapre S, Sharma A, Tiwari S (2016) Amelioration of drought tolerance in wheat by the interaction of plant growth-promoting rhizobacteria. Plant Biol 18:992–1000
Gontia-Mishra I, Sasidharan S, Tiwari S (2014) Recent developments in use of 1-amino cyclopropane-1-carboxylate (ACC) deaminase for conferring tolerance to biotic and abiotic stress. Biotechnol Lett 36:889–898
Gusain YS, Singh US, Sharma AK (2015) Bacterial mediated amelioration of drought stress in drought tolerant and susceptible cultivars of rice (*Oryza sativa* L.). Afr J Biotechnol 14:764–773
Hasegawa PM, Bressan RA, Zhu JK, Bohnert HJ (2000) Plant cellular and molecular responses to high salinity. Ann Rev Plant Physiol Plant Mol Biol 51:463–499
Hashem A, Kumar A, Al-Dbass AM et al (2019) Arbuscular mycorrhizal fungi and biochar improves drought tolerance in chickpea. Saudi J Biol Sci 26:614–624
Hepper CM (1975) Extracellular polysaccharides of soil bacteria. In: Walker N (ed) Soil microbiology, a critical review. Wiley, New York, pp 93–111
Hussain M, Malik MA, Farooq M, Ashraf MY, Cheema MA (2008) Improving drought tolerance by exogenous application of glycine betaine and salicylic acid in sunflower. J Agron Crop Sci 194:193–199

Ilhan S, Ozdemir F, Bor M (2015) Contribution of trehalose biosynthetic pathway to drought stress tolerance of *Capparis ovata* Desf. Plant Biol 17:402–407

Jaleel CA, Manivannan P, Sankar B, Kishorekumar A, Gopi R, Somasundaram R, Panneerselvam R (2007) Induction of drought stress tolerance by ketoconazole in *Catharanthus roseus* is mediated by enhanced antioxidant potentials and secondary metabolite accumulation. Colloids Surf, B 60:201–206

Jha B, Gontia I, Hartmann A (2012) The roots of the halophyte Salicornia brachiata are a source of new halotolerant diazotrophic bacteria with plant growth-promoting potential. Plant Soil 356(1–2):265–277

Jing H, Li C, Ma F, Ma JH, Khan A, Wang X, Zhao LY, Gong ZH, Chen RG (2016) Genome-wide identification, expression diversication of dehydrin gene family and characterization of CaDHN3 in pepper (*Capsicum annuum* L.). PloS ONE. 11:e0161073

Kang SM, Radhakrishnan R, Khan AL, Kim MJ, Park JM, Kim BR, Shin D-H, Lee I-J (2014) Gibberellin secreting rhizobacterium *Pseudomonas putida* H-2-3 modulates the hormonal and stress physiology of soybean to improve the plant growth under saline and drought conditions. Plant Physiol Biochem 84:115–124

Kashiwagi J, Krishnamurthy L, Crouch JH, Serraj R (2006) Variability of root length density and its contributions to seed yield in chickpea (*Cicer arietinum* L.) under terminal drought stress. Field Crops Res 95:171–181

Kaur G, Asthira B (2017) Molecular responses to drought stress in plants. Biol Plantarum 61:201–209

Kaushal M (2019) Microbes in cahoots with plants: MIST to hit the jackpot of agricultural productivity during drought. Int J Mol Sci 20:1769

Kaushal M, Wani SP (2016) Plant-growth-promoting rhizobacteria: drought stress alleviators to ameliorate crop production in drylands. Ann Microbiol 66:35–42

Khan N, Bano A, Rahman MA, Guo J, Kang Z, Babar MA (2019) Comparative physiological and metabolic analysis reveals a complex mechanism involved in drought tolerance in chickpea (*Cicer arietinum* L.) induced by PGPR and PGRs. Sci Rep 9:2097

Khan N, Zandi P, Ali S, Mehmood A, Adnan Shahid M (2018) Impact of salicylic acid and PGPR on the drought tolerance and phytoremediation potential of *Helianthus annus*. Front Microbiol 9:2507

Kloepper JW, Schroth M (1978) Plant growth promoting rhizobacteria on radishes. In: Proceedings of the 4th international conference on plant pathogenic bacteria, Angers, vol 2, 879–882

Knight H, Knight M (2001) Abiotic stress signalling pathways: specificity and cross-talk. Trends Plant Sci 6:262–267

Kour D, Rana KL, Kumar A, Rastegari AA, Yadav N, Yadav AN, Gupta VK (2019a) Extremophiles for hydrolytic enzymes productions: biodiversity and potential biotechnological applications. In: Molina G, Gupta VK, Singh BN, Gathergood N (eds) Bioprocessing for biomolecules production. Wiley, USA, pp 321–372

Kour D, Rana KL, Yadav AN, Yadav N, Kumar V, Kumar A, Sayyed RZ, Hesham AE-L, Dhaliwal HS, Saxena AK (2019b) Drought-tolerant phosphorus-solubilizing microbes: biodiversity and biotechnological applications for alleviation of drought stress in plants. In: Sayyed RZ, Arora NK, Reddy MS (eds) Plant growth promoting rhizobacteria for sustainable stress management, Volume 1: Rhizobacteria in Abiotic Stress Management. Springer, Singapore, pp 255–308. https://doi.org/10.1007/978-981-13-6536-2_13

Kour D, Rana KL, Yadav N, Yadav AN, Kumar A, Meena VS, Singh B, Chauhan VS, Dhaliwal HS, Saxena AK (2019c) Rhizospheric microbiomes: biodiversity, mechanisms of plant growth promotion, and biotechnological applications for sustainable agriculture. In: Kumar A, Meena VS (eds) Plant growth promoting rhizobacteria for agricultural sustainability: from theory to practices. Springer, Singapore, pp 19–65. https://doi.org/10.1007/978-981-13-7553-8_2

Kour D, Rana KL, Yadav N, Yadav AN, Singh J, Rastegari AA, Saxena AK (2019d) Agriculturally and industrially important fungi: current developments and potential biotechnological applications. In: Yadav AN, Singh S, Mishra S, Gupta A (eds) Recent advancement in white biotechnology through fungi, Volume 2: Perspective for Value-Added Products and Environments. Springer International Publishing, Cham, pp 1–64. doi:https://doi.org/10.1007/978-3-030-14846-1_1

Kumar A, Patel JS, Meena VS, Ramteke PW (2019a) Plant growth-promoting rhizobacteria: strategies to improve abiotic stresses under sustainable agriculture. J Plant Nutr 42:1402–1415

Kumar M, Saxena R, Rai PK, Tomar RS, Yadav N, Rana KL, Kour D, Yadav AN (2019) Genetic diversity of methylotrophic yeast and their impact on environments. In: Yadav AN, Singh S, Mishra S, Gupta A (eds) Recent advancement in white biotechnology through fungi: Volume 3: Perspective for Sustainable Environments. Springer International Publishing, Cham, pp 53–71. https://doi.org/10.1007/978-3-030-25506-0_3

Lakshmanan V, Castaneda R, Rudrappa T, Bais HP (2013) Root transcriptome analysis of *Arabidopsis thaliana* exposed to beneficial *Bacillus subtilis* FB17 rhizobacteria revealed genes for bacterial recruitment and plant defense independent of malate efflux. Planta 238:657–668

Levitt J (1980) Response of plants to environmental stress, vol 2. Academic Press, New York

Li H, Guo Q, Jing Y et al (2019a) Application of *Streptomyces pactum* Act12 enhances drought resistance in wheat. J Plant Growth Regul. https://doi.org/10.1007/s00344-019-09968-z

Li J, Meng B, Chai H, Yang X, Song W, Li S, Lu A, Zhang T, Sun W (2019b) Arbuscularmycorrhizal fungi alleviate drought stress in C3 (*Leymus chinensis*) and C4 (*Hemarthria altissima*) grasses via altering antioxidant enzyme activities and photosynthesis. Front Plant Sci 10:499

Li Y, Shi H, Zhang H, Chen S (2019c) Amelioration of drought effects in wheat and cucumber by the combined application of super absorbent polymer and potential biofertilizer. Peer J 7:e6073

Lim JH, Kim SD (2013) Induction of drought stress resistance by multi-functional PGPR *Bacillus licheniformis* K11 in pepper. Plant Pathol J 29:201–208

Lipiec J, Doussan C, Nosalewicz A, Kondracka K (2013) Effect of drought and heat stresses on plant growth and yield: a review. Int Agrophys 27:463–477

Liu CY, Zhang F, Zhang DJ, Srivastava AK, Wu QS, Zou YN (2018) Mycorrhiza stimulates root-hair growth and IAA synthesis and transport in trifoliate orange under drought stress. Sci Rep 8:1978

Liu F, Xing S, Ma H, Du Z, Ma B (2013) Cytokinin producing, plant growth promoting rhizobacteria that confer resistance to drought stress in *Platycladus orientalis* container seedlings. Appl Microbiol Biotechnol 97:9155–9164

Lu X, Liu S-F, Yue L, Zhao X, Zhang Y-B, Xie Z-K, Wang R-Y (2018) *Epsc* involved in the encoding of exopolysaccharides produced by *Bacillus amyloliquefaciens* FZB42 act to boost the drought tolerance of *Arabidopsis thaliana*. Int J Mol Sci 19:3795

Mikami K, Katagiri T, Iuchi S, Yamaguchi-Shinozaki K, Shinozaki K (1998) A gene encoding phosphatidylinositol-4-phosphate-5- kinase is induced by water stress and abscisic acid in *Arabidopsis thaliana*. Plant J 15:563–568

Miller G, Susuki N, Ciftci-Yilmaz S, Mittler R (2010) Reactive oxygen species homeostasis and signalling during drought and salinity stresses. Plant Cell Environ 33:45–467

Nahar K, Hasanuzzaman M, Fujita M (2016) Roles of osmolytes in plant adaptation to drought and salinity. Osmolytes and plants acclimation to changing environment: emerging omics technologies. Springer, New Delhi, pp 37–68

Naseem H, Ahsan M, Shahid MA, Khan N (2018) Exopolysaccharides producing rhizobacteria and their role in plant growth and drought tolerance. J Basic Microbiol 58:1009–1022

Nautiyal CS, Srivastava S, Chauhan PS, Seem K, Mishra A, Sopory SK (2013) Plant growth-promoting bacteria *Bacillus amyloliquefaciens* NBRISN13 modulates gene expression profile of leaf and rhizosphere community in rice during salt stress. Plant Physiol Biochem 66:1–9

Naveed M, Hussain MB, Zahir ZA, Mitter B, Sessitsch A (2014) Drought stress amelioration in wheat through inoculation with *Burkholderia phytofirmans* strain PsJN. Plant Growth Regul 73:121–131

Niu X, Song L, Xiao Y, Ge W (2018) Drought-tolerant plant growth-promoting rhizobacteria associated with foxtail millet in a semi-arid agroecosystem and their potential in alleviating drought stress. Front Microbiol 8:2580

Nocelli N, Bogino PC, Banchio E, Giordano W (2016) Roles of extracellular polysaccharides and biofilm formation in heavy metal resistance of rhizobia. Materials 9:418

Patten CL, Glick BR (2002) Role of *Pseudomonas putida* indoleacetic acid in development of host plant root system. Appl Environ Microbiol 48:3795–3801

Paul MJ, Primavesi LF, Jhurreea D, Zhang Y (2008) Trehalose metabolism and signaling. Ann Rev Plant Biol 59:417–441

Posas F, Wurgler-Murphy SM, Maeda T, Witten EA, Thai TC, Saito H (1996) Yeast HOG1 MAP kinase cascade is regulated by a multistep phosphorelay mechanism in the SLN1-YPD1-SSK1 "two-component" osmosensor. Cell 86:865–875

Quiroga G, Erice G, Aroca R, Chaumont F, Ruiz-Lozano JM (2017) Enhanced drought stress tolerance by the arbuscular mycorrhizal symbiosis in a drought-sensitive maize cultivar is related to a broader and differential regulation of host plant aquaporins than in a drought-tolerant cultivar. Front Plant Sci 8:1056

Quiroga G, Erice G, Aroca R, Zamarreño AM, García-Mina JM, Ruiz-Lozano JM (2018) Arbuscular mycorrhizal symbiosis and salicylic acid regulate aquaporins and root hydraulic properties in maize plants subjected to drought. Agric Water Manag 202:271–284

Rekha K, Kumar RM, Ilango K, Rex A, Usha B (2018) Transcriptome profiling of rice roots in early response to *Bacillus subtilis* (RR4) colonization. Botany 96:749–765

Rodriguez EM, Svensson JT, Malatrasi M, Choi DW, Close TJ (2005) Barley Dhn13 encodes a KS-type dehydrin with constitutive and stress responsive expression. Theo Appl Genet 110:852–858

Rodríguez-Salazar J, Suárez R, Caballero-Mellado J, Iturriaga G (2009) Trehalose accumulation in *Azospirillum brasilense* improves drought tolerance and biomass in maize plants. FEMS Microbiol Lett 296:52–59

Rolli E, Marasco R, Vigani G, Ettoumi B, Mapelli F, Deangelis ML, Gandolfi C, Casati E, Previtali F, Gerbino R et al (2014) Improved plant resistance to drought is promoted by the root-associated microbiome as a water stress-dependent trait. Environ Microbiol 17:316–331

Ruzzi M, Aroca R (2015) Plant growth-promoting rhizobacteria act as biostimulants in horticulture. Sci Hort 196:124–134

Saakre M, Baburao TM, Salim AP, Ffancies RM, Achuthan VP, Thomas G, Sivarajan SR (2017) Identification and characterization of genes responsible for drought tolerance in rice mediated by *Pseudomonas fluorescens*. Rice Sci 24:291–298

Saikia J, Sarma RK, Dhandia R, Yadav A, Bharali R, Gupta VK, Saikia R (2018) Alleviation of drought stress in pulse crops with ACC deaminase producing rhizobacteria isolated from acidic soil of Northeast India. Sci Rep 8(1)

Saleem AR, Brunetti C, Khalid A, Della Rocca G, Raio A, Emiliani G et al (2018) Drought response of *Mucuna pruriens* (L.) DC. inoculated with ACC deaminase and IAA producing rhizobacteria. PLoS ONE 13:e0191218

Sanchez-Romera B, Ruiz-Lozano JM, Zamarreno AM, Garcia-Mina JM, Aroca R (2016) Arbuscular mycorrhizal symbiosis and methyl jasmonate avoid the inhibition of root hydraulic conductivity caused by drought. Mycorrhiza 26:111–122

Sandhya V, SkZ Ali, Grover M, Reddy G, Venkateswaralu B (2010) Effect of plant growth promoting *Pseudomonas* spp. on compatible solutes antioxidant status and plant growth of maize under drought stress. Plant Growth Regul 62:21–30

Sandhya V, Ali SKZ, Grover M, Reddy G, Venkateswarlu B (2009) Alleviation of drought stress effects in sunflower seedlings by the exopolysaccharides producing *Pseudomonas putida* strain GAP-P45. Biol Fertil Soils 46:17–26

Sapre S, Gontia-Mishra I, Tiwari S (2019) ACC deaminase producing bacteria: a key player in alleviating abiotic stresses in plants. In: Kumar A, Meena VS (eds) Plant growth promoting rhizobacteria for agricultural sustainability-from theory to practices. Springer Nature, pp 267–291

Sarma RK, Saikia R (2014) Alleviation of drought stress in mung bean by strain *Pseudomonas aeruginosa* GGRJ21. Plant Soil 377:111–126

Selvakumar G, Panneerselvam P, Ganeshamurthy AN (2012) Bacterial mediated alleviation of abiotic stress in crops. In: Maheshwari DK (ed) Bacteria in agrobiology: stress management. Springer-Verlag, Berlin, pp 205–224

Shaharoona B, Bibi R, Arshad M, Zahir ZA, Hassan Z (2006) 1-aminocylopropane-1-carboxylate (ACC)-deaminase rhizobacteria extenuates ACC-induced classical triple response in etiolated pea seedlings. Pak J Bot 38:1491–1499

Sharifi R, Ryu C-M (2018) Revisiting bacterial volatile-mediated plant growth promotion: lessons from the past and objectives for the future. Ann Bot 122:349–358

Smirnoff N (1993) The role of reactive oxygen in the response of plants to water deficit and desiccation. J New Phytol 125:27–30

Srivastava S, Chaudhry V, Mishra A, Chauhan PS, Rehman A, Yadav A, Tuteja N, Nautiyal CS (2012) Gene expression profiling through microarray analysis in *Arabidopsis thaliana* colonized by *Pseudomonas putida* MTCC5279, a plant growth promoting rhizobacterium. Plant Sig Behav 7:235–245

Streeter JG (1985) Accumulation of alpha, alpha-trehalose by *Rhizobium* bacteria and bacteroids. J Bacteriol 164:78–84

Suárez R, Wong A, Ramírez M, Barraza A, Orozco MD, Cevallos MA, Lara M, Hernández G, Iturriaga G (2008) Improvement of drought tolerance and grain yield in common bean by overexpressing trehalose-6-phosphate synthase in rhizobia. Mol Plant-Microbe Interact 21:958–966

Sukkasem P, Kurniawan A, Kao TC, Chuang HW (2018) A multifaceted rhizobacterium *Bacillus licheniformis* functions as a fungal antagonist and a promoter of plant growth and abiotic stress tolerance. Environ Exper Bot 155:541–551

Sun X, Shi J, Ding G (2017) Combined effects of arbuscular mycorrhiza and drought stress on plant growth and mortality of forage sorghum. Appl Soil Ecol 119:384–391

Symanczik S, Lehmann MF, Wiemken A, Boller T, Courty P (2018) Effects of two contrasted arbuscular mycorrhizal fungal isolates on nutrient uptake by *Sorghum bicolor* under drought. Mycorrhiza 28:779–785

Timmusk S, El-Daim IAA, Copolovici L, Tanilas T, Kännaste A, Behers L, Nevo E, Seisenbaeva G, Stenström E, Niinemets Ü (2014) Drought-tolerance of wheat improved by rhizosphere bacteria from harsh environments: enhanced biomass production and reduced emissions of stress volatiles. PLoS ONE 9:e96086

Timmusk S, Wagner EGH (1999) The plant-growth-promoting rhizobacterium *Paenibacillus polymyxa* induces changes in *Arabidopsis thaliana* gene expression a possible connection between biotic and abiotic stress responses. Mol Plant-Microbe Interact 12:951–959

Tiwari S, Lata C, Chauhan PS, Nautiyal CS (2016) *Pseudomonas putida* attunes morphophysiological, biochemical and molecular responses in *Cicer arietinum* L. during drought stress and recovery. Plant Physiol Biochem 99:108–117

Tiwari S, Prasad V, Chauhan PS, Lata C (2017) *Bacillus amyloliquefaciens* confers tolerance to various abiotic stresses and modulates plant response to phytohormones through osmoprotection and gene expression regulation in rice. Front Plant Sci 8:1510

Ullah A, Manghwar H, Shaban M, Khan AH, Akbar A, Ali U, Fahad S (2019a) Phytohormones enhanced drought tolerance in plants: a coping strategy. Environ Sci Pollut Res 25:33103–33118

Ullah A, Nisar M, Ali H, Hazrat A, Hayat K, Keerio AA, Ihsan M, Laiq M, Ullah S, Fahad S, Khan A (2019b) Drought tolerance improvement in plants: an endophytic bacterial approach. Appl Microbiol Biotechnol 103:7385–7397

Urao T, Yakubova B, Satoha R, Yamaguchi-Shinozakia K, Sekib M, Hirayamab T, Shinozakib K (1999) A transmembrane hybrid-type histidine kinase in *Arabidopsis* functions as an osmosensor. Plant Cell 11:1743–1754

Vaishnav A, Choudhary DK (2019) Regulation of drought-responsive gene expression in *Glycine max* L. (merrill) is mediated through *Pseudomonas simiae* strain AU. J Plant Grow Regul 38:333–342

Vardharajula S, Zulfikar Ali S, Grover M, Reddy G, Bandi V (2011) Drought-tolerant plant growth promoting *Bacillus* spp.: effect on growth, osmolytes, and antioxidant status of maize under drought stress. J Plant Interact 6:1–14

Vargas L, Santa Brigida AB, MotaFilho JP et al (2014) Drought tolerance conferred to sugarcane by association with *Gluconacetobacter diazotrophicus*: a transcriptomic view of hormone pathways. PLoS ONE 9:e114744

Vejan P, Abdullah R, Khadiran T, Ismail S, Nasrulhaq Boyce A (2016) Role of plant growth promoting rhizobacteria in agricultural sustainability-a review. Molecules 21:573

Verma P, Yadav AN, Khannam KS, Panjiar N, Kumar S, Saxena AK, Suman A (2015a) Assessment of genetic diversity and plant growth promoting attributes of psychrotolerant bacteria allied with wheat (*Triticum aestivum*) from the northern hills zone of India. Ann Microbiol 65:1885–1899

Verma P, Yadav AN, Kumar V, Singh DP, Saxena AK (2017) Beneficial plant-microbes interactions: biodiversity of microbes from diverse extreme environments and its impact for crop improvement. In: Singh DP, Singh HB, Prabha R (eds) Plant-microbe interactions in agro-ecological perspectives: Volume 2: Microbial Interactions and Agro-Ecological Impacts. Springer, Singapore, pp 543–580.

Verma P, Yadav AN, Shukla L, Saxena AK, Suman A (2015b) Alleviation of cold stress in wheat seedlings by *Bacillus amyloliquefaciens* IARI-HHS2-30, an endophytic psychrotolerant K-solubilizing bacterium from NW Indian Himalayas. Natl J Life Sci 12:105–110

Vigani G, Rolli E, Marasco R, Dell'Orto M, Michoud G, Soussi A, Raddadi N, Borin S, Sorlini C, Zocchi G, Daffonchio D (2018) Root bacterial endophytes confer drought resistance and enhance expression and activity of a vacuolar H^+ -pumping pyrophosphatase in pepper plants. Environ Microbiol 21:3212–3228.

Vurukonda SSKP, Vardharajula S, Shrivastava M, SkZ A (2016) Enhancement of drought stress tolerance in crops by plant growth promoting rhizobacteria. Microbiol Res 184:13–24

Wu QS, He JD, Srivastava AK, Zou YN, Kuča K (2019) Mycorrhiza enhances drought tolerance of citrus by altering root fatty acid compositions and their saturation levels. Tree Physiol 39:1149–1158

Wu HH, Zou YN, Rahman MM, Ni QD, Wu QS (2017) Mycorrhizas alter sucrose and proline metabolism in trifoliate orange exposed to drought stress. Sci Rep 7:42389

Xie W, Hao Z, Zhou X, Jiang X, Xu L, Wu S, Zhao A, Zhang X, Chen B (2018) Arbuscular mycorrhiza facilitates the accumulation of glycyrrhizin and liquiritin in *Glycyrrhiza uralensis* under drought stress. Mycorrhiza 28:285–300

Xie Z, Chu Y, Zhang W, Lang D, Zhang X (2019) *Bacillus pumilus* alleviates drought stress and increases metabolite accumulation in *Glycyrrhiza uralensis* Fisch. Environ Exper Bot 158:99–106

Xu L, Li T, Wu Z, Feng H, Yu M, Zhang X, Chen B (2018) Arbuscular mycorrhiza enhances drought tolerance of tomato plants by regulating the 14-3-3 genes in the ABA signaling pathway. Appl Soil Ecol 125:213–221

Yadav AN, Kour D, Sharma S, Sachan SG, Singh B, Chauhan VS, Sayyed RZ, Kaushik R, Saxena AK (2019a) Psychrotrophic microbes: biodiversity, mechanisms of adaptation, and biotechnological implications in alleviation of cold stress in plants. In: Sayyed RZ, Arora NK, Reddy MS (eds) Plant growth promoting rhizobacteria for sustainable stress management: Volume 1: Rhizobacteria in Abiotic Stress Management. Springer, Singapore, pp 219–253. https://doi.org/10.1007/978-981-13-6536-2_12

Yadav AN, Kumar V, Prasad R, Saxena AK, Dhaliwal HS (2018a) Microbiome in crops: diversity, distribution and potential role in crops improvements. In: Prasad R, Gill SS, Tuteja N (eds) Crop improvement through microbial biotechnology. Elsevier, USA, pp 305–332

Yadav AN, Mishra S, Singh S, Gupta A (2019b) Recent advancement in white biotechnology through fungi Volume 1: Diversity and Enzymes Perspectives. Springer International Publishing, Cham

Yadav AN, Singh S, Mishra S, Gupta A (2019c) Recent advancement in white biotechnology through fungi. Volume 2: Perspective for Value-Added Products and Environments. Springer International Publishing, Cham

Yadav AN, Singh S, Mishra S, Gupta A (2019d) Recent advancement in white biotechnology through fungi. Volume 3: Perspective for Sustainable Environments. Springer International Publishing, Cham

Yadav AN, Verma P, Kumar M, Pal KK, Dey R, Gupta A, Padaria JC, Gujar GT, Kumar S, Suman A, Prasanna R, Saxena AK (2015) Diversity and phylogenetic profiling of niche-specific Bacilli from extreme environments of India. Ann Microbiol 65:611–629

Yadav AN, Verma P, Kumar S, Kumar V, Kumar M, Singh BP, Saxena AK, Dhaliwal HS (2018b) Actinobacteria from rhizosphere: molecular diversity, distributions and potential biotechnological applications. In: Singh B, Gupta V, Passari A (eds) New and future developments in microbial biotechnology and bioengineering. USA, pp 13–41. https://doi.org/10.1016/b978-0-444-63994-3.00002-3

Yadav AN, Verma P, Kumar V, Sachan SG, Saxena AK (2017a) Extreme cold environments: a suitable niche for selection of novel psychrotrophic microbes for biotechnological applications. Adv Biotechnol Microbiol 2:1–4

Yadav AN, Verma P, Sachan SG, Saxena AK (2017b) Biodiversity and biotechnological applications of psychrotrophic microbes isolated from Indian Himalayan regions. EC Microbiol ECO 01:48–54

Yadav N, Yadav A (2018) Biodiversity and biotechnological applications of novel plant growth promoting methylotrophs. J Appl Biotechnol Bioeng 5:342–344

Yasmin H, Nosheen A, Naz R, Bano A, Keyani R (2017) l-tryptophan-assisted PGPR-mediated induction of drought tolerance in maize (*Zea mays* L.). J Plant Interact 12:567–578

Zade NSE, Sadeghi A, Moradi P (2019) *Streptomyces* strains alleviate water stress and increase peppermint (*Mentha piperita*) yield and essential oils. Plant Soil 434:441–452

Zhang H, Murzello C, Sun Y, Kim MS, Xie X, Jeter RM, Zak JC, Dowd SE et al (2010) Choline and osmotic-stress tolerance induced in Arabidopsis by the soil microbe *Bacillus subtilis* (GB03). Mol Plant Microbe Interact 23:1097–1104

Zhang S, Moyne AL, Reddy MS, Kloepper JW (2002) The role of salicylic acid in induced systemic resistance elicited by plant growth-promoting rhizobacteria against blue mold of tobacco. Biol Cont 25:288–296

Zhao R, Guo W, Bi N, Guo J, Wang L, Zhao J, Zhang J (2015) Arbuscular mycorrhizal fungi affect the growth, nutrient uptake and water status of maize (*Zea mays* L.) grown in two types of coalmine spoils under drought stress. Appl Soil Ecol 88:41–49

Zhu JK (2002) Salt and drought stress signal transduction in plants. Ann Rev Plant Biol 53:247–273

Chapter 13
Microbial Consortium as Biofertilizers for Crops Growing Under the Extreme Habitats

Chuks Kenneth Odoh, Kabari Sam, Nenibarini Zabbey, Chibuzor Nwadibe Eze, Amechi S. Nwankwegu, Charity Laku and Boniface Barinem Dumpe

Abstract Biofertilizers are typically microbial formulations in organic carrier materials that improve soil health and crop growth and development. Of late the use of biofertilizers has gained much acceptance and research interest especially in the developed countries due to ecological impacts associated with the use of synthetic inorganic fertilizers in farming. Microbial formulations could be organism-specific or a consortium of organisms. Microbial consortium biofertilizers, the main focus of this chapter, have been reported as contributing significantly to plant adaptation to various abiotic stressors in "extreme" habitats. Many soil microorganisms are endowed with an array of capabilities ranging from production of growth-enhancing substances to the release of substances which ameliorate the effects of various abiotic stress conditions such as drought, salinity, pH stress, heat stress, pollutants, and nutrient deficiency. Besides exploring the MC biofertilizer operations and mechanisms (neutral and niche), it also relies on a network of intraspecific and interspecific

C. K. Odoh (✉) · K. Sam · N. Zabbey · C. Laku
Environment and Conservation Unit, Centre for Environment, Human Rights and Development (CEHRD), Legacy Centre, 6 Abuja Lane, D-Line, Port Harcourt, Rivers State, Nigeria
e-mail: kenchuks974@gmail.com; k.odoh@cehrd.org.ng

C. K. Odoh
Department of Microbiology, University of Nigeria, Nsukka 410001, Enugu State, Nigeria

K. Sam
Faculty of Marine Environmental Management, Department of Marine Environment and Pollution Control, Nigeria Maritime University, Warri, Delta State, Nigeria

N. Zabbey
Faculty of Agriculture, Department of Fisheries, University of Port Harcourt, PMB, 5323, East-West Road, Choba, Rivers State, Nigeria

C. N. Eze
Department of Microbiology, University of Nigeria, Nsukka 410001, Enugu State, Nigeria

A. S. Nwankwegu
Department of Environmental Science and Engineering, Hohai University, Nanjing, China

B. B. Dumpe
Faculty of Agriculture, Department of Crop and Soil, University of Port Harcourt, PMB, 5323, East-West Road, Choba, Rivers State, Nigeria

A. N. Yadav et al. (eds.), *Plant Microbiomes for Sustainable Agriculture*, Sustainable Development and Biodiversity 25, https://doi.org/10.1007/978-3-030-38453-1_13

interactions for sustainable growth and development of crops in challenged environments. Among these organisms are plant growth-promoting rhizobacteria (PGPR), arbuscular mycorrhizal fungi (AMF), mycorrhizal-helping bacteria (MHB), and a host of others that work together in consortium biofertilizer formulations.

Keywords Biofertilizers · Microbial consortium · Extreme habitat · Plant growth-promoting rhizobacteria (PGPR)

13.1 Introduction

The role of soil microorganisms in sustainable agriculture has gained prominence over the last decades. Soil microorganisms and their activities have contributed to all spheres of biotechnological advances such as molecular understudy of litter decomposition (Tamayo-Vélez and Nelson 2018), ecosystem processes and planetary earth systems (Lladó et al. 2017), and sustainable development-driven studies (Timmis et al. 2017). Among practitioners, policy-makers, academia, and international organizations, there has been increased calls for research to unravel more benefits of soil microbes particularly in food and agriculture (Compant et al. 2010; Clark et al. 2009; Odoh et al. 2019a), environmental remediation (Sam et al. 2017; Zabbey et al. 2017; Eze et al. 2018), and industrial applications (Odoh 2017; Zuroff and Curtis 2012).

Sustainable agriculture is a unique modern farming practice that promotes soil health, wholesome agricultural yield, and reduced pollution of cultivable soils. In addition, it is a strategic agrobiotechnology approach where the present societal food demands are met without compromising future generation's food security.

As the quest to feed the ever-growing human populations (7.7 billion) gets global momentum and the push to curb food insecurity intensifies (Glick 2012), there has been an un-quenching desire for increased yield per unit area production capacity. This is paramount so as to meet the growing demand for food, while also subverting pressures arising from the natural ecosystem (Pindi and Satyanarayana 2012; Chatzipavlidis et al. 2013). According to Food and Agriculture Organization (FAO) 2010 report, 60% increase in agricultural commodities demand is expected by 2030. More fascinating is the fact that more than 85% of this estimated yield will be emanating from developing countries whose economic growth is mainly dependent on agriculture (Mia and Shamsuddin 2010).

Microorganisms constitute one of the most diverse biological communities in the soil ecosystem. The interactions and activities of soil microbes aid increase in food production, earth geochemical stability, climatic, and biogeochemical cycles (Tringe et al. 2005; Hansel, et al. 2008). According to Odoh et al. (2019a), they live in complex biological settings within which exist interactions and influence of living and nonliving parameters. Advanced studies have, however, demonstrated the applicability of engineered species in medicine such as antibiotic resistance genes (Cycon et al. 2019) and microbial pigments in food industry (Sen et al. 2019) through understanding their behavioral pattern, complexity and mechanisms of actions (Kumar 2016; Ahmad et al. 2011).

In agriculture (Table 13.1), microbial consortium (MC) involves the symbiotic interactions of two or more microbial groups (Clark et al. 2009), for improved crop growth. Association of microbial groups enhances turnover of soil organic matters and mobilizes nutrients for plants growth and fixation of nitrogen (N) in leguminous crops (Nuti and Giovannetti 2015). As a result of the intrinsic advantages, MC acceptability and applicability by practitioners have increased unlike the use of a single strain, as demonstrated by Sarkar et al. (2011) who exploited metabolic versatility of microbial community for the treatment of organic wastes. In the bulk soil, MC lives in close interaction in the plant root canopy as phyllosphere bacillus, endophytes, and as members of a complex microbiota (Hacquard et al. 2015), thus justifying their seldom existence as a single strain. In spite of the individualistic characterization of the single cells, MC often responds to environmental stress as a unique organism. This is because they have more chances than any single strain among the population to adapt and take advantage of their internal beneficial interactions (Nuti and Giovannetti 2015). On regular intervals, their activities consist of a continuous shift between viable and culturable, and viable but non-culturable cells of the diverse components of the total population. Also, via unique chemical signaling mechanism ("quorum sensing"), MC detects and responds to cell population density and nutrient gradient through gene regulation. This mechanism thus helps in expressing appealing biochemical properties that enable their stability, functionality, robustness, and capacity to perform complex biochemical tasks.

With the increase in the global use of chemical fertilizer often beyond crop and soil's requirement thresholds (Sun et al. 2015; Liu et al. 2017), researchers and agriculturists are finding ways to limit the intense applications of agrochemicals and to significantly reduce their impacts on the soil ecosystem. Consequently, integration of natural biotechnological advantages of soil microbes in crop production systems would undoubtedly mitigate plants growth challenges, while also brightening the global desire for sustainable agriculture (Manindra et al. 2013; Odoh 2017). The process of co-metabolism is another interactive advantage of MC over single inoculum. This usually manifests when a specific microbial group gratuitously produces a specific metabolite which often serves as a potential limiting nutrient for another population within the consortium. This essentially helps to ensure complete mineralization of by-products which could constitute environmental nuisance to both plants and microbiota while optimizing soil arable functions.

Table 13.1 Microbial alleviation of extreme conditions associated with crop growth

Organisms	Stress conditions	Crop	Outcome	Mode and tolerant strategies	References
Rhizobium etli	Drought	*Phaseolus vulgaris*	Reduces the sensitivity of drought in plant and nodule dry weight, nitrogen content, and nodule functionality	Overexpressing this oxidase in bacteroides	Talbi et al. (2012)
Azospirillum brasilense	Drought	*Glycine max*	Improves plant traits that can help tolerance of water deficiency	Indole-3-acetic acid synthesis and nitrogen fixation	Hungria et al. (2015)
Azospirillum brasilense	Salinity	*Zea mays*	Enhance plant growth promotion and reduction in proline content	Alteration of the selectivity of Na^{+}, K^{+}, and Ca^{++} ions	*Fukami* et al. (2017) Fukami et al. (2018)
Achromobacter piechaudii	Heavy metal and salt stress	*Populus* Species *Lycopersicon esculentum*	Increase shoot and root growth and enhance root hair formation	IAA biosynthesis and overexpression of abiotic stress response gene	Fahad et al. (2015) Carmen and Roberto (2011)
Sinorhizobium arboris	Heat stress	Acacia Senegal Cajanus cajan	Maintain basal metabolic activity under adverse heat conditions	Enzymatic activity (Esterase, Chitinase, and glucanase production)	Leena et al. (2001) Kumar et al. (2010)
Bacillus megaterium *B. subtilis* *Bacillus thuringiensis*	Drought, pH, temperature	Triticum aestivum L Cicer arietinum	Enhanced leaf relative water content (RWC), greater biomass of shoot and root, and higher accumulation of protein, sugar, and phenolic compounds	Production of phytohormones	Khan et al. (2019)

(continued)

Table 13.1 (continued)

Organisms	Stress conditions	Crop	Outcome	Mode and tolerant strategies	References
Pantoea agglomerans	Heavy metals	Avena sativa	Ameliorate heavy metals stress on plant development		Pishchik et al. (2009)
Brevibacillus Brevis	Heat stress	Cotton crop	Enhance development of various plant growth parameters	Production of plant growth promoters such as IAA, ARA, antifungal activity, and ammonia synthesis	Nehra, et al. (2016)
Curvularia protuberata	Heat and drought	Dichanthelium lanuginosum		Colonization of roots	de Zelicourt et al. (2013)
Pseudomonas aeruginosa	Zn toxicity	Triticum aestivum		Improved biomass, N and P uptake, and total soluble protein	Islam et al. (2014)
Pseudomonas koreensis AGB-1	Cd, AS, Cu, Pb toxicity	Miscanthus sinensis		ACC deaminase, IAA production	Babu et al. (2015)
Photobacterium spp.	Hg toxicity	Phragmites australis		IAA, mercury reductase activity	Mathew et al. (2015)
Bacillus thuringiensis AZP2	Drought	Triticum aestivum		Production of volatile organic compounds	Timmusk et al. (2014)
Burkholderia sp., *Streptomyces platensis*	Nutrient deficiency	*Zea mays*			Oliveira et al. (2009)
Phyllobacterium		Strawberries	Aid in phosphate solubilization, plants protection against pathogens	Potassium and phosphate solubilization	Flores-Felix et al. (2015)
Rhizobium		*lettuce*	Increase plants biomass due to enhanced nutrient uptake	Indole acetic acid synthesis	Flores-Felix et al. (2013)
Penicillium janthinellum LK5		Solanum lycopersicum		Gibberellin synthesis	Khan et al. (2015)

(continued)

Table 13.1 (continued)

Organisms	Stress conditions	Crop	Outcome	Mode and tolerant strategies	References
Chryseobacterium		Solanum lycopersicum	Increase soil microbial biomass vis-à-vis soil nutrient.	Siderophore production	Radzki et al. (2013)
Rhizobium and Root-associated plant growth-promoting rhizobacteria (PGPR)	Salinity	Oryza sativa		Expression of salt stress-related RAB18 plant gene	Jha et al. (2014)
Pseudomonas koreensis strain AK-1	Salt	*Glycine max*		Reduction in Na + level and increase in K + level	Kasotia et al. (2015)
Rhizofungal flora	Hydrocarbon pollution	*Zea mays* and *Sorghum bicolor*	Enhanced germination, shoot growth		Eze et al. (2014)
Rhizobacteria	Chromium toxicity	*Vigna unguiculata* and *Arachis hypogea*	General growth performance		Eze et al. (2018)
PGPR and AMF	Hydrocarbon pollution and saline–alkali soil	Avena sativa	Improves the soil quality and degradation of total petroleum hydrocarbon	Augment the activities of essential enzymes, e.g., urease, sucrase, and dehydrogenase	Xun et al. (2015)
AMF	Drought	Leymus chinensis, Hemarthria altissima	Enhanced plant biomass photosynthetic rate, stomatal conductance	Enhance antioxidant enzyme activities	Li et al. (2019)

13.2 Microbial Consortium: Interactions, Operations, and Mechanism

Bacteria account for over 95% of the total microbial activities in the soil. It is primarily supported by their fast proliferation and ability to utilize wide range of nutrients (Odoh et al. 2019a). In the rhizosphere, rhizobacterial concentration in the soil is estimated to be 10^{12} CFU/g, while the rhizobacterial flora of the bulk soil is 10^{8}–10^{9} CFU/g (Compant et al. 2010; Foster 1988). During stress conditions, these rhizobacterial population structures get altered, thus affecting activities of the general soil ecosystem. These bacteria constantly interact with fungi and other associated organisms, thereby gaining adaptive capacity for complex mechanisms.

Arbuscular mycorrhizal fungi (AMF) also have developed aptitude to fix nitrogen, obtain iron using siderophores, and increase bioavailability of phosphorus via solubilization by its organic acid (Hardoim et al. 2015; Yadav et al. 2019a, b, c). According to Kim et al. (2008), MC biotechnology involves the use of more than two microbial species that are exposed to same environmental conditions be it nutrients, temperature, pH, or oxygen.

Ecological research has unveiled fascinating groups of microorganisms that live in close interactions. In the rhizosphere, diverse microbial populations acquire their nutrient via the root exudates (Philippot et al. 2013), which aid in regulating rhizospheric biotic and abiotic functions such as propagation, space competition, and soil physicochemical properties (Igiehon and Babalola 2018). This compound is made up of monosaccharides (glucose), disaccharides (sucrose), polysaccharides (arginine and benzoic acids), higher molecular-weight compounds (fatty acids, nucleotides, tannins), and vitamins (Rasmann and Turlings 2016). Technically, MC aggregation is driven by biofilm-mediated process, a biologically active body formed on/in surfaces in contact with water consisting of organic and inorganic minerals and microorganisms held together by matrix of organic polymers often produced and secreted by the microorganisms themselves (Nwankwegu and Onwosi 2017). Igiehon and Babalola (2018) suggest that these exudates play a huge role in facilitating symbiotic interactions particularly those involving rhizobacteria and arbuscular mycorrhizal fungi (AMF). This MC communicates and utilizes pheromone, a unique signaling molecule, for effective selection of partners, monitoring and maintaining integrity of their population densities (Sivasakthi et al. 2014). In addition, these signaling/sensing molecules coordinate microbial activities, which are essential for accomplishment of complex exploration of biological processes (Bernstein and Carlson 2012; Hays et al. 2015).

Harnessing this biotechnology advances in agriculture is of essence. This is because it improves crop growth and food production. Recent omics studies have thrown light on microorganisms in their natural habit, e.g., (rhizosphere), leading to the understanding of plant–microbial cooperation. In addition to providing clarity on the complexity of microbial structures in their natural environment, multi-omics research also reveals weakness of genetically engineered pure cultures or single strain in biotechnological applications (Jagmann and Philipp 2014; Igiehon and Babalola 2018; Kumar et al. 2019b; Rana et al. 2019; Yadav 2017b; Yadav et al. 2019d). There are two mechanisms that explain microbial coexistence in soil, namely, neutral/equalizing and niche/stabilizing mechanisms.

13.2.1 Neutral or Equalizing Mechanism

Neutralization theory also called equalizing mechanism involves minimizing differences among varying species in a growing population while considering other demographic parameters (Zhang et al. 2009; Ale et al. 2019). It is a mathematical tool and model program, which captures epiphenomena and broad-scale patterns of ecological communities. This theory predicts that changes in species composition are related to geographic distance between samples as a result of dispersal limitation. It also suggests that species abundance follows a zero-sum multinomial (ZSM) distribution (McGill et al. 2006), a unique species abundance distribution (Hubbell 2001; Fisher and Mehta 2014). This theory suggests that microbial community structure comes solely from stochastic processes and dispersal limitation, ensuring equalizing biodiversity of all species and a uniform population growth rate at an identical ecological zone. Integral microbial biodiversity characterizations such as birth, death, colonization, immigration, speciation, and dispersal limitations are linked to neutral mechanism.

According to Rosindell et al. (2011), organisms at the same trophic level are equivalent with respect to fitness in a specific environment; as illustrated in the Vellend's processes of microbial community assembly (Nemergut et al. 2013). Through heritable genetic barcodes, community compositions are tracked in a repeatedly subculture samples. In a study carried out by Cira et al. (2018), they revealed a transition between neutral and selective regimes, with a crossover point that is dependent on the fraction of immigrants and the magnitude of fitness differences. During exclusive competition, this intrinsic mechanism regulates activities via reduction of competitive speeds resulting in decrease in population growth usually at rare conditions. Zhang et al. (2009) observed a discrepancy between stabilizing (niche mechanism) and equalizing (neutral mechanisms) and identified that the former presumes negative frequency dependence in population growth of species while the latter does not. Their result, however, justifies Zhou and Zhang (2008), who earlier opined that neutral theory presupposes that all species are functionally equivalent with diversity having a little or no effect on the ecosystem.

13.2.2 Niche or Stabilizing Mechanisms

This mechanism suggests increasing importance of deterministic processes. Fundamentally, it explains the ecological traits differentiation among species within a community. Ale et al. (2019) argued that the niche-based mechanism characterized a robust and progressive activity, which deals with disconfirming data by generating new testable predictions. It is regarded as a negative intraspecific interaction relative to negative interspecific interactions in the soil ecosystem. Niche theories predict that a change in species composition is related to changes in environmental variables, and that species abundance follows a log-normal distribution (Leibold and McPeek 2006). Unlike in neutral theory where only random processes, such as birth, death, colonization, immigration, speciation, and dispersal limitations (Vanwonterghem et al. 2014), are considered, niche stabilizing mechanism utilizes environmental condition (abiotic and biotic factors), habitat heterogeneity, species interactions, and species relative abundances in shaping bacterial community structure (Dumbrell et al. 2010; Gilbert et al. 2012).

Literature in microbial ecology support the eminent contributions of niche-based and neutral processes in microbial community assembly (Burke et al. 2011; Logares et al. 2013). Worthy of note is that stabilizing mechanisms result in negative frequency-dependent selection where each species enjoys an advantage in population growth at rare condition. This bio-technique is paramount for long-term stable coexistence among species through resource partitioning, frequency-dependent predation (Chase et al. 2011), and fluctuations in population density and environmental factors (Letten et al. 2017). Aside from niche partitioning primarily regulating the composition and diversity of natural arbuscular mycorrhizal (AM), stochastic-neutral processes (Dumbrell et al. 2010) also influence them. Liao et al. (2016) suggest that habitat specialists (niche process) were significantly strongly shaped by environment selection, whereas habitat generalists (neutral process) were strongly assembled via neutral bioprocesses. They identified salinity, dissolved oxygen, water transparency, total phosphorus, ammonium nitrogen, temperature, and total nitrogen as the significant habitat specialists, whereas habitat generalists are driven by natural processes. The theory also suggest that organisms do not just come to existence all at once, rather through typically assemble and sequential species colonization events (Jiang and Patel 2008; Nemergut et al. 2013; Verbruggen et al. 2012). This ultimately gives rise to ecological succession.

13.3 Microbial Consortium as Biofertilizer

Belowground microbial interactions in soil ecology are dynamic and complex. It is a determinant of the increase in yield and productivity of the aboveground crop parts (Philippot et al. 2013). This occurs through a process called biofertilization—a phenomenon where microbial inoculants are seeded on plant surfaces, seeds, and/or soil to colonize root rhizosphere. This condition enhances growth through the supply and availability of primary nutrients to the plant (Odoh et al. 2019a; Odoh 2017). Biofertilizer (Table 13.2) is a culture of bacteria, fungi, and algae either alone or in combination that is packed in carrier materials to enhance plant growth. Several support materials preferably of organic origin (notably chitosan, gelatin, sawdust, k-carrageenan, zeolite, activated carbon, etc.) are stable bio-carriers used to immobilize these biofertilizers while substantially eliminating environmental perturbations (Nwankwegu and Onwosi 2017). In a comparative research evaluating microbial consortia versus single-strain inoculants, Bradáčová et al. (2019) suggest that microbial consortia increase the efficiency of crop production, particularly under challenging environmental conditions. Microbial fertilizer plays a critical role in atmospheric nitrogen fixation and mineralization of organic compounds. In arable agricultural application, it is considered an essential component for long-term soil fertility and sustainability. Nuti and Giovannetti (2015) in their view suggested that biofertilizers act by nourishing and fortifying the host plant, and inducing general pathogenic resistance, irrespective of its origin and nature.

Table 13.2 Biofertilizer and their formulating microbial strain

Product name	Organismal consortium	Target crops	Manufacturers
Amnite A 100®	Azotobacter, Bacillus, Rhizobium, Cheatonium, Pseudomonas	Cucumber, lettuce tomato, pepper	Cleveland biotech, United Kingdom (UK)
Armour-Zen®	Chitosan. An elicitor against Botrytis cinerea (gray mold), Sclerotinia scherofiorum (white rot)	Grapevine, ornamentals	Borty-Zen 2010 Ltd., New Zealand
Bioativo®	PGPR consortia, organic matters	Bean, maize, sugarcane, rice, carrot, cotton	Embrafos Ltd., Brazil
BactofilA10®	*A. brasilense, A. vinelandii, B. megaterium, P. fluorescens*	Cereals	Agro bio Hungary kft, Hungary
Biomix®, Biozink®, Biodine®	Azotobacter, *P. fluorescens*, phosphobacteria	Wide range of plant varieties, e.g., field crops	GreenMax Agrotech, India
Ceres®	P. fluorescens	Horticultural crop	Biovitis, France
Complete® plus	*B. pumilus, B subtilis, B. licheniformis,*	Nursery trees and field crop	Plant Health Care, United States of America (USA)
FZB 24® fl	B. amyloliquefaciens sp, planetarium	Vegetables	AbiTEP GmbH, Germany
Gmax PGPR	PGPR consortia	Field crops	GreenMax Agrotech, India
Galtrol®	*Agrobacterium radiobacter* strain 84	Ornamentals, Fruits, Nuts	AgBioChem, USA
Hyper Coating Seeds®	Rhizobium and legume seed	Legume	Tokachi Federation of Agricultural Cooperatives (TFAC), Japan
Inomix® biostimulant	B. polymyxa (LAB/BP/01), B. subtilis (LAB/BS/F1)	Cereals	LAB (Labiotech), Spain
Mycostop®	Streptomycin griseoviridis	Ornamentals, Tree Seedlings	Kemira Agro Oy, Finland

(continued)

Table 13.2 (continued)

Product name	Organismal consortium	Target crops	Manufacturers
Micosat F® cereal	B. subtilis BR62, Paenibacillus durus PD74, Streptomyces sp ST60	Tomato, soybean	CCS Aosta Srl, Italy
Mamezo®	Rhizobium-based formulation in peat	Legumes	TFAC, Japan
Nodulator®	Bradyrhizobium japonicum	Cereals and **horticultural plants**	Lallen and plant care BASF Inc. Canada
Nitrofix®	Azospirillum sp	Wheat, barley, carrot, maize, cabbage	Labiafam S.A, Cuba
Processing Seeds®	Rhizobium	Legumes	TFAC, Japan

Modified from Odoh et al. (2019a)

Due to excessive application of chemical fertilizers, leaching and runoff of essential minerals "phosphorus (P) and nitrogen (N)" occur leading to loss of soil nutrient. With the overwhelming importance of biofertilization in modern agriculture, research has focused on halting the overdependence on synthetic fertilizers coupled with the rising depletion of soil functionality (Bhardwaj et al. 2014). Chatzipavlidis et al. (2013) is of the view that biofertilizing system requires adequate preparation of the inoculants, selection of carrier, and designing of accurate delivery system. This bioprocess, however, requires optimization to support increased yield and economic viability of small and marginal farmers. MC, apart from being able to mobilize nutritionally important elements from non-usable form through biological process (Mazid et al. 2012), secretes fascinating bioactive ligands (Myc and Nod factors) using a transduction pathway (Roberts et al. 2013) for the release of Ca^{2+} in the cytosol (Sieberer et al. 2009).

13.3.1 Nitrogen Fixation as a Form of Biofertilizer

Nitrogen (N) is one of the major limiting nutrients constantly required for crop growth. It is a common and essential element occurring in all organisms. As a precursor of amino acid and a major constituent of protein and nucleic acid (DNA and RNA), it constitutes 3% of body mass index and fourth most important plants dry mass. Besides circulating the atmosphere, lithosphere, and biosphere through biogeochemical cycle, it serves as a nourishing nutrient and mineral for agricultural services. In plants, this element is fixed through nitrogen fixation. Here, atmospheric nitrogen gets converted into ammonia (NH_3) and/or nitrogenous compound. The conversion of dinitrogen (N_2) into NH_3 also called biological nitrogen fixation is important as it enables broad utilization by a number of microorganisms. During this process, soil free-living symbiotic diazotrophs, e.g., *Azotobacter, rhizobium,* and *spirilla,* as well as cyanobacteria (blue-green algae such as species of *Aphanizomenon, Anabaena, Nostoc,* etc. in aquatic ecosystems) produce a highly complex oxygen labile enzyme called nitrogenase (Verma et al. 2015; Franche et al. 2009; Simone et al. 2018). This enzyme aids the reduction of nitrogen (N_2) to ammonia (NH_3).

A combination of recent nitrogenase engineering biotechnology using advances in synthetic biology to broaden understanding of the enzyme biosynthesis and biochemistry by plant scientists has been innovated (Burén and Rubio 2018). This technology enables the engineering of plants to express their own specific nitrogenase enzymes, thus overcoming negative natural pressures (increased use and availability of reactive nitrogen) and apparent economic benefits and opportunities it presents. During biological process of nitrogen fixation, a two-form (symbiotic and nonsymbiotic) process occurs owing to their associated plants and group of microorganisms. It has, however, been established that nonsymbiotic processes fix less amount of nitrogen when compared to the rhizobia association with root nodule (symbiotic) (Sippel et al. 2018). Considering the overarching importance of this event and the role played by nitrogen in agriculture cum food production, PGPR have intrinsically developed capacity in augmenting this process through diverse strategies that would improvise and support availability of the nitrogen nutrient (Odoh et al. 2019a). Elsewhere, nitrogen fixation as a biofertilizer has been documented and demonstrated with a major contribution of PGPR and AMF as seen in the suppression of major biotic and abiotic stresses and threat (Majeed et al. 2018).

13.4 Plants–Microbial Interactions

Soil is the loose material of the earth's surface consisting of a mixture of organic matter, minerals, gases, liquids, and organisms that jointly support life. It is the natural component of the earth crust with proven biological, chemical, and physical properties. One of its rich nutritional components essential for plant crop growth is "soil organic matter" (SOM). It consists largely of residue of plants and animals usually at various stages of decomposition. These substances help in sustaining soil fauna and floras. Soil microorganism contributes over "8%" of the total SOM, while the nonliving remains, and humic substances in the soil account for about 60% of SOM (Varanini and Pinton 2001; Liste 2003; Htwe et al. 2019). SOM besides being an important portion of soil with pool of nutrient supporting the propagation of soil organisms and plants, it is also vital for cation exchange and sorption of contaminants (Eze et al. 2018). Its roles in erosion control, water, and air circulation as well as soil aggregation have also been documented (Guo et al. 2019a). However, with the dominance of SOM in soil due to plant derivatives, it thus supports the premise that accumulation of organic matter in soil profile (horizon A) occurs most where the number of plant roots is greatest.

Owing to the inability to physically understudy soil–plant–microbial mechanisms, omics molecular tool has revealed levels of interaction in the soil ecosystem. This advanced technique has helped to identify and quantify the microbial diversity associated with specific plants while giving clarity to their immense interactions to which plants are exposed. Schirawski and Perlin (2018) disclosed that plants constantly relate with microorganisms via a diverse number of mechanisms necessary for their survival. Through this association, plant benefits directly or by indirect effects of the associated microbes which composition around the root zones includes rhizobacteria and mycorrhizal fungi (Nadeem et al. 2014; Hamilton et al. 2016; Yadav et al. 2015a, b). Since multidimensional interactions occur in all plant organs, total microbiomes' evaluation could help provide evidence of plants' part specificity and its identifiable organism for agrobiology. Notably, plant root, apart from serving as a host for organisms, releases compounds and also serves as nutrient upon their death. These molecules induce more resistance to abiotic or biotic stress and defend species against malignant microbes.

Due to the rich microbial diversity and low nutrient composition in the soil, competition for dominance, adaptation to stress, and capacity to enhance crop growth are prevalent (Ngumbi and Kloepper 2016). Consequently, beneficial microorganisms interact with plant roots, thus supporting plant health via a myriad of mechanisms, e.g., biocontrol, biofertilization, and biostimulation (Glick 2014; Rashid et al. 2016; Odoh et al. 2019a; Yadav et al. 2016, c). Fungal network, according to Fabbro and Prati (2014), also gives protection to plants root zones against various phytopathogens, while helping in phosphorus acquisition and water availability during drought (Barnawal et al. 2014).

13.4.1 Forms of Interactions Among Microbial Consortium

13.4.1.1 Bacterial–Bacterial Interaction

Plant growth-promoting rhizobacteria (PGPR) includes all rhizobacteria capable of directly or indirectly enhancing crop growth. Example of these organisms includes *Alcaligenes, Pseudomonas, Azospirillum, Bacillus, Klebsiella, Azotobacter, Enterobacter, Burkholderia, Arthrobacter, and Serratia*—they facilitate crop development through a number of mechanisms (Saharan and Nehra 2011; Jambon et al. 2018; Odoh et al. 2019a). Bacteria–bacteria associations in plants are often exploited to enhance efficiency of pollutants sequestration (Eze et al. 2018; Odoh et al. 2019a). Owing to high nutrient availability in the rhizosphere, unlike in the rhizoplane (surface of the roots) and phyllosphere (surface of leaves), there tend to be heightened microbial and biochemical activity in the rhizoenvironment (Venturi and Keel 2016). PGPR also exhibit special role by hindering plant infections, increasing nutrient absorption, root and shoot formation, and improving seed germination and tolerance to environmental stress (Lugtenberg and Kamilova 2009; Odoh 2015). These functions are dependent on the recruitment of rhizospheric and rhizoplane microbes by plants from the bulk soil. Bulgarelli et al. (2012) noted that PGPR recruitment is dependent on the community structure of the bulk soil. Lundberg et al. (2012) opined that different plant genotypes select different rhizospheric communities, thus implying that genetic variation across plant species drives differential recruitment of beneficial microbes. These employed bacterial consortiums engage in fascinating roles such as nitrogen fixation, solubilization of phosphates, phytohormones production, and plant development (Ma et al. 2009; Odoh 2017; Htwe et al. 2019).

During cell–cell interaction in bacteria, members of associated community converse via signaling chemical process. Notable among these sensing mechanisms is *quorum sensing*. Quorum sensing is a microbial communication and regulation of gene expression mediated by small diffusible molecules called autoinducers or quorum-sensing molecules (QSM) (Barriuso 2015). It is described as a regulatory response for transcription of specific genes in response to the detected compound (Venturi and Keel 2016). As a self-regulatory innate mechanism, the accumulation of quorum-sensing molecules occurs throughout microbial growth. When microbial concentration reaches a threshold, regulatory response control by gene expression is initiated to control cell density and population outburst (Albuquerque and Casadevall 2012). This cell-to-cell communication signal is always specific and coordinate pathogenic activities by helping bacteria acclimatize to the disadvantages in the environment when activated (Qian et al. 2019). The QS signals in bacteria consist of acyl-homoserine lactone, autoinducing peptide, and autoinducer-2. They also regulate biochemical processes such as motility, biofilm formation, sporulation, and antibiotic production and play significant role in the secretion of virulence factors (Barriuso 2015; Fleitas-Martínez et al. 2019). Through this efficient cell–cell interaction, energetically cost-effective activities are only undertaken when bacteria population size is high enough to successfully accomplish a specific task (Clinton and Rumbaugh 2016).

In addition, volatile organic compounds (VOCs) and nodulation (Nod) factors of rhizobia have also been identified with properties aiding bacterial interactions (Jambon et al. 2018; Hung et al. 2015). VOCs aid long-distance interactions between microbes, microbes and plants, control symbiotic associations, and the distribution of saprophytic, mycorrhizal, and pathogenic organisms (Hung et al. 2015; Tyc et al. 2017; Brilli et al. 2019). Through this bio-technique, plant health is guaranteed as they act as biocide against plant pathogenic bacteria and fungi. Furthermore, bacterial VOCs promote plant growth through the use of acetoin, a chemical compound that induces systemic resistance and interference with plant gene expression (Bennett et al. 2012). However, responses to flavonoids and strigolactones from plant roots are recognized as host plant symbiosis and signaling molecules (Venturi and Keel 2016). Unlike the Myc factors produced by a specific mycorrhizal fungus, Nod factors by nodulation rhizobia and VOC are cell signaling secondary compounds found in root exudates aiding specificity of rhizobial interactions with their host plant (Oldroyd 2013).

13.4.1.2 Bacteria–Fungi Interaction

Historical ecological studies have revealed that bacteria and fungi often cohabit and share common ancestral origin where assemblage and dynamic co-evolving communities occur. Due to long years of microbiological research that distinguish bacteriology and mycology as two separate disciplines, many scientists and literatures have overlooked in reality, the coexistence of these organisms in the same ecosystem. This, however, has hampered the understanding of the interactions and biochemical processes exerted by the combination of their mutual partnership. Deveau et al. (2018) disclosed that bacteria–fungi interaction (BFI) is intrinsically modulated by behavioral properties of either or both of the interacting partners. Usually, during their coexistence, there exist intimate biophysical and metabolic associations leading to the development of bacterial–fungi interdependency. BFI research has in the last decade metamorphosed into interdisciplinary studies integrating molecular biology, genomics, chemical and microbial ecology, biophysics, and ecological modeling. Through the characterization of BFI, understanding of microbiomes (e.g., arabidopsis root microbiome) has been unraveled (Bergelson et al. 2019). This is primarily through the application of molecular tools where description of biomes and ecological habitat highlight the diversity of the microbes (Thompson et al. 2017). Considering the physical complexities that exist between bacteria and fungi and their applicability in agriculture and ecological studies, Frey-klett et al. (2011) suggested a transit from disordered poly-microbial communities to highly specific symbiotic associations of fungal hyphae and bacterial cells.

Arbuscular mycorrhizal fungi (AMF) and bacteria (PGPR) association have been reported to promote crop growth (Pathak et al. 2017). Besides this association having positive influence on crop yield, it also enhances soil nutritional status and soil microbial biodata. According to Pathak et al. (2017) and Franco et al. (2011), PGPR and AMF are major bio-inoculants with potential for halting dependence on agrochemicals, thus aiding in sustainable agricultural practices when serving as biofertilizer and biocontrol agents. PGPR are classified according to host and intra- and extracellular plant growth-promoting rhizobacteria, and boost plants through direct (growth-promoting hormones) and indirect (antimicrobial substances) mechanisms (Deshwal and Kumar 2013; Zheng et al. 2018). As aid to the process of mycorrhization, mycorrhizal-helping bacteria (MHB) and PGPR symbiotically interact with mycorrhizal fungi and mycorrhizal roots for nutrient uptake. Studies have revealed that rhizospheric AMF and PGPR elicit systemic host immune responses for plant resistance (Zamioudis and Pieterse 2012; Singh 2018). Experimental evidence has shown that co-inoculants of AMF and PGPR offer synergistic advantage to crop especially in nutrient-limited agricultural soil (Gouda et al. 2018). *Bacillus* sp., *Pseudomonas* sp. (PGPR), and AMF interaction proffer viability (Philippot et al. 2013) and show significant improvement in various field applications when used singly or in combined application (Pathak et al. 2017).

13.5 Microbial Consortium as Biofertilizers in Extreme Habitat

Microbial association and adaptation in extreme environments involve progressions characterized by genetic variations and distributions in the population through natural selection or drift. Changes in microbial community composition due to any implication aid shift in the behavioral occupation of organisms inhabiting an ecological niche. These implicating conditions are either newly introduced species, fluctuating environmental conditions or low fitness genotype variants within the host's microbiota (Yadav 2017a; Yadav et al. 2017). In ecological biotic or abiotic variation, tendency of adjustments of microbial physiology and metabolic homeostasis occurs. Liu et al. (2016) in their submission suggest that rapid adaptation of microbes occurs through genetic fluctuations at the level of individual bacterial cells. They further illustrated that intra-genomic recombination processes and epigenetic switches are precise phenomena in phase variation. Furthermore, integration of beneficial plant–microbes and microbiome interactions through agricultural microbial biotechnology has proven to be a sustainable solution for species adaptation and crop production in extreme habitat (Timmusk et al. 2017). In understanding this section, we focused and laid much emphasis on habitat under abiotic stress (heat, pH, salinity, and drought) and key environmental disturbances such as heavy metal and crude oil pollutants.

13.5.1 Heat Stress

Heat stress implies increase or decrease in a temperature more than the critical edge, at a particular time. Usually, this condition is adequate enough to cause irretrievable damage to plant growth and development (Ripa et al. 2019). Temperature stress also leads to soil fertility loss, microbial diversity, loss of nutrient resources, and a series of morphological, physiological, biochemical, and molecular changes with adverse effects on plant growth (Hayat et al. 2013; Chodak et al. 2015). Documented evidence has shown that temperature-related stress either high or low severely restricts crop production and most importantly, with the upset in the global earth temperature and upsurge in human activities. In the tropics, this abiotic stress has hindered agricultural developments particularly in climate-impacted community leading to weather change and alteration of farming seasons. Besides crop productivity being tempered by change in the plant metabolism during high temperature stress, there also occur cellular changes such as reactive oxygen species (ROS) (Hasanuzzaman et al. 2013). According to Akter and Islam (2017), heat stress significantly reduces seed germination and seedling growth, cell turgidity, and plant water-use efficiency. They further explained its role in the disturbance of cellular functions, enhancement of leaf senescence, deactivation of photosynthetic enzymes, and generation of oxidative damages to the chloroplasts.

In contrast, cold and frost (low temperatures) cause decreased kinetics of biomolecules leading to reduced cell membrane fluidity and a decreased rate of enzymatic reactions. It results in cell division, impairment of water transport, photosynthesis default, alteration of crop development, and growth in arctic environment (cold). This condition leads to the formation of ice crystals in soil due to reduction in water uptake by roots resulting in cellular dehydration. As a response to this stress condition, plants over the years have developed strategies to induce accumulation of several osmolytes and hydrophilic proteins such as dehydrins. According to Kosová et al. (2018), heat and cold temperature stress exacerbate imbalance between photosynthetic electron transport processes and carbon assimilation processes, thus resulting in enhanced photo-inhibition and thermal energy dissipation.

Different researches have reported the ability of some beneficial heat- and cold-tolerant bacteria to induce stability in plants (Chang et al. 2007; Chakraborty et al. 2018; Lamaoui et al. 2018; Yadav et al. 2015a, b). This microbial-tolerant synthesis often depends on the plant genus, stress type, microbial species, and the plants' microbial relationship. An example of this is seen in phyllosphere bacteria with ice-nucleating activity that damage plant in temperate region. This can be harnessed for onward application as foliar spray to suppress the ice nucleation effect in temperate region (Selvakumar et al. 2012). Consequently, these abiotic effects even at mildest state affect the smooth growth and development of plant at all states. With several independent researches, it has also been demonstrated that increase or decrease in temperature has the propensity of reducing crop yields by over 50% (Lamaoui et al. 2018). Through advances in biotechnology such as genomics and information technology, mitigation of these abiotic stresses through the use of agronomic management practices would be sound if integrated with tolerant MC, so as to aid the development of crop varieties while boosting harvest amidst rising stress (Fig. 13.1).

13.5.2 Drought Stress

One of the documented stresses that affect crop health and growth is drought. It is an abiotic condition where plants experience deficit supply of their water needs. When rate of transpiration exceeds rate of root water uptake, drought is said to occur—this leads to reduction in the cells' relative water content and development (Utsumi et al. 2019). Research on drought tolerance has become an important field of study owing to rising global climatic state. This, however, has cut the interest of most agriculturists and scientists to probe drought challenge and its related consequences on crop breeding (De Oliveira et al. 2017). The impact of this drought condition has necessitated the exploration of belowground microbial potentials and interactions to proffer ways of curtailing the rising menace (Igiehon and Babalola 2018). Characterized by cellular water deficiency, drought is a severe condition where plants get dehydrated and ultimately die off (plant dieback) due to limited water supply. This could lead to shedding leaves, breaking branches, weak root formation, and color alteration. According to Odoh et al. (2019b), this plant dieback occurs as a result

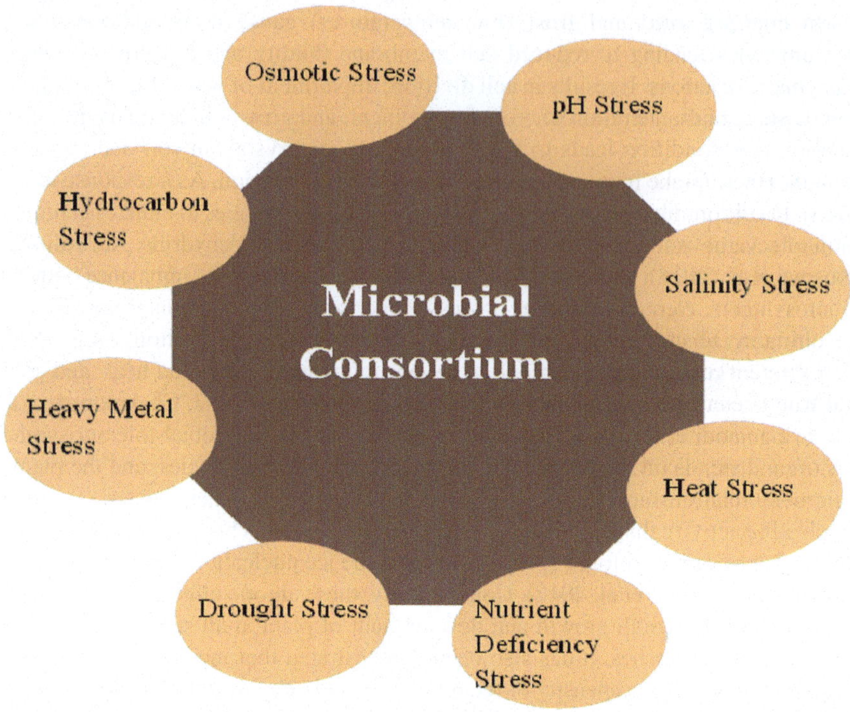

Fig. 13.1 Schematic representations of extreme conditions upturned by activities of microbial consortium

of drought, pathogens, parasite, soil acidity, and soil pollution. In general, drought stress reduces cell size, membrane integrity, produce reactive oxygen species, and promotion of leaves senescence, thus leading to decreased crop production (Tiwari et al. 2016). However, seed germination, seedling development, and morphological and molecular changes are all related developmental challenges impeded by crop exposure to drought (Nezhadahmadi et al. 2013; Varshney et al. 2018; Utsumi et al. 2019; Kour et al. 2019a, b; Rana et al. 2019; Verma et al. 2016, 2017; Yadav and Yadav 2018).

With the recent advances in omics studies, plant–microbial processes and species have showcased promising character in ameliorating agrobiotechnology-related stress (drought). At the transcriptomic level, bacterium *Paenibacillus polymyxa* B2 enhances drought tolerance in *Arabidopsis thaliana*. When PGPR (*A. thaliana*) is inoculated, it expresses genes for the overexpression of drought tolerance unlike when compared to the uninoculated plants (Glick et al. 2012; Alavilli et al. 2017). Lim and Kim (2013) in their studies using 2-D polyacrylamide gel electrophoresis and differential display polymerase chain reaction identified drought stress gene expressed in plants amended with *Bacillus licheniformis* in affected soils. Also through quantitative polymerase chain reaction (qPCR) *Bacillus amyloliquefaciens* 5113 and *Azospirillum brasilense* have shown priming effects on the expression of drought-responsive genes. Some of the genes usually expressed for drought abiotic stress include ascorbate peroxidase (APX1), S-adenosyl-methionine synthetase (SAMS1), heat shock protein (HSP17), and enzyme-enhancing drought stress in wheat leaves (Kasim et al. 2013). PGPR produce phytohormones or induce plants to synthesize molecules such as indole acetic acid, cytokinins, gibberellins, and abscisic acid (Hayat et al. 2010; Spaepen et al. 2008; Odoh et al. 2019a). These molecules trigger adaptation of plant in stress habitat through varying mechanisms (Spaepen and Vanderleyden 2011), thus influencing changes such as increase in root growth, length, surface area, and formation of lateral roots and root hairs for the acquisition of water (Egamberdieva and Kucharova 2009; Paul and Lade 2014; Manaf and Zayed 2015).

During flooding stress, anaerobiosis occurs due to excessive accumulation of water in the root; this leads to induced fermentation processes in the root region. It also results in enhanced accumulation of organic acids resulting in acidity of the cell cytoplasm affecting cellular enzymes (Kosová et al. 2018). Also using cellular organelles (nucleus, nucleolus, mitochondria, endoplasmic reticulum, plasma membrane, cell wall), a novel protein biosynthesis was found in nuclear proteome of soybean root tips (Yin and Komatsu 2016), suggesting expression of adaptive capacity in stress-driven environment.

13.5.3 Salt Stress

Salinity stress is a major abiotic factor commonly seen in most agricultural soil. It causes inhibition and impairment of crop growth and development via water stress and cytotoxicity by excessive uptake of ions (Isayenkov and Maathuis 2019). Salt stress has been reported to be detrimental on nitrogenase—an enzyme responsible for the fixation of nitrogen. Some of these ions causing salinity are responsible for changes in the ratio of ion homeostasis of plant system. This, however, leads to undue uptake of Na^+ and Cl^-, and reduction in K^+ and Ca^{2+} transport in the growing plant (Giri et al. 2007; Paul and Lade 2014). One of the significant effects of salt stress is low water budding; this leads to inability of plant to take up nutrients and water from the soil due to osmotic pressure. Álvarez-Aragón et al. (2016) in their studies revealed that overaccumulation of Na^+ and K^+ might be responsible for triggering growth reduction in NaCl-treated Arabidopsis plants. In their work, they demonstrated how stomatal regulation and/or systemic stress responses occur owing to salt concentrations. Typically, salinity stress is accompanied by oxidative stress caused by reactive oxygen species (ROS) (Isayenkov 2012). During soil salinity, electrically charged ions accumulate due to insufficient water distribution, hindering germination and uptake of nutrients. During this condition, growth responses in plant occur either as ion-independent growth reduction or by cytotoxic ion build-up. In the former, crop growth takes place within minutes to days—causing stomatal closure and inhibition of cell expansion (Rajendran et al. 2009; Isayenkov and Maathuis 2019), while the latter takes days or even weeks for development leading to slow metabolic processes, premature senescence, and cell death (Munns and Tester 2008; Roy et al. 2014).

However, harnessing PGPR potentials could directly and indirectly influence the promotion of plant growth under increased salinity concentration. Studies by Kasim et al. (2016) have it that when PGPR and endophytic microbes are inoculated, they mitigate effects of soil salt on plants using various mechanisms. In a related experiment, Bacilio et al. (2016) in their work revealed that plants inoculated with salt-tolerant bacteria (*Pseudomonas stutzeri*) reduce in high extent the negative impact of soil salinity on crop growth. Rhizobacteria sp. with capacity at reducing salt toxicity in several plants through the reduction of Na^+ concentration and increasing the K^+ and Ca^{2+} are reported (Bano and Fatima 2009; Hamdia et al. 2004; Kohler et al. 2009). This is achieved principally by altering host physiology, reducing foliar accumulation of ions (Na^+ and Cl^-) and improving the nutritional condition of macro (P and K), and micronutrients (Zn, Fe, Cu, and Mn) (Bano and Fatima 2009). In plant roots, endophytic fungi *Piriformospora indica and* AMF show ability in ameliorating ions toxicity by inducing host defense against salt stress (Ansari et al. 2013).

Cell homeostasis is the ability of a living cell to maintain internal ions concentration or remain nearly constant even in environmental stresses (Rombola-Caldentey 2019; Chérel et al. 2019). Under this ionic homeostasis stress, beneficial rhizobacteria are harnessed, e.g., *Pseudomonas* (Fu et al. 2010), *Azospirillum* (Ashraf et al. 2004), and *Bacillus subtilis* GB03 with *Arabidopsis* (Zhang et al. 2008) to maintain intracellular ionic homeostasis balance in crops such as eggplant and maize. These PGPRs have shown capacity to trigger all round physiological functions and development through the accumulation of osmolytes and several low-molecular-weight compounds such as methylated tertiary N compounds and amino acids (Guo et al. 2019). Primarily, these osmolytes operate through the modulation of osmotic pressure in the cytoplasm and cell membrane, thus stabilizing plant growth when water, salt, and/or temperatures seem unfavorable (Tiwari et al. 2010; Sze and Chanroj 2018; Chérel et al. 2019).

13.5.4 Soil Reaction (Acidity and Alkalinity)

Soil pH is an important abiotic condition that determines the growth, survival, and productivity of any given plant. In considering physicochemical properties, pH is a fundamental determinant of soil health. Some of the most important bioprocesses and microbial activities include nutrient solubility and availability to the host plant (Gentili et al. 2018). Many plant genetic traits, such as height, lateral spread, biomass, flower size and number, pollen production, etc., are influenced by soil pH (Jiang et al. 2013). Elsewhere, studies have shown that most micronutrients are more available to plants in acid soils compared to neutral–alkaline soils (Lončarić et al. 2008). In alkaline soils also, availability of most macronutrients increase with concomitant reduction of phosphorus, thus generally affecting crop growth. However, as these micronutrient accumulations increase along with other nonessential elements, they become poisonous to plants.

Alien invasive plants have usually been documented to have tolerance to environmental stress, including pH (Dassonville et al. 2008; Hao et al. 2017). They are species that upon introduction spread outside their natural habitat and threaten biological diversity. Unlike the cultivated or native plants whose optimum performing pH mostly ranges from 5.5 to 6.5 (Köpp et al. 2011), they propagate at extreme conditions. This characteristic allows them to adapt to a great variety of soil types and thus spread vigorously, while also colonizing environments not suitable for native species (Sărăteanu et al. 2010). In a study done by Gentili et al. (2018) on the effects of pH on germination, growth-related traits, reproduction, pollen production, and allergenicity of *Ambrosia artemisiifolia,* they reported that soil pH greatly affects the growth and development of *A. artemisiifolia* and may have contributed in limiting the distribution and growth of the plant. Despite the impact of this abiotic stress on plants, there has been relatively rare studies of its role in modern agro practices (Caplan and Yeakley 2006; Zeng and Clark 2013).

Some important plants' biosynthetic molecules and compounds such as proline, glycine betaine, and soluble sugars have proven to possess the capacity of ameliorating these abiotic stresses (Ranganayakulu et al. 2013; Goswami et al. 2015). Using PGPRs Zea mays exposed to pH stress showed increment in proline production upon inoculation with *Rhizobium* sp. and *Pseudomonas* sp. (Bano and Fatima 2009; Grover et al. 2011). Endomycorrhizal fungi have also shown capacity in inducing biosynthesis of glycine betaine and proline accumulation in plants subjected to abiotic stress (e.g., salt and pH) when compared to non-inoculated plants (Al-Garni 2006; Manaf and Zayed 2015). Different microorganisms have the ability to support accumulation of soluble sugar (trehalose) in plants. These organisms such as endomycorrhizal fungi, symbiotic bacteria such as Rhizobium spp., and free-living PGPRs suppress plant abiotic stresses (Grover et al. 2011 Suárez et al. 2008). Through genetic engineering, PGPRs could be designed for the overproduction of trehalose as biofertilizer, just as *Rhizobium etli* are used to overcome drought stress (Suárez et al. 2008). Similarly, genetically engineered *Azospirillum brasilense* when inoculated on maize plants overproduce trehalose which induce more resistance to varying abiotic conditions (Rodríguez-Salazar et al. 2009).

13.5.5 Heavy Metals and Hydrocarbon Stress

Heavy metals and crude oil hydrocarbon are both environmental pollutants with deleterious effects on crop food production. They are classified as environmental pollutants of major global concern due to inherent challenges associated with natural resource mining (Odoh et al. 2019b; Kumar et al. 2019a). Basically, hydrocarbon pollutants exist in the environment as total petroleum hydrocarbons (TPHs) and this according to Nwankwegu et al. (2018) are hydrocarbon derivatives/congeners, which bioaccumulate and bio-concentrate in food chain through soil ecosystem. Their presence in the ecosystem is attributed to industrialization, urbanization, and civilization (Odoh et al. 2017; Sam et al. 2017). In the developing economy, illegal activities such as artisanal refining and exploration are the basic routes of entering the soil ecosystem. It also has been traced to practices such as excessive fertilizer application, indiscriminate disposal of sewage, power plants/fossil fuel, municipal waste, and pesticides/insecticide usage among others. Examples of some heavy metal species commonly found in the soil are copper (Cu), zinc (Zn), nickel (Ni), lead (Pb), cadmium (Cd), cobalt (Co), mercury (Hg), chromium (Cr), and arsenic (As). Because they are nondegradable, they consistently bioaccumulate in the environment (Walker et al. 2003; Eze et al. 2018), thus impairing food production and posing health threat to man and animals. In agriculture, these heavy metal pollutants lead to decrease in crop yields and further economic loss.

In hydrocarbon stress, soil toxicity occurs leading to distorted microbial population. It causes severe ecological damage, loss of biodiversity structure, and climatic impact. In Nigeria Niger Delta region, hydrocarbon spills have left footprints with devastating impact on farming activities. This significantly alters the health indices in the region with huge ecological imbalance (Ite et al. 2013). Across the tropics, crude oil exploration has threatened food security leading to loss in soil fertility status owing to technical failures, sabotage and artisanal activities (Zabbey et al. 2017; Sam et al. 2017; Odoh et al. 2019b). Ajai (2010) stated that hydrocarbon stress has a direct impact on the total environment especially the food chain and other life-supporting entities.

In the rhizosphere, bacteria, fungi, protozoa, and algae coexist and exert multifunctional strategies in the utilization of mineral and organic wastes. These organisms mostly plant growth-promoting rhizobacteria (PGPR), phosphorus solubilizing bacteria, mycorrhizal-helping bacteria (MHB), and arbuscular mycorrhizal fungi (AMF) play a critical role through bioremediation and phytoremediation technology (Ahemad 2015; Stambulaka et al. 2018; Yadav and Yadav 2019a, b). These microorganisms alleviate the pollutants' noxious effects on plants through secretion of acids, proteins, phytoantibiotics, and other chemical molecules (Denton 2007; Wei et al. 2017; Pettit et al. 2019). Microorganisms, namely, bacteria, fungi, protozoa, and algae coexist in the soil especially within the rhizosphere region and serve as effective metal sequestering and growth-promoting bio-inoculants for plants in metal-stressed soils (Rajkumar and Freitas 2008; Stambulaka et al. 2018).

13.5.6 Osmotic Stress

Reactive oxygen species (ROS) are a bioproduct of metabolic activities and pathways localized in different cellular compartments (Apel and Hirt 2004). They are a group of very reactive, short-lived chemicals often produced during metabolic processes or after an oxidative reaction. Examples of ROS include superoxide ($\cdot O2^-$), hydroxyl radical ($\cdot OH$), hydrogen peroxide (H_2O_2), and singlet oxygen ($1O_2$) (Iqbal 2018).

These chemical molecules have the capacity to impair a number of physiological developments such as proliferation, differentiation, senescence, and apoptosis when found in minimal concentrations (Lai et al. 2007; Iqbal et al. 2016; Iqbal 2018). This occurs owing to their involvement in oxidative damage on proteins, DNA, and lipids (Carmen and Roberto 2011). Studies have shown that accumulation of ROS occurs most during stress condition unlike in normal growth conditions. This is a result of the disparity in the production and the number of scavenging molecules of ROS. According to Apel and Hirt (2004), ROS-scavenged molecules are antioxidative defense components that act to restore the physiological growth conditions. Intrinsically, this antioxidant defense system is synthesized in aerobic cells to offset the damaging effects of ROS (Ishizawa et al. 2017; Utami et al. 2018).

In plants, enzymatic and non-enzymatic components such as superoxide dismutase (SOD), catalases (CAT), ascorbate peroxidase (APX), and low-molecular-mass antioxidants play a key role in mopping up different types of ROS (Akram et al. 2017; Utami et al. 2018). Also, in the chloroplast and other cellular compartments, ascorbic acid and glutathione, which are found in high concentrations, play crucial roles in plant defense against oxidative stress (Miller et al. 2010). A combination and synergistic role of MC containing fungi and bacteria can be explored to improve plant growth and ecophysiological responses in extreme conditions. Here, the interactive effects of phytohormones-producing endophytic fungal and bacterial symbionts in plant growth and stress tolerance are harnessed. According to Bilalet al. (2018), phytohormones-producing endophytic *Paecilomyces formosus* LHL10 and *Sphingomonas* sp. modulated the stress state through reduced hydrogen peroxide, lipid peroxidation, and antioxidant enzymes (catalase and superoxide dismutase) when compared to the non-inoculated plants. Inoculation of PGPR strains has been attributed to the reduction of osmotic-related conditions in lettuce plants (Kohler et al. 2010). Apart from plant inoculated with *Pseudomonas mendocina* having the capacity to alleviate salt-related stress, it also reduces oxidative damage (reduced chlorosis, necrosis, and drying) in plants, thus enhancing activity related to antioxidant enzymes such as SOD, APX, GR, and POX (Bianco and Defez 2009). *P. entomophila*, *P. stutzeri*, *P. putida*, *P. syringae*, and *P. montelli* are some of the promising MC with significant capacity in osmotic and oxidative activity, drought, and salinity stress (Sandhya et al. 2010; Carmen and Roberto 2011).

13.5.7 Nutrient Deficiency

One of the fundamental targets of crop especially when exposed to harsh environmental condition is to survive its present threat. It does this through the activation of tolerant traits and adaptations using a number of mechanisms. It has been established that nutritional status of plants greatly affects their ability to adapt to adverse environmental conditions especially abiotic stress. A plethora of literature have reported exacerbated adverse effects of abiotic stresses on plants cultivated in nutrient-deficient agricultural soil (Munns and Tester 2008; García-Martí et al. 2019), which are alleviated by exogenous addition of macronutrients (Khoshgoftarmanesh et al. 2010; Turan et al. 2016; Meena et al. 2017). In a study investigating N deficiency and compensation with focus on its uptake, utilization, and the physiological characteristics in rice, Xiong et al. (2018) reported double-cropping super hybrid late rice and N compensation at young panicle differentiation stage after N deficiency at tillering stage. This according to their submission resulted in yield compensation in plants. Its deficiency in plants owing to rise in soil salinity has been reported to be detrimental (Carstensen et al. 2018). P deficiency has also been shown to limit plant productivity. Recent advances have shown that P deficiency affects electron transport to photosystem I (PSI), but the underlying mechanisms are still unknown (Mehra et al. 2018; Carstensen et al. 2018). Invariably, P reduction in saline soils has been linked to ionic strength effects which reduce the activity of phosphate and the tight control of P concentrations by sorption processes, and by low solubility of Ca-P minerals (Carmen and Roberto et al. 2011). With the concentration of these essential minerals (nitrogen and phosphorus), usually very low and insufficient for optimum crop growth, some PGPR have shown promising effects in augmenting this process by making the scarce nutrient readily available to plants (Odoh et al. 2019a). They exhibit this property by taking up several P forms, while the remaining part is adsorbed in the forms of HPO4-2 or H2PO4-1. PGPR play a leading role in mobilizing these nutrients through their participation in biogeochemical cycle (nitrogen and phosphorus cycle) where rhizobia species directly solubilize and mineralize inorganic phosphorus and facilitate the mobility of the organic forms (Richardson and Simpson 2011).

The conversion of some phosphate compounds, e.g., tricalcium phosphate, dicalcium phosphate, hydroxyapatite, and rock phosphate, is predominantly carried out by phosphate-solubilizing bacteria (PSB) such as *Arthrobacter, Pseudomonas, Alcaligenes, Bacillus, Burkholderia, Serratia, Enterobacter, Acinetobacter, Azospirillum, Azotobacter, Flavobacterium, Rhizobium,* and *Erwinia* (Zaidi et al. 2009, Odoh et al. 2019a). They exert this by the secretion of organic acids, e.g., carboxylic acid, formic acid, propionic acid, lactic acid, glycolytic acid, succinic, and fumaric acid. These acids, however, lower the pH of the rhizosphere, thus causing the release of the bound forms of phosphate like $Ca_3\ (PO4)_2$ in the calcareous soils (Kaur et al. 2016). Phosphate-solubilizing bacteria (PSB) also contribute to the mineralization of insoluble organic phosphate via the excretion of enzymes such as phytases, C-P lyases, and phosphatases (Weyens et al. 2010, Daur et al. 2018). *Medicago sativa L, Zea mays, Glycine max*, and a number of plants have been reported with increased yield through the inoculated PSB either singly or in combination with a number of rhizobacteria (Daur et al. 2018; Tagele et al. 2019).

13.6 Production and Commercialization Biofertilizer

With recent advances in agrobiotechnology, a wider coverage, application, production, and commercialization of MC biofertilizers holds promises. This is, however, welcoming as the world over is driving toward smart and sustainable agriculture. Consortium of PGPR is currently the most sought-after strain with multifunctional potentials. Their formulation and application in agriculture are currently advocated to reduce the overreliance on synthetic fertilizers and other agrochemicals. Backer et al. (2018) in their latest studies agree that members of phytomicrobiome (PGPR) offer huge potential in sustainable crop production, thus necessitating the need for more studies to unravel their potentials and challenges. Although the use of MC as an inoculant has been on for centuries, it has mainly focused on legumes and cereals in the past years (Sessitsch and Mitter 2015). As the world human population continuously gets on the rise in geometric fashion, there is an increasing need to meet the growing food demand through development of new agro technologies such as species engineering and screening for biomolecules production to enhance crop growth. PGPR, for instance, is designed to improve nitrogen fixation, ACC deaminase activity, auxin synthesis, and calcium phosphate solubilization when inoculated (Backer et al. 2018). In situ research approach should be adopted in carrying out research using PGPR MC to ascertain the most suitable strain and appropriate biotic condition needed for their growth, while paying attention on the soil quality and season of optimum performance (Odoh et al. 2019a).

For effective long shelf-life microbial inoculant to be developed and commercialized, field trial and conditions must be properly ascertained and approved. This is vital to curtail the release of strain in the environment. Here, PGPR are inoculated in plant material without an appropriate carrier or in quantities that do not allow for efficient rhizosphere colonization under field conditions (Backer et al. 2018). This is primarily due to competition with resident soil micro- and macro-fauna. Also, in the case of soils cultivated with value, they are often fumigated with broad-spectrum biocidal fumigants that alter the bio-community structure of the soil (Dangi et al. 2017). This fumigation system (short-term) is carefully done not to upset soil microbes and their interactions which help in nutrient acquisition and mobilization. In designing microbial consortia, their role specificity (e.g., bioremediation and plant growth potential) must be clearly defined (Macouzet 2016; Baez-Rogelio et al. 2017), as their bioprocess when inoculated will be based on the specific soil conditions. Combining effects of bioremediation and related abiotic condition ameliorations along with plant growth promotion would be essential in addressing some of the global agricultural problems. This, however, must be accompanied with training of staff, farmers, and associate users on efficient application of the bio-inoculants bearing in mind their soil specificity, environmental condition, and complexity or constraints for optimum impact (Bashan 2016; Parnell et al. 2016; Itelima et al. 2018). To develop a MC inoculant (Table 13.2), the following basic steps must be followed (Backer et al. 2018):

- Isolation of the bacterial strain.
- Screening in laboratory and controlled growth environment.
- Field assessment for a range of crops, geographic locations, planting dates, and soil types.
- Evaluation of the possible combinations of strains.

- Consideration of the management practices (e.g., agrochemical use and rotation).
- Refinement of the product.
- Experiments confirming absence of eco-toxicological effects.
- Product delivery formulation—e.g., peat, granular, liquid, or wettable powder.
- Registration and regulatory approval of the product.
- Product available on the market.

13.7 Regulation of MC Biofertilizer

In the developed economy, there are strict regulations and guidelines moderating the use and application of microbial-based nutrient (bio-inoculant). The first point of call of this product after its due formulation and successful testing is registration where the product must meet specific regulatory requirements. Prior to this, the product must be established in a carrier such as alginate (Bashan 2016) or biochar (Głodowska et al. 2016) through which the cells or inoculants are adhered to seeds using sticking agent at the time of sowing. In the case of liquid inoculants, they are spread on seeds prior to sowing or dripped into the seed furrow at the time of sowing. Importantly also are the storage and product lifespan so as to ensure microbial viability, survival, and/or bioactivity of the strain. There should also be clarity on acute versus chronic application of the biomolecule. In most cases, acute application occurs in a limited number of times during a growing season; it can also be on a target stage of crop development, or in response to environmental and abiotic conditions (drought), while in chronic application, the product could be applied at regular time sprays interval or as a slow-release seed treatment (Backer et al. 2018). With ambiguity on a clear-cut regulatory definition of plant biostimulants (MC), there has been complexity in the regulatory procedure and registration cum commercialization of these products across Europe and America. This, however, necessitates the need for unified standard, characterization and regulation of MC, genetically modified species (GMO), and other biostimulants across the globe especially in Africa and Asia which have huge agricultural potentials and high uneducated local engage in agropractices.

13.8 Prospects and Challenges of Biofertilizer Application

MC has recently been gaining public acceptance and recommendation for applicability in agrifood production. Even though its prominence has been in Asia, America, and Europe, there is still laxity of its growth in other regions especially in Africa. This is because of a number of factors ranging from lack of awareness, skilled man-power, and infrastructures to supportive regulatory framework. These identifiable constraints have militated against sustainable agricultural practices in the regions, thereby missing the benefits accruing to biofertilizer application "yield increase, biological nitrogen fixation, cost saving, nutrient uptake" unlike in some Asian countries (e.g., India, Thailand, and China).

13.8.1 Policy Definition and Enforcement

For any nation to produce, commercialize, and use her formulated MC biofertilizer successfully, she must have instituted a sound regulatory framework. This is necessary to curtail excesses that could emanate from application of engineered species into her environment. In Europe, United States, Canada, Argentina, Brazil, and India, for instance, there seems to be an existing policy and guideline governing application of biological products (biofertilizer, biostimulants, and biocontrol). There have also been discussions on science-based standards and policies that will clearly state the definitions and regulations of these bioproducts in India, European Union, Latin America, and USA. According to International Biocontrol Management Association (IBMA), these products are currently regulated by state governments in USA, thus necessitating discussions by the US Environmental Protection Agency (EPA), to establish unified federal regulations. The food and agricultural organization (FOA) should initiate and enforce these policies among existing member states so that they can harness the gains associated with MC. Also, governments in the developing regions in Africa should improve research in the agricultural sector as biofertilizers' application is still at the infant stage. This is primarily derailed by lack of awareness, human capacity, and infrastructure, thus not tapped the potentials of biofertilizers. A regional and national policy regulatory framework that will boost establishment of indigenous manufacturing firms bearing in mind the regions' specificity in terms of her inherent bioagents (microbes and plants), climatic conditions, soil quality, and complexity is also encouraged.

13.8.2 Global Action on Sustainable Agriculture

With the rising environmental stresses which have become unprecedented in recent time and impediment to agricultural productivity across the globe, a clear-cut global action is necessary to curtail possible monumental impact (food insecurity). In the 2014, the food and agricultural organization (FAO) and the world health organization jointly released the International Code of Conduct on Pesticide Management following series of death recorded from users of agrochemicals across the globe. In India, death associated with chemical pesticides used by farmers has continued to rise following flouting of the global norm and best practices, and also due to ill-informed users (farmers). These agrochemicals which are mostly classified as Class 1 are consistently released into the market by merchants and giant multinationals with no recourse to the harmful effects they have on the end users. These practices flout international regulations guiding a number of these products (e.g., monocrotophos and oxydemetonmethyl).

In Nigeria also, there is an upsurge in the use of agrochemicals in the last decades. The situation is indeed worrisome as citizens whose knowledge on the hazardous effects and impacts of these products (minimal concentration) have embraced in totality the use of agrochemicals (pesticides and herbicides) in every farming exercise. The government push to diversify the economy through agriculture has also worsened the situation owing to the sporadic campaign on the quest to rejuvenate the agricultural sector. Available statistics has shown that over 25% of the globally produced pesticides are used in Nigeria with 99% of the death associated with pesticide occurring in developing countries (Ojo 2016). These are linked to factors such as lack of education, non-use of safety procedures, use of cheaper but lethal chemicals, poor legislation and enforcement, and improper handling.

Also, in China, its water body (Lake Taihu) has been polluted by years of continuous seepage of runoff pesticides, herbicides, and fertilizers from nearby agricultural farms. Despite the prohibition of organochlorine pesticides by Chinese government in 1983, traces of HCH and DDT residues are still easily detected in its sediments (Feng et al. 2003). These happenings across the globe, however, pose serious threat not just to human health but to our ecosystem and biodiversity structures. It therefore calls for a global framework most importantly from FOA and WHO alongside regional and national governments to prohibit classified chemicals and properly regulate hazardous products, while also sensitizing the public especially in developing countries on the advantages of microbial-based formulated bioproducts and the need to hold onto green technology. This move will ultimately help in preventing and controlling the associated environmental problems such as air, water (eutrophication and fertilizer seepage), and soil pollutions linked to synthetic agrochemicals application.

13.8.3 Clarity of the Benefit of MC

It has become clear that the benefit associated with MC application is so enormous. This in comparison with single strain has recorded positive milestone in areas such as food production in industries, use in agricultural application, medicine, and environmental remediation. Through metabolic modeling and reconstruction of individual strains, a formidable and complex community of model microbial agents are formed for optimum performance and production of needed biochemical agents and biomass (Faust 2018). More metabolic research needs to be done on plants and microbial species for more understanding of genetic characters and genes needed to perform complex functions. There is need for farmers and practitioners in agricultural sectors to be abreast of this biotechnology owing to the growing abiotic conditions and alteration of climatic conditions. Proteomics and transcriptomic approaches and genetic engineering of species provide suitable alternatives to these challenges.

13.8.4 Lack of Skilled Manpower and Innovative Research

The thin line between successful utilization of biofertilizer in some Asian countries, Europe, America, and Africa is technical knowhow. The developed regions have sound research-based biotechnological approach for the formulation of bioproducts, increased awareness on their usage while fighting for corresponding decline on the use of chemical fertilizers. Developing countries in Africa are yet to pay adequate attention on the advantages of biofertilizer in agricultural system. This is evident in the skeletal application of biofertilizers by very few farmers across the region. This is in contrast to what is obtainable in Brazil, where almost all the crop protein produced is through BNF; in east and southern Africa, not up to 1% use any form of bio-inoculant. Most often these imported biofertilizers are formulated in conformity or tailored to the countries of origin's standard bearing in mind their local conditions (e.g., climatic conditions, storage condition). These parameters play a huge role in determining their shelf life and viability. With improved manpower development through training, and increased awareness, research, and innovations, locally influencing conditions will become a considerable factor when producing biofertilizer that will be indigenous to a particular region (African climatic conditions). It will also help limit loss of viability observed in some biofertilizers in the market across Africa (Jefwa et al. 2014) where storage conditions and handling play a major role. Through adequate training, knowledge gap usually witnessed in developing economy could be upturned as their trained agriculturists and scientists will see the need to localize their products for the overarching need of the populace. However, with the lack of far-reaching research to develop formulations that could cater for the spatial crop responses, Africa will not be able to benefit from the full potential of biofertilizers. It will also help improve qualitative product delivery in countries such as India, where significant government support has boosted biofertilizers production.

13.9 Conclusion and Future Prospects

Many arable lands are in urgent need of natural and eco-friendly alternatives to synthetic fertilizers for crop production and also to help cushion the shock emanating from abiotic stresses on crops growing in extreme habitats. Microbial consortium biofertilizers have been developed as dependable solution to this mayhem, and some parts of the developed world are already harnessing the benefits of the green technology. However, as appealing and promising as this agrotechnology can be, it is still replete with challenges especially in the undeveloped world where illiteracy and lack of skilled manpower impede proper implementation of its use in crop production.

Acknowledgements The authors are grateful to the management of Centre for Environment Human Right and Development (CEHRD), for providing the needed facilities and constant encouragement all through the period of preparing this manuscript.

References

Ahemad M (2015) Phosphate solubilising bacteria assisted phytoremediation of metalliferous soil: a review. Biotech 5:111–121. https://doi.org/10.1007/s13205.014.0206.0

Ahmad I, Khan MSA, Aqil F, Singh M (2011) Microbial applications in agriculture and the environment: a broad perspective. In: Ahmad I, Ahmad F, Pichtel J (eds) Microbes and Microbial Technology: Agricultural and Environmental Applications (New York, NY: Springer), 1–27

Ajai O (2010) Balancing of interests in environmental law in Nigeria. Environ Law Africa 379

Akram NA, Shafiq F, Ashraf M (2017) Ascorbic acid-a potential oxidant scavenger and its role in plant development and abiotic stress tolerance. Front Plant Sci 8:613. https://doi.org/10.3389/fpls.2017.00613

Akter N, Islam MR (2017) Heat stress effects and management in wheat. A review Agron Sustain Dev 37:37

Alavilli H, Lee H, Park M, Lee BH (2017) Antarctic moss multiprotein bridging factor 1c overexpression in Arabidopsis resulted in enhanced tolerance to salt stress. Front Plant Sci 8:1206. https://doi.org/10.3389/fpls.2017.01206

Albuquerque P, Casadevall A (2012) Quorum sensing in fungi—a review. Med Mycol 50:337–345. https://doi.org/10.3109/13693786.2011.652201

Ale SB, Halloway A, Mitchell WA, Whelan CJ (2019) Does God roll dice? Neutrality and determinism in evolutionary ecology. Biol Philos34: 3. https://doi.org/10.1007/s10539-018-9657-8

Al-Garni SMS (2006) Increasing NaCl-salt tolerance of a halophytic plant Phragmites australis by mycorrhizal symbiosis. Am-Eurasian J Agric Environ Sci 1:119–126

Álvarez-Aragón R, Haro R, Benito B, Rodríguez-Navarro A (2016) Salt intolerance in Arabidopsis: shoot and root sodium toxicity, and inhibition by sodium-plus-potassium overaccumulation. Planta 243:97–114

Ansari MW, Trivedi DK, Sahoo RK (2013) A critical review on fungi mediated plant responses with special emphasis to piriformosporaindica on improved production and protection of crops. Plant Physiol Bioch 70:403–10

Apel K, Hirt H (2004) Reactive oxygen species: metabolism, oxidative stress, and signal transduction. Annu Rev Plant Biol 55:373–399

Ashraf M, Hasnain S, Berge O, Mahmood T (2004) Inoculating wheat seedlings with exopolysaccharide-producing bacteria restricts sodium uptake and stimulates plant growth under salt stress. Biol Fertil Soils 40(3):157–162. https://doi.org/10.1007/s00374-004-0766-y

Babu AG, Shea PJ, Sudhakar D, Jung IB, Oh BT (2015) Potential use of *Pseudomonas koreensis* AGB-1 in association with *Miscanthus sinensis* to remediate heavy metal (loid) contaminated mining site soil. J Environ Manage 151:160–166. https://doi.org/10.1016/j.jenvman.2014.12.045

Bacilio M, Moreno M, Bashan Y (2016) Mitigation of negative effects of progressive soil salinity gradients by application of humic acids and inoculation with Pseudomonas stutzeri in a salt-tolerant and a salt-susceptible pepperAppl Soil Ecol 107:394–404

Backer R, Rokem JS, Ilangumaran G, Lamont J, Praslickova D, Ricci E, Subramanian S, Smith DL (2018) Plant growth-promoting rhizobacteria: context, mechanisms of action, and roadmap to commercialization of biostimulants for sustainable agriculture. Front Plant Sci 9:1473. https://doi.org/10.3389/fpls.2018.01473

Baez-Rogelio A, Morales-García YE, Quintero-Hernández V, Muñoz-Rojas J (2017) Next generation of microbial inoculants for agriculture and bioremediation. Microb Biotechnol 10:19–21. https://doi.org/10.1111/1751-7915.12448

Bano A, Fatima M (2009) Salt tolerance in Zea mays (L.) following inoculation with Rhizobium and Pseudomonas. Biol Fert Soils 45:405–413

Barnawal D, Bharti N, Maji D, Chanotiya CS, Kalra A (2014) ACC deaminase-containing Arthrobacter protophormiaein-duces NaCl stress tolerance through reduced ACC oxidase activity and ethylene production resulting in improved no-dulation and mycorrhization inPisum sativum. J Plant Physiol 171:884–894

Barriuso J (2015) Quorum sensing mechanisms in fungi. AIMS Microbiol 1:37–47.https://doi.org/10.3934/microbiol.2015.1.37

Bashan N (2016) Inoculant formulations are essential for successful inoculation with plant growth-promoting bacteria and business opportunities. Indian Phytopathol 69:739–743

Bennett JW, Hung R, Lee S, Padhi S (2012) Fungal and bacterial volatile organic compounds: an overview and their role as ecological signaling agents. In: Hock B (ed) Fungal Assoc, 2nd edn. Springer, Berlin Heidelberg (Germany), pp 373–393

Bergelson J, Mittelstrass J, Horton MW (2019) Characterizing both bacteria and fungi improves understanding of the Arabidopsis root microbiome. Sci Rep 9:24

Bernstein HC, Carlson RP (2012) Microbial consortia engineering for cellular factories: in vitro to in silico systems. Comput Struct Biotechnol J 3:1–8. https://doi.org/10.5936/csbj.201210017

Bhardwaj D, Ansari MW, Sahoo RK, Tuteja N (2014) Biofertilizers function as key player in sustainable agriculture by improving soil fertility, plant tolerance and crop productivity. Microb Cell Fact 13:1–10. https://doi.org/10.1186/1475-2859-13-66

Bianco C, Defez R (2009) *Medicago truncatula* improves salt tolerance when nodulated by an indole-3-acetic acid-overproducind *Sinorhizobium meliloti* strain. J Exp Bot. https://doi.org/10.1093/jxb/erp140

Bilal S, Shahzad R, Khan AL, Kang SM, Imran QM, Al-Harrasi A, Yun BW, Lee IJ (2018) Endophytic Microbial Consortia of Phytohormones Producing Fungus Paecilomyces formosus LHL10 and Bacteria Sphingomonas sp. LK11 to Glycine max L. Regulates Physio-hormonal Changes to Attenuate Aluminum and Zinc Stresses. Front Plant Sci 9:1273. https://doi.org/10.3389/fpls.2018.01273

Bradáčová K, Florea AS, Bar-Tal A, Minz D, Yermiyahu U, Shawahna R, Kraut-Cohen J, Zolti A, Erel R, Dietel K, Weinmann M, Zimmermann B, Berger N, Ludewig U, Neumann G, Poşta G (2019) Microbial consortia versus single-strain inoculants: an advantage in PGPM-assisted tomato production? Agronomy 9:105

Brilli F, Loreto F, Baccelli I (2019) Exploiting plant volatile organic compounds (VOCs) in agriculture to improve sustainable defense strategies and productivity of crops. Front Plant Sci 10:264. https://doi.org/10.3389/fpls.2019.00264

Bulgarelli D, Rott M, Schlaeppi K, Loren V, van Themaat E, Ahmadinejad N, Assenza F, Rauf P, Huettel B, Reinhardt R, Schmelzer E, Peplies J, Gloeckner FO, Amann R, Eickhorst T, Schulze-Lefert P (2012) Revealing structure and assembly cues for Arabidopsis root-inhabiting bacterial microbiota. Nature 488:91–95. https://doi.org/10.1038/nature11336

Burén S, Rubio LM (2018) State of the art in eukaryotic nitrogenase engineering. FEMS Microbiol Lett 365(2), fnx274. https://doi.org/10.1093/femsle/fnx274

Burke C, Steinberg P, Rusch D, Kjelleberg S, Thomas T (2011) Bacterial community assembly based on functional genes rather than species. Proc Natl Acad Sci U S A. 108:14288–14293

Caplan JS, Yeakley JA (2006) *Rubus armeniacus* (Himalayan blackberry) occurrence and growth in relation to soil and light conditions in western Oregon. Northwest Sci 80:9–17

Carmen B, Roberto D (2011) Soil bacteria support and protect plants against abiotic stresses A. Shan er (Ed.), Abiotic Stress in Plants mechanisms and Adaptations, Pub. InTech, pp 143–170

Carstensen A, Herdean A, Schmidt SB, Sharma A, Spetea C, Pribil M, Husted S (2018) The impacts of phosphorus deficiency on the photosynthetic electron transport chain. Plant Physiol 177:271–284

Chakraborty A, Ellefson E, Li C, Gittins D, Brooks JM, Bernard BB, Hubert CRJ (2018) Thermophilic endospores associated with migrated thermogenic hydrocarbons in deep Gulf of Mexico marine sediments. ISME J 12:1895–1906. https://doi.org/10.1038/s41396-018-0108-y

Chang WS, van de Mortel M, Nielsen L, de Guzman GN, Li X, Halverson LJ (2007) Alginate production by Pseudomonas putida creates a hydrated microenvironment and contributes to biofilm architecture and stress tolerance under water-limiting conditions. J Bacteriol 189:8290–8299

Chase JM, Myers JA (2011) Disentangling the importance of ecological niches from stochastic processes across scales. Philos T Roy Soc B 366:2351–2363

Chatzipavlidis I, Kefalogianni I, Venieraki A, Holzapfel W (2013) Status and trends of the conservation and sustainable use of microorganisms in agroindustrial processes. Commission on Genetic Resources for Food and Agriculture, Background Study Paper No, p 64

Chérel I, Gaillard I (2019) The complex fine-tuning of K^+ fluxes in plants in relation to osmotic and ionic abiotic stresses. Int J Mol Sci 20:715

Chodak M, Golebiewski M, Morawska-Ploskonka J, Kuduk K, Niklinska M (2015) Soil chemical properties affect the reaction of forest soil bacteria to drought and rewetting stress. Ann Microbiol 65(3):1627–1637

Cira NJ, Pearce MT, Quake SR (2018) Neutral and selective dynamics in a 22 synthetic microbial community. Proc Natl Acad Sci U S A 115, E9842-E9848, 23 https://doi.org/10.1073/pnas.1808118115

Clark DP, Dunlap PV, Madigan MT, Martinko JM (2009) Brock. *Biology of Microorganisms*. San Francisco: Pearson. p 485

Clinton A, Rumbaugh KP (2016) Interspecies and interkingdom signaling via quorum signals. Isr J Chem 56:265–272. https://doi.org/10.1002/ijch.201400132

Compant S, Clément C, Sessitsch A (2010) Plant growth promoting bacteria in the rhizo and endosphere of plants: Their role, colonization, mechanisms involved and prospects for utilization. Soil Biol Biochem 42:669–678

Cycon M, Mrozik A, Piotrowska-Seget Z (2019) Antibiotics in the soil environment degradation and their impact on microbial activity and diversity. Front Microbiol 10:338. https://doi.org/10.3389/fmicb.2019.00338

Dangi S, Tirado-Corbalá R, Gerik J, Hanson B (2017) Effect of longterm continuous fumigation on soil microbial communities. Agronomy 7:37. https://doi.org/10.3390/agronomy7020037

Dassonville N, Vanderhoeven S, Vanparys V, Hayez M, Gruber W, Meerts P (2008) Impacts of alien invasive plants on soil nutrients are correlated with initial site conditions in NW Europe. Oecologia 157:131–140. https://doi.org/10.1007/s00442-008-1054-6

Daur I, Saad MM, Eida AA, Ahmad S, Shah ZH, Ihsan MZ, Muhammad Y, Sohrab SS and Hirt H (2018) Boosting Alfalfa (*Medicago sativa* L.) Production with Rhizobacteria from Various Plants in Saudi Arabia. Front Microbiol 9:477. https://doi.org/10.3389/fmicb.2018.00477

De Oliveira EJ, Morgante CV, De Tarso AS, De Melo Chaves AR, Antonio RP, Cruz JL et al (2017) Evaluation of cassava germplasm for drought tolerance under field conditions. Euphytica 213:188. https://doi.org/10.1007/s10681-017-1972-7

de Zelicourt A, Al-Yousif M, Hirt H (2013) Rhizosphere microbes as essential partners for plant stress tolerance. Mol Plant 6:242–245. https://doi.org/10.1093/mp/sst028

Denton B (2007) Advances in phytoremediation of heavy metals using plant growth promoting bacteria and fungi. MMG 445 Basic Biotechnol 3:1–5

Deshwal VK, Kumar P (2013) Production of Plant growth promoting substance by Pseudomonads. J Academia Indust Res 2(4):221–225

Deveau A, Bonito G, Uehling J, Paoletti M, Becker M, Bindschedler S, Hacquard S, Hervé V, Labbé J, Lastovetsky OA, Mieszkin S, Millet LJ, Vajna B, Junier P, Bonfante P, Krom BP, Olsson S, van Elsas JD, Wick LY (2018) Bacterial-fungal interactions: ecology, mechanisms and challenges. FEMS Microbiol Rev 42:335–352

Dumbrell AJ, Nelson M, Helgason T, Dytham C, Fitter AH (2010) Relative roles of niche and neutral processes in structuring a soil microbial community. ISME J 4:337–345

Egamberdieva D, Kucharova Z (2009) Selection for root colonizing bacteria stimulating wheat growth in saline soils. Biol Fert Soil. https://doi.org/10.1007/s00374-009-0366-y

Eze CN, Odoh CK, Eze EA, Enemuor SC, Orjiakor IP, Okobo UJ (2018) Chromium (III) and its effects on soil microbial activities and phytoremediation potentials of Arachis hypogea and Vigna unguiculata. Afr J Biotechnol 17(38):1207–1214. https://doi.org/10.5897/ajb2018.16566

Eze CN, Ugwu CC, Eze EA, Eze US (2014) Evaluation of germination, shoot growth and rhizofungal flora of Zea mays and Sorghum bicolor in soil contaminated with varying levels of Bonny light crude oil. Int J Curr Microbiol App Sci 3(1):253–263

Fabbro CD, Prati D (2014) Early responses of wild plant see-dlings to arbuscular mycorrhizal fungi and pathogens. Basic Appl Ecol 15:534–542

Fahad S, Hussain S, Bano A, Saud S, Hassan S, Shan D, Khan FA, Khan F, Chen Y, Wu C, Tabassum MA (2015) Potential role of phytohormones and plant growth-promoting rhizobacteria in abiotic stresses: consequences for changing environment. Environ Sci Pollut Res Int 22:4907–4921

FAO (2010) Current status and options for biotechnologies in food processing and in food safety in developing countries. In: Proceedins of the FAO international technical conference Guadalajara, Mexico, pp 1–37

Faust K (2018) Microbial consortium design benefits from metabolic modeling. Trends in Biotechnol. https://doi.org/10.1016/j.tibtech.2018.11.004

Feng K, Yu BY, Ge DM, Wong MH, Wang XC, Cao ZH (2003) Organo-chlorine pesticide (DDT and HCH) residues in the Taihu Lake region and its movement in soil-water system. I. Field survey of DDT and HCH residues in ecosystem of the region. Chemosphere 50:683–687

Fisher CK, Mehta P (2014) The transition between the niche and neutral regimes in ecology. Proc Natl Acad Sci USA 111:13111–13116

Fleitas-Martínez O, Rigueiras PO, Pires ÁS, Porto WF, Silva ON, de la Fuente-Nunez C, Franco OL (2019) Interference with quorum-sensing signal biosynthesis as a promising therapeutic strategy against multidrug-resistant pathogens. Front Cell Infect Microbiol 8:444. https://doi.org/10.3389/fcimb.2018.00444

Flores-Félix JD, Menéndez E, Rivera LP, Marcos-García M, Martínez-Hidalgo P, Mateos PF, Martínez-Molina E, Velázquez MDLE, García-Fraile P, Rivas R (2013) Use of *Rhizobium leguminosarum* as a potential biofertilizer for *Lactuca sativa* and *Daucus carota* crops. J Plant Nutr Soil Sci 176:876–882

Flores-Felix JD, Silva LR, Rivera LP (2015) Plants probiotics as a tool to produce highly functional fruits: the case of Phyllobacterium and vitamin C in strawberries. PLoS ONE

Foster RC (1988) Microenvironments of soil organisms. Biol Fertility Soils 6:189–203. https://doi.org/10.1007/bf00260816

Franche C, Lindstrom K, Elmerich C (2009) Nitrogen-fixing bacteria associated with leguminous and non-leguminous plants. Plant Soil 321:35–59. https://doi.org/10.1007/s11104-008-9833-8

Franco JA, Bañón S, Vicente MJ, Miralles J, Martínez-Sánchez JJ (2011) Root development in horticultural plants grown under abiotic stress conditions-a review. J Hortic Sci Biotechnol 86:543–556. https://doi.org/10.1080/14620316.2011.11512802

Frey-Klett P, Burlinson P, Deveau A, Barret M, Tarkka M, Sarniguet A (2011) Bacterial-fungal interactions: hyphens between agricultural, clinical, environmental, and food microbiologists. Microbiol Mol Biol Rev 75:583–609

Fu QL, Liu C, Ding NF, Lin YC, Guo B (2010) Ameliorative effects of inoculation with the plant growth-promoting rhizobacterium Pseudomonassp. DW1 on growth of eggplant (Solanum melongena L.) seedlings under salt stress. Agr Water Manag 97(12):1994–2000. https://doi.org/10.1016/j.agwat.2010.02.003

Fukami J, Cerezini P, Hungria M (2018) Azospirillum: benefits that go far beyond biological nitrogen fixation. AMB Express 8(1):73. https://doi.org/10.1186/s13568-018-0608-1

Fukami J, Ollero FJ, Megías M, Hungria M (2017) Phytohormones and induction of plant-stress tolerance and defense genes by seed and foliar inoculation with Azospirillum brasilense cells and metabolites promote maize growth. AMB Express 7(1):153

García-Martí M, Piñero MC, García-Sanchez F, Mestre TC, López-Delacalle M, Martínez V, Rivero RM (2019) Amelioration of the oxidative stress generated by simple or combined abiotic stress through the K^+ and Ca^{2+} supplementation in tomato plants. Antioxidants 8:81

Gentili R, Ambrosini R, Montagnani C, Caronni S, Citterio S (2018) Effect of soil pH on the growth, reproductive investment and pollen allergenicity of ambrosia artemisiifolia L. Front Plant Sci 9:1335. https://doi.org/10.3389/fpls.2018.01335

Gilbert JA, Steele JA, Caporaso JG, Steinbrück L, Reeder J, Temperton B, Huse S, McHardy AC, Knight R, Joint I, Somerfield P, Fulrman JA, Field D (2012) Defining seasonal marine microbial community dynamics. ISME J 6(2):298–308. https://doi.org/10.1038/ismej.2011.107

Giri B, Kapoor R, Mukerji KG (2007) Improved tolerance of Acacia nilotica to salt stress by arbuscular mycorrhiza, Glomus fasciculatum may be partly related to elevated K/Na ratios in root and shoot tissues. Microb Ecol 54:753–760. https://doi.org/10.1007/s00248-007-9239-9

Glick BR (2012) Plant Growth promoting bacteria: mechanisms and applications. Scientifica Article ID 963401, 15 p, https://doi.org/10.6064/2012/963401

Glick BR (2014) Bacteria with ACC deaminase can promote plant growth and help to feed the world. Microbiol Res 169:30–39

Głodowska M, Husk B, Schwinghamer T, Smith D (2016) Biochar is a growth-promoting alternative to peat moss for the inoculation of corn with a pseudomonad. Agron Sustain Dev 36:1–10. https://doi.org/10.1007/s13593-016-0356-z

Goswami D, Thakker JN, Dhandhukia PC (2015) Simultaneous detection and quantification of indole-3-acetic acid (IAA) and indole-3-butyric acid (IBA) produced by Rhizobacteria from l-tryptophan (Trp) using HPTLC. J Microbiol Methods 110:7–14

Gouda S, Kerry RG, Das G, Paramithiotis S, Shin HS, Patra JK (2018) Revitalization of plant growth promoting rhizobacteria for sustainable development in agriculture. Microbiol Res 206:131–140

Grover M, Ali V, Sandhya SKZ, Venkateswarlu B (2011) Role of microorganisms in adaptation of agricultural crops to abiotic stressesWorld. J Microbiol Biotechnol 27:1231–1240

Guo J, Dong X, Han G, Wang B (2019) Salt-Enhanced Reproductive Development of Suaeda salsa L. Coincided With Ion Transporter Gene Upregulation in Flowers and Increased Pollen K + Content. Front Plant Sci 10:333 https://doi.org/10.3389/fpls.2019.00333

Guo Z, Zhang L, Yang W, Hua L, Cai C (2019b) Aggregate stability under long-term fertilization practices: the case of eroded ultisols of South Central China. Sustainability 11:1169

Hacquard S, Garrido-Oter R, Gonzalez A, Spaepen S, Ackermann G, Lebeis S, McHardy AC, Dangl JL, Knight R, Ley R, Schulze-Lefert P (2015) Microbiota and host nutrition across plant and animal kingdoms. Cell Host Microbe 17(5):603–616

Hamdia MAES, Shaddad MAK, Doaa MM (2004) Mechanisms of salt tolerance and interactive effects of Azospirillum brasilense inoculation on maize cultivars grown under salt stress conditions. Plant Growth Regul 44:165–174

Hamilton CE, Bever JD, Labb´e J, Yang XH, Yin HF (2016) Mitigating climate change through managing constructed-microbial communities in agriculture. Agr Ecosyst Environ 216:304–308

Hansel CM, Fendorf S, Jardine PM, Francis CA (2008) Changes in bacterial and archaeal community structure and functional diversity along a geochemically variable soil profile. Appl Environ Microbiol 74(5):1620–1633

Hao JH, Lv SS, Bhattacharya S, Fu JG (2017) Germination response of four alien congeneric *Amaranthus* species to environmental factors. PLoS ONE 12:e0170297. https://doi.org/10.1371/journal.pone.0170297

Hardoim PR, van Overbeek LS, Berg G, Pirttilä AM, Compant S, Campisano A, Döring M, Sessitsch A (2015) The hidden world within plants: Ecological and evolutionary considerations for defining functioning of microbial endophytes. Microbiol Mol Biol Rev 79(3):293–320. https://doi.org/10.1128/MMBR.00050-14

Hasanuzzaman M, Nahar K, Alam MM, Roychowdhury R, Fujita M (2013) Physiological, biochemical, and molecular mechanisms of heat stress tolerance in plants. Int J Mol Sci 14:9643–9684

Hayat R, Ali S, Amara U, Khalid R, Ahmed I (2010) Soil beneficial bacteria and their role in plant growth promotion: a review. Ann Microbiol 60:579–598

Hayat R, Khalid R, Ehsan M, Ahmed I, Yokotaand A, Ali S (2013) Molecular characterization of soil bacteria for improving crop yield in Pakistan. Pak J Bot 45:1045–1055

Hays SG, Patrick WG, Ziesack M, Oxman N, Silver PA (2015) Better together: engineering and application of microbial symbioses. Curr Opin Biotechnol 36:40–49. https://doi.org/10.1016/j.copbio.2015.08.008

Htwe AZ, Moh SM, Soe KM, Moe K, Yamakawa T (2019) Effects of Biofertilizer Produced from *Bradyrhizobium* and *Streptomyces griseoflavus* on Plant Growth, Nodulation, Nitrogen Fixation, Nutrient Uptake, and Seed Yield of Mung Bean, Cowpea, and Soybean. Agronomy 9:77

Hubbell SP (2001) The Unified neutral theory of biodiversity and biogeography. Monographs in Population Biology. Vol 32. Princeton University Press: Princeton, USA

Hung R, Lee S, Bennett JW (2015) Fungal volatile organic compounds and their role in ecosystems. Appl Microbiol Biotechnol 99:3395–3405. https://doi.org/10.1007/s00253-015-6494-4

Hungria M, Nogueira MA, Araujo RS (2015) Soybean seed co-inoculation with *Bradyrhizobium* spp. and *Azospirillum brasilense*: a new biotechnological tool to improve yield and sustainability. Am J Plant Sci. 6:811–817. https://doi.org/10.4236/ajps.2015.66087

Igiehon NO, Babalola OO (2018) Below-ground-above-ground plant-microbial interactions: focusing on soybean, rhizobacteria and mycorrhizal fungi. Open Microbiol J 12:261–279. https://doi.org/10.2174/1874285801812010261

Iqbal MJ (2018) Role of osmolytes and antioxidant enzymes for drought tolerance in wheat. Global Wheat Production, Shah Fahad, Abdul Basir and Muhammad Adnan, IntechOpen. https://doi.org/10.5772/intechopen.75926

Iqbal MJ, Maqsood Y, Abdin ZU, Manzoor A, Hassan M, Jamil A (2016) SSR markers associated with Proline in drought tolerant wheat germplasm. Appl Biochem Biotech 178:1042–1052

Isayenkov SV (2012) Physiological and molecular aspects of salt stress in plants. Cytol Genet 46:302–318. https://doi.org/10.3103/s0095452712050040

Isayenkov SV, Maathuis FJM (2019) Plant salinity stress: many unanswered questions remain. Front Plant Sci 10:80. https://doi.org/10.3389/fpls.2019.00080

Ishizawa H, Kuroda M, Morikawa M, Ike M (2017) Differential oxidative and antioxidative response of duckweed *Lemna minor* toward plant growth promoting/inhibiting bacteria. Plant Phys Biochem 118:667–673. https://doi.org/10.1016/j.plaphy.2017.08.006

Islam F, Yasmeen T, Ali Q, Ali S, Arif MS, Hussain S, Rizvi H (2014) Influence of *Pseudomonas aeruginosa* as PGPR on oxidative stress tolerance in wheat under Zn stress. Ecotoxicol Environ Saf 104:285–293. https://doi.org/10.1016/j.ecoenv.2014.03.008

Ite AE, Ibok UJ, Ite MU, Petters SW (2013) Petroleum exploration and production: past and present environmental issues in the Nigeria's Niger Delta. Am J Environ Prot 1(7):8–90

Itelima JU, Bang WJ, Onyimba IA, Oj E (2018) A review: biofertilizer; a key player in enhancing soil fertility and crop productivity. J Microbiol Biotechnol Rep 2:22–28

Jagmann N, Philipp B (2014) Reprint of design of synthetic microbial communities for biotechnological production processes. J Biotechnol 192:293–301. https://doi.org/10.1016/j.jbiotec.2014.11.005

Jambon I, Thijs S, Weyens N, Vangronsveld J (2018) Harnessing plant-bacteria-fungi interactions to improve plant growth and degradation of organic pollutants. J Plant Interact 13:119–130

Jefwa JM, Pypers P, Jemo M, Thuita M, Mutegi E, Laditi MA, Faye A, Kavoo A, Munyahali W, Herrmann L, Atieno M, Okalebo JR, Yusuf A, Ibrahim A, Ndung'u-Magiroi KW, Asrat A, Muletta D, Ncho C, Kamaa M, Lesueur D (2014) Do Commercial Biological and Chemical Products Increase Crop Yields and Economic Returns Under Smallholder Farmer Conditions?, In Challenges and opportunities for agricultural intensification of the humid highland systems of subSaharan Africa, B., Vanlauwe, P., van Asten, and G., Blomme, (eds.), Springer International Publishing, Switzerland, 81–96

Jha M, Sanjeet C, Sonia S (2013) Microbial consortium for sustainable rice production. Agroecology and Sustainable Food Systems 37(3):340–362

Jha Y, Sablok G, Subbarao N, Sudhakar R, Fazil MHUT, Subramanian RB et al (2014) Bacterial-induced expression of RAB18 protein in Orzya sativa salinity stress and insights into molecular interaction with GTP ligand. J Mol Recognit 27:521–527. https://doi.org/10.1002/jmr.2371

Jiang L, Patel SN (2008) Community assembly in the presence of dis-turbance: a microcosm experiment. Ecology 89:1931–1940

Jiang S, Zhang D, Wang L, Pan J, Liu Y, Kong X, Zhou Y, Li D (2013) A maize calcium-dependent protein kinase gene, ZmCPK4, positively regulated abscisic acid signaling and enhanced drought stress tolerance in transgenic Arabidopsis. Plant Phys Biochem 71:112–120

Kasim W, Osman M, Omar M, Abd El-Daim I, Bejai S, Meijer J (2013) Control of drought stress in wheat using plant-growth promoting rhizobacteria. J Plant Growth Regul 32:122–130

Kasim WA, Gaafar RM, Abou-Ali RM, Omar MN, Hewait HM (2016) Effect of biofilm forming plant growth promoting rhizobacteria on salinity tolerance in barley. Ann Agric Sci 61(2):217–227

Kasotia A, Varma A, Choudhary DK (2015) *Pseudomonas*-mediated mitigation of salt stress and growth promotion in *Glycine max*. Agric Res 4:31–41. https://doi.org/10.1007/s40003-014-0139-1

Kaur H, Kaur J, Gera R (2016) Plant growth promoting rhizobacteria: A boon to agriculture. Int J Cell Sci Biotech 5:17–22

Khan AL, Waqas M, Hussain J, Al-Harrasi A, Hamayun M, Lee IJ (2015) Phytohormones enabled endophytic fungal symbiosis improve aluminum phytoextraction in tolerant *Solanum lycopersicum*: an examples of *Penicillium janthinellum* LK5 and comparison with exogenous GA 3. J Hazard Mate. 295:70–78. https://doi.org/10.1016/j.jhazmat.2015.04.008

Khan N, Bano A, Babar MA (2019) Metabolic and physiological changes induced by plant growth regulators and plant growth promoting rhizobacteria and their impact on drought tolerance in *Cicer arietinum L*. PLoS ONE 14(3):e0213040. https://doi.org/10.1371/journal.pone.0213040

Khoshgoftarmanesh AH, Schulin R, Claney RL, Daneshbakhsh B Afyuni M (2010) Micronutrient efficient genotypes for crop yield and nutritional quality in sustainable agriculture. A review. Agron Sustain Dev 30, 83–107

Kim HJ, Boedicker JQ, Choi JW, Ismagilov RF (2008) Defined spatial structure stabilizes a synthetic multispecies bacterial community. Proc Natl Acad Sci USA 105:18188–18193. https://doi.org/10.1073/pnas.0807935105

Kohler J, Caravaca F, Roldàn A (2010) An AM fungus and a PGPR intensify the adverse effects of salinity on the stability of rhizosphere soil aggregates of *Lactuca sativa*. SoilBiol Bioch 42:429–434

Kohler J, Hernandez JA, Caravaca F, Roldan A (2009) Induction of antioxidant enzymes is involved in the greater effectiveness of a PGPR versus AM fungi with respect to increasing the tolerance of lettuce to severe salt stress. Environ Exp Bot 65:245–252

Köpp MM, Passos LP, da Silva VR, da Silva Lédo FJ, Meirelles Coimbra JL, Costa de Oliveira A (2011) Effects of nutrient solution pH on growth parameters of alfalfa (*Medicago sativa* L.) genotypes. Commun Sci 2:135–141

Kosová K, Vítámvás P, Urban MO, Prášil IT, Renaut J (2018) Plant abiotic stress proteomics: the major factors determining alterations in cellular proteome. Front Plant Sci 9:122. https://doi.org/10.3389/fpls.2018.00122

Kour D, Rana KL, Yadav AN, Yadav N, Kumar V, Kumar A, Sayyed RZ, Hesham AE-L, Dhaliwal HS, Saxena AK (2019a) Drought-Tolerant Phosphorus-Solubilizing Microbes: Biodiversity and Biotechnological Applications for Alleviation of Drought Stress in Plants. In: Sayyed RZ, Arora NK, Reddy MS (eds) Plant Growth Promoting Rhizobacteria for Sustainable Stress Management: Volume 1: Rhizobacteria in Abiotic Stress Management. Springer Singapore, Singapore, pp 255–308. https://doi.org/10.1007/978-981-13-6536-2_13

Kour D, Rana KL, Yadav N, Yadav AN, Singh J, Rastegari AA, Saxena AK (2019b) Agriculturally and industrially important fungi: current developments and potential biotechnological applications. In: Yadav AN, Singh S, Mishra S, Gupta A (eds) Recent Advancement in White Biotechnology through Fungi, Volume 2: Perspective for Value-Added Products and Environments. Springer International Publishing, Cham, pp 1–64. https://doi.org/10.1007/978-3-030-14846-1_1

Kumar A (2016) Role of microbes in food and industrial microbiology. J Food Ind Microbiol 2:e101. https://doi.org/10.4172/2572-4134.1000e101

Kumar A, Chaturvedi AK, Yadav K, Arunkumar KP, Malyan SK, Raja P, Kumar R, Khan SA, Yadav KK, Rana KL, Kour D, Yadav N, Yadav AN (2019a) Fungal Phytoremediation of Heavy Metal-Contaminated Resources: Current Scenario and Future Prospects. In: Yadav AN, Singh S, Mishra S, Gupta A (eds) Recent advancement in white biotechnology through Fungi: Volume 3: Perspective for Sustainable Environments. Springer International Publishing, Cham, pp 437–461. https://doi.org/10.1007/978-3-030-25506-0_18

Kumar H, Bajpai VK, Dubey RC (2010) Wiltdisease management and enhancement of growthand yield of Cajanus cajan (L) var. Manak bybacterial combinations amended with chemicalfertilizer. Crop Prot 29:591–598
Kumar M, Kour D, Yadav AN, Saxena R, Rai PK, Jyoti A, Tomar RS (2019b) Biodiversity of methylotrophic microbial communities and their potential role in mitigation of abiotic stresses in plants. Biologia 74:287–308. https://doi.org/10.2478/s11756-019-00190-6
Lai QX, Bao ZY, Zhu ZJ, Qian QQ, Mao BZ (2007) Effects of osmotic stress on antioxidant enzymes activities in leaf discs of PSAG12-IPT modified Gerbera. J Zhejiang Univ Sci B. 8(7):458–464. https://doi.org/10.1631/jzus.2007.b0458
Lamaoui M, Jemo M, Datla R, Bekkaoui F (2018) Heat and drought stresses in crops and approaches for their mitigation. Front Chem 6:26. https://doi.org/10.3389/fchem.2018.00026
Leena AR, Annelie ME, Janet J, Kristina L (2001) Effect of heat stress on cell activity and cell morphology of the tropical rhizobium. Sinorhizobium arboris, FEMS Microbiology Ecology 34(3):267–278. https://doi.org/10.1111/j.1574-6941.2001.tb00777.x
Leibold MA, McPeek MA (2006) Coexistence of the niche and neutral perspectives in community ecology. Ecology 87:1399–1410
Letten AD, Ke PJ, Fukami T (2017) Linking modern coexistence theory and contemporary niche theory. Ecol Monogr 87:161–177
Li J, Meng B, Chai H, Yang X, Song W, Li S, Lu A, Zhang T, Sun W (2019) Arbuscular Mycorrhizal Fungi Alleviate Drought Stress in C3 (Leymus chinensis) and C4 (Hemarthria altissima) Grasses via Altering Antioxidant Enzyme Activities and Photosynthesis. Front Plant Sci 10:499. https://doi.org/10.3389/fpls.2019.00499
Liao J, Cao, X, Zhao, L, Wang, J, Gao, Z, Wang MC, Huang Y (2016) The importance of neutral and niche processes for bacterial community assembly differs between habitat generalists and specialists. FEMS Microbiol Ecol 92: fiw174. https://doi.org/10.1093/femsec/fiw174
Lim JH, Kim SD (2013) Induction of drought stress resistance by multi-functional PGPR Bacillus licheniformis K11 in pepper. Plant Pathol J 29(2):201–208
Liste, HH (2003) Soil plant microbe interactions and their implications for agriculture and environment (Doctoral dissertation, Habilitation thesis, Humboldt University, Berlin)
Liu W, Zhang Y, Jiang S, Deng Y, Christie P, Murray PJ (2016) Arbuscular mycorrhizal fungi in soil and roots respond differently to phosphorus inputs in an intensively managed calcareous agricultural soil. Science 6:24902
Liu Z, Rong Q, Zhou W, Liang G (2017) Effects of inorganic and organic amendment on soil chemical properties, enzyme activities, microbial community and soil quality in yellow clayey soil. PLoS ONE 12(3):e0172767. https://doi.org/10.1371/journal.pone.0172767
Lladó S, López-mondéjar R, Baldrian P (2017) Forest soil bacteria: diversity, involvement in ecosystem processes, and response to global change. Microbiol Mol Biol Rev 81:1–27
Logares R, Lindstr ̈om ES, Langenheder S, et al. (2013) Biogeography of bacterial communities exposed to progressive long term environmental change. ISME J 7:937–48
Lončarić Z, Karalić K, Popović B, Rastija D, Vukobratović M (2008) Total and plant available micronutrients in acidic and calcareous soils in Croatia. Cereal Res Commun 36:331–334. https://doi.org/10.1556/crc.36
Lugtenberg B, Kamilova F (2009) Plant growth promoting rhizobacteria. Ann Rev Microbiol. 63:541–556
Lundberg DS, Lebeis SL, Paredes SH, Yourstone S, Gehring J, Malfatti S, Tremblay J, Engelbrektson A, Kunin V, Engelbrektson A, et al. (2012) Defining the core Arabidopsis thaliana root microbiome. Nature 488:86–90.https://doi.org/10.1038/nature11237
Ma Y, Rajkumar M, Freitas H (2009) Improvement of plant growth and nickel uptake by nickel resistant-plant-growth promoting bacteria. J Hazard Mater 166:1154–1161. https://doi.org/10.1016/j.jhazmat.2008.12.018
Macouzet M (2016) Critical aspects in the conception and production of microbial based plant biostimulants (MBPB). Probiotic Intelligentsia 5:29–38

Majeed A, Muhammad Z, Ahmad H (2018) Plant Cell Rep (2018) Plant growth promoting bacteria: role in soil improvement, abiotic and biotic stress management of crops 37:1599. https://doi.org/10.1007/s00299-018-2341-2

Manaf HH, Zayed MS (2015) Productivity of cowpea as affected by salt stress in presence of endomycorrhizae and Pseudomonas fluorescens. Ann Agric Sci https://doi.org/10.1016/j.aoas.2015.https://doi.org/10.013

Mathew DC, Ho YN, Gicana RG, Mathew GM, Chien MC, Huang CC (2015) A rhizosphere-associated symbiont, *Photobacterium* spp. strain MELD1, and its targeted synergistic activity for phytoprotection against mercury. PLoS ONE 10:e0121178. https://doi.org/10.1371/journal.pone.0121178

Mazid M, Khan TA, Mohammad F (2012) Role of nitrate reductase in nitrogen fixation under photosynthetic regulation. World J Pharm Res 1:386–414

McGill BJ, Maurer BA, Weiser MD (2006) Empirical evaluation of neutral theory. Ecology 87:1411–1423

Meena KK, Sorty AM, Bitla UM, Choudhary K, Gupta P, Pareek A, Singh DP, Prabha R, Sahu PK, Gupta VK, Singh HB, Krishanani KK, Minhas PS (2017) Abiotic stress responses and microbe-mediated mitigation in plants: the omics strategies. Front Plant Sci 8:172. https://doi.org/10.3389/fpls.2017.00172

Mehra P, Pandey BK, Verma L, Giri JA (2018) Novel Glycerophosphodiester Phosphodiesterase Improves Phosphate Deficiency Tolerance. Plant Cell Environ

Mia MAB, Shamsuddin ZH (2010) Rhizobium as a crop enhancer and biofertilizer for increased cereal production. Afr J Biotechnol 9:6001–6009

Miller G, Susuki N, Ciftci-Yilmaz S, Mittler R (2010) Reactive oxygen species homeostasis and signalling during drought and salinity stresses. Plant, Cell Environ 33:453–467

Munns R, Tester M (2008) Mechanisms of salinity tolerance. Annu Rev Plant Biol 59:651–681. https://doi.org/10.1146/annurev.arplant.59.032607.092911

Nadeem SM, Ahmad M, Zahir ZA, Javaid A, Ashraf M (2014) The role of mycorrhizae and plant growth promoting rhizobacteria (PGPR) in improving crop productivity under stressful environments. Biotechnol Adv 32:429–448

Nehra V, Saharan BS, Choudhary M (2016) Evaluation of Brevibacillus brevis as a potential plant growth promoting rhizobacteria for cotton (Gossypium hirsutum) crop. Springer Plus 5(1):948. https://doi.org/10.1186/s40064-016-2584-8

Nemergut DR, Schmidt SK, Fukami T, O'Neill SP, Bilinski TM, Stanish LF, Knelman JE, Darcy JL, Lynch RC, Wickey P, Ferrenberg S (2013) Patterns and processes of microbial community assembly. Microbiol Mol Biol Rev 77:342–356

Nezhadahmadi A, Prodhan ZH, Faruq G (2013) Drought tolerance in wheat. Sci World J 610721. https://doi.org/10.1155/2013/610721

Ngumbi E, Kloepper J (2016) Bacterial-mediated drought tolerance: Current and future prospect. Applied Soil Ecol 105: 109–125. https://doi.org/10.1016/j.apsoil.2016.04.009

Nuti M, Giovannetti G (2015) borderline products between bio-fertilizers/bio-effectors and plant protectants: the role of microbial consortia. J Agricult Sci Technol A 5:305–315 https://doi.org/10.17265/2161-6256/2015.05.001

Nwankwegu AS, Li Y, Jiang L, Lai Q, Weng S, Wei J, Acharya K (2018) Kinetic modelling of total petroleum hydrocarbon in spent lubricating petroleum oil impacted soil under different treatments. Environ Technol 1–10. https://doi.org/10.1080/09593330.2018.1498543

Nwankwegu AS, Onwosi CO (2017) Microbial cell immobilization: a renaissance to bioaugmentation inadequacies. A review. Environ Technol Rev 6(1):186–198

Odoh CK (2015) Effects of Some Heavy Metals on Soil Bacteria, Shoot Growth and Nodulation of Cowpea (*Vigna unguiculata*) and Groundnut (*Arachis hypogea*) Grown in Sandy Loam Soil. M.Sc. Thesis submitted to Department of Microbiology, University of Nigeria Nsukka. https://oer.unn.edu.ng/read/effects-of-some-heavy-metals-on-soil-bacteria-shoot-growth-and-nodulation-of-cowpea-vigna-unguiculata-and-groundnut-arachis-hypogea-grown-in-sandy-loam-soil

Odoh CK (2017) Plant growth promoting rhizobacteria (PGPR): a bioprotectant bioinoculant for sustainable agrobiology. A Review. Int J Adv Res Biol Sci 4 (5):123–142. http://dx.doi.org/10.22192/ijarbs.2017.04.05.014
Odoh CK, Akpi UK, Anya F (2017) Environmental impact of mineral exploration in Nigeria and their phytoremediation strategies for sustainable ecosystem. Global J Sci Front Res 17 (3):19–27 https://doi.org/10.17406/gjsfr
Odoh CK, Eze CN, Akpi UK, Unah VU (2019a) Plant growth promoting rhizobacteria (PGPR): a novel agent for sustainable food production. Am J Agri Biol Sci 14(35):54. https://doi.org/10.3844/ajabssp.2019.35.54
Odoh CK, Zabbey N, Sam K, Eze CN (2019b) Status, progress and challenges of phytoremediation—An African scenario. J Environ Manag 237:365–378. https://doi.org/10.1016/j.jenvman.2019.02.090
Ojo J (2016) Pesticides use and health in Nigeria. Ife J. Sci. 8:981–991
Oldroyd GED (2013) Speak, friend, and enter: signalling systems that promote beneficial symbiotic associations in plants. Nat Rev Microbiol 11:252–263. https://doi.org/10.1038/nrmicro2990
Oliveira CA, Alves VMC, Marriel IE, Gomes EA, Scotti MR, Carneiro NP, Guimaraes CT, Schaffert RE, Sá NMH (2009) Phosphate solubilizing microorganisms isolated from rhizosphere of maize cultivated in an oxisol of the Brazilian Cerrado Biome. Soil Biol Biochem 41:1782–1787. https://doi.org/10.1016/j.soilbio.2008.01.012
Parnell JJ, Berka R, Young HA, Sturino JM, Kang Y, Barnhart DM, DiLeo MV (2016) From the lab to the farm: an industrial perspective of plant beneficial microorganisms. Front Plant Sci 7:1110. https://doi.org/10.3389/fpls.2016.01110
Pathak D, Lone R, Koul KK (2017) Arbuscular Mycorrhizal Fungi (AMF) and Plant Growth-Promoting Rhizobacteria (PGPR) Association in Potato (Solanum tuberosum L.): A Brief Review. In: Kumar V, Kumar M, Sharma S, Prasad R (eds). Probiotics and Plant Health. Springer. Singapore, pp 401–420
Paul D, Lade H (2014) Plant growth promoting rhizobacteria to improve crop growth in saline soils: a review. Agron Sustain Dev, Springer/EDP Sciences/INRA, 34 (4), 737–752. ff10.1007/s13593-014-0233-6ff. ffhal-01234840f
Pettit T, Irga PJ, Torpy FR (2019) The in situ pilot-scale phytoremediation of airborne VOCs and particulate matter with an active green wall. Air Qual Atmos Health 12(1):33–44
Philippot L, Raaijmakers JM, Lemanceau P, van der Putten WH (2013) Going back to the roots: the microbial ecology of the rhizosphere. Nat Rev Microbiol 11(11):789–799. https://doi.org/10.1038/nrmicro3109
Pindi PK, Satyanarayana SDV (2012) Liquid microbial consortium- a potential tool for sustainable soil health. J Biofertil Biopestici 3:124. https://doi.org/10.4172/2155-6202.1000124
Pishchik VN, Provorov NA, Vorobyov NI, Chizevskaya EP, Safronova VI, Tuev AN Kozhemyakov AP (2009) Interactions between plants and associated bacteria in soils contaminated with heavy metals. Microbiology 78:785–793
Qian J, Jiashun C, Chengbo Y, Yulong Y, Kang Y (2019) Quorum sensing: a prospective therapeutic target for bacterial diseases. BioMed Research International Article ID 2015978, 15 p, https://doi.org/10.1155/2019/2015978
Radzki W, Gutierrez MFJ, Algar E, Lucas GJA, GarcíaVillaraco A, Ramos Solano B (2013) Bacterial siderophores efficiently provide iron to iron-starved tomato plants in hydroponics culture. Antonie Van Leeuwenhoek 104:321–330. https://doi.org/10.1007/s10482-013-9954-9
Rajendran K, Tester M, Roy SJ (2009) Quantifying the three main components of salinity tolerance in cereals. Plant, Cell Environ 32(3):237–249
Rajkumar M, Freitas H (2008) Influence of metal resistant plant growth-promoting bacteria on the growth of *Ricinus communis* in soil contaminated with heavy metals. Chemosphere 71:834–842
Rana KL, Kour D, Yadav AN (2019) Endophytic microbiomes: biodiversity, ecological significance and biotechnological applications. Res J Biotechnol 14:142–162

Ranganayakulu GS, Veeranagamallaiah G, Sudhakar C (2013) Effect of salt stress on osmolyte accumulation in two groundnut cultivars (*Arachis hypogaea* L.) with contrasting salt tolerance. Afr J Plant Sci 7:586–592

Rashid MA, Mujawar LH, Shahzad T, Almeelbi T, Ismail IMI, Oves M (2016) Bacteria and fungi can contribute to nutrients bioavailability and aggregate formation in degraded soils. Microbiol Res 183:26–41

Rasmann S, Turlings TC (2016) Root signals that mediate mutualistic interactions in the rhizosphere. Curr Opin Plant Biol 32:62–68. https://doi.org/10.1016/j.pbi.2016.06.017

Richardson AE, Simpson RJ (2011) Soil microorganisms mediating phosphorus availability. Plant Physiol 156:989–996. https://doi.org/10.1104/pp.111.175448

Ripa FA, Cao W, Tong S, Sun J (2019) Assessment of plant growth promoting and abiotic stress tolerance properties of wheat endophytic fungi. BioMed Res Int, Article ID 6105865, 12 p, https://doi.org/10.1155/2019/6105865

Roberts NJ, Morieri G, Kalsi G, Rose A, Stiller J, Edwards A, Xie F, Gresshoff PM, Oldroyd GE, Downie JA, Etzler ME (2013) Rhizobial and mycorrhizal symbioses in Lotus japonicus require lectin nucleotide phosphohydrolase, which acts upstream of calcium signaling. Plant Physiol 161:556–567

Rodrıguez-Salazar J, Suarez R, Caballero-Mellado J, Iturriaga G (2009) Trehalose accumulation in *Azospirillum brasilense* improves drought tolerance and biomass in maize plants. FEMS Microbiol Lett 296:52–59

Rombola-Caldentey MB (2019) Structure and regulation of NHX exchangers in the uptake of potassium into the vacuoles of arabidopsis thaliana (Tesis Doctoral Inédita). Universidad de Sevilla, Sevilla

Rosindell J, Hubbell SP, Etienne RS (2011) The unified neutral theory of biodiversity and biogeography at age ten. Trends Ecol Evol 26(7):340–348

Roy SJ, Negrao S, Tester M (2014) Salt resistant crop plants Curr Opin Biotechnol 26:115–124

Saharan BS, Nehra V (2011) Plant growth promoting rhizobacteria: A critical review. Life Sci Med Res 21:1–30

Sam K, Coulon F, Prpich G (2017) A multi-attribute methodology for the prioritization of oil contaminated sites in the Niger Delta. Sci Total Environ 579:1323–1332. https://doi.org/10.1016/j.scitotenv.2016.11.126

Sandhya V, Ali SKZ, Grover M, Reddy G, Venkateswarlu B (2010) Effect of plantgrowth promoting Pseudomonas spp. on compatible solutes, antioxidant status and plant growth of maize under drought stress.Plant Growth Regul 62: 21–30

Sărăteanu V, Moisuc A, Cotuna O (2010) *Ambrosia artemisiifolia* L. an invasive weed from ruderal areas to disturbed grasslands. Lucrări Stiinifice 53:28–31

Sarkar P, Meghvanshi M, Singh R (2011) Microbial consortium: a new approach in effective degradation of organic kitchen wastes. Int J Environ Sci Dev 2:3

Schirawski J, Perlin M (2018) Plant microbe interaction—the good, the bad and the diverse. Int J Mol Sci 19:1374

Selvakumar G, Panneerselvam P, Ganeshamurthy AN (2012) Bacterial mediated alleviation of abiotic stress in crops. In: Maheshwari DK (ed) Bacteria in Agrobiology: Stress Manag Springer, 205–224

Sen T, Barrow CJ, Deshmukh SK (2019) Microbial pigments in the food industry challenges and the way forward. Front Nutr 6:7. https://doi.org/10.3389/fnut.2019.00007

Sessitsch A, Mitter B (2015) 21st century agriculture: integration of plant microbiomes for improved crop production and food security. Microb Biotechnol 8:32–33. https://doi.org/10.1111/1751-7915.12180

Sieberer BJ, Chabaud M, Timmers AC, Monin A, Fournier J, Barker DG (2009) A nuclear targeted cameleon demonstrates intranuclear Ca^{2+} spiking in Medicago truncatula root hairs in response to rhizobial nodulation factors. Plant Physiol 151:1197–1206

Simone R, Lance CS, Brian MH (2018) Critical computational analysis illuminates the reductive-elimination mechanism that activates nitrogenase for N_2 reduction. Proc Natl Acad Sci 115(45):E10521–E10530. https://doi.org/10.1073/pnas.1810211115

Singh I (2018) Plant Growth Promoting Rhizobacteria (PGPR) and their various mechanisms for plant growth enhancement in stressful conditions: a review. *Eur J* Biol Res8 (4):191–213. http://www.journals.tmkarpinski.com/index.php/ejbr/article/view/90

Sippel D, Rohde M, Netzer J, Trncik C, Gies J, Grunau K, Djurdjevic I, Decamps L, Andrade SLA, Einsle O (2018) A bound reaction intermediate sheds light on the mechanism of nitrogenase. Science 359:1484–1489. https://doi.org/10.1126/science.aar2765

Sivasakthi S, Usharani G, Saranraj P (2014) Biocontrol potentiality of Plant Growth Promoting Bacteria (PGPR) Pseudomonas fluorescens and Bacillus subtilis: a review. Afr J Agri Res 9:1265–1277. https://doi.org/10.1007/bf00249074

Spaepen S, Boddelaere S, Croonenborghs A, Vanderleyden J (2008) Effect of *Azospirillum brasiliense* indole-3-acetic acid production on inoculated wheat plants. Plant Soil 312:15–23

Spaepen S, Vanderleyden J (2011) Auxin and plant-microbe interactions. Cold Spring Harbor Perspectives in Biology, https://doi.org/10.1101/cshperspect.a001438

Stambulska UY, Bayliak MM, Lushchak VI (2018) Chromium(VI) toxicity in legume plants: modulation effects of rhizobial rymbiosis. BioMed Res Int article ID 8031213, 13 p

Suarez R, Wong A, Ramirez M, Barraza A, OrozcoMdel C, Cevallos MA, Lara M, Hernandez G, Iturriaga G (2008) Improvement of drought tolerance and grain yield in common bean by overexpressing trehalose-6-phosphate synthase in rhizobia. Mol Plant Microbe Interact 21(7):958–966

Sun RB, Guo XS, Wang DZ, Chu HY (2015) Effects of long-term application of chemical and organic fertilizers on the abundance of microbial communities involved in the nitrogen cycle. Appl Soil Ecol 95:171–178

Sze H, Chanroj S (2018) Plant endomembrane dynamics: studies of K^+/H^+ antiporters provide insights on the effects of pH and ion homeostasis. Plant Physiol 177:875–895. https://doi.org/10.1104/pp.18

Tagele SB, Kim SW, Lee HG, Lee YS (2019) Potential of Novel Sequence Type of *Burkholderia cenocepacia*for Biological Control of Root Rot of Maize (*Zea mays* L.) Caused by *Fusarium temperatum*. Int J Mol Sci 20:1005

Talbi C, Sánchez C, Hidalgo-Garcia A, González EM, Arrese-Igor C, Girard L, Bedmar EJ, Delgada MJ (2012) Enhanced expression of *Rhizobium etli cbb3* oxidase improves drought tolerance of common bean symbiotic nitrogen fixation. J Exp Bot 63:5035–5043. https://doi.org/10.1093/jxb/ers101

Tamayo-Vélez A, Nelson WO (2018) Soil Fertility Improvement by Litter Decomposition and Inoculation with the Fungus *Mortierella* sp. Avocado Plantations of Colombia, Commun Soil Sci Plan 49(2):139–147. https://doi.org/10.1080/00103624.2017.1417420

Thompson LR, Sanders JG, McDonald D et al (2017) A communal catalogue reveals Earth's multiscale microbial diversity. Nature 551:457–463

Timmis K, de Vos WM, Ramos JL, Vlaeminck SE, Prieto A, Danchin A, Verstraete W, de Lorenzo V, Lee SY, Brüssow H, Timmis JK, Singh BK (2017) The contribution of microbial biotechnology to sustainable development goals. Microb Biotechnol 10(5):984–987. https://doi.org/10.1111/1751-7915.12818

Timmusk S, Abd El-Daim IA, Copolovici L, Tanilas T, Kännaste A, Behers L, Nevo E, Seisenbaeva G, Stenstrom E, Niinemets U (2014) Drought-tolerance of wheat improved by rhizosphere bacteria from harsh environments: enhanced biomass production and reduced emissions of stress volatiles. PLoS ONE 9(5):e96086. https://doi.org/10.1371/journal.pone.0096086

Timmusk S, Behers L, Muthoni J, Muraya A, Aronsson AC (2017) Perspectives and challenges of microbial application for crop improvement. Front Plant Sci 8:49

Tiwari JK, Munshi AD, Kumar R, Pandey RN, Arora A, Bhat JS, Sureja AK (2010) Effect of salt stress on cucumber: Na^+/K^+ ratio, osmolyte concentration, phenols and chlorophyll content. Acta Physiol Plant 32:103–114

Tiwari S, Abhisheki SC, Puneet SC, Chandra N (2016) Pseudomonas putida attunes morphophysiological: biochemical and molecular responses in Cicer arietinum L. during drought stress and recovery. Plant Physiol Biochem 99:108–117

Tringe SG, von Mering C, Kobayashi A, Salamov AA, Chen K, Chang HW, Podar M, Short JM, Mathur EJ, Detter JC, Bork P, Hugenholtz P, Rubin EM (2005) Comparative metagenomics of microbia communities. Science 308:554–557. https://doi.org/10.1126/science.1107851

Turan M, Kıtır N, Alkaya Ü, Günes A, Tüfenkçi Ş, Yıldırım E, Nikerel E (2016) Making soil more accessible to plants: the case of plant growth promoting rhizobacteria. IntechOpen 5:61–69. https://doi.org/10.5772/64826

Tyc O, Song C, Dickschat JS, Vos M, Garbeva P (2017) The ecological role of volatile and soluble secondary metabolites produced by soil bacteria. Trends Microbiol 25:280–292. https://doi.org/10.1016/j.tim.2016.12.002

Utami D, Kawahata A, Sugawara M, Jog RN, Miwa K, Morikawa M (2018) Effect of exogenous general plant growth regulators on the growth of the Duckweed *Lemna minor*. Front Chem 6:251. https://doi.org/10.3389/fchem.2018.00251

Utsumi Y, Utsumi C, Tanaka M, Ha CV, Takahashi S, Matsui A, Matsunaga TM, Matsunaga S, Kanno Y, Seo M, Okamoto Y, Moriya E, Seki M (2019) Acetic Acid treatment enhances drought avoidance in cassava (Manihot esculenta Crantz). Front Plant Sci 10:521. https://doi.org/10.3389/fpls.2019.00521

Vanwonterghem I, Jensen PD, Dennis PG, Hugenholtz P, Rabaey K, Tyson GW (2014) Deterministic processes guide long term synchronised population dynamics in replicate anaerobic digesters. ISME J 8:28

Varanini Z, Pinton R (2001) Direct versus indirect effects of soil humic substances on plant growth and nutrition. In: Pinton R, Varanini Z, Nannipieri P (eds) The rhizosphere: biochemistry and organic substances at the soil plant interface. Marcel Dekker, New York, NY, pp 141–157

Varshney RK, Tuberosa R, Tardieu F (2018) Progress in understanding drought tolerance: from alleles to cropping systems. J Exp Bot 69:3175–3179. https://doi.org/10.1093/jxb/ery187

Venturi V, Keel C (2016) Signaling in the rhizosphere. Trends Plant Sci 21:187–198. https://doi.org/10.1016/j.tplants.2016.01.005

Verbruggen E, Van Der Heijden MG, Weedon JT, Kowalchuk GA, Röling WF (2012) Community assembly, species richness and nestedness of arbuscular my-corrhizal fungi in agricultural soils. Mol Ecol 21:2341–2353

Verma P, Yadav AN, Khannam KS, Mishra S, Kumar S, Saxena AK, Suman A (2016) Appraisal of diversity and functional attributes of thermotolerant wheat associated bacteria from the peninsular zone of India. Saudi J Biol Sci. https://doi.org/10.1016/j.sjbs.2016.01.042

Verma P, Yadav AN, Khannam KS, Panjiar N, Kumar S, Saxena AK, Suman A (2015) Assessment of genetic diversity and plant growth promoting attributes of psychrotolerant bacteria allied with wheat (*Triticum aestivum*) from the northern hills zone of India. Ann Microbiol 65:1885–1899

Verma P, Yadav AN, Kumar V, Singh DP, Saxena AK (2017) Beneficial Plant-Microbes Interactions: Biodiversity of Microbes from Diverse Extreme Environments and Its Impact for Crop Improvement. In: Singh DP, Singh HB, Prabha R (eds) plant-microbe interactions in agro-ecological perspectives: Volume 2: Microbial Interactions and Agro-Ecological Impacts. Springer Singapore, Singapore, pp 543–580. https://doi.org/10.1007/978-981-10-6593-4_22

Walker DJ, Clemente R, Roig A, Bernal MP (2003) The effect of soil amendments on heavy metal bioavailability in two contaminated Mediterranean soils. Environ Pollution 22:303–312

Wei X, Lyu S, Yu Y, Wang Z, Liu H, Pan D, Chen J (2017) Phylloremediation of air pollutants: exploiting the potential of plant leaves and leaf-associated microbes. Front Plant Sci 8:1318. https://doi.org/10.3389/fpls.2017.01318

Weyens N, Truyens S, Dupae J, Newman L, Taghavi S, van der Lelie D, Carleer R, Vangronsveld J (2010) Potential of the TCE-degrading endophyte Pseudomonas putida W619-TCE to improve plant growth and reduce TCE phytotoxicity and evapotranspiration in poplar cuttings. Environ Pol 158:2915–2919. https://doi.org/10.1016/j.envpol.2010.06.004

Xiong Q, Tang G, Zhong L, He H, Chen X (2018) Response to nitrogen deficiency and compensation on physiological characteristics, yield formation, and nitrogen utilization of rice. Front Plant Sci 9:1075. https://doi.org/10.3389/fpls.2018.01075

Xun F, Xie B, Liu S, Guo C (2015) Effect of plant growth-promoting bacteria (PGPR) and arbuscular mycorrhizal fungi (AMF) inoculation on oats in saline alkali soil contaminated by petroleum to enhance phytoremediation. Environ Sci Pollut Res Int 22:598–608. https://doi.org/10.1007/s11356-014-3396-4

Yadav AN (2017a) Agriculturally important microbiomes: biodiversity and multifarious PGP attributes for amelioration of diverse abiotic stresses in crops for sustainable agriculture. Biomed J Sci Tech Res 1:1–4

Yadav AN (2017b) Beneficial role of extremophilic microbes for plant health and soil fertility. J Agric Sci 1:1–4

Yadav AN, Mishra S, Singh S, Gupta A (2019a) Recent Advancement in White Biotechnology Through Fungi Volume 1: Diversity and Enzymes Perspectives. Springer International Publishing, Cham

Yadav AN, Sachan SG, Verma P, Saxena AK (2015a) Prospecting cold deserts of north western Himalayas for microbial diversity and plant growth promoting attributes. J Biosci Bioeng 119:683–693

Yadav AN, Sachan SG, Verma P, Saxena AK (2016) Bioprospecting of plant growth promoting psychrotrophic Bacilli from cold desert of north western Indian Himalayas. Indian J Exp Biol 54:142–150

Yadav AN, Sachan SG, Verma P, Tyagi SP, Kaushik R, Saxena AK (2015b) Culturable diversity and functional annotation of psychrotrophic bacteria from cold desert of Leh Ladakh (India). World J Microbiol Biotechnol 31:95–108

Yadav AN, Sharma D, Gulati S, Singh S, Dey R, Pal KK, Kaushik R, Saxena AK (2015c) Haloarchaea endowed with phosphorus solubilization attribute implicated in phosphorus cycle. Sci Rep 5:12293

Yadav AN, Singh S, Mishra S, Gupta A (2019b) Recent advancement in white biotechnology through Fungi. Volume 2: Perspective for Value-Added Products and Environments. Springer International Publishing, Cham

Yadav AN, Singh S, Mishra S, Gupta A (2019c) Recent advancement in white biotechnology through Fungi. Volume 3: Perspective for Sustainable Environments. Springer International Publishing, Cham

Yadav AN, Verma P, Kaushik R, Dhaliwal HS, Saxena AK (2017) Archaea endowed with plant growth promoting attributes. EC Microbiol 8:294–298

Yadav AN, Yadav N (2018) Stress-adaptive microbes for plant growth promotion and alleviation of drought stress in plants. Acta Sci Agric 2:85–88

Yadav AN, Yadav N, Sachan SG, Saxena AK (2019b) Biodiversity of psychrotrophic microbes and their biotechnological applications. J Appl Biol Biotechnol 7:99–108

Yadav N, Yadav AN (2019a) Actinobacteria for sustainable agriculture. J Appl Biotechnol Bioeng 6:38–41

Yadav N, Yadav AN (2019b) Biodegradation of biphenyl compounds by soil microbiomes. Biodivers Int J 3:37–40

Yin X, Komatsu S (2016) Nuclear proteomics reveals the role of protein synthesis and chromatin structure in root tip of soybean during the initial stage of flooding stress. J Proteome Res 15:2283–2298. https://doi.org/10.1021/acs.jproteome.6b00330

Zabbey N, Sam K, Onyebuchi AT (2017) Remediation of contaminated lands in the Niger Delta, Nigeria: prospects and challenges. Sci Total Environ 586:952–965, https://doi.org/10.1016/j.scitotenv.2017.02.075

Zaidi A, Khan MS, Ahemad M, Oves M, Wani PA (2009) Recent Advances in Plant Growth Promotion by Phosphate-Solubilizing Microbes. In: Khan MS (ed) Microbial strategies for crop improvement. Springer, Berlin Heidelberg, pp 23–50

Zamioudis C, Pieterse CM (2012) Modulation of host immunity by beneficial microbes. Mol Plant Microbe Interact 25(2):139–150

Zeng Y, Clark MJ (2013) Optimal growing substrate pH for five *Sedum* species. HortScience 48:448–452

Zhang H, Kim MS, Sun Y, Dowd SE, Shi H, Pare PW (2008) Soil bacteria confer plant salt tolerance by tissue-specific regulation of the sodium transporter HKT1. Mol Plant Microbe Interact 21(6):737–744. https://doi.org/10.1094/mpmi-21-6-0737

Zhang QG, Buckling A, Godfray HCJ (2009) Quantifying the relative importance of niches and neutrality for coexistence in a model microbial system. Funct Ecol 23:1139–1147. https://doi.org/10.1111/j.1365-2435.2009.01579.x

Zheng W, Zeng S, Bais H, LaManna JM, Hussey DS, Jacobson DL et al. (2018) Plant Growth-Promoting Rhizobacteria (PGPR) reduce evaporation and increase soil water retention. Water Resour Res 178, 821–815. https://doi.org/10.1029/2018wr022656

Zhou SR, Zhang DY (2008) A nearly neutral model of biodiversity. Ecology 89:248–258

Zuroff TR, Curtis WR (2012) Developing symbiotic consortia for lignocellulosic biofuel production. Appl Microbiol Biotechnol 93:1423–1435. https://doi.org/10.1007/s00253-011-3762-9

Chapter 14
Global Scenario of Plant–Microbiome for Sustainable Agriculture: Current Advancements and Future Challenges

Simranjeet Singh, Vijay Kumar, Satyender Singh, Daljeet Singh Dhanjal, Shivika Datta and Joginder Singh

Abstract Phyto-microbiome are the microorganisms (fungi and bacteria) associated with all major plant components such as flowers, stems, roots, leaves, and fruits. They form symbiotic association with the plant, inhabit the intra- and intercellular positions without harming the host and frequently profit the host plant. They indicate the complex connections within the host plants involving the symbiotic, mutualistic relationship, and rarely the parasitism relationship. They are omnipresent and are known to improve the nutrient enrichment and growth of the plant. They not only produce root exudates but also release signal molecules which regulate various biochemical and genetic activities. They provide the immunity to plants from pests and insects and enhance the ability of plants to tolerate the impacts of abiotic and biotic stress and also produce bioactive compounds and phytohormones of biotechnological interest. In this book chapter, we will review the advent role of microbiome in plant growth and development. Efforts have been made to summarize the utilization of various hormones to mitigate the effects of various environmental stresses on cultivated plant communities. The final sections of the book chapter describe the applications of phyto-microbiome in twenty-first century and the clear out cut to commercialize of a phyto-microbiome-based technology.

Simranjeet Singh, Daljeet Singh Dhanjal, Satyender Singh, Vijay Kumar—Equally contributed.

S. Singh
Punjab Biotechnology Incubators, Mohali 160059, Punjab, India

S. Singh · D. S. Dhanjal · J. Singh (✉)
Department of Biotechnology, Lovely Professional University, Phagwara 144411, Punjab, India
e-mail: joginder.15005@lpu.co.in

S. Singh · S. Singh
Regional Advanced Water Testing Laboratory, Mohali 160059, Punjab, India

V. Kumar
Regional Ayurveda Research Institute for Drug Development, Gwalior 474009, Madhya Pradesh, India

S. Datta
Department of Zoology, Doaba College Jalandhar, Jalandhar, Punjab, India

A. N. Yadav et al. (eds.), *Plant Microbiomes for Sustainable Agriculture*, Sustainable Development and Biodiversity 25, https://doi.org/10.1007/978-3-030-38453-1_14

Keywords Biodiversity · Plant growth promotion · Plant–microbe interactions · Plant microbiome · Sustainable agriculture

14.1 Introduction

Soil is considered to be the reservoir for various microbes and organisms like ants, moles, and nematode (Dhanjal et al. 2017). Recent advances in sequencing techniques and elevation in number of microbial libraries have expanded the aura of the tree of life, as it is dominated by microbes (Anand et al. 2019). The horizontal gene transfer and mutation events have evolved wide array of variation among the microbial community. This significantly increases the diversity within species and endows them their functional characteristic (Bengtsson-Palme et al. 2017a). Here, soil plays the major role in recycling the nutrients (phosphorus and nitrogen) and imparting protection against abiotic and biotic stress (Dhanjal et al. 2018). Although the agricultural activities have increased the yield of crop, simultaneously it has also deteriorated the biological and physical properties of the soil (Gomiero 2017). Even usage of fertilizer aids in maintaining the soil fertility, but on tillage microbial communities get disrupted (Dong et al. 2012). The degradation of soil due to anthropogenic activities has emerged as a global concern and sustainable agriculture has become the need of the time for sustaining the life of humans on this earth. Thus, to sustain the life, plant–microbiome plays significant role in improving soil quality and plant growth, and providing resistance from stress (Rashid et al. 2016; Kumar et al. 2019a).

Plant–microbiome is essential as they contain distinct properties like production of secondary metabolites as well as phytohormone and nitrogen fixation and many more. Therefore, it represents the microbial community which is directly or indirectly associated with plants. Hence, they have been generally characterized into epiphytic, endophytic, and rhizosphere microbiome (Igiehon and Babalola 2018a; Kour et al. 2019; Kumar et al. 2019b; Rana et al. 2019a; Rana et al. 2019b). The soil contains both types of microbes, i.e., pathogenic and nonpathogenic in nature (Finkel et al. 2017). Nonpathogenic involves the symbiotic and neutral microbes which plays variety of roles in diverse fields like biodegradation, biofuel production, biocontrol, biotransformation, seed production, phytoremediation, and many more (Dwibedi and Saxena 2019). These potentials of microbes prompt us to understand the hyperdiversity of these unexplored plant microbial communities, not only for sustaining the ecosystem but also to preserve these biodiverse microbial communities beneficial for mankind (Braga et al. 2016). Hence, untapping and deep understanding of these plant microbes as whole have become important to explore the positive interactions for sustainable agriculture, especially during microbiome-dependent cropping approach (Busby et al. 2017). In this chapter, we will discuss the challenges and efforts put forward to advance our knowledge about different properties of microbes and how these properties affect plants. Further, we will also discuss about the soil microbiome improving the crop production.

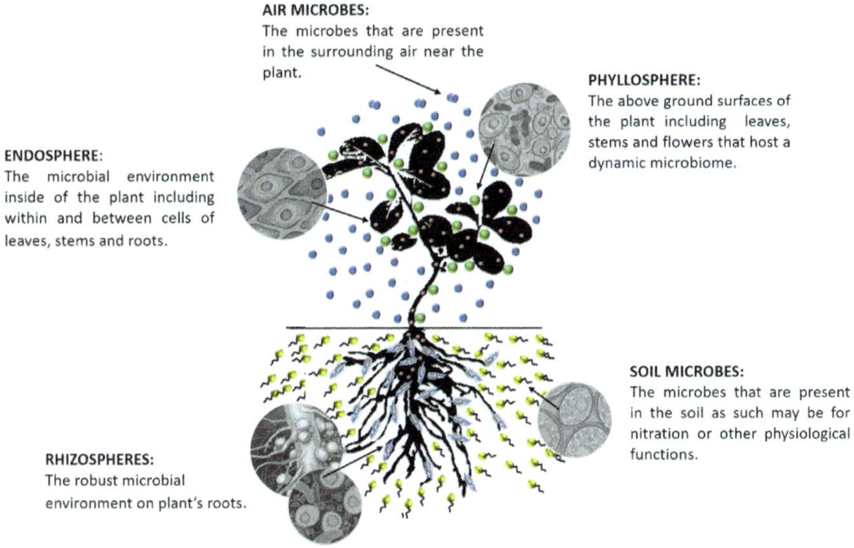

Fig. 14.1 Compartmentalization of plant microbiome

14.2 Composition and Function of Plant Microbiome

Plants are surrounded by a variety of microbes and vary according to their location like anthosphere, carposphere, phyllosphere, rhizosphere, and spermosphere (Shade et al. 2017) (Fig. 14.1). Here, we categorized the bacteria into three broad categories as follows.

14.3 Plant Microflora Below the Ground

Root microflora generally get horizontally transferred as they are predominantly present in soil (Lareen et al. 2016). Most dominating microbes belong to the family of *Acidobacteria, Actinobacteria, Bacteroidetes, Planctomycetes, Proteobacteria,* and *Verrucomicrobia* (Mendes et al. 2013; Kour et al. 2019; Suman et al. 2016; Verma et al. 2017b). There is a possibility of vertical transmission through seeds. Seeds also serve as reservoir for microbes, which allows them to multiply in the roots during plant development (Shahzad et al. 2018). Root system of plant provides distinctive niche to soil microbes residing in the rhizosphere (roots and certain portion above the ground) (Raaijmakers et al. 2009). Recently, Donn with his colleagues reported about the changes in bacterial community surrounding the roots of wheat and found 10-fold increase in population of actinobacteria, copiotrophs, oligotrophs, and pseudomonads at rhizosphere. Moreover, they also reported that over time, variation takes place in rhizoplane and rhizosphere community but there is no variation in soil microbial population (Donn et al. 2015). Similar results were also seen for *Brachypodium distachyon* rhizosphere which reported about the predominance of species of *Burkholderiales, Sphingobacteriales,* and *Xanthomonadales* family (Kawasaki et al. 2016). There have been reports which claimed that root exudates like amino acids, fatty acids, organic acids, plant growth regulators, phenolics, putrescine, sterols, sugars, and vitamins also affect microbes present in the rhizosphere, proclaimed as rhizosphere effect (Hunter et al. 2014). For example, secondary metabolite benzoxazinoids (BXs) synthesized by the roots of maize amends the composition of root-associated microbes, out of which members of *Actinobacteria* and *Proteobacteria* family were highly affected (Neal et al. 2012; Kudjordjie et al. 2019). Moreover, other researchers are investigating the mechanisms responsible for the assembly of microbial community and effects of substrate and root exudation on microbial community. Hence, this confirms that composition of rhizosphere microbes gets influenced by plants species and root exudates (Jacoby et al. 2017; Yadav et al. 2015a; Yadav et al. 2015b).

Various bacterial endophytes have colonized the roots of plants internally (Santoyo et al. 2016). These bacterial endophytes enter the root tissues via passive processes or root abrasion and active mechanisms (Santos et al. 2018). The transmission and colonization of these bacterial endophytes inside the plant depend on various factors like ability of endophytes to colonize plant and distribution of plant resources (Kandel et al. 2017). Various microbes have gained access to root tissues, for instance, *Acidobacteria, Actinobacteria, Bacteroidetes, Chloroflexi, Firmicutes, Gemmatimonadetes, Planctomycetes, Proteobacteria*, and *Verrucomicrobia* are the bacterial taxa which have gained access to grapevine roots (Morgan et al. 2017; Verma et al. 2016; Yadav et al. 2016). On the other hand, members of *Bradyrhizobiaceae, Comamonadaceae, Rhizobiaceae,* and *Streptomycetaceae* family are the one predominantly found in rice roots (Edwards et al. 2015).

14.4 Plant Microflora Above the Ground

Floral parts, leaves, and vegetative parts present above the ground provides distinctive environment for epiphyte and endophyte microbes (Frank et al. 2017). The endophytes predominantly translocate themselves through xylem to different parts of plants, which can either be fruit, leaves, or stem (Hardoim et al. 2015). With respect to location on the plant, aboveground microbes distribute themselves with other members of endophytic community (Nair and Padmavathy 2014). It has been observed that phyllosphere microbes obtained from the soil get influenced by external stimuli and later shows the profound effect (Carvalho and Castillo 2018). Subsequently, various microbes have been found in the phyllosphere and endosphere up to species level (Schlaeppi and Bulgarelli 2015). For example, on analyzing the carposphere and phyllosphere microbes of grapevine uncovered the presence of *Acinetobacter, Bacillus, Citrobacter, Curtobacterium, Enterobacter, Erwinia, Frigoribacterium, Methylobacterium, Pantoea*, and *Pseudomonas* species (Kecskeméti et al. 2016; Verma et al. 2017a; Verma et al. 2017b; Yadav et al. 2019; Yadav et al. 2018a; Yadav et al. 2018b; Yadav et al. 2018c). On the other hand, endophyte analysis of grape berries uncovered that *Bacillus, Burkholderia, Dyella, Mesorhizobium, Propionibacterium, Pseudomonas,* and *Ralstonia* species are the dominant (Campisano et al. 2014).

There is a study conducted on 300 different varieties of maize to analyze the leaf microbiome that revealed the predominance of *methylobacteria* and *sphingomonads* taxa (Wagner et al. 2019), whereas in apple flowers there is predominance of *Enterobacteriaceae* and *Pseudomonas* species (Steven et al. 2018). Various other studies conducted on almond, apple, grapefruit, pumpkin flower, and pumpkin also revealed the abundance of *Pseudomonas* species (Aleklett et al. 2014). Lately, seed-associated microbes have highlighted the abundance of *Actinobacteria, Bacteroidetes, Firmicutes,* and *Proteobacteria* (Qiao et al. 2017). Usually, microbes found on aboveground have originated from air, seed, and soil and adapted themselves to sustain their life on or inside the tissue of plant, where various factors like environmental condition, soil, and management influence microbial composition (Raaijmakers et al. 2009). Compartment-specific assembly with the host illustrates the strong relationship among the aboveground microbes and plant on the basis of functionality, but still there is a need to understand this association deeply (Garcia and Kao-Kniffin 2018). Aboveground microbes and endophytes are recognized for their ability to alleviate stress tolerance, increase disease resistance, and promote plant growth (Kumar and Verma 2018).

14.5 Satellite and Core Microflora

Core microflora or core plant–microbiome (CPM) refers to those microbes that closely linked with particular type of plant and are independent of soil and environmental conditions (Lakshmanan et al. 2014). *Bradyrhizobium, Microvirga,* and *Sphingobium* were found to be the core microbiome of *Solanum tuberosum* (potato) (Pfeiffer et al. 2016). Another study revealed that *Hyphomicrobiaceae*, *Micrococcaceae*, and *Pseudomonadaceae* sps. are the core microbes found in grapevine (Zarraonaindia et al. 2015). The CPM contains those microbial taxa that are essential for plant and have established themselves in plant during the course of evolution (Jacoby et al. 2017). The enhancement of microbial taxa through evolutionary selection process contains the genes responsible for survival and holobiont of the plant (Rosenberg and Zilber-Rosenberg 2018).

Distinctively, the microbial taxa found in less affluence and fewer locations are referred to as satellite taxa. They are described based on their habitat, geography of habitat, and their regional population. These taxa are perceived as the regulators of important mechanisms in an ecosystem (Banerjee et al. 2018). Few researchers found that microbial taxa having the scarce population play a crucial role in obstructing the entrance of undesirable microbial taxa in soil. Parallelly, the scarce bacterial population produces different antifungal chemicals which guard the plants against soil-induced infectious organisms (Ciancio et al. 2016). Various researches demonstrate that forfeiture of scarce microbial population compromises with plants' efficiency (Garcia and Kao-Kniffin 2018). Researchers also found that microbial taxa regulate vital mechanisms that can be unequal to their population number (Pineda et al. 2017). Numerous ecological factors should be taken into account to comprehend mechanisms of plant–microbiome taken in response to certain environmental tensions (Braga et al. 2016).

14.6 Factors Affecting Plant Microbiota

The microbial organization in plant's tissues is linked with various organic and inorganic constituents. These constituents can be type, pH, salinity, structure, moisture, organic matter, and exudates of soil, respectively (Jacoby et al. 2017; Yadav et al. 2015c; Yadav and Yadav 2018). These constituents above are important for parts of the plants which are present beneath the ground level. Other constituents such as infectious agents, environmental conditions, and human activities affect the microbial community of above- and belowground plant parts (Mendes et al. 2013). Using high-throughput approaches like a shotgun and 16S rRNA sequencing, Bulgarelli with his colleagues examined the root microbial community of various barley species and concluded that root exudates and innate immune system determine the organization of root microbiota (Bulgarelli et al. 2015). Various host-influence factors such as growth stage, age, and wellness of plants affect organization of plant microbiota by the formulation of root exudates and plant signaling processes (like induced and acquired systemic resistance, respectively) (Ortíz-Castro et al. 2009).

14.7 Function and Role of Plant Microbiome

Plant–microbiome consists of all neutral, beneficial as well as pathogenic microbes. There are few plant growth-promoting (PGP) bacteria which promote the growth of plant by synthesizing different phytohormones such as gibberellin, cytokinin, and auxin (Egamberdieva et al. 2017; Yadav 2017a, b; Yadav 2019; Yadav et al. 2017a). However, few PGPB produce ACC deaminase (1-aminocyclopropane-1-carboxylate deaminase) which regulate the level of ethylene (stress hormone) in plant (Glick 2014). Literature survey revealed that *Arthrobacter* spp., *Bacillus* spp., and *Pseudomonas* spp. are the ACC deaminase producer, which enables them to enhance the growth of plant (Souza et al. 2015). On analyzing the root microflora of soybean and wheat, it showed the dominance of *Pantoea* spp., *Paraburkholderia* spp., and *Pseudomonas* spp. These species promote the plant growth because of the properties like phytohormones (ACC deaminase, indole acetic acid), nitrogen fixation, and phosphate solubilization mechanism to enhance the stress tolerance

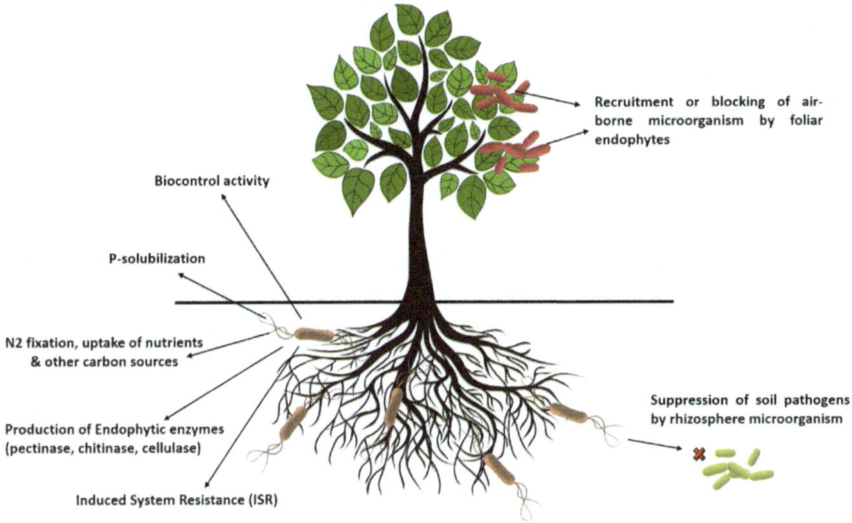

Fig. 14.2 Schematic overview of function and impact of plant–microbiome on the host plant

and improve nutrient uptake (Rascovan et al. 2016). Schematic representation of the function performed by microbiome is illustrated in Fig. 14.2.

On the other hand, there are few bacteria which show disease symptoms because of the production of phytohormones and phytotoxic compounds (Olanrewaju et al. 2017). For instance, *Pseudomonas syringae* is a plant pathogen which affects various plants like green bean, olive, tobacco, and tomato (Hirano and Upper 2000). *Erwinia amylovora*, another pathogenic bacterium, causes fire blight disease in ornamental plants. *Ralstonia solanacearum, Xylella fastidiosa,* and *Xanthomonas* sps. are the bacterial species that are associated with various diseases of banana and potato (Vrancken et al. 2013). There are certain factors which determines the severity as well as outcome of disease like biotic factors (like plant microflora), favorable environmental factor, host susceptibility, and population size of pathogens (Lareen et al. 2016). Both aboveground plant microflora and belowground microflora have been found to increase the resistance in host via commensal interactions of pathogen or due to modification in plant defense system (Igiehon and Babalola 2018b).

The plant microbiota uses diverse biocontrol mechanism like antibiotic production, siderophore production, lytic enzyme production, and pathogen-inhibiting volatile compound production to prevent the disease and pathogenic invasion (Compant et al. 2005). Few plants modulate the hormone level and induce resistance in plant to protect them from plant pathogens. Therefore, agricultural lands containing disease-suppressing microbes build pressure on pathogens and make it the disease-suppressive soil (Gómez Expósito et al. 2017). Particularly, *Bacillus, Burkholderia, Enterobacter, Paenibacillus, Pantoea, Paraburkholderia, Pseudomonas,* and *Streptomyces* have been found to play the main role in pathogen suppression (Yadav et al. 2017b). Carrión and his colleagues revealed that *Paraburkholderia graminis* PHS1 produces cysteine desulfurase and dimethyl sulfoxide reductase (sulfurous volatile compounds), which provide suppressive ability against fungal root pathogens (Carrión et al. 2018). On the other hand, Durán and his colleagues highlighted the role of endospheric bacteria belonging to *Enterobacter* and *Serratia* family, which have the biocontrol activity against *Gaeumannomyces graminis* (Durán et al. 2017).

14.8 Advancements in Plant–Microbiome Research (PMR)

Prediction about the advances in plant–microbiome research is quite difficult (Thijs et al. 2016). But following are the three points which help us in shaping the future of both applied and basic studies of soil microbiome.

14.8.1 Upgradation of Culturing Techniques

The advent of molecular techniques has rapidly increased the genomic data of uncultured microbial taxa persisting in the soil, for which closely related strains are currently unknown (Warinner et al. 2017). There is an exponential increase in data because of culture-independent approach like single-cell genomics or by the assembling of individual genome obtained via metagenome analysis (Blainey 2013). Although we have the genomic information of all the microbes present in the soil, still we lack to understanding about their functionality. Therefore, categorizing them in ecological taxa considering their genomic data is risky (Peršoh 2015). Additionally, the slow-growing microbe is another challenge. Therefore, approaches which use genomic data to intercept the information about ideal cultivation and isolation of uncultured microbes will unquestionably be helpful for the field to me in forward direction (Streit and Schmitz 2004).

14.8.2 Role of Viruses in Soil Microbiome

Approximately, 10^7–10^9 virus particles are found in one gram of soil, and this number is less than the number of viruses found in the aquatic ecosystem (Jacquet et al. 2010). Apparently, these viruses found in soil are highly diverse and remained unexplored (Williamson et al. 2017). Marine studies have revealed that phages play chief role in nutrient dynamics, as they maintain the nutrient dynamics by killing the 20–40% microbial population present in the water column (Sime-Ngando 2014). Various researches have demonstrated soil to be the reservoir for phages which specifically targets bacteria like *Rhizobium* spp. and *Xanthomonas* spp., still the effect of these viruses on activity and composition of soil remains comprehended (Buttimer et al. 2017). According to statistics, > 90% of viruses found in soil mainly get absorbed by clay as well as other soil surfaces, which leaves very low number of viruses in soil. Additionally, it is also uncertain that these viruses are capable enough to infect the plant–microbiome (Andika et al. 2016). Due to advancement in viral metagenomic field and development of new techniques, there is an enumeration in viral population, and it has prompted us to explore new viral community as well as study the effect of these viruses on microbial processes and populations (Jacquet et al. 2010). Basically, it clear to develop a holistic understanding of how these microbes (directly or indirectly) interact with each other instead of studying the individual microbial group (Tshikantwa et al. 2018).

14.8.3 Importance of Horizontal Gene Transfer

There are three main mechanisms of horizontal gene transfer, i.e., conjugation, transduction, and transformation, by which microbes their gene from one generation to another (Von Wintersdorff et al. 2016). Genes contains the sequence of various different traits like arsenic detoxification, antibiotic resistance, and xenobiotic degradation. Therefore, mobile genes may cause evolution of new phenotypic characteristic and may generate close relation with dissimilar genomes (Janssen et al. 2005). However, horizontal gene transfer can impose dilemma during the attempt of linking particular gene to targeted phylogenetic lineages. This definite control of horizontal gene transfer, its frequency in soil microbial communities, and their effect on the soil population are the topics which nurtures this expedition and motivate us to work in this direction (Andrews et al. 2018).

14.9 Use of Plant–Microbiome as a Biocontrol Agent

Worldwide, plant diseases have been held responsible for farmers' losses. FAO stated that 25% of crop loss occurs due to plant diseases and pests. Therefore, there is need for control plant disease (Savary et al. 2012). To achieve this, specific plants which are resistant to various diseases are bred along with different cultivation techniques like use of pathogen-free seeds, crop rotation, or by using biological and chemical agents (Cheng and Cheng 2015). All these plant protection methods influence the microbiome of plant; hence, information gathered from plant protection strategies should be taken into consideration during cultivation (Berg et al. 2014). In various cases, the onset of disease takes place due to imbalance of plant–microbiome and obstructs the desired objective of whole microbiome. Hence, analyzing metabolic pathway and plant–microbiome of the host plant unlock new opportunities to advance biocontrol approaches (Thijs et al. 2016).

Overall, better understanding unfolds the information about the diverse microbes which aid in plant growth and produce antagonistic chemicals against phytopathogens (Bargaz et al. 2018). Mosses are known to have unique microbial diversity and harbor microbes with exceptional antagonist potential because of their ecology (Bragina et al. 2012). Moreover, medicinal plants are also to be known to nurture diverse microbes as their secondary metabolites determine the composition of microbiome (Berg et al. 2015). Because of large diversity among microbes, it is believed that endophytes, especially seed endophytes, can emerge as the unique biocontrol agent (Ryan et al. 2008). Till now, largely bacteria as well as fungi have been used for biocontrol purposes (Whipps 2001). Recently, Archaea have been exploited to become the member of plant–microbiome but its biocontrol potential is still under evaluation (Müller et al. 2015).

As stated above, microbes linked with plant develops a network which gets influence during pathogen invasion. Therefore, these networks will enable us to develop new model and enhance disease management of the plant–soil microbes (Hassani et al. 2018). Even researchers have developed a framework to decipher these microbiome networks and showed the benefit of these frameworks to test their hypothesis for targeted microbes responsible for particular plant disease. They proposed four types of network analysis: (a) first type involves the general analysis of network to discover the aspiring taxa to maintain the existing microbial community; (b) second type involves the analysis majorly focusing on host as well as plant response; (c) third type involves the identification of pathogenic taxa which are directly or indirectly associated with prior known taxa of pathogens; and (d) last type involves the identification of those taxa which are associated with disease (Poudel et al. 2016).

Now, most of the biocontrol agents are screened through in vitro antagonisms effective against particular pathogen. As this screening allows the identification of potent biocontrol agent, still intense debate is going on about this screening approach (Larran et al. 2016). The chief reason is that biocontrol and plant growth-promoting (PGPR) microbes during in vivo evaluation show no antagonisms, whereas in the case of in vitro studies it shows modest result (Beneduzi et al. 2012). Different high-throughput techniques have been developed for plant assays but were found to be ineffective as includes artificial characters (Rasheed et al. 2017). For instance, rapeseed treated with *Paenibacillus* showed plant growth in natural soil, whereas no negative effect was observed in sterile soil under gnotobiotic conditions. Therefore, there is need for testing the potential of biocontrol and plant growth-promoting strains in soil as well as evaluation of their effect on soil type and plant species also be done (Bashan et al. 2014). Another research study showed the presence of similar genera of *Pseudomonas* in lettuce rhizosphere in three different soil samples from the field. This confirms that soil types do not play any role in biocontrol activity and rhizo-competence (Schreiter et al. 2018).

Additionally, targeted enhanced diversity is also being exploited as biomarker for screening purpose by researchers. These applicability and development in biological control is attaining attention globally (Atanasov et al. 2015), whereas previously only one microbe was used in biocontrol strategies, which explains about inconsistency.

But recent advancements have enabled us to develop microbiome-dependent biocontrol strategy (Ab Rahman et al. 2018). In 1999, Emmert and Handelsman stated Gram-positive bacteria to be potential candidate in biocontrol practices. On assessing the registered products for biocontrol and along with those that are in pipeline, almost all the products are bacillus based (Emmert and Handelsman 1999). There is technical reason for using *Bacillus* sps. as the base product in contrast to other Gram-negative bacteria, as form spores under unfavorable condition for survival. This spore formation increases their shelf-life (Berg et al. 2017). On examining the plant-associated microbes which harbor large number of bacterial species, it make us realize that we are seeing the very small portion of taxonomic diversity for biocontrol purpose (Chowdhury et al. 2017).

Moreover, the researchers have reported that 2% mean population is occupied by Bacillus species from the total bacterial community in the soil (Li et al. 2017). As an exception, another researcher published a paper in which they stated that 37% of *Bacillus* and *Paenibacillus* species are found in arid soil (Köberl et al. 2013). Still, the debate is going on comprehending the role of *Bacillus* on the plant and against its pathogen, or does it trigger the plant growth hormones or induce resistance as they are involved in plant–microbe interaction. It is already known that *Bacillus* and *Paenibacillus* strains synthesize various antibiotics and secondary metabolites (Mhlongo et al. 2018). As per studies, strains and spores of both *Bacillus* and *Paenibacillus* are predominantly found in the environment and this might be reason for reducing number of plant-associated microbial community. Hence, more emphasis is given on increasing the microbial diversity within products during biocontrol activity. This prompts to develop new strategies to explore the antagonistic potential of microbes associated with plants (Yadav et al. 2017c).

Nowadays, antibiotic resistance is becoming the global concern and making our treatment ineffective as a result people is suffering from disability, prolonged illness, and at last death (Laxminarayan et al. 2013). Regrettably, strategies used in agriculture like use of antibiotics in livestock are increasing the resistance level. These livestock aid in transferring the resistance in plant production system, which causes the enrichment of resistant bacteria in the rhizosphere (Founou et al. 2016), even though many microbes have developed resistance and are highly prevalent in the environment. Therefore, risk management studies to suppress the emergence of resistant microbes have become our first priority (Larsson et al. 2018). Though the different biocontrol products have shown lower persistence on soil and plant, usage of spore-forming bacteria will change the whole scenario (Paul et al. 2019). Additionally, soil microbes are found to be bacterial antagonists in nature. In particular, *Paenibacillus* have been found to contain not only the antibiotic synthesis gene but also contain antibiotic gene, justifying about their antagonistic ability within the microbiome (Cycoń et al. 2019). Therefore, further investigation is required to understand the mobilization of resistant gene through different ways of horizontal gene transfer (Bengtsson-Palme et al. 2017b).

14.10 Studies Related to Plant Microbiomes

Even though plant–microbiome is a broad topic, microbial group associated with plant is an active research area. The major reason is the exponential increase in publications in recent year on this topic as well as research targeting the niche of specific plant and how interaction takes place among particular microbial community (Naylor and Coleman-Derr 2018). Additionally, it has been observed that microbial assembly varies according to environmental conditions and plant species. This creates curiosity to learn more about how microbes are acquired by plant, either the microbes are subsequently passed to next generation or recruited from the environment (Yu and Hochholdinger 2018).

In spite of the role of the microbes in promoting the plant growth and nutrient cycle, we still require to gain more insight about microbial interaction which makes up the plant microbiome. This will allow us to comprehend the functioning of plant microbiome to changing agricultural practices and environmental conditions (Garcia and Kao-Kniffin 2018).

Various studies are being conducted to extend the link the shift in environmental condition with microbiome response (Purahong et al. 2018). For instance, the researcher conducted on the grape vineyard using culture-independent method revealed the presence of dynamic microbial community on different stages of plant lifecycle, enlightening the effect of agrochemicals on the plant (Morgan et al. 2017). Numerous studies have associated microbiome with physiology of plant, as microbes found in plant–microbiome in desert aid in providing the resistance to plant to survive in unfavorable conditions (Vurukonda et al. 2016). In 2012, Marasco and his colleague proposed that plant microbiomes are essential to develop resistance against drought stresses (Marasco et al. 2012). Another effective interaction was reported by another research group, where the described role of microbiome determines the composition of plant root exudates. These examples highlight the role of soil microbiome not only in plants but also in their evolutionary mechanism, regulated by host–microbe interaction (Lareen et al. 2016).

14.11 Conclusion and Future Prospects

The microbial communities possessing beneficial traits act as a powerful tool to elevate the sustainable agriculture by reducing the usage of fertilizes and combatting plant diseases to enhance the crop yield. The potential of microbiome is to elicit the plant growth, generate stress resistance, and improve plant health. Various microbial inoculants have been developed to achieve field success by either designing smart microbial consortia or engineering the microbiota with beneficial characteristics. In order to upsurge the crop yield and support the enhanced technologies, new plant breeding practices and suitable formulations are required. To achieve these goals, there is need to gather insight about the plant–microbiome interaction, microbial community of soil, and their stability during environmental changes. As single microbes are the key regulators within microbial community, therefore comprehensive investigation on these microbes along with microbial community of soil can assist us in expanding the horizon of this field. Moreover, the knowledge gained will enable us to completely comprehend the impact of these microbes on disease resistance, nutrient cycles, and yield of the crop. Furthermore, it will prompt us to explore novel strategies for microbiome engineering to move toward the sustainable agriculture.

References

Ab Rahman SF, Singh E, Pieterse CM, Schenk PM (2018) Emerging microbial biocontrol strategies for plant pathogens. Plant Sci 267:102–111

Aleklett K, Hart M, Shade A (2014) The microbial ecology of flowers: an emerging frontier in phyllosphere research. Botany 92(4):253–266

Anand P, Chopra RS, Dhanjal DS, Chopra C (2019) Isolation and characterization of microbial diversity of soil of Dhanbad coal mines using molecular approach. Res J Pharm Technol 12(3):1137–1140

Andika IB, Kondo H, Sun L (2016) Interplays between soil-borne plant viruses and RNA silencing-mediated antiviral defense in roots. Front Microbiol 7:1458

Andrews M, De Meyer S, James E, Stępkowski T, Hodge S, Simon M, Young J (2018) Horizontal transfer of symbiosis genes within and between rhizobial genera: occurrence and importance. Genes 9(7):321

Atanasov AG, Waltenberger B, Pferschy-Wenzig EM, Linder T, Wawrosch C, Uhrin P et al (2015) Discovery and resupply of pharmacologically active plant-derived natural products: a review. Biotechnol Adv 33(8):1582–1614

Banerjee S, Schlaeppi K, van der Heijden MG (2018) Keystone taxa as drivers of microbiome structure and functioning. Nat Rev Microbiol 16(9):567

Bargaz A, Lyamlouli K, Chtouki M, Zeroual Y, Dhiba D (2018) Soil microbial resources for improving fertilizers efficiency in an integrated plant nutrient management system. Front Microbiol 9:1606

Bashan Y, Bashan LE, Prabhu SR, Hernandez JP (2014) Advances in plant growth-promoting bacterial inoculant technology: formulations and practical perspectives (1998–2013). Plant Soil 378(1–2):1–33

Beneduzi A, Ambrosini A, Passaglia LM (2012) Plant growth-promoting rhizobacteria (PGPR): their potential as antagonists and biocontrol agents. Genet Mol Biol 35(4):1044–1051

Bengtsson-Palme J, Kristiansson E, Larsson DJ (2017a) Environmental factors influencing the development and spread of antibiotic resistance. FEMS Microbiol Rev 42(1):fux053

Bengtsson-Palme J, Kristiansson E, Larsson DJ (2017b) Environmental factors influencing the development and spread of antibiotic resistance. FEMS Microbiol Rev 42(1):fux053

Berg G, Grube M, Schloter M, Smalla K (2014) The plant microbiome and its importance for plant and human health. Front Microbiol 5:1

Berg G, Köberl M, Rybakova D, Müller H, Grosch R, Smalla K (2017) Plant microbial diversity is suggested as the key to future biocontrol and health trends. FEMS Microbiol Ecol 93(5):fix050

Berg G, Rybakova D, Grube M, Köberl M (2015) The plant microbiome explored: implications for experimental botany. J Exp Bot 67(4):995–1002

Blainey PC (2013) The future is now: single-cell genomics of bacteria and archaea. FEMS Microbiol Rev 37(3):407–427

Braga RM, Dourado MN, Araújo WL (2016) Microbial interactions: ecology in a molecular perspective. Braz J Microbiol 47:86–98

Bragina A, Berg C, Cardinale M, Shcherbakov A, Chebotar V, Berg G (2012) Sphagnum mosses harbour highly specific bacterial diversity during their whole lifecycle. ISME J 6(4):802–813

Bulgarelli D, Garrido-Oter R, Münch PC, Weiman A, Dröge J, Pan Y et al (2015) Structure and function of the bacterial root microbiota in wild and domesticated barley. Cell host Microb 17(3):392–403

Busby PE, Soman C, Wagner MR, Friesen ML, Kremer J, Bennett A et al (2017) Research priorities for harnessing plant microbiomes in sustainable agriculture. PLoS Biol 15(3):e2001793

Buttimer C, McAuliffe O, Ross RP, Hill C, O'Mahony J, Coffey A (2017) Bacteriophages and bacterial plant diseases. Front Microbiol 20(8):34

Campisano A, Antonielli L, Pancher M, Yousaf S, Pindo M, Pertot I (2014) Bacterial endophytic communities in the grapevine depend on pest management. PLoS ONE 9(11):e112763

Carrión VJ, Cordovez V, Tyc O, Etalo DW, de Bruijn I, de Jager VC (2018) Involvement of Burkholderiaceae and sulfurous volatiles in disease-suppressive soils. The ISME J 12(9):2307

Carvalho SD, Castillo JA (2018) Influence of light on plant-phyllosphere interaction. Front Plant Sci 9:1482

Cheng F, Cheng Z (2015) Research progress on the use of plant allelopathy in agriculture and the physiological and ecological mechanisms of allelopathy. Front Plant Sci 6:1020

Chowdhury ME, Jeon J, Rim SO, Park YH, Lee SK, Bae H (2017) Composition, diversity and bioactivity of culturable bacterial endophytes in mountain-cultivated ginseng in Korea. Sci Rep 7(1):10098

Ciancio A, Pieterse CM, Mercado-Blanco J (2016) Harnessing useful rhizosphere microorganisms for pathogen and pest biocontrol. Front Microbiol 7:1620

Compant S, Duffy B, Nowak J, Clément C, Barka EA (2005) Use of plant growth-promoting bacteria for biocontrol of plant diseases: principles, mechanisms of action, and future prospects. Appl Environ Microbiol 71(9):4951–4959

Cycoń M, Mrozik A, Piotrowska-Seget Z (2019) Antibiotics in the soil environment-degradation and their impact on microbial activity and diversity. Front Microbiol 10:338

Dhanjal DS, Chopra C, Anand P, Chopra RS (2017) Accessing the microbial diversity of sugarcane fields from Gujjarwal village, Ludhiana and their molecular identification. Res J Pharm Technol 10(10):3439–3442

Dhanjal DS, Sharma D (2018) Microbial metagenomics for industrial and environmental bioprospecting: the unknown envoy. Microbial bioprospecting for sustainable development. Springer, Singapore, pp 327–352

Dong W, Zhang X, Wang H, Dai X, Sun X, Qiu W, Yang F (2012) Effect of different fertilizer application on the soil fertility of paddy soils in red soil region of southern China. PLoS ONE 7(9):e44504

Donn S, Kirkegaard JA, Perera G, Richardson AE, Watt M (2015) Evolution of bacterial communities in the wheat crop rhizosphere. Environ Microbiol 17(3):610–621

Durán P, Jorquera M, Viscardi S, Carrion VJ, Mora MD, Pozo MJ (2017) Screening and characterization of potentially suppressive soils against Gaeumannomyces graminis under extensive wheat cropping by Chilean indigenous communities. Front Microbiol 8:1552

Dwibedi V, Saxena S (2019) Diversity and phylogeny of resveratrol-producing culturable endophytic fungi from Vitis species in India. 3 Biotech 9(5):182

Edwards J, Johnson C, Santos-Medellín C, Lurie E, Podishetty NK, Bhatnagar S (2015) Structure, variation, and assembly of the root-associated microbiomes of rice. Proc Natl Acad Sci 112(8):E911–920

Egamberdieva D, Wirth SJ, Alqarawi AA, Abd_Allah EF, Hashem A (2017) Phytohormones and beneficial microbes: essential components for plants to balance stress and fitness. Front Microbiol 8:2104

Emmert EA, Handelsman J (1999) Biocontrol of plant disease: a (Gram-) positive perspective. FEMS Microbiol Lett 171(1):1–9

Finkel OM, Castrillo G, Paredes SH, González IS, Dangl JL (2017) Understanding and exploiting plant beneficial microbes. Curr Opin Plant Biol 38:155–163
Founou LL, Founou RC, Essack SY (2016) Antibiotic resistance in the food chain: a developing country-perspective. Front Micro 23(7):1881
Frank A, Saldierna Guzmán J, Shay J (2017) Transm Bact Endophytes Microorg 5(4):70
Garcia J, Kao-Kniffin J (2018) Microbial group dynamics in plant rhizospheres and their implications on nutrient cycling. Front Micro 11(9):1516
Glick BR (2014) Bacteria with ACC deaminase can promote plant growth and help to feed the world. Microbiol Res 169(1):30–39
Gómez Expósito R, de Bruijn I, Postma J, Raaijmakers JM (2017) Current insights into the role of rhizosphere bacteria in disease suppressive soils. Front Micro 8:2529
Gomiero T (2017) Soil degradation, land scarcity and food security: Reviewing a complex challenge. Sustainability 8(3):281
Hardoim PR, Van Overbeek LS, Berg G, Pirttilä AM, Compant S, Campisano A (2015) The hidden world within plants: ecological and evolutionary considerations for defining functioning of microbial endophytes. Microbiol Mol Biol Rev 79(3):293–320
Hassani MA, Durán P, Hacquard S (2018) Microbial interactions within the plant holobiont. Microbiome 6(1):58
Hirano SS, Upper CD (2000) Bacteria in the leaf ecosystem with emphasis on *Pseudomonas syringae*—a pathogen, ice nucleus, and epiphyte. Microbiol Mol Biol Rev 64(3):624–653
Hunter PJ, Teakle G, Bending GD (2014) Root traits and microbial community interactions in relation to phosphorus availability and acquisition, with particular reference to Brassica. Front Plant Sci 5:27
Igiehon N, Babalola O (2018a) Rhizosphere microbiome modulators: contributions of nitrogen fixing bacteria towards sustainable agriculture. Int J Environ Res Public Health 15(4):574
Igiehon NO, Babalola OO (2018b) Below-ground-above-ground plant-microbial interactions: focusing on soybean, rhizobacteria and mycorrhizal fungi. Open Microbiol J 12:261
Jacoby R, Peukert M, Succurro A, Koprivova A, Kopriva S (2017) The role of soil microorganisms in plant mineral nutrition—current knowledge and future directions. Front Plant Sci 8:1617
Jacquet S, Miki T, Noble R, Peduzzi P, Wilhelm S (2010) Viruses in aquatic ecosystems: important advancements of the last 20 years and prospects for the future in the field of microbial oceanography and limnology. Adv Ocean Limnol 1(1):97–141
Janssen DB, Dinkla IJ, Poelarends GJ, Terpstra P (2005) Bacterial degradation of xenobiotic compounds: evolution and distribution of novel enzyme activities. Environ Microbiol 7(12):1868–1882
Kandel S, Joubert P, Doty S (2017) Bacterial endophyte colonization and distribution within plants. Microorganisms 5(4):77
Kawasaki A, Donn S, Ryan PR, Mathesius U, Devilla R, Jones A et al (2016) Microbiome and exudates of the root and rhizosphere of *Brachypodium distachyon*, a model for wheat. PLoS ONE 11(10):e0164533
Kecskeméti E, Berkelmann-Löhnertz B, Reineke A (2016) Are epiphytic microbial communities in the carposphere of ripening grape clusters (*Vitis vinifera* L.) different between conventional, organic, and biodynamic grapes? PloS One 11(8):e0160852
Köberl M, Schmidt R, Ramadan EM, Bauer R, Berg G (2013) The microbiome of medicinal plants: diversity and importance for plant growth, quality and health. Front Micro 20(4):400
Kour D, Rana KL, Yadav N, Yadav AN, Singh J, Rastegari AA, et al (2019) Agriculturally and industrially important fungi: current developments and potential biotechnological applications. In: Yadav AN, Singh S, Mishra S, Gupta A (eds) Recent advancement in white biotechnology through fungi, volume 2: perspective for value-added products and environments. Springer International Publishing, Cham, pp 1–64. https://doi.org/10.1007/978-3-030-14846-1_1
Kudjordjie EN, Sapkota R, Steffensen SK, Fomsgaard IS, Nicolaisen M (2019) Maize synthesized benzoxazinoids affect the host associated microbiome. Microbiome 7(1):59

Kumar A, Chaturvedi AK, Yadav K, Arunkumar KP, Malyan SK, Raja P, Kumar R, Khan SA, Yadav KK, Rana KL, Kour D, Yadav N, Yadav AN (2019) Fungal phytoremediation of heavy metal-contaminated resources: current scenario and future prospects. In: Yadav AN, Singh S, Mishra S, Gupta A (eds) Recent advancement in white biotechnology through fungi: volume 3: perspective for sustainable environments. Springer International Publishing, Cham, pp 437–461. https://doi.org/10.1007/978-3-030-25506-0_18

Kumar A, Verma JP (2018) Does plant—microbe interaction confer stress tolerance in plants: a review? Microbiol Res 207:41–52

Kumar M, Saxena R, Rai PK, Tomar RS, Yadav N, Rana KL, Kour D, Yadav AN (2019) Genetic diversity of methylotrophic yeast and their impact on environments. In: Yadav AN, Singh S, Mishra S, Gupta A (eds) Recent advancement in white biotechnology through fungi: volume 3: perspective for sustainable environments. Springer International Publishing, Cham, pp 53–71. https://doi.org/10.1007/978-3-030-25506-0_3

Lakshmanan V, Selvaraj G, Bais HP (2014) Functional soil microbiome: belowground solutions to an aboveground problem. Plant Physiol 166(2):689–700

Lareen A, Burton F, Schäfer P (2016) Plant root-microbe communication in shaping root microbiomes. Plant Mol Biol 90(6):575–587

Larran S, Simon MR, Moreno MV, Siurana MS, Perelló A (2016) Endophytes from wheat as biocontrol agents against tan spot disease. Biol Control 92:17–23

Larsson DJ, Andremont A, Bengtsson-Palme J, Brandt KK, de Roda Husman AM, Fagerstedt P (2018) Critical knowledge gaps and research needs related to the environmental dimensions of antibiotic resistance. Environ Int 117:132–138

Laxminarayan R, Duse A, Wattal C, Zaidi AK, Wertheim HF, Sumpradit N, Greko C (2013) Antibiotic resistance—the need for global solutions. Lancet Infect Dis 13(12):1057–1098

Li T, Liu T, Zheng C, Kang C, Yang Z, Yao X (2017) Changes in soil bacterial community structure as a result of incorporation of Brassica plants compared with continuous planting eggplant and chemical disinfection in greenhouses. PLoS ONE 12(3):e0173923

Marasco R, Rolli E, Ettoumi B, Vigani G, Mapelli F, Borin S, Abou-Hadid AF (2012) A drought resistance-promoting microbiome is selected by root system under desert farming. PLoS ONE 7(10):e48479

Mendes R, Garbeva P, Raaijmakers JM (2013) The rhizosphere microbiome: significance of plant beneficial, plant pathogenic, and human pathogenic microorganisms. FEMS Microbiol Rev 37(5):634–663

Mhlongo MI, Piater LA, Madala NE, Labuschagne N, Dubery IA (2018) The chemistry of plant–microbe interactions in the rhizosphere and the potential for metabolomics to reveal signaling related to defense priming and induced systemic resistance. Front Plant Sci 9:112

Morgan HH, Du Toit M, Setati ME (2017) The grapevine and wine microbiome: insights from high-throughput amplicon sequencing. Front Micro 11(8):820

Nair DN, Padmavathy S (2014) Impact of endophytic microorganisms on plants, environment and humans. Sci World J. https://doi.org/10.1155/2014/250693

Naylor D, Coleman-Derr D (2018) Drought stress and root-associated bacterial communities. Front Plant Sci 8:2223

Neal AL, Ahmad S, Gordon-Weeks R, Ton J (2012) Benzoxazinoids in root exudates of maize attract Pseudomonas putida to the rhizosphere. PLoS ONE 7(4):e35498

Olanrewaju OS, Glick BR, Babalola OO (2017) Mechanisms of action of plant growth promoting bacteria. World J Microbiol Biotechnol 33(11):197

Ortíz-Castro R, Contreras-Cornejo HA, Macías-Rodríguez L, López-Bucio J (2009) The role of microbial signals in plant growth and development. Plant Signal Behav 4(8):701–712

Paul C, Filippidou S, Jamil I, Kooli W, House GL, Estoppey A (2019) Bacterial spores, from ecology to biotechnology. Adv Appl Microbiol 106:79–111

Peršoh D (2015) Plant-associated fungal communities in the light of meta'omics. Fungal Divers 75(1):1–25

Pfeiffer S, Mitter B, Oswald A, Schloter-Hai B, Schloter M, Declerck S, Sessitsch A (2016) Rhizosphere microbiomes of potato cultivated in the High Andes show stable and dynamic core microbiomes with different responses to plant development. FEMS Microbiol Ecol 93(2):fiw242

Pineda A, Kaplan I, Bezemer TM (2017) Steering soil microbiomes to suppress aboveground insect pests. Trends Plant Sci 22(9):770–778

Poudel R, Jumpponen A, Schlatter DC, Paulitz TC, Gardener BM, Kinkel LL, Garrett KA (2016) Microbiome networks: a systems framework for identifying candidate microbial assemblages for disease management. Phytopathology 106(10):1083–1096

Purahong W, Orrù L, Donati I, Perpetuini G, Cellini A, Lamontanara A, (2018) Plant microbiome and its link to plant health: tree species, organs and Pseudomonas syringae pv. actinidiae infection shaping bacterial phyllosphere communities of kiwifruit plants. Front Plant Sci 9:1563

Qiao Q, Wang F, Zhang J, Chen Y, Zhang C, Liu G (2017) The variation in the rhizosphere microbiome of cotton with soil type, genotype and developmental stage. Sci Rep 7(1):3940

Raaijmakers JM, Paulitz TC, Steinberg C, Alabouvette C, Moënne-Loccoz Y (2009) The rhizosphere: a playground and battlefield for soilborne pathogens and beneficial microorganisms. Plant Soil 321(1–2):341–361

Rana KL, Kour D, Sheikh I, Dhiman A, Yadav N, Yadav AN, Rastegari AA, Singh K, Saxena AK (2019a) Endophytic fungi: biodiversity, ecological significance and potential industrial applications. In: Yadav AN, Mishra S, Singh S, Gupta A (eds) Recent advancement in white Biotechnology through fungi, vol 1. Diversity and Enzymes Perspectives. Springer, Switzerland, pp 1–62

Rana KL, Kour D, Sheikh I, Yadav N, Yadav AN, Kumar V, Singh BP, Dhaliwal HS, Saxena AK (2019b) Biodiversity of endophytic fungi from diverse niches and their biotechnological applications. In: Singh BP (ed) Advances in Endophytic Fungal Research: Present Status and Future Challenges. Springer International Publishing, Cham, pp 105–144. https://doi.org/10.1007/978-3-030-03589-1_6

Rascovan N, Carbonetto B, Perrig D, Díaz M, Canciani W, Abalo M, Alloati et al (2016) Integrated analysis of root microbiomes of soybean and wheat from agricultural fields. Sci Rep 6:28084

Rasheed A, Hao Y, Xia X, Khan A, Xu Y, Varshney RK, He Z (2017) Crop breeding chips and genotyping platforms: progress, challenges, and perspectives. Mol Plant 10(8):1047–1064

Rashid MI, Mujawar LH, Shahzad T, Almeelbi T, Ismail IM, Oves M (2016) Bacteria and fungi can contribute to nutrients bioavailability and aggregate formation in degraded soils. Microbiol Res 183:26–41

Rosenberg E, Zilber-Rosenberg I (2018) The hologenome concept of evolution after 10 years. Microbiome 6(1):78

Ryan RP, Germaine K, Franks A, Ryan DJ, Dowling DN (2008) Bacterial endophytes: recent developments and applications. FEMS Microbiol Lett 278(1):1–9

Santos ML, Berlitz DL, Wiest SL, Schünemann R, Knaak N, Fiuza LM (2018) Benefits associated with the interaction of endophytic bacteria and plants. Braz Arch Biol Technol 61:e18160431

Santoyo G, Moreno-Hagelsieb G, del Orozco-Mosqueda M C, Glick BR (2016) Plant growth-promoting bacterial endophytes. Microbiol Res 183:92–99

Savary S, Ficke A, Aubertot JN, Hollier C (2012) Crop losses due to diseases and their implications for global food production losses and food security Food Security 4(4):519–537

Schlaeppi K, Bulgarelli D (2015) The plant microbiome at work. Mol Plant-Microbe Interact 28(3):212–217

Schreiter S, Babin D, Smalla K, Grosch R (2018) Rhizosphere competence and biocontrol effect of Pseudomonas sp. RU47 independent from plant species and soil type at the field scale. Front Micro 9:97

Shade A, Jacques MA, Barret M (2017) Ecological patterns of seed microbiome diversity, transmission, and assembly. Curr Opin Microbiol 37:15–22

Shahzad R, Khan AL, Bilal S, Asaf S, Lee IJ (2018) What is there in seeds? Vertically transmitted endophytic resources for sustainable improvement in plant growth. Front Plant Sci 9:24

Sime-Ngando T (2014) Environmental bacteriophages: viruses of microbes in aquatic ecosystems. Front Micro 5:355
Souza RD, Ambrosini A, Passaglia LM (2015) Plant growth-promoting bacteria as inoculants in agricultural soils. Genet Mol Biol 38(4):401–419
Steven B, Huntley RB, Zeng Q (2018) The influence of flower anatomy and apple cultivar on the apple flower phytobiome. Phytobiomes 2(3):171–179
Streit WR, Schmitz RA (2004) Metagenomics–the key to the uncultured microbes. Curr Opin Microbiol 7(5):492–498
Suman A, Yadav AN, Verma P (2016) Endophytic Microbes in Crops: Diversity and Beneficial impact for Sustainable Agriculture. In: Singh D, Abhilash P, Prabha R (eds) Microbial inoculants in sustainable agricultural productivity, research perspectives. Springer-Verlag, India, pp 117–143. https://doi.org/10.1007/978-81-322-2647-5_7
Thijs S, Sillen W, Rineau F, Weyens N, Vangronsveld J (2016) Towards an enhanced understanding of plant–microbiome interactions to improve phytoremediation: engineering the metaorganism. Front Micro 16(7):341
Tshikantwa TS, Ullah MW, He F, Yang G (2018) Current trends and potential applications of microbial interactions for human welfare. Front Micro 9:1156
Verma P, Yadav AN, Khannam KS, Kumar S, Saxena AK, Suman A (2016) Molecular diversity and multifarious plant growth promoting attributes of Bacilli associated with wheat (*Triticum aestivum* L.) rhizosphere from six diverse agro-ecological zones of India. J Basic Microbiol 56:44–58
Verma P, Yadav AN, Khannam KS, Saxena AK, Suman A (2017a) Potassium-solubilizing microbes: diversity, distribution, and role in plant growth promotion. In: Panpatte DG, Jhala YK, Vyas RV, Shelat HN (eds) Microorganisms for green revolution: volume 1: microbes for sustainable crop production. Springer Singapore, Singapore, pp 125–149. https://doi.org/10.1007/978-981-10-6241-4_7
Verma P, Yadav AN, Kumar V, Singh DP, Saxena AK (2017b) Beneficial plant-microbes interactions: biodiversity of microbes from diverse extreme environments and its impact for crop improvement. In: Singh DP, Singh HB, Prabha R (eds) Plant-microbe interactions in agro-ecological perspectives: volume 2: microbial interactions and agro-ecological impacts. Springer Singapore, Singapore, pp 543–580. https://doi.org/10.1007/978-981-10-6593-4_22
Von Wintersdorff CJ, Penders J, van Niekerk JM, Mills ND, Majumder S, van Alphen LB (2016) Dissemination of antimicrobial resistance in microbial ecosystems through horizontal gene transfer. Front Micro 7:173
Vrancken K, Holtappels M, Schoofs H, Deckers T, Valcke R (2013) Pathogenicity and infection strategies of the fire blight pathogen Erwinia amylovora in Rosaceae: state of the art. Microbiology 159(5):823–832
Vurukonda SS, Vardharajula S, Shrivastava M, SkZ A (2016) Enhancement of drought stress tolerance in crops by plant growth promoting rhizobacteria. Microbiol Res 184:13–24
Wagner MR, Busby PE, Balint-Kurti P (2019) Breeding for broad-spectrum disease resistance alters the maize leaf microbiome. bioRxiv 1:647446
Warinner C, Herbig A, Mann A, Fellows Yates JA, Weiß CL, Burbano HA (2017) A robust framework for microbial archaeology. Annu Rev Genomics Hum Genet 31(18):321–356
Whipps JM (2001) Microbial interactions and biocontrol in the rhizosphere. J Exp Bot 1:487–511
Williamson KE, Fuhrmann JJ, Wommack KE, Radosevich M (2017) Viruses in soil ecosystems: an unknown quantity within an unexplored territory. Annu Rev Virol 4:201–219
Yadav AN (2017a) Agriculturally important microbiomes: biodiversity and multifarious PGP attributes for amelioration of diverse abiotic stresses in crops for sustainable agriculture. Biomed J Sci Tech Res 1:1–4
Yadav AN (2017b) Beneficial role of extremophilic microbes for plant health and soil fertility. J Agric Sci 1:1–4
Yadav AN (2019) Microbiomes of Wheat (Triticum aestivum L.) Endowed with Multifunctional Plant Growth Promoting Attributes. EC Microbiol 15:1–6

Yadav AN, Kour D, Sharma S, Sachan SG, Singh B, Chauhan VS, Sayyed RZ, Kaushik R, Saxena AK (2019) Psychrotrophic microbes: biodiversity, mechanisms of adaptation, and biotechnological implications in alleviation of cold stress in plants. In: Sayyed RZ, Arora NK, Reddy MS (eds) Plant growth promoting rhizobacteria for sustainable stress management: volume 1: rhizobacteria in abiotic stress management. Springer Singapore, Singapore, pp 219–253. https://doi.org/10.1007/978-981-13-6536-2_12

Yadav AN, Kumar V, Prasad R, Saxena AK, Dhaliwal HS (2018a) Microbiome in crops: diversity, distribution and potential role in crops improvements. In: Prasad R, Gill SS, Tuteja N (eds) Crop improvement through microbial biotechnology. Elsevier, USA, pp 305–332

Yadav AN, Sachan SG, Verma P, Saxena AK (2015a) Prospecting cold deserts of north western Himalayas for microbial diversity and plant growth promoting attributes. J Biosci Bioeng 119:683–693

Yadav AN, Sachan SG, Verma P, Saxena AK (2016) Bioprospecting of plant growth promoting psychrotrophic Bacilli from cold desert of north western Indian Himalayas. Indian J Exp Biol 54:142–150

Yadav AN, Sachan SG, Verma P, Tyagi SP, Kaushik R, Saxena AK (2015b) Culturable diversity and functional annotation of psychrotrophic bacteria from cold desert of Leh Ladakh (India). World J Microbiol Biotechnol 31:95–108

Yadav AN, Verma P, Kaushik R, Dhaliwal HS, Saxena AK (2017a) Archaea endowed with plant growth promoting attributes. EC Microbiol 8:294–298

Yadav AN, Verma P, Kour D, Rana KL, Kumar V, Singh B, Chauahan VS, Sugitha T, Saxena AK, Dhaliwal HS (2017b) Plant microbiomes and its beneficial multifunctional plant growth promoting attributes. Int J Environ Sci Nat Resour 3(1):1–8

Yadav AN, Verma P, Kumar M, Pal KK, Dey R, Gupta A, Padaria JC, Gujar GT, Kumar S, Suman A, Prasanna R, Saxena AK (2015c) Diversity and phylogenetic profiling of niche-specific Bacilli from extreme environments of India. Ann Microbiol 65:611–629

Yadav AN, Verma P, Kumar S, Kumar V, Kumar M, Singh BP, Saxena AK, Dhaliwal HS (2018b) Actinobacteria from Rhizosphere: molecular diversity, distributions and potential biotechnological applications. In: Singh B, Gupta V, Passari A (eds) New and future developments in microbial biotechnology and bioengineering. USA, pp 13–41. https://doi.org/10.1016/b978-0-444-63994-3.00002-3

Yadav AN, Verma P, Sachan SG, Kaushik R, Saxena AK (2018c) Psychrotrophic Microbiomes: Molecular Diversity and Beneficial Role in Plant Growth Promotion and Soil Health. In: Panpatte DG, Jhala YK, Shelat HN, Vyas RV (eds) Microorganisms for green revolution-volume 2: microbes for sustainable agro-ecosystem. Springer, Singapore, pp 197-240. https://doi.org/10.1007/978-981-10-7146-1_11

Yadav AN, Verma P, Singh B, Chauhan VS, Suman A, Saxena AK (2017c) Plant growth promoting bacteria: biodiversity and multifunctional attributes for sustainable agriculture. Adv Biotechnol Microbiol 5(5):1–6

Yadav AN, Yadav N (2018) Stress-adaptive microbes for plant growth promotion and alleviation of drought stress in plants. Acta Sci Agric 2:85–88

Yu P, Hochholdinger F (2018) The role of host genetic signatures on root-microbe interactions in the rhizosphere and endosphere. Front Plant Sci 9:1896

Zarraonaindia I, Owens SM, Weisenhorn P, West K, Hampton-Marcell J, Lax S (2015) The soil microbiome influences grapevine-associated microbiota. MBio 6(2):e02527–14

Chapter 15
Current Aspects and Applications of Biofertilizers for Sustainable Agriculture

Modhurima Misra, Ashish Sachan and Shashwati Ghosh Sachan

Abstract There has been a sharp increase in the world's population over the past few decades which can be threatening in terms of the food security of the people. Thus, to cater to the huge demand of food, agricultural production should be increased within a short span of time and with limited worldwide agricultural land resources. This situation has driven the farmers all over the world to rely heavily on the commercially available chemical fertilizers for enhanced agricultural productivity. Though there has been a significant rise in the production of crops, these fertilizers have proved to be detrimental for our ecosystem as well as animal and human health. The deteriorative effects of the large chemical inputs in the agricultural systems have not only challenged the sustainability of crop production but also the maintenance of the environment quality. Using biofertilizers is a natural, low-cost, environment-friendly way out to this problem. Biofertilizers comprise living microorganisms capable of supplying sufficient nutrients to the plants, while maintaining high yield. The present chapter aims at describing in brief the requirement of biofertilizers in the first place, as well as their choice over the conventional synthetic ones, the different types, their roles in agriculture, their production and mode of action and most importantly advantages and demerits. The inputs from various literature mentioned herein will help in understanding the immense importance of their usage in modern-day farming with a cleaner ecosystem.

Keywords Biofertilizers · Consortium · Microbiome · Plant growth promotion · Sustainable agriculture

M. Misra · S. G. Sachan (✉)
Department of Bio-Engineering, Birla Institute of Technology, Mesra, Ranchi, Jharkhand, India
e-mail: ssachan@bitmesra.ac.in

A. Sachan
Centre for Life Sciences, Central University of Jharkhand, Brambe, Ranchi 835205, Jharkhand, India

A. N. Yadav et al. (eds.), *Plant Microbiomes for Sustainable Agriculture*, Sustainable Development and Biodiversity 25, https://doi.org/10.1007/978-3-030-38453-1_15

15.1 Introduction

In the recent years, environmental pollution has become a matter of concern worldwide, with various industries and agricultural systems being the two most significant contributors. The components of agriculture which affect the environment are the different chemicals used for better production of crops that includes fertilizers, pesticides, insecticides, etc. Even though these can cause rapid production of crops (Santos et al. 2012) and put an end to the scarcity of food, their retention in the ecosystem, once introduced, can have deleterious effects on civilization in the long run. These are industrially manipulated compounds, having known quantities of different elements like nitrogen, phosphorus, sulphur, potassium, etc. Groundwater, contaminated from leaching of nitrates can cause a severe disease called Blue Baby Syndrome (Knobeloch et al. 2000). In this regard, safer, eco-friendly approaches are being sought after as potential alternatives to synthetic fertilizers. Thus, keeping these problems in mind, biofertilizers have been developed, which are often regarded as one of the best gifts to modern agricultural science. These have been proved to be quite successful in increasing the crop yield without disturbing the ecological balance. Europe and Latin America have started using biofertilizers significantly, due to the strict regulations on their consumption of chemical fertilizers (Raja 2013).

Numerous definitions of biofertilizers are available which can be attributed to the improved understanding of these compounds from time to time. Biofertilizers contain various microbes which mobilize nutritional elements through their regular metabolic activities (Vessey 2003), thereby enhancing their accessibility and uptake by the host plants. Mazid et al. (2011) defined them as "substances which contain living microorganisms that colonize the rhizosphere or the interior of the plants and promote growth by increasing the supply or availability of primary nutrients to the target crops, when applied to soils, seeds or plant surfaces". They are renewable, "eco-friendly" agro input of organic origin and consist of microorganisms which can be nitrogen fixers, phosphate solubilizers, sulphur oxidizers or organic matter decomposers (Vessey 2003). The application of microbial inoculum dates long back and started with small scale compost production that showed potential as a biofertilizer (Khosro and Yousef 2012).

The commercialization of biofertilizer dates back to 1895 when Nobbe and Hilther launched the product "Nitragin" (Yimer and Abena 2019). As of today, other microorganisms like *Azotobacter*, Blue-green algae are being used as potential biofertilizers. With the passage of time, these biofertilizers have become an indispensable part of modern-day organic farming, an environmentally sustainable form of agriculture that emphasizes protection and preservation of natural resources along with preventing the use of chemical fertilizers and pesticides (Barman et al. 2017).

15.2 Biofertilizers—A Natural Gift for Sustainable Agriculture

Indiscriminate use of chemical fertilizers for crop production can have hazardous effects on our environment which is of great concern in the context of sustainable agriculture as well as environmental remediation. These synthetic molecules, apart from being costly, contaminate the air, water and soil as well as cause eutrophication of surface water bodies (Youssef and Eissa 2014). These agrochemicals are capable of increasing the crop yield to a certain level and their excessive use can have long-lasting effects and negatively affect human health. Their overuse can lead to the decrease in soil fertility by increasing the salt content (Aggani 2013) and may have a role to play in greenhouse effect, depletion of ozone layer and acidification of soil and water (Kumar et al. 2018) (Fig. 15.1). They make the plants more susceptible to infections by weakening their roots.

In this regard, the exploitation of various microorganisms in augmenting nutrients to the host plants has gained significant importance as they have shown great potential as a substitute for the chemical fertilizers. The important benefits of these biofertilizers can be listed as follows:

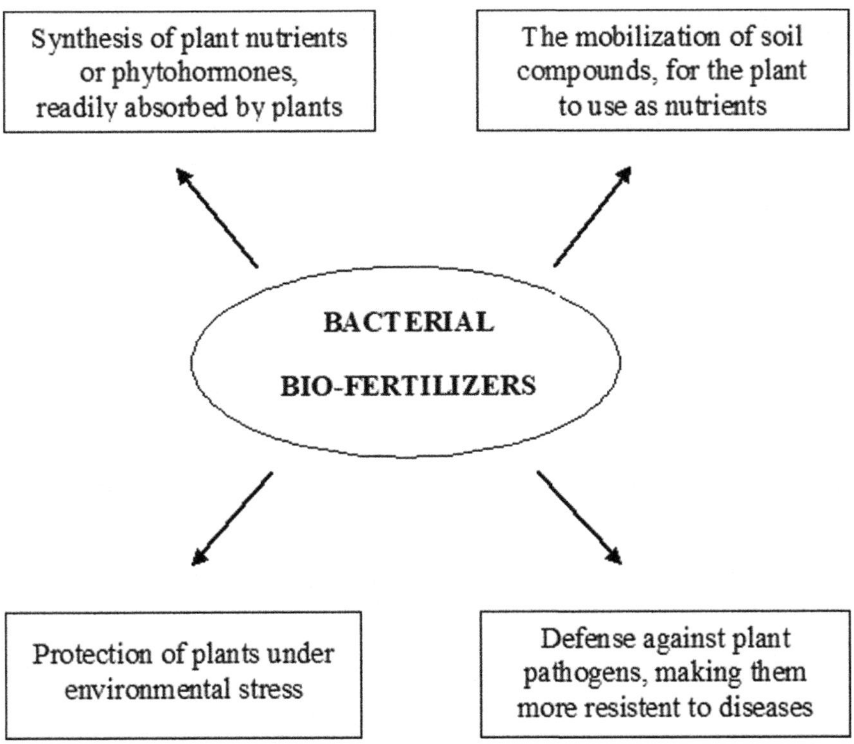

Fig. 15.1 Major functions of bacterial biofertilizers

- Being cost effective and ecosystem friendly (Khosro and Yousef 2012).
- They enrich the soil environment by accumulating various micro- and macronutrients through nitrogen fixation, phosphate solubilization and decomposition of organic matter (Sinha et al. 2010).
- They increase crop productivity by releasing different growth-regulating substances and synthesis of phytohormones.
- They help to increase the biodiversity of the soil where they are applied by inviting the localization of different beneficial bacteria and fungi.
- Biofertilizers form a part of nutrient cycling when applied as seed or soil inoculants (Singh et al. 2011).
- These beneficial microorganisms make the host plants more resistant towards environmental stress, like drought, salinity, etc., (Bhardwaj et al. 2014).
- The microbes are vital constituents of the Integrated Nutrient Management (INM) system by continuously and slowly releasing metabolic intermediates (Adesemoye and Kloepper 2009).

15.3 Types of Biofertilizers

Both rhizospheric and endophytic microbial communities can significantly determine crop vigor in agricultural systems through interactions with the host plants. The microorganisms which are important for agriculture can be distinguished as plant growth-promoting rhizobacteria (PGPRs), nitrogen-fixing cyanobacteria, phosphate-solubilizing microorganisms (PSMs), potassium-solubilizing microorganisms (KSMs), siderophore and phytohormone producing organisms, mycorrhiza, plant disease-suppressive bacteria, stress-tolerant endophytes and biodegrading microbes (Singh et al. 2011). The rhizosphere can have up to 10^{11} microbial cells per gram of root (Egamberdieva et al. 2008) and above 30,000 prokaryotic species (Mendes et al. 2013) that may automatically increase the crop productivity by enhancing the soil physicochemical properties, biodiversity of soil microbiome, plant growth and development (Sahoo et al. 2014). Figure 15.2 gives a brief idea about the major groups of biofertilizers (Table 15.1).

15.3.1 Nitrogen-Fixing Microorganisms

Nitrogen is considered to be one of the fundamental elements for plant development. The crops are dependent on bioavailable form of nitrogen for their growth, which is quite limited in nature. The process by which atmospheric nitrogen gets converted to organic forms like ammonia and nitrate, for ready utilization by the plants, is known as biological nitrogen fixation (BNF) (Gothwal et al. 2007). The different microorganisms capable of fixing atmospheric nitrogen can be grouped under three

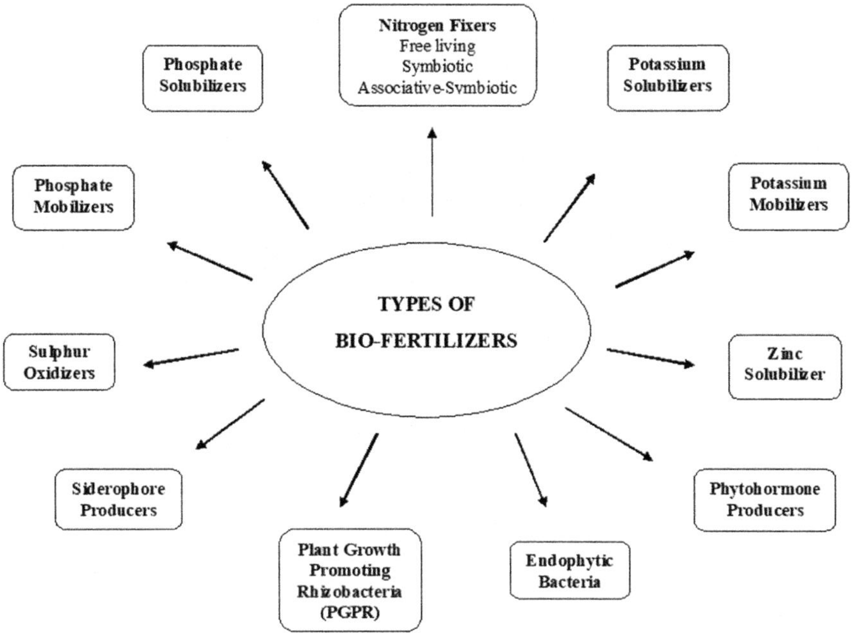

Fig. 15.2 Major groups of biofertilizers

categories—symbiotic, free living and associative (Gupta 2004). Symbiotic associations account for majority of BNF in nature. Nitrogen-fixing biofertilizers are crop specific in nature (Choudhury and Kennedy 2004; Yadav 2017a, b). Some of the important bacteria and algae which play a crucial part in this process are discussed below (Table 15.2).

15.3.1.1 *Rhizobium*

It is a symbiotic bacterium that colonizes the roots of the leguminous plants forming tumourous out growths known as root nodules, which helps in ammonia production. It fixes atmospheric nitrogen even in symbiotic association with a non-legume called *Parasponia* sp (Mahdi et al. 2010). They utilize the photosynthetic products of the plants as their carbon source and in return fix atmospheric nitrogen for their benefactor. The morphological and physiological characteristics of *Rhizobium* at free-living condition are different from those of the bacteroid form in nodules (Kumar et al. 2018). These are considered to be the most proficient biofertilizer for legumes as far as the amount of nitrogen fixed is concerned (Jehangir et al. 2017; Yadav et al. 2019a, b, c). The nodulation and nitrogen fixation by the bacterium is attributed to the nod, nif and fix genes.

Table 15.1 Major groups of biofertilizers with their types and examples

Major groups	Microorganisms	Examples
N_2—Fixers	Free living	*Azotobacter sp., Beijerinkia sp., Clostridium sp., Klebsiella sp., Anabaena sp., Nostoc sp.*
	Symbiotic	*Rhizobium sp., Frankia sp., Anabaena azollae*
	Associative Symbiotic	*Azospirillum sp., Anabaena sp.*
P—Solubilizers	Bacteria	*Bacillus sp., Pseudomonas sp., Rhizobium sp., Burkholderia sp., Microccocus sp., Achromobacter sp., Agrobacterium sp., Aereobacter sp., Flavobacterium sp., Erwinia sp.*
	Fungi	*Penicillium sp., Aspergillus awamori, Aspergillus niger.*
P—Mobilizers	Arbuscular mycorrhiza	*Glomus sp., Gigaspora sp., Acaulospora sp., Scutellospora sp., Sclerocystis sp.*
	Ectomycorrhiza	*Laccaria sp., Pisolithus sp., Boletus sp., Amanita sp.*
	Ericoid mycorrhizae	*Pezizella ericae*
	Orchid mycorrhiza	*Rhizoctonia solani*
K—Solubilizers	Bacteria	*Bacillus mucilaginosus, Bacillus edaphicus, Bacillus circulans, Acidithiobacillus ferrooxidans, Paenibacillus* spp
	Fungi	*Aspergillus sp*
K—Mobilizers	Bacteria	*Bacillus* Spp
S—Oxidising	Bacteria	*Thiobacillus sp*
Zn—Solubilizers	Bacteria	*Bacillus subtilis, Thiobacillus thioxidans*
	Yeast	*Saccharomyces sp*
Phytohormone producers	Bacteria	*Galactomyces sp., Pseudomonas sp., Azospirillum sp., Bacillus sp., Arthrobacter sp., Acinetobacter sp.*
	Fungi	*Gibberella fujikuroi*
Endophytic microorganisms	Bacteria	*Azoarcus* sp., *Gluconoacetobacter diazotrophicus, Herbaspirillum seropedicae, Serratia marcesens*
	Fungi	*Trichoderma stromaticum, T. evansi, T. amazonicum, T. taxi, T. martiale, T. theobromicola*
Plant Growth-Promoting Rhizobacteria (PGPR)	Bacteria	*Bacillus spp., Pseudomonas fluorescence, Azospirillum sp., Azotobacter sp., Burkholderia sp., Enterobacter sp., Klebsiella sp.*

Source Modified from Barman et al. (2017)

Table 15.2 A comparative study of the different N_2 fixers

N_2 fixers	Type	Family	Important species	N_2 fixed (Kg N/Ha)	Recommended crops
Rhizobium	Symbiotic	Rhizobiaceae	*R. meliloti* *R. leguminosarium* *R. ciceri* *R. etli* *R. tropici* *R fredii*	50–100	Pulse legumes like chickpea, red-gram, pea, lentil, black gram, etc., oil-seed legumes like soybean, groundnut and forage legumes like berseem and lucerne
Azotobacter	Free living	Azotobacteriaceae	*A. chroococcum,* *A. vinelandii,* *A. beijerinckii,* *A. insignis* *A. macrocytogenes*	20	Rice, wheat, barley, oat, sunflowers, maize, line, beetroot, tobacco, tea, coffee and coconuts
Cyanobacteria (BGA)	Symbiotic, Free living and Symbiotic-Associative	–	*Nostoc sp.,* *Anabaena sp.,* *Gloetrichia sp.,* *Tolypothrix sp.,* *Aulosira sp.,* *Aphanotheca sp.*	20–30	Submerged rice, barley, oats, tomato, radish, cotton, chilli, sugarcane, maize, lettuce
Azolla	Symbiotic-Associative	Salviniaceae	*A.caroliniana,* *A. microphylla,* *A. filiculoides* *A. Mexicana*	30–50	Submerged rice with maximum temperature
Azospirillum	Symbiotic-Associative	Spirilaceae	*A.lipoferum,* *A.brasilensehave* *A.amazonense,* *A.halopraeferens,* *A.brasilense,* *A. trakense*	20–40	Maize, sugarcane, sorghum, pearl millet

Source Compiled from Kumar et al. (2017) and Kumar et al. (2018)

15.3.1.2 *Azotobacter*

It is saccharophilic, aerobic, heterotrophic in nature and commonly an inhabitant of the neutral to alkaline soils. The most common species found in arable soils is *A.chroococcum* (Wani et al. 2013). It produces abundant slime helping in soil aggregation. *Azotobacter* sp is a potential nitrogen fixer in rice crops (Sahoo et al. 2014). Besides fixing atmospheric nitrogen, this bacterium is known to produce some antifungal substance which lessens the seedling mortality (Subba Rao 2001), as well as some growth-promoting substances (Mahdi et al. 2010), phytohormones like indole acetic acid, gibberellin, cytokinin (El-Fattah et al. 2013) and group B vitamins like thiamine and riboflavin (Revillas et al. 2000), antibiotics, exo-polysaccharides and pigments (Jimenez et al. 2011).

15.3.1.3 Cyanobacteria

Cyanobacteria or Blue-Green Algae (BGA) are photosynthetic prokaryotes found in snow and hot springs. Besides fixing atmospheric nitrogen, they induce growth of plants by producing auxins and gibberellins. It has been reported that the production of rice increased from 15% to 38% with the supplementation of BGA inoculants (Kumar et al. 2018). In India, free-living and symbiotic cyanobacteria have been applied in the cultivation of rice (Thajuddin and Subramanian 2005).

15.3.1.4 *Azolla-Anabaena* Symbiosis

BGA fixes nitrogen by forming symbiotic association with different organisms like fungi, ferns, flowering plants, etc. The most common association forms between *Azolla* and *Anabaena azollae*—a blue-green alga. *Azolla*, a free-floating aquatic fern, finds extensive use in the cultivation of rice as a substitute for synthetic nitrogen fertilizers (Yao et al. 2018). Moreover, they have a role in maintaining the soil quality by enhancing the fertility of the soil. It quickly decomposes in the soil (Kannaiyan 1990) and can also be supplied as a green manure before planting of rice saplings (Mahdi et al. 2010). Other advantages include supplementation of significant amount of elements like phosphorus, potassium, sulphur, zinc, iron, molybdenum and other micronutrients. The most common species found in India is *A. pinnata* (Mahdi et al. 2010).

15.3.1.5 *Azospirillum*

It is Gram-negative and heterotrophic in nature. It mainly forms associative symbiosis with plants which use the C4-dicarboxylic acid pathway or Hatch and Slack pathway for photosynthesis as it grows and fixes nitrogen on salts of malic acid and aspartic acid (Mishra and Dash 2014; Arun 2007). These are non-nodule forming and not only colonize around the roots of the host plants but also reside inside the root tissues (Mahdi et al. 2010; Verma et al. 2016a, b, 2015; Yadav et al. 2016). Currently, it is used as a pre-sowing seed treatment (Kumar et al. 2018). Mostly, *A. lipoferum* and *A. brasilense* have proved to be the successful inoculants worldwide. These are also noted for phytohormones and/or siderophores production and phosphate solubilization (Puente et al. 2004). Other advantages with *Azospirillium* sp inoculation include disease resistance, biocontrol activity (Bashan and De-bashan 2010) and unwanted stress (drought, salinity, toxic compounds) tolerance (Creus et al. 1997).

15.3.2 Phosphate Solubilizing Microorganisms

The plants are unable to utilize the inorganic form of phosphorus present in the soil as they are only capable of absorbing its two soluble forms—monobasic and dibasic. Regular usage of phosphate fertilizers is disadvantageous to the environment, and often it is too expensive for the farmers of the developing nations. The phosphate solubilizing microorganisms (PSMs) consist of different bacteria and fungi which are capable of fulfilling about 20–25% phosphorus necessity of the plants which also helps to lessen the manufacturing costs of phosphate fertilizers (Chang and Yang 2009; Verma et al. 2017b, 2015, 2018c). Phosphate solubilizing bacteria (PSBs) make up to about 1–50%, while phosphate solubilizing fungi (PSFs) are only 0.1–0.5% among the P-solubilizers (Chen et al. 2006). Recently, actinomycetes are also being experimented upon as potential phosphate solubilizing organisms as they can tolerate extreme environments like drought, antibiotics and phytohormones. Hamdali et al. 2008 had stated that approximately 20% of actinomycetes can solubilize phosphorus. These PSMs employ different mechanisms for phosphate solubilization like chelation, lowering of soil pH, mineralization, etc.

15.3.2.1 Phosphate Solubilizing Bacteria

Phosphate solubilizing bacteria consist of both aerobic and anaerobic strains. *Bacillus megaterium* is a Gram positive, rod, also known as "Phosphobacterium", has been reported to increase inorganic phosphorus (P) solubilization (Lach et al. 1990) as well as zinc, potassium, iron and manganese solubilizing potential (Amalraj et al. 2012). It helps the plants to avail the phosphorus up to the extent of 10–15 kg/ha and also causes proliferation of roots by releasing growth-promoting hormones. A novel strain, Micrococcus sp NII-0909 has been found to exhibit phosphate solubilizing capability as well as siderophores production (Dastager et al. 2010).

15.3.2.2 Phosphate Solubilizing Fungi

The important phosphate solubilizing fungi include *Aspergillus* sp and *Pencillium* sp. It has been reported that *Aspergillus niger* improved the quantity of dry biomass of chickpea plants by 22–33% with respect to the non-inoculated control (Kapri and Tewari 2010). Apart from these species, symbiotic nitrogenous rhizobia (Khan et al. 2009) and nematofungus *Arthrobotrys oligospora* (Khan et al. 2009; Thakur et al. 2014) have also shown phosphate solubilizing activity.

15.3.3 Phosphate Mobilizing Biofertilizers

These are microorganisms which scavenge phosphorus from a phosphorus enriched environment and channelize it to the plants. They can even solubilize organic phosphorus with the help of hyphal enzymes like phosphatase (Ezawa et al. 1995). Arbuscular mycorrhizal fungi (AMF) are significant contributors of phosphate mobilization. The potential of these fungi can be fully realized when the soil is deficient in phosphorus (Ghorbanian et al. 2012). This group of fungi remains symbiotically associated with the root of the plants in the rhizosphere and enhance phosphorus uptake by the extensive external fine absorbing hyphae that can exploit a larger soil volume (Jakobsen et al. 1992). These in turn receives the carbon requirements from the host plants. Studies by Yao et al. (2001), reveal that the AMF not only mobilize soil phosphates when its availability is low but also provide a greater surface area which permits increased contact with the phosphates with respect to the non-mycorrhizal root systems. The supply of phosphorus to the host plants depends upon the amounts of external mycelium produced by the fungi. Transfer of nutrients from the fungus to the plant mostly occurs at the arbuscular interface (Karandashov et al. 2004). The kinetics of nutrient uptake by the mycorrhizal roots are quite different from that of the non-mycorrhizal ones (Cress et al. 1979). Colonization of the roots by AMF thus helps in enhanced growth of the plants in low fertility soils or in soils with uneven distribution of nutrients. The PMBs are generally considered as broad-spectrum fertilizers (Itelima et al. 2018).

15.3.4 Potassium-Solubilizing Microorganisms

Only 1–2% potassium is available for uptake by the plants, though soil contains sufficient quantity of it. It is one of the fundamental elements for the plants and occurs primarily as silicate minerals in the soil which are inaccessible to plants. The microorganisms solubilize the silicates with the help of organic acids thus making them readily available to plants (Itelima et al. 2018). Different bacteria, fungi and actinomycetes have been reported for solubilizing soil potassium. Bacteria solubilize potassium by producing various organic and inorganic acids, acidolysis, chelation and exchange reactions (Archana et al. 2013; Meena et al. 2015; Verma et al. 2017a, b).

15.3.5 Potassium Mobilizing Biofertilizers

An example of potassium mobilizing biofertilizers (KMB) is *Bacillus* sp which mobilizes inaccessible forms of potassium in the soil and makes them accessible to the plants. Some established phosphate solubilizers like *Bacillus* sp and *Aspergillus* sp are known to mobilize potassium as well, along with solubilizing phosphate (Itelima et al. 2018). These microbes thus perform dual function.

15.3.6 Plant Growth-Promoting Rhizobacteria

These microorganisms reside in the rhizospheric soil and exhibit beneficial effects on crop productivity. They help to maintain the plant health and development in a holistic way. Directly they aid in the uptake of macro- and micronutrients as well as by modulating the levels of different phytohormones. Indirectly, they improve plant health by eliminating the harmful effects of plant pathogens (Kumar et al. 2018; Yadav et al. 2018a, b) (Fig. 15.3). Plant growth-promoting rhizobacteria (PGPR) not only promotes plant growth but also enhances the yield through different mechanisms. Several research works in the past as well as in the present have been devoted to the better understanding of the diverse nature, dynamics, significance and role of these PGPR in agriculture. Using PGPR species as biofertilizers has some definite economic and ecological advantages; still their administration should be carefully monitored as they behave as opportunistic pathogens in nosocomial infections (Tomar et al. 2016) (Table 15.3).

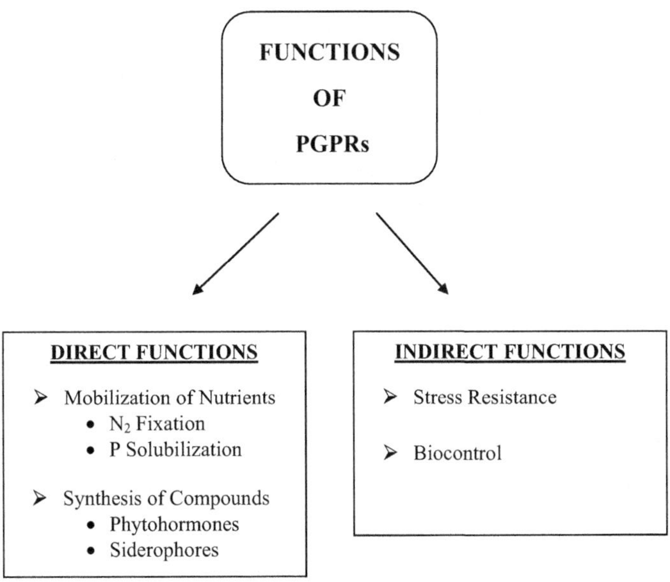

Fig. 15.3 Schematic representation of different functions of PGPRs

Table 15.3 Commercial biofertilizer products of plant growth-promoting rhizobacteria

Product	Company	Bacterial strains
Cell-Tech®	Novozymes	rhizobia
Nitragin Gold®	Novozymes	rhizobia
TagTeam®	Novozymes	rhizobia + *Penicillium bilaii*
Accomplish®	Loveland Products, Inc.	PGPR + enzymes + organic acids + chelators
Nodulator®	BASF Canada Inc.	*Bradyrhizobium japonicum*
Nodulator® N/T	BASF Canada Inc.	*Bacillus subtilis MBI 600* + *Bradyrhizobium japonicum*
Nodulator® PRO	BASF Canada Inc.	*Bacillus subtilis* + *Bradyrhizobium japonicum*
Nodulator® XL	BASF Canada Inc.	*Rhizobium leguminosarum biovar viceae 1435*
Bioboots®	Brett-Young Seeds	*Delftia acidovorans*
Bioboots® (soybean)	Brett-Young Seeds	*Delftia acidovorans* + *Bradyrhizobium* sp.
EVL coating®	EVL Inc.	PGPR consortia
Nitrofix®	Labiofam S. A.	*Azospirillum* sp.
Bioativo®	Instituto de Fosfato Biológico (IFB) Ltda.	PGPR consortia
VitaSoil®	Symborg	PGPR consortia
Azotobacterin®	JSC "Industrial Innovations"	*Azospirillum brasilense* B-4485
Mamezo®)	Tokachi Federation of Agricultural Cooperatives (TFAC)	rhizobia (in peat)
R-Processing Seeds®	Tokachi Federation of Agricultural Cooperatives (TFAC)	rhizobia (coated legume seeds)
Hyper Coating Seeds ®	Tokachi Federation of Agricultural Cooperatives (TFAC)	rhizobia (coated grass legume seeds)
Life®	Biomax	PGPR consortia
Biomix®	Biomax	PGPR consortia
Biodine®	Biomax	PGPR consortia
Biozink®	Biomax	PGPR consortia

Source García-Fraile et al. (2015)

15.3.7 Vesicular-Arbuscular Mycorrhizae

In agricultural soils, the fungi which are most commonly encountered are vesicular-arbuscular mycorrhizae (VAM) which play a significant role in sustainable agriculture. These are usually intracellular, endosymbiotic fungi belonging to the genera *Glomus, Gigaspora, and Sclerocysts* (Kumar et al. 2017; Rana et al. 2019a, b). This inoculant is generally recommended to different fruit plants, like banana, papaya, pineapple, etc., and ornamental plants. Besides, VAM has also been reported to decrease metal toxicity like aluminium. Their major roles are as follows:

- Helps in improved uptake of elements like potassium, zinc, iron, copper, etc.
- They secrete different organic acids which support enhanced availability of both macro- and micronutrients by the plants.
- These fungi increase the conduction of water by the roots of the plants even when the level of water is low in the soil (Bhattacharjee and Dey 2014).
- The mesh of hyphae helps in arresting the soil particles to form bigger aggregates as well as better carbon uptake by the plants. They also maintain the soil quality by synthesizing specific protein known as 'Glomulin' (Mahdi et al. 2010).
- Plants supplied with VAM exhibit elevated phytohormone (IAA, cytokinin) synthesis, thus indirectly helping if plant growth and development.
- VAM-inoculation causes a significant rise in phenolic and phytoalexin activity which renders the plant more resistance towards stress (Bhattacharjee and Dey 2014) (Fig. 15.4).

15.3.8 Endophytic Microorganisms

These microorganisms stimulate the development of non-leguminous crops and cause their nutritional enrichment through nitrogen fixation, phosphate solubilization and production of siderophore (Szilagyi-Zecchin et al. 2014) (Fig. 15.5). Besides they are also involved in the production of phytostimulators (phytohormones), cofactor pyrroloquinoline quinone (PQQ) and volatile acetoin (Tomar et al. 2016). They also help in combating stress by secreting stress modulators. Some species of *Trichoderma* make the plants resistant towards diseases and other abiotic stresses (Bae et al. 2009). These endophytic organisms are known to avoid any sort of competition with their rhizospheric counterparts (Yadav 2019; Yadav et al. 2017a, b, c; Yadav and Yadav 2019).

15.3.9 Sulphur Oxidizing Biofertilizers (SOBs)

These microorganisms oxidize sulphur to sulphates so that the plants can utilize it. *Thiobacillus* sp is one of the sulphur oxidizing biofertilizers (Itelima et al. 2018).

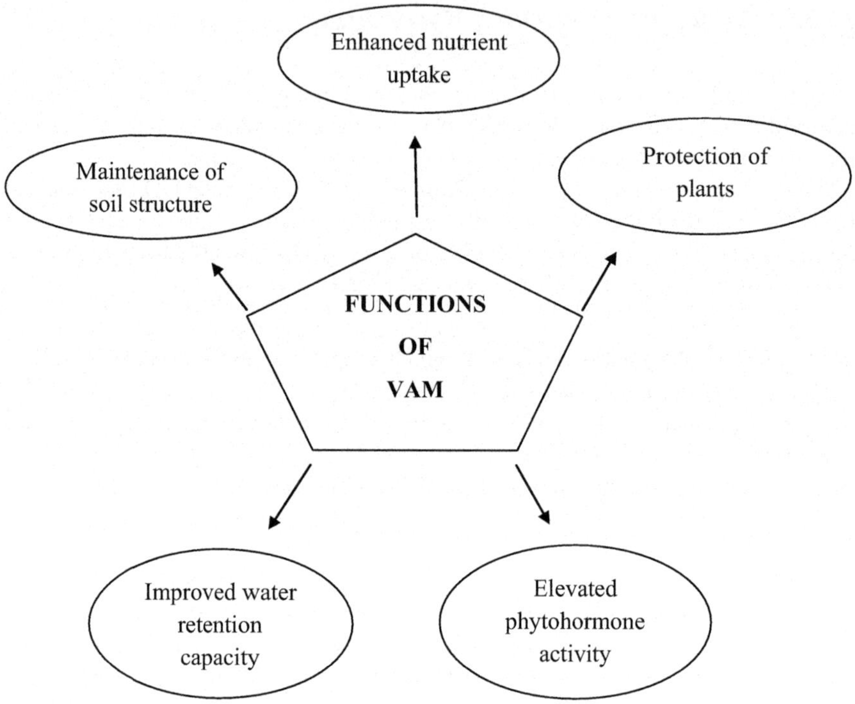

Fig. 15.4 Schematic representation of various functions of VAM

15.3.10 Zinc Solubilizing Microorganism

A wide array of microorganisms is there which help the host plants in the uptake of various micronutrients like zinc, iron, copper, etc. More than half (50%) of the Indian soils are deficient in zinc (Katyal and Rattan 1993). This is compensated by adding soluble zinc sulphate ($ZnSO_4$) from external sources of which only 1–4% is utilized by the crop and the rest 75% gets converted to different mineral fractions, not absorbed by the plants. Two mechanisms have been proposed for zinc fixation—one is applicable in acidic soils and is related with cation exchange and other occurs by chemisorptions of zinc on calcium carbonate with simultaneous complexation by organic ligands under alkaline conditions (Kumar et al. 2018). *Bacillus* sp has been conjugated with cheap, insoluble zinc compounds like zinc oxide, zinc carbonate and zinc sulphide in place of costly zinc sulphate and successfully applied for zinc solubilization (Mahdi et al. 2010; Yadav and Saxena 2018) (Kumar et al. 2019; Yadav et al. 2019d).

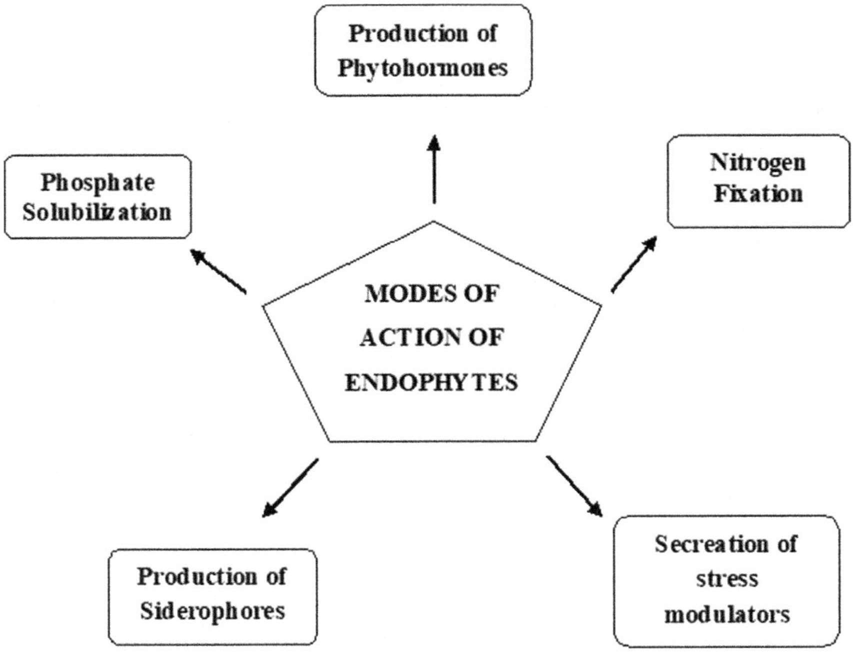

Fig. 15.5 Important functions of endophytic microorganisms

15.3.11 Siderophores Producers

Iron is quite abundantly available in nature; still its accessibility is limited. Iron is accumulated in the form of oxides and oxyhydroxides which make it more difficult for organisms to avail it. Microorganisms acquire Fe^{3+} by active transport through siderophores. These siderophores are low molecular weight compounds secreted by the microbes which act as iron scavengers. These compounds lessen plant stress due to high concentrations of heavy metals in soil. Plants can assimilate iron from bacterial siderophores by special mechanisms like ligand exchange, chelation or direct acquirement of iron siderophore complexes (Colo et al. 2014).

15.3.12 Phytohormone Producing Microorganisms

Phytohormones are organic compounds which are in charge of the overall development of the plant. They are synthesized in two ways—either endogenously by plant tissues or exogenously by PGPRs. The microorganisms inhabiting the soil and responsible for the production of phytohormones are *Galactomyces* sp., *Pseudomonas* sp., *Azospirillum* sp. and *Bacillus* sp. (Waqas et al. 2014; Lei and Ya-qing 2015; Rana et al. 2019c; Yadav et al. 2018a, b, c).

Microbial auxins help in the elongation of root as well as increase in its surface area, thus helping in better utilization of the soil resources. PGPR secreted auxins play a crucial role in plant–microbe interactions and phytostimulation (Spaepen and Vanderleyden 2011). The most common fungus which synthesizes almost 20 types of gibberellins is *Gibberella fujikuroi*. Recent studies by Ullah and his coworkers confirmed that *Photorhabdus temperate*—an entomopathogenic bacterium can produce unlike forms of gibberellic acid (Ullah et al. 2014). Another important plant hormone which helps in cellular differentiation, auxillary bud development, etc., is cytokinin and its most common microbial sources are *Azotobacter chroococcum, Azotobacter beijerinckii, Paenibacillus polymyxa, Pseudomonas putida* and *Pseudomonas fluorescens* (Kudoyarova et al. 2014; Arkhipova et al. 2007).

ACC-deaminase, the enzyme responsible for maintaining the ethylene levels in plants can also be synthesized by the PGPRs. It has been found to make the plants more tolerant towards high concentration of salts (Zahir et al. 2008, 2009) and provide resistance against radiation, poly-aromatic hydrocarbons, insect predation, wounding, etc. (Glick 2012). Rhizobacterial inoculants, synthesizing the hormone, induce plant shoot growth, root elongation, increase in nodulations in *Rhizobium* and promote mycorrhizal colonization (Glick 2012). Several rhizobacteria, like *Acinetobacter* sp., *Azospirillum* sp., *Agrobacterium* sp., *Achromobacte* sp., *Ralstonia* sp., *Enterobacter* sp., *Bacillus* sp., *Rhizobium* sp., *Serratia* sp., *Pseudomonas* sp., etc., have been reported for ACC-deaminase activity (Kang et al. 2010). Recently, *Pseudomonas stutzeri* A1501 capable of synthesizing ACC-deaminase has been reported to assist the cultivation of rice in the presence of salts and heavy metals (Han et al. 2015).

15.4 Production, Storage and Quality Control of Biofertilizers

Several parameters should be kept in mind during the production of biofertilizers, such as growth summary of the microbe(s) used, their types, optimum conditions of the microorganism(s), formulation of inoculums, etc. The formulation of inoculum method of administration and maintenance of the products plays an important role in the overall success of biofertilizer production. Biofertilizers can be amended with a carrier molecule (solid biofertilizer) or it can be produced in the form of liquid inoculants (liquid biofertilizers), both having their advantages and demerits. According to Somasegaran and Springer (1994), few things should be kept in mind while choosing the carrier molecule like

- It must be cheap and readily available when required
- The material should be such that bacterial nutrients can be added easily
- It should support easy sterilization by autoclaving or gamma irradiation
- Can be processed easily and there should not be any lump forming substances
- It should be non-toxic to the microorganisms and the plants,
- Moisture absorption capacity should be high for the material to be a good carrier.

- Water retention capacity of 50% or above should be preferable.
- It must stick to the seeds properly.
- The pH buffering capacity must be high.
- High organic matter content in the material is usually preferred.

The solid biofertilizers are associated with carriers to enhance the effectiveness and water ration capacity. Integration of microorganisms into carrier materials helps in easy handling as well as long term storage (Khosro and Yousef 2012). These are cost effective and easy to produce. Common carrier molecules used are saw dust, talcum dust, clay, peat, vermiculite, bentonite, diatomaceous earth, manure, rice or wheat bran, zeolite, rock phosphate pellets, charcoal, compost, etc. Though it is difficult to find a substance having all the characteristics of a suitable carrier, researches are still going on to develop one satisfying most of the criteria. Polymer-based carriers are close to the suitable ones. Usually, these are alginate beads, which immobilize the bacteria in their matrix, gradually releasing them in the soil environment after application. *Azospirillum brasilense* (> 10^{11} cfu/g inoculant) has been successfully encapsulated using alginate beads with a diameter in the micro range (100–200 μm) and this biofertilizer enhanced the production of wheat and tomato (Bashan et al. 2002).

In India, mostly carrier-based biofertilizers are manufactured and the microorganisms used usually have a shelf life of only 6 months (Rana et al. 2013). They are susceptible to UV and not thermostable beyond 30°C. At the time of production, the microbial density is only 10^8 cfu/ml, which reduces with time and the count is practically negligible at the end. This is one of the main reasons these biofertilizers could not be of much use and failed to draw the farmers, while another problem is that these are prone to contamination.

To overcome these problems, liquid biofertilizers were developed and marketed, which is quite a breakthrough in the history of biofertilizer production and truly a novel, innovative approach. These are specific microbial formulation supplied with certain additional chemicals or special cell protectants that stimulate the formation of resting spores or cysts for better shelf life and tolerance towards different abiotic stress. Some of the commonly used cell protectants are glycerol, molasses and sugars like glucose, maltose, sucrose, trehalose, etc. (Brar et al. 2012).

The microbes used in the preparation of liquid biofertilizers usually have a shelf life of 2 years. They are thermostable (can tolerate temperature as high as 55 °C), tolerant to UV radiations and contamination free. The count is around 10^9 cfu/ml, which remain constant up to 24 months. So, application of 1 ml of liquid biofertilizer is equivalent to that of 1 kg of 5 months old carrier-based counterparts (1000 times) (Mahdi et al. 2010). These are easy to handle and apply in the fields by the farmers. They are applied in various ways. Using hand sprayers, power sprayers or fertigation tanks are quite common. It can also be used basal manure and mixed along with farm yard manure. However, these biofertilizers are costly which hinders its widespread usage (Mahdi et al. 2010).

These bacterial products can be stored in water-in-oil emulsions. They are generally administered to the crops through irrigation systems (VanderGheynst et al. 2006).

The packets of formulated biofertilizers should provide certain pertinent information for the marketing of the products. They are (i) product name, (ii) the particular microorganism or consortia used in its formulation, (iii) specific crops to which it is applicable and is appropriate, (iv) the manufacture and expiry dates and (v) relevant instructions and directions for its administration.

Quality control of the entire production process is extremely important. Quality of the bioproduct determines the success and its acceptance by the farmers. The key steps in the production process that require quality control are (i) the fermentation, (ii) preparation of carrier, (iii) preparing a mixture of bacterial broth and the carrier, (iv) packaging and (v) storage. The carrier must be sterilized and aseptic conditions should be maintained at each step to avoid any contamination. Finally, the product should be stored specific temperature and humidity to ensure the viability of the microorganism during the expected lifespan of the biofertilizer. The final product should be checked for several parameters at different environments to explore its efficiency and limitations in order to guarantee its quality (Fig. 15.6).

15.5 Application of Biofertilizers

The application of solid and liquid biofertilizers in the field is different and is based on their formulations. In ideal situations, the farmer should be able to apply it with his own setup, such that no additional infrastructure is required. Liquid formulations can be applied to the soil using irrigation systems or with the help of different sprayers. For peach orchards and plants, like citrus, mango, vines, guava, apple, etc., spraying is usually recommended. Spreader centrifuges can be used to spread powdered or encapsulated products over the fields. In other cases, the seeds may be coated with the inoculants. The solid ones may be applied and mentioned below.

15.5.1 Seed Treatment

It is the most extensively used method. For every 5 kg of seeds, 100 gm fertilizer is applied. The amount of seed to be sown in a field is the determining factor behind the quantity of fertilizer required to apply per acre of agricultural land. A mixture of the biofertilizer and water in the ratio of 1:2 is prepared to form slurry which is poured in the container along with the seeds before its application. The combination is subjected to a gentle mixing using an adhesive like gum acacia, jaggery solution, etc. such that the seeds are uniformly coated with the biofertilizer (Kumar et al. 2017). The seeds are then spread on a clean sheet or cloth, dried under the shade for about 30 min to avoid direct sunlight and then sown immediately. For crops like pulses, oilseeds and fodder this method is usually suggested (García-Fraile et al. 2015).

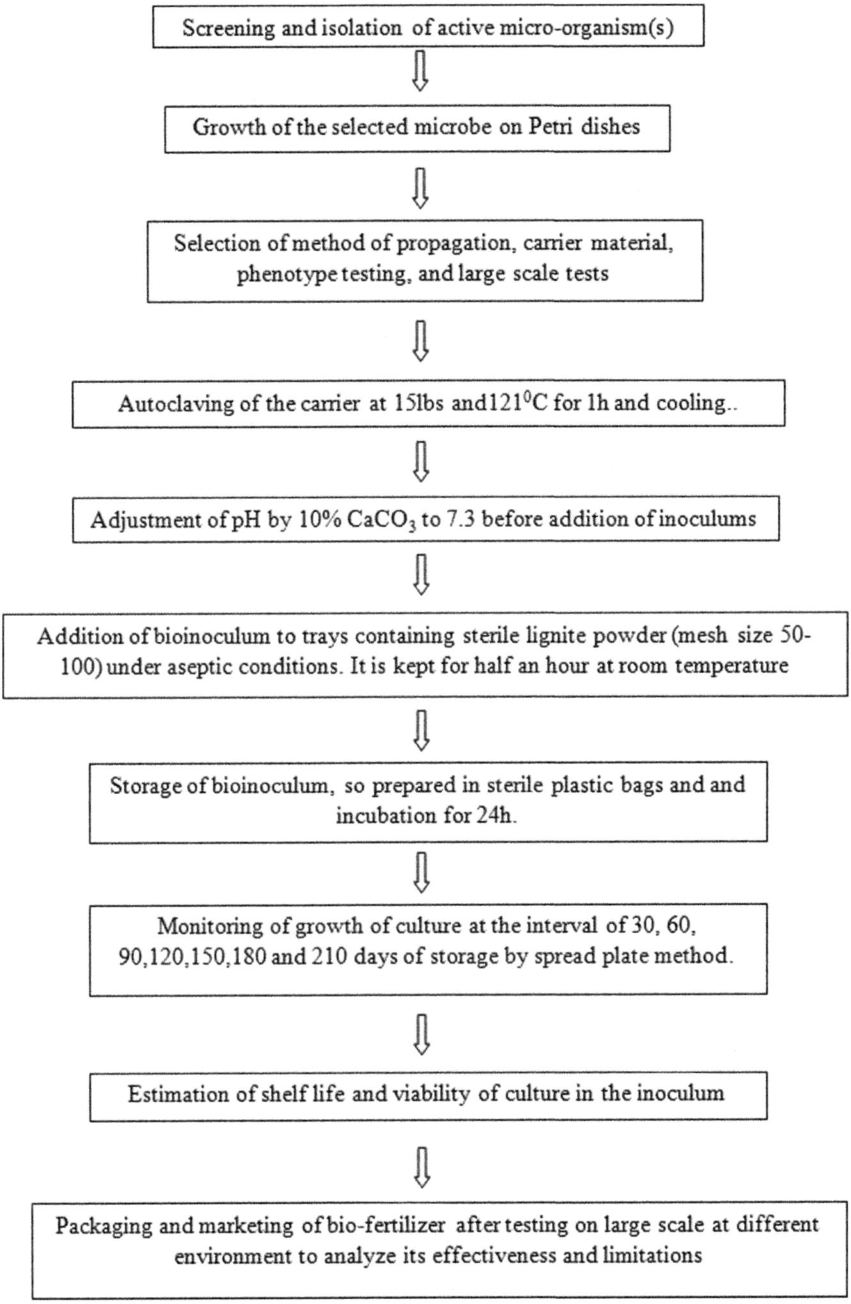

Fig. 15.6 Production of solid biofertilizers

15.5.2 Seedling Root Dip

A diluted formulation of about 1 part of biofertilizer in 10 parts of water is required for this type of treatment. This method is generally recommended for transplanted crops. For rice, a bed is made in the field and filled with water mixed with the specific biofertilizers. The roots of seedlings are dipped into this solution for 8–10 h and then transplanted. This method is suggested for crops which are replanted at seedling stage like paddy, tomato, potato, cabbage, onion, brinjal, chilly, etc. It can also be applied for treating ornamental plants like rose, jasmine, dahlia, marigold, chrysanthemum and many more (García-Fraile et al. 2015).

15.5.3 Soil or Main Field Treatment

The recommended biofertilizers (4 packets) are mixed with specific quantity of compost or dried, powdered manure from the farmyard (20 kg) and kept overnight. This mixture is then applied in the soil at the time of sowing of seeds or just before transplanting (Rana et al. 2013).

15.5.4 Set Treatment

For this type of treatment, biofertilizer and water are mixed in the ratio of 1:50. The explants are dipped into the solution containing the biofertilizer for 30 min with subsequent drying under the shade and planting in the field. This technique is generally applicable for crops like sugarcane, banana, grapes and strawberries (Rana et al. 2013). In a more direct approach, the biofertilizer is usually mixed with different carriers, like soil, compost, manure, rice husks, etc., in the ratio of 1:25 (1 kg per 25 kg of carrier) and applied in the soil.

15.6 Precautions in Using Biofertilizers

One should be cautious of the following factors while applying biofertilizers (Itelima et al. 2018).

- It should be never mixed with nitrogen fertilizers.
- Biofertilizers should not be administered along with any fungicides.
- The seeds soaked in liquid formulations, if required may be dried under the shade. Exposure of the biofertilizer to direct sunlight must be avoided.
- The optimum temperature for storing the biofertilizers is generally room temperature.

15.7 Constraints Regarding Utilization of Different Biofertilizers

Though the biofertilizer technology is a low cost, eco-friendly one, there are several restraints that have limited the widespread application or implementation of this technology. The restrictions may be:

15.7.1 Production Constraints

Though there is demand of biofertilizers over the conventional synthetic ones, the progress in the field of biofertilizer production is quite limited because of the following reasons:

Unavailability of appropriate and efficient strains—The biofertilizers are crop as well as soil specific in nature but the availability of region-specific strains is often a problem (Itelima et al. 2018).

Unavailability of suitable carrier—The search for a proper carrier molecule for the solid biofertilizers is still on which will maintain the microbe throughout its shelf life. Peat is a good option but in India poor quality peat is available which is low in carbon content (Kumar et al. 2017).

Unavailability of infrastructure—the lack of proper set-up, equipments, etc. limits its production.

Mutation during fermentation—Biofertilizers comprises microbes which are very much prone to mutation during the process of fermentation, which ultimately increases the production cost as well as makes it difficult for quality control.

15.7.2 Market Level Constraints

One of the important problems that hinder the use of these biofertilizers is lack of proper marketing strategies, retail outlets or commercial network for the producers (Mahdi et al. 2010).

Lack of awareness of farmers—Inspite of the considerable efforts in recent years to raise awareness regarding the use of biofertilizers, most of the farmers are not well versed in their application. They are not properly instructed about the benefits and often they are skeptical to accept any new development. The farmers are often seen using the chemical ones knowing very well about their hazardous effects.

Lack of quality assurance—corruption in marketing sector leads to the sale of inferior quality biofertilizers which results in loss of faith among farmers.

Seasonal and unassured demand—less demand of these fertilizers with simultaneous cropping operations has led to their occasional production which is not a continuous one.

15.7.3 Resource Constraint

Limited resource generation for biofertilizer production—it is a low investment industry and the private sectors dealing with this rely on demand based productions as the shelf life of these bioproducts is short. Sufficient funds are not available for a steady production.

15.7.4 Field Level Constraints

Soil and climatic factors—extreme environmental conditions like unfavorable pH, high temperature, drought, high nitrate level, and deficiency of important elements like phosphorus, copper, cobalt, molybdenum or presence of toxic compounds in the soil affect the growth of the microbes and crop response (Bhattacharjee and Dey 2014).

Native microbial population—Antagonistic microorganism already residing in soil may be dominating in nature and they may restrict the effective establishment of the microbial inoculants by outcompeting them (Mahdi et al. 2010).

15.8 Rejuvenation Strategies for Biofertilizers

The following strategies can be used for invigorating the sustainable agricultural practices:

- Identification and selection of suitable locations, crops, soil and specific strains for nitrogen and phosphorus fixation and zinc solubilization for different climatic conditions.
- Application of biotechnological methods for strain improvement
- Observing the activity of the different microbial cultures during storage to circumvent natural mutants.
- Development of suitable substitute formulations like liquid inoculants or granular formulations for all bio-inoculants, standardization of the media preparation, inoculation technique etc., for the new formulations.
- Employing experts like microbiologists in production units to supervise the production of the biofertilizers.
- Development of cold storage facilities in production units.

- Providing technical guidance on the production of biofertilizers as well as quality control of the products and rendering advice and projects to manufacturers.
- Providing organizational training to the extension workers and farmers for the widespread acceptance of the technology.
- Circulating information through mass media, publications and bulletins to make the users and consumers aware of these bioproducts (Gupta 2004).

15.9 Tripartite Relationship Among Different Biofertilizers

Knowledge regarding the synchronized administration of phosphate solubilizers and diazotrophs to the plants in presence of AMF is inadequate and there lies the scope of further research. However, it is seen that when there is dearth of both the essential elements—phosphorous and nitrogen, AMF aid in assimilating phosphates, thus helping in the growth of the plants. Elevated phosphorous content supports the diazotrophs in plants which further boosts up the nitrogenase activity. This helps in improving the rate of nitrogen fixation, which ultimately assists in the growth of roots and development of mycorrhiza. This is the exquisiteness of intergeneric interaction which makes further research all the more interesting.

15.10 Conclusion and Future Prospects

With the ever-growing population, there seems no end to the demand for food but with the availability of the chemical fertilizers, it was thought that the problem could be tackled. But the abundance in crop productivity came with a huge price, as in due course of time, these compounds started to affect our ecosystem in a negative way. On one hand, when one could be rest assured that the availability of food is ensured but at the same time, alleviation of the hazardous effects of the chemical fertilizers from the environment surfaced as a huge concern. These problems necessitated the search for safer alternatives. Amidst all these confusion, biofertilizers were developed and these seemed to be a viable option to put an end to the problem. They are able to enhance the productivity of crops without harming the environment thus playing a significant role in sustainable agriculture. Moreover, these compounds have the potential to minimize the ecological damages, to some extent, that have already been caused by the excessive use of the chemical fertilizers. Even if they cannot fully compensate for the chemical fertilizers, they can still be used as a supplement with the synthetic ones. This will help to reduce the total amount of chemical fertilizers to be used. Thus, the biofertilizer industry, in spite of having some limitations, does not fail to garner interest among recent researchers. Human population is getting more aware of the benefits of these biofertilizers with the passage of time and people all over the world are interested in consuming food grown with organic fertilizers rather than crops grown with chemical fertilizers.

References

Adesemoye AO, Kloepper JW (2009) Plant–microbes interactions in enhanced fertilizer-use efficiency. Appl Microbiol Biotechnol 85:1–12

Aggani SL (2013) Development of bio-fertilizers and its future perspective. Sch Acad J Pharm 4:327–332

Amalraj ELD, Maiyappan S, Peter AJ (2012) In vivo and in vitro studies of *Bacillus megaterium* var. *phosphaticum* on nutrient mobilization, antagonism and plant growth promoting traits. J Ecobiotechnol 1:35–42

Archana D, Nandish M, Savalagi V, Alagawadi A (2013) Characterization of potassium solubilizing bacteria (KSB) from rhizosphere soil. Bioinfolet—A Quarterly J Life Sci 10:248–257

Arkhipova TN, Prinsen E, Veselov SU (2007) Cytokinin producing bacteria enhance plant growth in drying soil. Plant Soil 292:305–315

Arun KS (2007) Bio-fertilizers for sustainable agriculture. Sixth edition, Agribios publishers, India, Mechanism of P-solubilization

Bae H, Sicher RC, Kim MS, Kim SH, Strem MD, Melnick RL, Bailey BA (2009) The beneficial endophyte *Trichoderma hamatum* isolate DS 219b promotes growth and delays the onset of the drought response in *Theobroma cacao*. J Exp Bot 60:3279–3295

Barman M, Paul S, Choudhury AG, Roy P, Sen J (2017) Biofertilizer as Prospective Input for Sustainable Agriculture in India. Int J Curr Microbiol App Sci 11:1177–1186

Bashan Y, de-Bashan LE (2010) How the plant growth-promoting bacterium *Azospirillum* promotes plant growth. A Critical Assess. Adv Agron 108:77–136

Bashan Y, Hernandez JP, Leyva LA (2002) Alginate microbeads as inoculant carriers for plant growth-promoting bacteria. Biol Fert Soils 35:359–368

Bhardwaj D, Ansari MW, Sahoo RK, Tuteja N (2014) Biofertilizers function as key player in sustainable agriculture by improving soil fertility, plant tolerance and crop productivity. Microb Cell Fact 13:66–76

Bhattacharjee R, Dey U (2014) Bio-fertilizer: a way towards organic agriculture: a review. Afr J Microbiol Res 8:2332–2342

Brar SK, Sarma SJ, Chaabouni E (2012) Shelf-life of Biofertilizers: an accord between formulations and genetics. J Biofertil Biopestici 3:5–6

Chang CH, Yang SS (2009) Thermo-tolerant phosphate-solubilizing microbes for multi-functional biofertilizer preparation. Bioresour Technol 100:1648–1658

Chen YP, Rekha PD, Arunshen AB, Lai WA, Young CC (2006) Phosphate solubilizing bacteria from subtropical soil and their tricalcium phosphate solubilizing abilities. Appl Soil Ecol 34:33–41

Choudhury MA, Kennedy IR (2004) Prospect and potentials for system of biological nitrogen fixation in sustainable rice production. Biol Fert Soils 39:219–227

Colo J, HajnaL-Jafari TI, Duric S (2014) Plant growth promotion rhizobacteria in onion production. Pol J Microbiol 6:83–88

Cress WA, Throneberry GO, Lindsey DL (1979) Kinetics of phosphorus absorption by mycorrhizal and nonmycorrhizal tomato roots. Plant Physiol 64:484–487

Creus C, Sueldo R, Barassi C (1997) Shoot growth and water status in *Azospirillum* inoculated wheat seedlings grown under osmotic and salt stresses. Plant Physiol Biochem 35:939–944

Dastager SG, Deepa CK, Pandey A (2010) Isolation and characterization of novel plant growth promoting Micrococcus sp NII-0909 and its interaction with cowpea. Plant Physiol Biochem 48:987–992

Egamberdieva D, Kamilova F, Validov S, Gafurova L, Kucharova Z, Lugtenberg B (2008) High incidence of plant growth stimulating bacteria associated with the rhizosphere of wheat grown on salinated soil in Uzbekistan. Environ Microbiol 10:1–9

El-Fattah DAA, Eweda WE, Zayed MS, Hassanein MK (2013) Effect of carrier materials, sterilization method, and storage temperature on survival and biological activities of *Azotobacter chroococcum* inoculant. Ann Agric Sci 58:111–118

Ezawa T, Saito M, Yoshida T (1995) Comparison of phosphatase localization in the intraradical hyphae of arbuscular mycorrhizal fungi, *Glomus* spp. and *Gigaspora* spp. Plant Soil 176:57–63
García-Fraile P, Menéndez E, Rivas R (2015) Role of bacterial biofertilizers in agriculture and forestry. Bioengineering 2:183–205
Ghorbanian D, Harutyunyan S, Mazaheri D, Rasoli V, Moheb IA (2012) Influence of Arbuscular mycorrhizal fungi and different levels of phosphorus on the growth of corn in water stress conditions. Afr J Agric Res 7:2575–2580
Glick BR (2012) Plant growth-promoting bacteria: mechanisms and applications. Scientifica 12:1–15
Gothwal RK, Nigam VK, Mohan MK, Sasmal D, Ghosh P (2007) Screening of nitrogen fixers from rhizospheric bacterial isolates associated with important desert plants. Appl Ecol Environ Res 6:101–109
Gupta AK (2004) The complete technology book on biofertilizers and organic farming. National Institute of Industrial Research Publishing, India
Hamdali H, Bouizgarne B, HafidI M, LebrihI A, Virolle MJ, Ouhdouch Y (2008) Screening for rock phosphate solubilizing Actinomycetes from Moroccan phosphate mines. Appl Soil Ecol 38:12–19
Han Y, Wang R, Yang Z (2015) 1-Aminocyclopropane-1-Carboxylate Deaminase from *Pseudomonas stutzeri* A1501 facilitates the growth of rice in the presence of salt or heavy metals. J Microbiol Biotechnol 25:1119–1128
Itelima JU, Bang WJ, Onyimba IA (2018) A review: biofertilizer; a key player in enhancing soil fertility and crop productivity. J Microbiol Biotechnol Rep 2:22–28
Jakobsen I, Abbott LK, Robson AD (1992) External hyphae of vesicular-arbuscular mycorrhizal fungi associated with *Trifolium subterrraneum* L. I. Spread of hyphae and phosphorus inflow into roots. New Phytol 120:371–380
Jehangir IA, Mir MA, Bhat MA, Ahangar MA (2017) Biofertilizers an approach to sustainability in agriculture: a review. Int J Pure App Biosci 5:327–334
Jiménez DJ, Montaña JS, Martínez MM (2011) Characterization of free nitrogen fixing bacteria of the genus Azotobacter in organic vegetable-grown Colombian soils. Braz J Microbiol 42:846–858
Kang BG, Kim WT, Yun HS, Chang SC (2010) Use of plant growth-promoting rhizobacteria to control stress responses of plant roots. Plant Biotechnol Rep 4:179–183
Kannaiyan S (1990) Blue green algal biofertilizers. In: Kannaiyan S (ed) The biotechnology of biofertilizers for rice crops. Narosa Publishing House, New Delhi, India, p 212
Kapri A, Tewari L (2010) Phosphate solubilization potential and phosphatase activity of rhizospheric *Trichoderma* spp. Braz J Microbiol 41:787–879
Karandashov V, Nagy R, Wegmuller S, Amrhein N, Bucher M (2004) Evolutionary conservation of a phosphate transporter in the arbuscular mycorrhizal symbiosis. Proc Natl Acad Sci USA 101:6285–6290
Katyal JC, Rattan RK (1993) Distribution of zinc in Indian soils. Fertilizer News 38:15–26
Khan AA, Jilani G, Akhtar MS, Naqvi SMS, Rasheed M (2009) Phosphorus solubilizing bacteria: occurrence, mechanisms and their role in crop production. ARPN J Agric Biol Sci 1:48–58
Khosro M, Yousef S (2012) Bacterial bio-fertilizers for sustainable crop production: a review. ARPN J Agric Biol Sci 7:307–316
Knobeloch L, Salna B, Hogan A, Postle J, Anderson H (2000) Blue babies and nitrate- contaminated well water. Environ Health Perspect 108:675–678
Kudoyarova GR, Melentiev AI, Martynenko EV (2014) Cytokinin producing bacteria stimulate amino acid deposition by wheat roots. Plant Physiol Biochem 83:285–291
Kumar M, Kour D, Yadav AN, Saxena R, Rai PK, Jyoti A, Tomar RS (2019) Biodiversity of methylotrophic microbial communities and their potential role in mitigation of abiotic stresses in plants. Biologia 74:287–308. https://doi.org/10.2478/s11756-019-00190-6
Kumar R, Kumawat N, Sahu YK (2017) Role of biofertilizers in agriculture. Pop Kheti 5:63–66
Kumar SM, Reddy GC, Phogat M, Korav S (2018) Role of bio-fertilizers towards sustainable agricultural development: a review. J Pharmacogn Phytochem 7:1915–1921

Lach DA, Sharma VK, Vary PS (1990) Isolation and characterization of a unique division mutant of *Bacillus megaterium*. J Gen Microbiol 136:545–553

Lei Z, Ya-qing Z (2015) Effects of phosphate solubilization and phytohormone production of *Trichoderma asperellum* Q1 on promoting cucumber growth under salt stress. J Integr Agric 14:1588–1597

Mahdi SS, Hassan GI, Samoon SA, Rather HA, Dar SA, Zehra B (2010) Bio fertilizers in organic agriculture. J Phytol 2:42–54

Mazid M, Khan TA, Mohammad F (2011) Potential of NO and H_2O_2 as signaling molecules in tolerance to abiotic stress in plants. J Ind Res Tech 1:56–68

Meena VS, Maurya BR, Verma JP, Aeron A, Kumar A, Kim K, Bajpai VK (2015) Potassium solubilizing rhizobacteria (KSR): Isolation, identification, and K-release dynamics from waste mica. Ecol Eng 81:340–347

Mendes R, Garbeva P, Raaijmakers JM (2013) The rhizosphere microbiome: significance of plant beneficial plant pathogenic and human pathogenic microorganisms. FEMS Microbiol Rev 37:634–663

Mishra P, Dash D (2014) Rejuvenation of biofertilizer for sustainable agriculture and economic development. Consilience: J Sustain Dev 11:41–61

Puente M, Li C, Bashan Y (2004) Microbial populations and activities in the rhizoplane of rock-weathering desert plants. II. Growth promotion of cactus seedlings. Plant Biol 6:643–650

Raja N (2013) Biopesticides and Biofertilizers: Ecofriendly sources for sustainable agriculture. J Biofertil Biopestici 4:1–2

Rana KL, Kour D, Sheikh I, Dhiman A, Yadav N, Yadav AN, Rastegari AA, Singh K, Saxena AK (2019a) Endophytic fungi: biodiversity, ecological significance and potential industrial applications. In: Yadav AN, Mishra S, Singh S, Gupta A (eds) Recent Advancement in White Biotechnology through Fungi, vol 1. Diversity and Enzymes Perspectives. Springer, Switzerland, pp 1–62

Rana KL, Kour D, Sheikh I, Yadav N, Yadav AN, Kumar V, Singh BP, Dhaliwal HS, Saxena AK (2019b) Biodiversity of endophytic fungi from diverse niches and their biotechnological applications. In: Singh BP (ed) advances in endophytic fungal research: present status and future challenges. Springer International Publishing, Cham, pp 105–144. https://doi.org/10.1007/978-3-030-03589-1_6

Rana KL, Kour D, Yadav AN (2019b) Endophytic microbiomes: biodiversity, ecological significance and biotechnological applications. Res J Biotechnol 14:142–162

Rana R, Ramesh KP (2013) Biofertilizers and their role in agriculture. Pop Kheti 1:56–61

Revillas JJ, Rodelas B, Pozo C, Martinez-Toledo MV, Gonzalez LJ (2000) Production of B-Group vitamins by two *Azotobacter* strains with phenolic compounds as sole carbon source under diazotrophic and adiazotrophic conditions. J Appl Microbiol 89:486–493

Sahoo RK, Ansari MW, Pradhan M, Dangar TK, Mohanty S, Tuteja N (2014) Phenotypic and molecular characterisation of native Azospirillum strains from rice fields to improve crop productivity. Protoplasma 251:943–953

Santos VB, Araújo ASF, Leite LFC, Nunes LAPL, Melo WJ (2012) Soil microbial biomass and organic matter fractions during transition from conventional to organic farming systems. Geoderma 170:227–231

Singh JS, Pandey VC, Singh DP (2011) Efficient soil microorganisms: a new dimension for sustainable agriculture and environmental development. Agric Ecosyst Environ 140:339–353

Sinha RK, Valani D, Chauhan K, Agarwal S (2010) Embarking on a second green revolution for sustainable agriculture by vermiculture biotechnology using earthworms: Reviving the dreams of Sir Charles Darwin. J Agric Biotech Sustain Dev 7:113–118

Somasegaran P, Springer H (1994) Carrier materials used in bio-fertilizer making. Nature publisher's 2:6

Spaepen S, Vanderleyden J (2011) Auxin and plant-microbe interactions. Cold Spring Harb Perspect Biol 3:1–13

Subba Roa NS (2001) An appraisal of biofertilizers in India. In: Kannaiyan S (ed) The biotechnology of biofertilizers for rice crops. Narosa Publishing House, New Delhi, India, pp 1–5
Szilagyi-Zecchin VJ, Ikeda AC, Hungria M, Adamoski D, Kava-Cordeiro V, Glienke C, Galli-Terasawa LV (2014) Identification and characterization of endophytic bacteria from corn (Zea mays L.) roots with biotechnological potential in agriculture. AMB Express 4:1–9
Thajuddin N, Subramanian G (2005) Cyanobacterial biodiversity and potential applications in biotechnology. Curr Sci 89:47–57
Thakur D, Kaushal R, Shyam V (2014) Phosphate solubilising microorganisms: role in phosphorus nutrition of crop plants a Review. Agric Rev 35:159–171
Tomer S, Suyal DC, Goel R (2016) Biofertilizers: A Timely Approach for Sustainable Agriculture. In: Choudhary DK, Varma A, Tuteja N (eds) Plant-microbe interaction: an approach to sustainable agriculture. Springer Nature, Singapore, pp 375–395
Ullah I, Khan AR, Jung BK, Khan AL, Lee IJ, Shin JH (2014) Gibberellins synthesized by the entomopathogenic bacterium, *Photorhabdus temperata* M1021 as one of the factors of rice plant growth promotion. J Plant Interact 9:775–782
VanderGheynst JS, Scher H, Guo HY (2006) Design of formulations for improved biological control agent viability and sequestration during storage. Ind Biotechnol 2:213–219
Verma P, Yadav AN, Khannam KS, Kumar S, Saxena AK, Suman A (2016a) Molecular diversity and multifarious plant growth promoting attributes of Bacilli associated with wheat (*Triticum aestivum* L.) rhizosphere from six diverse agro-ecological zones of India. J Basic Microbiol 56:44–58
Verma P, Yadav AN, Khannam KS, Mishra S, Kumar S, Saxena AK, Suman A (2016b) Appraisal of diversity and functional attributes of thermotolerant wheat associated bacteria from the peninsular zone of India. Saudi J Biol Sci. https://doi.org/10.1016/j.sjbs.2016.01.042
Verma P, Yadav AN, Khannam KS, Panjiar N, Kumar S, Saxena AK, Suman A (2015) Assessment of genetic diversity and plant growth promoting attributes of psychrotolerant bacteria allied with wheat (*Triticum aestivum*) from the northern hills zone of India. Ann Microbiol 65:1885–1899
Verma P, Yadav AN, Khannam KS, Saxena AK, Suman A (2017a) Potassium-Solubilizing Microbes: Diversity, Distribution, and Role in Plant Growth Promotion. In: Panpatte DG, Jhala YK, Vyas RV, Shelat HN (eds) Microorganisms for Green Revolution: Volume 1: Microbes for Sustainable Crop Production. Springer Singapore, Singapore, pp 125–149. https://doi.org/10.1007/978-981-10-6241-4_7
Verma P, Yadav AN, Kumar V, Singh DP, Saxena AK (2017b) Beneficial Plant-Microbes Interactions: Biodiversity of Microbes from Diverse Extreme Environments and Its Impact for Crop Improvement. In: Singh DP, Singh HB, Prabha R (eds) Plant-Microbe Interactions in Agro-Ecological Perspectives: Volume 2: Microbial Interactions and Agro-Ecological Impacts. Springer Singapore, Singapore, pp 543–580. https://doi.org/10.1007/978-981-10-6593-4_22
Vessey JK (2003) Plant growth promoting rhizobacteria as biofertilizers. Plant Soil 255:571–586
Wani SA, Chand S, Ali T (2013) Potential Use of *Azotobacter chroococcum* in Crop Production: An Overview. Curr Agric Res J 1:35–38
Waqas M, Khan AL, Kang SM, Kim YH, Lee IJ (2014) Phytohormone producing fungal endophytes and hardwood-derived biochar interact to ameliorate heavy metal stress in soybeans. Biol Fert Soils 50:1155–1167
Yadav AN (2017a) Agriculturally important microbiomes: biodiversity and multifarious PGP attributes for amelioration of diverse abiotic stresses in crops for sustainable agriculture. Biomed J Sci Tech Res 1:1–4
Yadav AN (2017b) Beneficial role of extremophilic microbes for plant health and soil fertility. J Agric Sci 1:1–4
Yadav AN (2019) Endophytic fungi for plant growth promotion and adaptation under abiotic stress conditions. Acta Sci Agric 3:91–93
Yadav AN, Kumar R, Kumar S, Kumar V, Sugitha T, Singh B, Chauhan V, Dhaliwal HS, Saxena AK (2017a) Beneficial microbiomes: biodiversity and potential biotechnological applications for sustainable agriculture and human health. J Appl Biol Biotechnol 5:45–57

Yadav AN, Kumar V, Prasad R, Saxena AK, Dhaliwal HS (2018a) Microbiome in Crops: Diversity, distribution and potential role in crops improvements. In: Prasad R, Gill SS, Tuteja N (eds) Crop Improvement through Microbial Biotechnology. Elsevier, USA, pp 305–332

Yadav AN, Mishra S, Singh S, Gupta A (2019a) Recent advancement in white biotechnology through Fungi Volume 1: diversity and enzymes perspectives. Springer International Publishing, Cham

Yadav AN, Sachan SG, Verma P, Saxena AK (2016) Bioprospecting of plant growth promoting psychrotrophic Bacilli from cold desert of north western Indian Himalayas. Indian J Exp Biol 54:142–150

Yadav AN, Saxena AK (2018) Biodiversity and biotechnological applications of halophilic microbes for sustainable agriculture. J Appl Biol Biotechnol 6:48–55

Yadav AN, Sharma D, Gulati S, Singh S, Dey R, Pal KK, Kaushik R, Saxena AK (2015) Haloarchaea endowed with phosphorus solubilization attribute implicated in phosphorus cycle. Sci Rep 5:12293

Yadav AN, Singh S, Mishra S, Gupta A (2019b) Recent advancement in white biotechnology through Fungi. Volume 2: Perspective for Value-Added Products and Environments. Springer International Publishing, Cham

Yadav AN, Singh S, Mishra S, Gupta A (2019c) recent advancement in white biotechnology through Fungi. Volume 3: Perspective for Sustainable Environments. Springer International Publishing, Cham

Yadav AN, Verma P, Kour D, Rana KL, Kumar V, Singh B, Chauahan VS, Sugitha T, Saxena AK, Dhaliwal HS (2017b) Plant microbiomes and its beneficial multifunctional plant growth promoting attributes. Int J Environ Sci Nat Resour 3:1–8 https://doi.org/10.19080/ijesnr.2017.03.555601

Yadav AN, Verma P, Kumar S, Kumar V, Kumar M, Singh BP, Saxena AK, Dhaliwal HS (2018b) Actinobacteria from Rhizosphere: Molecular Diversity, Distributions and Potential Biotechnological Applications. In: Singh B, Gupta V, Passari A (eds) New and Future Developments in Microbial Biotechnology and Bioengineering. USA, pp 13–41. https://doi.org/10.1016/b978-0-444-63994-3.00002-3

Yadav AN, Verma P, Sachan SG, Kaushik R, Saxena AK (2018c) Psychrotrophic Microbiomes: Molecular Diversity and Beneficial Role in Plant Growth Promotion and Soil Health. In: Panpatte DG, Jhala YK, Shelat HN, Vyas RV (eds) Microorganisms for Green Revolution-Volume 2: Microbes for Sustainable Agro-ecosystem. Springer, Singapore, pp 197–240. https://doi.org/10.1007/978-981-10-7146-1_11

Yadav AN, Verma P, Singh B, Chauhan VS, Suman A, Saxena AK (2017b) Plant growth promoting bacteria: biodiversity and multifunctional attributes for sustainable agriculture. Adv Biotechnol Microbiol 5:1–16

Yadav AN, Yadav N, Sachan SG, Saxena AK (2019b) Biodiversity of psychrotrophic microbes and their biotechnological applications. J Appl Biol Biotechnol 7:99–108

Yadav N, Yadav AN (2019) Actinobacteria for sustainable agriculture. J Appl Biotechnol Bioeng 6:38–41

Yao Q, Li X, Feng G, Christie P (2001) Mobilization of sparingly soluble inorganic phosphates by external mycelium of an arbuscular mycorrhizal fungus. Plant Soil 230:279–285

Yao YB, Zhanga B, Yuhua T, Miao Z, Ke ZB, Bowen Z (2018) Azolla biofertilizer for improving low nitrogen use efficiency in an intensive rice cropping system. Field Crop Res 216:158–164

Yimer D, Abena T (2019) Components, mechanisms of action, success under greenhouse and field condition, market availability, formulation and inoculants development on biofertilizer. Biomed J Sci and Tech Res 12:9366–9371

Youssef MMA, Eissa MFM (2014) Biofertilizers and their role in management of plant parasitic nematodes: a review. J Biotechnol Pharm Res 5:1–6

Zahir ZA, Ghani U, Naveed M (2009) Comparative effectiveness of *Pseudomonas* and *Serratia* sp. containing ACC-deaminase for improving growth and yield of wheat (*Triticum aestivum* L.) under salt-stressed conditions. Arch Microbiol 191:415–424

Zahir ZA, Munir A, Asghar HN (2008) Effectiveness of rhizobacteria containing ACC-deaminase for growth promotion of pea (*Pisum sativum*) under drought conditions. J Microbiol Biotechnol 18:958–963

Chapter 16
Plant Microbiomes for Sustainable Agriculture: Current Research and Future Challenges

Ajar Nath Yadav

Abstract The plant microbiomes play important role in plant growth promotion and soil fertility for sustainable agriculture. Plant and soil are valuable natural resource harbouring hotspots of microbes. The soil microbiomes play critical roles in the maintenance of global nutrient balance and ecosystem function. The microbes associated with plant as rhizospheric, endophytic and epiphytic with plant growth-promoting (PGP) attributes have emerged as an important and promising tool for sustainable agriculture. PGP microbes promote plant growth directly or indirectly, either by releasing plant growth regulators; solubilization of phosphorus, potassium and zinc; biological nitrogen fixation or by producing siderophore, ammonia, HCN and other secondary metabolites which are antagonistic against pathogenic microbes. The PGP microbes belonged to genera such as *Achromobacter, Arthrobacter, Aspergillus, Azospirillum, Azotobacter, Bacillus, Burkholderia, Gluconoacetobacter, Methylobacterium, Paenibacillus, Pantoea, Penicillium, Piriformospora, Planomonospora, Pseudomonas, Rhizobium, Serratia* and *Streptomyces*. These PGP microbes could be used as biofertilizers/bio-inoculants at place of chemical fertilizers for sustainable agriculture. This chapter exclusively concluded the horizon covered book content of plant microbiomes for sustainable agriculture. The concluding remark envisioned the future beneficial role of plant microbiomes in plant growth promotion and soil fertility.

Keywords Endophytic · Epiphytic · Microbial diversity · Plant microbiomes · Rhizospheric · Sustainable agriculture

This book contains current knowledge about plant microbiomes. The diverse groups of microbes are the key components of soil–plant systems, where they are engaged in an intense network of rhizosphere/endophytic/phyllosphere interactions. The rhizospheric, endophytic and epiphytic microbes with plant growth-promoting (PGP) attributes have emerged as an important and promising tool for sustainable agriculture. PGP microbes promote plant growth directly or indirectly, by releasing

A. N. Yadav (✉)
Department of Biotechnology, Dr. KSG Akal College of Agriculture, Eternal University, Baru Sahib, Sirmour, Himachal Pradesh, India
e-mail: ajar@eternaluniversity.edu.in; ajarbiotech@gmail.com

A. N. Yadav et al. (eds.), *Plant Microbiomes for Sustainable Agriculture*, Sustainable Development and Biodiversity 25, https://doi.org/10.1007/978-3-030-38453-1_16

plant growth regulators; solubilization of phosphorus, potassium and zinc; biological nitrogen fixation or by producing siderophores, ammonia, HCN and other secondary metabolites which are antagonistic against pathogenic microbes. These PGP microbes could be used as biofertilizers/bio-inoculants in place of chemical fertilizers for sustainable agriculture. The aim of the present book is to collect and compile the current developments in the understanding of the rhizospheric, endophytic and epiphytic microbial diversity associated with plants. The book encompasses current knowledge of plant microbiomes and their potential biotechnological applications for plant growth, crop yield and soil health for sustainable agriculture. The book will be highly useful to the faculty, researchers and students associated with microbiology, biotechnology, agriculture, molecular biology, environmental biology and related subjects.

Rhizosphere harbours potential microbiomes which play a pivotal role in nutrient cycling, enhancing soil fertility, maintaining plant health and productivity. Specific microbiomes that are assembled near roots are considered to be some of the most complex ecosystems on the Earth. Heterogeneous microbial communities of rhizospheric microbiomes considerably vary by soil type, land use pattern, plant species and host genotype. It is demonstrated that root exudates act as substrates and signalling molecules which are required for establishing plant–rhizobacterial interactions (Kour et al. 2019b; Mendes et al. 2013). These research priorities may enable us to manipulate agricultural microbiomes and thereby to develop management strategies for increased production and productivity of global agriculture in a sustainable manner. One of the challenges for future research work includes protection and conservation of rhizosphere biodiversity and their potential application in agricultural soils. Figure 16.1 represents the isolation, characterization and application of plant microbiomes for sustainable agriculture.

Endophytes are the microorganisms that live in the internal tissues of plants. Endophytic microbes hold great importance for the roles that they play in association with the host plants. Endophytes are known to promote the growth of the host plants by various activities such as detoxification of toxic compounds, protection against pathogens and production of plant growth-promoting hormones (Rana et al. 2019b; Suman et al. 2016). Many biotechnologically important metabolites are also produced by the endophytes such as anticancer and antimicrobial compounds. There is a rich diversity of endophytes that needs to be explored for biotechnological purposes. Such endophytes play an important role in plant growth promotion as these provide resistance to plant against different environmental stresses and toxic compounds, protect host plants against several pathogens, and produce many plant growth-promoting hormones. Endophytic microbes are also significantly important as biotransformers of different chemicals and help in recycling of nutrients. The endophytes also find many industrial usages as they are known for the production of many important enzymes and metabolites (Yadav et al. 2019a, b, c).

Endophytes constitute an important component of microbial diversity since 20 years, remarkable progress in the field revealed the significance of endophytic microorganisms. Endophytic microbes are unexplored group of organisms that has

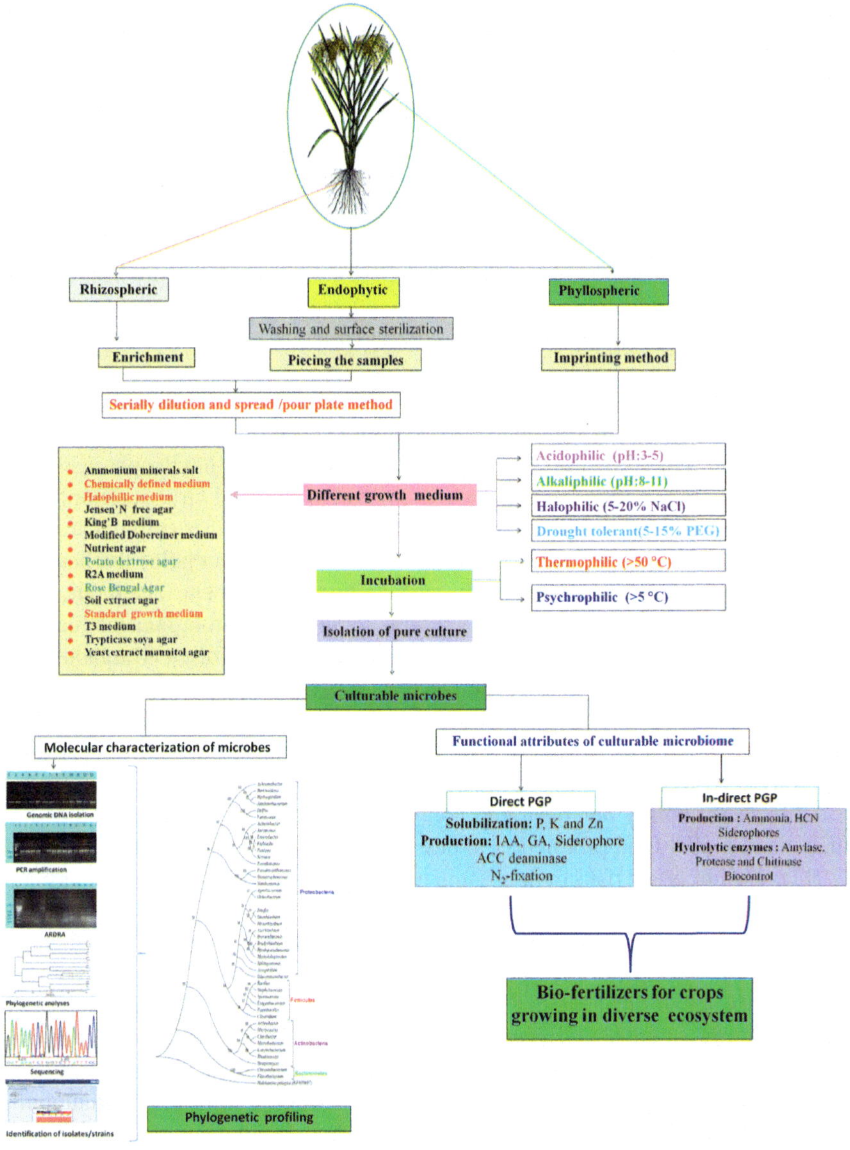

Fig. 16.1 A schematic representation of the isolation, characterization, identification and potential application of culturable and un-culturable microbiomes of crops. Adapted with permission from Verma et al. (2017)

huge potential for innovative pharmaceutical substances; they are established as anti-cancer, antioxidants, antifungal and antiinflammatory. Likewise in recent years, an incredible progress was made in developing them as therapeutic molecules against diverse ailments. In recent years more studies are warranted in bioprospecting new endophytic microorganisms and their applications. Bacterial and fungal endophytes are ubiquitous reside in the internal tissue of living plants. Endophytic fungi distributed out from tropical region to arctic region possess vast potential in terms of secondary metabolite production. It is pertinent to know that the various bioactive indispensable compounds evaluated by these endophytic fungi are host-specific. They are very significant to augmenting the adaptability of the endophyte and its host plants, for instance, biotic and abiotic stress tolerance (Rana et al. 2019a; Yadav 2018).

The phyllosphere referred to the total aerial plant surfaces (above-ground portions), as habitat for microorganisms. Microorganisms establish compositionally complex communities on the leaf surface. The microbiome of phyllosphere is rich in diversity of bacteria, fungi, actinomycetes, cyanobacteria and viruses (Kumar et al. 2019; Müller et al. 2016). Microbes commonly established either epiphytic or endophytic mode of life cycle on phyllosphere environment, which helps the host plant and functional communication with the surrounding environment. The phyllosphere is a unique environment colonized by a wide variety of microorganisms including epiphytes, beneficial and pathogenic, bacteria, fungi and viruses (Bargabus et al. 2002). Understanding the phyllosphere community structure, networking and physiology is a great challenge. However, extensive research on phyllosphere microbiota gives great potential for the applications in economic plant productivity specifically, agriculture and forestry, ecosystem cleaning and health.

Climate variability has been and continues to be, the principal source of fluctuations in global food production in developing countries Oseni and Masarirambi (2011). The important risks of increasing warming of globe are variable and untimely rainfall events, unstable winter seasons, more disease occurrences and crop failures (Adger et al. 2005). Extreme environments represent unique ecosystems which harbour novel biodiversity. Microbial communities associated with plant growing in most diverse conditions, including extremes of temperature, salinity, water deficiency and pH. In order to survive under such extreme conditions, these organisms referred to as extremophiles, have developed adaptive features, which permits them to grow optimally under one or more environmental extremes, while polyextremophiles grow optimally under multiple conditions. These extremophiles can grow optimally in some of the earth's most hostile environments of temperature (−2°–20 °C—psychrophiles; 60°–115 °C—thermophiles), salinity (2–5M NaCl—halophiles) and pH (<4 acidophiles and >9—alkaliphiles) (Yadav et al. 2015c). Microbes associated with crops are able to promote the plant growth. Several microbes have been reported that they can promote plant growth either directly or indirectly. Microbes have been shown to promote plant growth directly, e.g. by fixation of atmospheric nitrogen, solubilization of minerals such as phosphorus, potassium and zinc; production of siderophores and plant growth hormones such cytokinin, auxin and gibberellins.

Several bacteria support plant growth indirectly, via production of antagonistic substances by inducing resistance against plant pathogens (Glick et al. 1999; Tilak et al. 2005).

Salinity of the agriculture soil is the serious issue all over the world and it is also an important environmental factor for reduction of growth and yield of agricultural crops. The density of more salt available in soil may alter the physiological and metabolic activities in the agricultural crops and reduces the growth and production of crops both qualitative and quantitative ways. For combating against soil salinity, many transgenic salt-tolerant crops have been developed but far too little is a success. For solution, In the soils, the use of plant growth-promoting rhizobacteria (PGPR) can be reduced soil salinity, load of chemical fertilizers and pesticide in the agricultural field and improve soil health, seed germination, crop growth and productivity under saline condition PGPR accepted as potential microbes that can tolerant various atmospheric circumstances like more temperature, pH, and saline soils (Yadav and Saxena 2018). Halophilic microbes are isolated from saline soils or rhizosphere of halophytic plants and show plant growth-promoting characters directly like the production of IAA, solubilization of phosphate, production of siderophore, fixation of N_2, deaminase ACC activity or indirectly ways by controlling of phytopathogens under saline condition (Verma et al. 2017). Knowledge of plant–microbe interactions facilitates policies for the protection of crops and saline soil remediation and this type of interaction also observed in the area for ecological appreciative of microbes and which promotes halophyte to adaptability in salinity rich environment.

Drought is a conspicuous stress causing deleterious effect on plant growth and productivity. In order to compensate the yield loss due to drought, efficient and sustainable strategies are required for its management. Drought stress tolerance is a complex trait involving clusters of genes; hence, genetic engineering to generate drought-resistant varieties is a challenging task. In this context, the application of plant growth-promoting microbes (PGPM) to mitigate drought stress is gaining attention as an attractive and cost-effective alternative strategy (Kour et al. 2019a, b, c).

Microorganisms capable of coping with low temperatures are widespread in these natural environments where they often represent the dominant flora and they should, therefore, be regarded as the most successful colonizers of our planet. Psychrophilic microorganisms are adapted to thrive well at low temperatures close to the freezing point of water (Yadav et al. 2015a, b, 2016). Microbial activity of psychrophiles has even been reported at subzero temperatures. In general, psychrophilic microorganisms exhibit higher growth yield and microbial activity at low temperatures compared to temperatures close to the maximum temperature of growth and have more often been put forth as an explanation to successful microbial adaptation to the natural cold environment. Prospecting the cold habitats has led to the isolation of a great diversity of psychrotrophic microorganisms. The cold-adapted microbes have potential biotechnological applications in agriculture, medicine and industry. The bacterial diversity from the cold environment could serve as a database for selection of bioinoculants with PGP ability and could be used for improving the growth and yield of crops grown at high altitudes with prevailing low temperatures (Yadav et al. 2018; 2019d).

Biofertilizers are typically microbial formulations in organic carrier materials that improve soil health and crop growth and development. Of late the use of biofertilizers has gained much acceptance and research interest especially in the developed countries due to ecological impacts associated with the use of synthetic inorganic fertilizers in farming. Microbial formulations could be organism-specific or a consortium of organisms. Many soil microorganisms are endowed with an array of capabilities ranging from production of growth-enhancing substances to the release of substances which ameliorate the effects of various abiotic stress conditions such as drought, salinity, pH stress, heat stress, pollutants and nutrient deficiency.

There has been a sharp increase in the world's population over the past few decades which can be threatening in terms of the food security of the people. Thus, to cater to the huge demand of food, agricultural production should be increased within a short span of time and with limited worldwide agricultural land resources. This situation has driven the farmers all over the world to rely heavily on the commercially available chemical fertilizers for enhanced agricultural productivity. Though there has been a significant rise in the production of crops, these fertilizers have proved to be detrimental for our ecosystem as well as animal and human health. The deteriorative effects of the large chemical inputs in the agricultural systems have not only challenged the sustainability of crop production but also the maintenance of the environment quality. Using biofertilizers is a natural, low-cost, environment-friendly way out to this problem. Biofertilizers comprise living microorganisms capable of supplying sufficient nutrients to the plants, while maintaining high yield. With the ever-growing population, there seems no end to the demand of food but with the availability of the chemical fertilizers, it was thought that the problem could be tackled.

This book will be useful not only to the researchers, but also to each and every stakeholder that contributes toward evergreen agriculture. The enriched efforts of contributors and editorial team have resulted in a volume, which systematically describes to different issues, and applications of plant microbiomes in sustainable agriculture and environments. Needless to mention, that maybe, such volumes will be needed to place the biofertilizers and biopesticides technology in the field to its potential, but this particular book stands on its merit for the information and contents, which will be useful to all.

Acknowledgements The authors are grateful to the Department of Biotechnology, Dr. Khem Singh Gill Akal College of Agriculture, Eternal University, Baru Sahib and Department of Environment, Science & Technology, Shimla, Himachal Pradesh, India funded project "Development of microbial consortium as bio-inoculants for drought and low temperature growing crops for organic farming in Himachal Pradesh" for providing the facilities and financial support, to undertake the investigations. There are no conflicts of interest.

References

Adger WN, Arnell NW, Tompkins EL (2005) Successful adaptation to climate change across scales. Global Environ Change 15:77–86

Bargabus R, Zidack N, Sherwood J, Jacobsen B (2002) Characterisation of systemic resistance in sugar beet elicited by a non-pathogenic, phyllosphere-colonizing Bacillus mycoides, biological control agent. Physiol Mol Plant P 61:289–298

Glick BR, Patten C, Holguin G, Penrose D (1999) Biochemical and genetic mechanisms used by plant growth promoting bacteria. World Sci. https://doi.org/10.1142/p130

Kour D, Rana KL, Yadav AN, Yadav N, Kumar V, Kumar A, Sayyed RZ, Hesham AE-L, Dhaliwal HS, Saxena AK (2019a) Drought-tolerant phosphorus-solubilizing microbes: biodiversity and biotechnological applications for alleviation of drought stress in plants. In: Sayyed RZ, Arora NK, Reddy MS (eds) Plant growth promoting rhizobacteria for sustainable stress management, Volume 1: rhizobacteria in abiotic stress management. Springer Singapore, Singapore, pp 255–308. https://doi.org/10.1007/978-981-13-6536-2_13

Kour D, Rana KL, Yadav N, Yadav AN, Kumar A, Meena VS, Singh B, Chauhan VS, Dhaliwal HS, Saxena AK (2019b) Rhizospheric microbiomes: biodiversity, mechanisms of plant growth promotion, and biotechnological applications for sustainable agriculture. In: Kumar A, Meena VS (eds) Plant growth promoting rhizobacteria for agricultural sustainability: from theory to practices. Springer Singapore, Singapore, pp 19–65. https://doi.org/10.1007/978-981-13-7553-8_2

Kour D, Rana KL, Yadav N, Yadav AN, Singh J, Rastegari AA, Saxena AK (2019c) Agriculturally and industrially important fungi: current developments and potential biotechnological applications. In: Yadav AN, Singh S, Mishra S, Gupta A (eds) Recent advancement in white biotechnology through fungi, Volume 2: perspective for value-added products and environments. Springer International Publishing, Cham, pp 1–64. https://doi.org/10.1007/978-3-030-14846-1_1

Kumar M, Kour D, Yadav AN, Saxena R, Rai PK, Jyoti A, Tomar RS (2019) Biodiversity of methylotrophic microbial communities and their potential role in mitigation of abiotic stresses in plants. Biologia 74:287–308. https://doi.org/10.2478/s11756-019-00190-6

Mendes R, Garbeva P, Raaijmakers JM (2013) The rhizosphere microbiome: significance of plant beneficial, plant pathogenic, and human pathogenic microorganisms. FEMS Microbiol Rev 37:634–663

Müller CA, Obermeier MM, Berg G (2016) Bioprospecting plant-associated microbiomes. J Biotechnol 235:171–180

Oseni T, Masarirambi M (2011) Effect of climate change on maize (*Zea mays*) production and food security in Swaziland. Change 2:3

Rana KL, Kour D, Yadav AN (2019a) Endophytic microbiomes: biodiversity, ecological significance and biotechnological applications. Res J Biotechnol 14:142–162

Rana KL, Kour D, Sheikh I, Yadav N, Yadav AN, Kumar V, Singh BP, Dhaliwal HS, Saxena AK (2019b) Biodiversity of endophytic fungi from diverse niches and their biotechnological applications. In: Singh BP (ed) Advances in endophytic fungal research: present status and future challenges. Springer International Publishing, Cham, pp 105–144. https://doi.org/10.1007/978-3-030-03589-1_6

Suman A, Yadav AN, Verma P (2016) Endophytic microbes in crops: diversity and beneficial impact for sustainable agriculture. In: Singh D, Abhilash P, Prabha R (eds) Microbial inoculants in sustainable agricultural productivity, research perspectives. Springer, India, pp 117–143. https://doi.org/10.1007/978-81-322-2647-5_7

Tilak K, Ranganayaki N, Pal K, De R, Saxena A, Nautiyal CS, Mittal S, Tripathi A, Johri B (2005) Diversity of plant growth and soil health supporting bacteria. Curr Sci 89:136–150

Verma P, Yadav AN, Kumar V, Singh DP, Saxena AK (2017) Beneficial plant-microbes interactions: biodiversity of microbes from diverse extreme environments and its impact for crop improvement. In: Singh DP, Singh HB, Prabha R (eds) Plant-microbe interactions in agro-ecological perspectives: Volume 2: Microbial interactions and agro-ecological impacts. Springer Singapore, Singapore, pp 543–580. https://doi.org/10.1007/978-981-10-6593-4_22

Yadav AN (2018) Biodiversity and biotechnological applications of host-specific endophytic fungi for sustainable agriculture and allied sectors. Acta Sci Microbiol 1:01–05
Yadav AN, Saxena AK (2018) Biodiversity and biotechnological applications of halophilic microbes for sustainable agriculture. J Appl Biol Biotechnol 6:48–55
Yadav AN, Sachan SG, Verma P, Saxena AK (2015a) Prospecting cold deserts of north western Himalayas for microbial diversity and plant growth promoting attributes. J Biosci Bioeng 119:683–693
Yadav AN, Sachan SG, Verma P, Tyagi SP, Kaushik R, Saxena AK (2015b) Culturable diversity and functional annotation of psychrotrophic bacteria from cold desert of Leh Ladakh (India). World J Microbiol Biotechnol 31:95–108
Yadav AN, Verma P, Kumar M, Pal KK, Dey R, Gupta A, Padaria JC, Gujar GT, Kumar S, Suman A (2015c) Diversity and phylogenetic profiling of niche-specific Bacilli from extreme environments of India. Ann Microbiol 65:611–629
Yadav AN, Sachan SG, Verma P, Saxena AK (2016) Bioprospecting of plant growth promoting psychrotrophic Bacilli from cold desert of north western Indian Himalayas. Indian J Exp Biol 54:142–150
Yadav AN, Verma P, Sachan SG, Kaushik R, Saxena AK (2018) Psychrotrophic microbiomes: molecular diversity and beneficial role in plant growth promotion and soil health. In: Panpatte DG, Jhala YK, Shelat HN, Vyas RV (eds) Microorganisms for green revolution. Volume 2: microbes for sustainable agro-ecosystem. Springer, Singapore, pp 197–240. https://doi.org/10.1007/978-981-10-7146-1_11
Yadav AN, Mishra S, Singh S, Gupta A (2019a) Recent advancement in white biotechnology through fungi. Volume 1: diversity and enzymes perspectives. Springer International Publishing, Cham
Yadav AN, Singh S, Mishra S, Gupta A (2019b) Recent advancement in white biotechnology through fungi. Volume 2: perspective for value-added products and environments. Springer International Publishing, Cham
Yadav AN, Singh S, Mishra S, Gupta A (2019c) Recent advancement in white biotechnology through fungi. Volume 3: perspective for sustainable environments. Springer International Publishing, Cham
Yadav AN, Yadav N, Sachan SG, Saxena AK (2019d) Biodiversity of psychrotrophic microbes and their biotechnological applications. J Appl Biol Biotechnol 7:99–108

The manufacturer's authorised representative in the EU is Springer Nature Customer Service Centre GmbH, Europaplatz 3, 69115 Heidelberg, Germany. If you have any concerns regarding our products, please contact ProductSafety@springernature.com

Printed and bound by CPI Group (UK) Ltd, Croydon, CR0 4YY

03/10/2026

02227493-0002